Solid

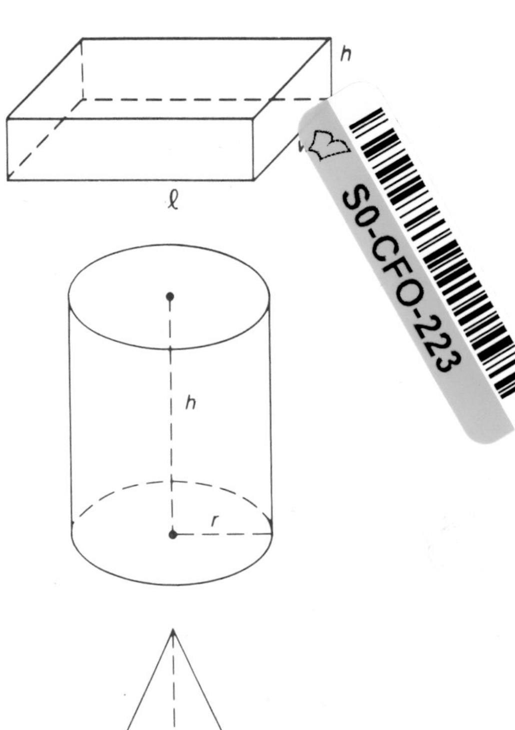

Rectangular Solid

Surface Area $S = 2\ell w + 2wh + 2\ell h$

Volume $V = \ell wh$

Right Circular Cylinder

Surface Area $S = 2\pi rh + 2\pi r^2$

Volume $V = \pi r^2 h$

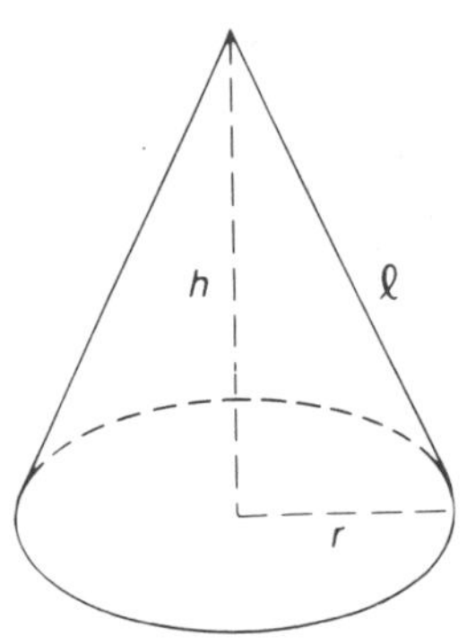

Right Circular Cone

Surface Area $S = \pi r\ell + \pi r^2$

Volume $V = \frac{1}{3}\pi r^2 h$

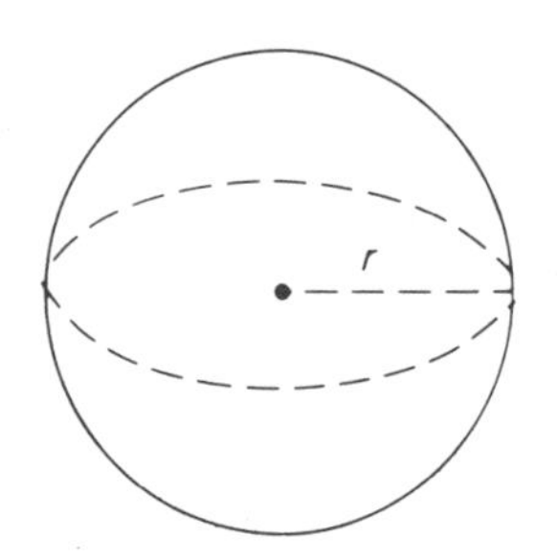

Sphere

Surface Area $S = 4\pi r^2$

Volume $V = \frac{4}{3}\pi r^3$

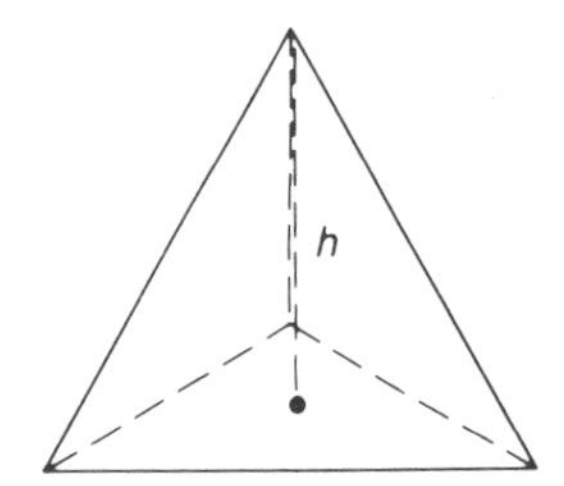

Right Pyramid

Volume $V = \frac{1}{3}bh$

(b is the area of the base)

ELEMENTARY ALGEBRA WITH APPLICATIONS

THIRD EDITION

ELEMENTARY ALGEBRA WITH APPLICATIONS

Terry H. Wesner
Henry Ford Community College

Harry L. Nustad

Wm. C. Brown Publishers

Book Team

Editor *Earl McPeek*
Developmental Editor *Theresa Grutz*
Production Editor *Eugenia M. Collins*
Designer *K. Wayne Harms*
Photo Editor *Carrie Burger*
Visuals Processor *Joseph P. O'Connell*

Wm. C. Brown Publishers

President *G. Franklin Lewis*
Vice President, Publisher *George Wm. Bergquist*
Vice President, Operations and Production *Beverly Kolz*
National Sales Manager *Virginia S. Moffat*
Group Sales Manager *Vincent R. Di Blasi*
Vice President, Editor in Chief *Edward G. Jaffe*
Marketing Manager *Elizabeth Robbins*
Advertising Manager *Amy Schmitz*
Managing Editor, Production *Colleen A. Yonda*
Manager of Visuals and Design *Faye M. Schilling*
Production Editorial Manager *Julie A. Kennedy*
Production Editorial Manager *Ann Fuerste*
Publishing Services Manager *Karen J. Slaght*

WCB Group

President and Chief Executive Officer *Mark C. Falb*
Chairman of the Board *Wm. C. Brown*

Library of Congress Catalog Card Number: 91-55591

ISBN 0-697-08580-5

Printed in the United States of America by Wm. C. Brown Publishers, 2460 Kerper Boulevard, Dubuque, IA 52001

10 9 8 7 6 5 4 3 2 1

To my parents
Harold and Nina,
for everything

Terry

To my parents
Harry and Cordelia,
for their guidance throughout my life

Harry

Contents

Chapter 1 ■ Operations with real numbers

Chapter 2 ■ Solving equations and inequalities

Chapter 3 ■ Polynomials and exponents

Chapter 4 ■ Factoring and solution of quadratic equations by factoring

Chapter 5 ■ Rational Expressions, Ratio and Proportion

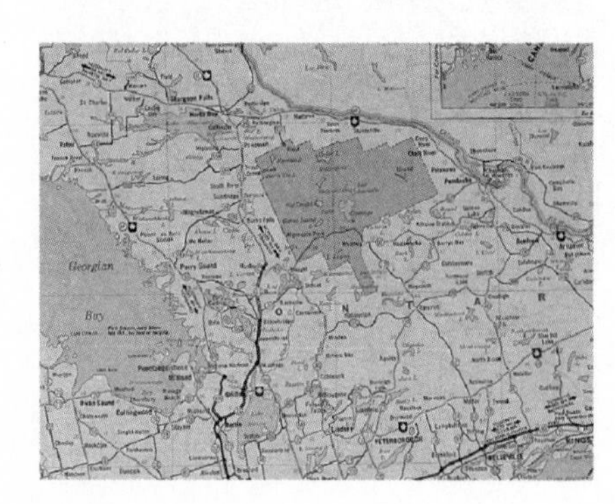

Chapter 6 ■ Operations with Rational Expressions

Chapter 7 ■ Linear Equations in Two Variables

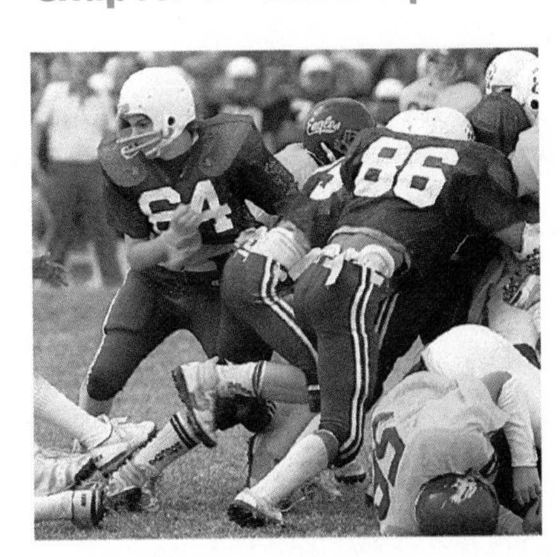

Chapter 8 ■ Systems of Linear Equations

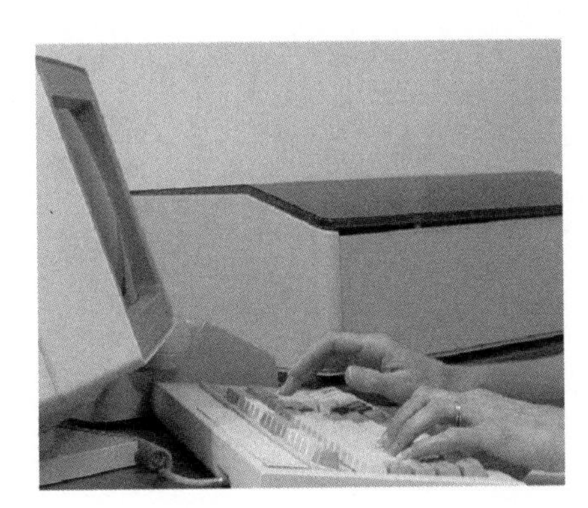

Chapter 9 ■ Roots and Radicals

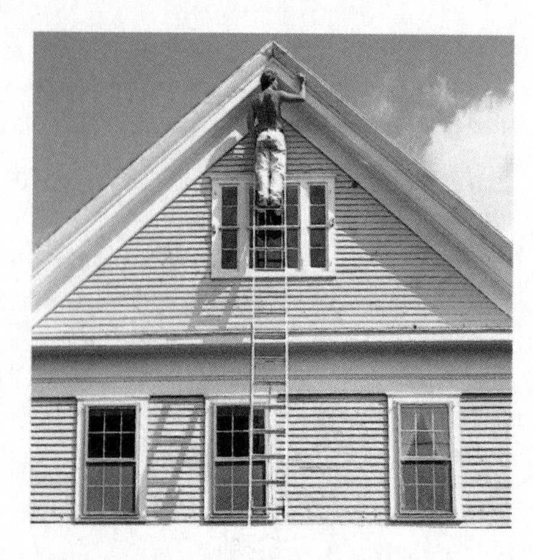

Chapter 10 ■ Solutions of Quadratic Equations

Wesner and Nustad

20 Point Learning System

Your students will count on Terry Wesner and Harry Nustad's integrated learning system. It is the product of over 50 years of combined teaching experience and has been developed with the help of feedback from users—both professors and students—through various texts and editions by this author team. The authors have fine-tuned and enhanced their learning system for this third edition of *Elementary Algebra with Applications*. A full-color design makes an already superb learning system even better. The pedagogical color scheme is used consistently throughout, providing a road map to guide students through the key points of each section. Much more than just adding visual appeal, the color in this text is an integral part of the learning system. Let's take a look at examples of the 20 points that make up the learning system.

1. Chapter Lead-in Problem and Solution
2. Explanations
3. Examples
4. Quick Checks
5. Procedure Boxes
6. Definitions
7. Concepts
8. Notes
9. Problem Solving
10. Mastery Points
11. Section Exercises
12. Quick-Reference Examples
13. Trial Problems
14. Core Exercise Problems
15. Section Review Exercises
16. Chapter Summary
17. Error Analysis
18. Critical Thinking
19. Chapter Review Exercises
20. Cumulative Test

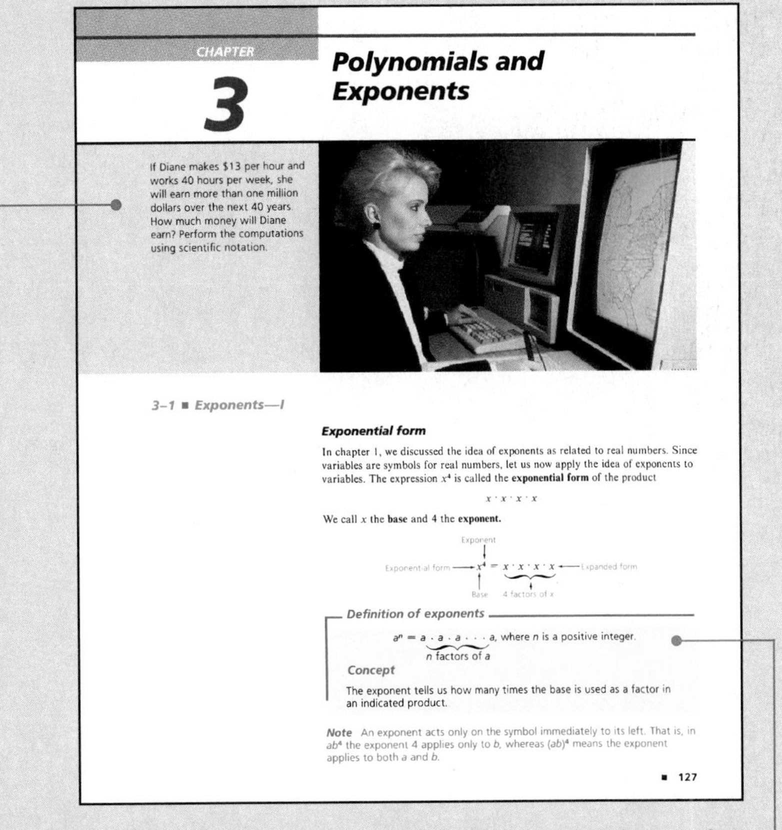

CHAPTER 3

Polynomials and Exponents

If Diane makes $13 per hour and works 40 hours per week, she will earn more than one million dollars over the next 40 years. How much money will Diane earn? Perform the computations using scientific notation.

3–1 ■ Exponents—I

Exponential form

In chapter 1, we discussed the idea of exponents as related to real numbers. Since variables are symbols for real numbers, let us now apply the idea of exponents to variables. The expression x^4 is called the **exponential form** of the product

$$x \cdot x \cdot x \cdot x$$

We call x the **base** and 4 the **exponent.**

Definition of exponents

$a^n = \underbrace{a \cdot a \cdot a \cdots a}_{n \text{ factors of } a}$, where n is a positive integer.

Concept

The exponent tells us how many times the base is used as a factor in an indicated product.

Note An exponent acts only on the symbol immediately to its left. That is, in ab^4 the exponent 4 applies only to b, whereas $(ab)^4$ means the exponent applies to both a and b.

■ 127

A **chapter-opening application** problem with full-color photo poses a problem that students will learn to solve as they progress through the chapter. Its step-by-step solution is shown before the chapter summary.

Definitions are stated precisely in easy-to-understand terms.

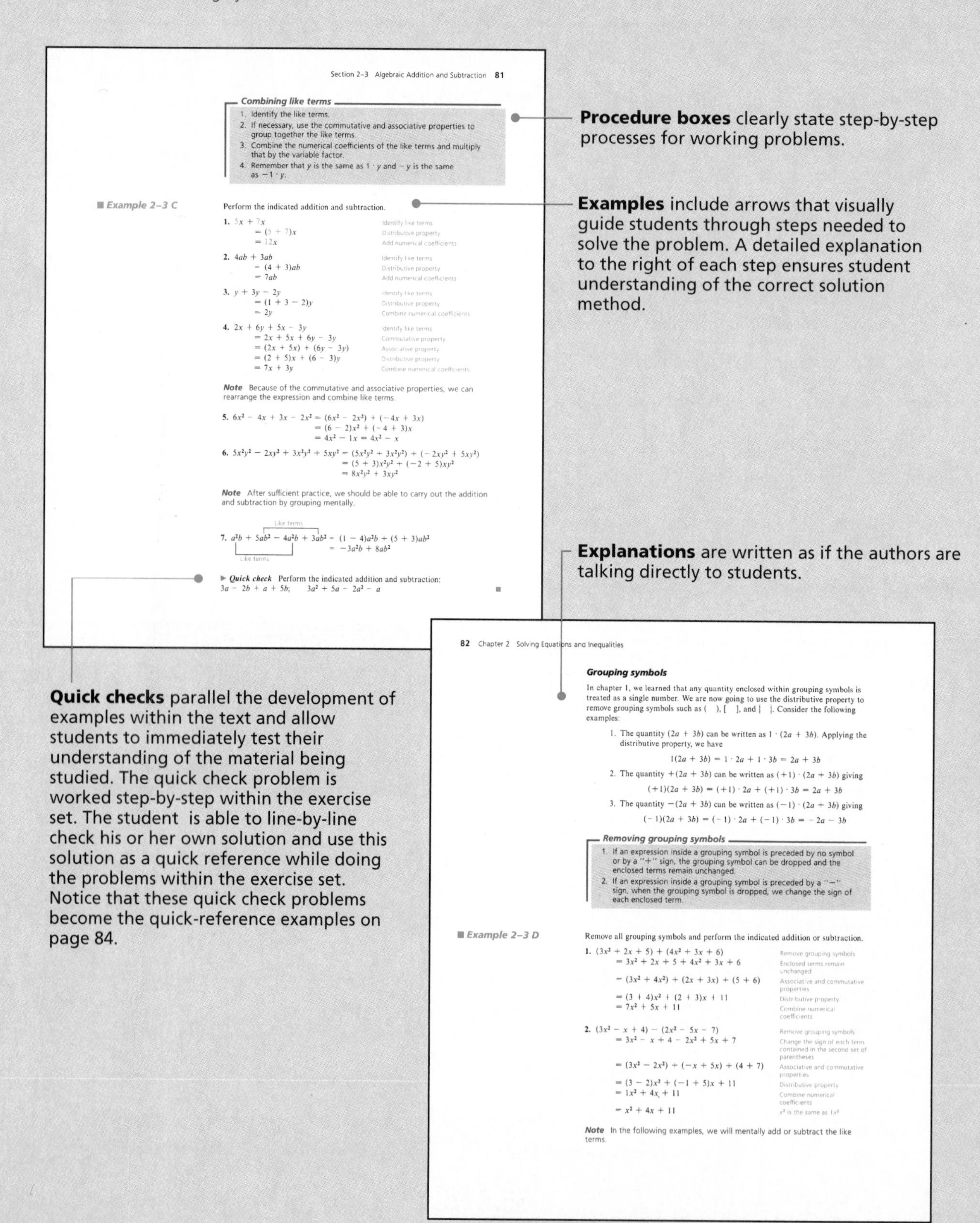
Procedure boxes clearly state step-by-step processes for working problems.
Examples include arrows that visually guide students through steps needed to solve the problem. A detailed explanation to the right of each step ensures student understanding of the correct solution method.
Explanations are written as if the authors are talking directly to students.
Quick checks parallel the development of examples within the text and allow students to immediately test their understanding of the material being studied. The quick check problem is worked step-by-step within the exercise set. The student is able to line-by-line check his or her own solution and use this solution as a quick reference while doing the problems within the exercise set. Notice that these quick check problems become the quick-reference examples on page 84.

Section 2–3 Algebraic Addition and Subtraction 81

Combining like terms

1. Identify the like terms.
2. If necessary, use the commutative and associative properties to group together the like terms.
3. Combine the numerical coefficients of the like terms and multiply that by the variable factor.
4. Remember that y is the same as $1 \cdot y$ and $-y$ is the same as $-1 \cdot y$.

■ *Example 2–3 C*

Perform the indicated addition and subtraction.

1. $5x + 7x$ — Identify like terms
$= (5 + 7)x$ — Distributive property
$= 12x$ — Add numerical coefficients

2. $4ab + 3ab$ — Identify like terms
$= (4 + 3)ab$ — Distributive property
$= 7ab$ — Add numerical coefficients

3. $y + 3y - 2y$ — Identify like terms
$= (1 + 3 - 2)y$ — Distributive property
$= 2y$ — Combine numerical coefficients

4. $2x + 6y + 5x - 3y$ — Identify like terms
$= 2x + 5x + 6y - 3y$ — Commutative property
$= (2x + 5x) + (6y - 3y)$ — Associative property
$= (2 + 5)x + (6 - 3)y$ — Distributive property
$= 7x + 3y$ — Combine numerical coefficients

Note Because of the commutative and associative properties, we can rearrange the expression and combine like terms.

5. $6x^2 - 4x + 3x - 2x^2 = (6x^2 - 2x^2) + (-4x + 3x)$
$= (6 - 2)x^2 + (-4 + 3)x$
$= 4x^2 - 1x = 4x^2 - x$

6. $5x^2y^2 - 2xy^2 + 3x^2y^2 + 5xy^2 = (5x^2y^2 + 3x^2y^2) + (-2xy^2 + 5xy^2)$
$= (5 + 3)x^2y^2 + (-2 + 5)xy^2$
$= 8x^2y^2 + 3xy^2$

Note After sufficient practice, we should be able to carry out the addition and subtraction by grouping mentally.

7. $a^2b + 5ab^2 - 4a^2b + 3ab^2 = (1 - 4)a^2b + (5 + 3)ab^2$ (Like terms)
$= -3a^2b + 8ab^2$

▶ ***Quick check*** Perform the indicated addition and subtraction: $3a - 2b + a + 5b$; $3a^2 + 5a - 2a^2 - a$ ■

82 Chapter 2 Solving Equations and Inequalities

Grouping symbols

In chapter 1, we learned that any quantity enclosed within grouping symbols is treated as a single number. We are now going to use the distributive property to remove grouping symbols such as ( ), [ ], and { }. Consider the following examples:

1. The quantity $(2a + 3b)$ can be written as $1 \cdot (2a + 3b)$. Applying the distributive property, we have
$$1(2a + 3b) = 1 \cdot 2a + 1 \cdot 3b = 2a + 3b$$
2. The quantity $+(2a + 3b)$ can be written as $(+1) \cdot (2a + 3b)$ giving
$$(+1)(2a + 3b) = (+1) \cdot 2a + (+1) \cdot 3b = 2a + 3b$$
3. The quantity $-(2a + 3b)$ can be written as $(-1) \cdot (2a + 3b)$ giving
$$(-1)(2a + 3b) = (-1) \cdot 2a + (-1) \cdot 3b = -2a - 3b$$

Removing grouping symbols

1. If an expression inside a grouping symbol is preceded by no symbol or by a "+" sign, the grouping symbol can be dropped and the enclosed terms remain unchanged.
2. If an expression inside a grouping symbol is preceded by a "−" sign, when the grouping symbol is dropped, we change the sign of each enclosed term.

■ *Example 2–3 D*

Remove all grouping symbols and perform the indicated addition or subtraction.

1. $(3x^2 + 2x + 5) + (4x^2 + 3x + 6)$ — Remove grouping symbols
$= 3x^2 + 2x + 5 + 4x^2 + 3x + 6$ — Enclosed terms remain unchanged
$= (3x^2 + 4x^2) + (2x + 3x) + (5 + 6)$ — Associative and commutative properties
$= (3 + 4)x^2 + (2 + 3)x + 11$ — Distributive property
$= 7x^2 + 5x + 11$ — Combine numerical coefficients

2. $(3x^2 - x + 4) - (2x^2 - 5x - 7)$ — Remove grouping symbols
$= 3x^2 - x + 4 - 2x^2 + 5x + 7$ — Change the sign of each term contained in the second set of parentheses
$= (3x^2 - 2x^2) + (-x + 5x) + (4 + 7)$ — Associative and commutative properties
$= (3 - 2)x^2 + (-1 + 5)x + 11$ — Distributive property
$= 1x^2 + 4x + 11$ — Combine numerical coefficients
$= x^2 + 4x + 11$ — x^2 is the same as $1x^2$

Note In the following examples, we will mentally add or subtract the like terms.

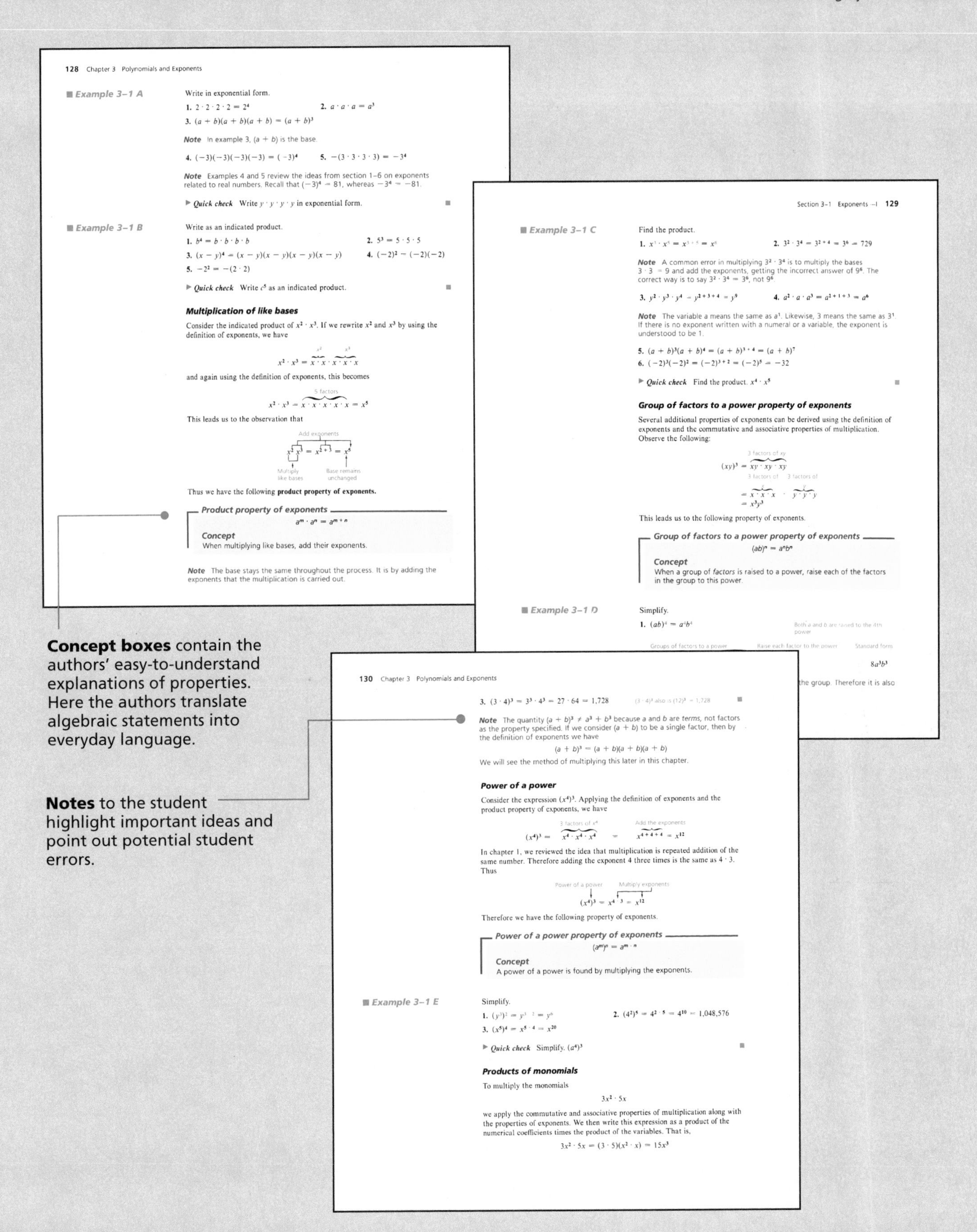

Concept boxes contain the authors' easy-to-understand explanations of properties. Here the authors translate algebraic statements into everyday language.
Notes to the student highlight important ideas and point out potential student errors.

128 Chapter 3 Polynomials and Exponents

■ Example 3–1 A

Write in exponential form.

1. $2 \cdot 2 \cdot 2 \cdot 2 = 2^4$
2. $a \cdot a \cdot a = a^3$
3. $(a + b)(a + b)(a + b) = (a + b)^3$

Note In example 3, $(a + b)$ is the base.

4. $(-3)(-3)(-3)(-3) = (-3)^4$
5. $-(3 \cdot 3 \cdot 3 \cdot 3) = -3^4$

Note Examples 4 and 5 review the ideas from section 1–6 on exponents related to real numbers. Recall that $(-3)^4 = 81$, whereas $-3^4 = -81$.

▶ *Quick check* Write $y \cdot y \cdot y \cdot y$ in exponential form. ■

■ Example 3–1 B

Write as an indicated product.

1. $b^4 = b \cdot b \cdot b \cdot b$
2. $5^3 = 5 \cdot 5 \cdot 5$
3. $(x - y)^4 = (x - y)(x - y)(x - y)(x - y)$
4. $(-2)^2 = (-2)(-2)$
5. $-2^2 = -(2 \cdot 2)$

▶ *Quick check* Write c^5 as an indicated product. ■

Multiplication of like bases

Consider the indicated product of $x^2 \cdot x^3$. If we rewrite x^2 and x^3 by using the definition of exponents, we have

$$x^2 \cdot x^3 = \underbrace{x \cdot x}_{x^2} \cdot \underbrace{x \cdot x \cdot x}_{x^3}$$

and again using the definition of exponents, this becomes

$$x^2 \cdot x^3 = \underbrace{x \cdot x \cdot x \cdot x \cdot x}_{5 \text{ factors}} = x^5$$

This leads us to the observation that

$$x^2 \cdot x^3 = x^{2+3} = x^5$$

(Add exponents; Multiply like bases; Base remains unchanged)

Thus we have the following **product property of exponents.**

Product property of exponents

$$a^m \cdot a^n = a^{m+n}$$

Concept
When multiplying like bases, add their exponents.

Note The base stays the same throughout the process. It is by adding the exponents that the multiplication is carried out.

Section 3–1 Exponents—I 129

■ Example 3–1 C

Find the product.

1. $x^3 \cdot x^5 = x^{3+5} = x^8$
2. $3^2 \cdot 3^4 = 3^{2+4} = 3^6 = 729$

Note A common error in multiplying $3^2 \cdot 3^4$ is to multiply the bases $3 \cdot 3 = 9$ and add the exponents, getting the incorrect answer of 9^6. The correct way is to say $3^2 \cdot 3^4 = 3^6$, not 9^6.

3. $y^2 \cdot y^3 \cdot y^4 = y^{2+3+4} = y^9$
4. $a^2 \cdot a \cdot a^3 = a^{2+1+3} = a^6$

Note The variable a means the same as a^1. Likewise, 3 means the same as 3^1. If there is no exponent written with a numeral or a variable, the exponent is understood to be 1.

5. $(a + b)^3(a + b)^4 = (a + b)^{3+4} = (a + b)^7$
6. $(-2)^3(-2)^2 = (-2)^{3+2} = (-2)^5 = -32$

▶ *Quick check* Find the product. $x^4 \cdot x^5$ ■

Group of factors to a power property of exponents

Several additional properties of exponents can be derived using the definition of exponents and the commutative and associative properties of multiplication. Observe the following:

$$(xy)^3 = \underbrace{xy \cdot xy \cdot xy}_{3 \text{ factors of } xy} = \underbrace{x \cdot x \cdot x}_{3 \text{ factors of } x} \cdot \underbrace{y \cdot y \cdot y}_{3 \text{ factors of } y} = x^3y^3$$

This leads us to the following property of exponents.

Group of factors to a power property of exponents

$$(ab)^n = a^nb^n$$

Concept
When a group of *factors* is raised to a power, raise each of the factors in the group to this power.

■ Example 3–1 D

Simplify.

1. $(ab)^4 = a^4b^4$ Both a and b are raised to the 4th power

Groups of factors to a power — Raise each factor to the power — Standard form

$8a^3b^3$

...the group. Therefore it is also

130 Chapter 3 Polynomials and Exponents

3. $(3 \cdot 4)^3 = 3^3 \cdot 4^3 = 27 \cdot 64 = 1{,}728$ $(3 \cdot 4)^3$ also is $(12)^3 = 1{,}728$ ■

Note The quantity $(a + b)^3 \neq a^3 + b^3$ because a and b are *terms*, not factors as the property specified. If we consider $(a + b)$ to be a single factor, then by the definition of exponents we have

$$(a + b)^3 = (a + b)(a + b)(a + b)$$

We will see the method of multiplying this later in this chapter.

Power of a power

Consider the expression $(x^4)^3$. Applying the definition of exponents and the product property of exponents, we have

$$(x^4)^3 = \underbrace{x^4 \cdot x^4 \cdot x^4}_{3 \text{ factors of } x^4} = \underbrace{x^{4+4+4}}_{\text{Add the exponents}} = x^{12}$$

In chapter 1, we reviewed the idea that multiplication is repeated addition of the same number. Therefore adding the exponent 4 three times is the same as $4 \cdot 3$. Thus

$$(x^4)^3 = x^{4 \cdot 3} = x^{12}$$

(Power of a power; Multiply exponents)

Therefore we have the following property of exponents.

Power of a power property of exponents

$$(a^m)^n = a^{m \cdot n}$$

Concept
A power of a power is found by multiplying the exponents.

■ Example 3–1 E

Simplify.

1. $(y^3)^2 = y^{3 \cdot 2} = y^6$
2. $(4^2)^5 = 4^{2 \cdot 5} = 4^{10} = 1{,}048{,}576$
3. $(x^5)^4 = x^{5 \cdot 4} = x^{20}$

▶ *Quick check* Simplify. $(a^4)^3$ ■

Products of monomials

To multiply the monomials

$$3x^2 \cdot 5x$$

we apply the commutative and associative properties of multiplication along with the properties of exponents. We then write this expression as a product of the numerical coefficients times the product of the variables. That is,

$$3x^2 \cdot 5x = (3 \cdot 5)(x^2 \cdot x) = 15x^3$$

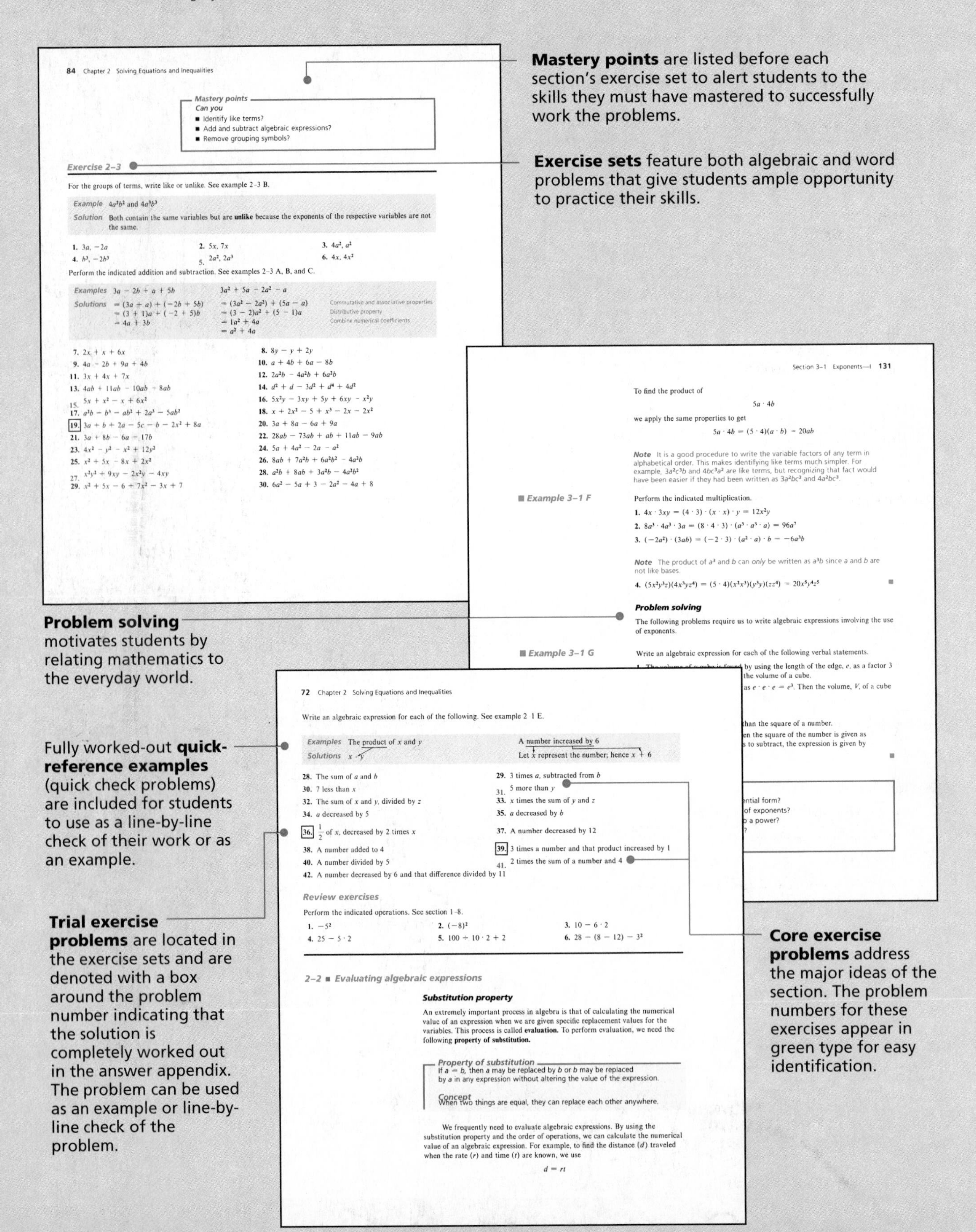
Mastery points are listed before each section's exercise set to alert students to the skills they must have mastered to successfully work the problems.
Exercise sets feature both algebraic and word problems that give students ample opportunity to practice their skills.
Problem solving motivates students by relating mathematics to the everyday world.
Fully worked-out **quick-reference examples** (quick check problems) are included for students to use as a line-by-line check of their work or as an example.
Trial exercise problems are located in the exercise sets and are denoted with a box around the problem number indicating that the solution is completely worked out in the answer appendix. The problem can be used as an example or line-by-line check of the problem.
Core exercise problems address the major ideas of the section. The problem numbers for these exercises appear in green type for easy identification.

Review exercises at the end of each section help students prepare for the following section and keep in touch with previous material.

A chapter **summary** synthesizes important concepts.

Chapter review exercises feature problems to help students determine if they need further work on a particular section. The problems are keyed to refer students back to the section from which they were drawn.

The completely worked-out solution for the **chapter-opening word problem** appears at the end of the chapter prior to the chapter summary.

Error analysis provides a group of problems where a common error has been made. The student is asked to correct the mistake. A page reference is provided so that the student can refer to examples and notes relative to the given problem.

Critical thinking provides special problems that the student must analyze and use their mathematical skills to solve. A series of hints are given in the Instructors Manual.

---

122 Chapter 2 Solving Equations and Inequalities

79. The perimeter (the sum of the sides) of a triangle is more than 52 cm. If two sides of the triangle are 18 cm and 16 cm, respectively, what are the possible values for the length of the third side?

80. Two sides of a triangle are 10 ft and 12 ft long, respectively. If the perimeter must be at least 31 ft, what are the possible values for the length of the third side?

Review exercises

Perform the indicated operations. See section 1–8.

1. -4^2 **2.** $(-4)^2$ **3.** -2^4 **4.** $(-2)^4$

Write an algebraic expression for each of the following. See section 2–1.

5. x raised to the fifth power **6.** A number cubed **7.** A number squared **8.** The product of x and y

Chapter 2 lead-in problem

Bonnie has $3,000 invested at 8% simple interest per year. How much more money must she invest at 7% simple interest if she wants an income of $660 per year ($55 per month) from her investments?

Solution

Let x = the number of dollars invested at 7%.

$$3{,}000(0.08) + x(0.07) = 660 \quad \text{Original equation}$$
$$240 + 0.07x = 660 \quad \text{Simplify}$$
$$0.07x = 420 \quad \text{Subtract 240}$$
$$x = 6{,}000 \quad \text{Divide by 0.07}$$

Therefore Bonnie needs to invest $6,000 at 7% so that her total income from both investments is $660 per year.

Chapter 2 summary

1. A **variable** is a symbol (generally a lowercase letter) that represents an unspecified number.
2. A **constant** is a symbol that does not change its value.
3. An **algebraic expression** is any meaningful collection of variables, constants, grouping symbols, and signs of operations.
4. The **terms** in an algebraic expression are any constants, variables, or products or quotients of these. They are separated by plus or minus signs.
5. In the expression $8x$, 8 is called the **numerical coefficient** or just the coefficient.
6. A **polynomial** is a special kind of algebraic expression. A **monomial** is a polynomial that contains one term; a **binomial** contains two terms; a **trinomial** contains three terms; a **multinomial** contains more than one term.
9. We can *add or subtract* only like, or similar, terms.
10. A **mathematical statement** can be labeled true or false.
11. An equation that is true for every permissible value of the variable is called an **identity.**
12. A replacement value for the variable that forms a true statement (satisfies that equation) is called a **root,** or a **solution,** of the equation.
13. The **solution set** is the set of all values for the variable that cause the equation to be a true statement.
14. A **linear equation** is an equation where the exponent of the unknown is 1.
15. The **addition and subtraction property of equality** enables us to add or subtract the same quantity in each member of an equation and the result will be an equivalent equation.

... **symmetric property of equality** allows us to ... change right and left members of an equation.

---

124 Chapter 2 Solving Equations and Inequalities

Chapter 2 review

[2–1]

Specify the number of terms in each expression.

1. $4x^2 + 3x + 2$ **2.** $5a^2b$ **3.** $7xy + 5$ **4.** $(ab + cd) + xy$

Determine which of the following algebraic expressions are polynomials. If they are not polynomials, state why not.

5. $\frac{x+y}{3} + z$ **6.** $x^3 - x^2$ **7.** $4a^2b^3c$ **8.** $\frac{a+b}{c}$

Write an algebraic expression for each of the following.

9. 5 times x **10.** 7 less than y **11.** 4 more than z **12.** 2 times a number, plus 6

[2–2]

Evaluate the following expressions if $a = 3$, $b = 4$, $c = -4$, and $d = -3$.

13. $3a - b + c$ **14.** $d - 2(a + c)$ **15.** $a^2b - a^2c$ **16.** $(2a - c)(b + 2d)$ **17.** $(c - 2d)^2$ **18.** $c^2 - d^2$

19. Evaluate R when $R = \frac{P \cdot L}{D^2}$ given (a) $P = 6$, $L = 8$, $D = 4$; (b) $P = 7$, $L = 3$, $D = \frac{2}{3}$.

20. The volume of a gas V_2 is given by $V_2 = \frac{P_1V_1}{P_2}$. Find V_2 when $P_1 = 780$, $V_1 = 80$, and $P_2 = 60$.

[2–3]

Remove all grouping symbols and combine like terms.

21. $(3x^2 + 2x - 1) + (x^2 - 5x + 4)$ **22.** $(a^2 - 3a + 4) - (2a^2 - 4a - 7)$
23. $(4a^2 - b^2) - (3a^2 + 2b^2) - (7a^2 - 3b^2)$ **24.** $(5x^3 - 2xy^2 + 3x^2y - 4y^3) - (4x^2 + 3x^2y - 2y^3 + 5xy^2)$
25. $(4ab + 7b^2c) - (15ab - 11bc)$ **26.** $(x - 2y + 7) - (x + 4y + 6)$
27. $(4ab - 2ac) - (6bc - 5ac) - (ab + 2bc)$ **28.** $3a - [4a - (a - 5)]$
29. $5x - [3x - (x - y)]$ **30.** $4x - (x - y) - [3x - y - (2x + 3y)]$
31. $x - \{5x - [3y - (2x - y)]\}$ **32.** $5a - \{6b + a - (5a - 4b)\}$

[2–4]

Determine whether the given statement is true or false when we replace the variable in each equation with the given number.

33. $x + 7 = 11; \{4\}$ **34.** $2x + 1 = 9; \{2\}$ **35.** $5x - 1 = 21; \{5\}$ **36.** $\frac{x}{2} + 5 = 12; \{14\}$

[2–4, 2–5, 2–6]

Find the solution set.

37. $x + 5 = 12$ **38.** $x - 4 = 17$ **39.** $a + 7 = -4$
40. $b - 3 = -9$ **41.** $5z + 3z - 7z + 3 = 7$ **42.** $2(3x - 4) - 5x = 11$
43. $3(2y + 3) = 7 + 5y$ **44.** $3(x - 1) - 2(x + 1) = 4$ **45.** $3x = 9$
46. $4x = 12$ **47.** $-2x = 14$ **48.** $-3x = 21$
49. $\frac{x}{3} = 4$ **50.** $\frac{x}{2} = 7$ **51.** $\frac{3x}{5} = 9$
52. $\frac{2x}{7} = 6$ **53.** $\frac{1}{3}x - 1 = \frac{3}{4}$ **54.** $\frac{1}{3}x + 1 = \frac{1}{6}x - 2$
55. $\frac{3}{4}x + 4 = \frac{5}{8}$ **56.** $\frac{3}{5}x + \frac{1}{2} = \frac{7}{10}x - 3$ **57.** $3x = 0$

Chapter 2 Critical Thinking 123

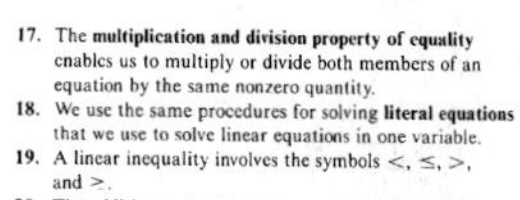

17. The **multiplication and division property of equality** enables us to multiply or divide both members of an equation by the same nonzero quantity.
18. We use the same procedures for solving **literal equations** that we use to solve linear equations in one variable.
19. A linear inequality involves the symbols $<$, $\leq$, $>$, and $\geq$.
20. The **addition and subtraction property of inequalities** states that the same number can be added to or subtracted from both members of an inequality without changing the direction (order) of the inequality symbol.
21. The **multiplication and division property of inequalities** states:
    a. The same *positive* number may be multiplied times or divided into both members of an inequality without changing the direction (order) of the inequality.
    b. When the same *negative* number is multiplied times or divided into both members of an inequality, the direction (order) of the inequality *must be changed.*

Chapter 2 error analysis

1. Degree of a polynomial
   Example: $2x - 3x^2 + x^3 - 1$ has degree 6.
   Correct answer: $2x - 3x^2 + x^3 - 1$ has degree 3.
   What error was made? *(see page 69)*
2. Terms in an algebraic expression
   Example: $x^2 + \frac{2x-3}{5}$ has 3 terms.
   Correct answer: $x^2 + \frac{2x-3}{5}$ has 2 terms.
   What error was made? *(see page 68)*
3. Applying the distributive property
   Example: $5(4 + b) = 20b$
   Correct answer: $5(4 + b) = 20 + 5b$
   What error was made? *(see page 80)*
4. Combining like terms
   Example: $3a^2 + 4a = 7a^3$
   Correct answer: $3a^2 + 4a = 3a^2 + 4a$
   What error was made? *(see page 81)*
5. Combining polynomials
   Example: $(3x^2 - 2x + 1) - (x^2 - x + 2)$ $= 3x^2 - 2x + 1 - x^2 - x + 2$ $= 2x^2 - 3x + 3$
   Correct answer: $(3x^2 - 2x + 1) - (x^2 - x + 2)$ $= 2x^2 - x - 1$
   What error was made? *(see page 82)*
6. Reciprocal of a number
   Example: The reciprocal of 0 is $\frac{1}{0}$.
   Correct answer: 0 has no reciprocal.
   What error was made? *(see page 94)*
7. Graphing linear inequalities
   Example: The graph of $x \leq 3$ is
   Correct answer: The graph of $x \leq 3$ is
   What error was made? *(see page 114)*
8. Multiplying members of an inequality
   Example: If $3 < 4$, then $3 \cdot -2 < 4 \cdot -2$.
   Correct answer: If $3 < 4$, then $3 \cdot -2 > 4 \cdot -2$.
   What error was made? *(see page 115)*
9. Multiplication of negative numbers
   Example: $(-5)(7) = 35$
   Correct answer: $(-5)(7) = -35$
   What error was made? *(see page 48)*
10. Division using zero
    Example: $\frac{7}{0} = 0$
    Correct answer: $\frac{7}{0}$ is undefined.
    What error was made? *(see page 53)*

Chapter 2 critical thinking

If you add any three consecutive odd integers, the sum will be a multiple of 3. Why is this true?

Chapter 2 Cumulative Test 125

58. $-x = -4$ 59. $3.7a = 22.2$ 60. $32.8 = -4.1x$
61. $2x + 5 = 11$ 62. $3b - 8 = 6$ 63. $y + (2y - 1) = 6$
64. $x + 3x = 5 + 7$ 65. $3(2a - 1) = 4a - 2$ 66. $5(x + 3) = 2x - 7$
67. $(3x - 2) - (4x - 1) = 3x$ 68. $2a + 5a - 4 = 3(1 - 2a)$ 69. $8 - 3x + 7 = 5(x + 7)$
70. $2y - 3(y + 1) = 11$ 71. $7x - 4(2x + 3) = 12$ 72. $8x - 14 = 14 - 8x$
73. $5b + 4 = 4 - 2b$ 74. $-3(2x + 1) = 4x - 5$
75. $3(c + 2) - 2(c + 1) = 5c + 11$ 76. $4x - 2(1 - 3x) = 8x + 2$

[2–7]

Solve for the specified variable.

77. $F = ma$, for a 78. $E = IR$, for I 79. $k = PV$, for P
80. $V = k + g + t$, for g 81. $A = \frac{1}{2}h(b + c)$, for c 82. $5x - y = 2x + 3y$, for x

[2–8]

Write an equation for the problem and solve for the unknown quantities.

83. The difference between two numbers is 23. Find the two numbers if their sum is 105.
84. If a number is divided by 9 and that result is then increased by 7, the answer is 11. Find the number.
85. The difference between one-third of a number and one-fifth of a number is 6. Find the number.
86. John invested part of $20,000 at 8% and the rest at 7%. If his income from the 8% investment was $250 more than that from the 7% investment, how much was invested at each rate?
87. Anne made two investments totaling $25,000. On one investment she made a 12% profit but on the other she took a 19% loss. If her net loss was $1,030, how much was in each investment?

[2–9]

Find the solution and graph the solution.

88. $3x > 12$ 89. $5x \leq 15$ 90. $-2x < 14$
91. $-4x > 16$ 92. $2x + 1 < 5$ 93. $7x - 4 > 11$
94. $3x + 7 < 5x - 2$ 95. $9x + 13 \geq 4x + 7$ 96. $6(2x - 1) \leq 3x - 4$
97. $-4 < 5x + 7 < 10$ 98. $0 \leq 1 - 5x < 6$ 99. $5 < 4x + 3 < 12$
100. $-8 \leq 3x + 5 \leq 4$

Chapter 2 cumulative test

Perform the indicated operations, if possible, and simplify.

[1–4] 1. $(-8) + (-4)$ [1–5] 2. $(-10) - (-14)$ [1–7] 3. $\frac{-24}{-8}$
[1–7] 4. $\frac{8}{0}$ [1–8] 5. -5^2 [1–2] 6. $\frac{4}{5} - \frac{3}{10}$
[1–4] 7. $\frac{2}{3} + \left(-\frac{5}{6}\right)$ [1–6] 8. $(-4)(0)(-2)$ [1–2] 9. $(2.3)(8.6)$
[1–8] 10. $5 + 6(8 - 2)$ [1–8] 11. $6 + 4(10 - 2)$ [1–8] 12. $10 - 2(15 - 3) - 5 \cdot 2$
[1–8] 13. $2[5(7 - 4) - 6 + 4]$ [1–8] 14. $5(-4 + 7) - 3(8 - 5)$ [1–8] 15. $14 + 2 \cdot 15 \div 6 - 3 + 4$
[2–3] 16. $5x + x - 2x$ [2–3] 17. $3x^2y^2 - 2xy - x^2y^2 + 5xy$ [2–3] 18. $(2a - b) - (a - 4b)$
[2–3] 19. $(3x^2y - 2xy^2) - (5xy^2 - x^2y)$ [2–3] 20. $5a + 3a^2 - 4a - a^2 + 5 + a^3 - 6$
[2–3] 21. $x - [3x - (y + x) + (2x - 3y)]$ [2–3] 22. $(3a - 2b) - [5a - (4b + 6a)]$

Cumulative tests emphasize the "building-block" nature of mathematics and help students retain knowledge and skills from previous chapters.

Equations and Inequalities

$= -2, b = -3, c = 4$, and $d = 5$.

[2–2] 24. $(3a - 2b) - (5c + d)$ [2–2] 25. $(a - 4c)(b - 2d)$

-12, and 6 [1–5] 27. Subtract -12 from -8.

ws of desks in a classroom. If each desks, how many desks are in the classroom?

[1–7] 29. A trip of 357 miles takes seven hours to complete. What was the average rate of speed?

Write an algebraic expression for each of the following.

[2–1] 30. x decreased by y [2–1] 31. A number increased by 6
[2–2] 32. Ann has d dimes, n nickels, and c cents. Express in cents the amount of money Ann has.

Find the solution set for 33–37 and the solution for 38–42.

[2–6] 33. $10x - 7 = 4x + 3$ [2–6] 34. $5x + 6 = 6$
[2–6] 35. $\frac{1}{3}x + 4 = \frac{5}{6}$ [2–6] 36. $3(2x - 1) + 2(5x - 3) = 8$
[2–6] 37. $16 - 2(4x - 1) = 3x - 12$ [2–9] 38. $-2x \geq 12$
[2–9] 39. $5x + 3x < 6x - 14$ [2–9] 40. $3x + (x - 1) > 7 - x$
[2–9] 41. $-1 < 2x + 3 < 11$ [2–9] 42. $-16 \leq 8 - 4x \leq 12$

Solve for the specified variable.

[2–7] 43. $P = a + b + c$, for b [2–7] 44. $x = a(y + z)$, for y

Solve the following word problems.

[2–8] 45. Phil has $10,000, part of which he invests at 6% and the rest at 5%. If his total income from the two investments was $560, how much did he invest at each rate?
[2–8] 46. The sum of three consecutive even integers is 48. Find the three integers.
[2–8] 47. If a number is increased by 9 and that result is divided by 3, the answer is 7. Find the number.
[2–8] 48. Dwala made two investments totaling $17,000. On one investment she made a 12% profit but on the other she took a 19% loss. If her net loss was $1,215, how much was in each investment?
[2–9] 49. Twice a number decreased by 2 is at most 10. Find all numbers that satisfy this condition.

Preface

Elementary Algebra with Applications is a beginning level text designed specifically for students who have not had a previous course in algebra. The book can be used in lecture-discussion classes or self-paced classes.

Problem-solving orientation The emphasis on problem solving begins in chapter 1 with word problems that have simple arithmetic solutions. The student also learns to change word phrases into algebraic expressions. In chapter 3 and throughout the rest of the text, the student is shown how to form and solve equations from word problems. **Diagrams** are used to show how the words are translated into mathematical symbols. **Tables** are provided to illustrate how several different word phrases become the same mathematical expression.

Critical thinking To encourage students to approach problems creatively in mathematics and the real world, we have included a critical thinking exercise in each chapter.

Error analysis Students can effectively increase their level of understanding of mathematical concepts by evaluating problems illustrating some of the most common mathematical errors. This strengthens the student's understanding of the concept and provides extra practice restating the concept in their own words.

Readability We have attempted to make the text as readable and accessible to students as possible by presenting the material in a manner similar to that which the instructor might use in the classroom.

Applications We have tried to provide a cross section of applications, mainly in the exercises. These are provided to help answer the perennial question "Why am I studying this stuff?" and to make the learning process itself more interesting. In particular, we have tried to show that algebra has become more important than ever in this age of the digital computer. Most ideas are supported by real-life applications relative to that concept.

Functional use of color In this third edition, color has been used to guide students through the text and clearly show the hierarchy of the text's elements. The effective use of color for each particular text element groups similar kinds of elements and helps students understand the relative importance of the elements.

- **Green** is reserved for the core ideas and core exercise problems presented in each chapter; it is used to highlight procedures, properties, definitions, notes, mastery points, and core exercise problems.
- **Blue** is used to emphasize explanations within the examples and exposition.
- **Red** is used to highlight extra practice opportunities for students within the development of each topic.

Highlights of the learning aids

Examples Examples present all aspects of the material being studied with a step-by-step development showing how the problem is worked. Examples have short phrase statements in **blue** type next to most steps stating exactly what has been done. The student is able to develop a clear understanding of how a problem is worked without having to guess what went on in a particular step.

Quick check exercises These exercises after a set of examples are designed to involve the student with the material while studying it. Quick check exercises directly parallel the development and examples in the text. As each new idea or procedure is illustrated with a set of examples, the student is asked to work a similar problem. A **red** triangle identifies each quick check exercise. Quick check exercises are worked step-by-step as quick-reference examples within the exercise set.

Procedure boxes Procedure boxes clearly state a step-by-step summary of the process by which types of problems are to be worked. **Green** has been consistently used for all procedure boxes to emphasize their importance to students.

Concept boxes Concept boxes include properties, theorems, or definitions along with an explanation in easy-to-understand language. **Green** is used to outline each concept box, emphasizing its importance.

Notes Notes to the student highlight important ideas and point out potential errors that students might make. The notes are printed in **green** type to attract the student's attention and emphasize their importance.

Mastery points Mastery points are listed before each exercise set. In essence, they are objectives for that section. They are specifically placed in this location to alert students to the particular skills they must know to successfully work the problems. When students have completed the section, the objectives have more meaning. The **green** outlined box is used to draw the students' attention to the mastery points before they begin the exercise set and to mark the mastery points as covering part of the main ideas of the section.

Exercise sets Exercise sets provide abundant opportunities for students to check their understanding of the concepts being presented. The problems in the exercise sets are carefully paired and graded by level of difficulty to guide the students easily from straightforward computations to more challenging, multi-step problems.

Green type problem numbers identify the core exercise problems in each exercise set.

The directions for each group of problems refer the student to a specific group of parallel examples. After each set of directions is a **quick-reference example**. This example is a specifically chosen quick check exercise from the section and is worked and explained step-by-step. The **red** shading over each quick reference example tells students it is related to the quick check exercises. Students can use this as a line-by-line check of their solution if they worked the problem while studying the material, or as an example they can refer to while working the exercise set.

Review exercises At the end of each section is a group of review problems. These exercises help reinforce the skills necessary for success in the following section. Answers are provided for all the review exercises.

Trial exercise problems Trial exercise problems appear throughout each exercise set and are denoted by a box around the problem number. This indicates that the solution is shown in its entirety in the answer appendix.

Chapter summaries End-of-chapter summaries synthesize the important ideas of each chapter.

Error analysis At the end of each chapter is a group of problems in which a typical error has been made. The student is asked to find and correct the mistake. If the student cannot find the error, a page reference is given which directs the student back to a specific note or group of examples that focus on this problem. Error analysis helps students increase their ability to find errors when checking their solutions, and encourages them to practice restating the important ideas of the chapter in their own words.

Critical thinking Following error analysis is a special problem that requires the student to analyze a problem and use their mathematical skills to answer it. The *Instructor's Resource Manual* contains a series of hints that can be used to guide the student through the analysis of the problem. The *Instructor's Resource Manual* also discusses various ways that critical thinking can be integrated into your course.

Chapter review A chapter review is placed at the end of each chapter. This problem set follows the same organization as the chapter. Each problem is keyed to the section from which it was drawn. Answers to all review problems are provided in the appendix.

Cumulative tests Cumulative tests give students the opportunity to work problems that are drawn from the chapter and from preceding chapters. If students need to review, they can use the section references to review the concept.

Answers Answers are given for all odd-numbered section exercise problems. The answers to all problems in the chapter reviews and cumulative tests are provided in the appendix.

New to this edition

Content

1. The arithmetic review has been moved from the appendix and is now in the opening chapter.
2. Solving equations, verbal problems, formulas, and linear inequalities have been moved from chapter 3 in the second edition to chapter 2 in the third edition to start the student on these concepts earlier.
3. Introduction to exponents has been moved to chapter 3 from chapter 2 in the first edition. In section 3–4,

special attention is now given to helping students effectively distinguish between the various operations with exponents and to apply the correct techniques to exponents.

4. Chapter 4 now includes a section on general strategies for factoring to improve the students' abilities to determine the correct factoring strategy to apply when encountering a variety of equation types to factor.
5. Addition and subtraction of rational expressions with like denominators is now covered in a separate section (6–2) before addition and subtraction of rational expressions with unlike denominators (6–3).
6. In chapter 7, the topic of functions of linear equations is now developed after equations in two variables.
7. Solutions of systems of linear equations is now developed with graphical solutions in section 8–1.
8. Systems of linear equations in three variables is not covered in the third edition.

Features

1. Quick-reference problems are worked out in the exercise sets to serve as further examples for the student.
2. Review exercises at the end of each section have been added to help prepare the student for the work of the following section.
3. Greater use of arrows to point up important steps taken in the development of an example.
4. *All* step-by-step procedures outlined for the major concepts are placed in boxes for emphasis.
5. Each chapter is introduced with an application problem (an accompanying related photo) that can be solved using the procedures studied in the ensuing chapter. The application problem is worked out in detail at the end of the chapter, just prior to the chapter summary for that chapter.
6. Error Analysis in each chapter helps students find errors and apply concepts in their own terms.
7. Critical Thinking activities in each chapter help students learn to address multi-step complex problems.
8. Color is used to clearly show which elements are related and to highlight the important concepts for the students.

For the instructor

The ***Instructor's Resource Manual*** has been expanded to include all critical thinking exercises from the text (with hints and solutions), a guide to the supplements that accompany *Elementary Algebra with Applications,* Third Edition, and reproducible quizzes, multiple chapter tests, and extension problems. Also included are a complete listing of all mastery points and suggested course schedules based on the mastery points. The final section of the *Instructor's Resource Manual* contains answers to the reproducible materials.

The ***Instructor's Solutions Manual*** contains completely worked-out solutions to all of the exercises in the textbook.

The ***Educator's Notebook*** is designed to assist you in formatting and presenting the concepts of *Elementary Algebra with Applications,* Third Edition to your students. Reproducible transparency masters are provided for each section of the textbook.

The ***Test Item File/Quiz Item File*** is a printed version of the computerized *TestPak* and *QuizPak* that allows you to choose test items based on chapter, section, or objective. The objectives are taken directly from the mastery points in *Elementary Algebra with Applications*, Third Edition. The items in the *Test Item File* and *Quiz Item File* are different from those in the prepared tests in the *Instructor's Manual.* Hence, you will have even more items to choose from for your tests.

WCB TestPak 3.0, our computerized testing service, provides you with a call-in/mail-in testing program and the complete *Test Item File* on diskette for use with IBM® PC, Apple®, or Macintosh® computers. *WCB TestPak* requires no programming experience. Tests can be generated randomly, by selecting specific test items or mastery points/objectives. In addition, new test items can be added and existing test items can be edited.

WCB GradePak, also a part of *TestPak 3.0*, is a computerized grade management system for instructors. This program allows you to track students' performance on examinations and assignments. It will compute each student's percentage and corresponding letter grade, as well as the class average. Printouts can be made utilizing both text and graphics.

WCB TestPak 3.0 disks and the WCB call-in service are available free to instructors adopting *Elementary Algebra with Applications*, Third Edition.

WCB QuizPak can be used to give your students on-line practice with the topics of elementary algebra. You can choose multiple-choice and true-false items from the Quiz Item File, edit items, or add your own items. Students' on-line test results are graded and scores then recorded in a GradePak file.

For the student

The ***Student's Solutions Manual*** contains overviews of every chapter of the text, chapter self-tests with solutions, and solutions to all proficiency checks, every other odd-numbered section exercise, and odd-numbered chapter review exercise problems. It is available for student purchase.

On the ***Videotapes***, the instructor introduces a concept, provides detailed explanations of example problems that illustrate the concept, including applications, and concludes with a summary. All of the topics presented in each section of *Elementary Algebra with Applications*, Third Edition are carefully reinforced by the comprehensive Wesner and Nustad Video series. The tapes are available free to qualified adopters.

The **Audiotapes** have also been developed specifically to accompany *Elementary Algebra with Applications,* Third Edition. They begin with a complete synopsis of the section, followed by clear discussions of examples with warning and hints where appropriate. Exercises are solved for each section of the text. Students are directed to turn off the tape and solve a specific problem and turn the tape on again for a complete explanation of the correct solution.

The concepts and skills developed in *Elementary Algebra with Applications*, Third Edition are reinforced though the interactive **Elementary Algebra Tutorial Practice Software.** Students practice solving section-referenced problems generated by the computer and review the major topics of elementary algebra. Step-by-step solutions with explanations guide students to mastery of the major concepts and skills of elementary algebra.

WCB QuizPak, a part of *TestPak 3.0*, provides students with true/false and matching questions from the *Quiz Item File* for each chapter in the text. Using this portion of the program will help your students prepare for examinations. Items in *QuizPak* are similar in level and coverage of concepts as the *TestPak* items. Also included with the *WCB QuizPak* is an on-line testing option that allows professors to prepare tests for students to take using the computer. The computer will automatically grade the test and update the gradebook file.

Acknowledgments

We wish to express our heartfelt thanks and grateful appreciation for the many comments and suggestions given to us during the preparation of the first edition. In particular, we wish to thank George Gullen III, Lynne Hensel, Terry Baker, Harry Datsun, and Robert Olsen for their excellent effort in reviewing each stage of the book and supplying us with the numerous valuable comments, suggestions, and constructive criticisms.

We also wish to thank Lisa Miyazaki for her superb help in preparing the manuscript and working all of the problems.

A very special thanks goes out to the memory of the former Mathematics Division Head, Raymond L. Spencer, for the frequent talks, advice, and encouragement.

The authors would like to acknowledge the contribution of Philip Mahler, who introduced to them the idea of using the tabular format to list all possible combinations of factoring in factoring trinomials. Mr. Mahler was also responsible for the idea of using the sign of the product "mn" as an operation in the second column of the table. The chief virtue of this method is that it is algorithmic. The authors have modified the method slightly by listing the greater factor first.

Vincent McGarry and Irene Doo of Austin Community College deserve special thanks for their careful preparation of the *Instructor's Solutions Manual.*

Throughout the development, writing, and production of this text, two people have been of such great value that we are truly indebted to them for their excellent work on our behalf. We wish to express our utmost thanks to Suresh Ailawadi and Eugenia M. Collins.

We would like to thank the following reviewers of the third edition of *Elementary Algebra with Applications:*

Marybeth Beno
South Suburban College

Sharlene Cadwallader
Mount San Antonio College

Pat Foard
South Plains College

Marty Hodges
Colorado Technical College

Wanda J. Long
St. Charles County Community College

Vincent McGarry
Austin Community College

J. Robert Malena
Community College of Allegheny County

Rita B. Sowell
Volunteer State Community College

Gerry C. Vidrine
Louisiana State University

Keith L. Wilson
Oklahoma City Community College

In addition, we would like to thank the reviewers of *Principles of Elementary Algebra with Applications*, First and Second Editions and *Elementary Algebra with Applications*, Second Edition, whose comments have positively influenced this edition.

Neil Aiken
Milwaukee Area Technical College

Joe Albree
Austin University at Montgomery

Ann Anderson
Broward Community College

Robert Baer
Miami University-Hamilton Branch

Pat Barbalich
Jefferson Community College

Charles Beals
Hartnell College

Don Bellairs
Grossmont College

John P. Bibbo
Southwestern College

Nancy Bray
San Diego Mesa College

Daniel Burns
Sierra College

P. M. Commons
Florida Junior College-South Campus

Ben Cornelius
Oregon Institute of Technology

Lena Dexter
Faulkner State Junior College

Louis Dyson
Clark College

Gail Earles
St. Cloud University

Alice Grandgeorge
Manchester Community College

Michele Greenfield
Middlesex County College

George Gullen III
Henry Ford Community College

Ray Haertel
Central Oregon Community College

Pam Hager
College of the Sequoias

Harry Hayward
Westmoreland County Community College

Lynne Hensel
Henry Ford Community College

Angela Hernandez
University of Montevallo

Tom Householder
Muskingham Area Technical College

Roe Hurst
Central Virginia Community College

Elizabeth Huttenlock
Pennsylvania State University

T. Henry Jablonski, Jr.
East Tennessee State University

Martha Jordan
Okaloosa-Walton Junior College

Glen Just
Mount St. Clare

Judy Kasabian
El Camino College

Margaret A. Kimbell
Texas State Technical Institute Waco Campus

Joanne F. Korsmo
New Mexico State University

Henry Kubo
West Los Angeles College

Theodore Lai
Hudson County Community College

Howard B. Lambert
East Texas State University

Calvin Latham
Monroe Community College

Jeri Vorwerk Love
Florida Junior College

Phil Mahler
Middlesex Community College

Gerald Marlette
Cuyahoga Community College

Hank Martel
Broward Community College Samuels Campus

Jerry J. Maxwell
Olney Central College

Donald Mazukelli
Los Angeles Valley College

Thomas McGannon
Chicago City College

Michael Montemuro
Westchester University

Robert Olsen
Dearborn Public Schools

Kelly Wyatt
Umpaqua Community College

Finally, we are grateful to our "book team," for without them there would be no book. In particular, we would like to express our sincere thanks to Earl McPeek, Gene Collins, Theresa Grutz, K. Wayne Harms, and Carrie Burger.

Study tips

When you work to your full capacity, you can hope to attain the knowledge and skills that will enable you to create your future and control your destiny. If you do not, you will have your future thrust upon you by others.

A Nation at Risk *

There are certain study skills that you as an algebra student need to have, or develop, to assure your success in this course. In addition to the following items listed, acquaint yourself with the text by reading the preface material that precedes these study tips. Then—

1. For every hour spent in class, plan to spend at least two hours studying outside class.
2. Before going to class, read the material to be covered. This will help you more easily understand the instructor's presentation.
3. Take time to become familiar with the learning aids in your textbook. This will allow you to get the maximum benefit from them. In *Elementary Algebra with Applications,* Third Edition color has been used to tie related features together.
   - **Green** is used for the core concepts, ideas, and exercises. Be sure you understand everything highlighted with green.
   - **Blue** indicates additional explanation and greater detail.
   - **Red** is used to identify quick checks and quick-reference examples, which give you greater opportunity to check your understanding of each problem type.
4. Review the material related to each exercise set *before* attempting to work the problems. Be sure you understand the underlying concepts in the worked-out examples and the reason for each step.
5. Carefully read the instructions to the exercise set. Look at the examples and determine what is being asked. Remember, these same instructions will most likely appear on tests.
6. When working the exercise set, take your time, think about what you are doing in each step, and ask yourself why you are performing that step. As you become more confident, increase your speed to better prepare yourself for test situations.
7. When working the exercise sets, compare examples to see in what ways they are alike and in what ways they are different. Problems often *look* similar but are not.

If you do not know how to begin a problem, or you get partway through and are unable to proceed, (a) look back through your notes or (b) look for an exercise you can do that has the answer given and try to analyze the similarities. If doing these things does not work, put the problem aside. Often getting away from it for a time will "open the door" when you try it again. Finally, if you need to, consult your instructor and show him/her the work you have done.

The fact that you will be "using tomorrow what you are doing today" makes it imperative that you learn each concept as you go along. Most concepts, especially the ones that give you the most difficulty, need constant review.

The practice of checking your work will aid you in two ways:

1. It will develop confidence, knowing you have done the problem correctly.
2. It will help you discover your errors on an exam that might otherwise have gone undetected had you not checked your work.

When checking your work, use a different method from the one you used to solve the problem. If the same procedure is used, a tendency to make the same mistake exists. Develop methods for checking your work as you do the practice exercises. This checking then becomes automatic when taking a test.

The following hints will aid you in preparing for an exam:

1. Begin studying and reviewing a number of days prior to the exam. This will enable you to contact your instructor for help if you need it. "All-night" sessions the night before the exam seldom (if ever) yield good results.

*The National Commission on Excellence in Education. *A Nation at Risk.* Washington, D.C.: U.S. Government Printing Office, 1983.

2. Take periodic breaks—10 to 15 minutes for each hour of study. Study for no longer than four hours at a time.
3. Work to develop understanding as well as skills. Memorization is seldom useful in an algebra course, so concentrate on understanding the methods and concepts. However do not ignore skill development, since doing so can often lead to what students call "stupid mistakes."

Prior to taking an exam, use the exercise sets, chapter reviews, and/or *Student's Solutions Manual* to make out a practice test, determine where your errors lie, and retake the test to be sure that you have corrected the mistakes. Allot the same amount of time you will be allowed on test day.

When taking the algebra exam you should:

1. Look over the exam to locate the easiest problems.
2. Work these problems first.
3. Work the more difficult and time-consuming problems next. Remember, when stuck on a problem, go on to other problems and return to those giving you difficulty *only after* completing all that you can.
4. Use what time remains to check your answers or to rework those problems that you found most difficult.

Don't panic should you "draw a blank." Avoid thoughts of failure. Should you feel this happening, relax and try to clear your mind. Search out the problems you feel most confident about and begin again. Should you be unable to complete the exam, be sure to check the problems that you have completed. Always be aware of the time remaining. Do not hurry and do not be intimidated by other students competing the exam early.

One final bit of advice. Show your work neatly. Develop this habit when working on your practice problems. There is a close correlation between neatly laid-out work and the correct answer. Your instructor will appreciate this and be more inclined to give you more credit if the answer is wrong.

ELEMENTARY

ALGEBRA

WITH APPLICATIONS

CHAPTER

1 Operations with Real Numbers

While on a trip to Canada, Tonya heard on the radio that the temperature today will be 20° Celsius. Will she need her winter coat? What will the temperature be in degrees Fahrenheit? We can determine what 20° Celsius is in degrees Fahrenheit by the following expression.

$$F = \frac{9}{5} \cdot 20 + 32$$

1–1 ■ Operations with fractions

Fractions

In day-to-day living, the numbers we use most often are the whole numbers,

$$0, 1, 2, 3, 4, \text{ and so on,}$$

for counting and the fractions, such as

$$\frac{1}{2}, \frac{3}{4}, \frac{9}{10}, \text{ and so on.}$$

In a fraction, the top number is called the **numerator** and the bottom number is called the **denominator.**

$$\frac{9}{10}$$

9 ← Numerator
10 ← Denominator

There are two types of fractions:

1. **Proper fractions** where the numerator is less than the denominator; for example, $\frac{9}{10}$.
2. **Improper fractions** where the numerator is greater than or equal to the denominator; for example, $\frac{10}{9}$ or $\frac{9}{9}$.

Prime numbers and factorization

Any whole number can be stated as a **product** of two or more whole numbers, called **factors** of the number. For example,

Product	Factors	
12	$= 2 \cdot 6$	2 and 6 are factors
12	$= 1 \cdot 12$	1 and 12 are factors
12	$= 4 \cdot 3$	4 and 3 are factors
12	$= 2 \cdot 2 \cdot 3$	2, 2, and 3 are factors

To factor a whole number is to write the number as a product of factors. In future work, it will be necessary to factor whole numbers such that the factors are **prime numbers.**

Prime numbers

A prime number is any whole number greater than 1 whose only factors are the number itself and 1.

The first ten prime numbers are

$$2, 3, 5, 7, 11, 13, 17, 19, 23, \text{and } 29.$$

Thus when we factored 12 as $12 = 2 \cdot 2 \cdot 3$, the number was stated as a product of prime factors.

■ *Example 1–1 A*

Write each number as a product of prime factors.

1. 36

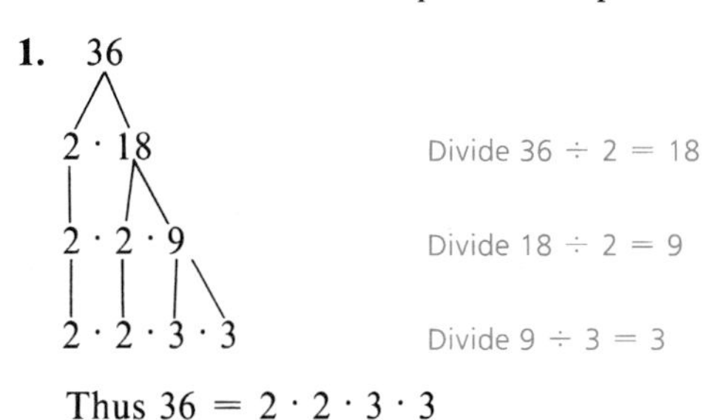

36
2 · 18
2 · 2 · 9
2 · 2 · 3 · 3
Divide 36 ÷ 2 = 18
Divide 18 ÷ 2 = 9
Divide 9 ÷ 3 = 3

Thus $36 = 2 \cdot 2 \cdot 3 \cdot 3$

Note This could have been done in the following way.

② 36	Divide 36 ÷ 2 = 18
② 18	Divide 18 ÷ 2 = 9
③ 9	Divide 9 ÷ 3 = 3
③	

We successively divide by prime numbers, starting with 2 if possible, until the quotient is a prime number.

2. 54

② 54	Divide 54 ÷ 2 = 27
③ 27	Divide 27 ÷ 3 = 9
③ 9	Divide 9 ÷ 3 = 3
③	

Thus $54 = 2 \cdot 3 \cdot 3 \cdot 3$. ■

Reducing fractions to lowest terms

A fraction is **reduced to lowest terms** when the only factor common to the numerator and the denominator is 1.

To reduce a fraction to lowest terms

1. Write the numerator and the denominator as a product of prime factors.
2. Divide the numerator and the denominator by all common factors.

Example 1–1 B

1. $\dfrac{14}{21} = \dfrac{2 \cdot 7}{3 \cdot 7}$ Write as a product of prime factors

$= \dfrac{2}{3}$ Divide numerator and denominator by common factor 7

$\dfrac{2}{3}$ is the answer since 2 and 3 have only 1 as a common factor.

2. $\dfrac{45}{60} = \dfrac{3 \cdot 3 \cdot 5}{2 \cdot 2 \cdot 3 \cdot 5}$ Write as a product of prime factors

$= \dfrac{3}{2 \cdot 2}$ Divide numerator and denominator by $3 \cdot 5$

$= \dfrac{3}{4}$ Multiply in denominator

▶ *Quick check* Reduce $\dfrac{25}{45}$ to lowest terms. ■

Products and quotients of fractions

To multiply two or more fractions, we use the following procedure.

To multiply fractions

1. Write the numerator and the denominator as an indicated product (do not multiply).
2. Reduce the resulting fraction to lowest terms.

Example 1–1 C

Multiply the following fractions and reduce to lowest terms.

1. $\dfrac{2}{3} \cdot \dfrac{5}{7} = \dfrac{2 \cdot 5}{3 \cdot 7}$ Multiply numerators / Multiply denominators

$= \dfrac{10}{21}$ Perform multiplications

2. $\dfrac{5}{6} \cdot \dfrac{3}{4} = \dfrac{5 \cdot 3}{6 \cdot 4}$ Multiply numerators / Multiply denominators

$= \dfrac{5 \cdot 3}{2 \cdot 3 \cdot 2 \cdot 2}$ Factor $6 = 2 \cdot 3$, $4 = 2 \cdot 2$

$= \dfrac{5}{2 \cdot 2 \cdot 2}$ Divide numerator and denominator by 3

$= \dfrac{5}{8}$ Multiply in the denominator ■

Suppose we multiply the fractions

$$\frac{5}{6} \cdot \frac{6}{5} = \frac{5 \cdot 6}{6 \cdot 5} = \frac{30}{30} = 1$$

When the product of two numbers is 1, we call each number the **reciprocal** of the other number. Thus

$\frac{5}{6}$ and $\frac{6}{5}$ are reciprocals,
$\frac{2}{7}$ and $\frac{7}{2}$ are reciprocals,
$\frac{14}{13}$ and $\frac{13}{14}$ are reciprocals.

We can see that the reciprocal of any fraction is obtained by interchanging the numerator and the denominator. The reciprocal of a fraction is used to divide fractions.

To divide two fractions

1. Multiply the first fraction by the **reciprocal** of the second fraction.
2. Reduce the resulting product to lowest terms.

■ ***Example 1–1 D***

Divide the following fractions and reduce to lowest terms.

1. $\frac{7}{8} \div \frac{6}{7} = \frac{7}{8} \cdot \frac{7}{6}$ Multiply by the reciprocal of $\frac{6}{7}$

$= \frac{7 \cdot 7}{8 \cdot 6}$ Multiply numerators / Multiply denominators

$= \frac{49}{48}$

Note The improper fraction $\frac{49}{48}$ can be written as the **mixed number** $1\frac{1}{48}$, which is the sum of a whole number and a proper fraction. This is obtained by dividing the numerator by the denominator.

$48\overline{)49}$ = $1\frac{1}{48}$ — quotient 1; 48 subtracted leaves 1 Remainder

Quotient (1); Remainder (1); Original denominator (48)

2. $\dfrac{\frac{4}{5}}{\frac{3}{7}} = \frac{4}{5} \div \frac{3}{7} = \frac{4}{5} \cdot \frac{7}{3}$ Multiply by the reciprocal of $\frac{3}{7}$

$= \frac{4 \cdot 7}{5 \cdot 3}$ Multiply numerators / Multiply denominators

$= \frac{28}{15}$ or $1\frac{13}{15}$ Perform indicated operations

The improper fraction answer is usually the one preferred in algebra. The mixed number form is usually preferred in an application problem.

3. $3\frac{1}{4} \div 5\frac{2}{3}$

We change the mixed numbers to improper fractions.

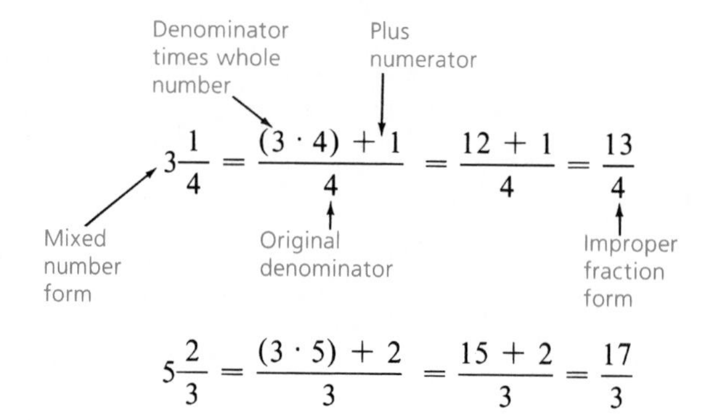

$$5\frac{2}{3} = \frac{(3 \cdot 5) + 2}{3} = \frac{15 + 2}{3} = \frac{17}{3}$$

We now divide as indicated.

$$3\frac{1}{4} \div 5\frac{2}{3} = \frac{13}{4} \div \frac{17}{3}$$

$$= \frac{13}{4} \cdot \frac{3}{17} \quad \text{Multiply by the reciprocal of } \frac{17}{3}$$

$$= \frac{13 \cdot 3}{4 \cdot 17} \quad \begin{array}{l}\text{Multiply numerators}\\ \text{Multiply denominators}\end{array}$$

$$= \frac{39}{68} \quad \text{Perform indicated operations}$$

▶ ***Quick check*** Divide $\dfrac{\frac{5}{6}}{\frac{5}{8}}$ and reduce to lowest terms.

4. The area of a rectangle is found by multiplying the length of the rectangle by the width of the rectangle. Find the area of a rectangle that is $2\frac{1}{2}$ feet long and $1\frac{5}{6}$ feet wide.

$$\text{Area} = 2\frac{1}{2} \cdot 1\frac{5}{6} \quad \text{Multiply the given dimensions}$$

$$= \frac{5}{2} \cdot \frac{11}{6} \quad \text{Change mixed numbers to improper fractions}$$

$$= \frac{5 \cdot 11}{2 \cdot 6} \quad \begin{array}{l}\text{Multiply numerators}\\ \text{Multiply denominators}\end{array}$$

$$= \frac{55}{12} \text{ or } 4\frac{7}{12} \quad \text{Perform indicated operations}$$

The area of the rectangle is $4\frac{7}{12}$ square feet. ■

Addition and subtraction of fractions

To add or subtract fractions, the fractions must have a *common* (same) *denominator.*

To add or subtract fractions with common denominators

1. Add or subtract the numerators.
2. Place the sum or difference over the common denominator.
3. Reduce the resulting fraction to lowest terms.

Example 1–1 E

Add or subtract the following fractions as indicated. Reduce to lowest terms.

1. $\frac{3}{8} + \frac{1}{8} = \frac{3+1}{8}$ Add numerators

$= \frac{4}{8}$ Combine in numerator

$= \frac{1}{2}$ Reduce to lowest terms

2. $\frac{7}{16} - \frac{5}{16} = \frac{7-5}{16}$ Subtract numerators

$= \frac{2}{16}$ Combine in numerator

$= \frac{1}{8}$ Reduce to lowest terms

When the fractions have different denominators, we must rewrite all of the fractions with a new common denominator. Many numbers can satisfy the condition for any set of denominators, but we want the *least* of these numbers, called the **least common denominator** (denoted by LCD). For example, 24 is the least common denominator of the fractions

$$\frac{7}{8} \text{ and } \frac{5}{6}$$

since it is the least (smallest) number that can be divided by 6 and 8 exactly. The procedure for finding the LCD is outlined next.

To find the least common denominator (LCD)

1. Express each denominator as a product of prime factors.
2. List all the *different* prime factors.
3. Write each prime factor the *greatest* number of times it appears in any of the prime factorizations in step 1.
4. The least common denominator is the product of all factors from step 3.

■ Example 1–1 F

Find the least common denominator (LCD) of the fractions with the following denominators.

1. 24 and 18

a. Write 24 and 18 as products of prime factors.
$24 = 2 \cdot 2 \cdot 2 \cdot 3$
$18 = 2 \cdot 3 \cdot 3$

b. The different prime factors are 2 and 3.

c. 2 is a factor three times in 24 and 3 is a factor two times in 18 (the greatest number of times).

d. The LCD is $2 \cdot 2 \cdot 2 \cdot 3 \cdot 3 = 72$.

2. 6, 8, and 14

a. $6 = 2 \cdot 3$
$8 = 2 \cdot 2 \cdot 2$
$14 = 2 \cdot 7$

b. The different prime factors are 2, 3, and 7.

c. 2 is a factor three times in 8, 3 is a factor once in 6, and 7 is a factor once in 14.

d. The LCD is $2 \cdot 2 \cdot 2 \cdot 3 \cdot 7 = 168$.

► ***Quick check*** Find the LCD for fractions with denominators of 6, 9, and 12. ■

Building fractions

To write the fraction $\frac{5}{6}$ as an equivalent fraction with new denominator 24, we find the number that is multiplied by 6 to get 24. Since

$$6 \cdot 4 = 24$$

we use the factor 4. Now multiply the given fraction $\frac{5}{6}$ by the fraction $\frac{4}{4}$. The fraction $\frac{4}{4}$ is equal to 1 and is called a unit fraction. Multiplying by the unit fraction $\frac{4}{4}$ will not change the value of $\frac{5}{6}$, only its form. Thus

$$\frac{5}{6} = \frac{5}{6} \cdot \frac{4}{4} = \frac{5 \cdot 4}{6 \cdot 4} = \frac{20}{24}$$

Multiplication by 1

We use the following procedure to write equivalent fractions.

To find equivalent fractions

1. Divide the original denominator into the new denominator.
2. Multiply the numerator and the denominator of the given fraction by the number obtained in step 1.

■ Example 1–1 G

Write equivalent fractions having the new denominator.

1. $\frac{3}{5} = \frac{?}{30}$

Since $30 \div 5 = 6$, multiply $\frac{3}{5}$ by $\frac{6}{6}$.

$\frac{3}{5} = \frac{3}{5} \cdot \frac{6}{6}$ Multiply by $\frac{6}{6}$

$= \frac{3 \cdot 6}{5 \cdot 6}$ Multiply numerators / Multiply denominators

$= \frac{18}{30}$

2. $\frac{7}{9} = \frac{?}{72}$

Since $72 \div 9 = 8$, multiply $\frac{7}{9}$ by $\frac{8}{8}$.

$\frac{7}{9} = \frac{7}{9} \cdot \frac{8}{8}$ Multiply by $\frac{8}{8}$

$= \frac{7 \cdot 8}{9 \cdot 8}$ Multiply numerators / Multiply denominators

$= \frac{56}{72}$ ■

To add or subtract fractions having different denominators, we use the following procedure.

To add or subtract fractions having different denominators

1. Find the LCD of the fractions.
2. Write each fraction as an equivalent fraction with the LCD as the new denominator.
3. Perform the addition or subtraction as before.
4. Reduce the resulting fraction to lowest terms.

■ Example 1–1 H

Add or subtract the following fractions as indicated. Reduce the resulting fraction to lowest terms.

1. $\frac{7}{8} + \frac{5}{6}$

a. The LCD of the fractions is 24.

b. Since $24 \div 8 = 3$, then

$\frac{7}{8} = \frac{7}{8} \cdot \frac{3}{3} = \frac{7 \cdot 3}{8 \cdot 3} = \frac{21}{24}$ Multiply by $\frac{3}{3}$

and since $24 \div 6 = 4$, then

$\frac{5}{6} = \frac{5}{6} \cdot \frac{4}{4} = \frac{5 \cdot 4}{6 \cdot 4} = \frac{20}{24}$ Multiply by $\frac{4}{4}$

c. $\frac{7}{8} + \frac{5}{6} = \frac{21}{24} + \frac{20}{24}$ Add fractions with LCD

$= \frac{21 + 20}{24}$ Add numerators

$= \frac{41}{24}$ or $1\frac{17}{24}$

2. $\frac{7}{8} - \frac{1}{3}$

a. The LCD of the fractions is 24.

b. Since $24 \div 8 = 3$, then

$\frac{7}{8} = \frac{7}{8} \cdot \frac{3}{3} = \frac{21}{24}$ Multiply by $\frac{3}{3}$

Since $24 \div 3 = 8$, then

$\frac{1}{3} = \frac{1}{3} \cdot \frac{8}{8} = \frac{8}{24}$ Multiply by $\frac{8}{8}$

c. $\frac{7}{8} - \frac{1}{3} = \frac{21}{24} - \frac{8}{24}$ Subtract fractions with LCD

$= \frac{21 - 8}{24} = \frac{13}{24}$ Subtract numerators

3. $3\frac{7}{8} - 2\frac{3}{4}$

Change each of the mixed numbers to an improper fraction.

$3\frac{7}{8} = \frac{(8 \cdot 3) + 7}{8} = \frac{31}{8}$; $2\frac{3}{4} = \frac{(4 \cdot 2) + 3}{4} = \frac{11}{4}$

a. The LCD of the fractions is 8.

b. $\frac{31}{8}$ already has the LCD in its denominator.

Since $8 \div 4 = 2$, then

$\frac{11}{4} = \frac{11 \cdot 2}{4 \cdot 2} = \frac{22}{8}$ Multiply by $\frac{2}{2}$

c. $3\frac{7}{8} - 2\frac{3}{4} = \frac{31}{8} - \frac{11}{4}$ Subtract improper fractions

$= \frac{31}{8} - \frac{22}{8}$ Subtract fractions with LCD

$= \frac{31 - 22}{8}$ Subtract numerators

$= \frac{9}{8}$ or $1\frac{1}{8}$

4. The perimeter (distance around) of a rectangle is found by *adding* the lengths of the four sides of the rectangle. Find the perimeter of a rectangle that is $1\frac{1}{4}$ yards long and $\frac{5}{6}$ of a yard wide.

Perimeter $= \frac{5}{6} + 1\frac{1}{4} + \frac{5}{6} + 1\frac{1}{4}$	Add all the sides
$= \frac{5}{6} + \frac{5}{4} + \frac{5}{6} + \frac{5}{4}$	Change to improper fractions
$= \frac{10}{12} + \frac{15}{12} + \frac{10}{12} + \frac{15}{12}$	LCD is 12
$= \frac{10 + 15 + 10 + 15}{12}$	Add numerators
$= \frac{50}{12} = \frac{25}{6}$	Reduce
or $4\frac{1}{6}$	Mixed number form

The perimeter of the rectangle is $4\frac{1}{6}$ yards.

▶ ***Quick check*** $4\frac{1}{2} - 2\frac{3}{4}$ ■

Mastery points

Can you

- Reduce a fraction to lowest terms?
- Multiply and divide fractions?
- Find the least common denominator (LCD) of two or more fractions?
- Add and subtract fractions?

Exercise 1–1

Reduce the following fractions to lowest terms. See example 1–1 B.

Example $\frac{25}{45}$

Solution

$\frac{25}{45} = \frac{5 \cdot 5}{3 \cdot 3 \cdot 5}$	Factor numerator Factor denominator
$= \frac{5}{3 \cdot 3}$	Divide numerator and denominator by 5
$= \frac{5}{9}$	Multiply in denominator

1. $\frac{4}{8}$ **2.** $\frac{3}{9}$ **3.** $\frac{10}{12}$ **4.** $\frac{8}{14}$ **5.** $\frac{16}{18}$ **6.** $\frac{14}{21}$

7. $\frac{28}{36}$ **8.** $\frac{50}{75}$ **9.** $\frac{64}{32}$ **10.** $\frac{96}{48}$ **11.** $\frac{100}{85}$ **12.** $\frac{120}{84}$

Multiply or divide the fractions as indicated. Reduce to lowest terms. See examples 1–1 C and D.

Example $\dfrac{\frac{5}{6}}{\frac{5}{8}}$

Solution $\dfrac{\frac{5}{6}}{\frac{5}{8}} = \dfrac{5}{6} \div \dfrac{5}{8} = \dfrac{5}{6} \cdot \dfrac{8}{5}$ Multiply by the reciprocal of $\frac{5}{8}$

$= \dfrac{5 \cdot 8}{6 \cdot 5}$ Multiply numerators, Multiply denominators

$= \dfrac{8}{6}$ Reduce by common factor 5

$= \dfrac{2 \cdot 2 \cdot 2}{2 \cdot 3}$ Write numerator and denominator as the product of prime factors

$= \dfrac{2 \cdot 2}{3}$ Reduce by common factor 2

$= \dfrac{4}{3}$ or $1\dfrac{1}{3}$ Multiply in denominator

13. $\dfrac{5}{6} \cdot \dfrac{3}{5}$

14. $\dfrac{2}{3} \cdot \dfrac{5}{6}$

15. $\dfrac{7}{8} \cdot \dfrac{7}{12}$

16. $\dfrac{7}{5} \cdot \dfrac{3}{2}$

17. $\dfrac{7}{9} \cdot \dfrac{3}{4}$

18. $\dfrac{3}{4} \cdot 6$

19. $\dfrac{3}{7} \div \dfrac{4}{5}$

20. $\dfrac{12}{25} \div \dfrac{8}{15}$

21. $\dfrac{6}{7} \div 3$

22. $4 \div \dfrac{3}{8}$

23. $\dfrac{15}{17} \div \dfrac{3}{5}$

24. $4 \div \dfrac{7}{2}$

25. $17 \div 2\dfrac{1}{3}$

26. $12 \cdot 1\dfrac{5}{6}$

27. $7\dfrac{1}{3} \cdot 2\dfrac{4}{7}$

28. $1\dfrac{1}{5} \cdot 2\dfrac{1}{2}$

29. $4\dfrac{4}{5} \cdot 2\dfrac{1}{2}$

30. $7\dfrac{1}{2} \div 5\dfrac{1}{4}$

31. $\dfrac{8}{\frac{2}{3}}$

32. $\dfrac{17}{\frac{3}{4}}$

33. $\dfrac{\frac{7}{8}}{\frac{4}{3}}$

34. $\dfrac{\frac{15}{64}}{\frac{45}{8}}$

35. $\dfrac{4}{5} \cdot \dfrac{2}{3} \cdot \dfrac{3}{8}$

36. $\dfrac{9}{8} \cdot \dfrac{2}{3} \cdot \dfrac{3}{8}$

37. $\dfrac{8}{3} \cdot \dfrac{4}{7}$

38. $\dfrac{8}{3} \div \dfrac{15}{14}$

39. $\dfrac{8}{3} \cdot \dfrac{15}{14}$

See example 1–1 D–4.

40. What is the total length of 25 pieces of steel, each $5\dfrac{1}{2}$ inches long?

41. The volume of a rectangular block is found by multiplying the length times the width times the height.

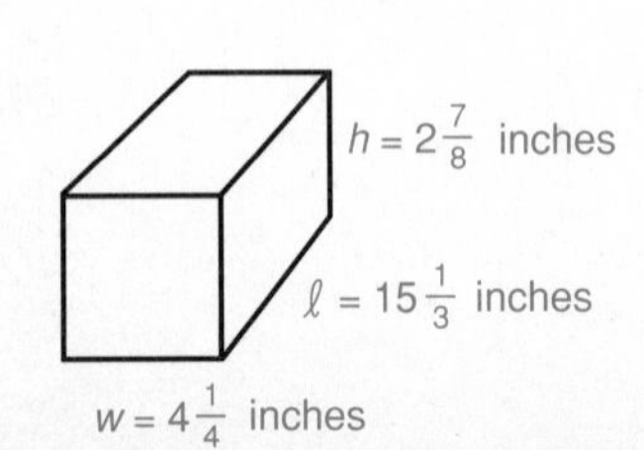

a. What is the volume in cubic inches of a rectangular block of wood $15\frac{1}{3}$ inches long, $4\frac{1}{4}$ inches wide, and $2\frac{7}{8}$ inches high?

b. What is the volume in cubic inches of a block of steel $8\frac{1}{2}$ inches long, $2\frac{1}{8}$ inches wide, and $1\frac{3}{4}$ inches high?

42. A wire $61\frac{1}{2}$ inches long is divided into 14 equal parts. What is the length of each part?

Find the LCD of the fractions with the following groups of denominators. See example 1–1 F.

Example 6, 9, and 12

Solution 1. State 6, 9, and 12 as a product of prime factors.

$6 = 2 \cdot 3$

$9 = 3 \cdot 3$

$12 = 2 \cdot 2 \cdot 3$

2. The different prime factors are 2 and 3.
3. 2 is a factor twice in 12 and 3 is a factor twice in 9.
4. The LCD is $2 \cdot 2 \cdot 3 \cdot 3 = 36$.

43. 3, 8, 10 **44.** 9, 15, 21 **45.** 6, 14, 18 **46.** 5, 10, 12

47. 16, 24, 36 **48.** 12, 16, 24 **49.** 5, 7, 11 **50.** 10, 20, 30

51. 68, 9, 12 **52.** 10, 14, 18 **53.** 10, 15, 20 **54.** 10, 15, 24

Add or subtract the following fractions as indicated. Reduce to lowest terms. See examples 1–1 E and H.

Example $4\frac{1}{2} - 2\frac{3}{4}$

Solution We first change the mixed numbers to improper fractions.

$$4\frac{1}{2} = \frac{(4 \cdot 2) + 1}{2} = \frac{9}{2}; \quad 2\frac{3}{4} = \frac{(4 \cdot 2) + 3}{4} = \frac{11}{4}$$

$$4\frac{1}{2} - 2\frac{3}{4} = \frac{9}{2} - \frac{11}{4}$$ Replace mixed numbers with improper fractions

The LCD is 4.

$$= \frac{18}{4} - \frac{11}{4}$$ Write $\frac{9}{2}$ as $\frac{18}{4}$

$$= \frac{18 - 11}{4}$$ Subtract numerators

$$= \frac{7}{4} \text{ or } 1\frac{3}{4}$$

55. $\frac{1}{3} + \frac{1}{3}$ **56.** $\frac{2}{5} + \frac{3}{10}$ **57.** $\frac{1}{3} + \frac{1}{4}$ **58.** $\frac{5}{6} - \frac{1}{6}$

59. $\frac{4}{5} - \frac{2}{10}$ **60.** $\frac{5}{6} - \frac{3}{8}$ **61.** $1 + \frac{5}{8}$ **62.** $3 + \frac{5}{6}$

63. $4 - \frac{3}{5}$
64. $\frac{2}{3} + \frac{3}{4}$
65. $\frac{3}{5} + \frac{7}{15}$
66. $\frac{5}{6} - \frac{1}{3}$
67. $\frac{3}{8} - \frac{1}{12}$
68. $\frac{7}{24} - \frac{3}{16}$
69. $\frac{7}{54} + \frac{19}{45}$
70. $\frac{1}{2} + \frac{1}{5} + \frac{1}{10}$
71. $\frac{7}{15} + \frac{5}{6} - \frac{3}{4}$
72. $\frac{9}{16} + \frac{5}{18} - \frac{2}{15}$
73. $8\frac{3}{16} - 4\frac{5}{8}$
74. $7\frac{1}{2} + 2\frac{3}{4}$
75. $\frac{2}{7} + \frac{2}{3} + \frac{5}{7}$

See example 1–1 H–4.

76. Jane owed Joan some money. If she paid Joan $\frac{1}{4}$ of the debt on June 15, $\frac{1}{3}$ of the original debt on July 1, and $\frac{3}{8}$ of the original debt on August 10, how much of her debt had Jane paid by August 10?
77. A flower garden in the form of a rectangle has two sides that are $24\frac{1}{2}$ feet long and two sides that are $18\frac{3}{4}$ feet long. Find the perimeter (total distance around) of the rectangle.
78. On a given day, Mrs. Jones purchased $\frac{5}{6}$ yard of one material, $\frac{3}{4}$ yard of another material, and $\frac{2}{3}$ yard of a third material. How many yards of material did she purchase altogether?
79. Butcher John has $32\frac{1}{4}$ pounds of pork chops. If he sells $21\frac{1}{3}$ pounds of the pork chops on a given day, how many pounds of pork chops does he have left?
80. A machinist has a piece of steel stock that weighs $12\frac{7}{8}$ ounces. If he cuts off $5\frac{1}{5}$ ounces, how many ounces does he have left?

1–2 ■ Operations with decimals and percents

In section 1–1, we studied fractions. A **decimal number** is a special fraction with a denominator that is 10, 100, 1,000, and so on.

In a number such as 23, the digits 2 and 3 have place value as follows:

$$23 = (2 \cdot 10) + (3 \cdot 1)$$

Now consider the same two digits with a dot, called the **decimal point,** in front of them, .23. We call this number a **decimal fraction,** or just plain **decimal.** (It is standard procedure to place a zero to the left of the decimal point if the decimal is less than 1. The zero helps the reader see the decimal point and emphasizes the fact that we are dealing with a decimal fraction.) In this new form, we have

$$0.23 = \left(2 \cdot \frac{1}{10}\right) + \left(3 \cdot \frac{1}{100}\right) = \frac{2}{10} + \frac{3}{100} = \frac{20}{100} + \frac{3}{100}$$

which can be written simply as

$$\frac{23}{100}$$

and read "twenty-three hundredths." The decimal point is placed so that the number of digits to the right of it indicates the number of zeros in the fraction's denominator. If there is one digit to the right of the decimal point, the denominator is 10, read "tenths." If there are two digits to the right of the decimal point, the denominator is 100, read "hundredths." If three digits are to the right of the decimal point, the denominator is 1,000, read "thousandths." Four digits to the right of the decimal point are read "ten-thousandths," five digits are read "hundred-thousandths," and so on.

To read a decimal fraction: Read the whole number (if any); next read "and" for the decimal point. Then read the portion after the decimal point as a whole number. Finally, read the name of the decimal place of the last digit on the right.

For example, 0.27, which is read "twenty-seven *hundredths,*" is written

$$0.27 = \frac{27}{100} \leftarrow \text{Hundred}$$

while 0.149, which is read "one hundred forty-nine *thousandths,*" is written

$$0.149 = \frac{149}{1{,}000} \leftarrow \text{Thousand}$$

The last digit in the number is the key to the denominator of the fraction.

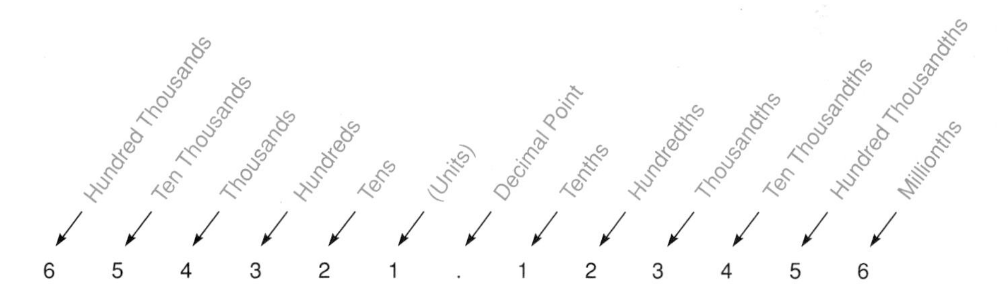

■ ***Example 1–2 A***

Write the following decimal numbers as fractions reduced to lowest terms.

1. 0.8 (read "eight *tenths*")

$$0.8 = \frac{8}{10} \leftarrow \text{Ten}$$

$$= \frac{4}{5} \quad \text{Reduce to lowest terms}$$

2. 0.57 (read "fifty-seven *hundredths*")

$$0.57 = \frac{57}{100} \leftarrow \text{Hundred}$$

3. 0.1234 (read "one thousand two hundred thirty-four *ten thousandths*")

$$0.1234 = \frac{1{,}234}{10{,}000} \leftarrow \text{Ten thousand}$$

$$= \frac{617}{5{,}000} \quad \text{Reduce to lowest terms}$$

Note A decimal number that is written as a fraction will reduce *only if* the numerator is divisible by 2 or 5. This was the case in examples 1 and 3.

▶ ***Quick check*** Write 0.42 as a fraction reduced to lowest terms. ■

Addition and subtraction of decimal numbers

To add or subtract decimal numbers, we place the numbers under one another so that the decimal points line up vertically and then proceed as in adding or subtracting whole numbers. The decimal point will appear in the answer directly below where it is lined up in the problem.

■ Example 1–2 B

Add or subtract the following numbers as indicated.

1. 5.67 + 32.046 + 251.7367 + 0.92

$$\begin{array}{r} 5.67 \\ 32.046 \\ 251.7367 \\ +\ 0.92 \\ \hline 290.3727 \end{array}$$

Decimal points aligned

Arrange numbers in columns

2. Subtract (a) 18.7 from 39.62, (b) 4.38 from 19.2

a.
$$\begin{array}{r} 3\overset{8}{\not 9}.62 \\ -18.7 \\ \hline 20.92 \end{array}$$

b.
$$\begin{array}{r} 1\overset{8}{\not 9}.\overset{1}{\not 2} \\ -\ 4.38 \\ \hline 14.82 \end{array}$$

3. What is the perimeter of the figure in the diagram? (Recall that the perimeter is the total distance around the figure.)

$$\begin{array}{r} 3.97 \text{ m} \\ 7.39 \text{ m} \\ 3.18 \text{ m} \\ +\ 7.83 \text{ m} \\ \hline 22.37 \text{ m} \end{array}$$

7.39 m
3.97 m
3.18 m
7.83 m

► ***Quick check*** Subtract 14.9 from 83.42 ■

To multiply decimal numbers

1. Multiply the numbers as if they are whole numbers (ignore the decimal points).
2. Count the number of decimal places in both factors. That is, count the number of digits to the right of the decimal point in each factor. This total is the number of decimal places the product must have.
3. Beginning at the right in the product, count off to the left the number of decimal places from step 2. Insert the decimal point. If necessary, zeros are inserted so there are enough decimal places.

■ Example 1–2 C

Multiply the following.

1. 2.36 × 0.403

$$\begin{array}{r} 2.36 \\ 0.403 \\ \hline 708 \\ 944 \\ \hline 0.95108 \end{array}$$

2.36 ← 2 decimal places
0.403 ← 3 decimal places (2 + 3 = 5)
0.95108 ← 5 decimal places

2. (18.14)(106.4)

```
     18.14 ←—— 2 decimal places
     106.4 ←—— 1 decimal place     (2 + 1 = 3)
      7256
    10884
   1814
 1,930.096 ←—— 3 decimal places
```

▶ ***Quick check*** (206.1)(9.36)

To divide decimal numbers, we must identify the divisor, the dividend, and the quotient in an indicated division.

```
      10 ←—— Quotient
  25)250 ←—— Dividend
   ↑
   └——————— Divisor
```

We now outline the procedure for dividing decimal numbers.

To divide decimal numbers

1. Change the *divisor* to a whole number by moving the decimal point to the *right* as many places as is necessary.
2. Move the decimal point in the *dividend* to the right this same number of places. If necessary, zeros are inserted so there are enough decimal places.
3. Insert the decimal point in the *quotient* directly above the new position of the decimal point in the dividend.
4. Divide as with whole numbers.

■ *Example 1–2 D*

Divide the following.

1. 360.5 ÷ 1.03

a. Write the problem 1.03)360.5

b. Move the decimal point *two* places to the right in 1.03 and 360.5

```
103)36,050. ←—— Zero inserted as placeholder
```

c. Now divide as with whole numbers.

```
       350.
103)36,050.
    30 9
     5 15
     5 15
        0
```

The quotient is 350.

2. If an automobile travels 429.76 miles and uses 15.8 gallons of gas, how many miles per gallon did the automobile achieve?

To determine the fuel economy, we divide the total number of miles traveled by the amount of gasoline used.

$429.76 \div 15.8$

```
        27.2
158.)4,297.6
     316
     1137
     1106
       316
       316
         0
```

The automobile achieved 27.2 miles per gallon.

▶ ***Quick check*** $4{,}950.3 \div 5.69$ ■

We can change a fraction into its decimal number equivalent using the following procedure.

To change a fraction to a decimal number

Divide the denominator into the numerator.

■ ***Example 1–2 E***

Convert each fraction to a decimal number.

1. $\frac{3}{4}$

We divide 3 by 4. To do this, we must add zero placeholders.

```
  0.75
4)3.00  ← Annex zeros
  2 8
    20
    20
     0
```

Thus $\frac{3}{4} = 0.75$

2. $\frac{1}{3}$

We divide $1 \div 3$.

```
        ┌── Continues indefinitely
  0.333̄
3)1.000  ← Add zeros
    9
    10
     9
    10
     9
    10
```

We can see that no matter how many zero placeholders we add, the quotient will continue to add digits of 3. This is called a **repeating decimal** (denoted by the bar placed over the last digit, or digits, that are repeating). Therefore, $\frac{1}{3} = 0.\overline{3}$. We can round a repeating decimal to as many places as are needed. We can say that $\frac{1}{3}$ is *approximately equal* to 0.333, denoted by

$$\frac{1}{3} \approx 0.333$$

▶ ***Quick check*** Convert $\frac{3}{8}$ to a decimal number. ■

Percent

We use decimal numbers extensively in our work with **percent.** The word percent means "per one hundred."

> Percent is defined to be parts per one hundred.

We use the symbol "%" to represent percent. Thus

3% means "three parts per one hundred"

or

3% means "three one hundredths."

From the above discussion,

$$3\% = \frac{3}{100} = 0.03$$

That is, we can write a percent as

1. a decimal number and
2. a fraction with denominator 100

■ ***Example 1–2 F***

Write each percent as a fraction and as a decimal number.

1. 39%
39% means "thirty-nine one hundredths."

$$39\% = \frac{39}{100} = 0.39$$

2. 123%
123% means "one hundred twenty-three one hundredths."

$$123\% = \frac{123}{100} = 1.23$$

▶ ***Quick check*** Write 241% as a fraction and as a decimal. ■

From these examples, we can see how to write a percent as a decimal number.

To write a percent as a decimal number

Move the decimal point two places to the *left* and drop the % symbol.

To write a fraction as a percent

Drop the % symbol and write the number over a denominator of 100.

To write a fraction or decimal number as a percent, we reverse the procedure.

To write a decimal number as a percent

Move the decimal point two places to the *right* and affix the % symbol.

To write a fraction as a percent

Find the decimal number equivalent of the fraction and change this decimal number to a percent.

■ *Example 1–2 G*

Write the following as decimal numbers, fractions, and percents.

1. 0.9

$0.9 = 90\%$ — Move the decimal point two places to the right and affix % symbol (Add a zero placeholder)

Since $0.9 = \frac{9}{10}$ — Write as a fraction

then $0.9 = \frac{9}{10} = 90\%$

2. 1.25

$1.25 = 125\%$ — Move the decimal point two places to the right and affix % symbol

Since $1.25 = \frac{125}{100}$ — Write as a fraction

$= \frac{5}{4}$ — Reduce to lowest terms

then $1.25 = \frac{5}{4} = 125\%$

3. $\frac{7}{8}$

Divide $7 \div 8$ to obtain the decimal equivalent. Doing this we find that

$$\frac{7}{8} = 0.875$$

Then $\frac{7}{8} = 0.875 = 87.5\%$ — Move the decimal point two places to the right and affix % symbol

▶ ***Quick check*** Write 1.75 as a fraction and a percent. ■

Percentage

When we find 60% of 500, we find the **percentage.** In the language of mathematics, "of" usually means the operation multiplication. Thus

$$60\% \text{ of } 500 \text{ means } 60\% \cdot 500$$

However, we cannot multiply 60% times 500. We must first change 60% to a decimal number (or a fraction) before we can perform the multiplication.

$$\begin{aligned} 60\% \text{ of } 500 &= 60\% \cdot 500 \\ &= 0.60 \cdot 500 && \text{Change 60\% to 0.60} \\ &= 300 && \text{Percentage} \end{aligned}$$

Therefore 60% · 500 = 300

(percent) · (base) (percentage)

Example 1–2 H

Find the following percentages.

1. 8% of 35

$8\% = 0.08$ — Change percent to a decimal number

$8\% \text{ of } 35 = 0.08 \cdot 35$ — Multiply
$= 2.8$

Thus 8% of 35 = 2.8

2. 224% of 50

$224\% = 2.24$ — Change percent to a decimal number

$224\% \text{ of } 50 = 2.24 \cdot 50$ — Multiply
$= 112$

Thus 224% of 50 = 112

3. $3\frac{1}{2}\%$ of 270

$3\frac{1}{2}\% = 3.5\%$ — $\frac{1}{2} = 0.5$ as a decimal number
$= 0.035$ — Change percent to a decimal number

$3\frac{1}{2}\% \text{ of } 270 = 0.035 \cdot 270$ — Multiply
$= 9.45$

Thus $3\frac{1}{2}\%$ of 270 = 9.45

▶ ***Quick check*** 236% of 20 ■

Mastery points

Can you

- Write decimal numbers as fractions?
- Add and subtract decimal numbers?
- Multiply and divide decimal numbers?
- Write fractions as decimal numbers?
- Change a percent to a decimal number?
- Change a decimal number to a percent?
- Change a fraction to a percent?
- Change a percent to a fraction?
- Find the percentage?

Exercise 1–2

Write each decimal number as a fraction reduced to lowest terms. See example 1–2 A.

Example 0.42

Solution 0.42 is read "forty-two *hundredths*."

$$0.42 = \frac{42}{100} \leftarrow \text{Hundred}$$
$$= \frac{21}{50} \quad \text{Reduce to lowest terms}$$

1. 0.4 **2.** 0.8 **3.** 0.15 **4.** 0.36
5. 0.125 **6.** 0.248 **7.** 0.875 **8.** 0.625

Add or subtract the following as indicated. See example 1–2 B.

Example Subtract 14.9 from 83.42

Solution We want 83.42 − 14.9

$$\begin{array}{r} \overset{7\,2}{83}.42 \\ -14.9 \\ \hline 68.52 \end{array}$$

9. 6.8 + 0.354 + 2.78 + 7.083 + 2.002
10. 4.76 + 0.573 + 3.57 + 40.09 + 13
11. 8.0007 + 360.01 + 25.72 + 6.362 + 140.2
12. 7.0001 + 8 + 7.067 + 803.1 + 5.25
13. 10.03 + 3.113 + 0.3342 + 0.0763 + 0.005
14. 27.376 − 14.007
15. 367.0076 − 210.02
16. 836 − 0.367
17. 1.07 − 0.00036
18. 4,563.2 − 274.063

Multiply the following. See example 1–2 C.

Example (206.1)(9.36)

Solution

$$\begin{array}{r} 206.1 \\ 9.36 \\ \hline 12366 \\ 6183 \\ 18549 \\ \hline 1{,}929.096 \end{array}$$

206.1 ← 1 decimal place
9.36 ← 2 decimal places (1 + 2 = 3)
1,929.096 ← 3 decimal places

19. (7.006)(1.36) **20.** (42.6)(73) **21.** (56.37)(0.0076) **22.** 703.6 × 1.7
23. 30.0303 × 0.030303 **24.** 2.456 × 0.00012

Divide the following. See example 1–2 D.

Example 4,950.3 ÷ 5.69

Solution 1. Write the problem $5.69\overline{)4,950.3}$
2. Move the decimal point *two* places in 5.69 and 4,950.3

$569\overline{)495,030.}$ ← Add zero placeholder

3. Divide as whole numbers.

$$\begin{array}{r} 870. \\ 569\overline{)495,030.} \\ \underline{4552} \\ 3983 \\ \underline{3983} \\ 0 \end{array}$$ ← Add zero placeholder

Thus 4,950.3 ÷ 5.69 = 870

25. 0.84 ÷ 0.7 **26.** 0.525 ÷ 0.5 **27.** 10.4 ÷ 0.26 **[28.]** 21.681 ÷ 8.03

29. 6,125.1 ÷ 60.05 **30.** 166.279 ÷ 64.7 **31.** 31.50 ÷ 0.0126 **32.** 2.9868 ÷ 0.057

Convert each fraction to a decimal number. See example 1–2 E.

Example $\frac{3}{8}$

Solution We divide 3 ÷ 8, adding zero placeholders where necessary.

$$\begin{array}{r} 0.375 \\ 8\overline{)3.000} \\ \underline{2\,4} \\ 60 \\ \underline{56} \\ 40 \\ \underline{40} \\ 0 \end{array}$$ ← Zero placeholders

The decimal equivalent of $\frac{3}{8}$ is 0.375

[33.] $\frac{3}{20}$ **34.** $\frac{5}{8}$ **35.** $\frac{13}{20}$ **36.** $\frac{17}{50}$ **[37.]** $\frac{2}{9}$ **38.** $\frac{5}{9}$

[39.] Heating oil costs 89.9 cents per gallon. What is the total cost of 14.36 gallons, correct to the nearest cent?

40. A carpenter has three pieces of wood which are 24.5 inches, 35.25 inches, and 62.375 inches long, respectively. How many inches of wood does she have all together?

41. A wood craftsman has 74.75 inches of a particular stock. He needs 5.75 inches of the stock to carve out a cardinal bird. How many cardinals can he make?

42. A rectangular field is 21.3 yards long and 15.75 yards wide. Find the area (length × width) of the field.

43. The 500-meter speed-skating event was won in a time of 43.33 seconds in the 1972 olympics. The winning time in 1976 was 42.76 seconds. How much faster was the 1976 time?

44. A student bought a book for $21.68. If she gave the cashier $25, how much change did she receive?

45. On a 4-day trip, the Adams family used 32.5 gallons, 28.36 gallons, 41.87 gallons, and 19.55 gallons of gasoline. How many gallons of gasoline did they use all together?

46. An airline pilot flew distances of 210.6 kilometers, 504.3 kilometers, 319.6 kilometers, 780.32 kilometers, and 421.75 kilometers on five flights. How many kilometers did he fly all together?

47. A rectangular field is 43.3 yards long and 25.34 yards wide. Find the area (length × width) of the field.

48. If a cubic foot of water weighs 62.5 pounds, how many pounds of water are there in a tank containing 10.4 cubic feet?

Write each percent as a fraction and as a decimal number. See example 1–2 F.

Example 241%

Solution 241% means "two hundred forty-one one hundredths."

$$241\% = \frac{241}{100} = 2.41$$

49. 5% **50.** 1% **51.** 12% **52.** 64%

53. 135% **54.** 150% **55.** 325% **56.** 570%

Write exercises 57–62 as a fraction and a percent and exercises 63–66 as a decimal and a percent. See example 1–2 G.

Example Write 1.75 as a fraction and a percent.

Solution $1.75 = 175\%$ Move decimal point two places to the right and affix % symbol

$$1.75 = \frac{175}{100}$$ Write as a fraction

$$= \frac{7}{4}$$ Reduce fraction to lowest terms

Thus $1.75 = \frac{7}{4} = 175\%$.

57. 0.8 **58.** 0.9 **59.** 0.54 **60.** 0.80 **61.** 1.15

62. 2.40 **63.** $\frac{3}{4}$ **64.** $\frac{5}{2}$ **65.** $\frac{3}{8}$ **66.** $\frac{5}{8}$

Find the following percentages. See example 1–2 H.

Example 236% of 20

Solution $236\% = 2.36$ Change percent to a decimal number

$236\% \text{ of } 20 = 2.36 \cdot 20$ Multiply

$= 47.2$

Thus 236% of 20 = 47.2

67. 5% of 40 **68.** 8% of 45 **69.** 26% of 130

70. 78% of 900 **71.** 110% of 500 **72.** 240% of 60

73. City Bank pays 5.7% interest per year on its savings accounts. What is the annual interest on a savings account that has $4,500? (*Hint:* Annual Interest = Percent · Amount in savings.)

74. The sales tax on retail sales in Michigan is 4%. How much sales tax does John pay on a purchase of $250?

75. If Jane pays 5% of her weekly salary in state income tax, how much state tax does she pay if her weekly salary is $460?

76. A local retailer predicts his profit in a given year will be 116% of the previous year. What is his predicted profit for this year if last year's profit was $42,500?

77. The local shoe store is giving a 25% discount on clearance items. How much discount is there on a pair of shoes costing $34? What is the price of the shoes *after* the discount excluding sales tax?

78. A company charges $4\frac{1}{2}$% shipping and handling charges on all items shipped. What are the shipping and handling charges on goods that cost $70? What is the total cost for the goods?

79. A bottle of solution is 4% salt. How much salt is there in a 24-fluidounce bottle of solution?

80. Self-employed persons must pay a Social Security tax of 12%. What is the Social Security tax on earnings of $25,000? If the person is in the 28% federal income tax bracket, how much federal income tax does the person pay?

1–3 ■ The set of real numbers and the real number line

Set symbolism

Algebra is often referred to as "a generalized arithmetic." The operations of arithmetic and algebra differ only in respect to the symbols we use in working with each of them. Therefore, we will begin our study of algebra by dealing first with numbers and their properties. From this work, we will see algebra develop naturally as a generalized arithmetic.

To begin our study, we will start with a very simple, but important, mathematical concept—the idea of the **set.*** **A set is any collection of things.** This may be a collection of books, people, coins, golf clubs, and so on. In mathematics, we use the idea of a set primarily to denote a group of numbers. Any one of the things that belong to the set is called a **member** or an **element** of the set. One way we write a set is by listing the elements, separating them by commas, and including this listing within a pair of braces, { }.

■ ***Example 1–3 A***

1. Using set notation, write the set of months that have exactly 30 days.
{April,June,September,November}

2. Using set notation, write the set of seasons of the year.
{spring,summer,fall,winter}

▶ ***Quick check*** Write the set of letters in the word "mathematics."
Write the set of odd numbers between 5 and 10. ■

Note When we form a set, the elements within the set are never repeated and they can appear in any order.

*Georg Cantor (1845–1918) is credited with the development of the ideas of set theory. He described a set as a grouping together of single objects into a whole.

We use capital letters $A, B, C, D,$ and so on, to represent a set. The symbol used to show that an element belongs to a set is the symbol $\in$, which we read "is an element of" or "is a member of." Consider the set $A = \{1,2,3,4\}$, which is read "the set A whose elements are 1, 2, 3, and 4." If we want to say that 2 is an element of the set A, this can be written symbolically as $2 \in A$.

A slash mark is often used in mathematics to negate a given symbol. Therefore if $\in$ means "is an element of the set," then $\notin$ would mean "is *not* an element of the set." To express the fact that 7 is not an element of set A, we could write $7 \notin A$.

Subset

Suppose that P is the set of people in a class and M is the set of men in the same class. It is obvious that the members of M are also members of P, so we say that M is a **subset** of P.

Definition

The set A is a subset of the set B if every element in A is also an element of B.

The symbol for subset is $\subseteq$, which we read "is a subset of." Therefore $A \subseteq B$ is read "the set A is a subset of the set B." Consider the following sets: $A = \{1,2,4\}$, $B = \{1,2,3,4,5\}$, and $C = \{1,3,5,7\}$. We observe that $A \subseteq B$ since every element in A is also an element of B. $C \nsubseteq B$ is read "the set C is *not* a subset of the set B," because not every element of C is an element of B. The set C contains the element 7, which is not an element of the set B.

Natural numbers and whole numbers

The most basic use of our number system is that of counting. We use 1, 2, 3, 4, 5, and so on, as symbols to represent the **natural** or **counting numbers.** The set of natural numbers will be denoted by N, as follows:

$$N = \{1,2,3,4,5, \cdots\}$$

The three dots tell us to continue this counting pattern indefinitely.

If we include 0 with the set of natural numbers, we have the set of **whole numbers,** W.

$$W = \{0,1,2,3,4,5, \cdots\}$$

We can use the set of whole numbers to represent physical quantities such as profit (100 dollars), room temperature (72 degrees), and distance (1,250 feet above sea level). However, with the set of whole numbers, we are not able to represent such things as losses of money, temperatures below zero, and distances below sea level. Therefore to represent such situations, we define a new set of numbers called the set of **integers.** We will denote this set by J.

Integers

We shall start by giving the natural numbers another name, the **positive integers.** We then form the *opposites,* or *negatives,* of the positive integers as follows: $-1, -2, -3, \cdots$. Combining the positive integers, the negative integers, and 0, we have the set of integers, J.

$$J = \{\cdots,-3,-2,-1,0,1,2,3, \cdots\}$$

■ ***Example 1–3 B***

Use integers to represent each of the following.

1. Bromine melts at seven degrees below zero Celsius.
   Answer: -7 degrees Celsius

   Less than zero is a negative value

2. Bromine boils at fifty-nine degrees Celsius.
   Answer: 59 degrees Celsius

   Greater than zero is a positive value

▶ ***Quick check*** Use integers to represent a debt of nine dollars. ■

Rational numbers

The set of integers is sufficient to represent many physical situations, but it is unable to provide an answer for the following problem. If we want to determine the miles per gallon (mpg) that our car is getting, and we find that 8 gallons of gas enable us to travel 325 miles, then our miles per gallon can be computed by dividing the number of miles by the number of gallons used.

$$\frac{325 \text{ miles}}{8 \text{ gallons}} = \frac{325}{8} \text{ mpg} = 40\frac{5}{8} \text{ mpg}$$

This value is not in the set of integers. Therefore to represent such situations, we define a new set of numbers called the set of **rational numbers,** which is denoted by Q. Recall that a quotient is an answer to a division problem. Hence $\frac{325}{8}$ is called a quotient of two integers since 325 and 8 are both integers.

Definition

A rational number is any number that can be expressed as a quotient of two integers in which the divisor is not zero.*

Other examples of rational numbers would be

$$\frac{2}{3}, -\frac{1}{2}, \frac{6}{1}, \frac{19}{5}, -\frac{23}{7}, \frac{15}{3}, \frac{0}{8}, \frac{-5}{1}$$

The decimal representation of a rational number is either a terminating or a repeating decimal. Some examples of terminating or repeating decimals are

$$\frac{1}{2} = 0.5, \quad \frac{1}{3} = 0.\overline{3}, \quad -\frac{1}{6} = -0.1\overline{6}, \quad -\frac{5}{4} = -1.25, \quad \frac{4}{33} = 0.\overline{12}$$

where a bar placed over a number or groups of numbers indicates that the number(s) repeat indefinitely.

Irrational numbers

At this point, we might feel that we now have numbers that will answer all possible physical situations. However that is not the case. Consider the following question.

What is the exact length of a side of a square whose area is 10 square units (figure 1–1)? To be able to answer this question, we need to find that number such that when it is multiplied with itself, the product is 10. If we use 3.16, the result would be $(3.16) \times (3.16) = 9.9856$. This is close to 10 but is not equal to 10.

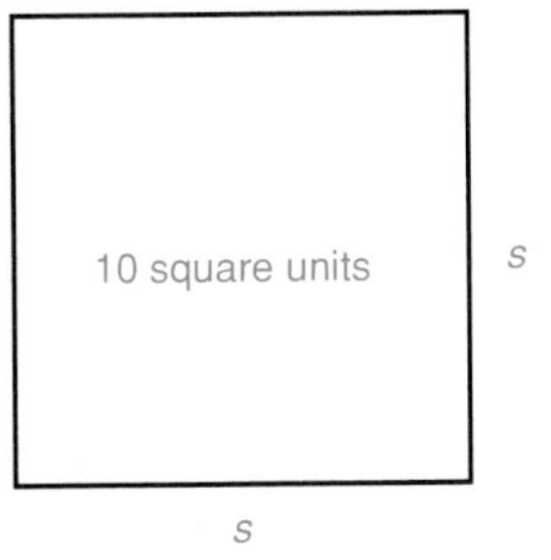

10 square units

s

s

Figure 1–1

*Division involving zero will be discussed in section 1–7.

It can be shown that there is no rational value that when multiplied with itself has a product of 10. The answer to this question and many others cannot be found in the set of rational numbers. In chapter 9, we will see that the answer to this question is $\sqrt{10}$ (read "the square root of 10"). Such numbers that cannot be expressed as the quotient of two integers belong to the set of **irrational numbers,** which is denoted by H

Real numbers

Since a rational number can be expressed as the quotient of two integers and an irrational number cannot, we should realize that a number can be rational or irrational, but it cannot be both. The set that contains all of the rational numbers and all of the irrational numbers is called the set of **real numbers,** which is denoted by R. Whenever we encounter a problem and a specific set of numbers is not indicated, it will be understood that we are dealing with real numbers.

All of the sets that we have examined thus far are subsets of the set of real numbers. Figure 1–2 shows the relationship.

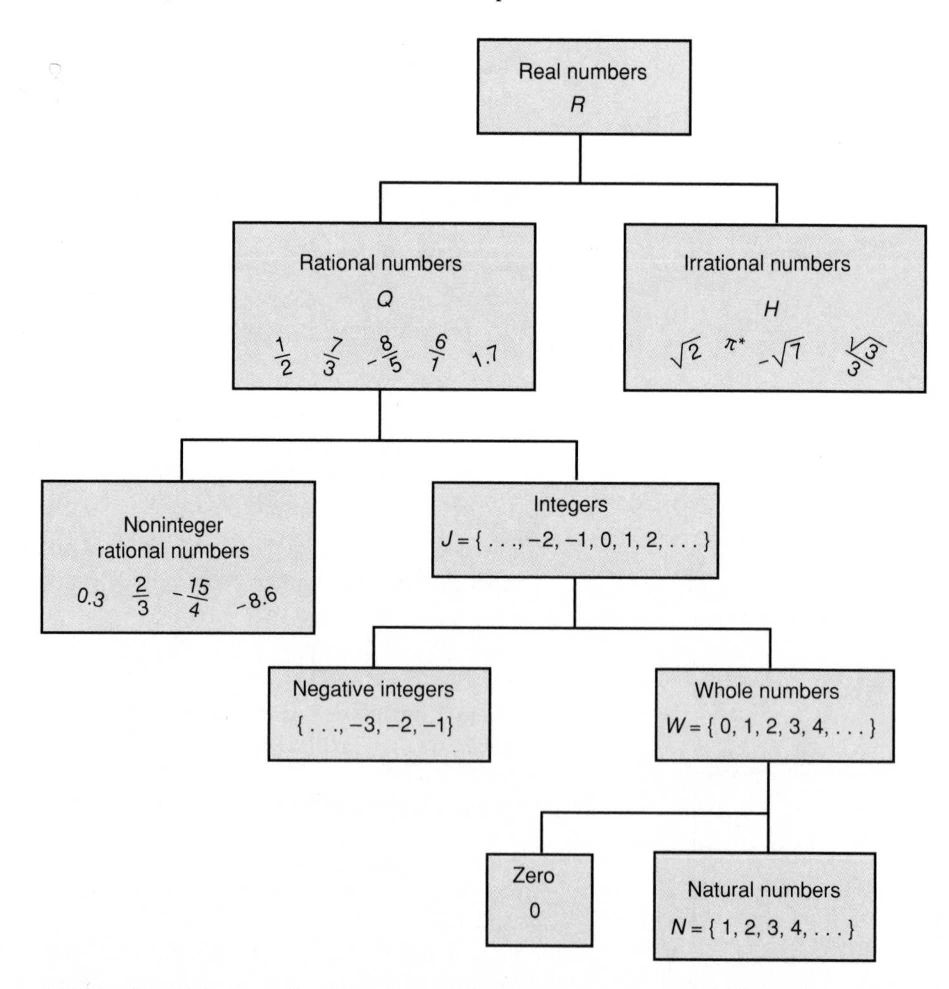

*π (pi) is the distance around a circle (circumference) divided by the distance across the circle through its center (diameter). Common approximations for π are 3.14 and $\frac{22}{7}$.

Figure 1–2

The real number line

To picture the set of real numbers, we shall use a real number line. We begin by drawing a line where the arrowhead at each end of the line indicates that the line continues on indefinitely in both directions. Next we choose any point on the line to represent 0. This point is called the **origin** of the number line. Numbers to the right of zero are positive and to the left of zero are negative (figure 1–3).

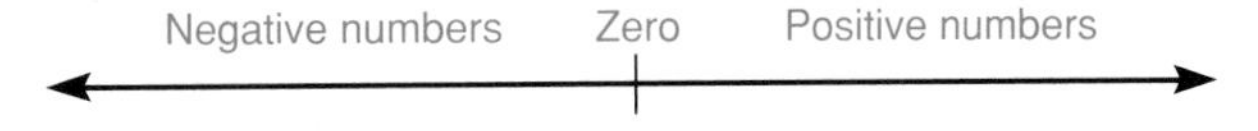

Figure 1–3

Any real number can now be located on the number line. Consider the number line in figure 1–4.

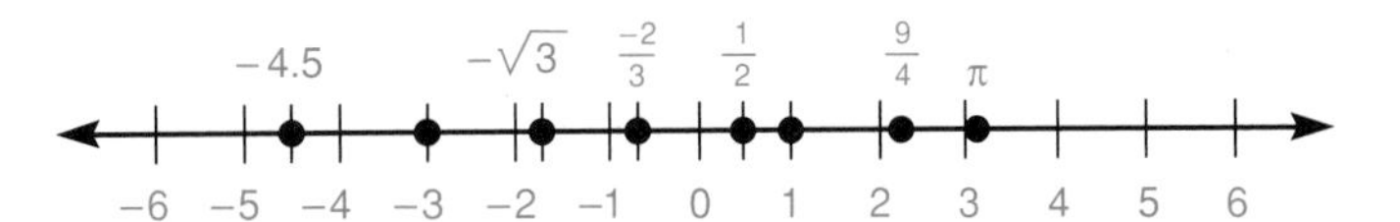

Figure 1–4

The number that is associated with each point on the line is called the **coordinate** of the point. The solid circle that is associated with each number is called the **graph** of that number. In figure 1–4, the numbers -4.5, -3, $-\sqrt{3}$, $\frac{-2}{3}$, $\frac{1}{2}$, 1, $\frac{9}{4}$, and π are the *coordinates* of the points indicated on the line by solid circles. The solid circles are the *graphs* of these numbers.

Note The coordinates $-\sqrt{3}$ and π represent irrational numbers. To graph these points, we would find a rational approximation using a calculator. $-\sqrt{3} \approx -1.732$, $\pi \approx 3.142$ ($\approx$ is read "is approximately equal to").

The direction in which we move on the number line is also important. If we move to the right, we are moving in a positive direction and the numbers are *increasing*. If we move to the left, we are moving in a negative direction and the numbers are *decreasing* (figure 1–5).

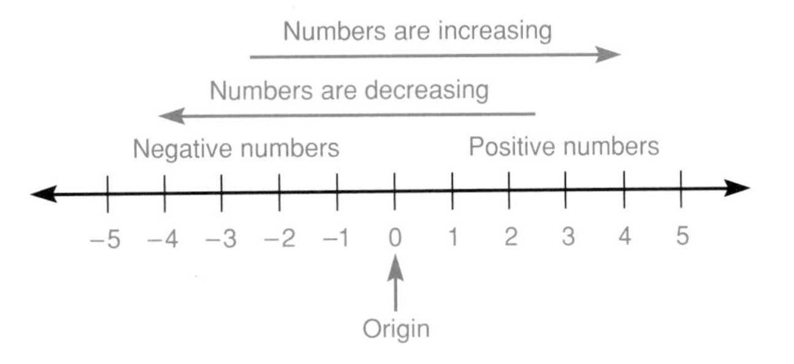

Figure 1–5

Order on the number line

So that the discussion can be more general, we will now introduce the concept of a **variable. A variable is a symbol (generally a lowercase letter) that represents an unspecified number.** A variable holds a position for a number.

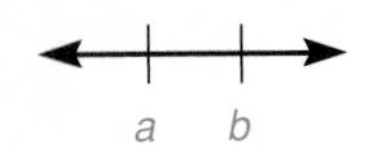

Figure 1–6

If we choose any two points on the number line and represent them by a and b, where a and b represent some *unspecified* numbers, we observe that there is an *order* relationship between a and b (figure 1–6). Since the point associated with a is to the *left* of the point associated with b, we say that a is **less than** b, which in symbols is $a < b$. We might also say that b is **greater than** a, which in symbols is $b > a$. The symbols $<$ (less than) and $>$ (greater than) are inequality symbols called **strict inequalities** and they denote an **order relationship** between numbers.

■ Example 1–3 C

Replace the ? with the proper inequality symbol ($<$ or $>$).

1. $-5 \ ? \ -3$ −6 −5 −4 −3 −2   Answer: $-5 < -3$ Because -5 is to the left of -3

Note If we have difficulty deciding which of two numbers is greater, think of the numbers as representing temperature readings. The -5 would be thought of as 5 degrees below zero and the -3 would be 3 degrees below zero. It is easy to realize that -3 is the greater (warmer) temperature and the inequality would be $-5 < -3$.

2. $0 \ ? \ -4$ −5 −4 −3 −2 −1 0 1   Answer: $0 > -4$ Because 0 is to the right of 4

3. $0 \ ? \ 3$ −1 0 1 2 3 4   Answer: $0 < 3$ Because 0 is to the left of 3

Note No matter which inequality symbol we use, the arrow *always* points at the lesser number.

▶ ***Quick check*** Replace ? with $<$ or $>$.

$2 \ ? \ 4$

$-3 \ ? \ -6$

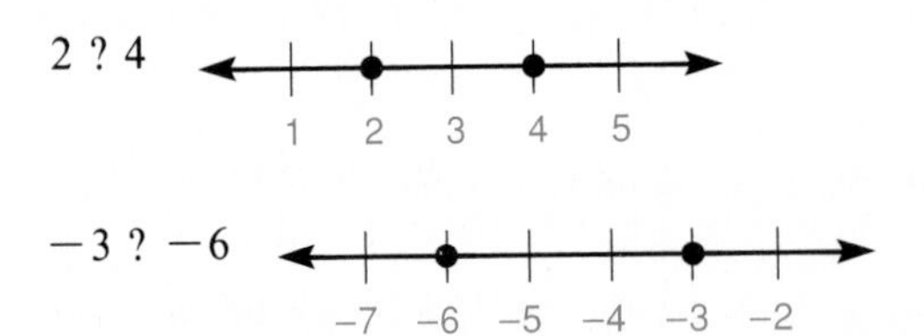

■

There are two other inequality symbols, called **weak inequalities.** They are **less than or equal to,** $\leq$, and **greater than or equal to,** $\geq$. The weak inequality symbol $\geq$, greater than or equal to, denotes that one number could either be greater than a second number or equal to that second number. The inequality $x \geq 3$ means that x is *at least* 3. That is, x represents all numbers that are 3 or more.

The other weak inequality symbol, $\leq$, less than or equal to, denotes that one number could either be less than a second number or equal to that second number. The inequality $x \leq 5$ means that x is *at most* 5. That is, x represents all numbers that are 5 or less.

Absolute value

As we study the number line, we observe a very useful property called **symmetry.** The numbers are symmetrical with respect to the origin. That is, if we go four units to the right of 0, we come to the number 4. If we go four units to the left of 0, we come to the *opposite* of 4, which is -4 (figure 1–7).

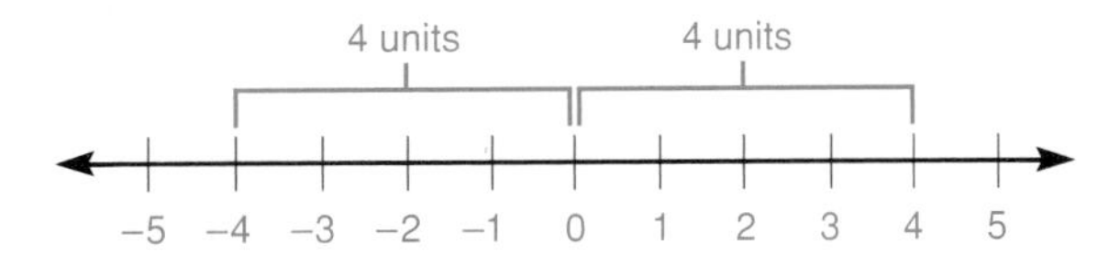

Figure 1–7

Each of these numbers is four units away from the origin. How far a given number is from the origin is called the **absolute value** of the number. **The absolute value of a number is the undirected distance that the number is from the origin.** The symbol for absolute value is $|\ \ |$.

■ ***Example 1–3 D***

Evaluate the following expressions.

1. $|-7| = 7$ **2.** $|3| = 3$ **3.** $|0| = 0$

4. $\left|\frac{2}{3}\right| = \frac{2}{3}$ **5.** $|-3.7| = 3.7$ **6.** $-|-6| = -6$

Note The absolute value of a number is *never* negative; that is, $|x| \geq 0$, for every $x \in R$, however the absolute value bars are only applied to the symbol contained within them. The $-$ sign in front of the absolute value bars is not affected by the absolute value bars. Example 6 would be read "the opposite of the absolute value of -6," and the answer is -6.

▶ ***Quick check*** Evaluate the following expressions.
$|-3|$; $|12|$; $-|-4|$ ■

Visualizing our number system on the number line demonstrates the fact that each number possesses two important properties.

1. The sign of the number denotes a direction from zero. (The absence of a sign indicates a positive number.)
2. The absolute value represents a distance from zero.

Mastery points

Can you

- Write sets?
- Draw a number line?
- Graph a number on the number line?
- Tell which of two real numbers is greater?
- Find absolute values?
- Approximate the value of the coordinate of a graph on the number line?

Exercise 1–3

Write each set by listing the elements. See example 1–3 A.

Examples The letters in the word "mathematics"

Solutions {m,a,t,h,e,i,c,s} The letters *m*, *a*, and *t* are not repeated within the set

The odd numbers between 5 and 10

{7,9} The word between means that we do not include the 5 or 10

1. The days of the week
2. The days of the week that begin with "T"
3. The first 3 months of the year
4. The months of the year that begin with "J"
5. The months with 31 days
6. The letters in the word "repeat"
7. The letters in the word "algebra"
8. The letters in the word "elementary"
9. The letters in the word "intermediate"
10. The even numbers between 5 and 11
11. The odd numbers between 2 and 10
12. The months of the year that begin with "A"
13. The days of the week that begin with "S"
14. The months of the year that begin with "M"

Use integers to represent each of the following. See example 1–3 B.

Example A debt of nine dollars

Solution -9 dollars A debt is a negative value

15. Ten dollars overdrawn in a checking account; 150 dollars in a savings account
16. Mercury's melting point is 39 degrees below zero Celsius. Its boiling point is 357 degrees Celsius.
17. A loss of 10 yards on a football play; a gain of 16 yards
18. Mt. Everest rises 29,028 feet above sea level; the Dead Sea has a depth of 1,290 feet below sea level.
19. The Dow Jones Industrial Stock Average fell 14 points; it rose 8 points.
20. Hydrogen's melting point is 259 degrees below zero Celsius. Water boils at 100 degrees Celsius. Water freezes at zero degrees Celsius.

Plot the graph of the following numbers, using a different number line for each exercise. See figure 1–4.

Example $-3, -2, 1, 1\frac{1}{2}, 3$

Solution

21. $-3, -1, \frac{1}{2}, 3, 4$
22. $-2, -1, \frac{3}{4}, 2, 4$
23. $-5, -2, 0, 1, 3\frac{1}{2}$
24. $-1, 0, \frac{2}{3}, 1, 3$
25. $-4, -2, 1, 2\frac{1}{2}, 4$
26. $-3, -1, \frac{1}{2}, 1, 2$
27. $-5, -1, 0, \frac{1}{2}, \sqrt{2}, 6$
28. $-4, -\sqrt{4}, \frac{1}{2}, \sqrt{3}, \pi$

Approximate the values of the set of coordinates of the graphs on the following number lines to the nearest $\frac{1}{4}$ of a unit. See figure 1–4.

Example

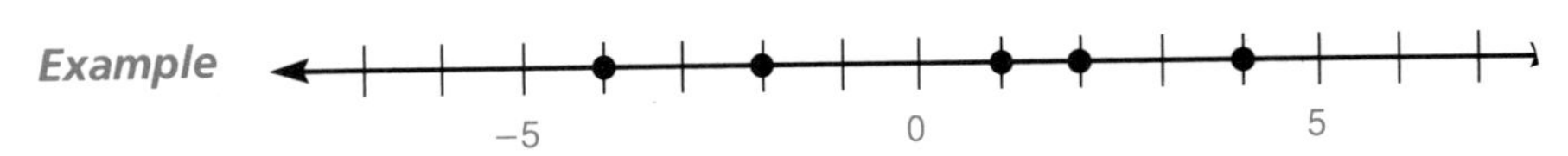

Solution $-4, -2, 1, 2, 4$

29.

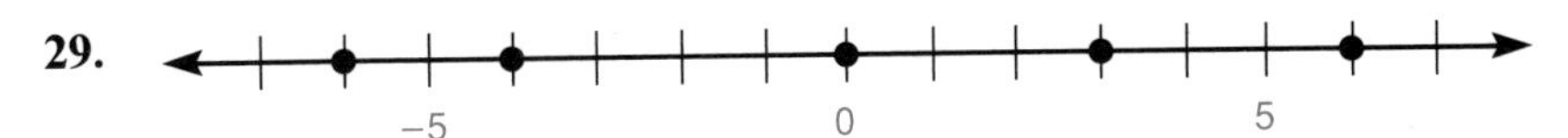

30.

31.

−10 −5 0

32.

−10 −5 0

33.

0 5 10

34.

−10 −5 0 5 10

35.

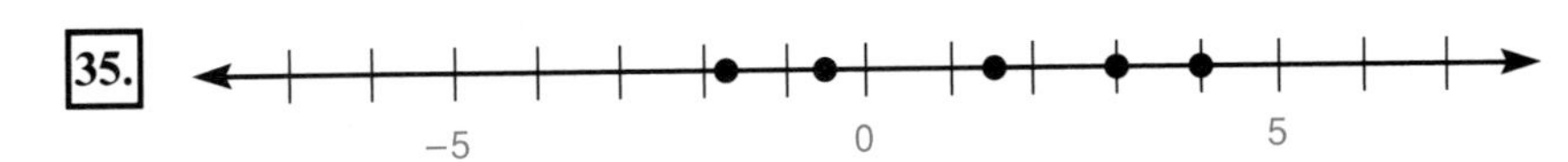

36.

−5 0 5

Replace the ? with the proper inequality symbol, $<$ or $>$. See example 1–3 C.

Examples	$2 \ ? \ 4$		$-3 \ ? \ -6$	
Solutions	$2 < 4$	Because 2 is to the left of 4 on the number line	$-3 > -6$	Because -3 is to the right of -6 on the number line

37. $4 \ ? \ 8$ **38.** $6 \ ? \ 3$ **39.** $9 \ ? \ 2$ **40.** $-2 \ ? \ -4$ **41.** $-3 \ ? \ -8$

42. $-9 \ ? \ -6$ **43.** $-10 \ ? \ -5$ **44.** $0 \ ? \ 2$ **45.** $0 \ ? \ 4$ **46.** $-3 \ ? \ 0$

47. $0 \ ? \ -6$

Evaluate the following expressions. See example 1–3 D.

Examples	$\lvert -3 \rvert$		$\lvert 12 \rvert$		$-\lvert -4 \rvert$	
Solutions	3	-3 is 3 units from the origin	12	12 is 12 units from the origin	-4	-4 is 4 units from the origin. The problem is to find the opposite of the absolute value

48. $\lvert 0 \rvert$ **49.** $\lvert 2 \rvert$ **50.** $\lvert 8 \rvert$ **51.** $\lvert -5 \rvert$ **52.** $\lvert -7 \rvert$

53. $|4|$ **54.** $\left|\frac{2}{3}\right|$ **55.** $\left|-\frac{1}{2}\right|$ **56.** $\left|-\frac{3}{4}\right|$ **57.** $\left|1\frac{1}{2}\right|$

58. $-\left|-2\frac{3}{4}\right|$ **59.** $-\left|\frac{5}{8}\right|$ **60.** $-|-2|$ **61.** $-|6|$

Replace the ? with the proper inequality symbol, $<$ or $>$. See examples 1–3 C and D.

Examples $|-6| \ ? \ |-3|$ $|4| \ ? \ |-7|$

Solutions $6 \ ? \ 3$ $|-6|$ is 6 and $|-3|$ is 3

$6 > 3$ 6 is to the right of 3 on the number line

then $|-6| > |-3|$

$4 \ ? \ 7$ $|4|$ is 4 and $|-7|$ is 7

$4 < 7$ 4 is to the left of 7 on the number line

then $|4| < |-7|$

62. $|-2| \ ? \ |-4|$ **63.** $|5| \ ? \ |-7|$ **64.** $|-3| \ ? \ |-4|$ **65.** $|0| \ ? \ |-2|$

66. $|-3| \ ? \ |4|$ **67.** $|-8| \ ? \ |-5|$ **68.** $|-9| \ ? \ |7|$ **69.** $|-6| \ ? \ |-2|$

70. $|-5| \ ? \ 3$ **71.** $7 \ ? \ |-2|$ **72.** $4 \ ? \ |-8|$ **73.** $|-4| \ ? \ 6$

Use absolute value to write each of the following. See figure 1–7.

Example The distance between -8 and 0

Solution $|-8|$ The absolute value of a number is the distance from the number to the origin.

74. The distance between 14 and 0

75. The distance between -27 and 0

76. The distance between 18 and 0

77. The distance between -9 and 0

78. The distance between -19 and 0

1–4 ■ Addition of real numbers

Addition of two positive numbers

When we perform the operations of addition and subtraction with integers, we will refer to this as operations with **signed numbers.** We use the minus sign ($-$) to indicate a negative number and the plus sign ($+$) to indicate a positive number. We should realize that the minus sign is identical to the symbol used for subtraction, and the plus sign is identical to the symbol used for addition. The meanings of these symbols will depend on their use in the context of the problem. In the case of a positive number, the plus sign need only be used if we wish to emphasize the fact that the number is positive. *When there is no sign, the number is understood to be positive.*

To visualize the idea of addition of signed numbers, we will use the number line to represent a checking account in which the origin represents a zero balance. We will let moves in the positive direction represent deposits and moves in the negative direction represent checks that we write, withdrawals. If we have a zero balance and deposit 5 dollars and 4 dollars, represented by $(+5)$ and $(+4)$, the balance in the account would be as shown in figure 1–8.

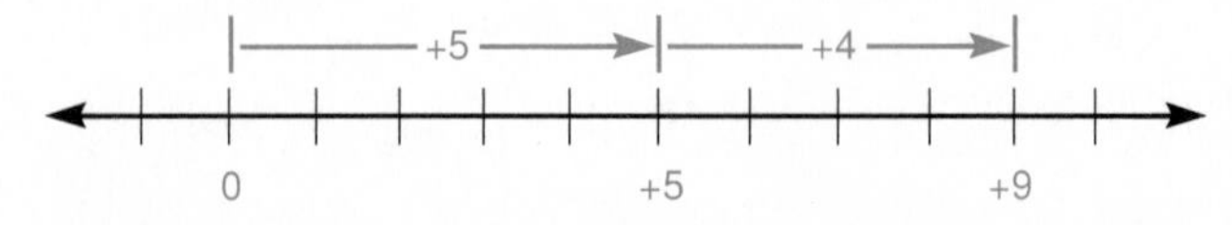

Figure 1–8

Performing the addition, we have $(+5) + (+4) = (+9)$. Notice that the sum, $(+9)$, has the same sign as the 5 and the 4.

So the discussion can be more general, we will use variables to state the process of addition on the number line. Recall that a variable represents an unspecified number. It is a placeholder for a number.

Addition on the number line

To add a and b, that is $a + b$, we locate a on the number line and move from there according to b.

1. If b is positive, we move to the right b units.
2. If b is negative, we move to the left the absolute value of b units.
3. If b is 0, we stay at a.

Example 1–4 A

Add the following numbers.

1. $(+4) + (+5) = +9$ — 4 and 5 are called addends, 9 is the sum

2. $(+3) + (+8) = +11$ — The sum of 3 and 8 is 11

3. $(+6) + (0) = +6$ — 6 plus 0 equals 6 ■

We have used the plus (+) sign in front of a number to emphasize the fact that the number was positive. In future examples, we will omit the plus sign from a positive number and it will be understood that 3 means $+3$.

From example 3, we see that when zero and a given number are added, the sum is the given number. For this reason, zero is called the **identity element of addition.** We now state this property.

Identity property of addition

For every real number a,

$$a + 0 = 0 + a = a$$

Concept

Adding zero to a number leaves the number unchanged.

Observe that our example in figure 1–8 and example 1 in example 1–4 A are the same addition problem with the order of the numbers reversed. That is, $(+5) + (+4) = +9$ and $(+4) + (+5) = +9$. This observation illustrates an important mathematical principle called the **commutative property of addition.**

Commutative property of addition

For every real number a and b,

$$a + b = b + a$$

Concept

This property says that when we are *adding* numbers, changing the order in which the numbers are added will not change the answer (sum).

Addition of two negative numbers

We now examine addition of two negative numbers. If we have a zero balance and write a check for 6 dollars, expressed as (-6), and another for 5 dollars, expressed as (-5), the loss to our checking account would be as shown in figure 1–9.

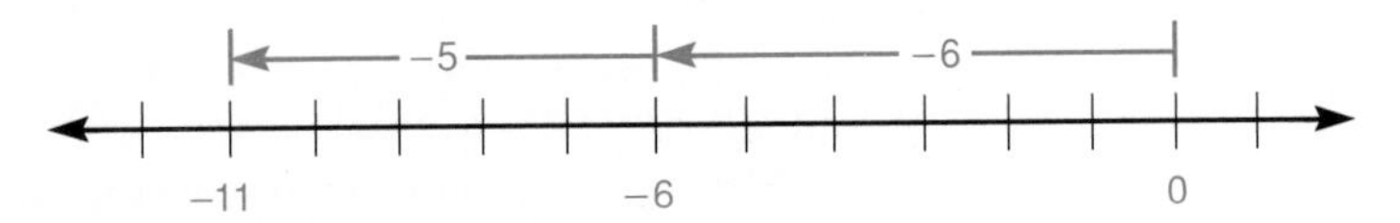

Figure 1–9

Our total withdrawal would be (-6) dollars $+$ (-5) dollars $=$ -11 dollars.

We can summarize what we have done by saying: *When we add two negative numbers, we add their absolute values and prefix the sum with their common sign,* $-$.

■ ***Example 1–4 B***

Add the following numbers.

1. $(-2) + (-7) = -9$

2. $(-9) + (-4) = -13$

3. $(-20) + (-30) = -50$

4. $(-6) + (-11) = -17$

▶ ***Quick check*** $(-5) + (-7)$ ■

We see from our examples that when we add two signed numbers and their signs are the same, we add their absolute values and prefix the sum with their common sign.

Addition of two numbers with different signs

To consider the addition of two numbers with different signs, we again refer to our checking account. Suppose we make a deposit to and a withdrawal from our checking account. If we deposit more money than we withdraw, we will have a positive balance in our account. For example, if we have a zero balance and deposit 15 dollars, represented by $(+15)$ dollars, and withdraw 10 dollars, represented by (-10) dollars, the result to our checking account will be as shown in figure 1–10.

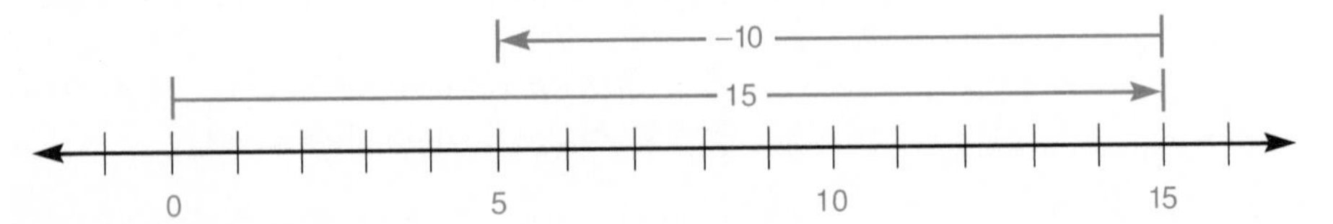

Figure 1–10

The balance in the checking account would be $(+15)$ dollars $+$ (-10) dollars $=$ $(+5)$ dollars.

We see that our answer is the absolute value of the difference of 15 and 10, prefixed by the sign of the number with the greater absolute value, $(+15)$.

Consider a second example in which we have a zero balance and deposit 10 dollars, $(+10)$ dollars, and write a check for 15 dollars, (-15) dollars. The result to our checking account this time would be as shown in figure 1–11.

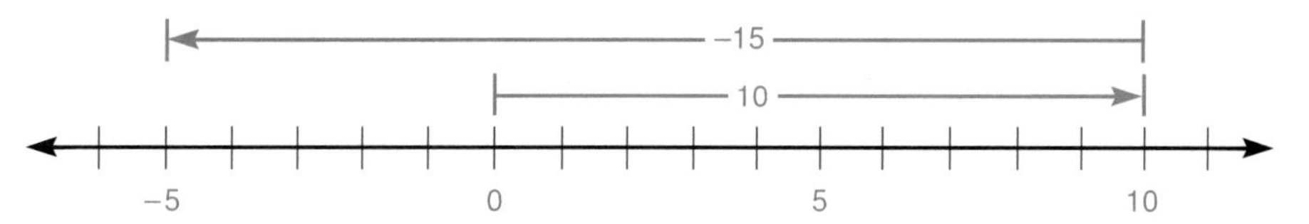

Figure 1–11

The balance would be $(+10)$ dollars $+$ (-15) dollars $=$ (-5) dollars.

We again see that our answer is the absolute value of the difference of the two numbers prefixed by the sign of the number with the greater absolute value, (-15).

We now summarize the procedure for adding two numbers of different signs. *The sum of a positive number and a negative number is found by subtracting the lesser absolute value from the greater absolute value. The answer has the sign of the number with the greater absolute value.*

■ ***Example 1–4 C***

Add the following numbers.

When numbers have the same sign
Sum has their common sign
Add their absolute values

1. $(-6) + (-10) = -16$

$|-6| + |-10|$: $6 + 10 = 16$

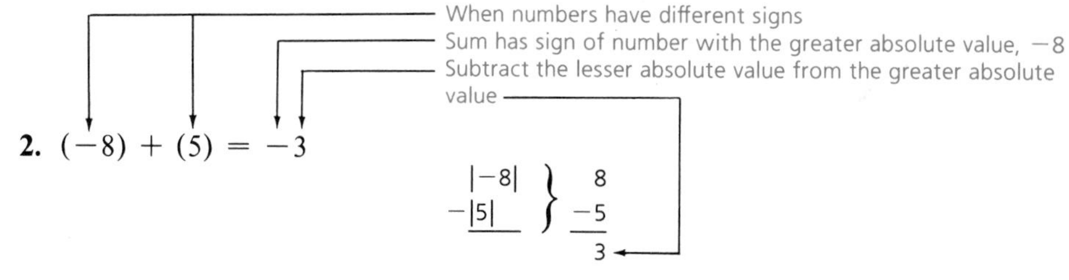

2. $(-8) + (5) = -3$

3. $(-4) + (6) = 2$ — Difference of the absolute values and the sign comes from the $+6$

4. $(10) + (-10) = 0$ — The sum of a number and its opposite is zero

▶ ***Quick check*** $(3) + (-8)$ ■

We observe from example 4 that $(10) + (-10) = 0$. That is, a number added to its opposite gives a sum equal to zero. The opposite of a number is also called the **additive inverse** of that number. We now state this property.

Additive inverse property

For every real number a,

$$a + (-a) = 0$$

and

$$(-a) + a = 0$$

Concept

The sum of a number and its opposite is zero. Its opposite is called its additive inverse.

We can summarize our procedure for addition of real numbers as follows:

Addition of two real numbers

1. If the signs are the same, we add their absolute values and prefix the sum by their common sign.
2. If the signs are different, we subtract the lesser absolute value from the greater absolute value. The answer has the sign of the number with the greater absolute value.
3. The sum of a number and its opposite (additive inverse) is zero.

In many problems, there will be more than two numbers being added together. In those situations, as long as the operation involved is *strictly addition,* we can add the numbers in any order we wish.

■ ***Example 1–4 D***

Find the sum.

1. $(-2) + (-7) + (12) = (-9) + (12)$ First $(-2) + (-7)$ is -9

$= 3$ Then $(-9) + (12)$ is 3

Note For convenience, we simply add the numbers as they appear, reading them from left to right.

2. $(14) + (-4) + (4) = (10) + (4)$ First $(14) + (-4)$ is 10

$= 14$ Then $(10) + (4)$ is 14 ■

Since the numbers in the preceding examples can be added in any order, we might feel in example 2 that it would be easier to add the (-4) and (4) first, as follows:

$$(14) + (-4) + (4) = (14) + 0 = 14$$

Therefore we observe the following:

$$[(14) + (-4)] + (4) = (14) + [(-4) + (4)] = 14$$

This illustrates a mathematical principle called the **associative property of addition.**

Associative property of addition

For every real number a, b, and c,

$$(a + b) + c = a + (b + c)$$

Concept

Changing the grouping of the numbers will not change the sum.

Problem solving

To solve the following word problems, we must find the sum of the given quantities. Represent gains by positive integers and losses by negative integers.

Example 1–4 E

Choose a variable to represent the unknown quantity and find its value.

1. A pipe that is 4 feet long is joined with a pipe that is 6 feet long. What is the total length of the pipe?

   Let ℓ = the total length of the pipe. To find the total length, we must *add* the individual lengths.

total length	is	4-foot pipe	joined with	6-foot pipe
ℓ	$=$	4	$+$	6

   $\ell = 4 + 6$
   $\ell = 10$

   The total length of the pipe is 10 feet.

2. The quarterback of the Detroit Lions attempted 4 passes with the following results: a 12-yard gain, an incomplete pass, a 5-yard loss (tackled behind the line), and a 15-yard gain. What was his total gain (or loss)?

   Let t = the total gain (or loss). To find the total gain (or loss), we must *add* the results of the 4 plays. Then

total gain (or loss)	is	12-yard gain		incomplete pass		5-yard loss		15-yard gain
t	$=$	(12)	$+$	(0)	$+$	(-5)	$+$	(15)

   $t = (12) + (0) + (-5) + (15)$
   $t = (12) + (-5) + (15) = (7) + (15) = 22$

   The total *gain* was 22 yards after the 4 plays. ■

Mastery points

Can you

- Add real numbers on the number line?
- Add real numbers mentally?
- Use the commutative and associative laws of addition?

Exercise 1–4

Find each sum. See examples 1–4 A, B, C, and D.

Examples $(-5) + (-7)$ | $(3) + (-8)$

Solutions -12 Signs are the same, add their absolute values and prefix the sum by their common sign | -5 Signs are different, subtract lesser absolute value from greater, sign comes from number with the greater absolute value

1. $(-9) + (-4)$
2. $(+3) + (-5)$
3. $(+7) + (-2)$
4. $(-14) + (-10)$
5. $(-8) + 3$
6. $(+12) + (-8)$
7. $4 + (-9)$
8. $(-11) + 7$
9. $4 + (-4)$
10. $(-3) + 3$
11. $(-8.7) + (-4.9)$
12. $(-12.1) + 8.6$

13. $(-3.7) + (-7.4)$

14. $(-8.3) + (15.8)$

15. $\left(-\frac{1}{6}\right) + \left(-\frac{1}{3}\right)$

16. $\frac{3}{4} + \left(-\frac{7}{8}\right)$

17. $\frac{1}{5} + \left(-\frac{1}{10}\right)$

18. $\left(-1\frac{1}{2}\right) + \left(-1\frac{1}{4}\right)$

19. $\left(2\frac{1}{2}\right) + \left(-3\frac{1}{4}\right)$

20. $\left(5\frac{3}{8}\right) + \left(-2\frac{1}{4}\right)$

21. $10 + (-5) + (-2)$

22. $3 + (-4) + 1$

23. $(-12) + (-10) + (+8) + (+24)$

24. $(-24) + (+12) + (+12)$

25. $(-30) + 14 + (-8) + (-20)$

26. $(-2) + (+3) + (-4) + (-5)$

27. $(-25) + 4 + (-32) + 28 + 3$

28. $(+11) + (-12) + (-14) + (-9)$

Find the sum.

Example The sum of -6, 2, and -9

Solution
$$(-6) + 2 + (-9) = (-4) + (-9) \quad \text{First add } -6 \text{ and } 2$$
$$= -13 \quad \text{Then add } -4 \text{ and } -9$$

29. The sum of 7, -11, and -6

30. The sum of -16, -6, and -5

31. 18 plus -14 plus -4

32. -9 plus 15 plus -17

33. The sum of -5 and 12 increased by 4

34. The sum of 15 and -18 increased by 10

35. 15 added to the sum of -9 and 9

36. 9 added to the sum of -6 and -11

See example 1–4 E.

Example If a man borrows \$1,800 and has \$700 in savings, what is his net worth?

Solution We represent the borrowed amount as $(-1,800)$ dollars and his savings as $(+700)$ dollars. His net worth is then represented by $(-1,800)$ dollars + $(+700)$ dollars = $(-1,100)$ dollars.

37. A temperature of $(-18)°$ C is increased by 25° C. What is the resulting temperature?

38. Tim Wesner received money for his birthday from 4 different people. He received \$10, \$8, \$7, and \$5. How much money did Tim receive for his birthday?

39. Jane Balch made profits of \$5, \$7, \$2, \$3, and \$15 in five days of selling Kool Aid. How much did she receive from her five-day sale?

40. The stock market rose by 23 points on Monday, fell by 10 points on Tuesday, rose by 8 points on Wednesday, rose by 31 points on Thursday, and fell by 19 points on Friday. What was the total gain (or loss) during the 5 days?

41. The barometric pressure rose 6 mb (millibars), then dropped 9 mb. Later that day the pressure dropped another 3 mb and then rose 8 mb. What was the total gain (or loss) in barometric pressure that day?

42. A small business showed profits of \$15, \$25, \$10, \$9, and \$27 on five consecutive days. What was the total profit?

43. John's blood pressure was 118. It changed by -19. Find his present blood pressure.

44. The temperature in Sault Ste. Marie was $-13°$ F at 8 A.M. By 1 P.M. that day, it had risen 39° F. What was the temperature at 1 P.M.?

45. The temperature on a given day in Anchorage, Alaska, was $-19°$ F. The temperature then went down 22° F. What was the final temperature that day?

46. A TWA plane is flying at an altitude of 33,000 feet. It suddenly hits an air pocket and drops 4,200 feet. What is its new altitude?

47. Mack Wooten has \$35 in his checking account. He deposits \$52, \$25, and \$32; he then writes checks for \$18 and \$62. What is the final balance in his checking account?

48. A football team has the ball on its 25-yard line. On three successive plays, the team gains 6 yards, loses 3 yards, and then gains 4 yards. Where does the ball rest for the fourth play?

1–5 ■ Subtraction of real numbers

Subtraction of two real numbers

We are already familiar with the operation of subtraction in problems such as $15 - 10 = 5$. From the definition of subtraction, we know that $15 - 10 = 5$ since $5 + 10 = 15$. In figure 1–6 of section 1–4, we see that $15 + (-10)$ is also equal to 5. That is,

Subtraction		Addition
$15 - 10 = 5$	and	$15 + (-10) = 5$

From this example, we see that we obtain the same results if we change the operation from subtraction to addition and change the sign of the number that we are subtracting. We would have an addition problem and could proceed as we did in section 1–4.

To summarize our procedure for subtracting real numbers, we can state algebraically:

Definition of subtraction

For any two real numbers, a and b,

$$a - b = a + (-b)$$

Concept

"a minus b" means the same as "a plus the opposite of b."

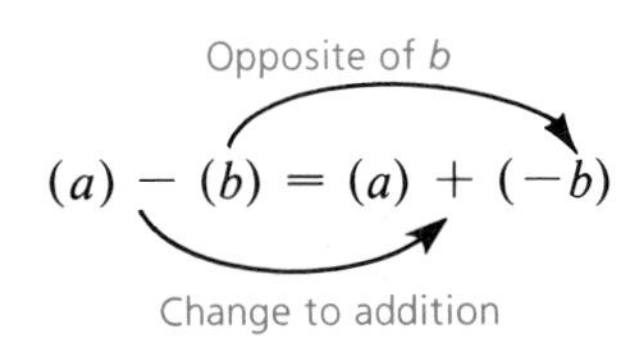

Our steps to carry out the subtraction would be as follows:

Subtraction of two real numbers

Step 1 We change the operation from subtraction to addition.
Step 2 We change the sign of the number that follows the subtraction symbol.
Step 3 We perform the addition, using our rules for adding signed numbers.

■ Example 1–5 A

Subtract the following numbers.

		Step 1 Subtraction to addition	***Step 2*** Change sign of number being subtracted		***Step 3*** Add
1. $(9) - (5) =$	(9)	$+$	(-5)	$=$	4
2. $(4) - (11) =$	(4)	$+$	(-11)	$=$	-7
3. $(-9) - (5) =$	(-9)	$+$	(-5)	$=$	-14

		Step 1 Subtraction to addition	Step 2 Change sign of number being subtracted		Step 3 Add
4. $(6) - (-8) =$	(6)	$+$	(8)	$=$	14
5. $(-12) - (-8) =$	(-12)	$+$	(8)	$=$	-4
6. $(-5) - (-14) =$	(-5)	$+$	(14)	$=$	9
7. $(8) - (5) =$	(8)	$+$	(-5)	$=$	3
8. $(5) - (8) =$	(5)	$+$	(-8)	$=$	-3

Note From examples 7 and 8, we see that the operation of subtraction is *not* commutative. That is,

$$(8) - (5) \neq (5) - (8)$$

▶ ***Quick check*** $(4) - (-6)$; $(-2) - (-8)$ ■

Addition and subtraction of more than two real numbers

When several numbers are being added and subtracted in a horizontal line, do the problem in order from left to right. For example, in

$$9 - 3 + 4 + 3 - 6 - 1 + 4$$

	Operation being performed
$= \underline{9 - 3} + 4 + 3 - 6 - 1 + 4$	$9 - 3 = 6$
$= \underline{6 + 4} + 3 - 6 - 1 + 4$	$6 + 4 = 10$
$= \underline{10 + 3} - 6 - 1 + 4$	$10 + 3 = 13$
$= \underline{13 - 6} - 1 + 4$	$13 - 6 = 7$
$= \underline{7 - 1} + 4$	$7 - 1 = 6$
$= \underline{6 + 4}$	$6 + 4 = 10$
$= 10$	

Note If we had changed each of the indicated subtractions to addition, $9 + (-3) + 4 + 3 + (-6) + (-1) + 4$, then the order in which the problem was carried out would not change the answer. For example, $9 - 3 \neq 3 - 9$, but $9 + (-3) = (-3) + 9$.

Grouping symbols

Many times, part of the problem will have a group of numbers enclosed with grouping symbols, such as parentheses ( ), brackets [ ], or braces { }. *If any quantity is enclosed with grouping symbols, we treat the quantity within as a single number.* Thus, in

$$9 - (3 + 2) + (6 - 2) - (5 - 4)$$

we perform operations within parentheses first to get

$$9 - 5 + 4 - 1$$

which gives

$$4 + 4 - 1 = 8 - 1 = 7$$

Then

$$9 - (3 + 2) + (6 - 2) - (5 - 4) = 7$$

■ **Example 1–5 B**

Perform the indicated operations.

1. $8 - 3 + 2 - 5 - 1 = 5 + 2 - 5 - 1$
$= 7 - 5 - 1$
$= 2 - 1$
$= 1$

2. $[6 + 1] - [2 - 5] + 7 + [9 - 6] = 7 - [-3] + 7 + 3$
$= 10 + 7 + 3$
$= 17 + 3$
$= 20$

3. $(14 - 7) - 2 = 7 - 2 = 5$

4. $14 - (7 - 2) = 14 - 5 = 9$

Note We observe from examples 3 and 4 that the operation of subtraction is *not* associative. That is, order does make a difference in subtraction.

$$(14 - 7) - 2 \neq 14 - (7 - 2)$$

▶ ***Quick check*** $14 - 11 + 18 - (7 - 12) + 2$ ■

Problem solving

To solve the following word problems, we must find the difference between two quantities. To find the difference we must *subtract.*

■ **Example 1–5 C**

Choose a letter for the unknown and find its value by subtracting.

1. On a given winter's day in Detroit, Michigan, the temperature was 31° in the afternoon. By 9 P.M. the temperature was −12°. How many degrees did the temperature drop from afternoon to 9 P.M.?

Let t = the number of degrees fall in temperature. We must find the difference between 31° and −12°. Thus

degrees fall	is	difference between 31° and −12°
t	$=$	$31 - (-12)$

$t = 31 - (-12) = 31 + 12 = 43$

There was a 43° drop in temperature.

2. From a board that is 16 feet long, John must cut a board that is 7 feet long. How much is left of the original board?

Let f = the number of feet of board left. We must find the difference between 16 and 7. Thus

feet left	is	difference between 16 and 7
f	$=$	$16 - 7$

$f = 16 - 7 = 9$

John has 9 feet of the original board left.

▶ ***Quick check*** A temperature of 14° Celsius is decreased by 18 degrees Celsius. What is the resulting temperature? ■

Mastery points

Can you

- Subtract real numbers?
- Add and subtract in order from left to right?
- Remember that subtraction is *not* commutative or associative?
- Remember that quantities within grouping symbols represent a single number?

Exercise 1–5

Find each sum or difference. See examples 1–5 A and B.

Examples		*Solutions* Step 1		Step 2		Step 3
$(4) - (-6)$	$=$	(4)	$+$	(6)	$=$	10
$(-2) - (-8)$	$=$	(-2)	$+$	(8)	$=$	6

1. $(4) - (5)$
2. $(-6) - (2)$
3. $(4) - (-2)$
4. $(-3) - (-7)$
5. $(-8) - (4)$
6. $(4) - (-8)$
7. $(4) - (9)$
8. $(-8) - (-5)$
9. $(-8) - (-4)$
10. $(-12) - (-16)$
11. $(8) - (-6)$
12. $(14) - (4)$
13. $(-6) + 0$
14. $(9) - (11)$
15. $(7) - 0$
16. $(6) + (-10)$
17. $\left(-\frac{1}{2}\right) - \left(-\frac{1}{4}\right)$
18. $\left(-\frac{2}{3}\right) - \left(-\frac{1}{4}\right)$
19. $1\frac{3}{8} - \left(-1\frac{1}{4}\right)$
20. $5\frac{5}{6} - \left(-2\frac{1}{3}\right)$
21. $-18.7 - (-9.3)$
22. $107.4 - (-12.6)$
23. $-215.8 - 96.2$
24. $-119.1 - 218.8$
25. $-512.7 - (-814.5)$
26. $(-12) - (-10) - (8)$
27. $(-30) + (14) - (8)$
28. $(-25) + (4) - (32) + (28)$
29. $(24) - (-12) - (12) + (-13)$
30. $(-2) - (3) + (-4) - (-5) + (-6)$
31. $(-15) - (13) - (-7) - (32)$
32. $(-17) - (11) - (-12) - (-5)$

Find each sum or difference. See example 1–5 B.

Example $14 - 11 + 18 - (7 - 12) + 2$

Solution
$$\begin{aligned} &= 14 - 11 + 18 - (-5) + 2 \\ &= 3 + 18 - (-5) + 2 \\ &= 21 - (-5) + 2 \\ &= 26 + 2 \\ &= 28 \end{aligned}$$

Perform operations within parentheses first, then add and subtract from left to right

33. $17 + 4 - (7 - 2)$
34. $(25 - 2) - (12 - 3)$
35. $(-6) - 4 + 8 - (8 - 7)$
36. $32 - 5 + 7 - 4 - (11 - 8)$
37. $10 - 10 + (10 + 10) - 10$
38. $12 + 3 - 16 - 10 - (12 + 5)$
39. $10 + (2 - 21) - (7 - 8)$
40. $(12 + 3) - 16 - 10 + (12 - 5)$
41. $(18 - 14) - (12 - 17) - 16$
42. $8 - 4 + 7 - (5 - 2) - 3$

See example 1–5 C.

Example A temperature of 14° C (Celsius) is decreased by 18 degrees Celsius. What is the resulting temperature?

Solution $14 - 18 = 14 + (-18)$
$= -4°$ C

43. A temperature of $(-6)°$ C is decreased by 32° C. What is the resulting temperature?

44. An electronics supply house has 432 resistors of a certain type. If 36 are sold during the first week, 72 during the second week, 29 during the third week, and 58 during the fourth week, how many are left at the end of the month?

45. Tim owes Tom and Rob $343 and $205, respectively, and Terry owes Tim $176. In terms of positive and negative symbols, how does Tim stand monetarily?

46. If a person has $78 after paying off a debt of $23, how much money did he have before paying off the debt? Write a statement involving the operation of subtraction of integers to show your answer.

47. A piece of wood 24 feet long is cut into three pieces so that two of the pieces measure 8 feet and 10 feet. What is the length of the third piece?

Find the difference.

Example 12 diminished by 20

Solution $12 - 20$ Diminished by 20 means subtract 20
$= 12 + (-20)$ Change to addition
$= -8$

48. -8 diminished by 11
49. -15 diminished by 7
50. -6 diminished by -21
51. -18 diminished by -9
52. Subtract -26 from -18.
53. Subtract -19 from 41.
54. Subtract -17 from 28.
55. From -26 subtract -45.
56. From -43 subtract -16.
57. 5 less than -8
58. 4 less than -12
59. 8 less than 3
60. 10 less than 5

Choose a letter for the unknown quantity and find the indicated difference. See example 1–5 C.

61. Rob has $23 in his savings account. If he spends $15 to buy a game, how much is left in his savings account?

62. Tom has $44 in his savings account. He wishes to buy a baseball glove for $21. How much is left in his savings account?

63. The temperature was $-15°$ at 6 A.M. but by noon the temperature has risen to 23°. How many degrees did it rise from 6 A.M. to noon?

64. The temperature dropped 22° from $-7°$ at midnight to just before daybreak at 7 A.M. What was the temperature at 7 A.M.?

65. Erin Nustad was born in 1986. How old will she be in the year 2000?

66. A chemist has 100 ml of acid and she needs 368 ml of the acid. How much more is needed?

67. Amy has $450 in assets and she wants to borrow enough money to buy a stereo system for $695. How much will she owe?

68. The top of Mt. Everest in Asia is 29,028 feet above sea level and the top of Mt. McKinley in Alaska is 20,320 feet above sea level. How much higher is the top of Mt. Everest than the top of Mt. McKinley?

69. Death Valley in California is 282 feet below sea level (-282). What is the difference in the altitude between Mt. McKinley and Death Valley? See exercise 68.

70. Mt. Whitney in California is 14,494 feet above sea level and the Salton Sea in California is 235 feet below sea level. What is the difference between the altitude of Mt. Whitney and Salton Sea?

71. Jim owes John \$25. He paid back \$13 and then had to borrow another \$7. How much does Jim owe John now?

72. On 4 successive hands in a poker game, Sheila won \$10, then lost \$9, then lost \$14, and finally lost \$6. What is her financial position after the 4 hands?

1–6 ■ Multiplication of real numbers

Multiplication of two positive numbers

We are already familiar with the fact that the product of two positive numbers is positive. We can see this fact by considering multiplication as repeated addition. For example, if we wish to add four 3s, then $3 + 3 + 3 + 3 = 12$. Another way of expressing this repeated addition is $4 \cdot 3 = 12$, in which case the raised dot, $\cdot$, means multiply or times. We could also have added three 4s: $4 + 4 + 4 = 12$, which could be written as $3 \cdot 4 = 12$. This observation illustrates an important mathematics principle called the **commutative property of multiplication.**

Commutative property of multiplication

For every real number a and b,

$$a \cdot b = b \cdot a$$

Concept

This property tells us that changing the order of the numbers when we multiply will not change the answer (product).

In the previous paragraph, the number 12 is called the **product** of 4 and 3, and 4 and 3 are called **factors** of 12. *The numbers or variables in an indicated multiplication are referred to as the factors of the product.*

In our example, we used a raised dot to indicate the operation of multiplication. The cross, $\times$, is used in arithmetic to indicate multiplication. We avoid using it in algebra because it may become confused with the variable x. Another way to indicate multiplication is the absence of any operation symbol between factors. The following are other examples of how we can express multiplication.

■ **Example 1–6 A**

1. $5 \cdot 7$ is read 5 times 7. The raised dot indicates multiplication.
2. $(4)(6)$ is read 4 times 6. The parentheses separate the numbers; the absence of any operation symbol between the numbers indicates multiplication.
3. $3a$ is read 3 times a.
4. ab is read a times b.
5. $6(8)$ is read 6 times 8. ■

Note 34 does not mean $3 \cdot 4$ and $3\frac{1}{2}$ does not mean $3 \cdot \frac{1}{2}$

Multiplication of two numbers with different signs

As an illustration of multiplying a positive number times a negative number, consider the following pattern:

$$\begin{aligned} 3 \cdot 3 &= 9 \\ 2 \cdot 3 &= 6 \\ 1 \cdot 3 &= 3 \\ 0 \cdot 3 &= 0 \\ (-1) \cdot 3 &= -3 \\ (-2) \cdot 3 &= -6 \\ (-3) \cdot 3 &= -9 \end{aligned}$$

The product decreases by 3

The product of a negative and a positive is a negative

We observe from this pattern that our product decreases by 3 each time. It logically follows that the product of a negative number and a positive number is a negative number.

A second observation from this pattern is that zero times a number has a product of zero. This is called the **zero factor property.**

Zero factor property

For every real number a,

$$a \cdot 0 = 0 \cdot a = 0$$

Concept

Multiplying any number by zero always gives zero as the answer. That is, whenever we are multiplying and zero is one of the factors, our product will be zero.

A third observation from this pattern is that 1 times a number is equal to the number. For this reason, 1 is called the **identity element of multiplication.**

Identity property of multiplication

For every real number a

$$a \cdot 1 = 1 \cdot a = a$$

Concept

Multiplying a number by 1 leaves the number unchanged.

Multiplication of two negative numbers

We will observe another pattern when we consider the following.

$$\begin{aligned} 3(-3) &= -9 \\ 2(-3) &= -6 \\ 1(-3) &= -3 \\ 0(-3) &= 0 \\ (-1)(-3) &= +3 \\ (-2)(-3) &= +6 \\ (-3)(-3) &= +9 \end{aligned}$$

The product increases by 3

The product of two negatives is a positive

From this pattern, we can see that our product increases by 3 each time. It logically follows that *the product of two negative numbers is a positive number.*

We can summarize our procedures for multiplication of real numbers as follows:

Multiplication of two real numbers

To multiply two real numbers, multiply their absolute values and

1. the product will be positive if the numbers have the same sign;
2. the product will be negative if the numbers have different signs.

■ Example 1–6 B

Multiply the following numbers.

1. $(-2) \cdot 3 = -6$

- Product of their absolute values: $2 \cdot 3 = 6$
- Product is negative because the numbers have different signs

2. $(-2)(-4) = 8$

- Product of their absolute values: $2 \cdot 4 = 8$
- Product is positive because the numbers have the same signs

3. $4(-4) = -16$ — Negative because signs are different

4. $(-5)(-5) = 25$ — Positive because signs are the same

▶ ***Quick check*** $(-4)(-3)$ ■

We can determine the sign of our answer when we multiply three or more real numbers. Consider the following examples.

1. $(-1)(2)(3)(4) = (-2)(3)(4) = (-6)(4) = -24$ — Odd number of negative factors
2. $(-1)(2)(-3)(4) = (-2)(-3)(4) = (6)(4) = 24$ — Even number of negative factors
3. $(-1)(2)(-3)(-4) = (-2)(-3)(-4) = (6)(-4) = -24$ — Odd number of negative factors
4. $(-1)(-2)(-3)(-4) = (2)(-3)(-4) = (-6)(-4) = 24$ — Even number of negative factors

Multiplication of two or more real numbers

1. If in the numbers being multiplied there is an **odd** number of negative factors, the answer will be negative.
2. If in the numbers being multiplied there is an **even** number of negative factors, the answer will be positive.

■ Example 1–6 C

Multiply the following numbers.

1. $(-7)(-2)(5) = (14)(5) = 70$ — Even number of negative factors

2. $(-6)(2)(-4) = (-12)(-4) = 48$ — Even number of negative factors

3. $[(-3)(5)](4) = (-15)(4) = -60$ — Odd number of negative factors

4. $(-3)[(5)(4)] = (-3)(20) = -60$ — Odd number of negative factors

▶ ***Quick check*** $(-3)(-2) \cdot 4$ ■

Examples 3 and 4 illustrate an important mathematical principle called the **associative property of multiplication.**

Associative property of multiplication

For every real number a, b, and c,

$$(a \cdot b)c = a(b \cdot c)$$

Concept

Changing the grouping of the numbers will not change the product.

Problem solving

To solve the following problems, we must multiply the quantities.

■ ***Example 1–6 D***

Choose a letter for the unknown and find the indicated product.

1. What is the cost of 7 VHS tapes if each tape costs \$5.95?

   Let c = the cost of the 7 tapes. We must multiply to find the total cost of all 7 tapes. Thus

total cost	is	7 tapes	at	\$5.95 each
c	$=$	7	$\cdot$	(5.95)

   $c = 7 \cdot (5.95) = 41.65$

   The 7 tapes cost \$41.65.

2. On 4 successive days, the stock market dropped 9 points (represented by a negative number) each day. How many points did the market change in the 4 days?

   Let d = the total drop. Represent 9-point drop by -9. We multiply to obtain

total drop	is	4 days	at	9-point drop each day
d	$=$	4	$\cdot$	(-9)

   $d = 4 \cdot (-9) = -36$

   The stock market dropped 36 points (-36) in the 4 days. ■

Mastery points

Can you

- Use the commutative and associative properties of multiplication, the zero factor property, and the identity property of multiplication?
- Multiply real numbers?

Exercise 1–6

Perform the indicated operations. See examples 1–6 B and C.

Examples $(-4)(-3)$

Solutions $= 12$ Product is positive because the numbers have like signs

$(-3)(-2) \cdot 4$

$= 6 \cdot 4$ $(-3)(-2) = 6$ because the numbers have like signs

$= 24$ 6 times 4 is 24

1. $(-3)(-5)$
2. $0 \cdot (-6)$
3. $4 \cdot (-7)$
4. $(-8) \cdot 3$
5. $4 \cdot (-3) \cdot 5$
6. $(-2)(2)(-2)$
7. $4 \cdot (-9)$
8. $(-3)(-2)(-8)$
9. $(-1)(-4)(5)$
10. $(-5)(2)(4)(3)$
11. $7 \cdot (-1)(-3)(-5)$
12. $2 \cdot (-3)(-1)(2)(-2)(3)$
13. $(-1.8)(2.4)$
14. $(-5.7)(-6.12)$
15. $(0.49)(-28.1)$
16. $(-8.9)(-8.9)$
17. $(-27)(0.08)$
18. $\left(-\frac{1}{3}\right)\left(\frac{3}{5}\right)$
19. $\left(-\frac{3}{4}\right)\left(-\frac{3}{4}\right)$
20. $\left(-\frac{3}{4}\right)\left(\frac{8}{9}\right)$
21. $\left(-\frac{5}{8}\right)\left(-\frac{2}{5}\right)$
22. $\left(\frac{5}{12}\right)\left(-\frac{9}{10}\right)$
23. $(-5)(-4)(-3)(2)$
24. $(-2)(-7)(7)(4)$
25. $(-3)(3)(-4)(4)$
26. $(-1)(-1)(-1)(-1)$
27. $(-2)(0)(3)(-4)$
28. $(-3)(-2)(4)(0)$
29. $(-5)(0)(-4)$

In exercises 30–46, two numbers are listed. Find two integers such that their product is the first number and their sum is the second number.

Examples 4, -4

Solutions Since $(-2)(-2) = 4$ and $(-2) + (-2) = -4$, then -2 and -2 are the integers.

-27, -6

Since $(-9)(3) = -27$ and $(-9) + (3) = -6$, then -9 and 3 are the integers.

30. -16, 0
31. -30, 1
32. 25, 10
33. 20, -9
34. -11, 10
35. 0, -7
36. -72, -21
37. -12, -1
38. 48, 16
39. 35, 12
40. 4, -4
41. -8, 7
42. -9, 0
43. -12, 1
44. -15, 2
45. -18, 3
46. -30, -1

Choose a letter for the unknown and multiply to find the value. See example 1–6 D.

47. Over a five-day period, the price of a particular stock suffered losses of \$3 on each of the first two days and \$2 on each of the last three days. If the stock originally sold for \$88, what was its price after the five-day period?
48. A man acquires a debt of \$6 each day for five days. If we represent a \$6 debt by (-6), write a statement of the change in his assets after five days. What is the change?
49. An auditorium contains 42 rows of seats. If each row contains 25 seats, how many people can be seated in the auditorium?
50. There are 7 rows of desks in a classroom. If each row contains 8 desks, how many students will the classroom hold?
51. A clothier ordered 15 suits, each costing him \$65. What was the total cost of the 15 suits?

52. Mrs. Jones purchased two dozen (24) cans of frozen orange juice concentrate that was on sale for 57¢ per can. How much did the orange juice cost her?

53. Jim Johnson lost an average of $23 in 4 successive poker games. What were his total losses? (Represent this by a negative answer.)

54. If a bank advertises that a person can double his investment in a savings account in 9 years, how much will $120 grow to in 9 years?

55. Joanie sold 35 glasses of lemonade at her corner stand. If she charged 15¢ a glass, how much did she make in sales?

56. Berkeley Gossett deposits his weekly allowance of $3 in a savings account at the bank. How much has he deposited in 3 years? (*Hint:* 52 weeks = 1 year.)

57. A grocer averages selling 25 gallons of milk each day. How many gallons of milk does he sell in 4 weeks? (Assume the grocery is open 7 days per week.)

58. A department store averages a loss of $75, due to thefts, each day. How much is lost in a month of 31 days?

1–7 ■ Division of real numbers

Division of two numbers

Recall that when we divide a number (called the *dividend*) by another number (called the *divisor*), we compute an answer (called the *quotient*). We define the operation of division as follows:

Definition of division

If $b \neq 0$*, $\frac{a}{b} = q$ provided that $b \cdot q = a$, where a is the dividend, b is the divisor, and q is the quotient.

The second part of our definition of division shows how to check the answer to the problem. We multiply the divisor by the quotient to get the dividend $(b \cdot q = a)$.

By the above definition, the quotient of the two negative numbers $(-20) \div (-5)$ or $\frac{-20}{-5}$ must be that number which multiplied by -5 gives -20. That number is 4, since $(-5)(4) = -20$. Therefore $\frac{-20}{-5} = 4$. We observe that *the quotient of two negative numbers is a positive number.*

To divide a positive number by a negative number, or a negative number by a positive number, consider the following divisions:

$$(-14) \div (2) = \frac{-14}{2} = -7$$

since

$$(2)(-7) = -14$$

and

$$(24) \div (-6) = \frac{24}{-6} = -4$$

because

$$(-6)(-4) = 24$$

We find that *the quotient of a positive number and a negative number is always a negative number.*

*The reason for this restriction will be explained on page 53.

We can summarize our procedures for multiplication and division of real numbers as follows:

Multiplication or division of two real numbers

To multiply or divide two real numbers, perform the operation (multiplication or division) using the absolute values of the numbers and

1. the quotient will be positive if the numbers have like signs;
2. the quotient will be negative if the numbers have different signs.

Example 1–7 A

Divide the following numbers.

1. $\dfrac{-14}{-7} = 2$, since $(-7)(2) = -14$

2. $\dfrac{-36}{-6} = 6$, since $(-6)(6) = -36$

3. $\dfrac{-24}{3} = -8$, since $(3)(-8) = -24$

4. $\dfrac{15}{-5} = -3$, since $(-5)(-3) = 15$

▶ ***Quick check*** $\dfrac{-18}{-9}$; $\dfrac{-15}{3}$ ■

In section 1–6, we developed a procedure for determining the sign of our answer when we multiply three or more real numbers. This same rule can be extended to apply to division. Consider the following examples.

Example 1–7 B

1. $\dfrac{(-1)(12)}{(2)(3)} = \dfrac{-12}{6} = -2$ Odd number of negative factors

2. $\dfrac{(-1)(12)}{(2)(-3)} = \dfrac{-12}{-6} = 2$ Even number of negative factors

3. $\dfrac{(-1)(12)}{(-2)(-3)} = \dfrac{-12}{6} = -2$ Odd number of negative factors

4. $\dfrac{(-1)(-12)}{(-2)(-3)} = \dfrac{12}{6} = 2$ Even number of negative factors ■

We can summarize the procedure for three or more real numbers in a multiplication or division problem as follows:

Multiplication or division of two or more real numbers

When we multiply or divide, if we have an odd number of negative factors, our answer will be negative; otherwise it will be positive.

Note Our procedure concerning multiplication or division of three or more signed numbers applies *only* when we are doing *strictly* the operations of multiplication and division. For example, in the problem $\frac{(-8) + (-4)}{-2}$, we have an odd number of negative numbers, but our solution would be as follows: $\frac{(-8) + (-4)}{-2} = \frac{-12}{-2} = 6$. We are not able to apply our procedure here because we are not performing strictly multiplication and division.

Division involving zero

In section 1–3, we defined a rational number to be any number that can be expressed as a quotient of two integers in which the divisor is not zero. The number zero, 0, is the only number that we cannot use as a divisor. To see why we exclude zero as a divisor, recall that we check a division problem by multiplying the divisor times the quotient to get the dividend. If we apply this idea in connection with zero as a divisor, we observe the following situations. Suppose there were a number q such that $3 \div 0 = q$. Then $q \cdot 0$ would have to be equal to 3 for our answer to check, but this product is zero regardless of the value of q. Therefore we cannot find an answer for this problem. We say that the answer is *undefined.* If we try to divide zero by zero and again call our answer q, we have $0 \div 0 = q$. When we check our work, $0 \cdot q = 0$, we see that any value for q will work. Since any value for q will work, we say our answer is *indeterminate.* We therefore decide that **division by zero is not allowed.**

It is important to note that although division by zero is not allowed, this does not extend to the division of zero by some other number. We can see that $\frac{0}{-4} = 0$ since $(-4) \cdot 0 = 0$. Thus, **the quotient of zero divided by any number other than zero is always zero.**

■ ***Example 1–7 C***

Perform the division, if possible.

1. $\frac{0}{5} = 0$

2. $\frac{2}{0}$ is undefined

3. $\frac{-7}{0}$ is undefined

4. $\frac{0}{-7} = 0$

5. $\frac{0}{0}$ is indeterminate

▶ ***Quick check*** Divide $\frac{11}{0}$, if possible. ■

Problem solving

To solve the following problems, we will have to divide the given quantities.

■ Example 1–7 D

Choose a letter for the unknown quantity and find the indicated quotient.

1. If \$7.68 is spent on 6 three-way light bulbs, how much did each light bulb cost?

   Let $c =$ the cost of each light bulb. We must divide \$7.68 by 6. Thus

cost per bulb	is equal to	total cost	divided	by 6 identical items
c	$=$	(7.68)	$\div$	6

   $c = (7.68) \div 6 = 1.28$

   Each light bulb cost \$1.28.

2. There are 400 people seated in a full auditorium. If there are 25 identical rows of seats, how many people are there in each row?

   Let $x =$ the number of people in each row. We must divide 400 by 25. Thus

people per row	is equal to	total number of people	divided	by 25 identical rows
x	$=$	400	$\div$	25

   $x = 400 \div 25 = 16$

   There are 16 people seated in each row. ■

Mastery points

Can you

- Perform division with real numbers?
- Remember the results of division involving zero?

Exercise 1–7

Perform the indicated operations, if possible. See examples 1–7 A, B, and C.

Examples	$\dfrac{-18}{-9}$		$\dfrac{-15}{3}$		$\dfrac{11}{0}$
Solutions	$= 2$	Quotient of two negatives is a positive	$= -5$	Quotient of a positive and a negative is a negative	is undefined

1. $\dfrac{-14}{-7}$ 2. $\dfrac{-15}{5}$ 3. $\dfrac{32}{-4}$ 4. $\dfrac{18}{3}$ 5. $\dfrac{-22}{-11}$

6. $\dfrac{18}{-3}$ 7. $\dfrac{-16}{2}$ 8. $\dfrac{-25}{-5}$ 9. $\dfrac{7}{0}$ 10. $\dfrac{-4}{0}$

11. $\dfrac{0}{-9}$ 12. $\dfrac{0}{5}$ 13. $\dfrac{0}{0}$ 14. $\dfrac{-24}{-6}$ 15. $\dfrac{49}{-7}$

16. $\dfrac{36}{-6}$ 17. $\dfrac{-25}{5}$ 18. $\dfrac{-64}{8}$ 19. $\dfrac{(-4)(-3)}{-6}$ 20. $\dfrac{(-18)(2)}{-4}$

21. $\frac{(16)(2)}{-8}$ **22.** $\frac{(-4)(0)}{-8}$ **23.** $\frac{(-16)(0)}{-8}$ **24.** $\frac{(-5)(-2)}{(-1)(-10)}$ **25.** $\frac{(-18)(3)}{(-2)(-9)}$

26. $\frac{(-2)(-4)}{(0)(4)}$ **27.** $\frac{(-3)(6)}{(0)(-2)}$ **28.** $\frac{8-8}{3+4}$ **29.** $\frac{(-6)(0)}{(-3)(0)}$ **30.** $\frac{6-6}{6-6}$

Example A football player carried the ball eight times, making the following yardages: gain of 6 yards (yd), loss of 3 yd, loss of 4 yd, gain of 4 yd, gain of 3 yd, loss of 1 yd, loss of 2 yd, gain of 5 yd. Show his gains and losses by positive and negative integers. What was his average gain or loss per carry?

Solution 6 yd, −3 yd, −4 yd, 4 yd, 3 yd, −1 yd, −2 yd, 5 yd. To find an average, we add together all of the values and divide by the total number of values.

$$\frac{6\text{ yd} + (-3)\text{ yd} + (-4)\text{ yd} + 4\text{ yd} + 3\text{ yd} + (-1)\text{ yd} + (-2)\text{ yd} + 5\text{ yd}}{8} = \frac{8\text{ yd}}{8} = 1\text{ yd}$$

31. The temperature at 1 P.M. for seven consecutive days in January was 5° C, −8° C, −7° C, −1° C, 10° C, −6° C, and 0° C. What was the average temperature for the seven days?

32. If the stock market showed the following gains and losses during six consecutive hours of trading on a given day, determine the average gain or loss during that six-hour period. Gain 36 points, loss 23 points, loss 72 points, gain 25 points, loss 31 points, loss 21 points.

33. Between Chicago and Detroit, a distance of 282 miles, a driver averages 47 miles per hour. How long will it take her to make the trip?

34. A trip of 369 miles takes nine hours to complete. What was the average rate of speed?

35. Light travels at a rate of 186,000 miles per second. How long will it take to travel 1,674,000 miles?

36. How long does it take light from the sun to reach earth if the sun is approximately 93,000,000 miles away? (Refer to exercise 35.)

Choose a letter for the unknown quantity and find the value by dividing. See example 1–7 D.

37. Mrs. Smith paid $36 for 9 crates of peaches for her fruit market. How much did each crate cost her?

38. A carpenter wishes to cut a 12-foot board into 3 pieces that are all the same length. Find the length of each piece.

39. A man drove 350 miles and used 14 gallons of gasoline. How many miles did he drive on each gallon of gasoline?

40. Irene drove 424 miles in 8 hours. How many miles did she travel each hour (in miles per hour) if she drove at a constant speed?

41. During a recent cold wave, the temperature fell 28° over a 7-day period. What was the average change per day?

42. Mary Ann typed 1,350 words in 30 minutes. How many words did she type per minute?

43. The college bookstore purchased 480 math textbooks. If the books came in 15 boxes of the same size, how many books were in each box?

44. Jim, John, Pete, and Mike worked together painting farmer Gene's barn. If he gave them $124 to split evenly among them, how much did each boy receive?

45. Alice took part in a 26-mile marathon run. If she ran the marathon in 5 hours and 12 minutes, how long did it take her to run 1 mile (in minutes) if she ran at a constant speed?

46. A farmer got 720 bushels of wheat from a 30-acre field. How many bushels did he get per acre?

1–8 ■ Properties of real numbers and order of operations

In the previous four sections, we introduced some of the properties of real numbers. We also saw how these properties are used when performing fundamental operations with numbers. Since variables represent numbers, we will be using these and other properties throughout our study of algebra. The properties that we have covered so far are listed and the page number where the property was first introduced is given for reference.

Properties of real numbers

If a, b, and c are any real numbers, then

$a + b = b + a$, commutative property of addition (page 35)
$a \cdot b = b \cdot a$, commutative property of multiplication (page 46)
$(a + b) + c = a + (b + c)$, associative property of addition (page 38)
$(a \cdot b)c = a(b \cdot c)$, associative property of multiplication (page 49)
$a \cdot 1 = 1 \cdot a = a$, identity property of multiplication (page 47)
$a + 0 = 0 + a = a$, identity property of addition (page 35)
$a + (-a) = 0$, additive inverse property (page 38)
$a \cdot 0 = 0 \cdot a = 0$, zero factor property (page 47)

Exponents

Consider the indicated products

$$4 \cdot 4 \cdot 4 = 64$$

and

$$3 \cdot 3 \cdot 3 \cdot 3 = 81$$

A more convenient way of writing $4 \cdot 4 \cdot 4$ is 4^3, which is read "4 to the third power" or "4 cubed." We call the number 4 the **base** of the expression and the number 3, to the upper right of 4, the **exponent.**

Thus

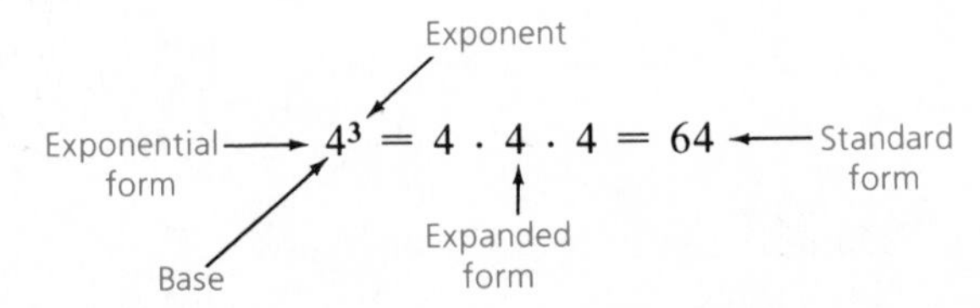

In like fashion, $3 \cdot 3 \cdot 3 \cdot 3$ may be written 3^4, where 3 is the base and 4 is the exponent. The expression is read "3 to the fourth power." Then

$$3 \cdot 3 \cdot 3 \cdot 3 = 3^4 = 81$$

Notice that *the exponent tells how many times the base is used as a factor in an indicated product.* We call this form of a product the **exponential form.** That is, the exponential form of the product $3 \cdot 3 \cdot 3 \cdot 3$ is 3^4.

Note The exponent is understood to be 1 when a number has no exponent. That is, $5 = 5^1$.

Remember that when we have a negative number, we place it inside parentheses. With this fact in mind, we can see that there is a definite difference between $(-2)^4$, which is read "-2 to the fourth power," and -2^4, which is read "the opposite of 2 to the fourth power." In the first case, the parentheses denote that this is a negative number to a power: $(-2)^4 = (-2)(-2)(-2)(-2) = +16$. In the second case, since there are no parentheses around the number, we understand that this is *not* (-2) to a power. It is, rather, the opposite of the answer when we raise 2^4: $-2^4 = -(2)^4 = -(2 \cdot 2 \cdot 2 \cdot 2) = -(16)$.

■ *Example 1–8 A*

Perform the indicated multiplication.

1. $(-3)^3 = (-3)(-3)(-3) = -27$
2. $-3^3 = -(3 \cdot 3 \cdot 3) = -27$
3. $(-3)^4 = (-3)(-3)(-3)(-3) = 81$
4. $-3^4 = -(3 \cdot 3 \cdot 3 \cdot 3) = -81$

▶ ***Quick check*** -3^2 ■

When we are performing several different types of arithmetic operations within an expression, we need to agree on an order in which the operations will be performed. To show that this is necessary, consider the following numerical expression.

$$3 + 4 \cdot 5 - 3$$

More than one answer is possible, depending on the order in which we perform the operations. To illustrate,

$$3 + 4 \cdot 5 - 3 = 7 \cdot 2 = 14$$

if we add and subtract as indicated before we multiply. However

$$3 + 4 \cdot 5 - 3 = 3 + 20 - 3 = 20^*$$

if we multiply before we add or subtract. A third possibility would be

$$3 + 4 \cdot 5 - 3 = 3 + 4 \cdot 2 = 3 + 8 = 11$$

if we subtract, then multiply, and finally add. To standardize the answer, we agree to the following order of operations, or priorities.

Order of operations, or priorities

1. **Groups:** Perform any operations within a grouping symbol such as ( ) parentheses, [ ] brackets, { } braces, | | absolute value, and above or below the fraction bar.
2. **Exponents:** Perform operations indicated by exponents.
3. **Multiply and divide:** Perform multiplication and division in order from left to right.
4. **Add and subtract:** Perform addition and subtraction in order from left to right.

Note

a. Within a grouping symbol, the order of operations will still apply.
b. If there are several grouping symbols intermixed, remove them by starting with the innermost one and working outward.

*This is the correct answer.

To illustrate this order, consider the numerical expression

$$6 + 5(7 - 3) - 2^2$$

We first evaluate within the grouping symbol, in this case parentheses, to get

$$6 + 5(4) - 2^2$$

We then perform the indicated power and have

$$6 + 5(4) - 4$$

Our third step is to carry out the multiplication, resulting in

$$6 + 20 - 4$$

Our last step is to perform the addition and subtraction in order from left to right, giving

$$\begin{aligned} &26 - 4 \\ &= 22 \end{aligned}$$

■ Example 1–8 B

Perform the indicated operations in the proper order and simplify.

1.
$$\begin{aligned} 7 + 8 \cdot 3 \div 2 &= 7 + 24 \div 2 && \text{Priority 3, multiply} \\ &= 7 + 12 && \text{Priority 3, divide} \\ &= 19 && \text{Priority 4, add} \end{aligned}$$

2.
$$\begin{aligned} (7 - 1) \div 2 + 3 \cdot 4 &= 6 \div 2 + 3 \cdot 4 && \text{Priority 1, parentheses} \\ &= 3 + 12 && \text{Priority 3, divide and multiply} \\ &= 15 && \text{Priority 4, add} \end{aligned}$$

3.
$$\begin{aligned} \frac{1}{2} + \frac{3}{4} \div \frac{5}{8} &= \frac{1}{2} + \frac{3}{\cancel{4}_{1}} \cdot \frac{\cancel{8}^{2}}{5} && \text{Priority 3, invert, divide out common factors} \\ &= \frac{1}{2} + \frac{6}{5} && \text{Priority 3, multiply} \\ &= \frac{5}{10} + \frac{12}{10} && \text{Least common denominator} \\ &= \frac{5 + 12}{10} && \text{Priority 4, add} \\ &= \frac{17}{10} \text{ or } 1\frac{7}{10} && \text{Priority 4, add} \end{aligned}$$

4.
$$\begin{aligned} 2^2 \cdot 3 - 3 \cdot 4 &= 4 \cdot 3 - 3 \cdot 4 && \text{Priority 2, exponent} \\ &= 12 - 12 && \text{Priority 3, multiply} \\ &= 0 && \text{Priority 4, subtract} \end{aligned}$$

5.
$$\begin{aligned} \frac{3}{4} - \frac{1}{2} \cdot \frac{2}{3} &= \frac{3}{4} - \frac{1}{\cancel{2}_{1}} \cdot \frac{\cancel{2}^{1}}{3} && \text{Priority 3, divide out common factors} \\ &= \frac{3}{4} - \frac{1}{3} && \text{Priority 3, multiply} \\ &= \frac{9}{12} - \frac{4}{12} && \text{Least common denominator} \\ &= \frac{9 - 4}{12} && \text{Priority 4, subtract} \\ &= \frac{5}{12} && \text{Priority 4, subtract} \end{aligned}$$

6. $(7.28 + 1.6) \div 2.4 - (6.1)(3.8)$

$= (8.88) \div 2.4 - (6.1)(3.8)$	Priority 1, parentheses
$= 3.7 - 23.18$	Priority 3, division and multiplication
$= -19.48$	Priority 4, subtract

7.

$\left(\frac{2}{3} + \frac{7}{8}\right) \div \frac{5}{6} = \left(\frac{16}{24} + \frac{21}{24}\right) \div \frac{5}{6}$	Priority 1, parentheses
$= \left(\frac{16 + 21}{24}\right) \div \frac{5}{6}$	Priority 1, parentheses
$= \frac{37}{24} \div \frac{5}{6}$	Priority 1, parentheses
$= \frac{37}{\cancel{24}_{4}} \cdot \frac{\cancel{6}^{1}}{5}$	Priority 3, invert, divide out common factors
$= \frac{37}{20}$ or $1\frac{17}{20}$	Priority 3, multiply

8. $(5.4)^2 - 4(3.1)(2.8)$

$= 29.16 - 4(3.1)(2.8)$	Priority 2, exponent
$= 29.16 - 34.72$	Priority 3, multiply
$= -5.56$	Priority 4, subtract

9.

$\frac{3(2 + 4)}{4 - 2} - \frac{4 + 6}{5} = \frac{3(6)}{4 - 2} - \frac{4 + 6}{5}$	Priority 1, groups: numerator and denominator
$= \frac{18}{2} - \frac{10}{5}$	Priority 1, numerator and denominator
$= 9 - 2$	Priority 3, divide
$= 7$	Priority 4, subtract

10. $5[7 + 3(10 - 4)]$

We first evaluate within the grouping symbol, applying the order of operations.

$5[7 + 3(6)] = 5[7 + 18]$	Priority 1, groups
$= 5[25]$	Priority 1, groups
$= 125$	Priority 3, multiply

▶ ***Quick check*** $18 \div 6 \cdot 3 + 10 - (4 + 5)$ ■

Problem solving

Solve the following word problems using the order of operations.

■ ***Example 1–8 C***

Choose a variable to represent the unknown quantity and find its value by performing the indicated operations.

1. Mrs. Hansen purchased 6 boxes of cereal at \$1.25 per box and 7 cans of tuna fish at 70¢ per can. What was her total bill?

Let t = Mrs. Hansen's total bill. 6 boxes at \$1.25 per box cost $6 \cdot \$1.25$; 7 cans at 70¢ per can cost $7 \cdot \$0.70$. The total bill is given by

total bill	is equal to	6 boxes of cereal	at	\$1.25 per box	and	7 cans of tuna	at	\$0.70 per can
t	$=$	6	$\cdot$	(1.25)	$+$	7	$\cdot$	(0.70)

$t = 6 \cdot (1.25) + 7 \cdot (0.70)$
$= 7.50 + 4.90$ Priority 3
$= 12.40$ Priority 4

Mrs. Hansen's total bill was \$12.40.

2. A man works a 40-hour week at \$12 per hour. If he works 11 hours of overtime at time and a half, how much will he receive for the 51 hours of work?

Let w = the man's total wages for the week. 40 hours at \$12 per hour is $40 \cdot \$12$. Hourly rate at time and a half is $\left(1\frac{1}{2} \cdot 12 = 18\right)$ and 11 hours at time and a half is $11 \cdot 18$. Thus

total wages	is equal to	40 hours	at	\$12 per hour	and	11 hours	at	\$18 per hour
w	$=$	40	$\cdot$	12	$+$	11	$\cdot$	18

$w = 40 \cdot 12 + 11 \cdot 18$
$w = 480 + 198$ Priority 3
$w = 678$ Priority 4

The man will receive \$678 for 51 hours of work. ■

Mastery points

Can you

- Perform multiple operations in the proper order?
- Use exponents?

Exercise 1–8

Perform the indicated operations. See example 1–8 A.

Example -3^2

Solution $= -(3^2)$
$= -9$

-3^2 is not the same as $(-3)^2$, it is the opposite of 3^2

1. $(-4)^2$
2. $(-5)^4$
3. $(-3)^3$
4. -4^2
5. -6^2
6. -2^4
7. -1^2
8. -2^2
9. $(-1)^2$
10. $(-2)^2$

Perform the indicated operations and simplify. See example 1–8 B.

Example $18 \div 6 \cdot 3 + 10 - (4 + 5)$

Solution

$= 18 \div 6 \cdot 3 + 10 - 9$	Priority 1, parentheses
$= 3 \cdot 3 + 10 - 9$	Priority 3, division
$= 9 + 10 - 9$	Priority 3, multiplication
$= 19 - 9$	Priority 4, addition
$= 10$	Priority 4, subtraction

11. $\dfrac{4 + 2}{3} + 2$

12. $-6 \cdot 7 + 8$

13. $6 + 5 \cdot 4$

14. $\dfrac{1}{5} \cdot 5 + 6$

15. $-2 + 10 \cdot \dfrac{1}{5}$

16. $4(3 - 2)(2 + 1)$

17. $0(5 + 2) + 3$

18. $\dfrac{24 \cdot 3}{9} - 6$

19. $(24 - 6) \div 3$

20. $(37 - 4) \div 11$

21. $\dfrac{2}{3} \div \left(\dfrac{5}{6} - \dfrac{4}{9}\right)$

22. $12 \cdot 4 + 2$

23. $2 + 3(8 - 5)$

24. $5 + 2(11 - 6)$

25. $6 + 4(8 + 2)$

26. $7 + 3(9 - 4)$

27. $8 - 3(6 - 4)$

28. $10 - 2(7 - 11)$

29. $15 \cdot 3^2 - 14$

30. $(8 - 3)(5 + 3)$

31. $\dfrac{7}{8} - \dfrac{1}{2} \div \dfrac{3}{4}$

32. $\dfrac{3}{8} + \dfrac{7}{12} \cdot \dfrac{3}{14}$

33. $3(6 - 2)(7 + 1)$

34. $12 + 3 \cdot 16 \div 4^2 - 2$

35. $9 - 3(12 + 3) - 4 \cdot 3$

36. $15 - 2(8 + 1) - 6 \cdot 4$

37. $50 - 4(6 - 8) + 5 \cdot 4$

38. $18 - 5(7 + 3) - 6$

39. $10 - 3 \cdot 4 \div 6 - 5$

40. $8 - (12 + 3) - 4 \cdot 3$

41. $4(2 - 5)^2 - 2(3 - 4)$

42. $6(-8 + 10) - 5(4 - 7)$

43. $\dfrac{5(3 - 5)}{2} - \dfrac{27}{-3}$

44. $\dfrac{3(8 - 6)}{2} - \dfrac{8}{-2}$

45. $\dfrac{5(6 - 3)}{3} - \dfrac{(-14)}{2}$

46. $(14.13 + 11.4) \div 3.7 - (2.4)(7.8)$

47. $(5.1 + 2.2)(4.8) - (6.3)(8.1)$

48. $(5.1)^2 \cdot 3 - (14.64) \div (6.1)$

49. $(1.9)^2 + 4(3.3)^2 - 8.7$

50. $5[10 - 2(4 - 3) + 1]$

51. $18 + [14 - 5(6 - 4) + 7]$

52. $(8 - 2)[16 + 4(5 - 7)]$

53. $(9 - 6)[21 + 5(4 - 6)]$

54. $\left(\dfrac{6 - 3}{7 - 4}\right)\left(\dfrac{14 + 2 \cdot 3}{5}\right)$

55. $\left(\dfrac{3}{12} - \dfrac{1}{6}\right)\left(\dfrac{2}{3} + \dfrac{1}{8}\right)$

56. $\left(\dfrac{1}{4} - \dfrac{1}{6}\right) \div \left(\dfrac{2}{3} - \dfrac{1}{8}\right)$

Perform the indicated operations and simplify. See example 1–8 B.

57. To convert 74° Fahrenheit (F) to Celsius (C), we use the expression

$C = \dfrac{5}{9}(F - 32)$; thus, in this case,

$C = \dfrac{5}{9}(74 - 32)$. Find C.

58. A Murray Loop is used to determine the point at which a telephone line is grounded. The unknown distance to the point of the ground, x, for a length of the loop of 32 miles and resistances of 222 and 384 ohms is given by

$$x = \frac{384}{222 + 384} \cdot 32$$

Find x.

59. The surface area in square inches of a flat ring whose inside radius is 2 inches and whose outside radius is 3 inches is approximately

$$\frac{22}{7} \cdot 3^2 - \frac{22}{7} \cdot 2^2$$

Find the area of this surface in square inches.

60. The surface area in square inches of a ring section whose inside diameter is 18 inches and whose outside diameter is 26 inches is approximately

$$\frac{22}{7} \cdot \frac{26 + 18}{2} \cdot \frac{26 - 18}{2}$$

Find the area of this surface in square inches.

61. To find the pitch diameter, D, of a gear with 36 teeth and an outside diameter of 8 inches, we use

$$D = \frac{36(8)}{38 + 3}$$

Find D in inches.

Choose a letter for the unknown quantity and use the order of operations to find its value. See example 1–8 C.

62. A woman purchased a case of soda (24 bottles) at 15¢ per bottle, 5 pounds of candy at 49¢ per pound, and 20 jars of baby food at 75¢ per jar. What was her total bill (a) in cents, (b) in dollars and cents?

63. Colleen Meadow is a typist in a law firm. Her base pay is \$7 per hour for a 40-hour week and she receives time and a half for every hour she works over 40 hours in a week. How much will she earn if she works 49 hours in one week?

64. In a series of poker games, Ace McGee won \$4,000 in each of 3 games, lost \$1,500 in each of 4 games, and won \$2,000 in each of 2 games. How much did Ace win (or lose) in the 9 games?

65. A carpenter must cut a 16-foot long board into 4-foot lengths and a 12-foot long board into 3-foot lengths. How many pieces of lumber will he have?

66. Jane, David, and Mary are typists in an office. David can type 75 words per minute, Jane can type 80 words per minute, and Mary can type 95 words per minute. How many words can they type together in 15 minutes?

67. The stock market opened at 2,725 points on a given day. If it lost 9 points per hour during the first 3 hours after opening and then gained 6 points per hour during the next 5 hours, what did the stock market close at?

Chapter 1 lead-in problem

While on a trip to Canada, Tonya heard on the radio that the temperature today will be 20° Celsius. Will she need her winter coat? What will the temperature be in degrees Fahrenheit? We can determine what 20° Celsius is in degrees Fahrenheit by the following expression:

$$F = \frac{9}{5} \cdot 20 + 32$$

Solution

$F = \frac{9}{5} \cdot 20 + 32$	Original expression
$F = 36 + 32$	Order of operations: multiply
$F = 68$	Add

The temperature is 68 degrees Fahrenheit. She will not need her winter coat.

Chapter 1 summary

1. A **prime number** is any whole number greater than 1 whose only factors are the number itself and 1.
2. To **reduce a fraction** to lowest terms
   a. Write the numerator and the denominator as a product of prime factors.
   b. Divide the numerator and the denominator by the common factors.
3. To **multiply** fractions
   a. Multiply the numerators and the denominators and place the product of the numerators over the product of the denominators.
   b. Reduce the resulting fraction to lowest terms.
4. To **divide** fractions, multiply the first fraction by the reciprocal of the second fraction and reduce to lowest terms.
5. To **add** or **subtract** fractions having the same denominator, add or subtract the numerators and place this sum or difference over the common denominator.
6. The **least common denominator** (LCD) of two or more fractions is the least (smallest) number that is exactly divisible by the denominators.
7. **Percent** is defined to be parts per one hundred.
8. To change a **percent to a decimal number,** move the decimal point two places to the left and drop the % symbol.
9. To change from a **decimal number to a percent,** move the decimal point two places to the right and affix the % symbol.
10. To change a **fraction to percent,** drop the % symbol and write the number over a denominator of 100.
11. A **set** is any collection of things.
12. The set A is a **subset** of the set B, which is denoted by $A \subseteq B$, if every element in A is also an element of B.
13. We use the following *sets of numbers:*
    N, natural numbers
    W, whole numbers
    J, integers
    Q, rational numbers
    H, irrational numbers
    R, real numbers
14. A **variable** is a symbol that represents an unspecified number.
15. The **number line** is a line on which we visually represent the set of real numbers.
16. The four **inequality symbols** that denote an order relationship between numbers are $<$ (less than), $>$ (greater than), $\leq$ (less than or equal to), $\geq$ (greater than or equal to).
17. The **absolute value,** $|\quad|$, of a number is the undirected distance that the number is from the origin.
18. For any two real numbers, a and b, $a - b = a + (-b)$.
19. If any quantity is enclosed with **grouping symbols,** we treat the quantity within as a single number.
20. The numbers or variables in an indicated multiplication are referred to as the **factors** of the **product.**
21. The **exponent** tells how many times the base is used as a factor in an indicated product.
22. If $b \neq 0$, $\dfrac{a}{b} = q$, provided that $b \cdot q = a$.
23. *Division by zero is not allowed.*
24. The quotient of *zero* divided by any number other than zero is always zero.
25. We use the following *properties of real numbers:*
    If a, b, and c are any real numbers, then
    $a + b = b + a$, commutative property of addition
    $a \cdot b = b \cdot a$, commutative property of multiplication
    $(a + b) + c = a + (b + c)$, associative property of addition
    $(a \cdot b)c = a(b \cdot c)$, associative property of multiplication
    $a \cdot 1 = 1 \cdot a = a$, identity property for multiplication
    $a + 0 = 0 + a = a$, identity property for addition
    $a + (-a) = (-a) + a = 0$, additive inverse property
    $a \cdot 0 = 0 \cdot a = 0$, zero factor property
26. **Operations with signed numbers**

    Addition

    Same signs: Add their absolute values and prefix the sum by their common sign.

    Different signs: Subtract the lesser absolute value from the greater absolute value. The result has the sign of the number with the greater absolute value.

    Additive inverse: The sum of a number and its additive inverse (opposite) is zero.

    Subtraction

    Change the sign of the number being subtracted and add that to the first number.

    Multiplication and division

    Perform the operation (multiplication or division) using the absolute value of the numbers.

    Same signs: Answer will be positive.

    Different signs: Answer will be negative.
27. *Order of Operations*
    a. Groups: Perform any operations within a grouping symbol such as ( ) parentheses, [ ] brackets, { } braces, | | absolute value, or in the numerator or the denominator of a fraction.
    b. Exponents: Perform operation indicated by exponents.
    c. Multiplication and Division: Perform multiplication and division in order from left to right.
    d. Addition and Subtraction: Perform addition and subtraction in order from left to right.

Chapter 1 error analysis

1. Determining order between numbers
   Example: $-3 < -5$
   Correct answer: $-3 > -5$
   What error was made? (*see page 30*)
2. Evaluate absolute value
   Example: $-|-3| = 3$
   Correct answer: $-|-3| = -3$
   What error was made? (*see page 31*)
3. Adding real numbers
   Example: $(-3) + 4 = 7$
   Correct answer: $(-3) + 4 = 1$
   What error was made? (*see page 37*)
4. Subtracting real numbers
   Example: $(-9) - (-4) = -13$
   Correct answer: $(-9) - (-4) = -5$
   What error was made? (*see page 41*)
5. Combining using grouping symbols
   Example: $4 - (5 - 2) = 4 - 5 - 2 = -3$
   Correct answer: $4 - (5 - 2) = 1$
   What error was made? (*see page 42*)
6. Exponents
   Example: $-3^2 = 9$
   Correct answer: $3\frac{1}{2} = -9$
   What error was made? (*see page 57*)
7. Multiplication of negative numbers
   Example: $(-2)(-6) = -12$
   Correct answer: $(-2)(-6) = 12$
   What error was made? (*see page 48*)
8. Division of real numbers
   Example: $\frac{-15}{3} = 5$
   Correct answer: $\frac{-15}{3} = -5$
   What error was made? (*see page 52*)
9. Division by zero
   Example: $\frac{-5}{0} = 0$
   Correct answer: $\frac{-5}{0}$ is undefined.
   What error was made? (*see page 53*)
10. Exponents
    Example: $3^3 = 9$
    Correct answer: $3^3 = 27$
    What error was made? (*see page 56*)

Chapter 1 critical thinking

A watch is started at 12 noon. Each time the watch reaches the next hour, it is stopped for 10 minutes. How long will it take the watch to go from 12 noon to 12 midnight?

Chapter 1 review

[1–1]

Reduce each fraction to lowest terms.

1. $\frac{10}{14}$
2. $\frac{36}{48}$
3. $\frac{120}{180}$

Multiply or divide the following as indicated. Reduce to lowest terms.

4. $\frac{6}{7} \cdot \frac{5}{3}$
5. $\frac{2}{3} \cdot \frac{9}{10}$
6. $\frac{7}{8} \div \frac{5}{6}$
7. $\frac{5}{12} \div \frac{10}{21}$
8. $3\frac{3}{4} \div 1\frac{1}{5}$
9. $2\frac{1}{2} \cdot 3\frac{1}{3}$
10. Hannah rents $\frac{3}{4}$ of a plot of land. If the plot is $\frac{5}{6}$ of an acre in size, how many acres does Hannah rent?
11. A recipe calls for $\frac{4}{5}$ of a cup of sugar. If Dene wishes to make $\frac{1}{2}$ of the recipe, how many cups of sugar should she use?

Add or subtract the following fractions as indicated. Reduce to lowest terms.

12. $\frac{3}{7} + \frac{5}{7}$

13. $\frac{5}{8} + \frac{1}{6}$

14. $\frac{11}{12} - \frac{1}{12}$

15. $\frac{8}{9} - \frac{2}{3}$

16. $4\frac{1}{4} + 2\frac{3}{5}$

17. $\frac{1}{5} - \frac{2}{3}$

18. Paula paid $\frac{1}{3}$ of her debt one week and $\frac{1}{4}$ of her debt the second week. At the end of the second week, how much of her debt had she paid off?

19. Bob Burger owns $3\frac{1}{8}$ acres of land. If he sells $2\frac{1}{4}$ acres to his friend Eric Hand, how many acres does he have left?

[1–2]

Perform the indicated operations on decimal numbers.

20. 20.6 + 1.373 + 210.42 + 0.027 + 31.09

21. 42.5 − 10.705

22. 213.4 × 6.35

23. 316.03 ÷ 22.1

24. Peter purchased an automobile for $3,450.63 and sold it for $4,016.12. How much profit did Peter make on the sale?

25. Linda owns 3 pieces of property 2.34, 3.61, and 1.91 acres in size. How many total acres of property does she own?

26. An automobile uses 15.2 gallons of gasoline to travel 188.8 miles. How many miles per gallon did the automobile average?

Find the following percentages.

27. 4% of 250

28. 57% of 120

29. 62.5% of 40

30. 131.2% of 60

[1–3]

List the elements of the following sets.

31. Integers between 49 and 56

32. Natural numbers less than 5

33. Whole numbers that are not natural numbers

34. Integers between −4 and 4

Plot the graphs of the following numbers, using a different number line for each problem.

35. $-2, -\frac{1}{2}, 0, 3$

36. $-\frac{3}{4}, 1, \frac{3}{2}, \frac{5}{2}$

37. $-4, -1, \sqrt{2}, 4$

38. $-3, -2, \frac{1}{2}, \pi$

Replace the ? with the proper inequality symbol (< or >) to get a true statement.

39. 4 ? 8

40. −5 ? 0

41. −10 ? −20

42. $|-10|$? $|-20|$

43. $|-5|$? $|0|$

44. $|-8|$? $|4|$

[1–4, 1–5]

Find the sum or difference.

45. (−1) + (−3)

46. 6 − 3

47. 7 − 13

48. (−4) + (5)

49. (−8) + (2)

50. 7 − (−8)

51. (−8) − (−4)

52. (−3) − (6)

53. 0 + (−3) − (−7) + 3 − (+4)

54. 4 − 3 + 7 − 8 + 12 − (−3)

[1–6, 1–7]

Find the product or quotient. If a quotient does not exist, so state.

55. $3 \cdot (-7)$

56. $(-4) \cdot (-3)$

57. $(-8) \cdot (3) \cdot (-1)$

58. $8 \cdot (-9) \cdot (-1) \cdot (-2)$

59. $(-4) \cdot (3) \cdot (-5) \cdot 0$

60. $\dfrac{-14}{2}$

61. $\dfrac{-8}{-4}$

62. $24 \div (-4)$

63. $\dfrac{7}{0}$

64. $\dfrac{0}{-8}$

65. $\dfrac{0}{0}$

66. $\dfrac{(-2)(-3)}{-6}$

67. Find two integer factors of 36 whose sum is -13.

68. A man suffers successive financial losses of $3,000, $2,560, and $3,300 on three business transactions. A loss is denoted by a negative number, and the man originally had $52,000.

a. Write a statement using negative numbers representing his assets after the losses.

b. Find the total assets after the losses.

69. At 7 A.M. the temperature was $-17°$. At noon that same day the reading was 23°. How much of a rise in temperature was there from 7 A.M. to noon?

70. The temperature readings during a five-hour period were 63°, 72°, 80°, 75°, and 69°.

a. Represent by positive and negative integers how much rise (+) and fall (−) there was from hour to hour.

b. Was the numerical value of the total rise greater than, equal to, or less than that of the total fall? How much? (Represent by a positive or negative integer.)

c. If the sixth hour showed a drop of 11°, what was the temperature during the sixth hour? Write a statement involving a negative integer representing this answer.

[1–8]

Perform the indicated operations and simplify.

71. $(-5)^2$

72. -4^3

73. -4^2

74. -3^3

75. $100 - 4 \cdot 5 + 18$

76. $-7 + 14 \div 7 + 2$

77. $18 + 3 \cdot 12 \div 2^2 - 7$

78. $19 - (14 - 6) + 7^2 - 11$

79. $\dfrac{8(2-4)}{4} - \dfrac{35}{7}$

80. $4[8 - 2(5 - 3) + 1]$

81. $\left[\dfrac{8 + (-2)}{-3}\right]\left[\dfrac{14 \div (-2)}{-1}\right]$

82. $\left[\dfrac{(-12) + (-6)}{4}\right]\left[\dfrac{(-18)(-3)}{-9}\right]$

CHAPTER

2 Solving Equations and Inequalities

Bonnie has $3,000 invested at 8% simple interest per year. How much more money must she invest at 7% simple interest if she wants an income of $660 per year ($55 per month) from her investments?

2–1 ■ Algebraic notation and terminology

Algebraic terminology

In section 1–3, we defined a **variable** to be a symbol that represents an unspecified number. A variable is able to take on any one of the different values that it represents. In the relationship

$$y = 2x$$

y and x are variables since they both can assume various numerical values.

A constant is a symbol that does not change its value. In the relationship

$$y = 2x$$

2 is a constant. A number is a constant. If a symbol represents only one value, that symbol is a constant.

Any meaningful collection of variables, constants, grouping symbols, and signs of operations is called an algebraic expression. Examples of algebraic expressions would be

$$5xy, \quad \frac{xy}{z}, \quad 2\ell + 2w, \quad \frac{x^2 - 1}{x^2 + 1}, \quad 3x^2 + 2x - 1, \quad 5(a + 2b).$$

In an algebraic expression, terms are any constants, variables, or products or quotients of these. *Terms are separated by plus or minus signs.*

Example 2–1 A

Determine the number of terms in the algebraic expression.

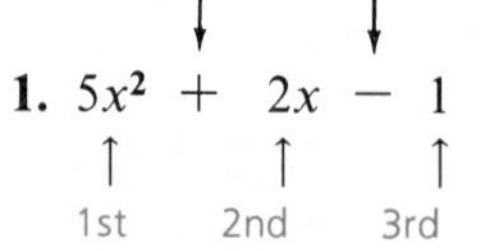

The plus and minus signs separate the algebraic expression into three terms

1. $5x^2 + 2x - 1$ (1st, 2nd, 3rd) — There are three terms

2. $x^2 + y^2$ (1st, 2nd) — There are two terms

3. $4x^5y^2z^4$ (1st) — There is one term

4. $a^2 + \dfrac{b + c^2}{d}$ (1st, 2nd) — There are two terms since the fraction bar forms a grouping. Observe that the second term has two terms in the numerator

▶ ***Quick check*** Determine the number of terms in $5 + x^2y - z$ and $4x^2 - \dfrac{2x + z}{y^2}$. ■

In the expression $5xy$, *each factor or grouping of factors is called the* **coefficient** *of the remaining factors.* That is, 5 is the coefficient of xy; x is the coefficient of $5y$; $5x$ is the coefficient of y; and so on. The 5 is called the **numerical coefficient,** and it tells us how many xy's we have in the expression.

Since we often talk about the numerical coefficients of a term, we will eliminate the word "numerical" and just say "coefficient." It will be understood that we are referring to the numerical coefficient. If no numerical coefficient appears in a term, the coefficient is *understood* to be 1.

Example 2–1 B

The algebraic expression $6x - 3y + z$ is thought of as the sum of terms $6x + (-3y) + z$, therefore 6 is the coefficient of x, -3 is the coefficient of y, and 1 is understood to be the coefficient of z.

▶ ***Quick check*** What are the coefficients in the algebraic expression $a^2 - 2a + 4b$? ■

A special kind of algebraic expression is a **polynomial.** The following are characteristics of a polynomial.

1. It has real number coefficients.
2. All variables in a polynomial are raised to only natural number powers.
3. The operations performed by the variables are limited to addition, subtraction, and multiplication.

A polynomial that contains just one term is called a **monomial;** a polynomial that contains two terms is called a **binomial;** and a polynomial that contains three terms is called a **trinomial.** Any polynomial that contains more than one term is called a **multinomial,** but no special names are given to polynomials that contain more than three terms.

■ ***Example 2–1 C***

Determine if each of the following algebraic expressions is a polynomial. If it is a polynomial, what name best describes it? If it is not a polynomial, state why it is not.

1. x, $4x$, 3, and $5x^2y$ are monomials.

2. $3x + 1$, $x + y$, and $81W^2 - 9T^2$ are binomials.

3. $5x^3 + 2y - 1$ and $z^2 + 9z - 10$ are trinomials.

4. $6x^3 - 2x^2 + 4x + 1$ is a polynomial of 4 terms.

5. $\dfrac{4}{x + 2}$ is not a polynomial since it contains a variable in the denominator.

Note We should simplify any expression, before identifying it. Also, in an expression, the combining of all of the constant terms is understood to be a single term. For example, $x + 3 + \pi$ is thought of as $x + (3 + \pi)$ and is a binomial.

▶ ***Quick check*** Determine if each is a polynomial. If it is, what name best describes it? If it is not, state why it is not.

$5x^2y + 2z$; $\quad 5x^2y + \dfrac{2}{z}$ ■

Another way that we identify different types of polynomials is by the degree of the polynomial. *The* **degree** *of a polynomial in one variable is the greatest exponent of that variable in any one term.*

■ ***Example 2–1 D***

Determine the degree of the polynomial.

1. $5x^3$ — Third degree because the exponent of x is 3

2. $x^4 - 2x^3 + 3x - 5$ — Fourth degree because the greatest exponent of x in any one term is 4

Note In example 2, the polynomial has been arranged in *descending powers* of the variable. This is the form that we will use when we write polynomials in one variable.

3. $4y^5 - 7y^2 + 3$ — Fifth degree because the greatest exponent of y in any one term is 5 ■

Algebraic notation

Many problems that we encounter will be stated verbally. These will need to be translated into algebraic expressions. While there is no standard procedure for changing a verbal phrase into an algebraic expression, the following guidelines should be of use.

1. Read the problem carefully, determining useful prior knowledge. Note what information is given and what information we are asked to find.
2. Let some letter represent one of the unknowns. Then express any other unknowns in terms of it.

3. Use the given conditions in the problem and the unknowns from step 2 to write an algebraic expression.

When translating verbal phrases into equations, we should be looking for phrases that involve the basic operations of addition, subtraction, multiplication, and division. Table 2–1 shows some examples of phrases that are commonly encountered. We will let x represent the unknown number.

Table 2–1

Phrase	Algebraic expression
Addition	
6 more than a number	$x + 6$
the sum of a number and 6	$x + 6$
6 plus a number	$x + 6$
a number increased by 6	$x + 6$
6 added to a number	$x + 6$
Subtraction	
6 less than a number	$x - 6$
a number diminished by 6	$x - 6$
the difference of a number and 6	$x - 6$
a number minus 6	$x - 6$
a number less 6	$x - 6$
a number decreased by 6	$x - 6$
6 subtracted from a number	$x - 6$
a number reduced by 6	$x - 6$
Multiplication	
a number multiplied by 6	$6x$
6 times a number	$6x$
the product of a number and 6	$6x$
Division	
a number divided by 6	$\frac{x}{6}$
the quotient of a number and 6	$\frac{x}{6}$
$\frac{1}{6}$ of a number	$\frac{x}{6}$

Example 2–1 E

Write an algebraic expression for each.

1. The product of a and b — $a \cdot b$
2. The sum of a and 4 — $a + 4$
3. x decreased by 9 — $x - 9$
4. y divided by 3 — $\frac{y}{3}$
5. A number increased by 6 — $n + 6$
6. Two times a number and that product decreased by 5 — $2n - 5$
7. A number divided by 3 and that quotient increased by 2 — $\frac{n}{3} + 2$

8. Twice the sum of x and 4 → $2(x + 4)$

▶ ***Quick check*** Write an algebraic expression for the product of x and y. Write an algebraic expression for a number increased by 6. ■

Mastery points

Can you

- Identify terms in an expression?
- Identify a polynomial?
- Write an algebraic expression?

Exercise 2–1

Specify the number of terms in each expression. See example 2–1 A.

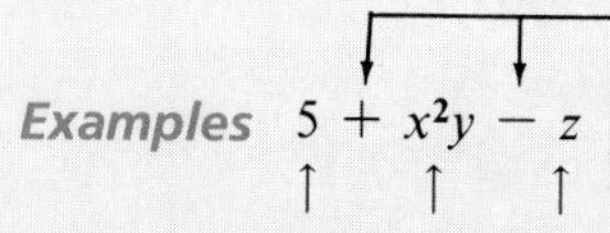

Examples $5 + x^2y - z$ $\qquad$ $4x^2 - \dfrac{2x + z}{y^2}$

Solutions 1st 2nd 3rd — Has three terms $\qquad$ 1st 2nd — Has two terms $\qquad$ Terms are separated by plus and minus signs

1. $3x + 4y$ **2.** $5xyz$ 3. $4x^2 + 3x - 1$ **4.** $x^3 - 4x + 7$

5. $\dfrac{6x}{5}$ **6.** $\dfrac{x}{3}$ 7. $8xy + \dfrac{5y}{2} - 6x$ **8.** $\dfrac{15x^2 + y}{8}$

9. $5x^3 + (3x^2 - 4)$ **10.** $x^2 + a^2(y^2 - z)$ **11.** $(x + y + z)$ **12.** 7

13. $a^2(b + c) - x^2(y + z)$ **14.** $x^2 + \dfrac{y - z}{a} + c$

Determine the numerical coefficients of the following algebraic expressions. See example 2–1 B.

Example $a^2 - 2a + 4b$

Solution 1 is understood to be the coefficient of a^2, -2 is the coefficient of a, 4 is the coefficient of b.

15. $5x^2 + x - 4z$ **16.** $a^2b + 4ab^2 - ab$ **17.** $x - y - 3z$

18. $3x^4 - x^2 + x^2$ **19.** $-2a - b + c$

Determine if each of the following algebraic expressions is a polynomial. If it is a polynomial, what name best describes it? If it is not a polynomial, state why it is not. See example 2–1 C.

Examples $5x^2y + 2z$ $\qquad$ $5x^2y + \dfrac{2}{z}$

Solutions It is a polynomial. Since there are two terms, it is a binomial. $\qquad$ Not a polynomial because a variable is used as a divisor (appears in the denominator)

20. $ax^2 + bx + c$ **21.** $mx + b$ **22.** $5x^2 + 2x$ 23. $y + \dfrac{1}{x}$

24. $\dfrac{a + b}{5} - c$ **25.** $\dfrac{a + b}{c} + d$ **26.** $4x^5 - 7x^3 + 3x - 2$ **27.** $9x^6 + 2x^2 + 4$

Write an algebraic expression for each of the following. See example 2–1 E.

Examples	The product of x and y	A number increased by 6
Solutions	$x \cdot y$	Let x represent the number; hence $x + 6$

28. The sum of a and b

29. 3 times a, subtracted from b

30. 7 less than x

31. 5 more than y

32. The sum of x and y, divided by z

33. x times the sum of y and z

34. a decreased by 5

35. a decreased by b

36. $\frac{1}{2}$ of x, decreased by 2 times x

37. A number decreased by 12

38. A number added to 4

39. 3 times a number and that product increased by 1

40. A number divided by 5

41. 2 times the sum of a number and 4

42. A number decreased by 6 and that difference divided by 11

Review exercises

Perform the indicated operations. See section 1–8.

1. -5^2

2. $(-8)^2$

3. $10 - 6 \cdot 2$

4. $25 - 5 \cdot 2$

5. $100 \div 10 \cdot 2 + 2$

6. $28 - (8 - 12) - 3^2$

2–2 ■ Evaluating algebraic expressions

Substitution property

An extremely important process in algebra is that of calculating the numerical value of an expression when we are given specific replacement values for the variables. This process is called **evaluation.** To perform evaluation, we need the following **property of substitution.**

Property of substitution

If $a = b$, then a may be replaced by b or b may be replaced by a in any expression without altering the value of the expression.

Concept

When two things are equal, they can replace each other anywhere.

We frequently need to evaluate algebraic expressions. By using the substitution property and the order of operations, we can calculate the numerical value of an algebraic expression. For example, to find the distance (d) traveled when the rate (r) and time (t) are known, we use

$$d = rt$$

If the rate is 45 miles per hour and the time is 3 hours, we can substitute these values into the expression as follows:

$$d = (45)(3) = 135$$

The distance traveled is 135 miles. We replaced the respective variables representing rate and time with their values. We then carried out the indicated arithmetic.

Note When replacing variables with the numbers they represent, it is a good procedure to put each of the numbers inside parentheses.

■ ***Example 2–2 A***

Evaluate the following expressions for the given real number replacement for the variable or variables.

1. $x^2 + 2x - 7$, when $x = 4$
The expression would be $(\quad)^2 + 2(\quad) - 7$ without the x.
Substituting 4 for each x, we have $(4)^2 + 2(4) - 7$. Using the order of operations, we have

$= 16 + 2(4) - 7$	Exponents
$= 16 + 8 - 7$	Multiply
$= 24 - 7$	Add
$= 17$	Subtract

Therefore the expression $x^2 + 2x - 7$ evaluated for $x = 4$ is 17.

2. $5a - b + 2(c + d)$, when $a = 2$, $b = 3$, $c = -2$, and $d = -3$.

$5a - b + 2(c + d)$	Original expression
$= 5(\quad) - (\quad) + 2[(\quad) + (\quad)]$	Expression ready for substitution
$= 5(2) - (3) + 2[(-2) + (-3)]$	Substitute
$= 5(2) - (3) + 2[-5]$	Order of operations, groups
$= 10 - (3) + (-10)$	Multiply
$= 7 + (-10)$	Subtract
$= -3$	Add

3. $4ab - c^2 + 3d$, when $a = 2$, $b = 3$, $c = -2$, and $d = -3$.

$4ab - c^2 + 3d$	Original expression
$= 4(\quad)(\quad) - (\quad)^2 + 3(\quad)$	Expression ready for substitution
$= 4(2)(3) - (-2)^2 + 3(-3)$	Substitute
$= 4(2)(3) - (4) + 3(-3)$	Order of operations, exponents
$= 8(3) - 4 + 3(-3)$	Multiply
$= 24 - 4 + (-9)$	Multiply
$= 20 + (-9)$	Subtract
$= 11$	Add

▶ ***Quick check*** Evaluate $3a - 2(c - d) + b$ when $a = 2$, $b = 3$, $c = -2$, and $d = -3$ ■

To evaluate an algebraic expression

1. Write parentheses in place of each variable.
2. Place the value that the variable is representing inside the parentheses.
3. Perform the indicated operations according to the order of operations.

Formulas

A **formula** expresses a relationship between quantities in the physical world, for example, $d = rt$.

■ Example 2–2 B

Evaluate the following formulas for the real number replacements of the variables.

1. The volume (V) of a rectangular solid is found by multiplying length (ℓ) times width (w) times height (h). The formula then reads $V = \ell wh$. Find the volume in cubic feet if $\ell = 12$ feet, $w = 4$ feet, and $h = 5$ feet.

$V = \ell wh$	Original formula
$V = (\quad)(\quad)(\quad)$	Formula ready for substitution
$V = (12)(4)(5)$	Substitute
$V = (48)(5)$	Order of operations, multiply
$V = 240$	Multiply

The volume is 240 cubic feet.

2. If we know the temperature in degrees Fahrenheit (F), the temperature in degrees Celsius (C) can be found by the formula $C = \frac{5}{9}(F - 32)$. Find the temperature in degrees Celsius if the temperature is 86 degrees Fahrenheit.

$C = \frac{5}{9}(F - 32)$	Original formula
$C = \frac{5}{9}[(\quad) - 32]$	Formula ready for substitution
$C = \frac{5}{9}[(86) - 32]$	Substitute
$C = \frac{5}{9}(54)$	Order of operations, groups first
$C = 30$	Multiply

The temperature is 30 degrees Celsius.

3. The perimeter* (P) of a rectangle is found by the formula $P = 2\ell + 2w$, where ℓ is the length of the rectangle and w is the width. Find the perimeter of the rectangle in meters if $\ell = 8$ meters and $w = 5$ meters.

$P = 2\ell + 2w$	Original formula
$P = 2(\quad) + 2(\quad)$	Formula ready for substitution
$P = 2(8) + 2(5)$	Substitute
$P = 16 + 10$	Order of operations, multiply
$P = 26$	Add

The perimeter of the rectangle is 26 meters.

*The perimeter is the distance around a closed geometric figure.

4. A formula in electricity is $I = \dfrac{E}{R}$, where I represents the current measured in amperes in a certain part of a circuit, E represents the potential difference in voltage across that part of the circuit, and R represents the resistance in ohms in that part of the circuit. Find I in amperes if $E = 110$ volts and $R = 44$ ohms.

$I = \dfrac{E}{R}$ Original formula

$I = \dfrac{(\ )}{(\ )}$ Formula ready for substitution

$I = \dfrac{(110)}{(44)}$ Substitute

$I = \dfrac{5}{2} = 2\dfrac{1}{2}$ Reduce and change to a mixed number

The current is $2\frac{1}{2}$ amperes.

▶ ***Quick check*** Evaluate $I = \dfrac{E}{R}$ when $E = 220$ volts and $R = 11$ ohms. ■

Subscripts

In some formulas, two or more measurements of the same unit may be given. It is customary to label these by using **subscripts.** To illustrate, given two different measurements of pressure in a science experiment, we might label them

$$P_1 \text{ and } P_2$$

The 1 and 2 are the subscripts. Subscripts are always written to the lower right of the letter. The symbols above are read "P sub-one" and "P sub-two."

Note Do not confuse a subscript, such as P_2, that helps distinguish between different measurements, and an exponent, such as P^2, that indicates the number of times a given base is used as a factor in an indicated product. Subscripts are written to the lower right of the symbol. Exponents are written to the upper right of the symbol.

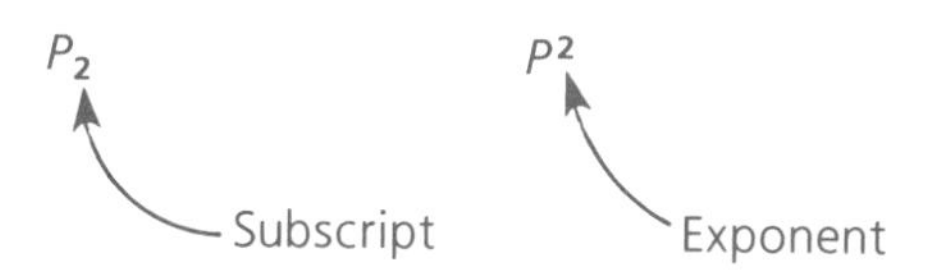

■ ***Example 2–2 C***

1. In a science problem $\dfrac{V_1}{V_2} = \dfrac{T_1}{T_2}$, where V_1 and V_2 represent different measurements of volume and T_1 and T_2 represent different measurements of temperature.

2. In $\frac{1}{R_t} = \frac{1}{R_1} + \frac{1}{R_2} + \frac{1}{R_3}$, R_1, R_2, and R_3 represent three different measurements of resistance and R_t represents the total resistance in the electrical circuit.

▶ ***Quick check*** Evaluate $C_t = \frac{C_1 \cdot C_2}{C_1 + C_2}$ when $C_1 = 4$ and $C_2 = 6$. ■

Problem solving

The following sets of word problems are designed to help us interpret word phrases and write expressions for them in algebraic symbols.

■ ***Example 2–2 D***

Write an algebraic expression for each of the following word phrases.

1. Nancy can type 90 words per minute. How many words can she type in n minutes?

   If Nancy can type 90 words in one minute, then we multiply

   $90 \cdot n$ or $90n$

   to obtain the number of words she can type in n minutes.

2. If John has n dollars in his savings account and on successive days he deposits \$15 and then withdraws \$34 to make a purchase, write an expression for the balance in his savings account.

   We *add* the deposits and *subtract* the withdrawals. Thus

   $n + 15 - 34$

   represents the balance in John's savings account after the two transactions.

3. A woman paid d dollars for a 30-pound bag of dog food. How much did the dog food cost her per pound?

   The price per pound is found by dividing the total cost by the number of pounds. Thus the price per pound of the dog food is represented by $\frac{d}{30}$ dollars.

▶ ***Quick check*** Express the cost in dollars of x cassette tapes if each tape costs \$2.95. ■

Mastery points

Can you

- Evaluate an algebraic expression?
- Evaluate a formula?
- Use subscripts?
- Write an algebraic expression?

Exercise 2–2

Evaluate the following expressions if $a = 2$, $b = 3$, $c = -2$, $d = -3$. See example 2–2 A.

Example $3a - 2(c - d) + b$

Solution

$= 3(\ \ ) - 2[(\ \ ) - (\ \ )] + (\ \ )$	Expression ready for substitution
$= 3(2) - 2[(-2) - (-3)] + (3)$	Substitute
$= 3(2) - 2(1) + 3$	Order of operations, groups
$= 6 - 2 + 3$	Multiply
$= 7$	Subtract and add

1. $2a + b - c$
2. $(a + b)$
3. $3a - 2b - (c + d)$
4. $a - 3(c + b)$
5. $(3a + 2b)(a - c)$
6. $2ab(c + d)$
7. $ac - bd$
8. $3c - 2(3a + b)$
9. $7a - d(6b + c)$
10. $(3a - 5c)(2b - 4d)$
11. $(5c - 3a)(4d - 2b)$
12. $(5c - d)4a$
13. $5a + 7b - 3c(a - d)$
14. $(4a + b) - (3a - b)(c + 2d)$
15. $a^2 - c^2$
16. $b^2 + 2d^2$
17. $3ab - 4c^2 + d$
18. $(c + d)^2$
19. $(c - d)^2$
20. $a^2b^2 + c^2d^2$
21. $3ac - 2a^2c^2$
22. $ab - ac$
23. $(ab)^2 - (ac)^2$
24. $(c - d)^2(a + b)$
25. $a^3b - c^3d$
26. $c^2 - d^3$
27. $3d^2 - 2c^3$
28. $(3d - 5c)^3$

Evaluate the following formulas. See examples 2–2 B and C.

Examples $I = \frac{E}{R}$, $E = 220$ and $R = 11$ $\qquad C_t = \frac{C_1 \cdot C_2}{C_1 + C_2}$, $C_1 = 4$ and $C_2 = 6$

Solutions

$I = \frac{(\ \ )}{(\ \ )}$	$C_t = \frac{(\ \ )(\ \ )}{(\ \ ) + (\ \ )}$	Formulas ready for substitution
$= \frac{(220)}{(11)}$	$= \frac{(4)(6)}{(4) + (6)}$	Substitute
$= 20$	$= \frac{24}{10}$	Order of operations
	$= \frac{12}{5}$	Reduce

29. $I = \frac{E}{R}$, $E = 220$ and $R = 33$
30. $V = \ell wh$, $\ell = 7$, $w = 5$, and $h = 6$
31. $I = prt$, $p = 1{,}000$; $r = 0.08$; and $t = 2$
32. $F = ma$, $m = 18$ and $a = 6$
33. $W = I^2R$, $I = 12$ and $R = 2$
34. $V = k + gt$, $k = 24$, $g = 9$, and $t = 4$
35. $A = \frac{1}{2}h(b_1 + b_2)$, $h = 6$, $b_1 = 8$, and $b_2 = 10$
36. $\ell = a + (n - 1)d$, $a = 2$, $n = 14$, and $d = 3$
37. $A = p + pr$, $p = 2{,}000$ and $r = 0.07$
38. $H = \frac{D^2N}{2}$, $D = 4$ and $N = 6$

39. $A = \dfrac{I^2R - 120E^2}{R}$, $E = 5$, $I = 12$, and $R = 100$

40. $C_t = \dfrac{C_1 \cdot C_2}{C_1 + C_2}$, $C_1 = 6$ and $C_2 = 12$

41. $S = \dfrac{1}{2}gt^2$, $g = 32$ and $t = 4$

42. $R_t = \dfrac{R_1 \cdot R_2}{R_1 + R_2}$, $R_1 = 8$ and $R_2 = 12$

43. $V_1 = \dfrac{V_2P_2}{P_1}$, $V_2 = 18$, $P_2 = 12$, and $P_1 = 36$

44. $V_2 = \dfrac{V_1T_2}{T_1}$, $V_1 = 12$, $T_1 = 4$, and $T_2 = 14$

Evaluate the following formulas. See examples 2–2 B and C.

45. Find the horsepower (h) required by a hydraulic pump when it needs to pump 10 gallons per minute (g) and the pounds per square inch (p) equal 3,000. Use $h = \dfrac{g \cdot p}{1{,}714}$.

46. The required ratio of gearing (R) of a milling machine is given by $R = (A - N) \cdot \dfrac{40}{A}$, where N = required number of divisions and A = approximate number of divisions. Find R when $A = 280$ and $N = 271$.

47. In a gear system, the velocity (V) of the driving gear is defined by $V = \dfrac{vn}{N}$, where v = velocity of follower gear, n = number of teeth of follower gear, and N = number of teeth of driving gear. Find V when $v = 90$ revolutions per minute, $n = 30$ teeth, $N = 65$ teeth.

48. The tap drill size (T) of a drill needed to drill threads in a nut is given by $T = D - \dfrac{1}{N}$, where D = diameter of the tap and N = number of threads per inch. Find T with a $\dfrac{1}{2}$-inch diameter tap and 13 threads per inch.

49. It is necessary to drag a box 600 feet across a level lot in 3 minutes. The force required to pull the box is 2,000 pounds. What is the horsepower (h) needed to do this if horsepower is defined by $h = \dfrac{\ell \cdot w}{33{,}000 \cdot t}$, where ℓ = length to be moved, w = force exerted through distance ℓ, and t = time in minutes required to move the box through ℓ?

50. A pulley 12 inches in diameter that is running at 320 revolutions per minute is connected by a belt to a pulley 9 inches in diameter. How many revolutions per minute will the smaller pulley make if $s = \dfrac{SD}{d}$, where s = speed of smaller pulley and d = diameter of smaller pulley, S = speed of larger pulley and D = diameter of larger pulley?

Write an algebraic expression for the following word statements. See example 2–2 D.

Example Express the cost in terms of dollars and in terms of cents for x cassette tapes if each tape costs $2.95.

Solution If we are buying x tapes at $2.95 each, then we must multiply x by $2.95. The algebraic expression in terms of dollars would be $2.95 \cdot x$ and in terms of cents, it would be $295 \cdot x$.

51. Jim enters 85 keystrokes per minute on the computer. How many keystrokes can he enter in m minutes?

52. Express the cost in dollars of h gallons of heating oil if each gallon costs $1.08.

53. A 10-pound box of candy costs y dollars. How much does the candy cost per pound?

54. Mike paid $25 for a ticket to a play. If the play lasted h hours, what did it cost him per hour to see the play?

55. Arlene has n nickels and d dimes in her purse. Express in cents the amount of money she has in her purse. (*Hint:* n nickels is represented by $5n$.)

56. Jack has q quarters, d dimes, and n nickels. Express in cents the amount of money Jack has.

57. Susan is p years old now. Express her age (a) 12 years from now, (b) 5 years ago.

58. Ann is 3 years old. If Jan is n times as old as Ann, express Jan's age. Express-Jan's age 8 years ago.

59. Bill's savings account has a current balance of \$258. He makes a withdrawal of n dollars and then makes a deposit of m dollars. Express his new balance in terms of n and m.

60. Paula has a balance of n dollars in her checking account. She makes a deposit of \$36 and then writes 3 checks for m dollars each. Express her new balance in terms of n and m.

61. Pete has c cents, all in half-dollars. Write an expression for the number of half-dollars Pete has.

62. If x represents a whole number, write an expression for the next greater whole number.

63. If y represents an even integer, write an expression for the next greater even integer.

64. If z represents an odd integer, write an expression for the next greater odd integer.

65. If Larry is f feet and t inches tall, how tall is Larry in inches?

66. John earns \$1,000 more than twice what Terry earns in a year. If Terry earns d dollars, write an expression for John's annual salary.

67. Jean's annual salary is \$2,000 less than n times Lisa's salary. If Lisa earns \$25,000 per year, express Jean's annual salary.

68. Express the total cost of purchasing x cans of tuna at 69¢ per can on Friday and y cans of the same tuna at 57¢ per can on Saturday.

69. A gallon of primer paint costs \$9.95 and a gallon of latex-base paint costs \$12.99. Express the cost in dollars of p gallons of primer and q gallons of latex-base paint.

70. Paula enters x calculations per minute on the calculator and Leigh enters 7 calculations per minute less than Paula. Write an expression for the number of calculations Leigh enters in 35 minutes.

Review exercises

Perform the indicated operations. See sections 1–4 and 1–5.

1. $(-12) + 6$

2. $4 + (-8)$

3. $10 - 18$

4. $9 - (-9)$

5. $-6 - (-6)$

6. $(-14) + (-7)$

2–3 ■ Algebraic addition and subtraction

Since algebraic expressions (including polynomials) represent real numbers when the variables are replaced by real numbers, the ideas and properties that apply to operations with real numbers also apply to algebraic expressions.

The distributive property

The distributive property is the only property that establishes a relationship between addition (or subtraction) and multiplication in the same expression. The distributive property allows us to change certain multiplication problems into sums or differences.

Distributive property

For every real number a, b, and c,

$$a(b + c) = ab + ac \text{ and } a(b - c) = ab - ac$$

Concept

If a number is being used to multiply the sum or difference of two others, it is "distributed" to them both. That is, it multiplies them both.

■ ***Example 2–3 A***

The following are applications of the distributive property.

Each term inside the parentheses is multiplied by 3

1. $3(4 + 5) = (3 \cdot 4) + (3 \cdot 5)$
$= 12 + 15$
$= 27$

Note Since we are able to add the numbers inside the parentheses, our solution without using the distributive property would be

$$3(4 + 5) = 3(9) = 27$$

2. $2(3 + a) = 2 \cdot 3 + 2 \cdot a$
$= 6 + 2a$

Note In this example, we could not add the numbers inside the parentheses. Therefore the multiplication could only be carried out by using the distributive property. ■

We are now going to use the distributive property to carry out addition and subtraction of algebraic expressions and to remove grouping symbols.

Like terms

We first need to define the types of quantities that can be added or subtracted. *We can add or subtract only like, or similar, quantities.* **Like terms or similar terms are terms whose variable factors are the same.**

Note For two or more terms to be called like terms, the variable factors of the terms, along with their respective exponents, must be identical. However the numerical coefficients of these identical variable factors may be different.

■ ***Example 2–3 B***

1. $3a^2b^3$ and $-2a^2b^3$ are like terms because the variables are the same (a and b) and the respective exponents are the same (a is to the second power in each term and b is to the third power).
2. $2x^2y$ and $2xy^2$ both contain the same variables but are *not* like terms because the exponents of the respective variables are not the same.

▶ ***Quick check*** Are $4a^2b^2$ and $4a^3b^3$ like terms? ■

Addition and subtraction

Using the definition of like terms and the distributive property, we are ready to carry out addition and subtraction of algebraic expressions. Consider the following example:

$$3a + 4a$$

Using the distributive property, the expression can be written

$$3a + 4a = (3 + 4)a = 7a$$

Note The process of addition or subtraction is performed only with the numerical coefficients. *The variable factor and its exponent remain unchanged.*

Combining like terms

1. Identify the like terms.
2. If necessary, use the commutative and associative properties to group together the like terms.
3. Combine the numerical coefficients of the like terms and multiply that by the variable factor.
4. Remember that y is the same as $1 \cdot y$ and $-y$ is the same as $-1 \cdot y$.

Example 2–3 C

Perform the indicated addition and subtraction.

1. $5x + 7x$ — Identify like terms
$= (5 + 7)x$ — Distributive property
$= 12x$ — Add numerical coefficients

2. $4ab + 3ab$ — Identify like terms
$= (4 + 3)ab$ — Distributive property
$= 7ab$ — Add numerical coefficients

3. $y + 3y - 2y$ — Identify like terms
$= (1 + 3 - 2)y$ — Distributive property
$= 2y$ — Combine numerical coefficients

4. $2x + 6y + 5x - 3y$ — Identify like terms
$= 2x + 5x + 6y - 3y$ — Commutative property
$= (2x + 5x) + (6y - 3y)$ — Associative property
$= (2 + 5)x + (6 - 3)y$ — Distributive property
$= 7x + 3y$ — Combine numerical coefficients

Note Because of the commutative and associative properties, we can rearrange the expression and combine like terms.

5. $6x^2 - 4x + 3x - 2x^2 = (6x^2 - 2x^2) + (-4x + 3x)$
$= (6 - 2)x^2 + (-4 + 3)x$
$= 4x^2 - 1x = 4x^2 - x$

6. $5x^2y^2 - 2xy^2 + 3x^2y^2 + 5xy^2 = (5x^2y^2 + 3x^2y^2) + (-2xy^2 + 5xy^2)$
$= (5 + 3)x^2y^2 + (-2 + 5)xy^2$
$= 8x^2y^2 + 3xy^2$

Note After sufficient practice, we should be able to carry out the addition and subtraction by grouping mentally.

Like terms

7. $a^2b + 5ab^2 - 4a^2b + 3ab^2 = (1 - 4)a^2b + (5 + 3)ab^2$
$= -3a^2b + 8ab^2$

Like terms

▶ ***Quick check*** Perform the indicated addition and subtraction:
$3a - 2b + a + 5b$; $3a^2 + 5a - 2a^2 - a$

Grouping symbols

In chapter 1, we learned that any quantity enclosed within grouping symbols is treated as a single number. We are now going to use the distributive property to remove grouping symbols such as ( ), [ ], and { }. Consider the following examples:

1. The quantity $(2a + 3b)$ can be written as $1 \cdot (2a + 3b)$. Applying the distributive property, we have

$$1(2a + 3b) = 1 \cdot 2a + 1 \cdot 3b = 2a + 3b$$

2. The quantity $+(2a + 3b)$ can be written as $(+1) \cdot (2a + 3b)$ giving

$$(+1)(2a + 3b) = (+1) \cdot 2a + (+1) \cdot 3b = 2a + 3b$$

3. The quantity $-(2a + 3b)$ can be written as $(-1) \cdot (2a + 3b)$ giving

$$(-1)(2a + 3b) = (-1) \cdot 2a + (-1) \cdot 3b = -2a - 3b$$

Removing grouping symbols

1. If an expression inside a grouping symbol is preceded by no symbol or by a "+" sign, the grouping symbol can be dropped and the enclosed terms remain unchanged.
2. If an expression inside a grouping symbol is preceded by a "−" sign, when the grouping symbol is dropped, we change the sign of each enclosed term.

■ Example 2–3 D

Remove all grouping symbols and perform the indicated addition or subtraction.

1. $(3x^2 + 2x + 5) + (4x^2 + 3x + 6)$ — Remove grouping symbols

$= 3x^2 + 2x + 5 + 4x^2 + 3x + 6$ — Enclosed terms remain unchanged

$= (3x^2 + 4x^2) + (2x + 3x) + (5 + 6)$ — Associative and commutative properties

$= (3 + 4)x^2 + (2 + 3)x + 11$ — Distributive property

$= 7x^2 + 5x + 11$ — Combine numerical coefficients

2. $(3x^2 - x + 4) - (2x^2 - 5x - 7)$ — Remove grouping symbols

$= 3x^2 - x + 4 - 2x^2 + 5x + 7$ — Change the sign of each term contained in the second set of parentheses

$= (3x^2 - 2x^2) + (-x + 5x) + (4 + 7)$ — Associative and commutative properties

$= (3 - 2)x^2 + (-1 + 5)x + 11$ — Distributive property

$= 1x^2 + 4x + 11$ — Combine numerical coefficients

$= x^2 + 4x + 11$ — x^2 is the same as $1x^2$

Note In the following examples, we will mentally add or subtract the like terms.

3. $(a^2 + 2ab + b^2) - (3a^2 - 4ab + b^2)$ — Remove grouping symbols

$= a^2 + 2ab + b^2 - 3a^2 + 4ab - b^2$ — Change the sign of each term in the second parentheses

$= (a^2 - 3a^2) + (2ab + 4ab) + (b^2 - b^2)$ — Associative and commutative properties

$= -2a^2 + 6ab + 0$ — Combine like terms

$= -2a^2 + 6ab$ — No b^2 is left

4. $(8R^2 - 2R + 3) - (6R^2 + 6R - 1)$ — Remove grouping symbols

$= 8R^2 - 2R + 3 - 6R^2 - 6R + 1$ — Change the sign of each term in the second parentheses

(Like terms: $8R^2$ and $-6R^2$; Like terms: $-2R$ and $-6R$; Like terms: 3 and 1)

$= 2R^2 - 8R + 4$ — Combine like terms

▶ ***Quick check*** Remove all grouping symbols and perform the indicated addition or subtraction: $(5x^2 + 2x - 1) - (3x^2 - 4x + 3)$ ■

There are many situations where there will be grouping symbols within grouping symbols. In these situations, *it is usually easier to remove the innermost grouping symbol first.*

■ Example 2–3 E

Remove all grouping symbols and perform the indicated addition or subtraction.

1. $2x - [y + (x - z)]$

$= 2x - [y + x - z]$ — Remove parentheses first

$= 2x - y - x + z$ — Next remove brackets

$= x - y + z$ — Combine like terms

2. $2a - [3b - (2a - b)]$

$= 2a - [3b - 2a + b]$ — Remove parentheses first

$= 2a - [4b - 2a]$ — Combine like terms within the brackets

$= 2a - 4b + 2a$ — Remove brackets

$= 4a - 4b$ — Combine like terms

Note After removing the parentheses, we added the like terms before removing the brackets. *Simplify inside grouping symbols as much as possible before going on.*

3. $(3R - 2S) - [5R - (R - S)]$

$= 3R - 2S - [5R - R + S]$ — Remove both sets of parentheses

$= 3R - 2S - [4R + S]$ — Combine like terms within brackets

$= 3R - 2S - 4R - S$ — Remove brackets

$= -R - 3S$ — Combine like terms

Note There were two separate sets of grouping symbols here. As long as they are separate, we may remove both sets at the same time.

▶ ***Quick check*** Remove all grouping symbols and perform the indicated addition or subtraction: $5a - \{2a + [5b - 3a]\}$ ■

Mastery points

Can you

- Identify like terms?
- Add and subtract algebraic expressions?
- Remove grouping symbols?

Exercise 2–3

For the groups of terms, write like or unlike. See example 2–3 B.

Example $4a^2b^2$ and $4a^3b^3$

Solution Both contain the same variables but are **unlike** because the exponents of the respective variables are not the same.

1. $3a, -2a$ **2.** $5x, 7x$ **3.** $4a^2, a^2$

4. $b^3, -2b^3$ **5.** $2a^2, 2a^3$ **6.** $4x, 4x^2$

Perform the indicated addition and subtraction. See examples 2–3 A, B, and C.

Examples $3a - 2b + a + 5b$ and $3a^2 + 5a - 2a^2 - a$

Solutions

$= (3a + a) + (-2b + 5b)$	$= (3a^2 - 2a^2) + (5a - a)$	Commutative and associative properties
$= (3 + 1)a + (-2 + 5)b$	$= (3 - 2)a^2 + (5 - 1)a$	Distributive property
$= 4a + 3b$	$= 1a^2 + 4a$	Combine numerical coefficients
	$= a^2 + 4a$	

7. $2x + x + 6x$

8. $8y - y + 2y$

9. $4a - 2b + 9a + 4b$

10. $a + 4b + 6a - 8b$

11. $3x + 4x + 7x$

12. $2a^2b - 4a^2b + 6a^2b$

13. $4ab + 11ab - 10ab - 8ab$

14. $d^2 + d - 3d^2 + d^4 + 4d^2$

15. $5x + x^2 - x + 6x^2$

16. $5x^2y - 3xy + 5y + 6xy - x^2y$

17. $a^2b - b^3 - ab^2 + 2a^3 - 5ab^2$

18. $x + 2x^2 - 5 + x^3 - 2x - 2x^2$

19. $3a + b + 2a - 5c - b - 2x^2 + 8a$

20. $3a + 8a - 6a + 9a$

21. $3a + 8b - 6a - 17b$

22. $28ab - 73ab + ab + 11ab - 9ab$

23. $4x^2 - y^2 - x^2 + 12y^2$

24. $5a + 4a^2 - 2a - a^2$

25. $x^2 + 5x - 8x + 2x^2$

26. $8ab + 7a^2b + 6a^2b^2 - 4a^2b$

27. $x^2y^2 + 9xy - 2x^2y - 4xy$

28. $a^2b + 8ab + 3a^2b - 4a^2b^2$

29. $x^2 + 5x - 6 + 7x^2 - 3x + 7$

30. $6a^2 - 5a + 3 - 2a^2 - 4a + 8$

Remove all grouping symbols and combine like terms. See examples 2–3 D and E.

Example $(5x^2 + 2x - 1) - (3x^2 - 4x + 3)$

Solution
$= 5x^2 + 2x - 1 - 3x^2 + 4x - 3$ Change the sign of each term in the second parentheses
$= (5x^2 - 3x^2) + (2x + 4x) + (-1 - 3)$ Commutative and associative properties
$= 2x^2 + 6x - 4$ Combine like terms

Example $5a - \{2a + [5b - 3a]\}$

Solution
$= 5a - \{2a + 5b - 3a\}$ Remove brackets
$= 5a - \{-a + 5b\}$ Combine like terms within braces
$= 5a + a - 5b$ Remove braces
$= 6a - 5b$ Combine like terms

31. $(2x + 3y) + (x + 5y)$

32. $(4a - b) + (3a + 2b)$

33. $(5x + y) - (3x - 2y)$

34. $(7x - 3y) - (5x - 6y)$

35. $(3a - b + 4c) - (a - 2b - c)$

36. $(4x - 3y - 2z) - (3x - 4y - z)$

37. $(8x + 3y - 4z) - (6x - y - 4z)$

38. $(7a - b - 3c) - (5a - 4b + 3c)$

39. $(2x^2y - xy^2 + 7xy) + (xy^2 - 5x^2y + 8xy)$

40. $(5x^2 - y^2) - (6x^2 - 3y^2) - (8x^2 + 2y^2)$

41. $(8a^3 - 2a^2b + 4ab^2 - 6b^3) - (4a^3 - 3a^2b - 2ab^2 - b^3)$

42. $(13a - 24bc) + (46bc - 16a - 26d)$

43. $(48a + 3b) - (-22a - 6b)$

44. $(3x^2y - 6xy + 32z) + (7xy - 3x^2y)$

45. $(8xy + 9y^2z) - (13xy - 14yz)$

46. $(18a + 31b) - (23a - 14bc)$

47. $(a - 3b + 2) - (a + 5b - 8)$

48. $(2x + 6z - 10y) - (8y + 3z - 6x + 4)$

49. $(3a - 2b) - (a + 4b) - (-a + 3b)$

50. $(5xy - y) - (3yz + 2xy) + (3y - 4xy)$

51. $(7x^2 - 2y) + (3z - 4y) - (4x^2 - 6y)$

52. $2x - [3x - (5x - 3)]$

53. $x - 1 + [2x - (x - 1)]$

54. $3a + [2a - (a - b)]$

55. $5x - [4a + 3b + (x - 2y)]$

56. $2a - [a - b - (3a + 2b)]$

57. $5a - (a + b) - [2a - b - (3a + 5b)]$

58. $x - [y + (2x - 3y)] + [2x - y]$

59. $-[4a + 7b - (3a + 5b)]$

60. $6x - \{5a + y + (4x - 7y)\}$

61. $2a + [a - (b - c)] - [2a - (b - c)]$

62. $3x - [6x - (4x - 3y)] - [4y - 3x]$

Review exercises

Write an algebraic expression for each of the following. See section 2–1.

1. The product of x and 3

2. 6 times the sum of a and 7

3. y decreased by 2 and that difference divided by 4

4. A number multiplied by 5

5. A number diminished by 12

6. A number divided by 8 and that quotient decreased by 9

2–4 ■ The addition and subtraction property of equality

Equations

An **equation** is a statement of equality. If two expressions represent the same number, then placing an equality sign, =, between them forms an equation. The following diagram is used to show the parts of an equation.

$$3x - 7 = 2x + 5$$

Left member of the equation — Equality sign — Right member of the equation

A **mathematical statement** is a mathematical sentence that can be labeled true or false. $2 + 3 = 5$ is a true statement, and $3 + 4 = 8$ is a false statement.

Other mathematical sentences, such as

$$2 + x = 8 \text{ and } 4 - x = 7,$$

cannot be labeled as true or false. Such sentences are called **open sentences.** The truth or falsity of the sentence is "open" since we do not know the value that the variable represents.

Solution set

A replacement value for the variable that forms a true statement is called a **root,** or a **solution,** of the equation. We say that a root of the given equation *satisfies* that equation. The **solution set** is the set of all values for the variable that cause the equation to be a true statement.

To check the solution of an equation

1. **Substitute:** Replace the variable in the original equation with the solution.
2. **Order of operations:** Perform the indicated operations.
3. **True statement:** If step 2 produces a true statement, the solution is correct.

■ Example 2–4 A

Determine if the given value is a solution of the equation.

1. $2 + x = 8$ when $x = 6$

If we replace x by 6 in the equation and simplify,

$$\begin{aligned} 2 + x &= 8 \\ 2 + (6) &= 8 \\ 8 &= 8 \end{aligned}$$

the equation is true. Then 6 is the root of the equation. The only solution to this equation is 6, and the solution set would then be $\{6\}$.

2. $4 - x = 7$ when $x = -3$

If we replace x by -3 in the equation and simplify,

$$\begin{aligned} 4 - x &= 7 \\ 4 - (-3) &= 7 \\ 7 &= 7 \end{aligned}$$

the equation is true, and -3 is the root of the equation. The solution set is $\{-3\}$.

▶ ***Quick check*** Determine if the equation $3x + 3 = 6$ is true when $x = 1$. Determine if the equation $2x - 1 = 3$ is true when $x = 4$. ■

Types of equations

An equation that is true for some values of the variable and false for other values of the variable is called a **conditional equation.** The equation $2 + x = 8$ is a conditional equation since it is true when $x = 6$ and false otherwise.

If the equation is true for every permissible value of the variable, it is called an **identical equation,** or **identity.** For example,

$$2(x + 3) = 2x + 6$$

is true for any real number replacement for x and is thus an identity. Properties such as

$$a + (b + c) = (a + b) + c \text{ and } a \cdot b = b \cdot a$$

are further examples of identities.

We will study conditional equations in this chapter. These equations will be *first-degree conditional* equations, also called **linear equations.** A linear equation is an equation where the exponent of the unknown is 1 and the solution set will contain at most one root. The equations $2 + x = 8$ and $4 - x = 7$ are linear equations since the variable, x, is to the first power. There is only one replacement value for the variable that will satisfy each equation. That is, for $2 + x = 8$, the solution set is $\{6\}$ since $2 + (6) = 8$, and for $4 - x = 7$, the solution set is $\{-3\}$ since $4 - (-3) = 7$.

Equivalent equations

So far we have looked at such linear equations as $2 + x = 8$ and $4 - x = 7$ for which the solution sets could be determined by inspection. Unfortunately, the majority of equations cannot be solved by inspection. We must develop a procedure for finding the roots.

If we wish to solve a more complicated equation, such as

$$2 + x + 3x - 4 = 2x + 4 + x,$$

the solution is not so obvious. To solve such an equation, we go through a series of steps whereby we form equations that are equivalent to the original equation until we have the equation in the form $x = n$, n being some real number. These equations that we form are called **equivalent equations.** *Equivalent equations are equations whose solution set is the same.*

■ Example 2–4 B

The following are equivalent equations whose solution set is $\{6\}$.

1. $2 + x + 3x - 4 = 2x + 4 + x$ **2.** $4x - 2 = 3x + 4$

3. $x - 2 = 4$ **4.** $x = 6$ ■

Addition and subtraction property of equality

Since an equation is a statement of equality between two expressions, identical quantities added to or subtracted from each expression will produce an equivalent equation. We can state this property as follows:

Addition and subtraction property of equality

For any algebraic expressions a, b, and c,

$$\text{if } a = b, \text{ then } a + c = b + c \text{ and } a - c = b - c$$

Concept

We can add or subtract the same quantity in each member of an equation and the result will be an equivalent equation.

Consider the equation $x - 2 = 4$. To determine the solution by means other than inspection, we want to form an equivalent equation of the form $x = n$. This can be done by applying the addition and subtraction property of equality. We add 2 to both members of the equation and then simplify each member separately.

$$x - 2 = 4$$
$$x - 2 + 2 = 4 + 2$$
$$x = 6$$

The root is 6, and the solution set is represented as $\{6\}$.

Check: $(6) - 2 = 4$ Substitute
$4 = 4$ True

Example 2–4 C

Find the solution set and check the answer.

1.
$x - 5 = 7$
$x - 5 + 5 = 7 + 5$ Add 5 to both members
$x + 0 = 12$ Additive inverse
$x = 12$ Solution

Check:
$(12) - 5 = 7$ Substitute
$7 = 7$ True

The solution set is $\{12\}$.

2.
$x + 4 = 12$
$x + 4 - 4 = 12 - 4$ Subtract 4 from both members
$x = 8$ Solution

Check:
$(8) + 4 = 12$ Substitute
$12 = 12$ True

The solution set is $\{8\}$.

Note Subtracting 4 is the same as adding -4. Either method may be used, but we must remember to perform the operation to *both* members of the equation.

3.
$x - 8 = -11$
$x - 8 + 8 = -11 + 8$ Add 8 to both members
$x = -3$ Solution

Check:
$(-3) - 8 = -11$ Substitute
$-11 = -11$ True

The solution set is $\{-3\}$.

4. $2 = x + 7$ | **Check:** $2 = (-5) + 7$ Substitute
$2 - 7 = x + 7 - 7$ Subtract 7 from both members | $2 = 2$ True
$-5 = x$ Solution

The solution set is $\{-5\}$.

▶ ***Quick check*** Find the solution set for $x - 7 = 12$ and check the answer. ■

Symmetric property of equality

The **symmetric property of equality** is also useful in finding the solution set of equations.

Symmetric property of equality

If $a = b$, then $b = a$

Concept

This property allows us to interchange the right and left members of the equation.

In example 2–4 C–4, instead of leaving the equation as $-5 = x$, we could use the symmetric property and write the equation as $x = -5$.

We can see that our goal in solving an equation is to isolate the unknown in one member of the equation and to place everything else in the other member. This forms an equation of the type $x = n$. When the unknown appears in both members of the equation, we use the addition and subtraction property of equality to form an equivalent equation where the unknown appears only in one member of the equation.

■ ***Example 2–4 D***

Find the solution set and check the answer.

1.
$5x - 4 = 4x + 3$
$5x - 4x - 4 = 4x - 4x + 3$ Subtract $4x$ from both members
$x - 4 = 3$
$x - 4 + 4 = 3 + 4$ Add 4 to both members
$x = 7$

Check: $5(7) - 4 = 4(7) + 3$ Substitute
$35 - 4 = 28 + 3$ Order of operations
$31 = 31$ (True) Solution checks

The solution set is $\{7\}$.

2.
$6x - 4 = 7x + 2$
$6x - 6x - 4 = 7x - 6x + 2$ Subtract $6x$ from both members
$-4 = x + 2$
$-4 - 2 = x + 2 - 2$ Subtract 2 from both members
$-6 = x$
$x = -6$ Symmetric property

Check: $6(-6) - 4 = 7(-6) + 2$ Substitute
$-36 - 4 = -42 + 2$ Order of operations
$-40 = -40$ (True) Checks

The solution set is $\{-6\}$.

3.
$$
\begin{aligned}
-2x - 5 &= -3x + 4 \\
-2x + 3x - 5 &= -3x + 3x + 4 && \text{Add } 3x \text{ to both members} \\
x - 5 &= 4 \\
x - 5 + 5 &= 4 + 5 && \text{Add 5 to both members} \\
x &= 9
\end{aligned}
$$

Note A good habit for us to develop is to form equivalent equations in which the unknown appears only in the member of the equation that has the greater coefficient of the unknown. This will ensure a positive coefficient for the unknown.

Check:
$$
\begin{aligned}
-2(9) - 5 &= -3(9) + 4 && \text{Substitute} \\
-18 - 5 &= -27 + 4 && \text{Order of operations} \\
-23 &= -23 \quad \text{(True)} && \text{Solution checks}
\end{aligned}
$$

The solution set is $\{9\}$.

▶ ***Quick check*** Find the solution set for $4x - 2 = 3x + 5$ and check the solution. ■

Sometimes it is necessary to use the associative, commutative, and distributive properties to perform indicated operations in one or both members of an equation. This will *simplify* the equation before the addition and subtraction property of equality is used. Consider the examples that follow.

■ Example 2–4 E

Find the solution set.

1.
$$
\begin{aligned}
5x - 4 + 2x &= 6x - 6 + 11 \\
7x - 4 &= 6x + 5 && \text{Simplify} \\
7x - 6x - 4 &= 6x - 6x + 5 && \text{Subtract } 6x \text{ from both members} \\
x - 4 &= 5 \\
x - 4 + 4 &= 5 + 4 && \text{Add 4 to both members} \\
x &= 9
\end{aligned}
$$

The solution set is $\{9\}$.

2.
$$
\begin{aligned}
3x &= 2(2x - 4) \\
3x &= 4x - 8 && \text{Simplify} \\
3x - 3x &= 4x - 3x - 8 && \text{Subtract } 3x \text{ from both members} \\
0 &= x - 8 \\
0 + 8 &= x - 8 + 8 && \text{Add 8 to both members} \\
8 &= x \\
x &= 8 && \text{Symmetric property}
\end{aligned}
$$

The solution set is $\{8\}$.

3.
$$
\begin{aligned}
3(3x - 1) + 4 &= 2(4x + 3) \\
9x - 3 + 4 &= 8x + 6 && \text{Simplify} \\
9x + 1 &= 8x + 6 && \text{Simplify} \\
9x - 8x + 1 &= 8x - 8x + 6 && \text{Subtract } 8x \text{ from both members} \\
x + 1 &= 6 \\
x + 1 - 1 &= 6 - 1 && \text{Subtract 1 from both members} \\
x &= 5
\end{aligned}
$$

The solution set is $\{5\}$.

▶ ***Quick check*** Find the solution set for $5x + 2x - 4 = 6x + 7$ and for $3(2x + 1) = x + 4x - 2$ ■

Problem solving

We are now ready to combine our ability to write an expression and our ability to solve an equation and apply them to solve a word problem. While there is no standard procedure for solving a word problem, the following guidelines should be useful.

Solving word problems

1. Read the problem carefully. Determine useful prior knowledge and note what information is given and what information we are asked to find.
2. Choose a variable to represent one of the unknowns and then express other unknowns in terms of it.
3. Use the given conditions in the problem and the unknowns from step 2 to write an algebraic equation.
4. Solve the equation for the unknown. Relate this answer to any other unknowns in the problem.
5. Check your results in the original statement of the problem.

■ ***Example 2–4 F***

Solve the following word problems by setting up an equation and solving it.

1. A number increased by 16 gives 24. Find the number.

Let $n =$ the number we are looking for. The key words to use are "increased by," which means *add,* and "gives," which means *equals.* The equation is then

a number	increased by	16	gives	24
n	$+$	16	$=$	24

$n + 16 = 24$
$n = 8$ Subtract 16 from each member

The number is 8.

2. Joan earned $15 less than Mary did last week. If Joan earned $342, how much did Mary earn?

Let $d =$ the amount that Mary earned last week. The key words are "less than." Since Joan earned $15 less than Mary, the equation is given by

Mary's salary	less $15	is	Joan's salary
d	-15	$=$	342

$d - 15 = 342$
$d = 357$ Add 15 to each member

Mary earned $357 last week. ■

Mastery points

Can you

- Determine if a given number is a root of an equation?
- Use the addition and subtraction property of equality?
- Simplify equations?
- Solve for an unknown?
- Check your answer?

Exercise 2–4

Determine if the given value is a solution of the equation. See example 2–4 A.

Examples $3x + 3 = 6; \{1\}$

Solutions $3(1) + 3 = 6$ Substitute
$3 + 3 = 6$ Order of operations
$6 = 6$ Checks

$2x - 1 = 3; \{4\}$

$2(4) - 1 = 3$ Substitute
$8 - 1 = 3$ Order of operations
$7 = 3$ Does not check

1. $4 + x = 8; \{4\}$
2. $3 - x = 4; \{-1\}$
3. $x + 7 = 10; \{3\}$
4. $3x - 2 = 4; \{2\}$
5. $8x + 6 = 2x - 6; \{-2\}$
6. $\frac{4}{5}x + 2 = 10; \{10\}$
7. $7x - 3 = 2x + 2; \{-2\}$
8. $3x + 2 = 5x - 1; \left\{\frac{3}{2}\right\}$
9. $2(x - 1) = 4x + 5; \left\{-\frac{7}{2}\right\}$
10. $5x - 1 = 11x - 1; \{0\}$
11. $\frac{x}{5} - 2 = 3x + 1; \{1\}$
12. $\frac{2x}{3} - 1 = \frac{x}{4} + 3; \{2\}$

Find the solution set by using the addition and subtraction property of equality. Check each solution. See examples 2–4 C and D.

Examples $x - 7 = 12$

Solutions $x - 7 + 7 = 12 + 7$ Add 7
$x = 19$

Check: $(19) - 7 = 12$ Substitute
$12 = 12$ (True) Checks

The solution set is $\{19\}$.

$4x - 2 = 3x + 5$

$4x - 3x - 2 = 3x - 3x + 5$ Subtract 3x
$x - 2 = 5$
$x - 2 + 2 = 5 + 2$ Add 2
$x = 7$

Check: $4(7) - 2 = 3(7) + 5$ Substitute
$28 - 2 = 21 + 5$ Order of operations
$26 = 26$ (True) Checks

The solution set is $\{7\}$.

13. $x - 4 = 12$
14. $y - 7 = 11$
15. $a + 5 = 2$
16. $b + 5 = 7$
17. $y - 6 = -8$
18. $5 = x + 7$
19. $9 = x + 14$
20. $a - 5 = -2$
21. $-10 = x - 4$
22. $a - 18 = -14$
23. $b + 7 = 0$
24. $y - 14 = 0$
25. $3x - 4 = 2x + 10$
26. $6x - 5 = 5x + 11$
27. $b + 4 = 2b + 5$
28. $-y - 6 = -2y + 1$
29. $-z - 8 = -2z - 4$
30. $5 - 3x = 7 - 4x$
31. $9 - 7a = 14 - 6a$
32. $3a - 5 = 2a - 2$

Find the solution set. See example 2–4 E.

Examples $5x + 2x - 4 = 6x + 7$

Solutions $7x - 4 = 6x + 7$ Combine like terms
$7x - 6x - 4 = 6x - 6x + 7$ Subtract 6x
$x - 4 = 7$
$x - 4 + 4 = 7 + 4$ Add 4
$x = 11$

The solution set is $\{11\}$.

$3(2x + 1) = x + 4x - 2$

$6x + 3 = 5x - 2$ Simplify
$6x - 5x + 3 = 5x - 5x - 2$ Subtract 5x
$x + 3 = -2$
$x + 3 - 3 = -2 - 3$ Subtract 3
$x = -5$

The solution set is $\{-5\}$.

33. $6a - 3a + 7 = 9a - 5a + 2$
34. $-4x - 2x + 1 = -5x + 7$
35. $7b - 2b + 5 - 4b = 11$
36. $12 = 6x + 3 - 4x - x$
37. $-4 - x = 4x + 2 - 6x$
38. $5(x + 2) = 4(x - 1)$
39. $2(2y - 1) = 3(y + 2)$
40. $5(3x + 2) = 7(2x + 3)$
41. $5x - 4 + x = 5(x - 2)$
42. $3(2x + 1) - 7 = 5x - 4$
43. $(4a + 5) - (2 + 3a) = 8$
44. $(9b + 7) - (8b + 2) = -4$
45. $3(z + 7) - (8 + 2z) = 6$
46. $4(x - 5) - (3x + 4) = -2$
47. $2(a - 3) - (a - 2) = 8$
48. $5(a + 1) - (4a + 3) = 14$
49. $2(3x - 1) + 3(x + 2) = 4(2x + 5)$
50. $3(4x - 5) + 2(x - 4) = 3(5x + 2)$
51. $-2(b + 1) + 3(b - 4) = -5$
52. $-3(x - 2) + 4(x - 5) = -7$

Solve the following problems by setting up an equation and solving for the unknown. See example 2–4 F.

53. A number increased by 11 yields 37. Find the number.

54. If a number is decreased by 16, the result is 52. Find the number.

55. If Gary's age is increased by 4 years, he is 37 years old. How old is Gary now?

56. Harry is 6 years older than Dene. If Dene is 54 years old, how old is Harry?

57. If Jake withdraws $340 from his savings account, his balance will be $395. How much does Jake have in his savings account now?

58. Pam deposits $42.50 in her checking account. If her new balance is $125.30, how much did she have in her account originally?

59. Mr. Johnson took in $560 on a given day in his grocery store. If he paid out $195 to his employees in wages, how much profit did he realize?

60. Marsha can groom 11 more dogs per day than Margaret can. If Marsha can groom 24 dogs per day, how many dogs per day can Margaret groom?

Review exercises

Perform the indicated operations. See sections 1–6 and 1–7.

1. $(-2)(-8)$ **2.** $(-4)(3)$ **3.** $\frac{-8}{-8}$ **4.** $\frac{6}{6}$ **5.** $\left(\frac{1}{3}\right)(3)$ **6.** $\left(-\frac{1}{4}\right)(-4)$

2–5 ■ The multiplication and division property of equality

Multiplication and division property of equality

In section 2–4, we used the associative, commutative, and distributive properties to simplify equations. We then used the addition and subtraction property of equality to solve for the unknown. These properties are sufficient to solve many of the equations that we encounter. However we cannot use them to solve such equations as

$$3x = 21 \quad \text{or} \quad \frac{2}{3}x = 12$$

Recall that we want our equation to be of the form $x = n$. This means that the coefficient of x must be 1. To achieve this, we make use of the multiplication and division property of equality.

The multiplication and division property of equality

For any algebraic expressions a, b, and c ($c \neq 0$)

$$\text{if } a = b, \text{ then } a \cdot c = b \cdot c \text{ and } a \div c = b \div c$$

Concept

An equivalent equation is obtained when we multiply or divide both members of an equation by the same nonzero quantity.

The multiplication and division property of equality enables us to multiply or divide both members of an equation by the same nonzero quantity. In the equation $3x = 21$, we use the multiplication and division property of equality to divide both members of the equation by 3. This forms an equivalent equation where x has a coefficient of 1, that is, $x = n$.

$$3x = 21$$

$$\frac{3x}{3} = \frac{21}{3} \quad \text{Divide both members by 3}$$

$$x = 7$$

The solution set is $\{7\}$.

For the equation $\frac{2}{3}x = 12$, recall that when we divide by a fraction, we invert and multiply. Therefore, if the coefficient is a fraction, we will multiply both members of the equation by the **reciprocal,** or the **multiplicative inverse,** of the coefficient.

Multiplicative inverse

The **multiplicative inverse** of a number, also called the **reciprocal** of the number, is such that when we multiply a number times its reciprocal, the answer will be 1.

Multiplicative inverse property

For every real number a, $a \neq 0$,

$$a \cdot \frac{1}{a} = 1$$

Concept

Every real number except zero has a multiplicative inverse, and the product of a number and its multiplicative inverse is always 1.

Note Zero is the only number that does not have a reciprocal. From the zero factor property, we know that zero times any number is zero. Therefore there can be no number such that zero times that number gives 1 as an answer.

$$a \cdot 0 = 0$$

■ ***Example 2–5 A***

The following examples are illustrations of the multiplicative inverse property, where the second number can be considered the reciprocal of the first, and the first can be considered the reciprocal of the second.

1. $5 \cdot \frac{1}{5} = 1$ **2.** $\frac{1}{2} \cdot 2 = 1$ **3.** $b \cdot \frac{1}{b} = 1, b \neq 0$

4. $\frac{3}{4} \cdot \frac{4}{3} = 1$ **5.** $\left(-\frac{5}{7}\right)\left(-\frac{7}{5}\right) = 1$ ■

We will now use the multiplicative inverse property to solve the equation $\frac{2}{3}x = 12$.

$$\frac{2}{3}x = 12$$
$$\frac{3}{2} \cdot \frac{2}{3}x = \frac{3}{2} \cdot 12$$ Multiply both members by the reciprocal of the coefficient
$$x = 18$$

The solution set is $\{18\}$.

Note In the earlier example $3x = 21$, we could have multiplied by the reciprocal of 3 to solve the equation. That is,

$$3x = 21$$
$$\frac{1}{3} \cdot 3x = \frac{1}{3} \cdot 21$$ Multiply both members by the reciprocal $\frac{1}{3}$
$$x = 7$$

Remember that to divide by a number is the same operation as to multiply by the reciprocal of that number.

■ ***Example 2–5 B***

Find the solution set.

1. $5x = 30$
$$\frac{5x}{5} = \frac{30}{5}$$ Divide both members by the coefficient 5
$$x = 6$$

The solution set is $\{6\}$.

2. $\frac{3}{4}x = 9$
$$\frac{4}{3} \cdot \frac{3}{4}x = \frac{4}{3} \cdot 9$$ Multiply both members by the reciprocal $\frac{4}{3}$
$$x = 12$$

The solution set is $\{12\}$.

3. $-x = -10$
$$-1 \cdot x = -10$$ -1 is the coefficient
$$\frac{-1 \cdot x}{-1} = \frac{-10}{-1}$$ Divide both members by the coefficient -1
$$x = 10$$

The solution set is $\{10\}$.

4. $6x = 10$
$$\frac{6x}{6} = \frac{10}{6}$$ Divide both members by the coefficient 6
$$x = \frac{5}{3}$$ Reduce the fraction

The solution set is $\left\{\frac{5}{3}\right\}$.

5. $\frac{x}{4} = 6$ — We can rewrite the left member to show that the coefficient is $\frac{1}{4}$

$\frac{1}{4}x = 6$

$4 \cdot \frac{1}{4}x = 4 \cdot 6$ — Multiply both members by the reciprocal 4

$x = 24$

The solution set is $\{24\}$.

6. $1.2x = 4.8$

$\frac{1.2x}{1.2} = \frac{4.8}{1.2}$ — Divide both members by 1.2

$x = 4$

The solution set is $\{4\}$.

▶ ***Quick check*** Find the solution set of the equations $5x = 35$, $\frac{2}{3}x = 8$, and $1.7x = 10.2$ ■

Problem solving

Now we will translate some word statements into equations and solve the resulting equations.

■ ***Example 2–5 C***

Write an equation for each problem and then solve the equation.

1. When a number is multiplied by -6, the result is 48. Find the number.

Let $n =$ the number for which we are looking. The formation of the equation would be as follows:

a number	multiplied by (-6)	result is	48
n	$\cdot(-6)$	$=$	48

$n \cdot (-6) = 48$

$\frac{n(-6)}{-6} = \frac{48}{-6}$ — Divide both members by -6

$n = -8$

The number is -8.

2. Alice makes \$4.50 per hour. If her pay was \$108, how many hours did she work?

Let $n =$ the number of hours that she worked. The formation of the equation would be as follows:

hourly rate	times	number of hours worked	gives	total pay
(4.50)		n	$=$	108

$(4.50) \cdot n = 108$

$\frac{(4.50)n}{4.50} = \frac{108}{4.50}$ — Divide both members by 4.50

$n = 24$

Alice worked 24 hours. ■

Mastery points

Can you

- Use the multiplication and division property of equality to form equivalent equations where the coefficient of the unknown is 1?
- Check your answer?

Exercise 2–5

Find the solution set by using the multiplication and division property of equality. Check each solution. See example 2–5 B.

Examples $5x = 35$

Solutions $\frac{5x}{5} = \frac{35}{5}$ Divide by 5

$x = 7$

The solution set is $\{7\}$.

Check:

$5(7) = 35$ Substitute

$35 = 35$ (True) Checks

Examples $\frac{2}{3}x = 8$

Solutions $\frac{3}{2} \cdot \frac{2}{3}x = \frac{3}{2} \cdot 8$ Multiply by $\frac{3}{2}$

$x = 12$

The solution set is $\{12\}$

Check:

$\frac{2}{3}(12) = 8$ Substitute

$8 = 8$ (True) Checks

Examples $1.7x = 10.2$

Solutions $\frac{1.7x}{1.7} = \frac{10.2}{1.7}$ Divide by 1.7

$x = 6$

The solution set is $\{6\}$.

Check:

$1.7(6) = 10.2$ Substitute

$10.2 = 10.2$ (True) Checks

1. $2x = 8$ **2.** $3x = 18$ **3.** $6x = 36$ **4.** $9x = 45$

5. $\frac{3}{4}x = 12$ **6.** $\frac{2}{5}x = 10$ **7.** $\frac{1}{7}x = 5$ **8.** $\frac{1}{5}x = 9$

9. $\frac{3}{2}x = 18$ **10.** $14 = \frac{7}{3}x$ **11.** $5x = -15$ **12.** $-8 = 2x$

13. $-24 = 6x$ **14.** $-5x = 30$ **15.** $-4x = -28$ **16.** $-30 = -6x$

17. $-x = 4$ **18.** $-x = -11$ **19.** $6x = 14$ **20.** $5x = 9$

21. $4x = 6$ **22.** $3x = -8$ **23.** $5x = 0$ **24.** $0 = 7x$

25. $-3x = 0$ **26.** $-2x = 0$ **27.** $\frac{x}{3} = 5$ **28.** $\frac{x}{4} = 8$

29. $\frac{x}{-2} = 7$ **30.** $-2 = \frac{x}{-3}$ **31.** $2.6x = 10.4$ **32.** $3.1x = 21.7$

33. $-4.8x = 33.6$ **34.** $-7.1x = 35.5$ **35.** $-42.9 = -3.9x$ **36.** $(0.4)x = 7.2$

37. $(0.3)x = -7.8$ **38.** $\frac{5}{7}x = 8$ **39.** $\frac{3}{8}x = 14$ **40.** $\frac{2}{9}x = 11$

Write an equation for each exercise and then solve the equation. See example 2–5 C.

41. When a number is multiplied by 6, the result is 54. Find the number.

42. When a number is multiplied by -4, the result is 36. Find the number.

43. When a number is divided by 9, the result is -7. Find the number.

44. When a number is divided by -8, the result is -8. Find the number.

45. Nancy worked for 30 hours and received $135. Find her hourly wage.

46. Adam worked for 14 hours and received $52.50. Find his hourly wage.

47. Four friends shared equally in the expenses for a party. If each person's share was $32.50, what was the total cost of the party?

48. Six friends shared equally in the cost of dinner. If the cost of the dinner was $51, what was each person's share?

49. If $\frac{3}{4}$ of a number is 48, find the number.

50. If $\frac{2}{3}$ of a number is 26, find the number.

Review exercises

Perform all indicated operations. See section 2–3.

1. $3x + 2x + 1 - 3$
2. $7x - 5x - 3 + 4$
3. $8x - 5 + 4x + 7$
4. $6x + 3 - 3x - 8$
5. $2(3x + 1) + 4x - 3$
6. $3(x - 1) + 2(x + 2)$

2–6 ■ Solving linear equations

Review of properties

We now are ready to combine the properties from the previous sections to help us solve more involved equations. The process consists of forming equivalent equations until we have our equation in the form of $x = n$. The properties that we will use are the following:

1. We can add or subtract the same number in both members of the equation.
2. We can multiply or divide both members of the equation by the same nonzero number.

If we use these two properties, making sure that both members of the equation are treated in exactly the same manner as we apply each of the properties, we will be forming equivalent equations.

Procedure for solving a linear equation

Using these properties, there are four basic steps to solve a linear equation. We shall now apply the properties to the equation $6(x + 1) = 4x + 10$.

Solving a linear equation

$$6(x + 1) = 4x + 10$$

Step 1 *Simplify each member of the equation.* Perform all indicated addition, subtraction, multiplication, and division. Remove all grouping symbols. In our example, step 1 would be to carry out the indicated multiplication in the left member as follows:

$$\begin{aligned} 6(x + 1) &= 4x + 10 \\ 6x + 6 &= 4x + 10 \end{aligned}$$

Step 2 *Use the addition and subtraction property of equality to form an equivalent equation where all the terms involving the unknown are in one member of the equation.* By subtracting $4x$ from *both* members of the equation, we have

$$\begin{aligned} 6x + 6 &= 4x + 10 \\ 6x - 4x + 6 &= 4x - 4x + 10 \\ 2x + 6 &= 10 \end{aligned}$$

Step 3 *Use the addition and subtraction property of equality to form an equivalent equation where all the terms not involving the unknown are in the other member of the equation.* Subtracting 6 from *both* members of the equation, we have

$$\begin{aligned} 2x + 6 &= 10 \\ 2x + 6 - 6 &= 10 - 6 \\ 2x &= 4 \end{aligned}$$

Step 4 *Use the multiplication and division property of equality to form an equivalent equation where the coefficient of the unknown is 1. That is, $x = n$.* By dividing *both* members of the equation by 2, we have

$$\begin{aligned} 2x &= 4 \\ \frac{2x}{2} &= \frac{4}{2} \\ x &= 2 \end{aligned}$$

The solution set is denoted by $\{2\}$.

Step 5 To check the solution, we substitute the solution in place of the unknown in the original equation. If we get a true statement, we say that the solution "satisfies" the equation.

In the equation $6(x + 1) = 4x + 10$, we found that $x = 2$. We can check the solution by substituting 2 in place of x in the original equation.

$$\begin{aligned} 6[(2) + 1] &= 4(2) + 10 && \text{Substitute} \\ 6[3] &= 8 + 10 && \text{Order of operations} \\ 18 &= 18 \quad \text{(True)} && \text{Solution checks} \end{aligned}$$

We see that $x = 2$ satisfies the equation.

■ Example 2–6 A

Find the solution set and check.

1. $6y + 5 - 7y = 10 - 2y + 3$

$$5 - y = 13 - 2y$$ Simplify each member by combining like terms
$$5 - y + 2y = 13 - 2y + 2y$$ Add $2y$ to both members
$$5 + y = 13$$
$$5 + y - 5 = 13 - 5$$ Subtract 5 from both members
$$y = 8$$

Check: $6(8) + 5 - 7(8) = 10 - 2(8) + 3$ Substitute
$$48 + 5 - 56 = 10 - 16 + 3$$ Order of operations
$$53 - 56 = -6 + 3$$
$$-3 = -3$$ (True) Solution checks

The solution set is $\{8\}$.

2. $8y + 5 - 7y = 10 - 2y + 3$

$$5 + y = 13 - 2y$$ Combine like terms
$$5 + y + 2y = 13 - 2y + 2y$$ Add $2y$
$$5 + 3y = 13$$ Combine like terms
$$5 + 3y - 5 = 13 - 5$$ Subtract 5
$$3y = 8$$ Combine like terms
$$\frac{3y}{3} = \frac{8}{3}$$ Divide by 3
$$y = \frac{8}{3}$$

Check: $8\left(\frac{8}{3}\right) + 5 - 7\left(\frac{8}{3}\right) = 10 - 2\left(\frac{8}{3}\right) + 3$ Substitute $\frac{8}{3}$ for y
$$\frac{64}{3} + \frac{15}{3} - \frac{56}{3} = \frac{30}{3} - \frac{16}{3} + \frac{9}{3}$$ Multiply, change to common denominator
$$\frac{64 + 15 - 56}{3} = \frac{30 - 16 + 9}{3}$$ Add and subtract in numerators
$$\frac{23}{3} = \frac{23}{3}$$ (True) Solution checks

The solution set is $\left\{\frac{8}{3}\right\}$.

3. $4(5x - 2) + 7 = 5(3x + 1)$

$$20x - 8 + 7 = 15x + 5$$ Distributive property
$$20x - 1 = 15x + 5$$ Combine like terms
$$20x - 15x - 1 = 15x - 15x + 5$$ Subtract $15x$
$$5x - 1 = 5$$
$$5x - 1 + 1 = 5 + 1$$ Add 1
$$5x = 6$$
$$\frac{5x}{5} = \frac{6}{5}$$ Divide by 5
$$x = \frac{6}{5}$$

Check:

$$4\left[5\left(\frac{6}{5}\right) - 2\right] + 7 = 5\left[3\left(\frac{6}{5}\right) + 1\right] \quad \text{Substitute } \frac{6}{5} \text{ for } x$$

$$4[6 - 2] + 7 = 5\left[\frac{18}{5} + 1\right] \quad \text{Order of operations}$$

$$4[4] + 7 = 5\left[\frac{18}{5} + \frac{5}{5}\right]$$

$$16 + 7 = 5\left[\frac{23}{5}\right]$$

$$23 = 23 \quad \text{(True)} \quad \text{Solution checks}$$

The solution set is $\left\{\frac{6}{5}\right\}$.

▶ ***Quick check*** Find the solution set for $5x + 2(x - 1) = 4 - 3x$ and check. ■

At this point, we will no longer show the check of our solution, but you should realize that a check of your solution is an important final step.

The following equations contain several fractions. When this occurs, it is usually easier to *clear the equation of all fractions.* We do this by multiplying both members of the equation by the least common denominator of all the fractions. Clearing all fractions is considered a means of simplifying the equation and will be done as a first step when necessary. Equations containing fractions will be studied more completely in chapter 6.

■ Example 2–6 B

Find the solution set.

1.

$$\frac{1}{4}x + 2 = \frac{1}{2}$$

$$4\left(\frac{1}{4}x + 2\right) = 4\left(\frac{1}{2}\right) \quad \text{The least common denominator of the fractions is 4, multiply both members by 4}$$

$$4\left(\frac{1}{4}x\right) + 4(2) = 4\left(\frac{1}{2}\right) \quad \text{Simplify (distributive property)}$$

$$x + 8 = 2 \quad \text{All fractions have been cleared}$$

$$x + 8 - 8 = 2 - 8 \quad \text{Subtract 8}$$

$$x = -6$$

The solution set is $\{-6\}$.

2.

$$\frac{5}{6}x - \frac{2}{3} = \frac{3}{4}x + 2$$

$$12\left(\frac{5}{6}x - \frac{2}{3}\right) = 12\left(\frac{3}{4}x + 2\right) \quad \text{The least common denominator of the fractions is 12; multiply by 12}$$

$$12\left(\frac{5}{6}x\right) - 12\left(\frac{2}{3}\right) = 12\left(\frac{3}{4}x\right) + 12(2) \quad \text{Simplify, distributive property}$$

$$10x - 8 = 9x + 24$$

$$10x - 9x - 8 = 9x - 9x + 24 \quad \text{Subtract } 9x$$

$$x - 8 = 24$$

$$x - 8 + 8 = 24 + 8 \quad \text{Add 8}$$

$$x = 32$$

The solution set is $\{32\}$. ■

Mastery points

Can you

- Solve linear equations?
- Check your answers?

Exercise 2–6

Find the solution set of the following equations, and check the solution. See examples 2–6 A and B.

Example $5x + 2(x - 1) = 4 - 3x$

Solution

$$5x + 2x - 2 = 4 - 3x \quad \text{Simplify (distributive property)}$$
$$7x - 2 = 4 - 3x \quad \text{Combine like terms}$$
$$7x + 3x - 2 = 4 - 3x + 3x \quad \text{Add } 3x$$
$$10x - 2 = 4 \quad \text{Combine like terms}$$
$$10x - 2 + 2 = 4 + 2 \quad \text{Add 2}$$
$$10x = 6 \quad \text{Combine like terms}$$
$$\frac{10x}{10} = \frac{6}{10} \quad \text{Divide by 10 and reduce}$$
$$x = \frac{3}{5}$$

Check:

$$5\left(\frac{3}{5}\right) + 2\left[\left(\frac{3}{5}\right) - 1\right] = 4 - 3\left(\frac{3}{5}\right) \quad \text{Substitute } \frac{3}{5} \text{ for } x$$
$$5\left(\frac{3}{5}\right) + 2\left[\frac{3}{5} - \frac{5}{5}\right] = 4 - \frac{9}{5} \quad \text{Order of operations}$$
$$5\left(\frac{3}{5}\right) + 2\left[\frac{-2}{5}\right] = \frac{20}{5} - \frac{9}{5}$$
$$\frac{15}{5} + \frac{-4}{5} = \frac{11}{5}$$
$$\frac{11}{5} = \frac{11}{5} \quad \text{(True)} \quad \text{Solution checks}$$

The solution set is $\left\{\frac{3}{5}\right\}$.

1. $2x = 4$ **2.** $3x = 11$ **3.** $5x = -10$ **4.** $-2x = 8$

5. $\frac{x}{2} = 18$ **6.** $\frac{x}{4} = 24$ **7.** $\frac{3x}{2} = 8$ **8.** $\frac{5x}{3} = 18$

9. $x + 7 = 11$ **10.** $x - 4 = 9$ **11.** $x + 5 = 5$ **12.** $x - 4 = -4$

13. $3x + 1 = 10$ **14.** $5x - 2 = 13$ **15.** $4x + 7 = 7$ **16.** $6x + 2 = 2$

17. $5x + 2x = x + 6$ **18.** $2x + (3x - 1) = 4 - x$ **19.** $2x + 3x - 6x = 4x - 8$

20. $\frac{x}{2} + 7 = 14$ **21.** $5 - \frac{3x}{5} = 11$ **22.** $\frac{5x + 2x}{6} = 10$

23. $\frac{1}{2}x + 3 = \frac{3}{4}$ **24.** $\frac{1}{5}x - 1 = \frac{7}{10}$ **25.** $\frac{1}{3}x + 2 = \frac{1}{2}x - 1$

26. $\frac{1}{4}x - 3 = \frac{1}{8}x + 1$

27. $\frac{2}{3}x + 5 = \frac{3}{4}$

28. $\frac{3}{5}x - 3 = \frac{3}{10}$

29. $\frac{3}{8}x + \frac{1}{2} = \frac{1}{4}x + 2$

30. $\frac{7}{12}x + 1 = \frac{2}{3}x - 1$

31. $3(2x - 1) = 4x + 3$

32. $5(7x - 3) = 30x + 11$

33. $12x - 8 = 5x + 2$

34. $3(2x + 5) = 4(x - 3)$

35. $8 - 2(3x + 4) = 5x - 16$

36. $(3x + 2) - (2x - 5) = 7$

37. $(7x - 6) - (4 - 3x) = 27$

38. $2(x - 4) - 3(5 - 2x) = 16$

39. $3(2x + 3) = 5 - 4(x - 2)$

40. $6(3x - 2) = 7(x - 3) - 2$

41. $2(x + 5) = 16$

42. $6 = 2(2x - 1)$

43. $2x - (3 - x) = 0$

44. $3(7 - 2x) = 30 - 7(x + 1)$

Find the solution set for the following equations, and check the answer. See examples 2–6 A and B.

45. To convert Celsius temperature to Fahrenheit, we use $F = \frac{9}{5}C + 32$. Find C when (a) $F = 18$, (b) $F = -27$, (c) $F = 2$.

46. The Stefan-Boltzmann Law in metallurgy, which is the temperature scale of radiation pyrometers, is given by $W = KT^4$. Find K when (a) $W = 36$, $T = 2$; (b) $W = 243$, $T = -3$.

47. The total creep of a metal (E_P) at time t is given by $E_P = E_0 + V_0t$, where E_0 = original creep, t = time, and V_0 = the original volume. Find V_0 when $E_P = 16$, $E_0 = 9$, and $t = 3$.

48. In a gear system, the speed, in number of revolutions, of two gears and the number of teeth in the gears are related by $S_D \cdot T_D = S_d \cdot T_d$, where S_D is the speed of the driver, T_D is the number of teeth in the driver, S_d is the speed of the driven gear, and T_d is the number of teeth in the driven gear. If
(a) $S_D = 240$, $T_D = 40$, and $S_d = 360$, find T_d;
(b) $S_D = 120$, $S_d = 90$, and $T_d = 18$, find T_D.

Review exercises

Evaluate the following formulas. See section 2–2.

1. $W = I^2R$, $I = 6$ and $R = 3$

2. $S = \frac{1}{2}gt^2$, $g = 32$ and $t = 3$

3. $A = \frac{1}{2}h(b + c)$, $h = 8$, $b = 10$, and $c = 12$

4. $I = prt$; $p = 2{,}000$; $r = 0.06$; and $t = 3$

5. $V = \ell wh$, $\ell = 10$, $w = 4$, and $h = 7$

6. $V = k + gt$, $k = 12$, $g = 16$, and $t = 5$

2–7 ■ Solving literal equations and formulas

Literal equations

Equations that contain two or more variables are called **literal equations.** In a literal equation, we generally solve the equation for one variable in terms of the remaining variables and constants. *The procedure for solving a literal equation is the same as the procedure for solving linear equations.*

Formulas

A formula is a mathematical equation that states the relationship between two or more physical conditions. Consider the formula $d = rt$, which expresses the fact that distance (d) is equal to rate (r) multiplied by the time (t). If we knew how far it was between two cities (d) and we wanted to travel this distance in a certain amount of time (t), then the equation could be solved for the necessary rate (r) to achieve this.

$$d = rt \qquad \text{Divide each member by } t$$
$$\frac{d}{t} = r$$

The equation is now solved for r in terms of d and t. If the distance and rate were known, the equation could be solved for time as follows:

$$d = rt \qquad \text{Divide each member by } r$$
$$\frac{d}{r} = t$$

The equation is now solved for t in terms of d and r.

We observe from this example that we may solve a literal equation or formula for a specified variable. The following list is a restatement of the procedure for solving linear equations that we will now apply to literal equations.

Solving a literal equation or a formula

Step 1 Simplify each member of the equation.

Step 2 Collect all the terms with the variable for which we are solving in one member of the equation. (Addition and subtraction property)

Step 3 Remove any term that is being added to or subtracted from the variable for which we are solving. (Addition and subtraction property)

Step 4 Divide each member of the equation by the coefficient of the variable for which we are solving. (Multiplication and division property)

■ Example 2–7 A

Solve for the specified variable.

1. The volume of a rectangular solid is found by multiplying length (ℓ) times width (w) times height (h), $V = \ell wh$. Solve the equation for h.

$V = \ell wh$	Original equation
$V = (\ell w)h$	Coefficient of h is ℓw
$\dfrac{V}{\ell w} = \dfrac{\ell wh}{\ell w}$	Divide by ℓw
$\dfrac{V}{\ell w} = h$	Equation is solved for h in terms of V, ℓ, and w
$h = \dfrac{V}{\ell w}$	Symmetric property

2. The simple interest (I) earned on the principal (P) over a time period (t) at an interest rate (r) is given by $I = Prt$. Solve for r.

$I = Prt$	Original equation
$I = (Pt)r$	Pt is the coefficient of r
$\dfrac{I}{Pt} = \dfrac{Ptr}{Pt}$	Divide by Pt
$\dfrac{I}{Pt} = r$	Equation is solved for r in terms of I, P, and t
$r = \dfrac{I}{Pt}$	Symmetric property

3. If we know the temperature in degrees Fahrenheit (F), the temperature in degrees Celsius (C) can be found by the equation $C = \dfrac{5}{9}(F - 32)$. Solve the formula for F.

$C = \dfrac{5}{9}(F - 32)$	
$9C = 9 \cdot \dfrac{5}{9}(F - 32)$	Clear the fraction
$9C = 5(F - 32)$	Apply the distributive property
$9C = 5F - 160$	
$9C + 160 = 5F$	Add 160
$\dfrac{9C + 160}{5} = F$	Divide by 5
$F = \dfrac{9C + 160}{5}$	Symmetric property

If the temperature is given in degrees Celsius, we use this form of the equation to determine the temperature in degrees Fahrenheit.

Note Although we have not stated any restrictions on the variables, it is understood that the values that the variables can take on must be such that no denominator is ever zero. That is, in example 1, $\ell \neq 0$ and $w \neq 0$; in example 2, $P \neq 0$, $t \neq 0$.

▶ ***Quick check*** Solve $P = 2\ell + 2w$ for ℓ. ■

Whether we are solving for x in a linear equation or a literal equation, the procedure is the same.

Linear equation	**Literal equation**	
$5(x + 1) = 2x + 7$	$5(x + y) = 2x + 7y$	Original equation
$5x + 5 = 2x + 7$	$5x + 5y = 2x + 7y$	Simplify (distributive property)
$3x + 5 = 7$	$3x + 5y = 7y$	All x's in one member
$3x = 2$	$3x = 2y$	Terms not containing x in the other member
$x = \frac{2}{3}$	$x = \frac{2y}{3}$	Divide by the coefficient

In the linear equation, we have a solution for x, and in the literal equation, we have solved for x in terms of y.

Mastery points

Can you

- Solve literal equations and formulas for a specified variable?

Exercise 2–7

Solve for the specified variable. See example 2–7 A.

Example $P = 2\ell + 2w$, for ℓ

Solution $P - 2w = 2\ell$ Subtract $2w$

$\frac{P - 2w}{2} = \ell$ Divide by 2

$\ell = \frac{P - 2w}{2}$ Symmetric property

1. $V = \ell wh$, for w
2. $V = \ell wh$, for ℓ
3. $I = Prt$, for P
4. $I = Prt$, for t
5. $F = ma$, for m
6. $E = IR$, for R
7. $K = PV$, for V
8. $E = mc^2$, for m
9. $W = I^2R$, for R
10. $A = \ell w$, for w
11. $P = 2\ell + 2w$, for w
12. $C = \pi D$, for π
13. $P = a + b + c$, for a
14. $A = \frac{1}{2}bh$, for b
15. $ay - 3 = by + c$, for a
16. $ay - 3 = by + c$, for b
17. $V = k + gt$, for k
18. $V = k + gt$, for t
19. $A = \frac{1}{2}h(b + c)$, for b
20. $A = \frac{1}{2}h(b + c)$, for h
21. $\ell = a + (n - 1)d$, for a
22. $\ell = a + (n - 1)d$, for d
23. $A = P(1 + r)$, for P
24. $\ell = a + (n - 1)d$, for n
25. $T = 2f + g$, for f
26. $i = \frac{prm}{12}$, for r
27. $D = dq + R$, for q
28. $M = -P(\ell - x)$, for x
29. $R = W - b(2c + b)$, for c
30. $F = k\frac{m_1m_2}{d^2}$, for k
31. $A = P(1 + rt)$, for r
32. $A = P(1 + rt)$, for P
33. $V = r^2(a - b)$, for a
34. $P = n(P_2 - P_1) - c$, for P_2
35. $3x - y = 4x + 5y$, for x
36. $3x - y = 4x + 5y$, for y
37. $2S = 2vt - gt^2$, for g
38. $ax + by = c$, for y

39. The distance s that a body projected downward with an initial velocity of v will fall in t seconds because of the force of gravity is given by $s = \frac{1}{2}gt^2 + vt$. Solve for g.

40. Solve the formula in exercise 39 for v.

41. The net profit P on sales of n identical tape decks is given by $P = n(S - C) - e$, where S is the selling price, C is the cost to the dealer, and e is the operating expense. Solve for S.

42. Solve the formula in exercise 41 for C.

43. Solve the formula in exercise 41 for e.

Review exercises

Perform the indicated operations. See section 1–8.

1. -5^2
2. $(-5)^2$
3. -3^4
4. $(-3)^3$

Write an algebraic expression for each of the following. See section 2–1.

5. x raised to the fourth power
6. A number squared
7. The product of a and b
8. x multiplied by y

2–8 ■ Word problems

Many problems that we will encounter will be verbally stated. These will need to be translated into algebraic equations. In chapter 1, we solved arithmetic word problems. In section 2–1, we saw how to take a word phrase and write an algebraic expression for it.

We are now ready to combine our arithmetic problem-solving skills, our ability to write an algebraic expression, and our ability to solve an equation and apply them to solve word problems. On page 91, some useful guidelines for solving word problems were given. You should review them at this time.

■ Example 2–8 A

Write an equation for the problem and solve for the unknown quantities.

1. One number is 4 more than a second number. If their sum is 38, find the two numbers.

Note In problems where we are finding more than one value, it is usually easiest to let the unknown represent the smallest unknown value.

Let $x =$ the smaller number (the second number). Then $x + 4$, which is 4 more than the smaller number, represents the other number. The parts that make up the equation are

smaller number	their sum	larger number	is	38
x	$+$	$(x + 4)$	$=$	38

$x + (x + 4) = 38$	Original equation
$x + x + 4 = 38$	Remove grouping symbol
$2x + 4 = 38$	Combine like terms
$2x = 34$	Subtract 4
$x = 17$	Divide by 2

Therefore the smaller number is 17 and the larger number is 4 more than the smaller number: $(x + 4)$ and $17 + 4 = 21$.

2. One number is 6 times a second number and their sum is 21. Find the numbers.

Let x = the second number. Then six times the second number or $6x$ = the other number. The parts that make up the equation are

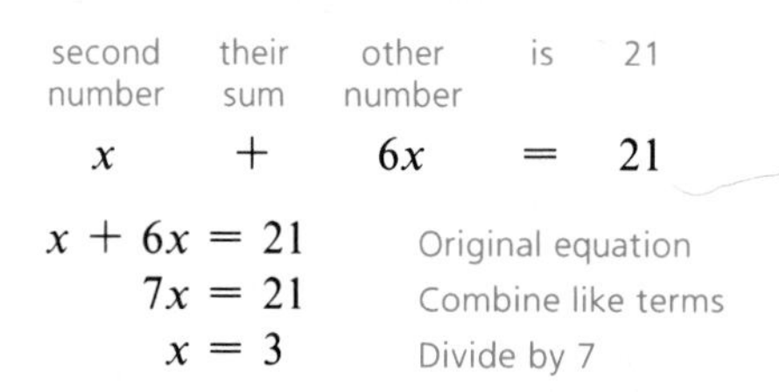

second number	their sum	other number	is	21
x	$+$	$6x$	$=$	21

$$x + 6x = 21 \quad \text{Original equation}$$
$$7x = 21 \quad \text{Combine like terms}$$
$$x = 3 \quad \text{Divide by 7}$$

Hence the second number is 3 and the other number is 6 times the second number and is $6 \cdot (3) = 18$.

3. If the first of two consecutive integers is multiplied by 3, this product is 4 more than the sum of the two integers. Find the integers.

Note Prior knowledge that is needed for this problem is that consecutive integers differ by 1. Therefore we add 1 to the first to get the second, we would add 2 to the first to get a third, and so on.

first	second	third	fourth	fifth
x	$x + 1$	$x + 2$	$x + 3$	$x + 4$

first integer	second integer
x	$x + 1$

The parts that make up the equation are

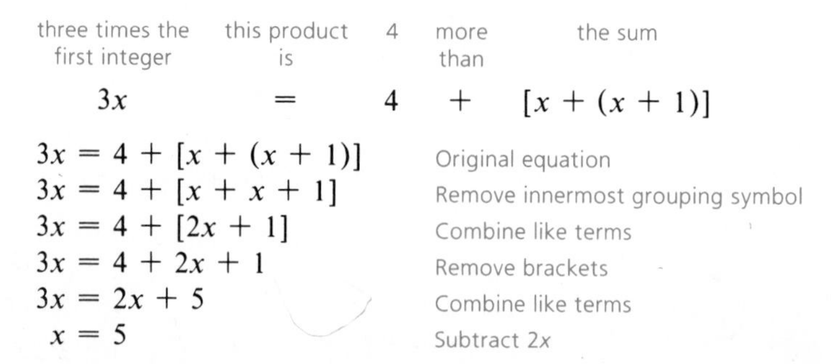

three times the first integer	this product is	4	more than	the sum
$3x$	$=$	4	$+$	$[x + (x + 1)]$

$$3x = 4 + [x + (x + 1)] \quad \text{Original equation}$$
$$3x = 4 + [x + x + 1] \quad \text{Remove innermost grouping symbol}$$
$$3x = 4 + [2x + 1] \quad \text{Combine like terms}$$
$$3x = 4 + 2x + 1 \quad \text{Remove brackets}$$
$$3x = 2x + 5 \quad \text{Combine like terms}$$
$$x = 5 \quad \text{Subtract } 2x$$

Therefore the first consecutive integer is 5 and the second integer is one more than the first $(x + 1)$ and is $5 + 1 = 6$.

▶ ***Quick check*** One natural number is 5 times another natural number and their sum is 36. Find the numbers.

The sum of three consecutive integers is 36. Find the integers. ■

Mastery points

Can you

- Write an equation for a word problem?
- Solve for the unknown quantities?

Exercise 2–8

Write an equation for the problem and solve for the unknown quantities. See example 2–8 A.

Example One natural number is 5 times another natural number and their sum is 36. Find the numbers.

Solution Let x = the smaller number. Then five times the smaller number or $5x$ = the other number. The parts that make up the equation are

smaller number	sum	larger number	is	36
x	$+$	$5x$	$=$	36

$x + 5x = 36$ Original equation

$6x = 36$ Combine like terms

$x = 6$ Divide by 6

Therefore the smaller number is 6 and the larger number is 5 times the smaller number $(5x)$ and is $5 \cdot (6) = 30$.

1. One number is 18 more than a second number. If their sum is 62, find the two numbers.
2. One number is 9 less than another number. If their sum is 47, find the two numbers.
3. The difference of two numbers is 17. Find the numbers if their sum is 87.
4. If three times a number is increased by 11 and the result is 47, what is the number?
5. If a number is divided by 4 and that result is then increased by 6, the answer is 13. Find the number.
6. If a number is decreased by 14 and that result is then divided by 5, the answer is 15. Find the number.
7. Nine times a number is decreased by 4, leaving 59. What is the number?
8. One-third of a number is 8 less than one-half of the number. Find the number.
9. The difference between one-half of a number and one-third of the number is 9. Find the number.
10. One-half of a number minus one-third of the number is 8. Find the number.
11. What number added to its double gives 63?
12. Six times a number, increased by 10, gives 94. Find the number.
13. Find a number such that twice the sum of that number and 7 is 44.
14. One number is seven times another. If their difference is 18, what are the numbers?
15. Find two numbers whose sum is 63 and whose difference is 5.
16. One number is 11 more than twice a second number. If their sum is 35, what are the numbers?
17. One number is 9 times a second number and their sum is 120. Find the numbers.

Example The sum of three consecutive integers is 36. Find the integers.

Solution

first integer	second integer	third integer
x	$(x + 1)$	$(x + 2)$

The parts that make up the equation are

the sum of three consecutive integers	is	36
$x + (x + 1) + (x + 2)$	$=$	36

$x + (x + 1) + (x + 2) = 36$	Original equation
$x + x + 1 + x + 2 = 36$	Remove grouping symbols
$3x + 3 = 36$	Combine like terms
$3x = 33$	Subtract 3
$x = 11$	Divide by 3

Hence the first integer is $x = 11$, the second integer is $x + 1 = (11) + 1 = 12$, and the third integer is $x + 2 = (11) + 2 = 13$.

18. The sum of three consecutive even integers is 72. Find the integers.

19. The sum of three consecutive odd integers is 51. Find the integers.

20. One number is 4 times a second number and their sum is 65. Find the numbers.

21. One number is 7 times a second number and their sum is 96. Find the numbers.

22. The sum of three consecutive integers is 69. Find the integers.

23. The sum of three consecutive even integers is 66. Find the integers.

24. The sum of three consecutive odd integers is 75. Find the integers.

25. The sum of three consecutive integers is 93. Find the integers.

26. The sum of three numbers is 44. The second number is three times the first number and the third number is 6 less than the first number. Find the three numbers.

27. The sum of three numbers is 63. The first number is twice the second number and the third number is three times the first number. Find the three numbers.

28. One number is 7 more than another number. Find the two numbers if three times the larger number exceeds four times the smaller number by 5.

29. One number is 4 more than another number. Find the two numbers if two times the larger number is 7 less than five times the smaller number.

30. One number is 33 more than another. The smaller number is one-fourth of the larger number. Find the numbers.

31. A number plus one-half of the number plus one-third of the number equal 44. Find the number.

32. A number is decreased by 7 and twice this result is 52. What is the number?

33. Four times the first of three consecutive integers is 27 less than three times the sum of the second and third. Find the three integers.

34. Five times the first of three consecutive even integers is 2 less than twice the sum of the second and third. Find the three integers.

35. One-fourth of the middle integer of three consecutive even integers is 27 less than one-half of the sum of the other two integers. Find the three integers.

Example The length of a rectangle is 3 times its width and its perimeter is 40 feet. Find the dimensions.

Solution Let x = the width, then 3 times the width or $3x$ = the length.

$P = 2w + 2\ell$	Formula for perimeter
$40 = 2(x) + 2(3x)$	Substitute
$40 = 2x + 6x$	Multiply
$40 = 8x$	Combine like terms
$5 = x$	Divide by 8

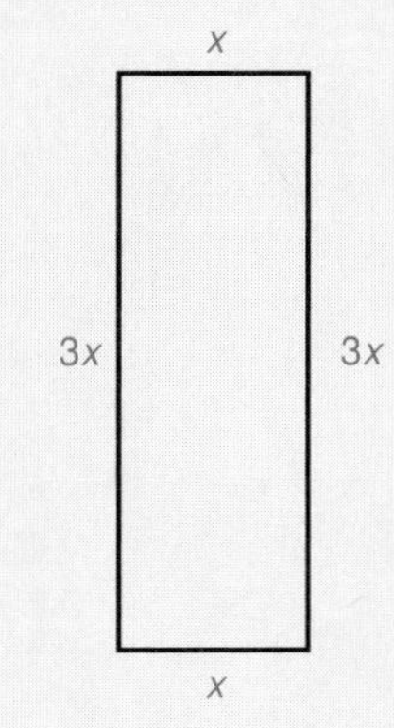

Therefore the width is $x = 5$ feet and the length is $3x = 3(5) = 15$ feet.

36. The length of a rectangle is 9 feet more than its width. The perimeter of the rectangle is 58 feet. Find the dimensions. (Prior knowledge: Perimeter = 2 times the length plus 2 times the width.)

37. The width of a rectangle is 3 feet less than its length. The perimeter of the rectangle is 70 feet. Find the dimensions. (See exercise 36.)

38. The width of a rectangle is $\frac{1}{3}$ of its length. If the perimeter is 96 feet, find the dimensions.

39. The length of a rectangle is 1 inch less than three times the width. Find the dimensions if the perimeter is 70 inches.

40. The width of a rectangle is 3 meters less than the length. If the perimeter of the rectangle is 142 meters, find the dimensions of the rectangle.

41. The length of a rectangle is 5 feet more than its width. If the perimeter is 82 feet, find the length and width.

42. The sum of the number of teeth on two gears is 74 and their difference is 22. How may teeth are on each gear?

43. A 12-foot board is cut into two pieces so that one piece is 4 feet longer than the other. How long is each piece?

44. A 24-foot rope is cut into two pieces so that one piece is twice as long as the other. How long is each piece?

45. A 50-foot extension cord is cut into two pieces so that one piece is 12 feet longer than the other piece. How long is each piece?

46. The sum of two currents is 80 amperes. If the greater current is 24 amperes more than the lesser current, find their values.

47. Two gears have a total of 59 teeth. One gear has 15 less teeth than the other. How many teeth are on each gear?

48. Two electrical voltages have a total of 156 volts (V). If one voltage is 32 V more than the other, find the two voltages.

49. The sum of two voltages is 89 and their difference is 32. Find the two voltages.

50. The sum of two resistances in a series is 30 ohms and their difference is 14 ohms. How many ohms are in each resistor?

Example A man has $10,000, part of which he invests at 11% and the rest at 8%. If his total income from the two investments for one year is $980, how much does he invest at each rate?

Solution **Note** All interest problems in this textbook will be simple interest. The prior knowledge that is needed for this problem is that Interest = Principal · Rate · Time. Time will be equal to 1 year in the following problems.

11% investment	8% investment
x	$10{,}000 - x$

equation

$$x(0.11) + (10{,}000 - x)(0.08) = 980$$

If we have a total amount of $10,000 to invest and we invest x dollars at 11%, then the amount left to invest at 8% would be the total amount minus what we have already invested, $10{,}000 - x$.

$$\begin{aligned} x(0.11) + (10{,}000 - x)(0.08) &= 980 && \text{Original equation} \\ 0.11x + 800 - 0.08x &= 980 && \text{Distributive property} \\ 0.03x + 800 &= 980 && \text{Combine like terms} \\ 0.03x &= 180 && \text{Subtract 800} \\ x &= 6{,}000 && \text{Divide by 0.03} \end{aligned}$$

Hence he has invested $x = 6{,}000$ dollars at 11% and $10{,}000 - x = 10{,}000 - (6{,}000) = 4{,}000$ dollars at 8%.

51. Phil has $20,000, part of which he invests at 8% interest and the rest at 6%. If his total income for one year was $1,460 from the two investments, how much did he invest at each rate?

52. Nancy has $18,000. She invests part of her money at 7½% interest and the rest at 9%. If her income for one year from the two investments was $1,560, how much did she invest at each rate?

53. Tammy has $15,000. She invest part of this money at 8% and the rest at 6%. Her income for one year from these investments totals $1,120. How much is invested at each rate?

54. Alanzo invested $26,000, part at 10% and the rest at 12%. If his income for one year from these investments is $2,720, how much was invested at each rate?

55. Rich has $18,000, part of which he invests at 10% interest and the rest at 8%. If his income from each investment was the same, how much did he invest at each rate?

56. Amy invests a total of $12,000, part at 10% and part at 12%. Her total income for one year from the investments is $1,340. How much is invested at each rate?

57. Barb has $30,000, part of which she invests at 9% interest and the rest at 7%. If her income from the 7% investment was $820 more than that from the 9% investment, how much did she invest at each rate?

58. Paul invested a total of $18,000, part at 5% and part at 9%. If his income for one year from the 9% investment was $200 less than his income from the 5% investment, how much was invested at each rate?

59. Lynne made two investments totaling $25,000. On one investment she made an 18% profit, but on the other investment she took an 11% loss. If her net profit was $2,180, how much was each investment?

60. Grace made two investments totaling $18,000. She made a 14% profit on one investment, but she took a 9% loss on the other investment. If her net profit was $220, how much was each investment?

61. Larry made two investments totaling $21,000. One investment made him a 13% profit, but on the other investment, he took a 9% loss. If his net loss was $196, how much was each investment?

62. Jeff made two investments totaling $34,000. One investment made him a 12% profit, but on the other investment he took a 21% loss. If his net loss was $2,940, how much was each investment?

63. Dale has invested $5,000 at an 8% rate. How much more must he invest at 10% to make the total income for one year from both sources a 9% rate?

64. Jeremy has $9,000 invested at 6%; how much more must he invest at 10% to realize a net return of 9%?

65. Jennifer has $14,000 invested at 7% and is going to invest an additional amount at 11% so that her total investment will make 9%. How much does she need to invest at 11% to achieve this?

Review exercises

Evaluate the following formulas. See section 2–2.

1. $I = prt$, $p = 2{,}000$; $r = 0.05$; $t = 1$

2. $V = \ell wh$, $\ell = 7$, $w = 4$, $h = 3$

3. $F = ma$, $m = 34$, $a = 6$

4. $V = k + gt$, $k = 12$, $g = 32$, $t = 3$

5. $A = p + pr$, $p = 3{,}000$; $r = 0.06$

6. $A = \frac{1}{2}(b + c)$, $b = 20$, $c = 12$

7. $S = \frac{1}{2}gt^2$, $g = 32$, $t = 4$

8. $\ell = a + (n - 1)d$, $a = 4$, $n = 10$, $d = 4$

2–9 ■ Solving linear inequalities

Inequality symbols

In chapter 1, we studied the meaning of the inequality symbols

$<$ "is less than"
$\leq$ "is less than or equal to"
$>$ "is greater than"
$\geq$ "is greater than or equal to."

These symbols define the *sense* or *order* of an inequality. Some examples of how we use these symbols would be:

1. If we want to state symbolically that 4 is less than 7, we write $4 < 7$.
2. If we wish to denote that the variable x represents 5 or any number greater than 5, we write $x \geq 5$.

Note $x \geq 5$ represents *any* real number that is greater than or equal to 5, and not just any integer greater than or equal to 5. Remember that 5.1, 5.004, and so on, are all greater than 5.

3. If we wish to denote that the variable T represents any number less than 3, but not 3 itself, we write $T < 3$.

Linear inequalities

When we replace the equal sign in a conditional linear equation with one of these inequality symbols, we form a *conditional linear inequality.*

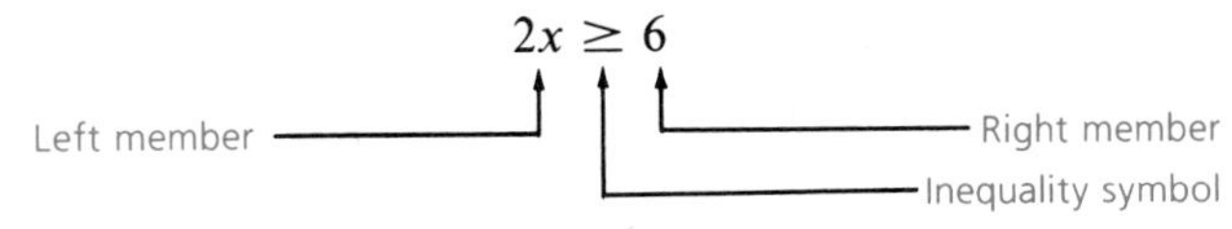

A major difference between the linear equation and the linear inequality is the solution. The solution of a linear equation has at most one solution, whereas the solution of a linear inequality may consist of an unlimited number of solutions. Consider the inequality $2x \geq 6$. We can, by inspection, see that if we substitute 3, $3\frac{1}{2}$, 4, or 5 for x, the inequality will be true. In fact, we see that if we were to substitute any number greater than or equal to 3, the inequality would be true. This demonstrates the fact that the inequality has an unlimited number of solutions. The values for x that would satisfy the inequality would be $x \geq 3$.

Another way to indicate the solution of an inequality is by graphing. To graph the solution, we simply draw a number line (as we did in chapter 1), place a solid circle at 3 on the number line to signify that 3 is in the solution, and draw an arrow extending from the solid circle to the right (figure 2–1). The solid line indicates that *all* numbers greater than or equal to 3 are part of the graph.

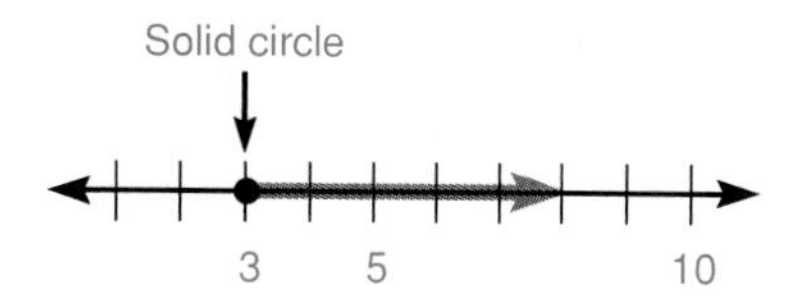

Figure 2–1

■ *Example 2–9 A*

Graph the following linear inequalities.

1. $x < 4$ Here x represents all real numbers less than 4, but not 4 itself. To denote the fact that x cannot equal 4, we put a **hollow circle** at 4.

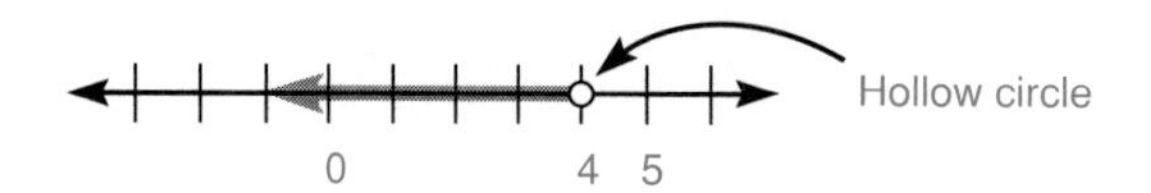

2. $x \geq -3$ The greater than or equal to symbol, $\geq$, indicates that the graph will contain the point -3, and we place a **solid circle** at -3.

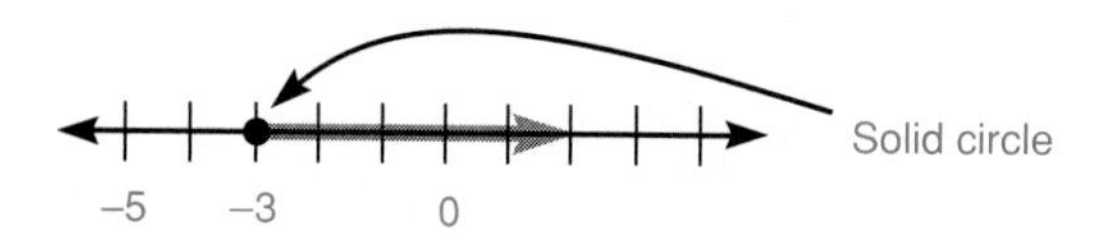

3. $-3 \leq x < 4$ This statement is called a *compound inequality*. It is read "-3 is less than or equal to x and x is less than 4." We place a solid circle at -3 to denote that -3 is included and place a hollow circle at 4 to show that 4 is not included. We then draw a line segment between the two circles.

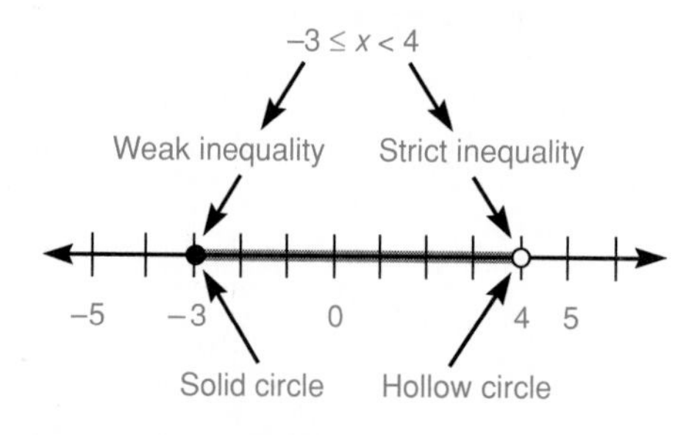

Note When we graph inequalities, a strict inequality ($<$ or $>$) is represented by a hollow circle at the number. A weak inequality ($\leq$ or $\geq$) is represented by a solid circle at the number.

▶ ***Quick check*** Graph $-3 < x \leq 2$ ■

Solving linear inequalities

The properties that we will be using to solve linear inequalities are similar to those that we used to solve linear equations.

Addition and subtraction property of inequalities

For all real numbers a, b, and c, if $a < b$, then

$$a + c < b + c \text{ and } a - c < b - c$$

Concept

The same number can be added to or subtracted from both members of an inequality without changing the direction of the inequality symbol.

Multiplication and division property of inequalities

For all real numbers a, b, and c, if $a < b$, and

1. If $c > 0$ (c represents a positive number), then

$$a \cdot c < b \cdot c \text{ and } \frac{a}{c} < \frac{b}{c}$$

Concept

We can multiply or divide *both* members of the inequality by the same positive number without changing the direction of the inequality symbol.

2. If $c < 0$ (c represents a negative number), then

$$a \cdot c > b \cdot c \text{ and } \frac{a}{c} > \frac{b}{c}$$

Concept

We can multiply or divide *both* members of an inequality by the same *negative* number, provided that we **reverse** the direction of the inequality symbol.

Note The two properties were stated in terms of the less than ($<$) symbol. The properties apply for any of the other inequality symbols ($>$, $\leq$, or $\geq$).

To demonstrate these operations, consider the inequality $8 < 12$.

1. If we add or subtract 4 in each member, we still have a true statement.

$8 < 12$	or	$8 < 12$	Original true statement
$8 + 4 < 12 + 4$		$8 - 4 < 12 - 4$	Add or subtract 4
$12 < 16$		$4 < 8$	New true statement

2. If we multiply or divide by 4 in each member, we still have a true statement.

$8 < 12$	or	$8 < 12$	Original true statement
$8 \cdot 4 < 12 \cdot 4$		$\frac{8}{4} < \frac{12}{4}$	Multiply or divide by 4
$32 < 48$		$2 < 3$	New true statement

3. But if we multiply or divide by -4 in each member, we have to reverse the direction of the inequality before we have a true statement.

$8 < 12$	or	$8 < 12$	Orignial true statement
$8(-4) > 12(-4)$		$\frac{8}{-4} > \frac{12}{-4}$	Multiply or divide by -4 and reverse direction of the inequality symbol
$-32 > -48$		$-2 > -3$	New true statement

Note When we reverse the direction of the inequality symbol, we say that we **reversed the sense or order** of the inequality.

To summarize our properties, we see that they are the same as the properties for linear equations, with one exception. **Whenever we multiply or divide both members of an inequality by a negative number, we must reverse the direction of the inequality symbol.**

We shall now solve some linear inequalities. The procedure for solving a linear inequality uses the same four steps that we used to solve a linear equation.

Solving a linear inequality

1. Simplify in each member, where necessary, by performing the indicated operations.
2. Add, or subtract, to get all terms containing the unknown in one member of the inequality.
3. Add, or subtract, to get all terms *not* containing the unknown in the other member of the inequality.
4. Multiply, or divide, to obtain a coefficient of 1 for the unknown. *Remember, when multiplying or dividing by a negative number, always change the direction (order) of the inequality symbol.*

■ *Example 2–9 B*

Find the solution.

1. $2x + 5x - 1 < 4x + 2$

Step 1 *We simplify* the inequality by carrying out the indicated addition in the left member.

$$2x + 5x - 1 < 4x + 2$$
$$7x - 1 < 4x + 2$$

Step 2 *We want all the terms containing the unknown,* x, *in one member of the inequality.* Therefore we subtract $4x$ from both members of the inequality.

$$7x - 1 < 4x + 2$$
$$7x - 4x - 1 < 4x - 4x + 2$$
$$3x - 1 < 2$$

Note A negative coefficient of the unknown can be avoided if we form equivalent inequalities where the unknown appears only in the member of the inequality that has the greater coefficient of the unknown.

Step 3 *We want all the terms not involving the unknown in the other member of the inequality.* Therefore we add 1 to both members of the inequality.

$$3x - 1 < 2$$
$$3x - 1 + 1 < 2 + 1$$
$$3x < 3$$

Step 4 *We form an equivalent inequality where the coefficient of the unknown is 1.* Hence we divide both members of the inequality by 3.

$$3x < 3$$
$$\frac{3x}{3} < \frac{3}{3}$$
$$x < 1$$

We can also graph the solution.

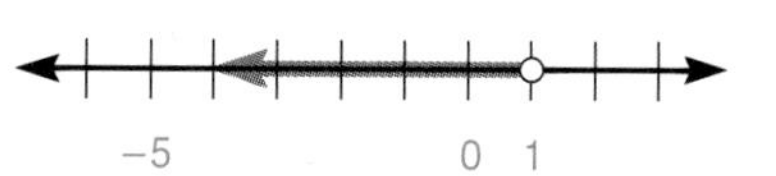

Note We should be careful to observe in step 4 whether we are multiplying or dividing by a positive or negative number so that we will form the correct inequality.

2. $-2x \leq 4$

The only operation we need to perform to solve the inequality is to divide by -2. Since we are dividing by a negative number, we must remember to **reverse** the direction of the inequality symbol.

$$-2x \leq 4$$
$$\frac{-2x}{-2} \geq \frac{4}{-2} \quad \text{Reverse the direction of the inequality symbol}$$
$$x \geq -2$$

Graph

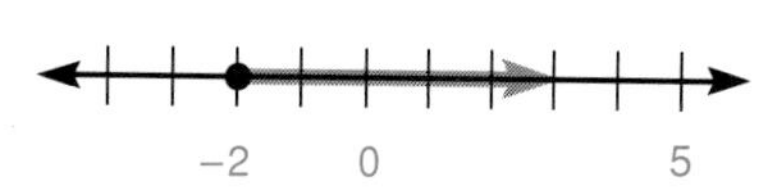

3.
$$5(2x + 1) \leq 7x - 4x + 3$$
$$10x + 5 \leq 3x + 3 \quad \text{Simplify by multiplying in left member and combining like terms in right member}$$
$$10x - 3x + 5 \leq 3x - 3x + 3 \quad \text{Subtract } 3x \text{ from both members}$$
$$7x + 5 \leq 3$$
$$7x + 5 - 5 \leq 3 - 5 \quad \text{Subtract 5 from both members}$$
$$7x \leq -2$$
$$\frac{7x}{7} \leq \frac{-2}{7} \quad \text{Divide both members by 7}$$
$$x \leq \frac{-2}{7}$$

Graph

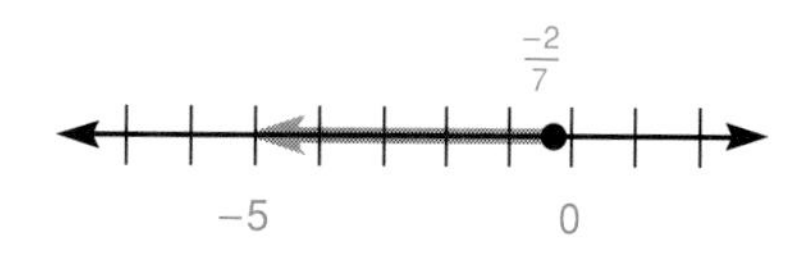

4. $-3 \le 2x + 1 < 5$

When solving a compound inequality, the solution must be such that the unknown appears only in the middle member of the inequality. We can still use all of our properties, if we apply them to all three members, and we must reverse the direction of *all* inequality symbols when multiplying or dividing by a negative number.

$$-3 \le 2x + 1 < 5$$

$$-3 - 1 \le 2x + 1 - 1 < 5 - 1 \quad \text{Subtract 1 from all three members}$$

$$-4 \le 2x < 4$$

$$\frac{-4}{2} \le \frac{2x}{2} < \frac{4}{2} \quad \text{Divide all three members by 2}$$

$$-2 \le x < 2$$

Graph 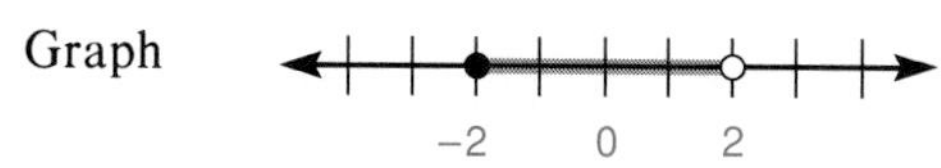

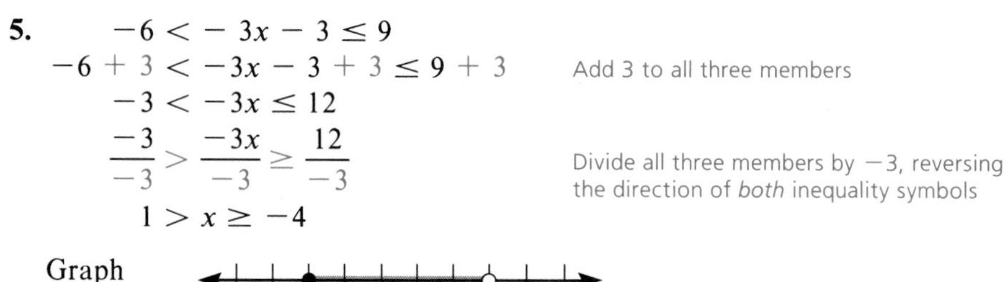

5. $-6 < -3x - 3 \le 9$

$$-6 + 3 < -3x - 3 + 3 \le 9 + 3 \quad \text{Add 3 to all three members}$$

$$-3 < -3x \le 12$$

$$\frac{-3}{-3} > \frac{-3x}{-3} \ge \frac{12}{-3} \quad \text{Divide all three members by } -3\text{, reversing the direction of } \textit{both} \text{ inequality symbols}$$

$$1 > x \ge -4$$

Graph 

Note From our discussions in chapter 1, we could have written the solution in the previous problem as $-4 \le x < 1$. This is usually the preferred form.

▶ ***Quick check*** Find the solution for $4x + 5x - 4 < 6x - 1$ ■

Problem solving

We are now ready to combine our abilities to write an expression and to solve an inequality and apply them to solve word problems. The guidelines for solving a linear inequality are the same as those for solving a linear equation in section 2–4. The following table shows a number of different ways that an inequality symbol could be written with words.

Symbol	$<$	$\le$	$>$	$\ge$
In words	is less than is fewer than is almost	is at most is no more than is no greater than is less than or equal to	is greater than is more than exceeds	is at least is no less than is no fewer than is greater than or equal to

■ Example 2–9 C

1. Write an inequality for the following statement: A student's test grade, G, must be at least 75 to have a passing grade.

   If the student's test grade must be *at least* 75, the grade must be 75 or greater. Thus,

   $G \geq 75$

2. Four times a number less 5 is to be no more than three times the number increased by 2. Find the number.

   Let x represent the number.

4 times a number	less	5	is no more than	3 times the number	increased by	2
$4x$	$-$	5	$\leq$	$3x$	$+$	2

   The inequality is $4x - 5 \leq 3x + 2$

$$4x - 5 \leq 3x + 2$$
$$4x - 5 - 3x \leq 3x + 2 - 3x \quad \text{Subtract } 3x \text{ from each member}$$
$$x - 5 \leq 2 \quad \text{Combine in each member}$$
$$x - 5 + 5 \leq 2 + 5 \quad \text{Add 5 to each member}$$
$$x \leq 7 \quad \text{Combine in each member}$$

   The number is any real number x such that $x \leq 7$.

▶ ***Quick check*** To complete an order for cement, a company will need at least 3 trucks. Write an inequality for the number of trucks needed.

If 4 is subtracted from three times a number, the result is greater than 2 more than twice the number. Find the number. ■

Mastery points

Can you

- Graph inequalities and compound inequalities?
- Solve linear inequalities and compound linear inequalities?
- Solve verbal inequalities?

Exercise 2–9

Graph the following. See example 2–9 A.

Example $-3 < x \leq 2$

Solution

1. $x > 2$ 2. $x > -2$ 3. $x \geq 1$ 4. $x \geq 4$ 5. $x < 0$
6. $x < -2$ 7. $x > 0$ 8. $x \leq 3$ 9. $x \leq -4$ 10. $-2 < x < 0$
11. $-1 < x < 2$ 12. $-3 \leq x \leq 4$ 13. $0 \leq x \leq 5$ 14. $1 \leq x < 4$ 15. $-1 < x \leq 3$

Find the solution and graph the solution. See example 2–9 B.

Example $4x + 5x - 4 < 6x - 1$

Solution

$9x - 4 < 6x - 1$ Combine like terms in left member
$9x - 4 - 6x < 6x - 1 - 6x$ Subtract 6x from each member
$3x - 4 < -1$ Combine like terms
$3x - 4 + 4 < -1 + 4$ Add 4 to each member
$3x < 3$ Combine like terms
$x < 1$ Divide each member by 3

Thus $x < 1$ and the solutions are all real numbers less than 1.

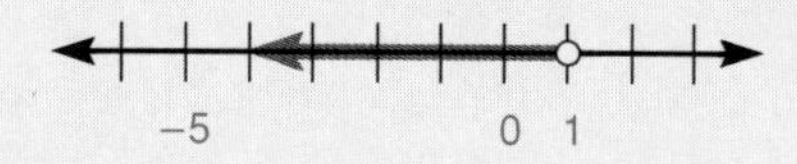

16. $4x > 10$
17. $2x \leq 5$
18. $3x \geq 15$
19. $5x < 30$
20. $\frac{3}{4}x < 9$
21. $\frac{2}{3}x \geq 12$
22. $-4x < 12$
23. $-3x \leq 27$
24. $-6x > 18$
25. $-2x < 10$
26. $4x + 3x \geq 2x + 7$
27. $8x - 2x > 4x - 5$
28. $3x + 2x < x + 6$
29. $\frac{3x}{2} > 8$
30. $\frac{4x}{3} > 12$
31. $x + 7 \leq 9$
32. $x + 4 \geq -12$
33. $x - 5 < 6$
34. $x - 12 < -9$
35. $2x + (3x - 1) > 5 - x$
36. $2(3x + 1) < 7$
37. $3(2x - 1) \geq 4x + 3$
38. $4(5x - 3) \leq 25x + 11$
39. $12x - 8 > 5x + 2$
40. $9x + 4 > x - 11$
41. $2 + 5x - 16 < 6x - 4$
42. $3(2x + 5) > 4(x - 3)$
43. $8 - 2(3x + 4) > 5x - 16$
44. $(3x + 2) - (2x - 5) > 7$
45. $(7x - 6) - (4 - 3x) \leq 27$
46. $2(x - 4) - 16 \leq 3(5 - 2x)$
47. $3(1 - 2x) \geq 2(4 - 4x)$
48. $4(5 - x) > 7(2 - x)$
49. $3(2x + 3) \geq 5 - 4(x - 2)$
50. $6(3x - 2) \leq 7(x - 3) - 2$
51. $-1 < 2x + 3 < 4$
52. $-3 < 3x - 4 < 6$
53. $-2 \leq 5x + 2 \leq 3$
54. $0 \leq 7x - 1 \leq 7$
55. $-5 < 4x + 3 \leq 8$
56. $-2 < -x \leq 3$
57. $-1 \leq -x < 4$
58. $-4 \leq 2 - x < 3$
59. $-3 < 4 - x \leq 5$
60. $1 < 3 - 4x < 6$
61. $0 \leq 1 - 3x < 7$
62. $-4 \leq 3 - 2x \leq 0$

Write an inequality to represent the following statements. See Example 2–9 C.

Example To complete an order for cement, a company will need at least 3 trucks.

Solution The words *at least* 3 trucks means that the company will need *3 or more* trucks. If x is the number of trucks needed, then

$x \geq 3$

63. Mark's score must be at least 72 on the final exam to pass the course.

64. The temperature today will be less than 38.

65. An automobile parts company needs to order at least 8 new lift trucks.

66. An accounting company will hire at least 2 new employees, but not more than 7.

67. The selling price (P) must be at least twice the cost (C).

Write an inequality using the given information and solve. See example 2–9 C.

Example If 4 is subtracted from three times a number, the result is greater than 2 more than twice the number. Find the number.

Solution Let x represent the number. Then

$3x - 4$ is "4 subtracted from three times the number,"
$2x + 2$ is "2 more than twice the number."

Since the two expressions are related by "is greater than," the inequality is $3x - 4 > 2x + 2$

$$3x - 4 > 2x + 2$$
$$3x - 4 - 2x > 2x + 2 - 2x \quad \text{Subtract } 2x \text{ from each member}$$
$$x - 4 > 2 \quad \text{Combine in each member}$$
$$x - 4 + 4 > 2 + 4 \quad \text{Add 4 to each member}$$
$$x > 6 \quad \text{Combine in each member}$$

The number is any number that is greater than 6.

68. When 7 is subtracted from two times a number, the result is greater than or equal to 9. Find all numbers that satisfy this condition.

69. Five times a number minus 11 is less than 19. Find all numbers that satisfy this condition.

70. The product of 6 times a number added to 2 is greater than or equal to 1 subtracted from five times the number. What are the numbers that satisfy this condition?

71. Twice a number increased by 7 is no more than three times the number decreased by 5. Find the numbers that satisfy this condition.

72. If one-third of a number is added to 23, the result is greater than 30. Find all numbers that satisfy this condition.

73. Eugenia has scores of 7, 6, and 8 on three quizzes. What must she score on the fourth quiz to have an average of 7 or higher?

74. Sam has scores of 72, 67, and 81 on three tests. If an average of 70 is required to pass the course, what is the minimum score he must have on the fourth test to pass?

75. Two times a number plus 4 is greater than 6 but less than 14. Find all numbers that satisfy these conditions.

76. Four times a number minus 7 is greater than 17 but less than 25. Find all numbers that satisfy these conditions.

77. The perimeter of a square must be greater than 20 inches but less than 108 inches. Find all values of a side that satisfy these conditions.

78. The perimeter of a rectangle must be less than 100 feet. If the length is known to be 30 feet, find all numbers that the width could be. (*Note:* The width of a rectangle must be a positive number.)

79. The perimeter (the sum of the sides) of a triangle is more than 52 cm. If two sides of the triangle are 18 cm and 16 cm, respectively, what are the possible values for the length of the third side?

80. Two sides of a triangle are 10 ft and 12 ft long, respectively. If the perimeter must be at least 31 ft, what are the possible values for the length of the third side?

Review exercises

Perform the indicated operations. See section 1–8.

1. -4^2 **2.** $(-4)^2$ **3.** -2^4 **4.** $(-2)^4$

Write an algebraic expression for each of the following. See section 2–1.

5. x raised to the fifth power **6.** A number cubed **7.** A number squared **8.** The product of x and y

Chapter 2 lead-in problem

Bonnie has $3,000 invested at 8% simple interest per year. How much more money must she invest at 7% simple interest if she wants an income of $660 per year ($55 per month) from her investments?

Solution

Let x = the number of dollars invested at 7%.

income from 8% investment + income from 7% investment = total income

$$
\begin{aligned}
3{,}000(0.08) + x(0.07) &= 660 && \text{Original equation}\\
240 + 0.07x &= 660 && \text{Simplify}\\
0.07x &= 420 && \text{Subtract 240}\\
x &= 6{,}000 && \text{Divide by 0.07}
\end{aligned}
$$

Therefore Bonnie needs to invest $6,000 at 7% so that her total income from both investments is $660 per year.

Chapter 2 summary

1. A **variable** is a symbol (generally a lowercase letter) that represents an unspecified number.
2. A **constant** is a symbol that does not change its value.
3. An **algebraic expression** is any meaningful collection of variables, constants, grouping symbols, and signs of operations.
4. The **terms** in an algebraic expression are any constants, variables, or products or quotients of these. They are separated by plus or minus signs.
5. In the expression $8x$, 8 is called the **numerical coefficient** or just the coefficient.
6. A **polynomial** is a special kind of algebraic expression. A **monomial** is a polynomial that contains one term; a **binomial** contains two terms; a **trinomial** contains three terms; a **multinomial** contains more than one term.
7. We use the *property of substitution* to *evaluate* algebraic expressions.
8. **Like terms** or **similar terms** are terms whose variable factors are the same.
9. We can *add or subtract* only like, or similar, terms.
10. A **mathematical statement** can be labeled true or false.
11. An equation that is true for every permissible value of the variable is called an **identity.**
12. A replacement value for the variable that forms a true statement (satisfies that equation) is called a **root,** or a **solution,** of the equation.
13. The **solution set** is the set of all values for the variable that cause the equation to be a true statement.
14. A **linear equation** is an equation where the exponent of the unknown is 1.
15. The **addition and subtraction property of equality** enables us to add or subtract the same quantity in each member of an equation and the result will be an equivalent equation.
16. The **symmetric property of equality** allows us to interchange right and left members of an equation.

17. The **multiplication and division property of equality** enables us to multiply or divide both members of an equation by the same nonzero quantity.
18. We use the same procedures for solving **literal equations** that we use to solve linear equations in one variable.
19. A linear inequality involves the symbols $<$, $\leq$, $>$, and $\geq$.
20. The **addition and subtraction property of inequalities** states that the same number can be added to or subtracted from both members of an inequality without changing the direction (order) of the inequality symbol.
21. The **multiplication and division property of inequalities** states:
    a. The same *positive* number may be multiplied times or divided into both members of an inequality without changing the direction (order) of the inequality.
    b. When the same *negative* number is multiplied times or divided into both members of an inequality, the direction (order) of the inequality *must be changed.*

Chapter 2 error analysis

1. Degree of a polynomial
   Example: $2x - 3x^2 + x^3 - 1$ has degree 6.
   Correct answer: $2x - 3x^2 + x^3 - 1$ has degree 3.
   What error was made? (*see page 69*)
2. Terms in an algebraic expression
   Example: $x^2 + \dfrac{2x - 3}{5}$ has 3 terms.
   Correct answer: $x^2 + \dfrac{2x - 3}{5}$ has 2 terms.
   What error was made? (*see page 68*)
3. Applying the distributive property
   Example: $5(4 + b) = 20b$
   Correct answer: $5(4 + b) = 20 + 5b$
   What error was made? (*see page 80*)
4. Combining like terms
   Example: $3a^2 + 4a = 7a^3$
   Correct answer: $3a^2 + 4a = 3a^2 + 4a$
   What error was made? (*see page 81*)
5. Combining polynomials
   Example: $(3x^2 - 2x + 1) - (x^2 - x + 2)$
   $= 3x^2 - 2x + 1 - x^2 - x + 2$
   $= 2x^2 - 3x + 3$
   Correct answer: $(3x^2 - 2x + 1) - (x^2 - x + 2)$
   $= 2x^2 - x - 1$
   What error was made? (*see page 82*)
6. Reciprocal of a number
   Example: The reciprocal of 0 is $\dfrac{1}{0}$.
   Correct answer: 0 has no reciprocal.
   What error was made? (*see page 94*)
7. Graphing linear inequalities
   Example: The graph of $x \leq 3$ is

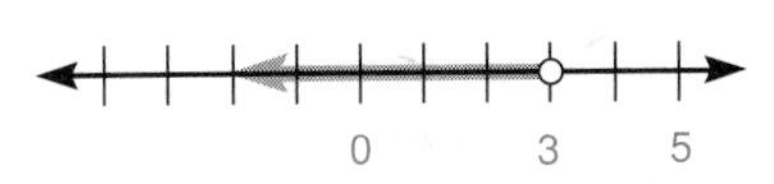

   Correct answer: The graph of $x \leq 3$ is

   What error was made? (*see page 114*)
8. Multiplying members of an inequality
   Example: If $3 < 4$, then $3 \cdot -2 < 4 \cdot -2$.
   Correct answer: If $3 < 4$, then $3 \cdot -2 > 4 \cdot -2$.
   What error was made? (*see page 115*)
9. Multiplication of negative numbers
   Example: $(-5)(7) = 35$
   Correct answer: $(-5)(7) = -35$
   What error was made? (*see page 48*)
10. Division using zero
   Example: $\dfrac{7}{0} = 0$
   Correct answer: $\dfrac{7}{0}$ is undefined.
   What error was made? (*see page 53*)

Chapter 2 critical thinking

If you add any three consecutive odd integers, the sum will be a multiple of 3. Why is this true?

Chapter 2 review

[2–1]

Specify the number of terms in each expression.

1. $4x^2 + 3x + 2$
2. $5a^2b$
3. $7xy + 5$
4. $(ab + cd) + xy$

Determine which of the following algebraic expressions are polynomials. If they are not polynomials, state why not.

5. $\frac{x + y}{3} + z$
6. $x^3 - x^2$
7. $4a^2b^3c$
8. $\frac{a + b}{c}$

Write an algebraic expression for each of the following.

9. 5 times x
10. 7 less than y
11. 4 more than z
12. 2 times a number, plus 6

[2–2]

Evaluate the following expressions if $a = 3$, $b = 4$, $c = -4$, and $d = -3$.

13. $3a - b + c$
14. $d - 2(a + c)$
15. $a^2b - a^2c$
16. $(2a + c)(b + 2d)$
17. $(c - 2d)^2$
18. $c^2 - d^2$
19. Evaluate R when $R = \frac{P \cdot L}{D^2}$ given (a) $P = 6$, $L = 8$, $D = 4$; (b) $P = 7$, $L = 3$, $D = \frac{2}{3}$.
20. The volume of a gas V_2 is given by $V_2 = \frac{P_1V_1}{P_2}$. Find V_2 when $P_1 = 780$, $V_1 = 80$, and $P_2 = 60$.

[2–3]

Remove all grouping symbols and combine like terms.

21. $(3x^2 + 2x - 1) + (x^2 - 5x + 4)$
22. $(a^2 - 3a + 4) - (2a^2 - 4a - 7)$
23. $(4a^2 - b^2) - (3a^2 + 2b^2) - (7a^2 - 3b^2)$
24. $(5x^3 - 2xy^2 + 3x^2y - 4y^3) - (4x^2 + 3x^2y - 2y^3 + 5xy^2)$
25. $(4ab + 7b^2c) - (15ab - 11bc)$
26. $(x - 2y + 7) - (x + 4y + 6)$
27. $(4ab - 2ac) - (6bc - 5ac) + (ab + 2bc)$
28. $3a - [4a - (a - 5)]$
29. $5x + [3x - (x - y)]$
30. $4x - (x - y) - [3x - y - (2x + 3y)]$
31. $x - \{5x - [3y - (2x - y)]\}$
32. $5a - \{6b + a - (5a - 4b)\}$

[2–4]

Determine whether the given statement is true or false when we replace the variable in each equation with the given number.

33. $x + 7 = 11; \{4\}$
34. $2x + 1 = 9; \{2\}$
35. $5x - 1 = 21; \{5\}$
36. $\frac{x}{2} + 5 = 12; \{14\}$

[2–4, 2–5, 2–6]

Find the solution set.

37. $x + 5 = 12$
38. $x - 4 = 17$
39. $a + 7 = -4$
40. $b - 3 = -9$
41. $5z + 3z - 7z + 3 = 7$
42. $2(3x - 4) - 5x = 11$
43. $3(2y + 3) = 7 + 5y$
44. $3(x - 1) - 2(x + 1) = 4$
45. $3x = 9$
46. $4x = 12$
47. $-2x = 14$
48. $-3x = 21$
49. $\frac{x}{3} = 4$
50. $\frac{x}{2} = 7$
51. $\frac{3x}{5} = 9$
52. $\frac{2x}{7} = 6$
53. $\frac{1}{3}x - 1 = \frac{3}{4}$
54. $\frac{1}{3}x + 1 = \frac{1}{6}x - 2$
55. $\frac{3}{4}x + 4 = \frac{5}{8}$
56. $\frac{3}{5}x + \frac{1}{2} = \frac{7}{10}x - 3$
57. $3x = 0$

58. $-x = -4$

59. $3.7a = 22.2$

60. $32.8 = -4.1x$

61. $2x + 5 = 11$

62. $3b - 8 = 6$

63. $y + (2y - 1) = 6$

64. $x + 3x = 5 + 7$

65. $3(2a - 1) = 4a - 2$

66. $5(x + 3) = 2x - 7$

67. $(3x - 2) - (4x - 1) = 3x$

68. $2a + 5a - 4 = 3(1 - 2a)$

69. $8 - 3x + 7 = 5(x + 7)$

70. $2y - 3(y + 1) = 11$

71. $7x - 4(2x + 3) = 12$

72. $8x - 14 = 14 - 8x$

73. $5b + 4 = 4 - 2b$

74. $-3(2x + 1) = 4x - 5$

75. $3(c + 2) - 2(c + 1) = 5c + 11$

76. $4x - 2(1 - 3x) = 8x + 2$

[2–7]

Solve for the specified variable.

77. $F = ma$, for a

78. $E = IR$, for I

79. $k = PV$, for P

80. $V = k + g + t$, for g

81. $A = \frac{1}{2}h(b + c)$, for c

82. $5x - y = 2x + 3y$, for x

[2–8]

Write an equation for the problem and solve for the unknown quantities.

83. The difference between two numbers is 23. Find the two numbers if their sum is 105.

84. If a number is divided by 9 and that result is then increased by 7, the answer is 11. Find the number.

85. The difference between one-third of a number and one-fifth of a number is 6. Find the number.

86. John invested part of $20,000 at 8% and the rest at 7%. If his income from the 8% investment was $250 more than that from the 7% investment, how much was invested at each rate?

87. Anne made two investments totaling $25,000. On one investment she made a 12% profit but on the other she took a 19% loss. If her net loss was $1,030, how much was in each investment?

[2–9]

Find the solution and graph the solution.

88. $3x > 12$

89. $5x \leq 15$

90. $-2x < 14$

91. $-4x > 16$

92. $2x + 1 < 5$

93. $7x - 4 > 11$

94. $3x + 7 < 5x - 2$

95. $9x + 13 \geq 4x + 7$

96. $6(2x - 1) \leq 3x - 4$

97. $-4 < 5x + 7 < 10$

98. $0 \leq 1 - 5x < 6$

99. $5 < 4x + 3 < 12$

100. $-8 \leq 3x + 5 \leq 4$

Chapter 2 cumulative test

Perform the indicated operations, if possible, and simplify.

[1–4] **1.** $(-8) + (-4)$

[1–5] **2.** $(-10) - (-14)$

[1–7] **3.** $\frac{-24}{-8}$

[1–7] **4.** $\frac{8}{0}$

[1–8] **5.** -5^2

[1–2] **6.** $\frac{4}{5} - \frac{3}{10}$

[1–4] **7.** $\frac{2}{3} + \left(-\frac{5}{6}\right)$

[1–6] **8.** $(-4)(0)(-2)$

[1–2] **9.** $(2.3)(8.6)$

[1–8] **10.** $5 + 6(8 - 2)$

[1–8] **11.** $6 + 4(10 - 2)$

[1–8] **12.** $10 - 2(15 - 3) - 5 \cdot 2$

[1–8] **13.** $2[5(7 - 4) - 6 + 4]$

[1–8] **14.** $5(-4 + 7) - 3(8 - 5)$

[1–8] **15.** $14 + 2 \cdot 15 \div 6 - 3 + 4$

[2–3] **16.** $5x + x - 2x$

[2–3] **17.** $3x^2y^2 - 2xy - x^2y^2 + 5xy$

[2–3] **18.** $(2a - b) - (a - 4b)$

[2–3] **19.** $(3x^2y - 2xy^2) - (5xy^2 - x^2y)$

[2–3] **20.** $5a + 3a^2 - 4a - a^2 + 5 + a^3 - 6$

[2–3] **21.** $x - [3x - (y + x) + (2x - 3y)]$

[2–3] **22.** $(3a - 2b) - [5a - (4b + 6a)]$

Evaluate the following if $a = -2$, $b = -3$, $c = 4$, and $d = 5$.

[2–2] **23.** $(a + 2d)^2$

[2–2] **24.** $(3a - 2b) - (5c + d)$

[2–2] **25.** $(a - 4c)(b - 2d)$

Find the answer.

[1–4] **26.** The sum of 8, -12, and 6

[1–5] **27.** Subtract -12 from -8.

[1–6] **28.** There are 6 rows of desks in a classroom. If each row contains 7 desks, how many desks are in the classroom?

[1–7] **29.** A trip of 357 miles takes seven hours to complete. What was the average rate of speed?

Write an algebraic expression for each of the following.

[2–1] **30.** x decreased by y

[2–1] **31.** A number increased by 6

[2–2] **32.** Ann has d dimes, n nickels, and c cents. Express in cents the amount of money Ann has.

Find the solution set for 33–37 and the solution for 38–42.

[2–6] **33.** $10x - 7 = 4x + 3$

[2–6] **34.** $5x + 6 = 6$

[2–6] **35.** $\frac{1}{3}x + 4 = \frac{5}{6}$

[2–6] **36.** $3(2x - 1) + 2(5x - 3) = 8$

[2–6] **37.** $16 - 2(4x - 1) = 3x - 12$

[2–9] **38.** $-2x \geq 12$

[2–9] **39.** $5x + 3x < 6x - 14$

[2–9] **40.** $3x + (x - 1) > 7 - x$

[2–9] **41.** $-1 < 2x + 3 < 11$

[2–9] **42.** $-16 \leq 8 - 4x \leq 12$

Solve for the specified variable.

[2–7] **43.** $P = a + b + c$, for b

[2–7] **44.** $x = a(y + z)$, for y

Solve the following word problems.

[2–8] **45.** Phil has $10,000, part of which he invests at 6% and the rest at 5%. If his total income from the two investments was $560, how much did he invest at each rate?

[2–8] **46.** The sum of three consecutive even integers is 48. Find the three integers.

[2–8] **47.** If a number is increased by 9 and that result is divided by 3, the answer is 7. Find the number.

[2–8] **48.** Dwala made two investments totaling $17,000. On one investment she made a 12% profit but on the other she took a 19% loss. If her net loss was $1,215, how much was in each investment?

[2–9] **49.** Twice a number decreased by 2 is at most 10. Find all numbers that satisfy this condition.

CHAPTER

3 Polynomials and Exponents

If Diane makes $13 per hour and works 40 hours per week, she will earn more than one million dollars over the next 40 years. How much money will Diane earn? Perform the computations using scientific notation.

3–1 ■ Exponents—I

Exponential form

In chapter 1, we discussed the idea of exponents as related to real numbers. Since variables are symbols for real numbers, let us now apply the idea of exponents to variables. The expression x^4 is called the **exponential form** of the product

$$x \cdot x \cdot x \cdot x$$

We call x the **base** and 4 the **exponent.**

$$\underbrace{x^4}_{\text{Exponential form; base } x\text{, exponent } 4} = \underbrace{x \cdot x \cdot x \cdot x}_{4 \text{ factors of } x} \leftarrow \text{Expanded form}$$

Exponent → 4; Exponential form → x^4; Base → x; 4 factors of x; Expanded form → $x \cdot x \cdot x \cdot x$

Definition of exponents

$$a^n = \underbrace{a \cdot a \cdot a \cdot \cdots a}_{n \text{ factors of } a}, \text{ where } n \text{ is a positive integer.}$$

Concept

The exponent tells us how many times the base is used as a factor in an indicated product.

Note An exponent acts only on the symbol immediately to its left. That is, in ab^4 the exponent 4 applies only to b, whereas $(ab)^4$ means the exponent applies to both a and b.

■ Example 3–1 A

Write in exponential form.

1. $2 \cdot 2 \cdot 2 \cdot 2 = 2^4$ **2.** $a \cdot a \cdot a = a^3$

3. $(a + b)(a + b)(a + b) = (a + b)^3$

Note In example 3, $(a + b)$ is the base.

4. $(-3)(-3)(-3)(-3) = (-3)^4$ **5.** $-(3 \cdot 3 \cdot 3 \cdot 3) = -3^4$

Note Examples 4 and 5 review the ideas from section 1–6 on exponents related to real numbers. Recall that $(-3)^4 = 81$, whereas $-3^4 = -81$.

▶ ***Quick check*** Write $y \cdot y \cdot y \cdot y$ in exponential form. ■

■ Example 3–1 B

Write as an indicated product.

1. $b^4 = b \cdot b \cdot b \cdot b$ **2.** $5^3 = 5 \cdot 5 \cdot 5$

3. $(x - y)^4 = (x - y)(x - y)(x - y)(x - y)$ **4.** $(-2)^2 = (-2)(-2)$

5. $-2^2 = -(2 \cdot 2)$

▶ ***Quick check*** Write c^5 as an indicated product. ■

Multiplication of like bases

Consider the indicated product of $x^2 \cdot x^3$. If we rewrite x^2 and x^3 by using the definition of exponents, we have

$$x^2 \cdot x^3 = \overbrace{x \cdot x}^{x^2} \cdot \overbrace{x \cdot x \cdot x}^{x^3}$$

and again using the definition of exponents, this becomes

$$x^2 \cdot x^3 = \overbrace{x \cdot x \cdot x \cdot x \cdot x}^{\text{5 factors}} = x^5$$

This leads us to the observation that

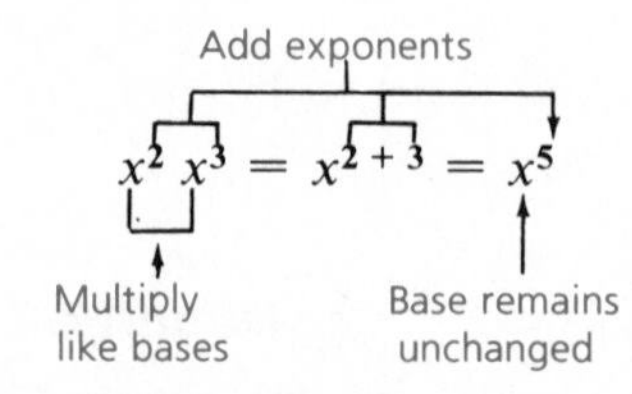

Thus we have the following **product property of exponents.**

> ***Product property of exponents***
>
> $$a^m \cdot a^n = a^{m+n}$$
>
> ***Concept***
>
> When multiplying like bases, add their exponents.

Note The base stays the same throughout the process. It is by adding the exponents that the multiplication is carried out.

■ *Example 3–1 C*

Find the product.

1. $x^3 \cdot x^5 = x^{3+5} = x^8$

2. $3^2 \cdot 3^4 = 3^{2+4} = 3^6 = 729$

Note A common error in multiplying $3^2 \cdot 3^4$ is to multiply the bases $3 \cdot 3 = 9$ and add the exponents, getting the incorrect answer of 9^6. The correct way is to say $3^2 \cdot 3^4 = 3^6$, not 9^6.

3. $y^2 \cdot y^3 \cdot y^4 = y^{2+3+4} = y^9$

4. $a^2 \cdot a \cdot a^3 = a^{2+1+3} = a^6$

Note The variable a means the same as a^1. Likewise, 3 means the same as 3^1. If there is no exponent written with a numeral or a variable, the exponent is understood to be 1.

5. $(a+b)^3(a+b)^4 = (a+b)^{3+4} = (a+b)^7$
6. $(-2)^3(-2)^2 = (-2)^{3+2} = (-2)^5 = -32$

▶ ***Quick check*** Find the product. $x^4 \cdot x^5$ ■

Group of factors to a power property of exponents

Several additional properties of exponents can be derived using the definition of exponents and the commutative and associative properties of multiplication. Observe the following:

$$(xy)^3 = \overbrace{xy \cdot xy \cdot xy}^{\text{3 factors of } xy}$$
$$= \overbrace{x \cdot x \cdot x}^{\text{3 factors of } x} \cdot \overbrace{y \cdot y \cdot y}^{\text{3 factors of } y}$$
$$= x^3y^3$$

This leads us to the following property of exponents.

Group of factors to a power property of exponents

$$(ab)^n = a^nb^n$$

Concept

When a group of *factors* is raised to a power, raise each of the factors in the group to this power.

■ *Example 3–1 D*

Simplify.

1. $(ab)^4 = a^4b^4$ Both a and b are raised to the 4th power

	Groups of factors to a power		Raise each factor to the power		Standard form
2.	$(2ab)^3$	$=$	$2^3a^3b^3$	$=$	$8a^3b^3$

Note In example 2, the number 2 is a factor in the group. Therefore it is also raised to the indicated power.

3. $(3 \cdot 4)^3 = 3^3 \cdot 4^3 = 27 \cdot 64 = 1{,}728$ $(3 \cdot 4)^3$ also is $(12)^3 = 1{,}728$ ■

Note The quantity $(a + b)^3 \neq a^3 + b^3$ because a and b are *terms,* not factors as the property specified. If we consider $(a + b)$ to be a single factor, then by the definition of exponents we have

$$(a + b)^3 = (a + b)(a + b)(a + b)$$

We will see the method of multiplying this later in this chapter.

Power of a power

Consider the expression $(x^4)^3$. Applying the definition of exponents and the product property of exponents, we have

$$(x^4)^3 = \overbrace{x^4 \cdot x^4 \cdot x^4}^{\text{3 factors of } x^4} = \overbrace{x^{4+4+4}}^{\text{Add the exponents}} = x^{12}$$

In chapter 1, we reviewed the idea that multiplication is repeated addition of the same number. Therefore adding the exponent 4 three times is the same as $4 \cdot 3$. Thus

Power of a power — Multiply exponents

$$(x^4)^3 = x^{4 \cdot 3} = x^{12}$$

Therefore we have the following property of exponents.

Power of a power property of exponents

$$(a^m)^n = a^{m \cdot n}$$

Concept
A power of a power is found by multiplying the exponents.

■ ***Example 3–1 E***

Simplify.

1. $(y^3)^2 = y^{3 \cdot 2} = y^6$

2. $(4^2)^5 = 4^{2 \cdot 5} = 4^{10} = 1{,}048{,}576$

3. $(x^5)^4 = x^{5 \cdot 4} = x^{20}$

▶ ***Quick check*** Simplify. $(a^4)^3$ ■

Products of monomials

To multiply the monomials

$$3x^2 \cdot 5x$$

we apply the commutative and associative properties of multiplication along with the properties of exponents. We then write this expression as a product of the numerical coefficients times the product of the variables. That is,

$$3x^2 \cdot 5x = (3 \cdot 5)(x^2 \cdot x) = 15x^3$$

To find the product of

$$5a \cdot 4b$$

we apply the same properties to get

$$5a \cdot 4b = (5 \cdot 4)(a \cdot b) = 20ab$$

Note It is a good procedure to write the variable factors of any term in alphabetical order. This makes identifying like terms much simpler. For example, $3a^2c^3b$ and $4bc^3a^2$ are like terms, but recognizing that fact would have been easier if they had been written as $3a^2bc^3$ and $4a^2bc^3$.

■ Example 3–1 F

Perform the indicated multiplication.

1. $4x \cdot 3xy = (4 \cdot 3) \cdot (x \cdot x) \cdot y = 12x^2y$
2. $8a^3 \cdot 4a^3 \cdot 3a = (8 \cdot 4 \cdot 3) \cdot (a^3 \cdot a^3 \cdot a) = 96a^7$
3. $(-2a^2) \cdot (3ab) = (-2 \cdot 3) \cdot (a^2 \cdot a) \cdot b = -6a^3b$

Note The product of a^3 and b can *only* be written as a^3b since a and b are not like bases.

4. $(5x^2y^3z)(4x^3yz^4) = (5 \cdot 4)(x^2x^3)(y^3y)(zz^4) = 20x^5y^4z^5$ ■

Problem solving

The following problems require us to write algebraic expressions involving the use of exponents.

■ Example 3–1 G

Write an algebraic expression for each of the following verbal statements.

1. The volume of a cube is found by using the length of the edge, e, as a factor 3 times. Write an expression for the volume of a cube.

   We write e as a factor 3 times as $e \cdot e \cdot e = e^3$. Then the volume, V, of a cube is given by

   $V = e^3$.

2. Write an expression for 5 less than the square of a number.

   Let n represent the number, then the square of the number is given as n^2, and since "less than" means to subtract, the expression is given by

   $n^2 - 5$. ■

Mastery points

Can you

- Write a product in exponential form?
- Use the product property of exponents?
- Raise a group of factors to a power?
- Raise a power to a power?
- Multiply monomials?

Exercise 3–1

Write the following expressions in exponential form. See example 3–1 A.

Example $y \cdot y \cdot y \cdot y$

Solution $= y^4$ *y* to the fourth power

1. $aaaaa$
2. $bbbb$
3. $(-2)(-2)(-2)(-2)$
4. $-(2 \cdot 2 \cdot 2 \cdot 2)$
5. $xxxxxx$
6. $(2a)(2a)(2a)$
7. $(xy)(xy)(xy)(xy)$
8. $(a + b)(a + b)$
9. $(x - y)(x - y)(x - y)$
10. $(2a - b)(2a - b)(2a - b)$

Write as an indicated product (expanded form). See example 3–1 B.

Example c^5

Solution $= c \cdot c \cdot c \cdot c \cdot c$ *c* written as a factor 5 times

11. x^4
12. y^5
13. $(-2)^3$
14. -2^4
15. 5^3
16. $(5x)^3$
17. $(4y)^4$
18. $(a + b)^3$
19. $(x - y)^2$
20. $(2x + y)^3$

Simplify by using the properties of exponents. See examples 3–1 C, D, E, and F.

Examples $x^4 \cdot x^5$

Solutions $= x^{4+5}$ Like bases

$= x^9$ Add exponents

$(a^4)^3$

$= a^{4 \cdot 3}$ Power of a power

$= a^{12}$ Multiply exponents

21. $x^4 \cdot x^7$
22. $a^5 \cdot a^5$
23. $R^2 \cdot R$
24. $a \cdot a^4$
25. $a^2 \cdot a^3 \cdot a^4$
26. $x^5 \cdot x \cdot x^3$
27. $5^2 \cdot 5^3$
28. $6 \cdot 6^3$
29. $4 \cdot 4^2 \cdot 4^4$
30. $(a + b)^2(a + b)^5$
31. $(x - 2y)^4(x - 2y)^6$
32. $(3a + b)^2(3a + b)^3$
33. $(a - b)^4(a - b)^7$
34. $(ab)^5$
35. $(xy)^4$
36. $(2abc)^3$
37. $(4xyz)^3$
38. $(a^2)^4$
39. $(x^5)^3$
40. $(y^2)^2$
41. $(b^5)^5$
42. $(c^9)^3$
43. $(2xy^2)(3x^3y)$
44. $(4x^2y^3)(5xy^4)$
45. $(a^2b^3)(a^5b^2)$
46. $(x^2y^2)(x^4y^3)$
47. $(6x^3)(5x^2)$
48. $(4a)(3a^4)$
49. $(2a^3b^4c)(6a^4b^3)$
50. $(5xy)(xy)$
51. $(3a^2b)(4a^3b^2)$
52. $(a^3b^4)(5a^2b^5)$
53. $(-2a^2b)(3ab^4)$
54. $(-5x^2y^5)(-2x^2y)$
55. The formula for finding the volume of a cube is $V = e^3$, where V represents volume in some cubic unit of measure and e represents the length of the edge of the cube. Write an expression for the volume in expanded form, and then determine the number of cubic units in the figure for each of the following values of e: (a) $e = 5$, (b) $e = 4$, (c) $e = 6$.

Using exponents, write an expression for each of the following verbal statements. See example 3–1 G.

56. The area, A, of a square is found by using the length of the side, s, as a factor twice. Write an expression for the area of a square.

57. The distance, s, a falling object will fall in time, t, seconds is found by multiplying $\frac{1}{2}$ times the gravity, g, times the square of t. Write an expression for the distance the object will fall.

58. The area of a circle is found by multiplying the constant π times the length of the radius, r, used as a factor 2 times. Write an expression for the area of a circle.

59. The volume, V, of a sphere is found by multiplying $\frac{4}{3}\pi$ times the radius, r, used as a factor 3 times. Write an expression for the volume of a sphere.

60. Johnny is n years old. His mother says that she is 6 years more than the cube of Johnny's age. Write an expression for his mother's age.

61. Jane is m years old. Her father is 8 years less than Jane's age used as a factor 4 times. Write an expression for her father's age.

62. Write an expression for 2 times the square of t.

63. Write an expression for twice the square of x less the cube of y.

64. A number can be written in the form a times 10 used as a factor 8 times, where a is a number between 1 and 10. Write an expression for the number in terms of a.

65. Write an expression for the quotient of the cube of p divided by the square of q.

Review exercises

Perform the indicated addition and subtraction. See section 2–3.

1. $2a + 3a + 4a$
2. $5x + x + 2x$
3. $3ab - 2ab + 5ab$
4. $9xy + 4xy - 6xy$
5. $4a^2 + 3a^2 - 2a + 7a$
6. $6x^2 + 3x - x^2 + 2x$
7. $2x^2y - x^2y + 3xy^2 + 4xy^2$
8. $5ab^2 + 3a^2b - 2ab^2 - a^2b$

3–2 ▪ Products of algebraic expressions

Product of a monomial and a multinomial

To multiply a monomial and a multinomial (a polynomial of more than one term), we use the distributive property. For example, to multiply

$$3x^2y(x^2 + xy + y^2)$$

we multiply each term in the trinomial by the monomial $3x^2y$ to get

$$(3x^2y \cdot x^2) + (3x^2y \cdot xy) + (3x^2y \cdot y^2)$$

which yields

$$3x^2y(x^2 + xy + y^2) = 3x^4y + 3x^3y^2 + 3x^2y^3$$

In each indicated product, note that we multiplied like bases by using the properties of exponents. For example, in the first term,

$$3x^2y \cdot x^2 = 3 \cdot (x^2 \cdot x^2) \cdot y = 3 \cdot x^{2+2} \cdot y = 3x^4y$$

Example 3–2 A

Perform the indicated multiplication.

1. $5y(2y + 3) = 5y \cdot 2y + 5y \cdot 3$ — Distribute $5y$ times each term in the parentheses
   $= 10y^2 + 15y$ — Multiply monomials

2. $x^3(x^2 + xy - y^2) = x^3 \cdot x^2 + x^3 \cdot xy - x^3 \cdot y^2$
   $= x^5 + x^4y - x^3y^2$

Note In example 2, when we multiplied x^3 times the third term of the trinomial, y^2, the subtraction sign remained, giving $-x^3y^2$.

3. $-5a^3(a^2 + 2ab - b^3) = -5a^3 \cdot a^2 - 5a^3 \cdot 2ab + 5a^3 \cdot b^3$
   $= -5a^5 - 10a^4b + 5a^3b^3$

4. $4x^2y(2x^3 - 3x^2y^2 + y^4) = 4x^2y \cdot 2x^3 - 4x^2y \cdot 3x^2y^2 + 4x^2y \cdot y^4$
   $= 8x^5y - 12x^4y^3 + 4x^2y^5$

▶ ***Quick check*** Perform the indicated multiplication. $3ab^2(2a - 3b)$ ■

Product of two multinomials

The product of two multinomials requires the use of the distributive property several times. That is, in the product

$$(x + 2y)(x + y)$$

we consider $(x + 2y)$ a single number and apply the distributive property.

$$(x + 2y)(x + y) = (x + 2y) \cdot x + (x + 2y) \cdot y$$

We now apply the distributive property again.

$$(x + 2y) \cdot x + (x + 2y) \cdot y = x \cdot x + 2y \cdot x + x \cdot y + 2y \cdot y$$
$$= x^2 + 2xy + xy + 2y^2$$

The last step in the problem is to combine like terms, if there are any.

$$x^2 + (2xy + xy) + 2y^2 = x^2 + 3xy + 2y^2$$

Notice that in this product, each term of the first factor is multiplied by each term of the second factor. We can generalize our procedure as follows:

Multiplying two multinomials

When we are multiplying two multinomials, we multiply each term in the first multinomial by each term in the second multinomial. We then combine like terms.

Example 3–2 B

Perform the indicated multiplication and simplify.

1. $(a + 3)(a - 4) = a \cdot a - a \cdot 4 + 3 \cdot a - 3 \cdot 4$ — Distribute multiplication
   (arrows labeled 1, 2, 3, 4 above the products)
   $= a^2 - 4a + 3a - 12$ — Multiply monomials
   $= a^2 - a - 12$ — Combine like terms

Note We have drawn arrows to indicate the multiplication that is being carried out. This should be a convenient way for us to indicate the multiplication to be performed.

2. $(2x + 3)(5x - 2) = 10x^2 - 4x + 15x - 6$ Distribute and multiply
$= 10x^2 + 11x - 6$ Combine like terms

Note A word that is useful for remembering the multiplication to be performed when multiplying two binomials is **FOIL.** Foil is an abbreviation signifying **F**irst times first, **O**uter times outer, **I**nner times inner, and **L**ast times last.

3. $(a + b)(a + 2b) = a^2 + 2ab + ab + 2b^2$ Distribute and multiply
$= a^2 + 3ab + 2b^2$ Combine like terms

▶ ***Quick check*** Perform the indicated multiplication and simplify.
$(2x + y)(x - 3y)$ ■

Special products

Three special products appear so often that the form of the answers can be written without computation. Consider the product

$$(x + 6)^2 = (x + 6)(x + 6)$$

which becomes

$$x^2 + 6x + 6x + 36$$

When we combine the second and third terms, we get

$$x^2 + 12x + 36$$

This is called the **square of a binomial** or a **perfect square trinomial** and has certain characteristics. Inspection shows us that in

$$(x + 6)^2 = x^2 + 12x + 36$$

the three terms of the product can be obtained in the following manner:

The square of a binomial

1. The first term of the product is the *square of the first term* of the binomial $[(x)^2 = x^2]$.
2. The second term of the product is *two times the product of the two terms of the binomial* $[2(x \cdot 6) = 12x]$.
3. The third term of the product is the *square of the second term* of the binomial $[(6)^2 = 36]$.

If we apply this to

we get

$$\begin{aligned} &(x - 7)^2 \\ &= [x + (-7)]^2 \\ &= x^2 + [2 \cdot x \cdot (-7)] + (-7)^2 \end{aligned}$$

and so

$$\begin{aligned} (x - 7)^2 &= x^2 + (-14x) + 49 \\ &= x^2 - 14x + 49 \end{aligned}$$

In general, for real numbers a and b,

$$(a + b)^2 = a^2 + 2ab + b^2$$

and

$$(a - b)^2 = a^2 - 2ab + b^2$$

Note $(a + b)^2 = a^2 + 2ab + b^2$, *not* $a^2 + b^2$. This is a common error. *The square of a binomial is always a trinomial.*

■ Example 3–2 C

Perform the indicated multiplication and simplify.

1. $(2x + 3)^2 = (2x)^2 + (2 \cdot 2x \cdot 3) + (3)^2$ — Apply special products property
   $= 4x^2 + 12x + 9$ — Multiply monomials
2. $(5a - 4b)^2 = (5a)^2 - [2 \cdot 5a \cdot (4b)] + (4b)^2$ — Special products property
   $= 25a^2 - [40ab] + 16b^2$ — Multipy monomials
   $= 25a^2 - 40ab + 16b^2$ — Standard form ■

The third special product is obtained by multiplying the sum and the difference of the same two terms. Consider the following:

$$(x + 3)(x - 3) = x^2 - 3x + 3x - 9$$
$$= x^2 - 9$$

Special characteristics are evident in this product also.

The difference of two squares

For real numbers a and b,

$$(a + b)(a - b) = a^2 - b^2$$

Concept

1. The product is obtained by first squaring the first term of the factors, and then
2. subtracting the square of the second term of the factors.

■ Example 3–2 D

Perform the indicated multiplication and simplify.

1. $(x + 7)(x - 7) = (x)^2 - (7)^2 = x^2 - 49$
2. $(a + 2b)(a - 2b) = (a)^2 - (2b)^2 = a^2 - 4b^2$
3. $(3x - 2y)(3x + 2y) = (3x)^2 - (2y)^2 = 9x^2 - 4y^2$ ■

In all the examples that we have looked at, whether they were special products or not, a single rule is sufficient.

When multiplying two multinomials together, we multiply each of the terms in the first multinomial times each of the terms in the second multinomial and then combine like terms.

■ ***Example 3–2 E***

Perform the indicated multiplication and simplify.

1. $(3x - y)(2x + 3y) = 6x^2 + 9xy - 2xy - 3y^2$ — Distribute multiplication

$= 6x^2 + 7xy - 3y^2$ — Combine like terms

2. $(a - 2)(2a^2 + 3a + 2) = 2a^3 + 3a^2 + 2a - 4a^2 - 6a - 4$ — Distribute multiplication

$= 2a^3 - a^2 - 4a - 4$ — Combine like terms

Note Although there are three terms in the second parentheses, we still follow the procedure of every term in the first parentheses times every term in the second parentheses.

3. $(x - y)(x^2 + 3xy - y^2)$ — Distribute multiplication

$= x^3 + 3x^2y - xy^2 - x^2y - 3xy^2 + y^3$ — Combine like terms

$= x^3 + 2x^2y - 4xy^2 + y^3$

4. $(a + 6)(a - 2)(a - 1)$. When there are three multinomials to be multiplied, we apply the associative property to multiply two of them together first and take that product times the third.

$$\begin{aligned}[(a + 6)(a - 2)](a - 1) &= [a^2 - 2a + 6a - 12](a - 1) \\ &= [a^2 + 4a - 12](a - 1) \\ &= a^3 - a^2 + 4a^2 - 4a - 12a + 12 \\ &= a^3 + 3a^2 - 16a + 12\end{aligned}$$

■

Mastery points

Can you

- Multiply a monomial and a multinomial?
- Multiply multinomials?
- Find the special products of the square of a binomial or the difference of two squares?

Exercise 3–2

Perform the indicated multiplication and simplify. See examples 3–2 A, B, C, D, and E.

Examples $3ab^2(2a - 3b)$

Solutions $= 3ab^2 \cdot 2a - 3ab^2 \cdot 3b$ Distributive property
$= 6a^2b^2 - 9ab^3$ Multiply monomials

$(2x + y)(x - 3y)$
$= 2x \cdot x - 2x \cdot 3y + y \cdot x - y \cdot 3y$ Distributive property
$= 2x^2 - 6xy + xy - 3y^2$ Multiply monomials
$= 2x^2 - 5xy - 3y^2$ Combine like terms

1. $2ab(a^2 - bc + c^2)$
2. $6x(4y + 7z)$
3. $3a(5b^2 - 7c^2)$
4. $-ab(a^4 - a^2b^2 - b^4)$
5. $-5ab^2(3a^2 - ab + 4b^2)$
6. $6x^2(4x^2 - 2x + 3)$
7. $3ab(a^2 - 2ab - b^2)$
8. $(2x)(x - y + 5)(5y)$
9. $(3a)(2a - b)(2b^2)$
10. $(x^2y)(x^2 + y^2)(xy^2)$
11. $(x + 3)(x + 4)$
12. $(a + 5)(a - 3)$
13. $(y - 9)(y - 4)$
14. $(z + 7)(z - 11)$
15. $(a + 1)(a + 1)$
16. $(b - 1)(b - 1)$
17. $(R - 3)^2$
18. $(R + 2)(R - 2)$
19. $(a + 3)(a - 3)$
20. $(3x + 2)(x - 4)$
21. $(3a - 5)(2a - 7)$
22. $(3 - 2y)(2 - y)$
23. $(7 + 2x)(2x - 7)$
24. $(4r + 3)(r - 12)$
25. $(3k + w)(k - 6w)$
26. $(a - 6bc)(5a + 4bc)$
27. $(a + 6b)^2$
28. $(2a + 3b)^2$
29. $(2a + 3b)(2a - 3b)$
30. $(4x - y)(4x + y)$
31. $(a + 4b)(a^2 - 2ab + b^2)$
32. $(x - 2y)(2x^2 - 3xy + y^2)$
33. $(x + 4)(6x^2 - 3x + 7)$
34. $(x - y)(x^2 - 2xy + y^2)$
35. $(x^2 - 2x - 3)(x^2 + x + 4)$
36. $(a^2 - 3a + 6)(a^2 + 2a - 5)$
37. $(a - 6)(a - 2)(a + 1)$
38. $(2b - 1)(b + 2)(2b + 1)$
39. $(a - b)(a + b)(2a - 3b)$
40. $(a + b)^3$
41. $(a - b)^3$
42. $(2a + b)^3$
43. $(a - 2b)^3$
44. The area of the shaded region between the two circles is $\pi(R + r)(R - r)$. Perform the indicated multiplication.

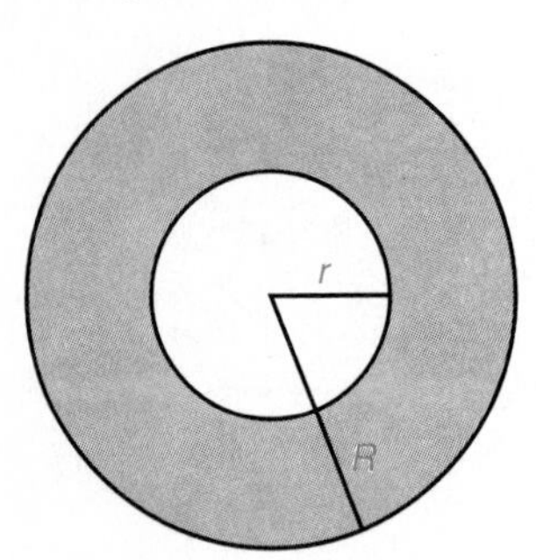

45. When squares of c units on a side are cut from the corners of a square sheet of metal x units on a side, and the metal sheet is then folded up into a tray, the volume is $c(x - 2c)(x - 2c)$. Perform the indicated multiplication.

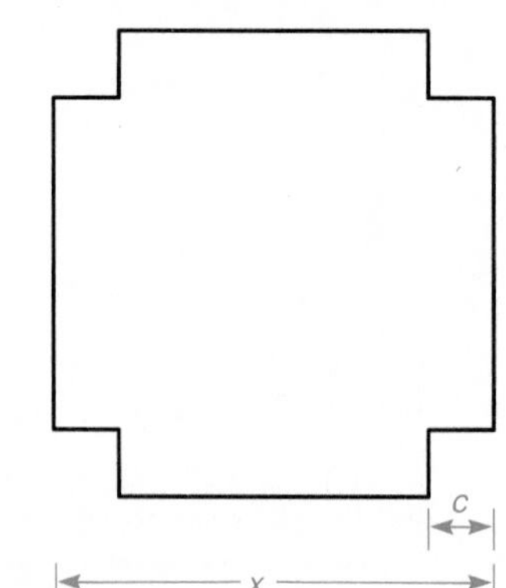

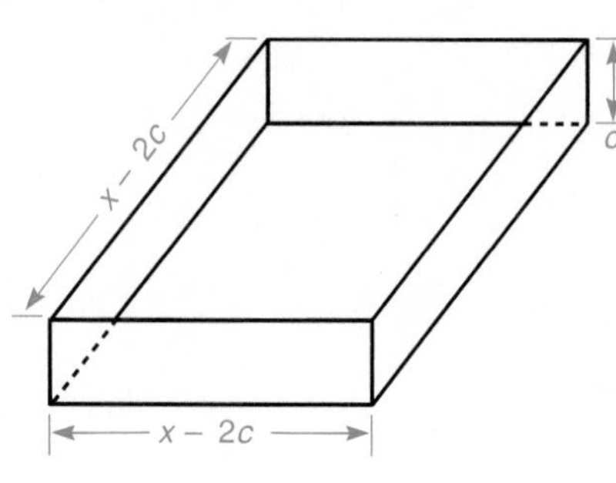

Review exercises

Perform the indicated addition or subtraction. See sections 1–4 and 1–5.

1. $(-3) + (-2)$ **2.** $8 - (-4)$ **3.** $(-7) - (-10)$ **4.** $2 + (-5) + (-6)$

Simplify by using the properties of exponents. See section 3–1.

5. $x^4 \cdot x^8$ **6.** $(a^3)^5$ **7.** $(3ab)^3$ **8.** $(2x)^3$

3–3 ■ Exponents–II

Fraction to a power property of exponents

In section 3–1, we learned several useful properties of exponents. Now we shall learn several more.

Our next property of exponents can be derived from the definition of exponents. Consider the expression $\left(\frac{a}{b}\right)^3$.

$$\left(\frac{a}{b}\right)^3 = \underbrace{\frac{a}{b} \cdot \frac{a}{b} \cdot \frac{a}{b}}_{\text{3 factors of } \frac{a}{b}} = \frac{\overbrace{a \cdot a \cdot a}^{\text{3 factors of } a}}{\underbrace{b \cdot b \cdot b}_{\text{3 factors of } b}} = \frac{a^3}{b^3}$$

Thus

Fraction raised to a power → $\left(\frac{a}{b}\right)^3 = \frac{a^3}{b^3}$ ← Numerator raised to the power; ← Denominator raised to the power

Fraction to a power property of exponents

$$\left(\frac{a}{b}\right)^n = \frac{a^n}{b^n},\ b \neq 0$$

Concept

Whenever a fraction is raised to a power, the numerator and the denominator are *both* raised to that power.

■ Example 3–3 A

Perform the indicated operations and simplify.

1. $\left(\frac{3}{4}\right)^3 = \frac{3^3}{4^3} = \frac{27}{64}$

2. $\left(\frac{a}{b}\right)^5 = \frac{a^5}{b^5}$

3. $\left(\frac{2a}{b}\right)^3 = \frac{(2a)^3}{b^3} = \frac{2^3a^3}{b^3} = \frac{8a^3}{b^3}$ ■

Division of expressions with like bases

Consider the expression

$$\frac{x^6}{x^2}$$

We can use the definition of exponents to write the fraction as

$$\frac{x^6}{x^2} = \frac{x \cdot x \cdot x \cdot x \cdot x \cdot x}{x \cdot x}$$

We reduce the fraction as follows:

$$\frac{x \cdot x \cdot x \cdot x \cdot x \cdot x}{x \cdot x} = \frac{x \cdot x \cdot x \cdot x}{1} = \frac{x^4}{1} = x^4$$

In our example, we reduced by two factors of x, leaving $6 - 2 = 4$ factors of x in the numerator. Therefore

$$\frac{x^6}{x^2} = x^{6-2} = x^4$$

Thus we have the following property of exponents.

Quotient property of exponents

$$a^m \div a^n = \frac{a^m}{a^n} = a^{m-n}, a \neq 0$$

Concept

To divide quantities having *like* bases, subtract the exponent of the denominator from the exponent of the numerator to get the power of the given base in the quotient.

Note If the base a is zero, $a = 0$, we have an expression that has no meaning. Therefore $a \neq 0$ indicates that we want our variables to assume no values that would cause the denominator to be zero.

Example 3–3 B

Simplify. Assume that no variable is equal to zero.

1. $\frac{x^7}{x^5} = x^{7-5} = x^2$
2. $a^{11} \div a^4 = a^{11-4} = a^7$
3. $\frac{5^4}{5} = 5^{4-1} = 5^3 = 125$

Note Remember that when we are dividing like bases, their exponents are subtracted, but *the base is not changed.*

4. $\frac{a^5 \cdot a^2}{a^4} = \frac{a^{5+2}}{a^4} = \frac{a^7}{a^4} = a^{7-4} = a^3$
5. $\frac{x^3}{y^2} = \frac{x^3}{y^2}$

Note In example 5, we cannot simplify. The bases must be the same in order to subtract exponents when we divide.

6. $\dfrac{2^5x^9y^{15}}{2^3x^5y^{12}} = 2^{5-3}x^{9-5}y^{15-12} = 2^2x^4y^3 = 4x^4y^3$

▶ ***Quick check*** Simplify. Assume that no variable is equal to zero. $a^{11} \div a^7$ ■

Negative exponents

To this point, we have considered only those problems where the exponent of the numerator is greater than the exponent of the denominator. Consider the example

$$\frac{x^2}{x^6}$$

By the definition of exponents, this becomes

$$\frac{x^2}{x^6} = \frac{x \cdot x}{x \cdot x \cdot x \cdot x \cdot x \cdot x}$$

and reducing the fraction,

$$\frac{x \cdot x}{x \cdot x \cdot x \cdot x \cdot x \cdot x} = \frac{1}{x \cdot x \cdot x \cdot x} = \frac{1}{x^4}$$

Again, we reduced by two factors of x, leaving $6 - 2 = 4$ factors of x in the denominator. Hence

$$\frac{x^2}{x^6} = \frac{1}{x^4}$$

However using the quotient property of exponents to carry out the division, we would have

$$\frac{x^2}{x^6} = x^{2-6} = x^{-4}$$

Since we should arrive at the same answer regardless of which procedure we use, then x^{-4} must be $\frac{1}{x^4}$, thus $x^{-4} = \frac{1}{x^4}$. This leads us to the definition of negative exponents.

Definition of negative exponents

$$a^{-n} = \frac{1}{a^n},\ a \neq 0$$

Concept

A negative exponent on any base (except zero) can be written as 1 over that base with a positive exponent.

■ ***Example 3–3 C***

Write the following problems with positive exponents. Assume that no variable is equal to zero.

1. $x^{-3} = \dfrac{1}{x^3}$ Rewritten as 1 over x to the positive 3rd

2. $a^{-9} = \dfrac{1}{a^9}$ Rewritten as 1 over a to the positive 9th

Note From the definition of negative exponents, if a **factor** is moved from either the numerator to the denominator or from the denominator to the numerator, the sign of its exponent will change. The sign of the base will not be affected by this change.

2. *Alternative procedure*

$$a^{-9} = \frac{a^{-9}}{1}$$ Rewrite as a fraction

$$= \frac{1}{a^9}$$ Sign of the exponent is changed as the **factor** is moved from the numerator to the denominator

3. $$\frac{1}{b^{-4}} = \frac{b^4}{1}$$ Sign of the exponent is changed as the **factor** is moved from the denominator to the numerator

$$= b^4$$ Standard form is to leave only positive exponents

4. $$(-3)^{-3} = \frac{1}{(-3)^3}$$ Sign of the exponent is changed as the factor is moved from the numerator to the denominator

$$= \frac{1}{-27} \text{ or } -\frac{1}{27}$$ Standard form

▶ ***Quick check*** Write b^{-2} with positive exponents. ■

Zero as an exponent

Now consider the situation involving the division of like bases that are raised to the same power.

$$\frac{x^3}{x^3}, x \neq 0$$

By the definition of exponents, we have

$$\frac{x^3}{x^3} = \frac{x \cdot x \cdot x}{x \cdot x \cdot x} = \frac{1}{1} = 1$$

By the quotient property of exponents,

$$\frac{x^3}{x^3} = x^{3-3} = x^0$$

Since $\frac{x^3}{x^3} = 1$ and $\frac{x^3}{x^3} = x^0$, then x^0 must be equal to 1. This leads us to the definition of zero as an exponent.

Definition of zero as an exponent

$$a^0 = 1, a \neq 0$$

Concept

Any number other than zero raised to the zero power is equal to 1.

■ Example 3–3 D

Simplify. Assume that no variable is equal to zero.

1. $b^0 = 1$ **2.** $r^0 = 1$ **3.** $5^0 = 1$ **4.** $(-2)^0 = 1$

5. $(a + b)^0 = 1$ **6.** $3x^0 = 3 \cdot 1 = 3$ **7.** $(3x)^0 = 1$

Note The exponent acts only on the symbol immediately to its left. In example 6, only the x is raised to the zero power. The exponent of 3 is understood to be 1. In example 7, the parentheses indicate that both the 3 and the x are raised to the zero power.

▶ ***Quick check*** Simplify. Assume that no variable is equal to zero. C^0 ■

■ Example 3–3 E

Simplify. Leave the answer with only positive exponents. Assume that no variable is equal to zero.

1. $\frac{x^5}{x^{11}} = x^{5-11}$ Division of like bases

$= x^{-6}$ Subtract exponents

$= \frac{1}{x^6}$ Standard form

2. $a^{-7} \cdot a^5 = a^{-7+5}$ Multiplication of like bases

$= a^{-2}$ Add exponents

$= \frac{1}{a^2}$ Standard form

3. $(b^{-2})^{-4} = b^{(-2)\cdot(-4)}$ Power of a power

$= b^8$ Multiply exponents

4. $\frac{a^3b^5}{a^7b^2} = a^{3-7}b^{5-2}$ Division of like bases

$= a^{-4}b^3$ Subtract exponents

$= \frac{b^3}{a^4}$ Standard form

5. $\frac{a^3b^2c^4}{ab^5c^4} = a^{3-1}b^{2-5}c^{4-4}$ Division of like bases

$= a^2b^{-3}c^0$ Subtract the exponents

$= \frac{a^2 \cdot 1}{b^3}$ The a's remain in the numerator, the b's drop to the denominator, and c^0 is 1

$= \frac{a^2}{b^3}$ Standard form

6. $\frac{a^{-2}b^4}{a^{-5}b^6} = a^{-2-(-5)}b^{4-6}$ Division of like bases

$= a^3b^{-2}$ Subtract exponents

$= \frac{a^3}{b^2}$ Standard form

▶ ***Quick check*** Simplify. Leave the answer with only positive exponents. Assume that no variable is equal to zero. $\frac{b^4}{b^{10}}$ ■

Mastery points

Can you

- Raise a fraction to a power?
- Perform division on expressions having like bases?
- Perform operations involving negative exponents?
- Perform operations involving zero as an exponent?

Exercise 3–3

Write each expression with only positive exponents. Assume that no variable is equal to zero. See examples 3–3 C and D.

Examples C^0 b^{-2}

Solutions $= 1$ By definition is equal to 1 $= \dfrac{1}{b^2}$ Rewritten as 1 over b to the positive 2nd

1. x^0
2. $(2y)^0$
3. $5a^0$
4. $7x^0$
5. $(3B)^0$
6. S^{-2}
7. R^{-5}
8. $(2x)^{-3}$
9. $(3P)^{-2}$
10. $4z^{-2}$
11. $9C^{-4}$
12. $\dfrac{5}{x^{-4}}$
13. $\dfrac{1}{2y^{-3}}$
14. $\dfrac{1}{3x^{-2}}$
15. $2x^{-4}y^2$
16. $x^{-2}y^4$
17. $p^0r^{-2}t^5$
18. $x^{-3}y^2z^{-4}$

Perform all indicated operations and leave your answer with only positive exponents. Assume that no variable is equal to zero. See examples 3–3 A, B, and E.

Examples $a^{11} \div a^7$ $\dfrac{b^4}{b^{10}}$

Solutions

$= a^{11-7}$ Division of like bases
$= a^4$ Subtract exponents

$= b^{4-10}$ Division of like bases
$= b^{-6}$ Subtract exponents
$= \dfrac{1}{b^6}$ Standard form

19. $\left(\dfrac{a}{b}\right)^6$
20. $\left(\dfrac{x}{y}\right)^4$
21. $\left(\dfrac{2}{3}\right)^3$
22. $\left(\dfrac{1}{2}\right)^4$
23. $\left(\dfrac{2x}{y}\right)^4$
24. $\left(\dfrac{2ab}{c}\right)^3$
25. $\left(\dfrac{3a}{b}\right)^3$
26. $x^{12} \div x^6$
27. $y^4 \div y^2$
28. $\dfrac{a^5}{a^3}$
29. $\dfrac{b^9}{b^7}$
30. $\dfrac{c^6}{c^9}$
31. $\dfrac{R^4}{R^8}$
32. $\dfrac{3^4}{3^2}$
33. $\dfrac{2^5}{2^3}$
34. $\dfrac{4^2}{4^5}$
35. $\dfrac{6}{6^3}$
36. $\dfrac{x^4x^3}{x^2}$
37. $\dfrac{y^5y}{y^2}$
38. $\dfrac{a^4a^2}{a^5}$
39. $\dfrac{a^4}{a^2a}$
40. $\dfrac{x^7}{x^2x^3}$
41. $\dfrac{y^3}{y^4y^5}$
42. $\dfrac{b^2}{bb^4}$
43. $\dfrac{a^7b^5}{a^4b^2}$
44. $\dfrac{x^9y^7}{x^4y}$
45. $\dfrac{2^3x^3y^7}{2xy^5}$
46. $\dfrac{3^3a^4b^5}{3^2a^2b^3}$
47. $\dfrac{3a^2b^5}{3^4a^5b^5}$
48. $\dfrac{5^2a^3b}{5^3a^7b^3}$

49. $x^{-4}x^7$ **50.** $y^{-2}y^{10}$ **51.** a^5a^{-11} **52.** $R^{-2}R^{-5}$ **53.** $x^{-2}x^4x^0$

54. $x^5x^0x^{-2}$ **55.** $a^0a^{-5}a^3$ **56.** $a^{-7}a^4a^0$ **57.** $(-5)^{-3}$ **58.** $(-2)^{-4}$

59. $\dfrac{3^{-2}}{3^{-5}}$ **60.** $\dfrac{2^{-6}}{2^{-3}}$ **61.** $(a^{-2})^3$ **62.** $(b^4)^{-4}$ **63.** $(x^5)^{-2}$

64. $(y^{-3})^4$ **65.** $(a^{-2})^{-3}$ **66.** $(z^{-4})^{-4}$ **67.** $(x^0)^{-2}$ **68.** $(a^{-3})^0$

69. $(x^{-2})^0$ **70.** $(b^0)^{-4}$ **71.** $\dfrac{R^2S^{-4}}{R^{-3}S^5}$ **72.** $\dfrac{4y^{-3}}{4^{-1}y^2}$ **73.** $\dfrac{4^{-1}a^{-2}b^3c^0}{2a^{-3}b^{-1}c^{-2}}$

Review exercises

Perform the indicated operations. See sections 1–4 to 1–8.

1. $(-4) + (-6)$ **2.** $(-3)(-7)$ **3.** $(-4) - (-8)$ **4.** -4^2

Simplify by using the properties of exponents. See section 3–1.

5. a^3a^5 **6.** $(x^3)^4$ **7.** xx^2x^3 **8.** $(2ab)^2$

3–4 ■ Exponents—III

Properties and definitions of exponents

The following is a summary of the properties and definitions of exponents that we have studied so far.

Definitions

$$a^n = \overbrace{a \cdot a \cdot a \cdots a}^{n \text{ factors}}, \text{ where } n \text{ is a positive integer}$$

$$a^{-n} = \frac{1}{a^n}, \ a \neq 0$$

$$a^0 = 1, \ a \neq 0$$

Properties

$$a^m \cdot a^n = a^{m+n}$$

$$(ab)^n = a^nb^n$$

$$(a^m)^n = a^{m \cdot n}$$

$$\left(\frac{a}{b}\right)^n = \frac{a^n}{b^n}, \ b \neq 0$$

$$a^m \div a^n = \frac{a^m}{a^n} = a^{m-n}, \ a \neq 0$$

The following examples illustrate some more problems in which more than one property of exponents is applied within the same problem.

■ Example 3–4 A

Simplify. Leave the answer with only positive exponents. Assume that no variable is equal to zero.

1. $(2a^2b^3)^3 = 2^3(a^2)^3(b^3)^3$ — Each factor in the group is raised to the 3rd power

$= 2^3a^6b^9$ — Power of a power, multiply exponents

$= 8a^6b^9$ — 8 is the standard form of 2^3

2. $(5a^4b^2)^4 = 5^4(a^4)^4(b^2)^4$ — Each factor is raised to the power

$= 5^4a^{16}b^8$ — Power of a power

$= 625a^{16}b^8$ — Standard form

3. $(3a^{-2}b^3)^{-3} = 3^{-3}(a^{-2})^{-3}(b^3)^{-3}$ — Each factor is raised to the power

$= 3^{-3}a^6b^{-9}$ — Power of a power

$= \dfrac{a^6}{3^3b^9}$ — The 3s and the b's drop to the denominator

$= \dfrac{a^6}{27b^9}$ — Standard form

4. $(-3a^2)(2ab^3)(-4a^3b^5)$

$= [(-3)(2)(-4)](a^2aa^3)(b^3b^5)$ — Multiply like bases using the commutative and associative properties

$= 24a^6b^8$ — Signed numbers and multiplication of like bases, add exponents

5. $(3a^4b)^2(3^2ab^5)^2 = 3^2(a^4)^2b^2 \cdot (3^2)^2a^2(b^5)^2$ — Group of factors to a power

$= 3^2a^8b^2 \cdot 3^4a^2b^{10}$ — Power of a power

$= (3^2 \cdot 3^4)(a^8a^2)(b^2b^{10})$ — Multiply like bases

$= 3^6a^{10}b^{12}$ — Add exponents

$= 729a^{10}b^{12}$ — Standard form

6. $\left(\dfrac{2a^2b^3}{c^5}\right)^3 = \dfrac{(2a^2b^3)^3}{(c^5)^3}$ — Both the numerator and the denominator are raised to the 3rd power

$= \dfrac{2^3(a^2)^3(b^3)^3}{(c^5)^3}$ — Each factor in the numerator is raised to the 3rd power

$= \dfrac{8a^6b^9}{c^{15}}$ — Power of a power, multiply exponents

7. $\dfrac{a^{-2}b^3}{a^{-4}b^6} = a^{(-2)-(-4)}b^{3-6}$ — Division of like bases

$= a^2b^{-3}$ — Subtract exponents

$= \dfrac{a^2}{b^3}$ — Standard form

8. $\left(\dfrac{a^{-2}b}{c^3}\right)^{-2} = \dfrac{(a^{-2}b)^{-2}}{(c^3)^{-2}}$ — Numerator and denominator are raised to the power

$= \dfrac{(a^{-2})^{-2}b^{-2}}{(c^3)^{-2}}$ — Numerator has a group of factors to a power

$= \dfrac{a^{(-2)(-2)}b^{-2}}{c^{(3)(-2)}}$ — Power of a power

$= \dfrac{a^4b^{-2}}{c^{-6}}$ — Multiply exponents

$= \dfrac{a^4c^6}{b^2}$ — Standard form, factors raised to a negative power are moved to the other side of the fraction bar

▶ ***Quick check*** Simplify. Leave the answer with only positive exponents. Assume that no variable is equal to zero. $(2a^{-2}b^3)^3$ ■

Mastery points

Can you

- Apply the definitions and properties of exponents?

Exercise 3–4

Simplify by using the properties and definitions of exponents. Leave the answer with only positive exponents. Assume that no variable is equal to zero. See example 3–4 A.

Example $(2a^{-2}b^3)^3$

Solution

$= 2^3(a^{-2})^3(b^3)^3$ Groups of factors to a power

$= 2^3a^{-6}b^9$ Power of a power

$= \dfrac{8b^9}{a^6}$ Standard form

1. $(2a^2)^3$
2. $(3x^4)^2$
3. $(2x^2y)^3$
4. $(4ab^3)^2$
5. $(x^4y^3z)^4$
6. $(2a^5b^2c)^3$
7. $(5a^5b^2c^4)^2$
8. $(a^3b^2)^3$
9. $(2a^2)^{-2}$
10. $(5x^{-3})^{-2}$
11. $(4^{-1}x^2)^{-2}$
12. $(2a^2b^{-3})^{-2}$
13. $(3xy^{-4})^{-3}$
14. $(3x^{-2}y^{-3})^2$
15. $(x^2y^{-5}z^3)^{-2}$
16. $(a^{-3}b^2c^{-4})^{-3}$
17. $(3x^2)(2x^0y^2)(x^5y)$
18. $(a^2b)(-3a^0b^2)(a^3b)$
19. $(-2x^2y)(3x^3y^2)(x^5y)$
20. $(a^2bc)(-2a^2b^2c^2)(3abc)$
21. $(x^3yz^4)(-3xyz^2)(-2x^2yz)$
22. $(a^2b)(-3b^2c^2)(2a^2c^2)$
23. $\left(\dfrac{2x}{y^2}\right)^3$
24. $\left(\dfrac{x^2y}{z^2}\right)^3$
25. $\left(\dfrac{3a^2c^0}{b^3}\right)^2$
26. $\left(\dfrac{x^3}{y^0z^4}\right)^3$
27. $\left(\dfrac{2x^2}{y^3}\right)^3$
28. $\left(\dfrac{ab^2}{c^4}\right)^4$
29. $\left(\dfrac{2x^3y^3}{z^5}\right)^2$
30. $\left(\dfrac{a^5bc^4}{d^2e}\right)^5$
31. $\dfrac{a^{-2}b^3}{a^3b^{-5}}$
32. $\dfrac{x^{-5}y^2}{x^3y^{-4}}$
33. $\dfrac{3R^{-1}S^{-2}}{9R^{-3}S^2}$
34. $\dfrac{R^2S^{-4}}{R^{-3}S^5}$
35. $\dfrac{8a^{-2}b^{-5}}{2a^{-1}b^4}$
36. $\dfrac{2x^{-1}y^{-2}}{3x^{-2}y^2}$
37. $\dfrac{6R^{-2}S^0}{2R^2S^{-3}}$
38. $\dfrac{2a^{-1}b^0c^2}{5a^3b^{-1}c^{-3}}$
39. $(a^2b^3)^3(ab^2)^4$
40. $(xy^2)^3(x^2y^2)^2$
41. $(2a^3)^2(2a^2)^3$
42. $(3x^5)^3(3x^3)^2$
43. $(2x^2y)^3(2x^4y^5)^2$
44. $(3r^2s^4)^3(r^5s^6)^2$
45. $\left(\dfrac{xy^{-2}}{z^{-4}}\right)^{-1}$
46. $\left(\dfrac{x^{-3}y}{z^5}\right)^{-2}$
47. $\left(\dfrac{2a^{-3}}{b^5}\right)^{-2}$
48. $\left(\dfrac{4^{-1}a^{-2}}{b^{-5}}\right)^{-2}$
49. $\left(\dfrac{ab^{-2}}{c^{-1}}\right)^{-3}$
50. $\left(\dfrac{2^{-2}x^3}{y^{-2}}\right)^{-3}$

Review exercises

Perform the indicated multiplication. See section 1–2.

1. $(6.2) \cdot (5.7)$
2. $(2.8) \cdot (3.7)$
3. $(1.9) \cdot (8.8)$
4. $(4.2) \cdot (6.9)$
5. $(9.9) \cdot (1.9)$
6. $(7.5) \cdot (6.6)$

3–5 ■ Scientific notation

Scientific notation

An important use of integer exponents is in scientific, engineering, and technical fields where we deal with very large or very small numbers. For example, the mass of a hydrogen atom is 0.000 000 000 000 000 000 000 001 67 gram; the mass of an electron is 0.000 000 000 000 000 000 000 000 000 91 gram; the half-life of lead-204 is 14,000,000,000,000,000,000 years. To work with such numbers on the calculator, they must often be entered in scientific notation. We define the scientific notation of a positive number X to be the product

$$X = a \times 10^n$$

where $1 \leq a < 10$ and n is an integer. To achieve this form of the decimal number X, use the following steps.

Scientific notation

Step 1 Move the decimal point to a position immediately following the first nonzero digit in X.

Step 2 Count the number of places the decimal point has been moved. This is the power, n, to which 10 is raised.

Step 3 If
a. the decimal point is moved to the *left*, n is *positive*.
b. the decimal point is moved to the *right*, n is *negative*.
c. the decimal point already follows the first nonzero digit, n is *zero*.

■ Example 3–5 A

Express the following numbers in scientific notation.

1. 250

$250 = 2.50. \times 10^2 = 2.5 \times 10^2$

2. 45,000,000

$45{,}000{,}000 = 4.5000000. \times 10^7 = 4.5 \times 10^7$

3. 5

$5 = 5 \times 10^0$

4. 0.000152

$0.000152 = 0.0001.52 \times 10^{-4} = 1.52 \times 10^{-4}$

Note To write a negative number in scientific notation, we use the same procedure as for a positive number except that a negative sign, −, is placed in front of a.

Example
To write −0.0234 in scientific notation, we proceed as follows:

$$-0.0234 = -0.02.34 \times 10^{-2} = -2.34 \times 10^{-2}$$

▶ ***Quick check*** Express the following numbers in scientific notation.
4,380 −0.00592 ■

Standard form

Sometimes it is necessary to convert a number in scientific notation to its standard form. To do this, we apply the rules in reverse.

> **Standard form**
>
> When the power of 10 is
> 1. *positive,* the decimal point is moved to the *right n* places.
> 2. *negative,* the decimal point is moved to the *left n* places.
> 3. *zero,* the decimal point is not moved.

Example 3–5 B

Express the following numbers in standard form.

1. 1.45×10^4
Since the exponent of 10 is positive 4, we move the decimal point 4 places to the *right* to get

$1.45 \times 10^4 = 1.4500. = 14{,}500$

2. 5.23×10^{-3}
The *negative* exponent, -3, tells us to move the decimal point 3 places to the *left* to get

$5.23 \times 10^{-3} = 0.005.23 = 0.00523$

Note In each example, it was necessary to insert zeros to properly locate the decimal point.

3. -4.07×10^{-2}
With a negative exponent, -2, move the decimal point 2 places to the *left* to get

$-4.07 \times 10^{-2} = -0.04.07 = -0.0407$

Note The negative sign preceding the number is carried along into the standard form.

▶ ***Quick check*** Express the following numbers in standard form.
9.98×10^{-4} $\quad$ -5.63×10^4 ■

Computation using scientific notation

Scientific notation can be used to simplify numerical calculations when the numbers are very large or very small. We first change the numbers to scientific notation and use the properties of exponents to help perform the indicated operations.

Example 3–5 C

Perform the indicated operations using scientific notation.

1. $(349{,}000{,}000)(0.0816)$

$= (3.49 \times 10^8)(8.16 \times 10^{-2})$	Scientific notation
$= (3.49 \cdot 8.16) \times (10^8 \cdot 10^{-2})$	Commutative and associative properties
$= 28.4784 \times 10^6$	Multiply
$= 28{,}478{,}400$	Standard form

2. $\dfrac{(102{,}000{,}000)(0.00105)}{(1{,}190)(0.012)}$

$= \dfrac{(1.02 \times 10^8)(1.05 \times 10^{-3})}{(1.19 \times 10^3)(1.2 \times 10^{-2})}$ Scientific notation

$= \dfrac{(1.02)(1.05)10^8 \cdot 10^{-3}}{(1.19)(1.2)10^3 \cdot 10^{-2}}$ Commutative and associative properties

$= \dfrac{(1.02)(1.05)}{(1.19)(1.2)} \times 10^4$ Properties of exponents

$= 0.75 \times 10^4$ Multiplication and division

$= 7{,}500$ Standard form ■

Mastery points

Can you

- Express a number in scientific notation?
- Convert a number from scientific notation to standard form?
- Do computations using scientific notation?

Exercise 3–5

Express the following numbers in scientific notation. See example 3–5 A.

Examples 4,380 | -0.00592

Solutions $= 4.38 \times 10^3$ Three places to the left, exponent is 3 | $= -5.92 \times 10^{-3}$ Three places to the right, exponent is -3

1. 255 **2.** 65,000,000 **3.** 12,345 **4.** 14,800 **5.** 155,000 **6.** 14.36 **7.** 855.076 **8.** 1,570.7 **9.** 1,007,600 **10.** 6,000,736 **11.** 0.00012 **12.** 0.0863 **13.** 0.0000081 **14.** 0.0000147 **15.** 0.0007 **16.** 0.12079 **17.** 0.000000000094 **18.** -456 **19.** $-4{,}500$ **20.** -0.00087 **21.** $-5{,}850{,}000$ **22.** -0.0567 **23.** -45.78 **24.** $-34{,}000{,}000$ **25.** -0.00000002985

Convert the following numbers in scientific notation to their standard form. See example 3–5 B.

Examples 9.98×10^{-4} | -5.63×10^4

Solutions $= 0.000998$ Exponent is -4, move 4 places to the left | $= -56{,}300$ Exponent is 4, move 4 places to the right

26. 2.07×10^3 **27.** 4.99×10^7 **28.** 5.061×10^5 **29.** 7.23×10^0 **30.** 1.073×10^4 **31.** 4.2×10^{-3} **32.** 7.611×10^{-7} **33.** 1.47×10^{-6} **34.** 5.0×10^{-2} **35.** 7.89×10^{-4} **36.** -2.3×10^5 **37.** -4.82×10^{-9} **38.** -2.61×10^2 **39.** -4.92×10^{-6} **40.** -9.3×10^8

Express the following numbers in scientific notation or in standard form. See examples 3–5 A and B.

41. A millicron equals 0.000000001 of a meter. Write this number in scientific notation.

42. The speed of light is approximately 30,000,000,000 centimeters per second. Write this in scientific notation.

43. The average human body manufactures approximately 2,000,000,000,000 red blood cells each day. Write this in scientific notation.

44. A person working for 30 years earning at least \$13 per hour will earn in excess of 7.6×10^5 dollars. Write this in standard form.

45. A person with a heartbeat of 72 beats per minute will experience more than 3.56×10^7 heartbeats in one year. Write this in standard form.

46. Light travels approximately 9,500,000,000,000 kilometers in one year. Write this number in scientific notation.

47. One atom of iron has a mass of 9.3×10^{-23}. Write this in standard form.

48. The planet Pluto is about 4,250 million miles from the Earth. Write this in scientific notation.

49. One gallon of oil releases 1.4×10^5 BTU of heat energy when burned. Write this in standard form.

50. There are 3.3×10^{22} molecules in one cubic centimeter of water. Write this in standard form.

Perform the indicated operations using scientific notation. Leave the answer in scientific notation. See example 3–5 C.

51. $(6{,}370{,}000) \cdot (19{,}200{,}000)$

52. $(18{,}700{,}000) \cdot (52{,}600{,}000)$

53. $(9.41 \times 10^{12}) \cdot (3.86 \times 10^{-14})$

54. $(4.49 \times 10^{-18}) \cdot (5.89 \times 10^{27})$

55. $(0.00341) \cdot (0.0000519)$

56. $(0.00827) \cdot (0.0196)$

57. $(5.93 \times 10^{-4}) \cdot (8.17 \times 10^{11})$

58. $(177{,}000) \div (0.15)$

59. $(1.344 \times 10^{-8}) \div (9.6 \times 10^{-12})$

60. $(1.036 \times 10^{15}) \div (3.7 \times 10^{-4})$

61. $\dfrac{(92{,}000{,}000) \cdot (0.0036)}{(0.018) \cdot (4{,}000)}$

62. $\dfrac{(39{,}600) \cdot (0.00264)}{(0.00000132) \cdot (66{,}000{,}000)}$

Review exercises

Perform the indicated addition or subtraction. See section 2–3.

1. $3x + 2x$

2. $6a^2 - 3a^2$

3. $5ab + 7ab$

Perform the indicated multiplication. See section 3–2.

4. $x^2(x + 2)$

5. $3a(2a - 5)$

6. $x^2y(3x + 2y - 7)$

Chapter 3 lead-in problem

If Diane makes \$13 per hour and works 40 hours per week, she will earn more than one million dollars over the next 40 years. How much money will Diane earn? Perform the computations using scientific notation.

Solution

hourly rate		hours worked per week		weeks per year		number of years worked	
13	·	40	·	52	·	40	Original problem

$= (1.3 \times 10)(4.0 \times 10)(5.2 \times 10)(4.0 \times 10)$ — Scientific notation

$= (1.3)(4.0)(5.2)(4.0) \times (10 \cdot 10 \cdot 10 \cdot 10)$ — Associative and commutative properties

$= 108.16 \times 10^4$ — Multiply

$= 1{,}081{,}600$ — Standard form

Diane will earn \$1,081,600 over the next 40 years.

Chapter 3 summary

1. In the expression x^6, x is called the **base** and 6 the **exponent.**
2. $a^n = \overbrace{a \cdot a \cdot a \cdots a}^{n \text{ factors}}$, where n is a positive integer.
3. **Properties and definitions of exponents**

   $a^m \cdot a^n = a^{m+n}$

   $(ab)^n = a^n b^n$

   $(a^m)^n = a^{m \cdot n}$

   $\left(\dfrac{a}{b}\right)^n = \dfrac{a^n}{b^n}, b \neq 0$

   $a^m \div a^n = \dfrac{a^m}{a^n} = a^{m-n}, a \neq 0$

   $a^{-n} = \dfrac{1}{a^n}, a \neq 0$

   $a^0 = 1, a \neq 0$
4. When *multiplying* two multinomials, we multiply each term in the first multinomial by each term in the second multinomial. We then combine like terms.
5. Three **special products** are:

   $(a + b)^2 = a^2 + 2ab + b^2$

   $(a - b)^2 = a^2 - 2ab + b^2$

   $(a + b)(a - b) = a^2 - b^2$
6. The scientific notation of a positive number X is $X = a \times 10^n$, where $1 \leq a < 10$ and n is an integer.

Chapter 3 error analysis

1. Exponents
   Example: $xy^3 = x^3y^3$
   Correct answer: $xy^3 = xy^3$
   What error was made? (*see page 127*)
2. Multiplication of like bases
   Example: $4^2 \cdot 4^3 = 16^5$
   Correct answer: $4^2 \cdot 4^3 = 4^5$
   What error was made? (*see page 129*)
3. Power of a power
   Example: $(a^3)^2 = a^5$
   Correct answer: a^6
   What error was made? (*see page 130*)
4. Multiplying unlike bases
   Example: $x^3 \cdot y = (xy)^4$
   Correct answer: $x^3 \cdot y = x^3y$
   What error was made? (*see page 131*)
5. Product of a monomial and a multinomial
   Example: $-x(x^2 - 2x + 1) = x^3 - 2x^2 + x$
   Correct answer: $-x^3 + 2x^2 - x$
   What error was made? (*see page 134*)
6. Squaring a binomial
   Example: $(4x + y)^2 = (4x)^2 + (y)^2 = 16x^2 + y^2$
   Correct answer: $16x^2 + 8xy + y^2$
   What error was made? (*see page 136*)
7. Dividing like bases
   Example: $\dfrac{x^3}{x} = x^4$
   Correct answer: x^2
   What error was made? (*see page 140*)
8. Negative exponents
   Example: $\dfrac{1}{a^{-3}} = -a^3$
   Correct answer: $\dfrac{1}{a^{-3}} = a^3$
   What error was made? (*see page 141*)
9. Zero exponent
   Example: $(x + y)^0 = x^0 + y^0 = 1 + 1 = 2$
   Correct answer: $(x + y)^0 = 1$
   What error was made? (*see page 143*)
10. Evaluate absolute value
    Example: $-|-4| = 4$
    Correct answer: $-|-4| = -4$
    What error was made? (*see page 31*)

Chapter 3 critical thinking

Given the number 52^2, determine a method by which you can square the 52 mentally.

Chapter 3 review

[3–1]

Simplify and leave the answers with only positive exponents.

1. $a^5 \cdot a^7$
2. $a \cdot a^4 \cdot a^9$
3. $4^3 \cdot 4^2$
4. $(xy)^4$
5. $(a^3)^5$
6. $(5ab^3)(4a^3b^2)$
7. $(3x^2y^3)(2xy^4)$
8. $(-5x^2)(3x^3)$
9. $(2a^2b)(3ab^4)$
10. $(5x^2y)(2x^3y^4)$
11. $(-3a^2b^3)(2a^4b^7)$

[3–2]

Perform the indicated multiplication and simplify.

12. $5x(3x - 2y)$
13. $-3a^2b(2a^2 - 3ab + 4b^2)$
14. $(5x)(3x - y)(2y^2)$
15. $(x + 3)(x - 4)$
16. $(x + 5)^2$
17. $(a - 7)(a + 7)$
18. $(5x - y)(3x + 2y)$
19. $(x - 2y)(x^2 + 3xy + y^2)$
20. $(3a - b)(a + b)(a - 2b)$
21. $(2a + b)^3$

[3–3]

Simplify and leave the answers with only positive exponents.

22. $\dfrac{b^5}{b^7}$
23. $5a^{-2}$
24. $a^{-5} \cdot a^9$
25. $\dfrac{x^3x^2}{x^8}$
26. $(3a^2b)^0$
27. $5x^{-3}y^{-2}$
28. $\dfrac{a^{-4}}{a^{-7}}$
29. $\left(\dfrac{a}{b}\right)^5$
30. $\left(\dfrac{2yz}{x}\right)^2$
31. $\dfrac{2a^2b^4}{2^3a^5b}$
32. $\dfrac{a^5b^{-2}}{a^{-4}b}$

[3–4]

Simplify and leave the answers with only positive exponents.

33. $(2a^2b^3)^3$
34. $(3^3x^4y^5)^4$
35. $(2xy^{-3})^{-2}$
36. $\dfrac{2x^{-1}y^0z^3}{4x^{-2}y^{-3}}$
37. $\dfrac{8a^{-5}b^{-4}c^0}{4a^{-7}b^2c^{-3}}$
38. $(x^5y^4)^4(2x^2y^3)^3$
39. $(2a^2b)^3(3a^4)^2$
40. $\left(\dfrac{a^3b^0}{c^4}\right)^3$
41. $\left(\dfrac{3a^3b^2}{c^5}\right)^2$
42. $\left(\dfrac{2xy^4}{z^6}\right)^5$

[3–5]

Express the following numbers in scientific notation.

43. 1,840
44. 0.00157
45. 107,000,000
46. 849,000,000,000
47. −37.5
48. −0.00543

Express the following numbers in standard form.

49. 5.04×10^5
50. 6.39×10^{-3}
51. -5.96×10^2
52. -8.86×10^{-3}
53. 7.35×10^{-7}
54. 8.12×10^8

Perform the indicated operations using scientific notation. Leave the answer in scientific notation.

55. $(456{,}000{,}000) \cdot (0.000587)$
56. $(0.0000183) \cdot (0.000846)$
57. $(756{,}000) \div (105{,}000{,}000)$
58. $(0.00525) \div (42{,}000)$

Chapter 3 cumulative test

Determine if the following statements are true or false.

[1–3] 1. $|-10| < 0$

[1–8] 2. $-3^2 = (-3)^2$

[1–3] 3. $|-3| < |-7|$

Perform the indicated operations, if possible, and simplify.

[1–7] 4. $\dfrac{(8)}{(-4)}$

[1–7] 5. $\dfrac{(-9)}{0}$

[1–5] 6. $(2) - (-6)$

[1–6] 7. $(-2)(4)(0)(-4)$

[1–8] 8. $(-2)^4$

[1–8] 9. $48 - 24 \div 8 - 3 - 2^2$

[2–3] 10. $(3x^2y - 4xy + 2xy^2) - (2x^2y - 4xy + 3x^2y^2)$

[1–6] 11. $(5)(-2)(4)$

[1–7] 12. $\dfrac{0}{-3}$

[3–2] 13. $(3x - y)^2$

[1–6] 14. $(2)(-7)(0)(3)$

[3–4] 15. $(2a^2b^5)^2$

[1–6] 16. -5^2

[1–8] 17. $10 - 10 \div 10 \cdot 10 - 10 + 10$

[3–2] 18. $(3x - 2y)(3x + 2y)$

[3–1] 19. $(3x^2y)(2xy^4)$

[1–8] 20. $2[5(7 - 4) - 6 + 4]$

[1–7] 21. $\dfrac{(-6)(-4)}{(5)(0)}$

[2–3] 22. $(3a - 2b) - [5a - (4b + 6a)]$

[3–2] 23. $(x + 1)(x^2 - x - 1)$

[3–3] 24. $x^{-3}x^5x^0$

[3–3] 25. $\dfrac{a^{-5}}{a^{-9}}$

Find the solution set.

[2–6] 26. $3x - 4 = x + 10$

[2–6] 27. $2(x - 4) + 7 = 8x - 11$

[2–6] 28. $\dfrac{2}{3}x + 4 = \dfrac{5}{6}$

Find the solution.

[2–9] 29. $8 - 3x < 9$

[2–9] 30. $6x + 5 - 4 > 2$

[2–6] 31. $3 - 2x = 6$

[2–9] 32. $-9 \leq 2x + 7 \leq 5$

[2–8] 33. If a number is decreased by 17 and that result is then divided by 5, the answer is 16. Find the number.

[2–8] 34. One-third of a number is 12 less than one-half of the number. Find the number.

[2–8] 35. Brenda invested part of \$30,000 at 8% and the rest at 7%. If her income for one year from the 8% investment was \$675 more than that from the 7% investment, how much was invested at each rate?

CHAPTER

4 Factoring and Solution of Quadratic Equations by Factoring

The formula $s = vt - 16t^2$ gives the height s in feet that an object will travel in t seconds if it is propelled directly upward at an initial velocity of v feet per second. If an object is thrown upward at 96 feet per second, how long will it take the object to reach a height of 144 feet?

4–1 ■ Common factors

Greatest common factor

To find the solution of certain equations that are not linear, we will need to study a technique called *factoring a polynomial.* Factoring polynomials will also be useful in dealing with algebraic fractions since, as we have seen with arithmetic fractions, we must have the numerator and the denominator of the fractions in a factored form to reduce or to find the least common denominator.

The first type of factoring that we will do involves finding the greatest common factor (GCF) of each term in the polynomial. Recall the statement of the distributive property.

$$a(b + c) = ab + ac$$

$a(b + c)$ is called the *factored form* of $ab + ac$. We now use the distributive property to write $3x + 6$ in the factored form. First we notice that $3x + 6$ can be written as

$$3 \cdot x + 3 \cdot 2 \quad \text{3 is a common factor in both terms}$$

By applying the distributive property, we have

$$3x + 6 = 3 \cdot x + 3 \cdot 2 = 3(x + 2)$$

$3(x + 2)$ is the factored form of $3x + 6$.

This type of factoring, as its name implies, involves looking for numbers or variables that are **common factors** in *all* of the original terms. In our example, the number 3 was common to all of the original terms, and we were able to factor it out.

When a polynomial is factored, we "factor out" the greatest common factor, GCF. The greatest common factor consists of the following:

Greatest common factor

1. The greatest integer that is a common factor of all the numerical coefficients and
2. the variable factor(s) raised to the least power to which they were raised in any of the terms

← Factoring

Polynomial *(terms)*	Distributive Property *(determine the GCF)*	Factored Form *(factors)*
$3x + 6$	$\mathbf{3} \cdot x + \mathbf{3} \cdot 2$	$3(x + 2)$
$10x^2 + 15y$	$\mathbf{5} \cdot 2x^2 + \mathbf{5} \cdot 3y$	$5(2x^2 + 3y)$
$12a - 42b$	$\mathbf{6} \cdot 2a - \mathbf{6} \cdot 7b$	$6(2a - 7b)$
$18xy + 12xz$	$\mathbf{6x} \cdot 3y + \mathbf{6x} \cdot 2z$	$6x(3y + 2z)$

Multiplying →

In the previous examples, we determined the greatest common factor by inspection. In some problems, this may not be possible, and the following procedure will be necessary. Factor the polynomial $12x^3y + 30x^2y^3$.

Step 1 Factor each term such that it is the product of primes* and variables to powers.

$12x^3y$	$30x^2y^3$
$2^2 \cdot 3 \cdot x^3 \cdot y$	$2 \cdot 3 \cdot 5 \cdot x^2 \cdot y^3$

Step 2 Write down all the numbers and variables that are common to *every* term.

$2 \cdot 3 \cdot x \cdot y$

Note We do not have 5 as part of our greatest common factor since it does not appear as a factor in *all* of the terms.

Step 3 Take the numbers and variables in step 2. Raise them to the *lowest* power to which they were raised in any of the terms.

$2^1 \cdot 3^1 \cdot x^2 \cdot y^1 = 6x^2y$

This is the greatest common factor.

Step 4 Find the multinomial factor, the polynomial within the parentheses, by dividing each term of the polynomial being factored by the GCF.

$$\frac{12x^3y}{6x^2y} = 2x \text{ and } \frac{30x^2y^3}{6x^2y} = 5y^2$$

$$6x^2y(2x + 5y^2)$$

Step 5 We can now write the polynomial in its factored form.

$$12x^3y + 30x^2y^3$$
$$= 6x^2y \cdot 2x + 6x^2y \cdot 5y^2$$
$$= 6x^2y(2x + 5y^2)$$

*Primes and prime factorization are covered in section 1–1.

Completely factored form

In the previous example, $12x^3y + 30x^2y^3$ could also be factored to $3xy(4x^2 + 10xy^2)$ or $12y\left(x^3 + \frac{5}{2}x^2y^2\right)$. This allows room for a given polynomial to be factored in many ways, unless some restrictions are placed on the procedure. We wish to factor each polynomial in a unique manner that will not permit such variations in the results. Thus it is customary to adopt the following criteria for a completely factored polynomial.

A polynomial with integer coefficients will be considered to be in **completely factored form** when it satisfies the following criteria:

Completely factored form

1. The polynomial is written as a product of polynomials with integer coefficients.
2. None of the polynomial factors other than the monomial factor can be further factored.

We see that $6x^2y(2x + 5y^2)$ is the completely factored form of the expression $12x^3y + 30x^2y^3$ since all of the coefficients are integers and, except for the monomial factor $6x^2y$, the remaining factor contains no other factor with integer coefficients.

In general, whenever we factor a monomial out of a polynomial, we factor the monomial so that it has a positive coefficient. Realize that we could also factor out the opposite, or negative, of this common factor. We could have factored out $-6x^2y$ from our example. The completely factored form would have been

$$-6x^2y(-2x - 5y^2)$$

Observe that the only change in our answer when we factor out the opposite of the common factor is that this changes the signs of all terms inside the parentheses.

■ *Example 4–1 A*

Write in completely factored form.

1. $7a^3 + 14a$

$= 7a(\quad + \quad)$ — Multinomial factor will have as many terms as the original expression; GCF is $7a$

$= 7a(a^2 + 2)$ — Completely factored form; $\frac{14a}{7a}$; $\frac{7a^3}{7a}$

If we want to check the answer, we apply the distributive property and perform the multiplication as follows:

$7a(a^2 + 2) = 7a \cdot a^2 + 7a \cdot 2$ — Distributive property

$= 7a^3 + 14a$ — Carry out the multiplication

2. $9x^5 + 6x^3 - 18x^2$

$= 3^2x^5 + 2 \cdot 3x^3 - 2 \cdot 3^2x^2$ — Factor each term

$= 3x^2(\quad + \quad - \quad)$ — Determine the GCF

$= 3x^2(3x^3 + 2x - 6)$ — Completely factored form

3. $72a^2b - 84a^3b^4 + 48a^4b^2$

$= 2^3 \cdot 3^2 \cdot a^2 \cdot b - 2^2 \cdot 3 \cdot 7 \cdot a^3 \cdot b^4 + 2^4 \cdot 3 \cdot a^4 \cdot b^2$ — Factor each term

$= 12a^2b(\quad - \quad + \quad)$ — Determine the GCF

$= 12a^2b(6 - 7ab^3 + 4a^2b)$ — Completely factored form

4. $3x^3y^2 + 15x^2y^4 + 3xy^2$

$= 3 \cdot x^3 \cdot y^2 + 3 \cdot 5 \cdot x^2 \cdot y^4 + 3 \cdot x \cdot y^2$ — Factor each term

$= 3xy^2(\quad + \quad + \quad)$ — Determine the GCF

$= 3xy^2(x^2 + 5xy^2 + 1)$ — Completely factored form

Note In example 4, the last term in the factored form is 1. This situation occurs when a term and the GCF are the same, that is, whenever we are able to factor all of the numbers and variables out of a given term. For example, $\frac{3xy^2}{3xy^2} = 1$. The number of terms inside the parentheses must be equal to the number of terms in the original polynomial.

▶ ***Quick check*** Write $9x + 6y + 3$ and $9x^3y^2 - 18x^2y^3$ in completely factored form. ■

Remember that when an expression is within a grouping symbol, we treat the quantity as just one number. Therefore if we have a quantity common to all of the terms, we can factor it out of the polynomial.

■ ***Example 4–1 B***

Factor completely.

1. $x(a - 2b) + y(a - 2b)$

The quantity $(a - 2b)$ is common to both terms. We then factor the common quantity out of each term and place the remaining factors from each term in the second parentheses.

$x(a - 2b) + y(a - 2b)$

$(a - 2b)(x + y)$

Common factor — Remaining factors

2. $x^2(a + b) + (a + b)$

$= (a + b)(\quad + \quad)$ — Determine the GCF

$= (a + b)(x^2 + 1)$ — Completely factored form

3. $3x^2(2a - b) - 9x(2a - b)$

$= 3x^2(2a - b) - 3^2x(2a - b)$ — Factor each term

$= 3x(2a - b)(\quad - \quad)$ — Determine the GCF

$= 3x(2a - b)(x - 3)$ — Completely factored form

▶ ***Quick check*** Factor $x(y + 5) - z(y + 5)$ ■

Four-term polynomials

Consider $ax + ay + bx + by$. We observe that this is a *four-term polynomial* and we will *try* to factor it by grouping.

$$ax + ay + bx + by = (ax + ay) + (bx + by)$$

There is a common factor of a in the first two terms and a common factor of b in the last two terms.

$$(ax + ay) + (bx + by) = a(x + y) + b(x + y)$$

The quantity $(x + y)$ is common to both terms. Factoring it out, we have

$$a(x + y) + b(x + y) = (x + y)(a + b)$$

Therefore we have factored the polynomial by grouping.

Factoring a four-term polynomial by grouping

1. Arrange the four terms so that the first two terms have a common factor and the last two terms have a common factor.
2. Determine the GCF of each pair of terms and factor it out.
3. If step 2 produces a common binomial factor in each term, factor it out.
4. If step 2 does not produce a common binomial factor in each term, try grouping the terms of the original polynomial in a different way.
5. If step 4 does not produce a common binomial factor in each term, the polynomial will not factor by this procedure.

■ ***Example 4–1 C***

Factor completely.

1. $ax + 2ay + bx + 2by$

$= (ax + 2ay) + (bx + 2by)$ Group in pairs

$= a(x + 2y) + b(x + 2y)$ Factor out the GCF

$= (x + 2y)(a + b)$ Factor out the common binomial

2. $3ac + 6ad - 2bc - 4bd$

$= (3ac + 6ad) - (2bc + 4bd)$ Group in pairs

$= 3a(c + 2d) - 2b(c + 2d)$ Factor out the GCF

$= (c + 2d)(3a - 2b)$ Factor out the common binomial

3. $2ax - 2ay + bx - by$

$= (2ax - 2ay) + (bx - by)$ Group in pairs

$= 2a(x - y) + b(x - y)$ Factor out the GCF

$= (x - y)(2a + b)$ Factor out the common binomial

4. $6ax + by + 3ay + 2bx$

$= 6ax + 3ay + 2bx + by$ Rearrange the terms

$= (6ax + 3ay) + (2bx + by)$ Group in pairs

$= 3a(2x + y) + b(2x + y)$ Factor out the GCF

$= (2x + y)(3a + b)$ Factor out the common binomial

Note As in example 4, sometimes the terms must be rearranged so that the pairs will have a common factor.

▶ ***Quick check*** Factor $3ax + 6bx + 2ay + 4by$ ■

It is important to remember to look for the greatest common factor first *when we attempt to determine the completely factored form of any polynomial. If we fail to do this, the answer may not be in a completely factored form or we may not see how to factor the problem by an appropriate procedure.*

Mastery points

Can you

- Determine the greatest common factor?
- Factor a four-term polynomial by grouping?

Exercise 4–1

Write in completely factored form. See example 4–1 A.

Examples $9x + 6y + 3$

Solutions
$= 3 \cdot 3x + 3 \cdot 2y + 3 \cdot 1$ Factor each term
$= 3(\quad + \quad + \quad)$ Determine the GCF
$= 3(3x + 2y + 1)$ Completely factored form

$9x^3y^2 - 18x^2y^3$
$= 9x^2y^2 \cdot x - 9x^2y^2 \cdot 2y$ Factor each term
$= 9x^2y^2(\quad - \quad)$ Determine the GCF
$= \quad (x - 2y)$ Completely factored form

1. $2y + 6$
2. $3a - 12$
3. $4x^2 + 8y$
4. $8y^2 + 10x^2$
5. $3x^2y + 15z$
6. $5r^2 + 10rs - 20s$
7. $7a - 14b + 21c$
8. $8x - 12y + 16z$
9. $15xy - 18z + 3x^2$
10. $18ab - 27a + 3ac$
11. $42xy - 21y^2 + 7$
12. $15a^2 - 27b^2 + 12ab$
13. $8x - 10y + 12z - 18w$
14. $15L^2 - 21W^2 + 36H$
15. $20a^2b - 60ab + 45ab^2$
16. $4x^2 + 8x$
17. $3x^2y + 6xy$
18. $8x^3 + 4x^2$
19. $2R^4 - 6R^2$
20. $3x^2 - 3xy + 3x$
21. $2x^3 - x^2 + x$
22. $24a^2 + 12a - 6a^3$
23. $15ab + 18ab^2 - 3a^2b$
24. $2x^4 - 6x^2 + 8x$
25. $xy^2 + xyz + xy$
26. $3R^2S - 6RS^2 + 12RS$
27. $2L^3 - 18L + 2L^2$
28. $V^2 + V^3 - V^4 + 2V$
29. $5p^2 + 10p + 15p^3$
30. $16x^3y - 3x^2y^2 + 24x^2y^3$

Supply the missing factor.

Example $-3a - 6b = -3(\quad)$

Solution Since $-3a - 6b = (-3) \cdot a + (-3) \cdot 2b$, then $-3a - 6b = -3(a + 2b)$, the missing factor is $(a + 2b)$.

Example $-a^2b^3 + a^2b^2 = -a^2b^2(\quad)$

Solution $-a^2b^3 + a^2b^2 = (-a^2b^2)(b) + (-a^2b^2)(-1)$ Divide each term by $-a^2b^2$ to find the missing factor
$= -a^2b^2(b - 1)$, the missing factor is $(b - 1)$.

31. $-6x - 9 = -3(\quad)$
32. $-5a + 10b = -5(\quad)$
33. $6x - 8z - 12w = 2(\quad)$
34. $-4a^3 - 36ab + 16ab^2 - 24b^3 = -4(\quad)$
35. $-12L + 15W - 6H = -3(\quad)$
36. $-3a + a^3b = -a(\quad)$
37. $-x + x^2 - x^3 = -x(\quad)$
38. $-x + 2xy + xy^2 = -x(\quad)$

39. $-xyz + x^2yz - xy^2z + xyz^2 = -xyz(\quad)$

40. $-4x^2 + 8x - 12x^3 = -4x(\quad)$

41. $-10a^2b^2 + 15ab - 20a^3b^3 = -5ab(\quad)$

42. $-24RS - 16R + 32R^2 = -8R(\quad)$

Write in completely factored form. See example 4–1 B.

Example $x(y + 5) - z(y + 5)$

Solution $= (y + 5)(\quad - \quad)$ Determine the GCF

$= (y + 5)(x - z)$ Completely factored form

43. $x(a + b) + y(a + b)$

44. $3a(x - y) + b(x - y)$

45. $15x(2a + b) + 10y(2a + b)$

46. $21R(L + 2N) - 35S(L + 2N)$

47. $3x(a + 4b) + 6y(a + 4b)$

48. $4RS(2P + q) - 8RT(2P + q)$

49. $8a(b + 6) - (b + 6)$

Write the following in completely factored form. See example 4–1 C.

Example $3ax + 6bx + 2ay + 4by$

Solution $= (3ax + 6bx) + (2ay + 4by)$ Group in pairs

$= 3x(a + 2b) + 2y(a + 2b)$ Factor out the GCF

$= (a + 2b)(3x + 2y)$ Factor out the common binomial

50. $rt + ru + st + su$

51. $ac + ad + bc + bd$

52. $5ax - 3by + 15bx - ay$

53. $6ax - 2by + 3bx - 4ay$

54. $2ax^2 - bx^2 + 6a - 3b$

55. $4ax + 2ay - 2bx - by$

56. $ac + 3ad - 4bc - 12bd$

57. $20x^2 + 5xz - 12xy - 3yz$

58. $a^2x + 3a^2y - 3x - 9y$

59. $4ax + 12bx - 3ay - 9by$

60. $ac + ad - 2bc - 2bd$

61. $2ac + 6bc - ay - 3by$

62. $2ac + bc - 4ay - 2by$

63. $2ac + 3bc + 8ay + 12by$

64. $5ac - 3by + 15bc - ay$

65. $6ax + by + 2ay + 3bx$

66. $2ax - ad + 4bx - 2bd$

67. $3ax - 2bd - 6ad + bx$

68. $6ax + 3bd - 2ad - 9bx$

69. $2a^3 + 15 + 10a^2 + 3a$

70. $3a^3 - 6a^2 + 5a - 10$

71. $8a^3 - 4a^2 + 6a - 3$

Write in completely factored form. See example 4–1 A.

72. The area of the surface of a cylinder is determined by $A = 2\pi rh + 2\pi r^2$. Factor the right member. (π is the Greek letter pi.)

73. The total surface area of a right circular cone is given by $A = \pi rs + \pi r^2$. Factor the right member.

74. The equation for the distance traveled by a rocket fired vertically upward into the air is given by $S = 560t - 16t^2$, where the rocket is S feet from the ground after t seconds. Factor the right member.

75. In engineering, the equation for deflection of a beam is given by

$$Y = \frac{2wx^4}{48EI} - \frac{3\ell wx^3}{48EI} - \frac{\ell^3wx}{48EI}$$

Factor the right member.

Review exercises

Two numbers are listed. Find two integers such that their product is the first number and their sum is the second number. See section 1–6.

1. 20, 9

2. 12, 7

3. −16, −6

4. −16, 6

5. 16, 10

6. 16, −10

7. 36, 12

8. 11, 12

4–2 ■ Factoring trinomials of the form $x^2 + bx + c$

Determining when a trinomial will factor

In section 3–2, we learned how to multiply two binomials as follows:

Factors — Terms

$$(x + 2)(x + 6) = x^2 + 6x + 2x + 12 = x^2 + 8x + 12$$

Multiplying →

In this section, we are going to reverse the procedure and factor the trinomial.

Terms — Factors

$$x^2 + 8x + 12 = (x + 2)(x + 6)$$

Factoring →

The following group of trinomials will enable us to see how a trinomial factors.

$12 = 2 \cdot 6$ — Product

1. $x^2 + 8x + 12 = (x + 2)(x + 6)$

$8 = 2 + 6$ — Sum

$12 = (-2) \cdot (-6)$ — Product

2. $x^2 - 8x + 12 = (x - 2)(x - 6)$

$-8 = (-2) + (-6)$ — Sum

$-12 = (-2) \cdot 6$ — Product

3. $x^2 + 4x - 12 = (x - 2)(x + 6)$

$4 = (-2) + 6$ — Sum

$-12 = 2 \cdot (-6)$ — Product

4. $x^2 - 4x - 12 = (x + 2)(x - 6)$

$-4 = 2 + (-6)$ — Sum

In general,

$$(x + m)(x + n) = x^2 + (m + n)x + m \cdot n$$

The trinomial $x^2 + bx + c$ will factor with integer coefficients only if there are two integers, which we will call m and n, such that $m + n = b$ and $m \cdot n = c$.

Sum: $m + n$ — Product: $m \cdot n$

$$x^2 + bx + c = (x + m)(x + n)$$

The signs (+ or −) for m and n

1. If c is positive, then m and n have the same sign as b.
2. If c is negative, then m and n have different signs and the one with the greater absolute value has the same sign as b.

■ *Example 4–2 A*

Factor completely each trinomial.

1. $a^2 + 11a + 18$ $\quad m + n = 11$ and $m \cdot n = 18$

Since $b = 11$ and $c = 18$ are both positive, then m and n are both positive.

List the factorizations of 18	Sum of the factors of 18	
$1 \cdot 18$	$1 + 18 = 19$	
$2 \cdot 9$	$2 + 9 = 11$	← Correct sum
$3 \cdot 6$	$3 + 6 = 9$	

The m and n values are 2 and 9. The factorization is

$$a^2 + 11a + 18 = (a + 2)(a + 9)$$

The answer can be checked by performing the indicated multiplication.

$$(a + 2)(a + 9) = a^2 + 9a + 2a + 18 = a^2 + 11a + 18$$

Note The commutative property allows us to write the factors in any order. That is, $(a + 2)(a + 9) = (a + 9)(a + 2)$.

2. $b^2 - 2b - 15$ $\quad m + n = -2$ and $m \cdot n = -15$

Since $b = -2$ and $c = -15$ are both negative, then m and n have different signs and the one with the greater absolute value is negative.

Factorizations of −15, where the negative sign goes with the factor with the greater absolute value	Sum of the factors of −15	
$1 \cdot (-15)$	$1 + (-15) = -14$	
$3 \cdot (-5)$	$3 + (-5) = -2$	← Correct sum

The m and n values are 3 and −5. The factorization is

$$b^2 - 2b - 15 = (b + 3)(b - 5)$$

3. $5x - 24 + x^2$

It is easier to identify b and c if we write the trinomial in descending powers of the variable, which is called **standard form.**

$x^2 + 5x - 24$ $\quad m + n = 5$ and $m \cdot n = -24$

Since $b = 5$ is positive and $c = -24$ is negative, m and n have different signs and the one with the greater absolute value is positive.

Factorizations of −24, where the positive factor is the one with the greater absolute value	Sum of the factors of −24	
$(-1) \cdot 24$	$(-1) + 24 = 23$	
$(-2) \cdot 12$	$(-2) + 12 = 10$	
$(-3) \cdot 8$	$(-3) + 8 = 5$	← Correct sum
$(-4) \cdot 6$	$(-4) + 6 = 2$	

The m and n values are −3 and 8. The factorization is

$$x^2 + 5x - 24 = (x - 3)(x + 8)$$

4. $c^2 - 9c + 14$ $\quad m + n = -9$ and $m \cdot n = 14$
Since $b = -9$ is negative and $c = 14$ is positive, m and n are both negative.

List the factorizations of 14	Sum of the factors of 14
$(-1)(-14)$	$(-1) + (-14) = -15$
$(-2)(-7)$	$(-2) + (-7) = -9$ ← Correct sum

The m and n values are -2 and -7. The factorization is

$$c^2 - 9c + 14 = (c - 2)(c - 7)$$

5. $x^2 + 5x + 12$ $\quad m + n = 5$ and $m \cdot n = 12$
Since $b = 5$ and $c = 12$ are both positive, m and n are both positive.

Factorizations of 12	Sum of the factors of 12
$1 \cdot 12$	$1 + 12 = 13$
$2 \cdot 6$	$2 + 6 = 8$
$3 \cdot 4$	$3 + 4 = 7$

No sum equals 5

Since none of the factorizations of 12 add to 5, there is no pair of integers (m and n) and the trinomial will not factor using integer coefficients. We call this a **prime polynomial.**

6. $x^4 - 4x^3 - 21x^2 = x^2(x^2 - 4x - 21)$ $\quad$ Common factor of x^2
To complete the factorization, we see if the trinomial $x^2 - 4x - 21$ will factor. We need to find m and n that add to -4 and multiply to -21. The values are 3 and -7. The completely factored form is

$$x^4 - 4x^3 - 21x^2 = x^2(x + 3)(x - 7)$$

Note A common error when the polynomial has a common factor is to factor it out but to forget to include it as one of the factors in the completely factored form.

7. $x^2y^2 + 9xy + 20$
Rewriting the polynomial as $(xy)^2 + 9xy + 20$, we want to find values for m and n that add to 9 and multiply to 20. The numbers are 4 and 5. The factorization is

$$x^2y^2 + 9xy + 20 = (xy + 4)(xy + 5)$$

8. $x^2 - 5ax + 6a^2$ $\quad m + n = -5a$ and $m \cdot n = 6a^2$
We need to find m and n that add to $-5a$ and multiply to $6a^2$. The values are $-2a$ and $-3a$. The factorization is

$$x^2 - 5ax + 6a^2 = (x - 2a)(x - 3a)$$

▶ ***Quick check*** Factor $z^2 + 8z - 20$ ■

Factoring a trinomial of the form $x^2 + bx + c$

1. Factor out the GCF. If there is a common factor, make sure to include it as part of the final factorization.
2. Determine if the trinomial is factorable by finding m and n such that $m + n = b$ and $m \cdot n = c$. If m and n do not exist, we conclude that the trinomial will not factor.
3. Using the m and n values from step 2, write the trinomial in factored form.

Mastery points

Can you

- Determine two integers whose product is one number and whose sum is another number?
- Recognize when the trinomial $x^2 + bx + c$ will factor and when it will not?
- Factor trinomials of the form $x^2 + bx + c$?
- Always remember to look for the greatest common factor before applying any of the factoring rules?

Exercise 4–2

Factor completely each trinomial. See example 4–2 A.

Example $z^2 + 8z - 20$

Solution Since $b = 8$ is positive and $c = -20$ is negative, m and n have different signs and the one with the greater absolute value is positive.

Factorizations of -20, where the positive factor is the one with the greater absolute value	Sum of the factors of -20
$(-1) \cdot 20$	$(-1) + 20 = 19$
$(-2) \cdot 10$	$(-2) + 10 = 8$ ← Correct sum
$(-4) \cdot 5$	$(-4) + 5 = 1$

The m and n values are -2 and 10. The factorization is

$$z^2 + 8z - 20 = (z - 2)(z + 10)$$

1. $a^2 + 9a + 18$
2. $c^2 + 9c + 20$
3. $x^2 + 11x - 12$
4. $x^2 + 13x + 12$
5. $y^2 + 13y - 30$
6. $a^2 + 9a + 14$
7. $x^2 - 14x + 24$
8. $b^2 - 10b + 21$
9. $a^2 + 5a - 24$
10. $y^2 + 9y - 36$
11. $x^2 + 8x + 12$
12. $c^2 + 8c + 15$
13. $a^2 - 2a - 24$
14. $z^2 - 5z - 36$
15. $2x^2 + 6x - 20$
16. $2a^2 + 26a + 24$
17. $3x^2 - 18x - 48$
18. $a^2 - 9a + 4$
19. $x^2 + 5x + 7$
20. $x^2 - 4x + 6$
21. $y^2 + 17y + 30$
22. $b^2 + 13b + 40$
23. $4x^2 - 4x - 24$
24. $5y^2 + 5y - 30$
25. $5a^2 - 15a - 50$
26. $x^2y^2 - 4xy - 21$
27. $x^2y^2 - 3xy - 18$
28. $x^2y^2 - xy - 30$
29. $x^2y^2 + 13xy + 12$
30. $4a^2b^2 - 32ab + 28$
31. $3x^2y^2 - 3xy - 36$
32. $3x^2y^2 + 21xy + 36$
33. $x^2 + 3xy + 2y^2$
34. $a^2 - ab - 2b^2$
35. $a^2 - 2ab - 3b^2$
36. $a^2 - 7ab + 10b^2$
37. $a^2 - ab - 6b^2$
38. $x^2 + 2xy - 8y^2$
39. $x^2 - 2xy - 15y^2$
40. $a^2 + 7ab + 12b^2$

Review exercises

Factor completely. See section 4–1.

1. $ax^2 + bx^2 + cx^2$
2. $3x^3 + 12x^2 - 6x$
3. $3x(2x + 1) + 5(2x + 1)$
4. $2x(3x - 2) + 3(3x - 2)$
5. $4x(5x + 1) + (5x + 1)$
6. $6x(2x + 3) - (2x + 3)$
7. $x(3x - 5) - 2(3x - 5)$
8. $7x(x - 9) - 3(x - 9)$

4–3 ■ Factoring trinomials of the form $ax^2 + bx + c$

How to factor trinomials

In this section, we are going to factor trinomials of the form $ax^2 + bx + c$. This is called the **standard form** of a trinomial, where we have a single variable and the terms of the polynomial are arranged in descending powers of that variable. The a, b, and c in our standard form represent integer constants. For example,

$$2x^2 + 9x + 9$$

is a trinomial in standard form, where $a = 2$, $b = 9$, and $c = 9$.

Consider the product

$$(2x + 3)(x + 3)$$

By multiplying these two quantities together, we get a trinomial.

$$\begin{aligned}(2x + 3)(x + 3) &= 2x^2 + 6x + 3x + 9\\ &= 2x^2 + 9x + 9\end{aligned}$$

To completely factor the trinomial $2x^2 + 9x + 9$ entails reversing this procedure to get

$$(2x + 3)(x + 3)$$

The trinomial will factor with integer coefficients if we can find a pair of integers (m and n) whose sum is equal to b, and whose product is equal to $a \cdot c$. In the trinomial $2x^2 + 9x + 9$, b is equal to 9, and $a \cdot c$ is $2 \cdot 9 = 18$. Therefore we want $m + n = 9$ and $m \cdot n = 18$. The values for m and n are 3 and 6.

If we observe the multiplication process in our example, we see that m and n appear as the coefficients of the middle terms that are to be combined for our final answer.

$$\begin{aligned}(2x + 3)(x + 3) &= 2x^2 + 6x + 3x + 9\\ &= 2x^2 + 9x + 9\end{aligned}$$

This is precisely what we do with the m and n values. We replace the coefficient of the middle term in the trinomial with these values. In our example, m and n are 6 and 3 and we replace the 9 with them.

$$2x^2 + 9x + 9 = 2x^2 + \overbrace{6x + 3x}^{9x} + 9$$

Our next step is to group the first two terms and the last two terms.

$$(2x^2 + 6x) + (3x + 9)$$

Now we factor out what is common in each pair. We see that the first two terms contain the common factor $2x$ and the last two terms contain the common factor 3.

$$2x(x + 3) + 3(x + 3)$$

When we reach this point, what is inside the parentheses in each term will be the same. Since the quantity $(x + 3)$ is common to both terms, we can factor it out.

$$2x\underbrace{(x + 3)} + 3\underbrace{(x + 3)}$$

Common to both terms

Having factored out what is common, what is left in each term is placed in the second parentheses.

$$2x(x + 3) + 3(x + 3)$$
$$(x + 3)(2x + 3)$$

Common factor   Remaining factors

The trinomial is factored.

A summary of the steps follows:

Factoring a trinomial of the form $ax^2 + bx + c$

Step 1 Determine if the trinomial $ax^2 + bx + c$ is factorable by finding m and n such that $m \cdot n = a \cdot c$ and $m + n = b$. If m and n do not exist, we conclude that the trinomial will not factor.

Step 2 Replace the middle term, bx, by the sum of mx and nx.

Step 3 Place parentheses around the first and second terms and around the third and fourth terms. Factor out what is common to each pair.

Step 4 Factor out the common quantity of each term and place the remaining factors from each term in the second parentheses.

We determine the signs (+ or −) for m and n in a fashion similar to that of section 4–2.

The signs (+ or −) for m and n

1. If $a \cdot c$ is positive, then m and n have the same sign as b.
2. If $a \cdot c$ is negative, then m and n have different signs and the one with the greater absolute value has the same sign as b.

Example 4–3 A

Factor completely the following trinomials. If a trinomial will not factor, so state.

1. $6x^2 + 13x + 6$

Step 1 $m \cdot n = 6 \cdot 6 = 36$ and $m + n = 13$

We determine by inspection that m and n are 9 and 4.

Step 2	$= 6x^2 + \overbrace{9x + 4x}^{13x} + 6$	Replace bx with mx and nx
Step 3	$= (6x^2 + 9x) + (4x + 6)$	Group the first two terms and the last two terms
	$= 3x(2x + 3) + 2(2x + 3)$	Factor out what is common to each pair
Step 4	$= (2x + 3)(3x + 2)$	Factor out the common quantity

Note The order in which we place m and n into the problem will not change the answer.

(Alternate)

Step 2	$= 6x^2 + \overbrace{4x + 9x}^{13x} + 6$	Replace bx with nx and mx
Step 3	$= (6x^2 + 4x) + (9x + 6)$	Group the first two terms and the last two terms
	$= 2x(3x + 2) + 3(3x + 2)$	Factor out what is common to each pair
Step 4	$= (3x + 2)(2x + 3)$	Factor out the common quantity

We see that the outcome in step 4 is the same regardless of the order of m and n in the problem.

Note The order in which the two factors are written in the answer does not matter. That is, $(2x + 3)(3x + 2) = (3x + 2)(2x + 3)$

2. $3x^2 + 5x + 2$

Step 1 $m \cdot n = 3 \cdot 2 = 6$ and $m + n = 5$

m and n are 2 and 3.

Step 2	$= 3x^2 + \overbrace{2x + 3x}^{5x} + 2$	Replace bx with mx and nx
Step 3	$= (3x^2 + 2x) + (3x + 2)$	Group the first two terms and the last two terms
	$= x(3x + 2) + 1(3x + 2)$	Factor out what is common to each pair

We observe in the last two terms that the greatest common factor is only 1 or -1. We factor out 1 so that we have the same quantity inside the parentheses.

Step 4	$= (3x + 2)(x + 1)$	Factor out the common quantity

3. $4x^2 - 11x + 6$

Step 1 $m \cdot n = 4 \cdot 6 = 24$ and $m + n = -11$

m and n are -3 and -8.

Step 2	$= 4x^2 \overbrace{- 3x - 8x}^{-11x} + 6$	Replace bx with mx and nx

Step 3 $= (4x^2 - 3x) + (-8x + 6)$ Group the first two terms and the last two terms

$= x(4x - 3) - 2(4x - 3)$ Factor out what is common to each pair

We have 2 or -2 as the greatest common factor in the last two terms. We factor out -2 so that we will have the same quantity inside the parentheses.

Step 4 $= (4x - 3)(x - 2)$ Factor out the common quantity

Note If the third term in step 2 is preceded by a minus sign, we will usually factor out the negative factor.

4. $12x^2 - 4x - 5$

Step 1 $m \cdot n = 12(-5) = -60$ and $m + n = -4$

m and n are 6 and -10.

Step 2 $= 12x^2 + \overbrace{6x - 10x}^{-4x} - 5$ Replace bx with mx and nx

Step 3 $= (12x^2 + 6x) + (-10x - 5)$ Group the first two terms and the last two terms

$= 6x(2x + 1) - 5(2x + 1)$ Factor out what is common to each pair

Step 4 $= (2x + 1)(6x - 5)$ Factor out the common quantity

5. $6x^2 - 9x - 4$

$m \cdot n = 6 \cdot (-4) = -24$ and $m + n = -9$

Our m and n values are not obvious by inspection.

Note If you cannot determine the m and n values by inspection, then you should use the following systematic procedure to list all the possible factorizations of $a \cdot c$. This way you will either find m and n or verify that the trinomial will not factor using integers.

1. Take the natural numbers 1,2,3,4, · · · and divide them into the $a \cdot c$ product. Those that divide into the product evenly we write as a factorization using the correct m and n signs.

Factorization of -24, where the negative sign goes with the factor with the greater absolute value

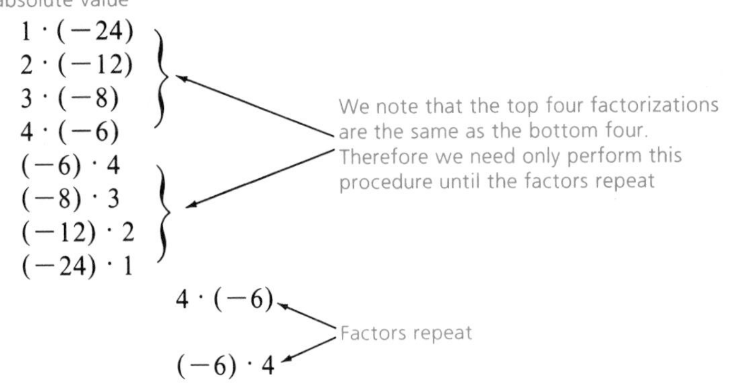

1 · (−24)
2 · (−12)
3 · (−8)
4 · (−6)
(−6) · 4
(−8) · 3
(−12) · 2
(−24) · 1
We note that the top four factorizations are the same as the bottom four. Therefore we need only perform this procedure until the factors repeat
4 · (−6)
(−6) · 4
Factors repeat

2. Find the sum of the factorizations of $a \cdot c$. If there is a sum equal to b, the trinomial will factor. If there is no sum equal to b, then the trinomial will not factor with integer coefficients.

Factorizations of -24	Sum of the factors of -24	
$1 \cdot (-24)$	$1 + (-24) = -23$	
$2 \cdot (-12)$	$2 + (-12) = -10$	
$3 \cdot (-8)$	$3 + (-8) = -5$	← Passed -9
$4 \cdot (-6)$	$4 + (-6) = -2$	
	No sum equals -9	

Since none of the factorizations of -24 add to -9, there is no pair of integers (m and n) and the trinomial will not factor.

Note Regardless of the signs of m and n, the column of values of the sum of the factors will either be increasing or decreasing. Therefore, once the desired value has been passed, the process can be stopped and the trinomial will not factor.

6. $24x^2 - 39x - 18$

Before we attempt to apply any factoring rule, recall that we must always factor out what is common to each term. Therefore we have

$24x^2 - 39x - 18 = 3(8x^2 - 13x - 6)$ Common factor of 3

Now we are ready to factor the trinomial $8x^2 - 13x - 6$.

Step 1 $m \cdot n = 8(-6) = -48$ and $m + n = -13$
m and n are 3 and -16.

$$\overbrace{}^{-13x}$$

Step 2 $= 3(8x^2 + 3x - 16x - 6)$ Replace bx with mx and nx

Step 3 $= 3[(8x^2 + 3x) + (-16x - 6)]$ Group the first two terms and the last two terms

$= 3[x(8x + 3) - 2(8x + 3)]$ Factor out what is common to each pair

Step 4 $= 3(8x + 3)(\quad - 2)$ Factor out the common quantity

Note In example 6, we factored out 3 that was common to all the original terms. A common error is to forget to include it as one of the factors in the answer.

▶ ***Quick check*** Factor $6x^2 + 23x + 15$ and $12x^2 + 12x - 9$ ■

Factoring by inspection—an alternative approach

In the beginning of this section, we studied a systematic procedure for determining if a trinomial will factor and how to factor it. In many instances, we can determine how the trinomial will factor by inspecting the problem rather than by applying this procedure.

Factoring by inspection is accomplished as follows: Factor $7x + 2x^2 + 3$.

Step 1 Write the trinomial in standard form.

$2x^2 + 7x + 3$ Arrange terms in descending powers of x

Step 2 Determine the possible combinations of first-degree factors of the first term.

$(2x \quad)(x \quad)$ — The only factorization of $2x^2$ is $2x \cdot x$

Step 3 Combine with the factors of step 2 all the possible factors of the third term.

$(2x \quad 3)(x \quad 1)$ — The only factorization of 3 is $3 \cdot 1$
$(2x \quad 1)(x \quad 3)$

Step 4 Determine the possible symbol (+ or −) between the terms in each binomial.

$(2x + 3)(x + 1)$
$(2x + 1)(x + 3)$

The rules of real numbers given in chapter 1 provide the answer to step 4.

1. If the third term is preceded by a + sign and the middle term is preceded by a + sign, then the symbols will be

   $(\quad + \quad)(\quad + \quad)$

2. If the third term is preceded by a + sign and the middle term is preceded by a − sign, then the symbols will be

   $(\quad - \quad)(\quad - \quad)$

3. If the third term is preceded by a − sign, then the symbols will be

   $(\quad + \quad)(\quad - \quad)$

   or

   $(\quad - \quad)(\quad + \quad)$

Note It is assumed that the first term is preceded by a + sign or no sign. If it is preceded by a − sign, these rules could still be used if (−1) is first factored out of all the terms.

Step 5 Determine which factors, if any, yield the correct middle term.

$(2x + 3)(x + 1)$

inner: $+3x$, outer: $+2x$ — $(+3x) + (+2x) = +5x$

$(2x + 1)(x + 3)$

inner: $+x$, outer: $+6x$ — $(+x) + (+6x) = +7x$ — Correct middle term

The second set of factors gives us the correct middle term. Therefore

$(2x + 1)(x + 3)$ is the factorization of $2x^2 + 7x + 3$.

Example 4–3 B

Factor completely the following trinomials by inspection.

1. $6x^2 + 17x + 5$

Step 1 $6x^2 + 17x + 5$ — Standard form

Step 2 $(6x \quad)(x \quad)$
$(3x \quad)(2x \quad)$ — $6x^2 = 3x \cdot 2x$ or $6x \cdot x$

Step 3 $(6x \quad 5)(x \quad 1)$
$(6x \quad 1)(x \quad 5)$
$(3x \quad 5)(2x \quad 1)$
$(3x \quad 1)(2x \quad 5)$ — The only factorization of 5 is $5 \cdot 1$

Step 4 $(6x + 5)(x + 1)$
$(6x + 1)(x + 5)$
$(3x + 5)(2x + 1)$
$(3x + 1)(2x + 5)$ — Using the rules of signed numbers, determine the possible signs between the terms

Step 5 $(6x + 5)(x + 1)$ — inner: $+5x$, outer: $+6x$ — $(+5x) + (+6x) = +11x$

$(6x + 1)(x + 5)$ — inner: $+x$, outer: $+30x$ — $(+x) + (+30x) = +31x$

$(3x + 5)(2x + 1)$ — inner: $+10x$, outer: $+3x$ — $(+10x) + (+3x) = +13x$

$(3x + 1)(2x + 5)$ — inner: $+2x$, outer: $+15x$ — $(+2x) + (+15x) = +17x$ — Correct middle term

The last set of factors gives us the correct middle term. Hence $(3x + 1)(2x + 5)$ is the factorization of $6x^2 + 17x + 5$.

2. $4x^2 - 5x + 1$

Step 1 $4x^2 - 5x + 1$ — Standard form

Step 2 $(4x \quad)(x \quad)$
$(2x \quad)(2x \quad)$ — $4x^2 = 2x \cdot 2x$ or $4x \cdot x$

Step 3 $(4x \quad 1)(x \quad 1)$
$(2x \quad 1)(2x \quad 1)$ — The only factorization of 1 is $1 \cdot 1$

Step 4 $(4x - 1)(x - 1)$
$(2x - 1)(2x - 1)$ — Determine the possible signs between the terms

Step 5 $(4x - 1)(x - 1)$

$-x$ (inner product), $-4x$ (outer product)

$(-x) + (-4x) = -5x$ Correct middle term

$(2x - 1)(2x - 1)$

$-2x$ (inner product), $-2x$ (outer product)

$(-2x) + (-2x) = -4x$

The first set of factors gives us the correct factorization.

$4x^2 - 5x + 1 = (4x - 1)(x - 1)$

3. $13x - 5 + 6x^2$

Step 1 $6x^2 + 13x - 5$ Write in standard form

Step 2 $(6x \quad)(x \quad)$ $6x^2 = 6x \cdot x$ or $2x \cdot 3x$

$(2x \quad)(3x \quad)$

Step 3 $(6x \quad 5)(x \quad 1)$

$(6x \quad 1)(x \quad 5)$ The only factorization of 5 is $5 \cdot 1$

$(2x \quad 5)(3x \quad 1)$

$(2x \quad 1)(3x \quad 5)$

Step 4 $(6x + 5)(x - 1)$ or $(6x - 5)(x + 1)$ Determine the possible signs between the terms

$(6x + 1)(x - 5)$ or $(6x - 1)(x + 5)$

$(2x + 5)(3x - 1)$ or $(2x - 5)(3x + 1)$

$(2x + 1)(3x - 5)$ or $(2x - 1)(3x + 5)$

Step 5 $(6x + 5)(x - 1)$ or $(6x - 5)(x + 1)$

$+5x$, $-6x$ $\quad$ $-5x$, $+6x$

$(+5x) + (-6x) = -x$ or $(-5x) + (+6x) = +x$

$(6x + 1)(x - 5)$ or $(6x - 1)(x + 5)$

$+x$, $-30x$ $\quad$ $-x$, $+30x$

$(+x) + (-30x) = -29x$ or $(-x) + (+30x) = +29x$

$(2x + 5)(3x - 1)$ or $(2x - 5)(3x + 1)$

$+15x$, $-2x$ $\quad$ $-15x$, $+2x$

$(+15x) + (-2x) = +13x$ or $(-15x) + (+2x) = -13x$

Correct middle term

$$(2x + 1)(3x - 5) \text{ or } (2x - 1)(3x + 5)$$

$+3x$ $\quad$ $-3x$

$-10x$ $\quad$ $+10x$

$$(+3x) + (-10x) = -7x \text{ or } (-3x) + (+10x) = +7x$$

The factorization of $6x^2 + 13x - 5$ is $(2x + 5)(3x - 1)$. ■

Mastery points

Can you

- Determine two integers whose product is one number and whose sum is another number?
- Recognize when the trinomial $ax^2 + bx + c$ will factor and when it will not?
- Factor trinomials of the form $ax^2 + bx + c$?
- Always remember to look for the greatest common factor before applying any of the factoring rules?

Exercise 4–3

Factor completely each trinomial. If a trinomial will not factor, so state. See examples 4–3 A and B.

Example $6x^2 + 23x + 15$

Solution $m \cdot n = 6 \cdot 15 = 90$ — Determine m and n

$m + n = 23$

m and n are 5 and 18.

$$= 6x^2 + \overbrace{5x + 18x}^{23x} + 15$$ Replace bx with mx and nx

$= (6x^2 + 5x) + (18x + 15)$ — Group first two terms and last two terms

$= x(6x + 5) + 3(6x + 5)$ — Factor out what is common to each pair

$= (6x + 5)(x + 3)$ — Factor out the common quantity

Example $12x^2 + 12x - 9$

Solution $= 3(4x^2 + 4x - 3)$ — Common factor of 3

$m \cdot n = 4(-3) = -12$

$m + n = 4$

m and n are -2 and 6. — Determine m and n

$$= 3[4x^2 \overbrace{- 2x + 6x}^{4x} - 3]$$ Replace bx with mx and nx

$= 3[2x(2x - 1) + 3(2x - 1)]$ — Group first two terms and last two terms, factor out what is common to each pair

$= 3(2x - 1)(2x + 3)$ — Factor out the common quantity

1. $2x^2 + x - 6$ **2.** $3x^2 + 7x - 6$ **3.** $2x^2 + 3x + 1$ **4.** $4x^2 - 5x + 1$

5. $2R^2 - 7R + 6$ **6.** $R^2 - 4R + 6$ **7.** $5x^2 - 7x - 6$ **8.** $2x^2 - x - 1$

9. $9x^2 - 6x + 1$
10. $8x^2 - 17x + 2$
11. $5x^2 + 4x + 6$
12. $2x^2 - 11x + 12$
13. $6x^2 + 13x + 6$
14. $2r^2 + 13r + 18$
15. $4x^2 + 20x + 21$
16. $7R^2 + 20R - 3$
17. $4x^2 - 2x + 5$
18. $4x^2 - 4x - 3$
19. $9y^2 - 21y - 8$
20. $6x^2 - 23x - 4$
21. $10x^2 + 7x - 6$
22. $10x^2 + 9x + 2$
23. $2x^2 - 9x + 10$
24. $7x^2 - 3x + 6$
25. $4x^2 + 14x + 12$
26. $5R^2 - 9R - 2$
27. $4x^2 + 10x + 6$
28. $6x^2 - 17x + 12$
29. $6x^2 + 7x - 3$
30. $3x^2 + 12x + 12$
31. $2x^2 + 6x - 20$
32. $3a^2 + 8a - 4$
33. $6x^2 + 5x - 6$
34. $3x^2 - 19x + 20$
35. $4x^2 + 12x + 9$
36. $9z^2 - 30z + 25$
37. $7x^2 - 36x + 5$
38. $3x^2 + 2x + 4$
39. $15P^2 + 2P - 1$
40. $12x^2 + 13x - 4$
41. $2x^3 - 6x^2 - 20x$
42. $4x^2 + 10x + 4$
43. $2a^3 + 15a^2 + 7a$
44. $9x^2 + 27x + 8$
45. $8x^2 - 14x - 15$
46. $8x^2 - 18x + 9$

47. When a stone is thrown vertically into the air, the height S of the stone at any instant in time t is given by $S = -16t^2 + 32t - 16$. Factor the right member.

Review exercises

Use the special product rules to carry out the indicated multiplication. See section 3–2.

1. $(x - y)(x + y)$
2. $(3a - 2b)(3a + 2b)$
3. $(x - y)^2$
4. $(5a + 4b)(5a - 4b)$
5. $(2a + b)^2$
6. $(4x - y)^2$
7. $(x^2 + 1)(x^2 - 1)$
8. $(a^2 - 4)(a^2 + 4)$

4–4 ■ *Factoring the difference of two squares and perfect square trinomials*

In section 3–2, we saw that the product of $(a + b)(a - b)$ was $a^2 - b^2$. We refer to the indicated product $(a + b)(a - b)$ as the *product* of the *sum* and *difference* of the same two terms. Notice that in one factor we *add* the terms and in the other we find the *difference* between these same terms. The product will *always* be the *difference of the squares* of the two terms. To factor the **difference of two squares,** we reverse the formula from section 3–2.

Factoring the difference of two squares

$$a^2 - b^2 = (a + b)(a - b)$$

Concept

(first term)² − (second term)² factors into
(first term + second term)(first term − second term)

To use this factoring technique, we must be able to recognize **perfect squares.**

Example 4–4 A

Write the following as a quantity squared, if possible.

1. $16 = 4 \cdot 4$ — 16 is a perfect square
   $= (4)^2$ — 16 can be written as 4 squared

2. $x^2 = x \cdot x$ — x is written as a factor twice
   $= (x)^2$ — Writing x^2 as $(x)^2$ shows this is a perfect square

3. $25a^2 = 5a \cdot 5a$ — 25 is $5 \cdot 5$ and a^2 is $a \cdot a$
   $= (5a)^2$ — It is now rewritten as a square

4. $9y^4 = 3y^2 \cdot 3y^2$ — 9 is $3 \cdot 3$ and y^4 could be written as $y^2 \cdot y^2$
   $= (3y^2)^2$ — It is now rewritten as a square

▶ ***Quick check*** Write 64 and $9x^4$ each as a quantity squared. ■

This is the procedure we use for factoring the difference of two squares.

Factoring the difference of two squares

Step 1 Identify that we have a perfect square *minus* another perfect square.

Step 2 Rewrite the problem as a first term squared minus a second term squared.

$$(\text{first term})^2 - (\text{second term})^2$$

Step 3 Factor the problem into the first term plus the second term times the first term minus the second term.

$$(\text{first term} + \text{second term})(\text{first term} - \text{second term})$$

Example 4–4 B

Write the following in completely factored form.

	Step 1 Identify		*Step 2* Rewrite		*Step 3* Factor
	$a^2 - b^2$	$=$	$(a)^2 - (b)^2$	$=$	$(a + b)(a - b)$
1.	$x^2 - 9$	$=$	$(x)^2 - (3)^2$	$=$	$(x + 3)(x - 3)$
2.	$4a^2 - b^2$	$=$	$(2a)^2 - (b)^2$	$=$	$(2a + b)(2a - b)$
3.	$4p^2 - 25v^2$	$=$	$(2p)^2 - (5v)^2$	$=$	$(2p + 5v)(2p - 5v)$
4.	$r^4 - 49$	$=$	$(r^2)^2 - (7)^2$	$=$	$(r^2 + 7)(r^2 - 7)$

▶ ***Quick check*** Write $t^2 - 64$ and $4a^2 - b^2c^2$ in completely factored form. ■

Our first step in any factoring problem is to look for any common factors. Often an expression that does not appear to be factorable becomes so by taking out the greatest common factor. When we have applied a factoring rule to a problem, *we must inspect all parts of our answer to make sure that nothing will factor further.*

Example 4–4 C

Write the following in completely factored form.

1. $2x^2 - 18y^2 = 2(x^2 - 9y^2)$ — Factor out what is common, 2
   $= 2[(x)^2 - (3y)^2]$ — Identify and rewrite as squares
   $= 2(x + 3y)(x - 3y)$ — Factor and inspect the factors

2. $3w^2 - 48 = 3(w^2 - 16)$ — Common factor of 3
$= 3[(w)^2 - (4)^2]$ — Identify and rewrite
$= 3(w + 4)(w - 4)$ — Factor and inspect the factors

3. $5a^4 - 45a^2b^2 = 5a^2(a^2 - 9b^2)$ — Common factor of $5a^2$
$= 5a^2[(a)^2 - (3b)^2]$ — Identify and rewrite
$= 5a^2(a + 3b)(a - 3b)$ — Factor and inspect the factors

4. $a^4 - 16 = (a^2)^2 - (4)^2$ — Identify and rewrite
$= (a^2 + 4)(a^2 - 4)$ — Factor and inspect the factors
$= (a^2 + 4)[(a)^2 - (2)^2]$ — Identify and rewrite
$= (a^2 + 4)(a + 2)(a - 2)$ — Factor and inspect

Note In example 4, $a^2 + 4$ is called the *sum of two squares*. This will *not* factor using integers.

5. $2x^4 - 162 = 2(x^4 - 81)$ — Common factor of 2
$= 2[(x^2)^2 - (9)^2]$ — Identify and rewrite
$= 2(x^2 + 9)(x^2 - 9)$ — Factor and inspect
$= 2(x^2 + 9)[(x)^2 - (3)^2]$ — Identify and rewrite
$= 2(x^2 + 9)(x + 3)(x - 3)$ — Factor and inspect

Note A common error in examples 1, 2, 3, and 5 is to factor out something that is common but forget to include it as a factor in the final answer. ■

Perfect square trinomials

In section 3–2, two of the special products that we studied were the squares of a binomial. We will now restate those special products.

$$a^2 + 2ab + b^2 = (a + b)^2$$

and

$$a^2 - 2ab + b^2 = (a - b)^2$$

The right members of the equations are called the squares of binomials, and the left members are called perfect square trinomials. Perfect square trinomials can always be factored by our factoring procedure. However if we observe that the first and last terms of a trinomial are perfect squares, we should see if the trinomial will factor as the square of a binomial. To factor a trinomial as a perfect square trinomial, the following three conditions need to be met.

Necessary conditions for a perfect square trinomial

1. The first term must have a positive coefficient and be a perfect square, a^2.
2. The last term must have a positive coefficient and be a perfect square, b^2.
3. The middle term must be twice the product of the bases of the first and last terms, $2ab$ or $-2ab$.

We observe that

$$9x^2 + 12x + 4$$
$$= (3x)^2 + 2(3x)(2) + (2)^2$$

Condition 1 Condition 3 Condition 2

Therefore it is a perfect square trinomial and factors into

$$(3x + 2)^2$$

■ Example 4–4 D

The following examples show the factoring of some other perfect square trinomials.

	Condition 1		*Condition 3*		*Condition 2*		*Square of a binomial*
1. $4x^2 + 20x + 25 =$	$(2x)^2$	$+$	$2(2x)(5)$	$+$	$(5)^2$	$=$	$(2x + 5)^2$
2. $9x^2 - 6x + 1 =$	$(3x)^2$	$-$	$2(3x)(1)$	$+$	$(1)^2$	$=$	$(3x - 1)^2$
3. $16x^2 + 24x + 9 =$	$(4x)^2$	$+$	$2(4x)(3)$	$+$	$(3)^2$	$=$	$(4x + 3)^2$
4. $9y^2 - 30y + 25 =$	$(3y)^2$	$-$	$2(3y)(5)$	$+$	$(5)^2$	$=$	$(3y - 5)^2$

■

Mastery points

Can you

- Identify and rewrite a perfect square?
- Factor the difference of two squares?
- Remember that the sum of two squares will not factor?
- Factor out any common factors before applying other factoring rules?
- Inspect all factors to make sure the problem is completely factored?
- Factor perfect square trinomials?

Exercise 4–4

Write the following as a quantity squared, if possible. See example 4–4 A.

Examples 64

Solutions $= 8 \cdot 8$ Identify
$= (8)^2$ Rewrite

$9x^4$

$= 3x^2 \cdot 3x^2$ Identify
$= (3x^2)^2$ Rewrite

1. 36 **2.** 25 **3.** c^2 **4.** e^2
5. $16x^2$ **6.** $49b^2$ **7.** $4z^4$ **8.** $25b^2$

Write in completely factored form. See examples 4–4 B, C, and D.

Examples $t^2 - 64$ Identify

Solutions $= (t)^2 - (8)^2$ Rewrite
$= (t + 8)(t - 8)$ Factor

$4a^2 - b^2c^2$ Identify

$= (2a)^2 - (bc)^2$ Rewrite
$= (2a + bc)(2a - bc)$ Factor

9. $x^2 - 1$
10. $x^2 - 25$
11. $a^2 - 4$
12. $r^2 - s^2$
13. $9 - E^2$
14. $49 - R^2$
15. $1 - k^2$
16. $4y^2 - 9$
17. $9b^2 - 16$
18. $x^2 - 16z^2$
19. $b^2 - 36c^2$
20. $16x^2 - y^2$
21. $4a^2 - 25b^2$
22. $16a^2 - b^2$
23. $25p^2 - 81$
24. $r^2 - 4s^2$
25. $8x^2 - 32y^2$
26. $3a^2 - 27b^2$
27. $5r^2 - 125s^2$
28. $20 - 5b^2$
29. $50 - 2x^2$
30. $x^2y^2 - 4z^2$
31. $r^2s^2 - 25t^2$
32. $a^4 - 25$
33. $x^4 - 9$
34. $x^4 - 1$
35. $r^4 - 81$
36. $16t^4 - 1$
37. $49x^2 - 64y^4$
38. $125p^2 - 20v^2$
39. $98x^2y^2 - 50p^2c^2$
40. $a^2 + 10a + 25$
41. $c^2 - 14c + 49$
42. $b^2 + 8b + 16$
43. $a^2 + 6a + 9$
44. $x^2 - 12x + 36$
45. $y^2 - 6y + 9$
46. $a^2 + 6ab + 9b^2$
47. $4a^2 - 12ab + 9b^2$
48. $x^2 - 16xy + 64y^2$
49. $9c^2 - 12cd + 4d^2$
50. $9a^2 - 30ab + 25b^2$
51. In engineering, the equation of transverse shearing stress in a rectangular beam is given by

$$T = \frac{V}{8I}(h^2 - 4v_1^2)$$

Factor the right member.

Review exercises

Factor completely. See sections 4–1, 4–2, 4–3, and 4–4.

1. $x^2 + 8x + 12$
2. $49a^2 - 81$
3. $3ax + bx - 12ay - 4by$
4. $2x^3 + 14x^2 + 24x$
5. $10a^2 + 21a + 9$
6. $4a^2 - 20a + 25$
7. $x^2y^2 + 8xy + 15$
8. $x^2 + 4xy + 4y^2$

4–5 ■ Other types of factoring

The difference of two cubes

In section 4–4, we factored expressions that involved the difference of two squares. To factor these types of expressions, we identified the two terms as perfect squares and applied the procedure. In this section, we will factor the *sum and difference of two cubes* in a similar fashion.

Consider the indicated product of $(a - b)(a^2 + ab + b^2)$. If we carry out the multiplication, we have

$$\begin{aligned}(a - b)(a^2 + ab + b^2) &= a^3 + a^2b + ab^2 - a^2b - ab^2 - b^3\\ &= a^3 - b^3\end{aligned}$$

Therefore $(a - b)(a^2 + ab + b^2) = a^3 - b^3$ and $(a - b)(a^2 + ab + b^2)$ is the factored form of $a^3 - b^3$.

The difference of two cubes factors as

$$a^3 - b^3 = (a - b)(a^2 + ab + b^2)$$

Concept

If we are able to write a two-term polynomial as a first term cubed minus a second term cubed, then it will factor as the difference of two cubes.

$$(\text{1st term})^3 - (\text{2nd term})^3$$
$$= (\text{1st term} - \text{2nd term})[(\text{1st term})^2 + (\text{1st term})(\text{2nd term}) + (\text{2nd term})^2]$$

To use this factoring technique, we must be able to recognize **perfect cubes.**

■ ***Example 4–5 A***

Write the following as a quantity cubed, if possible.

1. $27 = 3 \cdot 3 \cdot 3$ — 27 is a perfect cube
   $= (3)^3$ — 27 can be written as 3 cubed
2. $a^3 = a \cdot a \cdot a$ — a is written as a factor three times
   $= (a)^3$ — Writing a^3 as $(a)^3$ shows this is a perfect cube
3. $8a^3 = 2a \cdot 2a \cdot 2a$ — 8 is $2 \cdot 2 \cdot 2$ and a^3 is $a \cdot a \cdot a$
   $= (2a)^3$ — It is now rewritten as a cube
4. $64x^6 = 4x^2 \cdot 4x^2 \cdot 4x^2$ — 64 is $4 \cdot 4 \cdot 4$ and x^6 is $x^2 \cdot x^2 \cdot x^2$
   $= (4x^2)^3$ — It is now rewritten as a cube

▶ ***Quick check*** Write the following as a quantity cubed, if possible. $27y^3$ ■

This is the procedure we use for factoring the difference of two cubes.

Step 1 Identify that we have a perfect cube minus another perfect cube.

Step 2 Rewrite the problem as a first term cubed minus a second term cubed.

$$(\text{1st term})^3 - (\text{2nd term})^3$$

Step 3 Factor the expression into the first term minus the second term, times the first term squared plus the first term times the second term plus the second term squared.

$$(\text{1st term} - \text{2nd term})[(\text{1st term})^2 + (\text{1st term} \cdot \text{2nd term}) + (\text{2nd term})^2]$$

■ ***Example 4–5 B***

Factor completely.

1. $x^3 - 27$ We rewrite x^3 as a cube and 27 as a cube.

   $x^3 - 27 = (x)^3 - (3)^3$

   The first term is x and the second term is 3. Then we write the procedure for factoring the difference of two cubes.

   $$(\quad - \quad)[(\quad)^2 + (\quad)(\quad) + (\quad)^2]$$
   $$\uparrow \text{1st} \quad \uparrow \text{2nd} \quad \uparrow \text{1st} \qquad \uparrow \text{1st} \quad \uparrow \text{2nd} \qquad \uparrow \text{2nd}$$

Now substitute x where the first term is in the procedure and 3 where the second term is.

$$(x - 3)[(x)^2 + (x)(3) + (3)^2]$$

↑ 1st ↑ 2nd ↑ 1st ↑ 1st ↑ 2nd ↑ 2nd

Finally we simplify.

$$(x - 3)(x^2 + 3x + 9)$$

Therefore $x^3 - 27 = (x - 3)(x^2 + 3x + 9)$.

2. $8x^3 - y^3 = (2x)^3 - (y)^3$ — First term is $2x$ and second term is y.
Then $(\quad - \quad)[(\quad)^2 + (\quad)(\quad) + (\quad)^2]$ — Factoring procedure ready for substitution
$= (2x - y)[(2x)^2 + (2x)(y) + (y)^2]$ — The first term is $2x$, the second term is y
$= (2x - y)(4x^2 + 2xy + y^2)$ — Simplify within the second group

3. $2a^3 - 54b^3 = 2(a^3 - 27b^3)$ — Factor out the common quantity of 2
$2(a^3 - 27b^3) = 2[(a)^3 - (3b)^3]$ — Rewrite as cubes
Then $2(\quad - \quad)[(\quad)^2 + (\quad)(\quad) + (\quad)^2]$ — Factoring procedure ready for substitution
$= 2(a - 3b)[(a)^2 + (a)(3b) + (3b)^2]$ — The first term is a, the second term is $3b$
$= 2(a - 3b)(a^2 + 3ab + 9b^2)$ — Simplify within the second group

4. $a^{15} - 64b^3 = (a^5)^3 - (4b)^3$ — Rewrite as cubes
Then $(\quad - \quad)[(\quad)^2 + (\quad)(\quad) + (\quad)^2]$ — Factoring procedure ready for substitution
$= (a^5 - 4b)[(a^5)^2 + (a^5)(4b) + (4b)^2]$ — The first term is a^5, the second term is $4b$
$= (a^5 - 4b)(a^{10} + 4a^5b + 16b^2)$ — Simplify within the second group

Note In example 4, we observe that a number raised to a power that is a multiple of 3 can be written as a cube by dividing the exponent by 3. The quotient is the exponent of the number inside the parentheses and the 3 is the exponent outside the parentheses. For example, $y^{12} = (y^4)^3$ or $z^{24} = (z^8)^3$.

▶ ***Quick check*** Factor $16R^3 - 54$ ■

The sum of two cubes

If we carry out the indicated multiplication in $(a + b)(a^2 - ab + b^2)$, we have

$$\begin{aligned}(a + b)(a^2 - ab + b^2) &= a^3 - a^2b + ab^2 + a^2b - ab^2 + b^3 \\ &= a^3 + b^3\end{aligned}$$

Therefore $(a + b)(a^2 - ab + b^2) = a^3 + b^3$ and $(a + b)(a^2 - ab + b^2)$ is the factored form of $a^3 + b^3$.

The sum of two cubes factors as

$$a^3 + b^3 = (a + b)(a^2 - ab + b^2)$$

Concept

If we are able to write a two-term polynomial as a first term cubed plus a second term cubed, then it will factor as the sum of two cubes.

$$(\text{1st term})^3 + (\text{2nd term})^3$$
$$= (\text{1st term} + \text{2nd term})[(\text{1st term})^2 - (\text{1st term})(\text{2nd term}) + (\text{2nd term})^2]$$

This is the procedure we use for factoring the sum of two cubes.

> ***Step 1*** Identify that we have a perfect cube plus another perfect cube.
> ***Step 2*** Rewrite the problem as a first term cubed plus a second term cubed.
>
> $$(\text{1st term})^3 + (\text{2nd term})^3$$
>
> ***Step 3*** Factor the expression into the first term plus the second term, times the first term squared minus the first term times the second term plus the second term squared.
>
> $$(\text{1st term} + \text{2nd term})[(\text{1st term})^2 - (\text{1st term})(\text{2nd term}) + (\text{2nd term})^2]$$

■ *Example 4–5 C*

Factor completely.

1. $a^3 + 8 = (a)^3 + (2)^3$

We now write the procedure for the sum of two cubes.

$$(\ \ +\ \ )[(\ \ )^2 - (\ \ )(\ \ ) + (\ \ )^2]$$

↑ 1st ↑ 2nd ↑ 1st ↑ 1st ↑ 2nd ↑ 2nd

Substituting,

$$(a + 2)[(a)^2 - (a)\,(2) + (2)^2]$$

↑ 1st ↑ 2nd ↑ 1st ↑ 1st ↑ 2nd ↑ 2nd

simplifying, $(a + 2)(a^2 - 2a + 4)$

therefore $a^3 + 8 = (a + 2)(a^2 - 2a + 4)$

2. $x^3 + 125 = (x)^3 + (5)^3$ — Rewrite as cubes

Then $(\ \ +\ \ )[(\ \ )^2 - (\ \ )(\ \ ) + (\ \ )^2]$ — Factoring procedure ready for substitution

$= (x + 5)[(x)^2 - (x)(5) + (5)^2]$ — The first term is x, the second term is 5

$= (x + 5)(x^2 - 5x + 25)$ — Simplify within the second group

3. $8a^3 + b^{21} = (2a)^3 + (b^7)^3$ — Rewrite as cubes

Then $(\ \ +\ \ )[(\ \ )^2 - (\ \ )(\ \ ) + (\ \ )^2]$ — Factoring procedure ready for substitution

$= (2a + b^7)[(2a)^2 - (2a)(b^7) + (b^7)^2]$ — The first term is $2a$, the second term is b^7

$= (2a + b^7)(4a^2 - 2ab^7 + b^{14})$ — Simplify within the second group

4. $x^3y^3 + z^3 = (xy)^3 + (z)^3$ — Rewrite as cubes

Then $(\ \ +\ \ )[(\ \ )^2 - (\ \ )(\ \ ) + (\ \ )^2]$ — Factoring procedure ready for substitution

$= (xy + z)[(xy)^2 - (xy)(z) + (z)^2]$ — The first term is xy, the second term is z

$= (xy + z)(x^2y^2 - xyz + z^2)$ — Simplify within the second group

▶ ***Quick check*** Factor $27x^3 + y^3$ ■

Mastery points

Can you

- Identify and rewrite a perfect cube?
- Factor the sum or difference of two cubes?

Exercise 4–5

Write the following as a quantity cubed, if possible. See example 4–5 A.

Example $27y^3$

Solution $= 3y \cdot 3y \cdot 3y$ — 27 is $3 \cdot 3 \cdot 3$ and y^3 is $y \cdot y \cdot y$
$= (3y)^3$ — It is now rewritten as a cube

1. 64 **2.** 8 **3.** 125 **4.** 1 **5.** $27x^3$
6. $64a^3$ **7.** a^6 **8.** x^9 **9.** $8b^{15}$ **10.** $64c^{21}$

Factor completely. If an expression will not factor, so state. See examples 4–5 B and C.

Example $16R^3 - 54$

Solution
$= 2[8R^3 - 27]$ — Common factor of 2
$= 2[(2R)^3 - (3)^3]$ — Rewrite as cubes
$2(\quad - \quad)[(\quad)^2 + (\quad)(\quad) + (\quad)^2]$ — Factoring procedure ready for substitution
$= 2(2R - 3)[(2R)^2 + (2R)(3) + (3)^2]$ — The first term is 2*R*, the second term is 3
$= 2(2R - 3)(4R^2 + 6R + 9)$ — Simplify within the second group

Example $27x^3 + y^3$

Solution
$= (3x)^3 + (y)^3$ — Rewrite as cubes
$(\quad + \quad)[(\quad)^2 - (\quad)(\quad) + (\quad)^2]$ — Factoring procedure ready for substitution
$= (3x + y)[(3x)^2 - (3x)(y) + (y)^2]$ — The first term is 3*x*, the second term is *y*
$= (3x + y)(9x^2 - 3xy + y^2)$ — Simplify within the second group

11. $r^3 + s^3$ **12.** $L^3 + 8$ **13.** $8x^3 + y^3$ **14.** $27r^3 + 8$
15. $h^3 - k^3$ **16.** $p^3 - q^3$ **17.** $a^3 - 8$ **18.** $b^3 + 64$
19. $x^3 - 8y^3$ **20.** $27a^3 - b^3$ **21.** $64x^3 - y^3$ **22.** $r^3 - 27$
23. $27x^3 - 8y^3$ **24.** $64a^3 - 8$ **25.** $8a^3 + 27b^3$ **26.** $64s^3 + 1$
27. $2a^3 + 16$ **28.** $3x^3 + 81$ **29.** $2x^3 - 16$ **30.** $81a^3 - 3b^3$
31. $x^5 + 27x^2y^3$ **32.** $16a^3 + 2b^3$ **33.** $x^6 + y^3$ **34.** $x^3 + y^9$
35. $a^9 - b^3$ **36.** $a^6 - 8$ **37.** $x^{12} - 27$ **38.** $x^{15} + 64$
39. $8a^2b^3 - a^5$ **40.** $2x^3 - 54y^3$ **41.** $54r^3 + 2s^3$ **42.** $b^5 + 64b^2c^3$
43. $x^3y^3 - z^3$ **44.** $x^3y^9 - 1$ **45.** $a^{15}b^6 - 8c^9$ **46.** $x^{18}y^9 - 27z^3$
47. $a^3b^3 + 8$ **48.** $x^3y^6 + z^3$ **49.** $x^9y^{12} + z^{15}$ **50.** $a^{12}b^{15} + c^{24}$

Review exercises

Write in completely factored form. See sections 4–1, 4–2, 4–3, and 4–4.

1. $a^2 - 7a + 10$
2. $6ax + 2bx - 3ay - by$
3. $x^2 + 4xy + 4y^2$
4. $6a^2 - ab - b^2$
5. $5a^3 - 40a^2 + 75a$
6. $6x^2 + x - 12$

4–6 ■ Factoring: A general strategy

In this section, we will review the different methods of factoring that we have studied in the previous sections. The following outline gives a general strategy for factoring polynomials.

I. Factor out any common factors.
Examples
1. $5a^3 - 25a^2 = 5a^2(a - 5)$
2. $c(a - 2b) + 2d(a - 2b) = (a - 2b)(c + 2d)$

II. Count the number of terms.

A. Two terms: Check to see if the polynomial is the difference of two squares, the difference of two cubes, or the sum of two cubes.
Examples
1. $a^2 - 16b^2 = (a - 4b)(a + 4b)$ — Difference of two squares
2. $8a^3 - b^3 = (2a - b)(4a^2 + 2ab + b^2)$ — Difference of two cubes
3. $m^3 + 64n^3 = (m + 4n)(m^2 - 4mn + 16n^2)$ — Sum of two cubes

B. Three terms: Check to see if the polynomial is a perfect square trinomial. If it is not, use one of the general methods for factoring a trinomial.
Examples
1. $a^2 + 6a + 9 = (a + 3)^2$ — Perfect square trinomial
2. $a^2 + 5a - 14 = (a + 7)(a - 2)$ — General trinomial, leading coefficient of 1
3. $6a^2 + 7a - 20 = (2a + 5)(3a - 4)$ — General trinomial, leading coefficient other than 1

C. Four terms: Check to see if we can factor by grouping.
Examples
1. $ac + 3a - 2bc - 6b = (a - 2b)(c + 3)$
2. $a^3 + 2a^2 - 3a - 6 = (a^2 - 3)(a + 2)$

III. Check to see if any of the factors we have written can be factored further. Any common factors that were missed in part I can still be factored out here.
Examples
1. $c^4 - 11c^2 + 28 = (c^2 - 4)(c^2 - 7)$
$= (c - 2)(c + 2)(c^2 - 7)$ — Difference of two squares
2. $4a^2 - 36b^2 = (2a - 6b)(2a + 6b)$
$= 2(a - 3b)2(a + 3b)$
$= 4(a - 3b)(a + 3b)$ — Overlooked common factor

The following examples illustrate our strategy for factoring polynomials.

■ Example 4–6 A

Completely factor the following polynomials.

1. $3x^3 - 3xy^2$

I. First we look for any common factors.

$3x^3 - 3xy^2 = 3x(x^2 - y^2)$ Common factor of $3x$

II. The factor $x^2 - y^2$ has two terms and is the difference of two squares.

$x^2 - y^2 = (x - y)(x + y)$ Factoring the binomial

III. After checking to see if any of the factors will factor further, we conclude that $3x(x - y)(x + y)$ is the completely factored form. Therefore

$3x^3 - 3xy^2 = 3x(x - y)(x + y)$

2. $3ax + bx + 6ay + 2by$

I. There is no common factor (other than 1 or -1).

II. The polynomial has four terms and we factor it by grouping.

$(3ax + bx) + (6ay + 2by)$ Group in pairs

$= x(3a + b) + 2y(3a + b)$ Factor out what is common to each pair

$= (3a + b)(x + 2y)$ Factor out the common quantity

III. None of the factors will factor further.

$3ax + bx + 6ay + 2by = (3a + b)(x + 2y)$

3. $3a^2 - 2a - 8$

I. There is no common factor (other than 1 or -1).

II. The polynomial has three terms and the coefficient of a^2 is not 1. Therefore we must find m and n and factor the trinomial. $m + n = -2$ and $m \cdot n = -24$, the values for m and n are -6 and 4.

$= 3a^2 - 6a + 4a - 8$ Replace $-2a$ with $-6a + 4a$

$= (3a^2 - 6a) + (4a - 8)$ Group the first two terms and the last two terms

$= 3a(a - 2) + 4(a - 2)$ Factor out what is common to each pair

$= (a - 2)(3a + 4)$ Factor out the common quantity

III. None of the factors will factor further.

$3a^2 - 2a - 8 = (a - 2)(3a + 4)$

▶ ***Quick check*** Factor $4x^2 - 36y^2$ ■

Mastery points

Can you

- Factor out the greatest common factor?
- Factor the difference of two squares?
- Factor the difference of two cubes?
- Factor the sum of two cubes?
- Factor trinomials?
- Factor a four-term polynomial?
- Use the general strategy for factoring polynomials?

Exercise 4–6

Completely factor the following polynomials. If a polynomial will not factor, so state. See the outline of the general strategy for factoring polynomials and example 4–6 A.

Example $4x^2 - 36y^2$

Solution

I.	$4x^2 - 36y^2 = 4(x^2 - 9y^2)$	Common factor of 4
II.	$= 4(x - 3y)(x + 3y)$	Factoring the difference of two squares
III.	$4x^2 - 36y^2 = 4(x - 3y)(x + 3y)$	Completely factored form

1. $n^2 - 49$
2. $a^2 + 6a + 5$
3. $7b^2 + 36b + 5$
4. $2x^2 + 15x + 18$
5. $x^2y^2 + 2xy - 8$
6. $y^2 + 11y + 10$
7. $36 - y^2$
8. $25a^2(3b + c) + 5a(3b + c)$
9. $10a^2 - 20ab + 10b^2$
10. $a^2b^2 - 5ab - 14$
11. $4a^2 - 16b^2$
12. $12x^3y^2 - 18x^2y^2 + 16xy^4$
13. $3ax + 6ay - bx - 2by$
14. $5x^2 + 18x - 60$
15. $6x^2 + 7x - 5$
16. $9x^5y - 6x^3y^3 + 3x^2y^2$
17. $6am + 4bm - 3an - 2bn$
18. $5x^2 - 32x - 21$
19. $7b^2 + 16b - 15$
20. $3a^2 + 13a + 4$
21. $4x^2 + 17x - 15$
22. $5y^2 + 16y + 12$
23. $6x^2 - 24xy - 48y^2$
24. $4ab(x + 3y) - 8a^2b^2(x + 3y)$
25. $3x^2y(m - 4n) + 15xy^2(m - 4n)$
26. $4x^2 - 20xy + 25y^2$
27. $9a^2 - 30ab + 25b^2$
28. $80y^4 - 5y$
29. $3a^5 - 48a$
30. $3a^5b - 18a^3b^3 + 27ab^5$
31. $3a^3b^3 + 6a^2b^4 + 3ab^5$
32. $3b^2 + 8b - 91$
33. $3b^2 - 32b - 91$
34. $b^4 - 81$
35. $3ax + 6bx + 2ay + 4by$
36. $12ax + 4bx - 3ay - by$
37. $6x^2 + 11x - 2$
38. $6x^2 - 17x - 3$
39. $3x^4 - 48x^2$
40. $3x^3 + 3x^2 - 18x$
41. $y^3 + 27z^3$
42. $8b^3 - c^3$
43. $x^3 - y^9$
44. $a^3b^3 + 64$

Review exercises

Find the solution set of the following equations. See section 2–6.

1. $2x + 6 = 0$
2. $4x - 12 = 0$
3. $3x - 18 = 0$
4. $5x + 3 = 0$
5. $6x + 4 = 0$
6. $3x + 1 = 0$
7. $4x - 1 = 0$
8. $3x = 0$

4–7 ■ Solving quadratic equations by factoring

The standard form of a quadratic equation

In chapter 2, we studied linear equations, also called first-degree equations. Recall that the **degree** of an equation in one variable is the greatest exponent of that variable in any one term. In this section, we will find the solutions to an equation that contains the second, but no higher, power of that variable. Such an equation is a **second-degree equation,** also called a **quadratic equation.**

Quadratic equations can be written in the form

$$ax^2 + bx + c = 0$$

where a, b, and c are constants, $a \neq 0$.

Note It is necessary that $a \neq 0$. If $a = 0$ and $b \neq 0$, then $0 \cdot x^2 + bx + c$ becomes $bx + c = 0$, which is a linear equation.

We call $ax^2 + bx + c = 0$ where $a > 0$ the **standard quadratic form** of a quadratic equation. Notice that in standard form the terms of the left member of the equation are written in descending powers of the variable. The right member contains *only zero*. In most of our work with quadratic equations, it will be necessary to write the equation in standard quadratic form.

Solution of quadratic equations in factored form

Now suppose we have the equation $x^2 - x - 6 = 0$ stated in the factored form, $(x - 3)(x + 2) = 0$. This equation states that the product of two factors, $x - 3$ and $x + 2$, is 0. To find the necessary numbers, we use the algebraic property called the **zero product property.**

Zero product property

Given real numbers p and q, if $pq = 0$, then $p = 0$ or $q = 0$.

Concept

If the product of two factors is zero, then at least one of the factors is zero.

Extending this property, if $(x + p)(x + q) = 0$, then $x + p = 0$ or $x + q = 0$. Therefore by this property, $(x - 3)(x + 2) = 0$ only if $x - 3 = 0$ or $x + 2 = 0$. Since each of these equations is linear, we use the methods for solving linear equations. We find $x = 3$ when $x - 3 = 0$ and $x = -2$ when $x + 2 = 0$. Then 3 and -2 are solutions of the equation $(x - 3)(x + 2) = 0$.

From this discussion, we can see that to solve any equation in *factored form* whose product is 0, we solve an equation of the form $p \cdot q = 0$.

Solving an equation of the form $p \cdot q = 0$

1. Set *each* factor equal to 0.
2. Solve the resulting equations for the variable.

Solution set

In the equation $x^2 - x - 6 = 0$, we saw that the solutions of the equation, $(x - 3)(x + 2) = 0$, were 3 and -2. To express this as a solution set, we would write the solutions in any order, separated by a comma and enclosed within a pair of braces. The solution set for the equation is $\{-2,3\}$.

Example 4–7 A

Find the solution set of the following equations.

1. $(x + 5)(x - 4) = 0$ — The equation is in factored form

$x + 5 = 0$ or $x - 4 = 0$ — Set each factor equal to 0 and solve

$x = -5$ $\quad$ $x = 4$ — The solutions

Check for $x = -5$ $\qquad$ Check for $x = 4$

$(-5 + 5)(-5 - 4) = 0$ $\quad$ $(4 + 5)(4 - 4) = 0$ — Substitute the solution for x

$0 \cdot (-9) = 0$ $\quad$ $9 \cdot 0 = 0$ — Order of operations

$0 = 0$ $\quad$ $0 = 0$ — True, both solutions check

The solution set is $\{-5,4\}$.

Note In future examples, we will not always show a check of the solutions, but checking your solutions is always an important part of the problem.

2. $(x - 3)(3x + 1) = 0$ — The equation is in factored form

$x - 3 = 0$ or $3x + 1 = 0$ — Set each factor equal to 0 and solve

$x = 3$ $\quad$ $3x = -1$ — Add 3, subtract 1

$x = -\frac{1}{3}$ — Divide by 3

$x = 3$ or $x = -\frac{1}{3}$ — The solutions

The solution set is $\left\{-\frac{1}{3},3\right\}$.

3. $3x(x - 7) = 0$ — The equation is in factored form

$3x = 0$ or $x - 7 = 0$ — Set each factor equal to 0 and solve

$x = 0$ $\quad$ $x = 7$ — Dividing by 3, adding 7 gives the solutions

The solution set is $\{0,7\}$.

▶ ***Quick check*** Find the solution set of $(2x + 3)(x + 1) = 0$ ■

Solving quadratic equations by factoring

In general, to find the solution set of a quadratic equation by factoring, we use the following procedure:

Solving a quadratic equation by factoring

Step 1 Write the equation in standard quadratic form.

Step 2 Completely factor the left member.

Step 3 Set each of the factors containing the variable equal to 0 and solve the resulting equations.

Step 4 Write the solutions in a solution set.

Step 5 Check your solutions by substituting into the original equation.

Example 4–7 B

Find the solution set of the following equations. Check your solutions.

1.

$$x^2 + 5x = -6$$

$x^2 + 5x + 6 = 0$ — Write the equation in standard form

$(x + 2)(x + 3) = 0$ — Factor $x^2 + 5x + 6 = (x + 2)(x + 3)$

$x + 2 = 0$ or $x + 3 = 0$ — Set each factor equal to 0

$x = -2$ $\quad$ $x = -3$ — Solve each equation, giving the solutions

Check:

(1) Let $x = -2$

$$\begin{aligned} x^2 + 5x &= -6 \\ (-2)^2 + 5(-2) &= -6 \\ 4 - 10 &= -6 \\ -6 &= -6 \quad \text{True} \end{aligned}$$

(2) Let $x = -3$

$$\begin{aligned} x^2 + 5x &= -6 \\ (-3)^2 + 5(-3) &= -6 \\ 9 - 15 &= -6 \\ -6 &= -6 \quad \text{True} \end{aligned}$$

The solution set is $\{-3,-2\}$.

2.

$$x^2 = 2x$$

$x^2 - 2x = 0$ — Write the equation in standard form

$x(x - 2) = 0$ — Factor completely

$x = 0$ or $x - 2 = 0$ — Set each factor equal to 0

$x = 0$ $\quad$ $x = 2$ — Solve each equation, giving the solutions

Check your solutions by substituting 0 and 2 for x in the original equation.

The solution set is $\{0,2\}$.

3.

$$x^2 = 16$$

$x^2 - 16 = 0$ — Write the equation in standard form.

$(x - 4)(x + 4) = 0$ — Factor $x^2 - 16 = (x - 4)(x + 4)$

$x - 4 = 0$ or $x + 4 = 0$ — Set each factor equal to 0

$x = 4$ $\quad$ $x = -4$ — Solve each equation, giving the solutions

Check your solutions by substituting 4 and -4 for x in the original equation.

The solution set is $\{-4,4\}$.

4.

$$4x^2 = 20x - 25$$

$4x^2 - 20x + 25 = 0$ — Write the equation in standard form

$(2x - 5)^2 = 0$ — $4x^2 - 20x + 25 = (2x - 5)^2$

$2x - 5 = 0$ — Set the repeated factor equal to 0

$x = \dfrac{5}{2}$ — Solve the equation for x, giving the solution

Check your solution by substituting $\dfrac{5}{2}$ for x in the original equation.

The solution set is $\left\{\dfrac{5}{2}\right\}$.

Note In example 4, we have *two* factors, $(2x - 5)$ and $(2x - 5)$, but since they are the same, the equation has only *one* distinct solution.

5.

$$3x^2 + 3 = -6x$$

$3x^2 + 6x + 3 = 0$	Write the equation in standard form
$3(x^2 + 2x + 1) = 0$	Factor $3x^2 + 6x + 3 = 3(x + 1)^2$
$3(x + 1)^2 = 0$	Set the only distinct factor with a variable equal to 0
$x + 1 = 0$	
$x = -1$	Solve the equation, giving the solution

Check your solution by substituting -1 for x in $3x^2 + 3 = -6x$.

The solution set is $\{-1\}$.

▶ ***Quick check*** Find the solution set of each of the following: $2a^2 = 5a$, $4y^2 = 9$, $x^2 - 7x + 12 = 0$, and $12(x^2 - 2x) = -9$ ■

In conclusion, let us compare the quadratic equation with the linear equation.

1. A *linear equation* is an equation of the form $ax + b = 0$, where $a \neq 0$. A *quadratic equation* is an equation of the form $ax^2 + bx + c = 0$, where a, b, and c are constants, $a > 0$.
2. A linear equation is solved by isolating the variable. Some quadratic equations are solved by factoring and setting the linear factors equal to zero.
3. A conditional linear equation has at most *one solution.* A quadratic equation has at most *two real solutions,* which may or may not be distinct.

Mastery points

Can you

- Find the solution set of an equation in factored form whose product is equal to zero?
- Find the solution set of a quadratic equation by factoring?

Exercise 4–7

Find the solution set of the following equations. See example 4–7 A.

Example $(2x + 3)(x + 1) = 0$ — The equation is in factored form

Solution Set each factor equal to 0 and solve the equations.

$2x + 3 = 0$	or	$x + 1 = 0$	The factors are set equal to 0
$2x = -3$			Solve each equation
$x = -\dfrac{3}{2}$		$x = -1$	The solutions

The solution set is $\left\{-\dfrac{3}{2}, -1\right\}$.

1. $(x + 5)(x - 5) = 0$ **2.** $(x - 1)(x + 1) = 0$ **3.** $x(x + 6) = 0$

4. $x(x - 8) = 0$ **5.** $3a(a - 7) = 0$ **6.** $5p(p + 9) = 0$

7. $(3x - 9)(2x + 3) = 0$ **8.** $(8x - 4)(5x + 10) = 0$ **9.** $(2x + 1)(3x - 2) = 0$

10. $(4y - 3)(5y + 2) = 0$ **11.** $(5x - 1)(5x + 1) = 0$ **12.** $(4a - 1)(4a + 1) = 0$

13. $(8 - y)(7 - y) = 0$ **14.** $(1 - x)(3 - x) = 0$ **15.** $(4 - 3u)(8 - 5u) = 0$

16. $(7 + 3y)(2 - 3y) = 0$ **17.** $(5 - x)(5 + x) = 0$ **18.** $(8 - x)(8 + x) = 0$

19. $x(2x - 3)(x + 1) = 0$ **20.** $3x(x - 12)(3x + 1) = 0$ **21.** $(5x + 3)(x - 10)(4x - 1) = 0$

22. $(8x - 1)(2x + 7)(x - 3) = 0$

Find the solution set of the following quadratic equations by factoring. Check the solutions. See example 4–7 B.

Example $2a^2 = 5a$

Solution

$2a^2 - 5a = 0$	Write the equation in standard form
$a(2a - 5) = 0$	Factor $2a^2 - 5a = a(2a - 5)$
$a = 0$ or $2a - 5 = 0$	Set each factor equal to 0
$a = 0$ $\quad 2a = 5$	Solve each equation
$a = 0$ $\quad a = \frac{5}{2}$	The solutions

Check:

(1) Let $a = 0$

$2a^2 = 5a$

$2(0)^2 = 5(0)$

$0 = 0$ (True) Checks

(2) Let $a = \frac{5}{2}$

$2\left(\frac{5}{2}\right)^2 = 5\left(\frac{5}{2}\right)$ Substitute

$2\left(\frac{25}{4}\right) = \frac{25}{2}$ Order of operations

$\frac{25}{2} = \frac{25}{2}$ (True) Checks

The solution set is $\left\{0, \frac{5}{2}\right\}$.

Example $4y^2 = 9$

Solution

$4y^2 - 9 = 0$	Write the equation in standard form
$(2y + 3)(2y - 3) = 0$	Factor $4y^2 - 9 = (2y + 3)(2y - 3)$
$2y + 3 = 0$ or $2y - 3 = 0$	Set each factor equal to 0
$2y = -3$ $\quad 2y = 3$	Solve each equation
$y = -\frac{3}{2}$ $\quad y = \frac{3}{2}$	The solutions

Check by replacing y with $-\frac{3}{2}$ and $\frac{3}{2}$ in the original equation.

The solution set is $\left\{-\frac{3}{2}, \frac{3}{2}\right\}$.

23. $x^2 + 4x = 0$ **24.** $y^2 - 4y = 0$ **25.** $3a^2 - 5a = 0$ **26.** $4x^2 + 7x = 0$

27. $2x^2 + 6x = 0$ **28.** $3b^2 = 9b$ **29.** $2y^2 - 18 = 0$ **30.** $6x^2 = 24x$

31. $10a^2 = -15a$ **32.** $4y^2 = -6y$ 33. $a^2 = 25$ **34.** $y^2 = 49$

35. $x^2 - 9 = 0$ **36.** $y^2 - 100 = 0$ **37.** $2x^2 = 32$ **38.** $3a^2 = 27$

39. $7x^2 - 28 = 0$ **40.** $5y^2 - 45 = 0$ 41. $8x^2 - 18 = 0$ **42.** $7x^2 - 7 = 0$

Find the solution set of the following quadratic equations by factoring. Check the solutions. See example 4–7 B.

Example $x^2 - 7x + 12 = 0$

Solution

$x^2 - 7x + 12 = 0$ The equation is in standard form

$(x - 3)(x - 4) = 0$ Factor $x^2 - 7x + 12 = (x - 3)(x - 4)$

$x - 3 = 0$ or $x - 4 = 0$ Set each factor equal to 0

$x = 3$ $x = 4$ Solve each equation, giving the solutions

Check:

(1) Let $x = 3$

$x^2 - 7x + 12 = 0$

$(3)^2 - 7(3) + 12 = 0$

$9 - 21 + 12 = 0$

$-12 + 12 = 0$

$0 = 0$ (True) Checks

(2) Let $x = 4$

$x^2 - 7x + 12 = 0$

$(4)^2 - 7(4) + 12 = 0$ Substitute

$16 - 28 + 12 = 0$ Order of operations

$-12 + 12 = 0$

$0 = 0$ (True) Checks

The solution set is $\{3,4\}$.

43. $y^2 + 6y - 16 = 0$ **44.** $x^2 - 3x - 4 = 0$ 45. $a^2 + 14a + 49 = 0$

46. $x^2 - 16x + 64 = 0$ **47.** $b^2 + 5b - 14 = 0$ **48.** $x^2 + x - 42 = 0$

49. $x^2 + 3x + 2 = 0$ **50.** $y^2 = -11y - 10$ 51. $a^2 - 11a = 12$

52. $x^2 - 14x = 15$ **53.** $y^2 - 32 = 4y$ **54.** $x^2 = 27 - 6x$

Example $3x^2 = 7x + 6$

Solution

$3x^2 - 7x - 6 = 0$ Write in standard form

$(3x + 2)(x - 3) = 0$ Factor $3x^2 - 7x - 6 = (3x + 2)(x - 3)$

$3x + 2 = 0$ or $x - 3 = 0$ Set each factor equal to 0

$3x = -2$ $x = 3$ Solve the equations

$x = -\frac{2}{3}$ $x = 3$ The solutions

Check by replacing x with $-\frac{2}{3}$ and 3 in the original equation.

The solution set is $\left\{-\frac{2}{3},3\right\}$.

55. $2x^2 - 7x - 9 = 0$ **56.** $2y^2 - y - 3 = 0$ 57. $6a^2 - 5a + 1 = 0$

58. $3p^2 + 10p - 8 = 0$ **59.** $6x^2 + x - 12 = 0$ **60.** $12x^2 - 25x + 12 = 0$

61. $4y^2 = 4y + 3$ **62.** $6a^2 + 3 = 11a$ **63.** $9x^2 = 8x + 1$

64. $-6x = -3x^2 - 3$ **65.** $6p^2 - 7p = 20$ **66.** $9y^2 + 20 = -27y$

67. $3x^2 + 6x = -3$ **68.** $15 - 2x^2 = -x$ **69.** $9x^2 + 30x = -25$

70. $12y^2 - 15 = 8y$ **71.** $3x^2 - 4x - 28 = x$ **72.** $6z^2 + z = -10z - 3$

73. $2b^2 + 2b = 9 - b$ **74.** $3a - 3 = 15a^2 + 2a - 5$

Example $12(x^2 - 2x) = -9$

Solution

$$12x^2 - 24x = -9$$ Multiply in the left member

$$12x^2 - 24x + 9 = 0$$ Write the equation in standard form

$$3(4x^2 - 8x + 3) = 0$$ Factor common factor 3

$$3(2x - 1)(2x - 3) = 0$$ Factor $4x^2 - 8x + 3 = (2x - 1)(2x - 3)$

$$2x - 1 = 0 \quad \text{or} \quad 2x - 3 = 0$$ Set each factor containing the variable equal to 0

Note The constant factor 3 cannot be zero, so we disregard it when finding the solutions.

$$2x = 1 \quad \text{or} \quad 2x = 3$$ Solve each equation

$$x = \frac{1}{2} \qquad x = \frac{3}{2}$$ The solutions

Check by replacing x with $\frac{1}{2}$ and $\frac{3}{2}$ in the original equation.

The solution set is $\left\{\frac{1}{2}, \frac{3}{2}\right\}$.

75. $x(x + 3) = -2$
76. $3x(3x + 2) = 24$
77. $2x(2x + 6) = -8$
78. $2(x^2 - 6) = -5x$
79. $x(x - 6) = 18 + x$
80. $x(x + 1) - 6 = 0$
81. $4 = 3x(4 - 3x)$
82. $2x(3 - x) = 4$
83. $5(x^2 - 5) = 20x$
84. $x(x + 7) = 36 - 2x$

Review exercises

Write an algebraic expression for each of the following. See section 2–1.

1. 7 more than a number
2. A number decreased by 11
3. 6 times the sum of x^2 and x
4. The sum of x^2 and $2x$, divided by 8

Solve the following word problems. See section 2–8.

5. Three times a number is increased by 12 and the result is 51. Find the number.
6. One number is two more than five times another number. If their sum is 38, find the numbers.
7. The sum of three consecutive even integers is 72. Find the integers.

4–8 ■ Applications of the quadratic equation

Many formulas used in the physical world are quadratic in nature since they become second-degree equations when solving for one of the variables. Likewise, many word problems require the use of quadratic equations for their solutions. We now consider some of these common uses of the quadratic equation.

Be aware that a check of the solutions by merely substituting them into the equation set up to solve the problem *will not* guarantee the correct answer. We should check the results to ensure that the physical conditions of the word problem are satisfied.

■ Example 4–8 A

1. The product of two consecutive even integers is 168. Find the integers.

Let x = the lesser even integer. Then $x + 2$ = the next consecutive even integer.

Note Consecutive even or odd integers are given by $x, x + 2, x + 4, \cdots$.

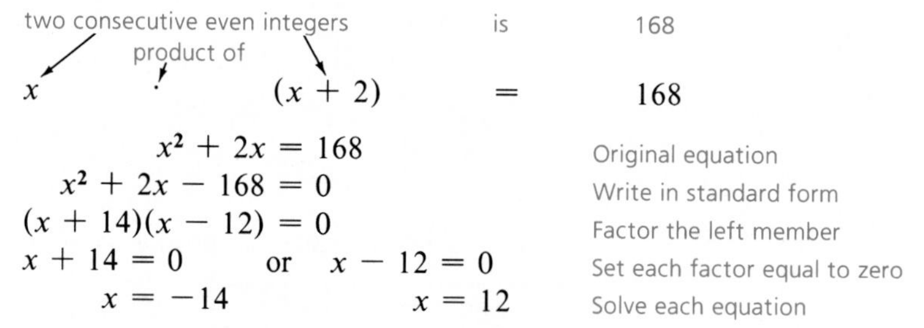

$x^2 + 2x = 168$	Original equation
$x^2 + 2x - 168 = 0$	Write in standard form
$(x + 14)(x - 12) = 0$	Factor the left member
$x + 14 = 0$ or $x - 12 = 0$	Set each factor equal to zero
$x = -14$ $\quad$ $x = 12$	Solve each equation

When $x = -14$, then $x + 2 = -14 + 2 = -12$.
When $x = 12$, then $x + 2 = 12 + 2 = 14$.
Check: Since $(-14)(-12) = 168$ and $(12)(14) = 168$, and both solutions are consecutive even integers, the conditions of the problem are met.
Therefore, the two integers are -14 and -12 or 12 and 14.

2. The area of a rectangle is $A = \ell w$, where ℓ is the length and w is the width of the rectangle. The length of a rectangle is 2 inches more than three times the width. If $A = 33$ square inches, find the length and width of the rectangle.

Let w = the width of the rectangle. Length $\ell = 3w + 2$

area of a rectangle	is	length	times	width
A	$=$	ℓ	$\cdot$	w
33	$=$	$(3w + 2)$	$\cdot$	w

$w(3w + 2) = 33$	Original equation
$3w^2 + 2w = 33$	Distribute multiplication
$3w^2 + 2w - 33 = 0$	Write in standard form
$(3w + 11)(w - 3) = 0$	Factor the left member
$3w + 11 = 0$ or $w - 3 = 0$	Set each factor equal to zero
$3w = -11$ $\quad$ $w = 3$	Solve each equation
$w = -\frac{11}{3}$ $\quad$ $w = 3$	Solutions of the equation

The solution set of the equation we wrote is $\left\{-\frac{11}{3}, 3\right\}$. Since the width cannot be negative, $w = -\frac{11}{3}$ is not a solution of the problem, even though it *is* a solution of the equation. So $w = 3$ is the only physical solution and

$\ell = 3w + 2 = 3(3) + 2 = 9 + 2 = 11$

Check: Since 11 is two more than 3 times 3 and $(3)(11) = 33$, the conditions of the problem are met.
The rectangle is 3 inches wide and 11 inches long.

3. Current in a circuit flows according to the equation $i = 16 - 16t^2$, where i is the current in amperes and t is the time in seconds. Find the time t when $i = 0$ amperes (no current).

Replacing i by 0, we have the equation

$0 = 16 - 16t^2$	Substitute 0 for i
$0 = 16(1 - t^2)$	Factor the common factor 16
$0 = 16(1 - t)(1 + t)$	Factor $1 - t^2 = (1 - t)(1 + t)$
$1 - t = 0$ or $1 + t = 0$	Set each factor containing the variable equal to 0
$t = 1$ $\quad$ $t = -1$	Solutions of the equation

Since time cannot be negative, $t = 1$ second. ■

Mastery points

Can you

- Set up a quadratic equation for a word problem and solve the equation?
- Substitute values into a formula and solve the resulting quadratic equation?

Exercise 4–8

Solve the following word problems by setting up a quadratic equation. See example 4–8 A–1.

1. The product of two consecutive odd integers is 143. Find the integers.
2. The product of two consecutive integers is 132. Find the integers.
3. The product of two consecutive even integers is 288. Find the integers.
4. The product of two consecutive integers is 306. Find the integers.
5. One integer is six more than a second integer. The product of the two integers is 91. Find the integers.
6. One integer is eight less than a second integer. The product of the two integers is 153. Find the integers.
7. The product of two consecutive even integers is four more than two times their sum. Find the integers.
8. The product of two consecutive odd integers is five more than six times the lesser integer. Find the integers.
9. The sum of two integers is -13 and their product is 36. Find the integers.
10. The sum of two integers is -3 and their product is -70. Find the integers.
11. One number is one more than three times the other. Their product is 14. Find the numbers.
12. One number is two more than the other and their product is -1. Find the numbers.

See example 4–8 A–2.

13. The length of a rectangle is 2 meters less than twice the width. If the area is 24 square meters, what are the dimensions of the rectangle?
14. The area of a rectangle is 21 square feet. What are its dimensions if the length is 5 feet less than four times the width?
15. The area of a rectangle is numerically equal to twice the length. If the length is 3 feet more than the width, what are the dimensions of the rectangle?
16. The length of a rectangle is three less than twice the width. If the area is numerically five times the length, find the dimensions of the rectangle.
17. The height of a page of a book is 3 inches more than the width. If the area of the page is six less than ten times the width, find the width of the page.

18. The height of a page of a book is 4 inches more than the width. If there is a margin of 1 inch all around the printed matter of the page and the area of the printed matter is 32 square inches, what is the height of the page?

See example 4–8 A–3.

19. An object with initial velocity v undergoes an acceleration a for time t. The displacement s of the object for this time is given by the equation

$$s = vt + \frac{1}{2}at^2$$

a. Find t when $s = 8$, $v = 2$, $a = 2$
b. Find t when $s = 6$, $v = 3$, $a = 6$

20. The current in a circuit flows according to the equation $i = 12 - 12t^2$, where i is the current in amperes and t is the time in seconds. Find t when i is (a) 0 amperes, (b) 9 amperes.

21. The power output of a generator armature is given by $P_0 = E_gI - r_gI^2$.
a. Find I when $P_0 = 120$, $E_g = 22$, $r_g = 1$
b. Find I when $P_0 = 120$, $E_g = 16$, $r_g = \frac{1}{2}$

22. The output power P of a 100-volt electric generator is defined by

$$P = 100I - 5I^2$$

where I is in amperes. Find I when (a) $P = 480$, (b) $P = 375$

23. A ball rolls down a slope and travels a distance $d = 6t + \frac{t^2}{2}$ feet in t seconds. Find t when (a) $d = 14$ feet, (b) $d = 32$ feet.

24. Because of gravity, an object falls a distance s feet according to the formula $s = 16t^2$, where t seconds is the time it falls. How long will it take the object to fall (a) 256 feet; (b) 49 feet; (c) $2\frac{1}{4}$ feet; (d) 1,024 feet?

The formula $s = v_0t - 16t^2$ gives the height s in feet that an object will travel in t seconds if it is propelled directly upward at an *initial* velocity of v_0 feet per second. (Use this formula in exercises 25 through 28.)

25. If an object is thrown upward at 96 feet per second, how long will it take the object to reach a height of 80 feet?

26. How long will it take before the object of exercise 25 hits the ground? (*Hint:* $s = 0$ when this happens.)

27. A projectile is fired upward with an initial velocity of 144 feet per second. How long will it take before the projectile strikes an object 288 feet directly overhead?

28. How long will it take before the projectile in exercise 27 falls back to the ground?

29. The formula for the area A of a trapezoid is $A = \frac{1}{2}h(b_1 + b_2)$, where h is the altitude (height) of the trapezoid and the parallel bases are b_1 and b_2 (see diagram). If the area of the trapezoid is 63 square inches, base b_1 is 10 inches long, and the altitude h is 1 inch less than the length of b_2, find the length of h and b_2.

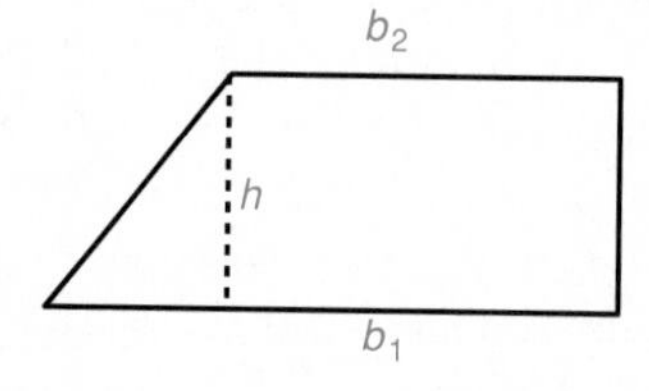

30. If the area of a trapezoid is 21 square feet, base b_2 is 5 feet long, and base b_1 is 6 feet longer than the altitude h, find the altitude of the trapezoid.

31. The altitude and base b_2 of a trapezoid have the same length. If the area of the trapezoid is 24 square meters and base b_1 is twice as long as base b_2, find the dimensions of the trapezoid.

32. One base of a trapezoid is three times the length of the other base. If the altitude is twice as long as the shorter base and the area is 36 square centimeters, find the dimensions of the trapezoid.

33. The volume of a box (rectangular solid) is given by $V = \ell wh$, where ℓ is the length, w is the width, and h is the height of the box (see diagram). If the box is 4 feet tall, the length is 1 foot longer than the width, and the volume is 224 cubic feet, find the length and the width of the box.

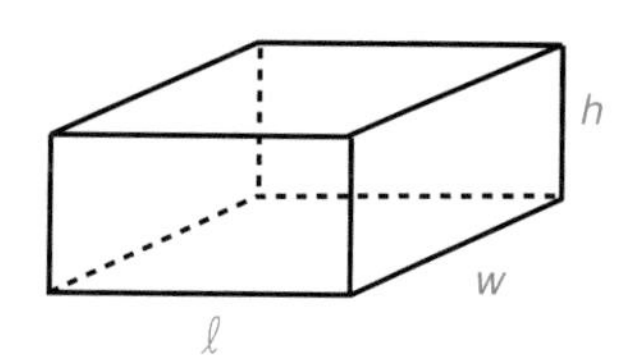

34. A box is 9 inches long and has a volume of 162 cubic inches. If the width of the box is twice the height, find the width and the height of the box.

35. A storage room is three times as long as it is wide. If the room contains 756 cubic feet of space and has a ceiling 7 feet high, find the length and width of the room.

36. A cardboard box has a volume of 108 cubic inches. If the length of the box is 1 inch more than two times its width and the box has a height of 3 inches, find the length and the width of the box.

Review exercises

Find the solution set. See sections 2–6 and 4–7.

1. $3x - 2 = 0$ **2.** $4a - 5 = 0$ **3.** $x^2 - 9 = 0$ **4.** $x^2 + 6x + 8 = 0$

Evaluate the following expressions if $a = 4$, $b = -3$, $c = -6$, and $d = 5$. See section 2–2.

5. $ab - cd$ **6.** $b^2 - c^2$ **7.** $\dfrac{c - 2b}{d - a}$ **8.** $\dfrac{c + d}{3a + 4b}$

Chapter 4 lead-in problem

The formula $s = vt - 16t^2$ gives the height s in feet that an object will travel in t seconds if it is propelled directly upward at an initial velocity of v feet per second. If an object is thrown upward at 96 feet per second, how long will it take the object to reach a height of 144 feet?

Solution

$$s = vt - 16t^2 \quad \text{Original equation}$$
$$(144) = (96)t - 16t^2 \quad \text{Substitute 144 for } s \text{ and 96 for } v$$
$$16t^2 - 96t + 144 = 0 \quad \text{Standard form}$$
$$16(t^2 - 6t + 9) = 0 \quad \text{Common factor}$$
$$16(t - 3)(t - 3) = 0 \quad \text{Factor the trinomial}$$
$$t - 3 = 0 \quad \text{Set the repeated factor equal to zero}$$
$$t = 3 \quad \text{Solve}$$

The object will reach a height of 144 feet in 3 seconds.

Chapter 4 summary

1. **Common factors** are factors that appear in all the original terms.
2. A polynomial with integer coefficients is in **completely factored form** when
   a. the polynomial is written as a product of polynomials with integer coefficients;
   b. none of the polynomial factors, other than the monomial factor, can be factored further.
3. The **difference of two squares** factors as

$$a^2 - b^2 = (a + b)(a - b)$$

4. The **trinomial** $ax^2 + bx + c$ will factor only if we can find a pair of integers, m and n, whose product is $a \cdot c$ and whose sum is b.
5. **Perfect square trinomials** factor as

   $a^2 + 2ab + b^2 = (a + b)^2$

   and

   $a^2 - 2ab + b^2 = (a - b)^2$
6. The **sum or difference of two cubes** factors as

   $a^3 + b^3 = (a + b)(a^2 - ab + b^2)$

   and

   $a^3 - b^3 = (a - b)(a^2 + ab + b^2)$
7. We try to factor **four-term polynomials** by grouping.
8. A **quadratic equation** is an equation that contains the second, but no higher, power of the variable.
9. The product of two factors $(x + p)(x + q) = 0$ only if $x + p = 0$ or $x + q = 0$

Chapter 4 error analysis

1. Factoring a common factor
   Example: $3x^2 - 6x - 3 = 3(x^2 - 2x)$
   Correct answer: $3(x^2 - 2x - 1)$
   What error was made? (*see page 158*)
2. Completely factoring a trinomial
   Example: $3x^2 - 6x + 3 = 3(x^2 - 2x + 1)$
   $= (x - 1)^2$
   Correct answer: $3(x - 1)^2$
   What error was made? (*see page 170*)
3. Factoring the sum of two squares
   Example: $4x^2 + 9y^2 = (2x + 3y)^2$
   Correct answer: $4x^2 + 9y^2$ is not factorable.
   What error was made? (*see page 177*)
4. Factoring the difference of two squares
   Example: $16x^2 - 4y^2 = (4x)^2 - (2y)^2$
   $= (4x + 2y)(4x - 2y)$
   Correct answer: $4(2x + y)(2x - y)$
   What error was made? (*see page 177*)
5. Factoring the sum of two cubes
   Example: $x^3 + y^3 = (x + y)(x^2 - 2xy + y^2)$
   Correct answer: $x^3 + y^3 = (x + y)(x^2 - xy + y^2)$
   What error was made? (*see page 182*)
6. Order between real numbers
   Example: $-15 > 4$
   Correct answer: $-15 < 4$
   What error was made? (*see page 30*)
7. Combining using grouping symbols
   Example: $6 - (10 + 3) = 6 - 10 + 3 = -1$
   Correct answer: -7
   What error was made? (*see page 57*)
8. Multiplication of real numbers
   Example: $5\frac{3}{4} = 5 \cdot \frac{3}{4} = \frac{15}{4}$
   Correct answer: $\frac{23}{4}$
   What error was made? (*see page 46*)
9. Division by zero
   Example: $\frac{0}{0} = 1$
   Correct answer: $\frac{0}{0}$ is indeterminate.
   What error was made? (*see page 53*)
10. Exponents
    Example: $-2^2 = 4$
    Correct answer: -4
    What error was made? (*see page 57*)

Chapter 4 critical thinking

If n is an integer, for what values of n will $4n^2 + 10n + 4$ represent an even number?

Chapter 4 review

[4–1]

Supply the missing factors.

1. $3x + 9 = 3(\quad)$
2. $9x^2 - 18x = 9x(\quad)$
3. $-4y^3 + 8y^2 = -4y^2(\quad)$
4. $7a + 14b - 28c = 7(\quad)$
5. $-5a^2 - 15a + 30a^3 = -5a(\quad)$

Write in completely factored form.

6. $3a^2 - 3ab + 3b$
7. $x^2y + xyz + xy^2z$
8. $a^3b + a^3b^2$
9. $3R^3 - R^2 + 5R^4$
10. $4y^2 + 8y + 12y^3$
11. $x^4 + 3x^3 + 9x^2$
12. $16R^3S^2 - 12R^4S^3 + 24R^2S^2$
13. $10a^4b^3 + 15a^2b^2 - 20a^3b^2$
14. $2(a + b) + x(a + b)$
15. $y(x - 3z) + 4(x - 3z)$
16. $a(3R + 1) + b(3R + 1)$
17. $2a(x - 3y) - 3b(x - 3y)$
18. $6ax - 3ay - 2bx + by$
19. $4ax + 6by + 8ay + 3bx$
20. $ax + 3bx - 4a - 12b$
21. $ax^2 - 2bx^2 + 4a - 8b$

[4–2]

Write in completely factored form.

22. $x^2 - 9x + 14$
23. $2a^3 - 8a^2 - 10a$
24. $a^2 + 14a + 24$
25. $x^2 - 4x - 32$
26. $a^2 - 16a - 36$
27. $3x^2 - 9x - 30$
28. $x^3 - x^2 - 6x$
29. $x^3 - 4x^2 - 21x$
30. $a^2b^2 + ab - 6$
31. $a^2b^2 + 10ab + 24$
32. $a^2b^2 - 9ab + 18$
33. $a^2b^2 - 8ab - 20$

[4–3]

Write in completely factored form.

34. $4x^2 + 4x + 1$
35. $9r^2 - 36r + 36$
36. $4x^2 - 5x + 1$
37. $9a^2 + 9a - 10$
38. $8a^2 - 2a - 3$
39. $24x^2 + 22x + 3$
40. $8a^2 - 18a + 9$
41. $2a^2 + 15a + 18$

[4–4]

Write in completely factored form.

42. $4a^2 - 9$
43. $36b^2 - c^2$
44. $25 - a^2$
45. $16x^2 - 4y^2$
46. $9x^2 - y^4$
47. $x^4 - 16$
48. $y^4 - 81$
49. $b^2 + 12b + 36$
50. $c^2 - 10c + 25$
51. $4x^2 - 12x + 9$
52. $9x^2 - 12x + 4$

[4–5]

Write in completely factored form.

53. $R^3 + 8S^3$
54. $16x^3 - 54$
55. $27a^3 + 125b^3$
56. $x^3y^3 - 1$
57. $2x^9 + 250$
58. $64x^{12} - y^{15}$
59. $a^3b^6 + c^9$

[4–6]

Write in completely factored form.

60. $12x^4 - 3x^3$
61. $a^2 - 3a - 10$
62. $4a^2 - 21a + 5$
63. $9y^2 - 4$
64. $6ax + 9bx - 4a - 6b$
65. $b^2 - b - 20$
66. $9x^2 + 21x + 10$
67. $a^2 + 14a + 49$
68. $12x^5 - 3x^3$
69. $c^3 + 9c^2 + 20c$
70. $16a^2 - 8a + 1$
71. $b^4 - 1$

[4–7]

Determine the solution set.

72. $(x-1)(x+3)=0$

73. $3x(x-8)=0$

74. $(5x+1)(3x-7)=0$

75. $(7x-1)(5x-8)=0$

76. $(4-3x)(9-x)=0$

77. $5x(x+9)(3x+12)=0$

78. $(5x-4)(5x+4)=0$

79. $(4x+4)(5x-15)(7x+14)=0$

Find the solution set of the following quadratic equations by factoring.

80. $4x^2-9x=0$

81. $y^2=1$

82. $2a^2=128a$

83. $3x^2-75=0$

84. $x^2-x-30=0$

85. $2y=y^2+1$

86. $4a^2+13a+3=0$

87. $5x^2-4=8x$

88. $x(x-8)=-16$

89. $3(x^2+3)=12x$

90. $8y^2+y-6=-y$

91. $4a-12=2a-2a^2$

[4–8]

Solve the following word problems.

92. If the product of two consecutive integers is ten less than the square of the greater integer, determine the integers.

93. The length of a rectangular flower garden is 5 feet more than its width. If the area of the garden is 104 square feet, find the dimensions of the flower garden.

94. A farmer has some cattle to sell to a slaughterhouse. When the manager of the slaughterhouse asked how many cattle he had to sell, the farmer replied, "If you triple the square of the number of cattle you get 1,200." How many cattle did the farmer have to sell?

95. The height h in feet of a projectile launched vertically upward from the top of a 96-foot tall tower when time $t = 0$ is given by $h = 96 + 80t - 16t^2$. How long will it take the projectile to strike the ground?

Chapter 4 cumulative test

Perform the indicated operations and simplify.

[1–8] 1. $40 - 2 \cdot 8 \div 4 - 6 + 3$

[3–4] 2. $(2a^2b)^3$

[3–2] 3. $(a+2b)^2$

[3–1] 4. $a^3 \cdot a^2 \cdot a$

[1–8] 5. $3[6-4(5-2)+10]$

[3–1] 6. $(x^2y^3)(x^3y)$

[2–3] 7. $(3x-y)-(4y-3x)-(x+2y)$

[3–2] 8. $(3x-2y)(3x+2y)$

[3–3] 9. $\left(\frac{2a^2}{b}\right)^3$

[3–3] 10. $x^3 \cdot x^{-5}$

[2–3] 11. $3x-[x-y-(2x+3y)]$

[3–4] 12. $(3x^{-3}y^2)^{-2}$

[2–2] 13. If $P(x) = x^2 - x - 6$, find (a) $P(-2)$, (b) $P(3)$, (c) $P(0)$.

Solve the inequalities and find the solution set for the equations.

[2–6] 14. $3(3x-2)=4x+3$

[2–9] 15. $\frac{3}{4}x+2<\frac{1}{2}x+5$

[2–6] 16. $2(2x-1)=6(x-2)-4$

[2–9] 17. $2x+7\geq 13$

[4–7] 18. $x^2-9=0$

[4–7] 19. $x^2-7x+10=0$

[2–9] 20. $3(2x+1)<4x+8$

[2–9] 21. $-2\leq 3x-8\leq 6$

Solve for the specified variable.

[2–7] **22.** $3x - y = x + 5y$ for x

[2–7] **23.** $2x + 5 - 3y = 5x + 3 - 8y$ for x

Write in completely factored form.

[4–1] **24.** $2ab - 4a^2b^2 - 8a^3b^5$

[4–4] **25.** $4a^2 + 12a + 9$

[4–4] **26.** $25c^2 - 9d^2$

[4–3] **27.** $4a^2 - 4a - 15$

[4–2] **28.** $x^2 + 9x + 18$

Set up an equation and solve for the unknown(s).

[2–8] **29.** One number is eleven more than twice a second number. If their sum is 53, what are the numbers?

[4–8] **30.** The product of two consecutive positive odd integers is 143. Find the integers.

[2–8] **31.** Terry has $15,000. He invests part of this money at 8% and the rest at 6%, and his income for one year from these investments totals $1,100. How much was invested at each rate?

[4–8] **32.** The area of a rectangle is six more than ten times the width. If the length is 5 meters longer than the width, what are the dimensions of the rectangle?

CHAPTER

5 Rational Expressions, Ratio and Proportion

On a road map, 3 inches represents a distance of 45 miles. If the distance between Detroit and Sault Ste. Marie is 23 inches on the map, how far is it from Detroit to Sault Ste. Marie?

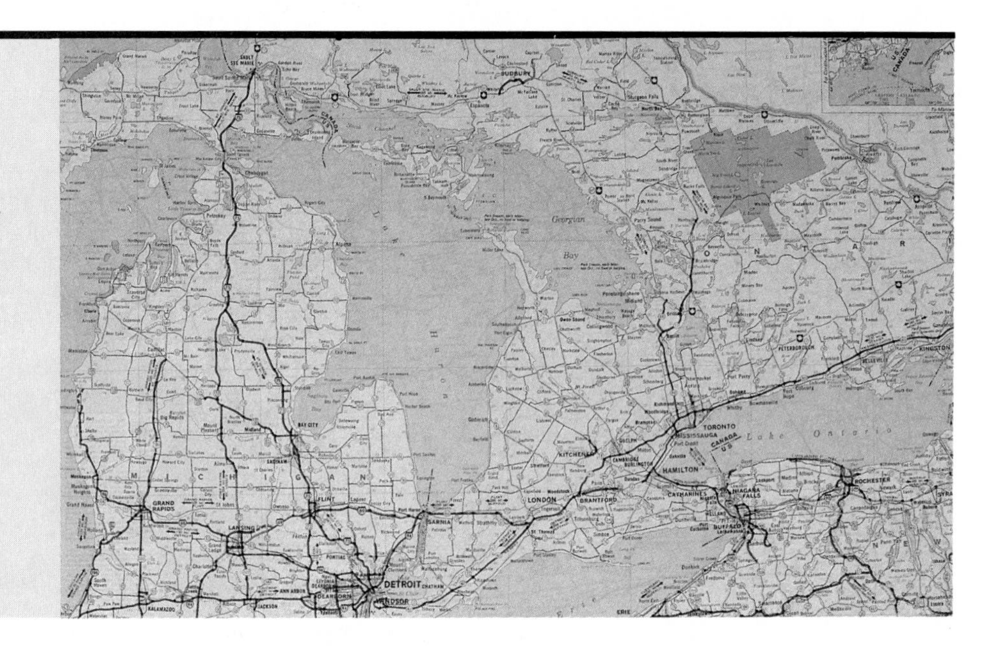

5–1 ■ Rational numbers and rational expressions

A rational expression

In chapter 1, we defined a rational number.

Rational numbers

A rational number is a number that can be written as the quotient of two integers with the divisor (denominator) not zero.

Examples of rational numbers are

$$6, \quad \frac{3}{4}, \quad \frac{-5}{6}, \quad \text{and } \frac{8}{7}$$

We extend this definition to involve the quotient of two polynomials and define a **rational expression.**

Rational expressions

A **rational expression** is an expression of the form

$$\frac{P}{Q}$$

where P and Q are polynomials, $Q \neq 0$.

Concept

A rational expression is an expression that can be written as the quotient of two polynomials with the denominator not zero.

For example,

$$\frac{2x}{x+1}, \quad \frac{x^2-2}{x^2-x-6}, \quad \text{and} \quad \frac{x^2+x}{5}$$

are all rational expressions.

Just as a rational number has a numerator and a denominator, so does a rational expression. In the rational expression

$$\frac{x^2-2}{x^2-x-6}$$

the polynomial on the top, $x^2 - 2$, is called the **numerator** and the polynomial on the bottom, $x^2 - x - 6$, is called the **denominator.**

Note Any polynomial is a rational expression since the denominator can be considered to be 1. For example, $x^2 - 2 = \dfrac{x^2-2}{1}$.

Evaluating a rational expression

In chapter 2, we evaluated algebraic expressions by substituting given values for the variables and performing the indicated operations. We follow the same procedure when evaluating rational expressions.

■ Example 5–1 A

Evaluate the following rational expressions for the given value of the variable.

1. $\dfrac{5x-2}{4x+3}, x = 2$

$$\frac{5x-2}{4x+3} = \frac{5(2)-2}{4(2)+3}$$ Replace x with 2

$$= \frac{10-2}{8+3}$$ Perform indicated operations using order of operations

$$= \frac{8}{11}$$

2. $\dfrac{x+2}{x^2-3x-10}, x = 5$

$$\frac{x+2}{x^2-3x-10} = \frac{5+2}{(5)^2-3(5)-10}$$ Replace x with 5

$$= \frac{7}{25-15-10}$$ Perform indicated operations

$$= \frac{7}{0} \quad \text{(undefined)}$$

Note Anytime zero is in the denominator, the expression is undefined. Therefore, for the rational expression to be defined, x cannot be 5.

3. $\dfrac{x-5}{2x^2+x-1}, x = -\dfrac{1}{2}$

$$\frac{x-5}{2x^2+x-1} = \frac{\left(-\frac{1}{2}\right) - 5}{2\left(-\frac{1}{2}\right)^2 + \left(-\frac{1}{2}\right) - 1}$$ Replace x with $-\frac{1}{2}$

$$= \frac{-\frac{1}{2} - \frac{10}{2}}{\frac{1}{2} - \frac{1}{2} - 1}$$ Perform indicated operations

$$= \frac{-\frac{11}{2}}{-1}$$

$$= \frac{11}{2}$$ Multiply numerator and denominator by -1

▶ ***Quick check*** Evaluate: $\dfrac{3y-2}{y+3}$ when $y = 4$ and $\dfrac{y^2-y-1}{y^2-4}$ when $y = -2$ ■

Domain of a rational expression

Notice in example 2, the answer was undefined since division by zero is not defined. The rational expression becomes *meaningless* for those values of the variable for which the denominator equals zero, as with $x = 5$ in example 2. Finding the value(s) of the variable that will make the denominator zero is called *finding the restrictions on the variable(s).* All other values of the variable for which the expression is defined make up the **domain** of the rational expression.

Domain of a rational expression

The set of all replacement values of the variable for which a rational expression is defined determines the **domain** of the rational expression.

In finding the domain of a rational expression, we use the following procedure.

Finding the domain of a rational expression

1. Factor the denominator into a product of prime polynomials, if possible.
2. Set each factor of the denominator containing the variable equal to zero (using the zero product property).
3. Solve the resulting equations. The solutions are the restrictions placed on the variable.

■ ***Example 5–1 B***

Determine the domain of each of the following rational expressions.

1. $\dfrac{a-3}{a-4}$

$a - 4 = 0$ — Set denominator equal to 0

$a = 4$ — Solve equation for a

The restriction is that $a \neq 4$. Domain is all real numbers except 4.

Note We look *only* at the denominator. The value(s) of the variable for which the numerator is zero is of no concern to us.

2. $\dfrac{3x^2}{x^2 - x - 6}$

$\dfrac{3x^2}{x^2 - x - 6} = \dfrac{3x^2}{(x-3)(x+2)}$ — Factor the denominator

$x - 3 = 0$ or $x + 2 = 0$ — Set each factor equal to 0

$x = 3$ $\quad x = -2$ — Solve each equation for x

The restrictions are $x \neq 3$ or $x \neq -2$. Domain is all real numbers except 3 or -2.

3. $\dfrac{x+3}{x^2+4}$

$x^2 + 4$ does not factor. If there is a restriction, it will occur when x^2 is -4. Since x^2 is never negative, the sum $x^2 + 4$ can never be zero and there are no restrictions on the variable. Thus, the domain is the set of all real numbers.

4. $\dfrac{x-3}{x^2-x}$

$\dfrac{x-3}{x^2-x} = \dfrac{x-3}{x(x-1)}$ — Factor the denominator

$x = 0$ or $x - 1 = 0$ — Set each factor equal to 0

$x = 0$ $\quad x = 1$ — Solve each equation for x

The restrictions are $x \neq 0$ or $x \neq 1$. Domain is all real numbers except 0 or 1.

Note In example 4, a common error is to forget to place restrictions on the factor x.

▶ ***Quick check*** Determine the domain: $\dfrac{x+3}{x^2+x-6}$ ■

Mastery points

Can you

- Evaluate a rational expression for a given value of the variable?
- Determine the restrictions on the variable in a rational expression?
- Determine the domain of a rational expression?

Exercise 5–1

Evaluate each given rational expression using the given value of the variable. If the given value is not in the domain of the rational expression, so state. See example 5–1 A.

Examples $\dfrac{3y-2}{y+3}$; $y = 4$ $\qquad$ $\dfrac{y^2+2y-1}{y^2-4}$; $y = -2$

Solutions

$\dfrac{3(4)-2}{(4)+3}$ Replace y with 4

$\dfrac{12-2}{4+3}$ Perform indicated operations

$\dfrac{10}{7}$

$\dfrac{(-2)^2+2(-2)-1}{(-2)^2-4}$ Replace y with -2

$\dfrac{4-4-1}{4-4}$ Perform indicated operations

$\dfrac{-1}{0}$ (Undefined)

-2 is not in the domain.

1. $\dfrac{x}{3x}$; $x = 2$
2. $\dfrac{2x+1}{5x-3}$; $x = 3$
3. $\dfrac{5a^2+2}{a-1}$; $a = 1$
4. $\dfrac{x^2-1}{x}$; $x = 0$
5. $\dfrac{-4p^2}{2p-3}$; $p = -1$
6. $\dfrac{-5b^3}{5-2b}$; $b = -2$
7. $\dfrac{x+9}{x^2+2x-1}$; $x = 4$
8. $\dfrac{2n-3}{3n^2+n-1}$; $n = -3$
9. $\dfrac{(-2x)^2}{x^2+3x+7}$; $x = 2$
10. $\dfrac{(-x)^3}{2x^2-5}$; $x = 3$
11. $\dfrac{3x^2+2x+1}{4-x-x^2}$; $x = 1$
12. $\dfrac{8-b-2b^2}{4-3b^2}$; $b = -2$
13. $\dfrac{3-4x}{x^2-x}$; $x = 0$
14. $\dfrac{x^3-2x^2+x}{x-2}$; $x = \dfrac{1}{2}$
15. $\dfrac{-x+3}{x^2+2x+1}$; $x = -\dfrac{1}{3}$

Determine the domain of the given rational expression. See example 5–1 B.

Example $\dfrac{x+3}{x^2+x-6}$

Solution $\dfrac{x+3}{(x+3)(x-2)}$ Factor the denominator

$x+3 = 0$ or $x-2 = 0$ Set each factor equal to zero

$x = -3 \qquad x = 2$ Solve for x

The restrictions are $x \neq -3$ or $x \neq 2$ and the domain is the set of all real numbers except -3 or 2.

16. $\dfrac{4}{3x}$
17. $\dfrac{5}{4x}$
18. $\dfrac{8}{x-2}$
19. $\dfrac{10}{x-5}$
20. $\dfrac{x}{x+7}$
21. $\dfrac{3x^2}{x+3}$
22. $\dfrac{x+1}{2x-1}$
23. $\dfrac{a+9}{4a-3}$
24. $\dfrac{p-3}{5-2p}$
25. $\dfrac{y+4}{8-3y}$
26. $\dfrac{x+7}{x^2+3x-18}$
27. $\dfrac{8b+1}{b^2-7b+6}$
28. $\dfrac{5s^2+7}{2s^2-s-3}$
29. $\dfrac{8z}{3z^2+2z-8}$
30. $\dfrac{4}{x^2-4}$

31. $\frac{5x}{9x^2 - 4}$

32. $\frac{a - 2}{4a^2 - 16}$

33. $\frac{b + 3}{5b^2 - 45}$

34. $\frac{7x}{3x - 15}$

35. $\frac{5b - 1}{9b - 21}$

36. $\frac{16x}{8x^2 - 18}$

37. $\frac{17q}{3q^2 - 3}$

38. $\frac{23}{x^2 + 1}$

39. $\frac{5x - 3}{2x^2 + 1}$

40. $\frac{2x^2}{x^2 + 9}$

41. In business, the acid-test ratio (A) is given by $A = \frac{C + R}{L}$, where C is the cash on hand, R represents receivables, and L is the current liability. For what value of L is A not defined?

42. In college algebra, the sum S of the terms in some infinite geometric sequence is given by $S = \frac{a}{1 - r}$, where a is the first term and r is the common ratio. For what value of r is S not defined?

43. In physics, the *ideal gas law* of a constant mass of gas is given by $\frac{P_1V_1}{T_1} = \frac{P_2V_2}{T_2}$, where P_1 = the pressure in the first state, V_1 = the volume in the first state, T_1 = the temperature in Kelvin degrees in the first state, P_2 = the pressure in the second state, V_2 = the volume in the second state, and T_2 = the temperature in Kelvin degrees in the second state. What restrictions on the variables must be placed in this formula?

Review exercises

1. The statement $4(5 + 6) = 4 \cdot 5 + 4 \cdot 6$ demonstrates what property of real numbers? See section 1–8.

2. Simplify the expression $2[3 - 4(1 + 6)]$ by performing the indicated operations. See section 1–8.

3. Evaluate the expression $a[b + (c - d)]$ when $a = 1$, $b = -2$, $c = 3$, and $d = -4$. See section 2–2.

4. Reduce the fraction $\frac{35}{63}$ to lowest terms. See section 1–1.

Factor the following expressions. See section 4–3.

5. $2x^2 - 9x - 5$

6. $4y^2 - 40y + 100$

5–2 ■ Simplifying rational expressions

The fundamental principle of rational expressions

One of the most important procedures we can use when we work with rational expressions is the simplification of the rational expression. To do this, we use a principle called the **fundamental principle of rational expressions.**

Fundamental principle of rational expressions

If P is any polynomial and Q and R are nonzero polynomials, then

$$\frac{PR}{QR} = \frac{P}{Q} \text{ and } \frac{P}{Q} = \frac{PR}{QR}$$

Concept

To change the appearance of a rational expression without changing its value, we may multiply or divide both the numerator and the denominator by the same nonzero polynomial.

This property is based on 1 being the identity element for multiplication. That is,

$$\frac{PR}{QR} = \frac{P}{Q} \cdot \frac{R}{R} = \frac{P}{Q} \cdot 1 = \frac{P}{Q}$$

This property permits us to **reduce** rational expressions to *lowest terms.* A rational expression is *completely reduced* if the greatest factor common to both the numerator and the denominator is 1 or -1. We can see that the key to reducing rational expressions is *finding* and dividing out *factors* that are common to both the numerator and the denominator.

To reduce a rational expression

1. Write the numerator and the denominator in factored form.
2. Divide the numerator and the denominator by all common factors.

Example 5–2 A

Simplify the following rational expressions by reducing to lowest terms. Assume that all denominators are nonzero.

1. $\frac{45}{60} = \frac{3 \cdot 3 \cdot 5}{2 \cdot 2 \cdot 3 \cdot 5}$ Factor numerator and denominator to prime factors

$= \frac{3 \cdot (3 \cdot 5)}{2 \cdot 2 \cdot (3 \cdot 5)}$ Group common factors $(3 \cdot 5)$

$= \frac{3}{2 \cdot 2}$ Divide numerator and denominator by $(3 \cdot 5)$

$= \frac{3}{4}$ Multiply remaining factors

2. $\frac{14x^2}{10x^3} = \frac{7 \cdot 2 \cdot x \cdot x}{5 \cdot 2 \cdot x \cdot x \cdot x}$ Factor numerator and denominator

$= \frac{7 \cdot (2 \cdot x \cdot x)}{5 \cdot x \cdot (2 \cdot x \cdot x)}$ Group common factors $(2 \cdot x \cdot x)$

$= \frac{7}{5x}$ Divide numerator and denominator by common factor $(2 \cdot x \cdot x)$

3. $\frac{5a - 15}{4a - 12} = \frac{5(a - 3)}{4(a - 3)}$ Factor numerator and denominator

$= \frac{5}{4}$ Divide numerator and denominator by common factor $(a - 3)$

4. $\frac{y - 7}{y^2 - 49} = \frac{y - 7}{(y + 7)(y - 7)}$ Factor denominator

$= \frac{1}{y + 7}$ Divide numerator and denominator by common factor $(y - 7)$

▶ ***Quick check*** Reduce $\frac{25z^4}{15z^5}$ and $\frac{a^2 - 36}{a^2 - a - 30}$ to lowest terms. ■

Note This reducing process is often called "cancelling" the common factors. It is important to remember that we are *dividing* both the numerator and the denominator by the same common factor; we are *not* just "crossing out" quantities. This leads us to our next topic.

Common errors when reducing

The fundamental property allows us to divide by common *factors only.* A common error in example 4 is to divide the numerator and the denominator by y and 7. These are *terms* and this cannot be done. For example,

$$\frac{9}{11} = \frac{8+1}{8+3} \neq \frac{\not{8}+1}{\not{8}+3} = \frac{1}{3}$$

This error can be avoided by always remembering that the *fundamental principle of rational expressions* allows us to *divide* the numerator and the denominator by common factors. y and 7 are *terms* and *not factors.*

Reducing $\frac{a-b}{b-a}$

Consider the rational expression $\frac{x-5}{5-x}$, which does not appear to be reducible by a common factor. However,

$$\begin{aligned} 5 - x &= -1(-5 + x) && \text{Factor out } -1 \\ &= -1(x - 5) && \text{Commute the terms and use the definition of subtraction} \end{aligned}$$

Thus,

$$\begin{aligned} \frac{x-5}{5-x} &= \frac{x-5}{-1(x-5)} \\ &= \frac{1}{-1} && \text{Reduce by common factor } (x-5) \\ &= -1 \end{aligned}$$

> In general, for all real numbers a and b, $a \neq b$,
>
> $$\frac{a-b}{b-a} = -1$$

■ Example 5–2 B

Simplify the following rational expressions by reducing to lowest terms. Assume that no denominator equals zero.

1. $\frac{4-x}{x^2-16}$

$$\begin{aligned} &= \frac{4-x}{(x-4)(x+4)} && \text{Completely factor the denominator} \\ &= \frac{4-x}{x-4} \cdot \frac{1}{x+4} && \text{Factor opposites } 4 - x \text{ and } x - 4 \\ &= -1 \cdot \frac{1}{x+4} && \frac{4-x}{x-4} = -1 \\ &= \frac{-1}{x+4} && \text{Multiply numerator by } -1 \end{aligned}$$

2. $\dfrac{1 - x^2}{2x^2 + x - 3}$

$= \dfrac{(1 - x)(1 + x)}{(x - 1)(2x + 3)}$ — Completely factor numerator and denominator

$= \dfrac{1 - x}{x - 1} \cdot \dfrac{1 + x}{2x + 3}$ — Factor opposites $1 - x$ and $x - 1$

$= -1 \cdot \dfrac{1 + x}{2x + 3}$ — $\dfrac{1 - x}{x - 1} = -1$

$= \dfrac{-1(1 + x)}{2x + 3}$ — Multiply numerator by -1

$= \dfrac{-1 - x}{2x + 3}$ or $\dfrac{-x - 1}{2x + 3}$ — Alternative forms of answer

▶ ***Quick check*** Reduce $\dfrac{16 - y^2}{3y^2 - 11y - 4}$ to lowest terms. ■

Mastery points

Can you

- Reduce a rational expression to lowest terms using the fundamental principle of rational expressions?
- Recognize factors $a - b$ and $b - a$ and use $\dfrac{a - b}{b - a} = -1$?

Exercise 5–2

Simplify the following rational expressions by reducing to lowest terms. Assume that no denominator equals zero. See example 5–2 A.

Examples $\dfrac{25z^4}{15z^5}$

Solutions

$= \dfrac{5 \cdot 5 \cdot z \cdot z \cdot z \cdot z}{5 \cdot 3 \cdot z \cdot z \cdot z \cdot z \cdot z}$ — Factor numerator and denominator

$= \dfrac{5 \cdot (5 \cdot z \cdot z \cdot z \cdot z)}{3 \cdot z \cdot (5 \cdot z \cdot z \cdot z \cdot z)}$ — Group common factors $(5 \cdot z \cdot z \cdot z \cdot z)$

$= \dfrac{5}{3z}$ — Divide numerator and denominator by $(5 \cdot z \cdot z \cdot z \cdot z)$

$\dfrac{a^2 - 36}{a^2 - a - 30}$

$= \dfrac{(a + 6)(a - 6)}{(a + 5)(a - 6)}$ — Factor numerator and denominator

$= \dfrac{a + 6}{a + 5}$ — Divide numerator and denominator by $(a - 6)$

1. $\dfrac{54}{72}$ **2.** $\dfrac{75}{145}$ **3.** $\dfrac{6x}{15}$ **4.** $\dfrac{8a}{10}$

5. $\dfrac{16x^2}{12x}$ **6.** $\dfrac{15b^3}{20b}$ **7.** $\dfrac{-8x^2}{6x^4}$ **8.** $\dfrac{3a^6}{-9a^3}$

9. $\dfrac{16a^2b}{20ab^2}$ **10.** $\dfrac{15a^2x^3}{35ax^2}$ 11. $\dfrac{20ab^2c^3}{-4ab^2c^3}$ **12.** $\dfrac{-72x^4y^3z^2}{9x^4y^3z^2}$

13. $\dfrac{10(x + 5)}{8(x + 5)}$ **14.** $\dfrac{24(x - 3)}{15(x - 3)}$ **15.** $\dfrac{6(x - 2)}{(x + 3)(x - 2)}$ **16.** $\dfrac{-8(x + 1)}{4(x + 1)(x - 6)}$

17. $\dfrac{a + b}{a^2 - b^2}$ 18. $\dfrac{x^2 - y^2}{x - y}$ 19. $\dfrac{3m - 6}{5m - 10}$ 20. $\dfrac{8b + 12}{10b + 15}$

21. $\dfrac{3x - 3}{6x + 6}$ 22. $\dfrac{6y - 6}{8y^2 - 8}$ 23. $\dfrac{x^2 - 9}{x^2 + 6x + 9}$ 24. $\dfrac{a^2 - 10a + 25}{a^2 - 25}$

25. $\dfrac{x^2 - 3x - 10}{x^2 - x - 6}$ 26. $\dfrac{y^2 - y - 42}{y^2 + 12y + 36}$ 27. $\dfrac{2y^2 - 3y - 9}{4y^2 - 13y + 3}$ 28. $\dfrac{4m^2 - 15m - 4}{8m^2 - 18m - 5}$

29. $\dfrac{x - 3}{x^3 - 27}$ 30. $\dfrac{x + 2}{x^3 + 8}$ 31. $\dfrac{a^2 - b^2}{a^3 + b^3}$ 32. $\dfrac{x^3 + y^3}{x^2 - y^2}$

Simplify by reducing to lowest terms. Assume that no denominator is equal to zero. See example 5–2 B.

Example $\dfrac{16 - y^2}{3y^2 - 11y - 4}$

Solution

$= \dfrac{(4 - y)(4 + y)}{(y - 4)(3y + 1)}$ Factor numerator and denominator

$= \dfrac{4 - y}{y - 4} \cdot \dfrac{4 + y}{3y + 1}$

$= -1 \cdot \dfrac{4 + y}{3y + 1}$ $\dfrac{4 - y}{y - 4} = -1$

$= \dfrac{-1(4 + y)}{3y + 1}$ Multiply the numerator by -1

$= \dfrac{-4 - y}{3y + 1}$ or $\dfrac{-y - 4}{3y + 1}$ Alternative forms of answer

33. $\dfrac{4x - 4y}{y - x}$ 34. $\dfrac{8b - 8a}{a - b}$ 35. $\dfrac{2x - 8}{12 - 3x}$ 36. $\dfrac{12a - 8b}{10b - 15a}$

37. $\dfrac{2y^2 - 2x^2}{x - y}$ 38. $\dfrac{3p - 3q}{6q^2 - 6p^2}$ 39. $\dfrac{(x - y)^2}{y^2 - x^2}$ 40. $\dfrac{a - b}{b^2 - a^2}$

41. $\dfrac{n^2 - m^2}{(m + n)^2}$ 42. $\dfrac{p^2 - q^2}{q^2 - p^2}$ 43. $\dfrac{4x - 4y}{y^2 - x^2}$ 44. $\dfrac{4 - y}{2y^2 - 7y - 4}$

45. $\dfrac{x - 3}{12 - x - x^2}$

Review exercises

1. Write the decimal number 0.000314 in scientific notation. See section 3–5.

2. Given $x = 2$, $y = -3$ and $z = -1$, evaluate the expression $\dfrac{4x - y}{2y + z}$. See sections 2–2 and 5–1.

3. A piece of lumber 16 feet long is to be divided into two pieces so that one piece is 1 foot longer than twice the length of the other piece. Find the lengths of the two pieces of lumber. See section 2–8.

Simplify the following expressions. Assume all denominators are nonzero. Express answers with positive exponents only. See section 3–4.

4. $(5x^{-2}y^3)^0$ 5. $\dfrac{3xy^2}{3^{-2}xy^{-1}}$ 6. $(-2x^3y^2)^3$

5–3 ▪ The quotient of two polynomials

In section 3–3, we observed the process of dividing a monomial by a monomial. We shall first review this process before we deal with other types of polynomial division. Recall the quotient property for division of expressions that have like bases, $a^m \div a^n = a^{m-n}, a \neq 0$.

▪ Example 5–3 A

Find the indicated quotients. Assume all variables are nonzero.

1. $x^7 \div x^4 = x^{7-4}$ — Subtract exponents when dividing
   $= x^3$

2. $\dfrac{2^2a^5}{2a^3} = 2^{2-1}a^{5-3}$ — Divide like bases by subtracting exponents
   $= 2^1a^2$ — Perform indicated subtractions
   $= 2a^2$ — $2^1 = 2$

▶ ***Quick check*** Find the quotient of $\dfrac{5^3x^4}{5x}$ ▪

Division of a polynomial by a monomial

Consider the indicated division.

$$\frac{3x^3 - 9x^2 + 15x}{3x}$$

To perform this division, we use a principle of fractions.

$$\frac{a}{c} + \frac{b}{c} = \frac{a+b}{c} \quad (c \neq 0)$$

By reversing this equation, the principle can be used to divide a polynomial by a monomial.

Division of a polynomial by a monomial

$$\frac{a+b}{c} = \frac{a}{c} + \frac{b}{c} \quad (c \neq 0)$$

Concept

To divide a polynomial by a monomial, divide each term of the polynomial by the monomial.

▪ Example 5–3 B

Find the indicated quotients. Assume all denominators are nonzero.

1. $\dfrac{3x^3 - 9x^2 + 15x}{3x} = \dfrac{3x^3}{3x} - \dfrac{9x^2}{3x} + \dfrac{15x}{3x}$ — Divide each term of the numerator by the monomial denominator
   $= \dfrac{3}{3}x^{3-1} - \dfrac{9}{3}x^{2-1} + \dfrac{15}{3}x^{1-1}$ — Quotient property of exponents
   $= 1 \cdot x^2 - 3 \cdot x^1 + 5 \cdot x^0$ — Subtract and divide as indicated
   $= x^2 - 3x + 5 \cdot 1$ — $x^0 = 1$
   $= x^2 - 3x + 5$

2. $\dfrac{8a^4 + 4a^2 - 12a}{4a} = \dfrac{8a^4}{4a} + \dfrac{4a^2}{4a} - \dfrac{12a}{4a}$ — Divide each term of numerator by the monomial denominator

$= 2a^3 + a - 3$ — Simplify each term by reducing

3. $\dfrac{5a^7 + 15a^5 - 10a}{5a^2} = \dfrac{5a^7}{5a^2} + \dfrac{15a^5}{5a^2} - \dfrac{10a}{5a^2}$ — Divide each term of numerator by the monomial denominator

$= a^5 + 3a^3 - \dfrac{2}{a}$ — Simplify each term by reducing

Recall that we can check our division by

(quotient)(divisor) = dividend

In example 2,

$(2a^3 + a - 3)(4a) = 2a^3 \cdot 4a + a \cdot 4a - 3 \cdot 4a$ — Distributive property

$= 8a^4 + 4a^2 - 12a$ — Dividend

▶ ***Quick check*** Find the quotient: $\dfrac{16x^5 + 20x^3 - 4x^2}{4x^2}$ ■

Note A common error in this type of problem is demonstrated in the following example.

$$\frac{x^3 + x^2}{x^2} \neq \frac{x^3 + 1}{1}$$

It is tempting to simply "cancel" the x^2 in the numerator with the x^2 in the denominator, but the correct procedure would be

$$\frac{x^3 + x^2}{x^2} = \frac{x^3}{x^2} + \frac{x^2}{x^2} = x + 1$$

Remember that the entire numerator and the entire denominator must be divided by the same quantity. In the example, only *part* of the numerator was divided by x^2. *Only factors* may be divided by factors.

Division of a polynomial by a polynomial

Consider a quotient in which the divisor is not a monomial. For example,

$$\frac{y^2 - y - 2}{y - 2} \quad \begin{matrix} \leftarrow \text{Dividend} \\ \leftarrow \text{Divisor} \end{matrix}$$

which involves the division of a trinomial by a binomial. We handle this just like long division with numbers. Set it up in the form

$$y - 2\overline{)y^2 - y - 2}$$

Note The divisor and dividend must be arranged in descending powers of one variable with zeros inserted to hold the position of any missing term.

The following table demonstrates writing a polynomial in descending powers of the variable and inserting zeros to hold the position of missing terms.

Dividend	**Dividend arranged in descending powers**
$x^3 + 2x + 3x^4 + 4x^2 - 1$	$3x^4 + x^3 + 4x^2 + 2x - 1$
$x^3 + x - 9$	$x^3 + 0x^2 + x - 9$
$x^4 - 1$	$x^4 + 0x^3 + 0x^2 + 0x - 1$

The method for dividing polynomials is similar to the long division used in dividing whole numbers. To demonstrate this, we divide 972 by 36 step-by-step as we divide $(y^2 - y - 2)$ by $(y - 2)$.

$$36\overline{)972} \qquad\qquad y - 2\overline{)y^2 - y - 2}$$

Step 1 Divide 36 into 97, which goes 2 times. Place 2 over 7 in the dividend.

$$\begin{array}{r} 2 \\ 36\overline{)972} \end{array}$$

Divide y into y^2, which goes y times. Place y over y in the dividend.

$$\begin{array}{r} y\phantom{{}-2} \\ y - 2\overline{)y^2 - y - 2} \end{array}$$

Step 2 Multiply 2 times 36, place 72 below 97 in the dividend.

$$\begin{array}{r} 2 \\ 36\overline{)972} \\ 72 \end{array}$$

Multiply y times $(y - 2)$, place $y^2 - 2y$ below $y^2 - y$ in the dividend.

$$\begin{array}{r} y\phantom{{}-2} \\ y - 2\overline{)y^2 - y - 2} \\ \underline{y^2 - 2y}\phantom{{}-2} \end{array}$$

Step 3 Subtract 72 from 97. The difference is 25.

$$\begin{array}{r} 2 \\ 36\overline{)972} \\ (-)\underline{72} \\ 25 \end{array}$$

Subtract $y^2 - 2y$ from $y^2 - y$.
$(y^2 - y) - (y^2 - 2y) =$
$y^2 - y - y^2 + 2y = y$

$$\begin{array}{r} y\phantom{{}-2} \\ y - 2\overline{)y^2 - y - 2} \\ (-)\underline{y^2 - 2y}\phantom{{}-2} \\ y\phantom{{}-2} \end{array}$$

Change signs and add

Step 4 Bring down the next digit of the dividend, 2.

$$\begin{array}{r} 2 \\ 36\overline{)972} \\ \underline{72} \\ 252 \end{array}$$

Bring down the next term of the dividend, -2.

$$\begin{array}{r} y\phantom{{}-2} \\ y - 2\overline{)y^2 - y - 2} \\ \underline{y^2 - 2y}\phantom{{}-2} \\ y - 2 \end{array}$$

Step 5 Divide 36 into 252, which goes 7 times. Place 7 over 2 in the dividend.

$$\begin{array}{r} 27 \\ 36\overline{)972} \\ \underline{72} \\ 252 \end{array}$$

Divide y into y, which goes 1 time. Place 1 over 2 in the dividend with a plus sign between y and 1.

$$\begin{array}{r} y + 1 \\ y - 2\overline{)y^2 - y - 2} \\ \underline{y^2 - 2y}\phantom{{}-2} \\ y - 2 \end{array}$$

Step 6 Multiply 7 times 36, which is 252. Place this product below 252 at the bottom.

$$\begin{array}{r} 27 \\ 36\overline{)972} \\ \underline{72} \\ 252 \\ (-)\underline{252} \\ 0 \end{array}$$

Multiply 1 times $(y - 2)$, which is $y - 2$. Place this below $y - 2$ at the bottom.

$$\begin{array}{r} y + 1 \\ y - 2\overline{)y^2 - y - 2} \\ \underline{y^2 - 2y} \\ y - 2 \\ (-)\underline{y - 2} \\ 0 \end{array}$$

Change signs and add

Step 7 Subtract $252 - 252 = 0$. There is no remainder.

$972 \div 36 = 27$

Subtract $(y - 2) - (y - 2) = 0$.

$(y^2 - y - 2) \div (y - 2)$
$= y + 1$

Step 8 Check your division by multiplying the quotient by the divisor to see if you get the original dividend.

$27 \cdot 36 = 972$ $\qquad$ $(y + 1)(y - 2) = y^2 - y - 2$

Note A common error is committed when we subtract polynomials as we did in step 3. Remember, to subtract two polynomials, *change the signs of the second polynomial and then add.*

$$\begin{array}{rcl} y^2 - y & \rightarrow & y^2 - y \\ (-)\underline{y^2 - 2y} & \rightarrow & \underline{-y^2 + 2y} \\ & & 0 + y = y \end{array}$$

The large majority of errors in this type of problem occur when polynomials are subtracted.

Example 5–3 C

Find the indicated quotient and check the answer.

1. $\dfrac{x^2 + 3x - 4}{x + 4}$

$$\begin{array}{r} x - 1 \\ x + 4\overline{)x^2 + 3x - 4} \\ \underline{x^2 + 4x} \\ -x - 4 \\ \underline{-x - 4} \\ 0 \end{array}$$

$x(x + 4) = x^2 + 4x$
Subtract to get $-x$ and bring down -4
$-1(x + 4) = -x - 4$
Subtract to get 0

Therefore, $\dfrac{x^2 + 3x - 4}{x + 4} = x - 1$.

Check: $(x - 1)(x + 4) = x^2 + 4x - x - 4$
$= x^2 + 3x - 4$

If we still have a remainder after "bringing down" all of the terms of the dividend, handle it as follows:

2. $\dfrac{a^2 + 5a + 6}{a - 2}$

$$\begin{array}{r} a + 7 \\ a - 2\overline{)a^2 + 5a + 6} \\ \underline{a^2 - 2a} \\ 7a + 6 \\ \underline{7a - 14} \\ 20 \end{array}$$

$a(a - 2) = a^2 - 2a$
Subtract to get $7a$. Bring down 6
$7(a - 2) = 7a - 14$
$(7a + 6) - (7a - 14) = 7a + 6 - 7a + 14 = 20$

Hence, $\dfrac{a^2 + 5a + 6}{a - 2} = a + 7 + \dfrac{20}{a - 2}$, where the remainder 20 is placed over the divisor $a - 2$.

To check our answer, we add the remainder of 20 to the product of $(a + 7)$ and $(a - 2)$.

$$\begin{aligned}(a + 7)(a - 2) + (20) &= a^2 - 2a + 7a - 14 + (20)\\ &= a^2 + 5a + 6\end{aligned}$$

3. $\dfrac{x^3 - x + 2}{x - 3}$

Note that there is no term in the dividend that contains x^2. The division will be easier to perform if the term $0x^2$ is inserted as a placeholder so that all powers of the variable x are present in descending order. Thus, we have

$$\frac{x^3 + 0x^2 - x + 2}{x - 3}$$

and the value of the dividend has not been changed since we have added $0x^2$, which is 0. Therefore, to perform the division, we get

$$\begin{array}{r l}
 & x^2 + 3x + 8 \\
x - 3 \,\big) & \overline{x^3 + 0x^2 - x + 2} \\
 & (-)\underline{x^3 - 3x^2} \\
 & \quad 3x^2 - x \\
 & \quad (-)\underline{3x^2 - 9x} \\
 & \qquad 8x + 2 \\
 & \qquad (-)\underline{8x - 24} \\
 & \qquad\qquad 26
\end{array}$$

$$\frac{x^3 - x + 2}{x - 3} = x^2 + 3x + 8 + \frac{26}{x - 3}$$

Check: $(x^2 + 3x + 8)(x - 3) + (26)$
$= x^3 - 3x^2 + 3x^2 - 9x + 8x - 24 + 26$
$= x^3 - x + 2$

Note When performing division, always count the number of terms in the denominator. If there is only one term, *do not use long division.*

▶ ***Quick check*** Find the quotient: $\dfrac{6x^2 - 7x - 3}{2x - 3}$ ■

Mastery points

Can you

- Divide a monomial by a monomial?
- Divide a polynomial by a monomial?
- Divide a polynomial by a polynomial?
- Check the answer?

Exercise 5–3

Perform the indicated divisions and check the answers. See example 5–3 A.

Example $\dfrac{5^3x^4}{5x}$

Solution $= 5^{3-1}x^{4-1}$ Divide like bases by subtracting exponents

$= 5^2x^3$ Perform subtractions

$= 25x^3$ $5^2 = 25$

1. $\dfrac{8x^3}{2x}$
2. $\dfrac{-15x^5}{3x^2}$
3. $\dfrac{-65x^4y^2z}{13xy}$
4. $\dfrac{-28a^3b}{-7ab}$
5. $\dfrac{3(a-b)^2}{a-b}$
6. $\dfrac{5(x+y)^3}{x+y}$
7. $\dfrac{6a^2(b-c)^2}{3a(b-c)}$
8. $\dfrac{-10x^3(y-z)^3}{2x^2(y-z)}$
9. $\dfrac{12a^3b^2c(x+y)^3}{-3abc(x+y)^2}$

See example 5–3 B.

Example $\dfrac{16x^5 + 20x^3 - 4x^2}{4x^2}$

Solution $= \dfrac{16x^5}{4x^2} + \dfrac{20x^3}{4x^2} - \dfrac{4x^2}{4x^2}$ Divide denominator into each term of numerator

$= 4x^3 + 5x - 1$ Divide constants and apply properties of exponents

10. $\dfrac{6x-9}{3}$
11. $\dfrac{24a^2 - 12a}{-6}$
12. $\dfrac{bx^2 - bx}{bx}$
13. $\dfrac{a^3 - 3a^2 + 2a}{a}$
14. $\dfrac{12x^3 - 8x^2 + 3x}{4x}$
15. $\dfrac{15a^3 - 9a^2 + 12a - 6}{3a}$
16. $\dfrac{13a - a^2b^2 + a^2b}{a^2b}$
17. $\dfrac{x^2y - xy^2 - 2xy^3}{-xy}$
18. $\dfrac{14a^2b^3 - 21a^2b^2 - 28ab}{7ab}$
19. $\dfrac{30x^3y^4 + 21x^2y^2 - 18x^2y^4}{3x^2y^2}$
20. $\dfrac{-21m^2n^5 + 35m^3n^2 - 14m^2n^2}{-7m^2n^2}$
21. $\dfrac{a(b-1) - c(b-1)}{b-1}$
22. $\dfrac{a(x-y) - b(x-y)}{x-y}$

See example 5–3 C.

Example $\dfrac{6x^2 - 7x - 3}{2x - 3}$

Solution

$$\begin{array}{r} 3x + 1 \\ 2x - 3\overline{)6x^2 - 7x - 3} \\ (-)\underline{6x^2 - 9x} \\ 2x - 3 \\ (-)\underline{2x - 3} \\ 0 \end{array}$$

$3x(2x - 3) = 6x^2 - 9x$

Subtract to get $2x$ and bring down -3

$1(2x - 3) = 2x - 3$

$(2x - 3) - (2x - 3) = 2x - 3 - 2x + 3 = 0$

$$\frac{6x^2 - 7x - 3}{2x - 3} = 3x + 1$$

The check is left to the student.

23. $\dfrac{a^2 + 7a + 10}{a - 2}$

24. $\dfrac{x^2 + 8x + 15}{x + 5}$

25. $\dfrac{a^2 + 5a + 10}{a + 3}$

26. $\dfrac{x^2 - x - 72}{x + 8}$

27. $(a^2 + 6a + 10) \div (a + 3)$

28. $(4a^2 + 1 + 4a) \div (2a + 1)$

29. $(9a^2 - 24a + 12) \div (3a - 4)$

30. $(27a^3 - 1) \div (3a - 1)$

31. $(x^3 - 8) \div (x - 2)$

32. $(x^4 - 14) \div (x - 2)$

33. $\dfrac{x^3 + 4x^2 + 7x + 6}{x + 2}$

34. $\dfrac{2a^3 - 3a^2 - 13a + 12}{a - 5}$

35. $\dfrac{b^3 + 6b^2 + 7b - 8}{b - 1}$

36. $\dfrac{6x^4 - x^3 - 2x^2 - 7x - 19}{2x - 3}$

37. $(15a^2 + 28a - 32) \div (5a - 4)$

38. $(x^4 - 2x^3 + 4x^2 - x + 3) \div (x^2 - x + 4)$

39. $(x^4 + 3x^3 - 6x^2 + 3x - 8) \div (x^2 + 3x - 5)$

40. $(y^4 + 2y^3 - 4y + 2) \div (y^2 - y + 1)$

41. $(y^4 + 2y - 3) \div (y^2 + 2y - 5)$

42. A contractor uses the expression $x^2 + 6x + 8$ to represent the area of the floor of a room. If she decides that the length of the room will be represented by $x + 4$, what will the width of the room be in terms of x?

43. An electrician uses the expression $4x^2 + 11x + 6$ to determine the amount of wire to order when wiring a house. If the expression comes from multiplying the number of rooms times the number of outlets and he knows the number of rooms to be $x + 2$, find the number of outlets in terms of x.

44. What polynomial when divided by $3x - 2$ yields the quotient $2x^2 + 3x - 5$?

45. What polynomial when divided by $-2x + 5$ yields the quotient $3x^3 - 2x + 6$?

Review exercises

1. Find the solution set of the quadratic equation $4y^2 + 9y + 2 = 0$. See section 4–7.

2. The area of a rectangle is 42 square feet. If the length is 1 foot longer than the width, what are the dimensions of the rectangle? See section 4–8.

Find the following products. See section 3–2.

3. $(4x - 3)^2$

4. $(x + 2)(x^2 + x - 1)$

5. $(5y - 1)(5y + 1)$

6. Reduce $\dfrac{3y^2 - 5y - 2}{2y^2 - y - 6}$ to lowest terms. See section 5–2.

5–4 ■ Ratio and proportion

A ratio

We learned the fraction $\frac{a}{b}$ represents the indicated quotient of a divided by b. A **ratio** compares two numbers, or quantities, in the same way.

Ratio

A ratio is the comparison of two numbers (or quantities) by division.

The ratio of the number a to the number b is written

$$a \text{ to } b, \quad \frac{a}{b}, \quad \text{or} \quad a : b$$

We read $a : b$ as "the ratio of a to b," where a and b are called the *terms* of the ratio. The first number given is always the numerator and the second number is the denominator of the fraction representing the ratio.

■ Example 5–4 A

Write each ratio statement in the forms $a : b$ and $\frac{a}{b}$ reduced to lowest terms.

1. The ratio of 3 to 4

$3:4$ — Form $a : b$

$\frac{3}{4}$ — Written as a fraction

2. The ratio of 15 to 9

$15 : 9$ or $5 : 3$ — Form $a : b$ (divide 15 and 9 by the common factor 3)

$\frac{15}{9} = \frac{5}{3}$ — Written as a fraction reduced to lowest terms

3. The ratio of $2\frac{1}{2}$ to $3\frac{1}{4}$

We first write each mixed number as an improper fraction.

$$2\frac{1}{2} = \frac{5}{2} \text{ and } 3\frac{1}{4} = \frac{13}{4}$$

$$\frac{2\frac{1}{2}}{3\frac{1}{4}} = \frac{5}{2} \div \frac{13}{4} = \frac{5}{2} \cdot \frac{4}{13} = \frac{10}{13}$$

Therefore, the ratio $2\frac{1}{2} : 3\frac{1}{4}$ becomes $10 : 13$ or $\frac{10}{13}$.

▶ ***Quick check*** Write the ratio of 18 to 21 in the forms $a : b$ and $\frac{a}{b}$ reduced to lowest terms.

If the quantities have the same unit of measure, the ratio will be expressed by a fraction without any unit designation required.

4. 45 minutes to 60 minutes

$$45 \text{ min} : 60 \text{ min} = \frac{45 \text{ min}}{60 \text{ min}} \qquad \text{Write as fraction } \frac{a}{b}$$

$$= \frac{3}{4} \text{ or } 3 : 4 \qquad \text{Reduce by dividing each term by 15}$$

When the compared quantities are not of the same unit of measure but *can be stated* in the same unit, it may be desirable to do so. The ratio again becomes only a fraction, as in example 4.

5. 3 feet to 4 inches

Since 1 ft = 12 in., then 3 ft = 36 in., we have

$$36 \text{ in.} : 4 \text{ in.} = \frac{36 \text{ in.}}{4 \text{ in.}} \qquad \text{Write as fraction } \frac{a}{b}$$

$$= \frac{9}{1} \text{ or } 9 : 1 \qquad \text{Reduce by dividing each term by 4}$$

Note We reduced to $\frac{9}{1}$ to demonstrate the comparison that is present.

6. 35 cents to 4 dollars

Since there are 400 cents in 4 dollars, we have

$$35 \text{ cents} : 400 \text{ cents} = \frac{35 \text{ cents}}{400 \text{ cents}} \qquad \text{Write as a fraction } \frac{a}{b}$$

$$= \frac{7}{80} \text{ or } 7 : 80 \qquad \text{Reduce by dividing each term by 5}$$

Note When we change to a common unit of measure, it is easiest to change to the *smaller* unit of measure, as we did in the previous examples. Changing to the larger unit of measure usually involves fractions that are more difficult to reduce.

▶ ***Quick check*** Write the ratio of 16 minutes to 2 hours in the forms $a : b$ and $\frac{a}{b}$ reduced to lowest terms. ■

Ratios are used to indicate relationships in many areas of the physical world.

1. The geographer makes maps and prints to scale, 20 miles to 1 inch.
2. The physicist measures air pressure and uses force per unit of area.

$$14.7 \text{ lb/in.}^2 = \frac{14.7 \text{ lb}}{1 \text{ in.}^2}$$

3. The auto mechanic interprets engine specifications by compression ratio, 9 to 1 or 9 : 1.
4. The machinist is concerned with gear ratio, 2 to 1 or 2 : 1.

When you compare two measurable quantities by ratio, it is not necessary for them to have the same unit of measure. If the units are not the same, you *must include the units* when you are expressing the ratio. These ratios represent rates of change.

■ *Example 5–4 B*

Express the following as a ratio in lowest terms.

1. 50 miles to 1 inch

$$50 \text{ mi} : 1 \text{ in.} = \frac{50 \text{ mi}}{1 \text{ in.}} \text{ or 50 miles per inch}$$

2. 350 miles to 7 hours

$$350 \text{ mi} : 7 \text{ hr} = \frac{350 \text{ mi}}{7 \text{ hr}}$$ Write as a fraction $\frac{a}{b}$

$$= \frac{50 \text{ mi}}{1 \text{ hr}} \text{ (stated 50 miles per hour)}$$ Reduce by dividing each term by 7

▶ ***Quick check*** Express 64 pounds to 8 square inches as a ratio in lowest terms. ■

A proportion

A **proportion** establishes a relationship between two ratios.

Definition of a proportion

A **proportion** is a statement of equality of two ratios.

Given the ratios a to b and c to d,

$$\frac{a}{b} = \frac{c}{d} \quad \text{or} \quad a : b = c : d$$

is a proportion. We read the statement $a : b = c : d$ "a is to b as c is to d." The numbers a, b, c, and d are called the *terms* of the proportion.

Given the proportion $\frac{a}{b} = \frac{c}{d}$,

$$bd \cdot \frac{a}{b} = bd \cdot \frac{c}{d}$$ Multiply each member by bd

$$ad = bc$$ Reduce by b on the left and by d on the right

Property of proportions

$$\text{If } \frac{a}{b} = \frac{c}{d}, \text{ then } ad = bc \ (b, d \neq 0)$$

Note The products ad and bc are found by multiplying diagonally.

$$\frac{a}{b} \times \frac{c}{d} \qquad \begin{matrix} bc \\ ad \end{matrix}$$

This process is frequently called *cross-multiplying,* especially by persons in applied fields, and ad and bc are called the *cross products.*

Example 5–4 C

Determine if the following statements form a proportion.

1. $\frac{3}{5} = \frac{12}{20}$

Using the property of proportions, we obtain

$5 \cdot 12 = 60$ and $3 \cdot 20 = 60$

The cross products are both 60, so we have a proportion.

2. $\frac{5}{6} = \frac{16}{18}$

Using the property of proportions, we obtain

$6 \cdot 16 = 96$ and $5 \cdot 18 = 90$

The cross products are not the same so we do not have a proportion. ■

We use the property of proportions to find the unknown term of a proportion if three of the four terms are known.

Example 5–4 D

Find the unknown term of the given proportion. Check your solution.

1. $\frac{x}{8} = \frac{16}{64}$

$64 \cdot x = 8 \cdot 16$	Property of proportions
$64x = 128$	Multiply as indicated
$x = 2$	Divide each member by 64

Check: $\frac{2}{8} = \frac{16}{64}$ Then

$$2 \cdot 64 = 8 \cdot 16$$
$$128 = 128$$

2. $\frac{49}{y} = \frac{35}{5}$

$49 \cdot 5 = 35 \cdot y$	Property of proportions
$35y = 245$	Multiply as indicated
$y = 7$	Divide each member by 35

Check: $\frac{49}{7} = \frac{35}{5}$ Then

$$5 \cdot 49 = 7 \cdot 35$$
$$245 = 245$$

▶ ***Quick check*** Find the value of z in the proportion $\frac{72}{z} = \frac{30}{6}$. ■

Proportions are used in solving many applied problems. Consider the following examples.

■ *Example 5–4 E*

Set up a proportion for each problem and solve.

1. Two gears are in the ratio of 4 : 5. If the smaller gear has 32 teeth, how many teeth are there in the larger gear?

Let $x =$ the number of teeth in the larger gear. Set up a proportion: one ratio involving 4 and the number of teeth in the smaller gear and the other involving 5 and the number of teeth in the larger gear. Corresponding numbers must be in the numerator and the denominator.

$\frac{4}{32} = \frac{5}{x}$	Set up a proportion
$4 \cdot x = 32 \cdot 5$	Property of proportions
$4x = 160$	Multiply as indicated
$x = 40$	Divide each member by 4

Thus, there are 40 teeth in the larger gear.

2. On a map, 1 inch represents 6 miles. How many inches are needed to represent 28 miles?

Let $x =$ the number of inches representing 28 miles. Now, 1 inch is to 6 miles as x inches is to 28 miles.

$\frac{1 \text{ in.}}{6 \text{ mi}} = \frac{x \text{ in.}}{28 \text{ mi}}$	Set up a proportion
$6 \cdot x = 1 \cdot 28$	Property of proportions
$x = \frac{1 \cdot 28}{6} = \frac{28}{6}$	Divide each member by 6
$= \frac{14}{3} \text{ or } 4\frac{2}{3}$	Reduce to lowest terms

Therefore, 28 miles are represented by $4\frac{2}{3}$ inches on the map.

Note In example 2, **the same units of measure are in the numerator of the ratios and the same units of measure are in the denominators.** That is, we placed inches in the numerator and miles in the denominator of each ratio. This step is important in setting up the proportion you will use to solve for the unknown.

3. Cheryl set aside \$20 per week for her savings program when her salary was \$200 per week. If her salary is now \$250 per week, how much should she set aside for her weekly savings to be proportional to what she saved before?

Let $x =$ the amount to be set aside when Cheryl earns \$250 per week. Then, \$20 is to \$200 as x is to \$250.

$\frac{20}{200} = \frac{x}{250}$	Set up a proportion
$200 \cdot x = 20 \cdot 250$	Property of proportions
$200x = 5{,}000$	Multiply as indicated
$x = \frac{5{,}000}{200}$	Divide each member by 200
$x = 25$	

Cheryl should set aside \$25 when making \$250 per week.

▶ ***Quick check*** On a map, 1 inch represents 9 miles. How many inches are needed to represent 42 miles? ■

Mastery points

Can you

- Write ratios?
- Reduce ratios?
- Set up proportions?
- Solve proportions for the unknown?
- Set up proportions to solve problems?

Exercise 5–4

Express the given ratios in two forms, $\frac{a}{b}$ and $a : b$, reduced to lowest terms. See example 5–4 A.

Example The ratio of 18 to 21

Solution $18 : 21$ or $6 : 7$ — Form $a : b$ (reduce by dividing each term by 3)

$\frac{18}{21}$ or $\frac{6}{7}$ — Write as a fraction reduced to lowest terms

1. 12 to 7 **2.** 8 to 19 **3.** 7 to 42 **4.** 32 to 60

5. 16 to 6 **6.** 24 to 9 **7.** 15 to 5 **8.** 90 to 40

9. 8 to $\frac{3}{4}$ **10.** 12 to $1\frac{3}{4}$ **11.** $2\frac{1}{2}$ to 10 **12.** $5\frac{5}{6}$ to 3

13. $\frac{5}{6}$ to $\frac{2}{3}$ **14.** $\frac{7}{5}$ to $\frac{3}{4}$ **15.** $3\frac{1}{4}$ to $2\frac{2}{3}$ **16.** $3\frac{4}{5}$ to $4\frac{1}{10}$

17. 4.2 to 2.1 **18.** 2.4 to 6.0 **19.** 1.02 to 2.38 **20.** 3.06 to 2.55

Find the indicated ratios reduced to lowest terms expressed in two forms. See examples 5–4 A and B.

Example 16 minutes to 2 hours

Solution Since 2 hours = 120 minutes (1 hour = 60 minutes),

$16 \text{ min} : 2 \text{ hr} = 16 \text{ min} : 120 \text{ min}$ — Replace 2 hr with 120 min

$= 16 : 120$ — Eliminate unit of measure

$= 2 : 15$ or $\frac{2}{15}$ — Reduce to lowest terms (Divide by 8)

The ratio of 16 min to 2 hr is $2 : 15$ or $\frac{2}{15}$.

Example 64 pounds to 8 square inches

Solution $64 \text{ lb to } 8 \text{ sq in.} = \frac{64 \text{ lb}}{8 \text{ sq in.}}$ — Write ratio as a fraction

$= \frac{8 \text{ lb}}{1 \text{ sq in.}}$ — Divide by 8 to reduce to lowest terms

$= 8 \text{ lb/sq in.}$ — Write as a rate

21. 6 in. to 14 in.
22. 4 ft to 18 ft
23. 25 cm to 10 cm
24. 35 lb to 5 lb
25. 36 km to 24 km
26. 48 lb to 16 lb
27. 15 in. to 3 ft
28. 10 ft to 4 yd
29. \$3 to 35¢
30. 5 days to 15 weeks
31. 30 min to 13 hr
32. 16 lb to 8 oz
33. 48 lb to 24 ft³
34. 50 cm to 5 in.³
35. 16 grams to 2 cm³
36. 300 mi to 10 gal
37. 1,020 mi to 17 hr
38. 105 kg to 35 m³

Solve the following applied problems.

39. The *output* in horsepower is the useful energy delivered *by an engine* and the *input* in horsepower is the amount of energy delivered *to an engine*. The *mechanical efficiency* of the engine is given by the ratio

$$\text{mechanical efficiency} = \frac{\text{output}}{\text{input}}$$

Find the mechanical efficiency of an engine rated to deliver 425 horsepower (input) when it delivers only 375 horsepower.

40. The *pitch* of a roof is the ratio of the *rise* of a rafter to the *span* of the roof.

$$\text{pitch} = \frac{\text{rise of rafter}}{\text{span of roof}}$$

Find the pitch if the roof rises 7 feet in a span of 21 feet.

41. The smaller of two belted pulleys makes 240 revolutions per minute and the larger one makes 100 revolutions per minute. What is the ratio of the speed of the larger pulley to the smaller pulley? Of the smaller pulley to the larger pulley?

42. The *mechanical advantage* (*MA*) of a machine is given by the ratio

$$\text{mechanical advantage } (MA) = \frac{\text{resistance to effort (output)}}{\text{input effort}}$$

If a machine has an effort of 30 pounds that results in a resistance of 120 pounds, what is the mechanical advantage of the machine?

43. The mechanical advantage of a hydraulic press can be given by the ratio

$$\text{mechanical advantage} = \frac{\text{area of the large piston}}{\text{area of the small piston}}$$

If the large piston has area 32 square centimeters and the small piston has area 12 square centimeters, find the mechanical advantage of the press.

44. Tool steel may be worked at a cutting speed of 20 feet per minute in a lathe, and cast iron may be worked at a cutting speed of 45 feet per minute. What is the ratio of the cutting speed of tool steel to cast iron?

45. An automobile engine is rated at 350 horsepower. When the engine is tested, it produces only 325 horsepower. What is the mechanical efficiency of the engine? (Refer to exercise 39.)

46. An electric motor uses 10 volts of electricity to produce an equivalent output of 8 volts. What is the mechanical efficiency of the motor? (Refer to exercise 39.)

47. A particular stock costing \$63 paid an earnings of \$6. What is the cost : earnings ratio?

48. A room is 24 feet long and 18 feet wide. What is the ratio of its length to its width?

49. A cement block weighs 120 pounds and a steel block weighs 1,860 pounds. What is the ratio of the weight of the steel block to the weight of the cement block?

50. A mathematics class contains 32 male students and 10 female students. What is the ratio of the male students to the female students?

51. A doctor having earnings of \$60,000 in a given year paid income taxes of \$8,400. What is the ratio of taxes to income?

52. The stress caused by a heavy load is defined by the ratio

$$\text{stress} = \frac{\text{distorting force } (F)}{\text{area } (A)}$$

measured in lb/in.² Find the stress of a force of a 4,400-pound load on an area of 1,200 square inches.

53. Power is defined by the ratio of the work done (F) to the time taken (t). Find the power if 42 ft-lb of work is done in 6 seconds, in ft-lb/sec.

54. The magnification (M) of an object by a lens is given by the ratio

$$M = \frac{q}{p}$$

where q = the image distance and p = the object distance from the lens. Find the magnification of an object whose image distance is 27 feet and object distance is 12 feet.

Find the value of the unknown that makes the proportion true. See example 5–4 D.

Example Find the value of z in the proportion $\frac{72}{z} = \frac{30}{6}$

Solution

$$\frac{72}{z} = \frac{30}{6}$$

$30 \cdot z = 72 \cdot 6$ Property of proportions

$30z = 432$ Perform indicated multiplication

$z = \frac{432}{30}$ Divide by 30

$= \frac{72}{5}$ or $14\frac{2}{5}$ Reduce to lowest terms

Thus, $z = 14\frac{2}{5}$ makes the equation a proportion.

55. $\frac{9}{x} = \frac{36}{5}$ **56.** $\frac{y}{7} = \frac{30}{42}$ **57.** $\frac{5}{9} = \frac{p}{20}$ **58.** $\frac{14}{10} = \frac{21}{z}$

59. $6 : 15 = x : 8$ **60.** $R : 12 = 15 : 100$ **61.** $1\frac{1}{2} : a = 4\frac{3}{4} : 2$ **62.** $\frac{3}{4} : 4 = \frac{1}{2} : b$

63. $1.2 : x = 3.6 : 9$ **64.** $4.5 : 3 = y : 2$ **65.** $\frac{3\frac{1}{4}}{5} = \frac{2\frac{1}{2}}{a}$ **66.** $\frac{2.4}{4.2} = \frac{b}{2.1}$

Solve the following problems by first choosing a letter to represent the unknown and then setting up the proper proportion. See example 5–4 E.

Example On a map, 1 inch represents 9 miles. How many inches are needed to represent 42 miles?

Solution Let x = the number of inches representing 42 miles. Then, 1 inch is to 9 miles as x inches is to 42 miles.

$\frac{1 \text{ inch}}{9 \text{ miles}} = \frac{x \text{ inches}}{42 \text{ miles}}$ Set up a proportion

$9 \cdot x = 1 \cdot 42$ Property of proportions

$9x = 42$ Multiply as indicated

$x = \frac{42}{9}$ Divide each member by 9

$= \frac{14}{3}$ or $4\frac{2}{3}$ Reduce to lowest terms

Thus, $4\frac{2}{3}$ inches represents 42 miles on the map.

67. A man earns \$180 per week. How many weeks must he work to earn \$1,260?

68. An automobile uses 8 liters of gasoline to travel 84 kilometers. How many liters are needed to travel 1,428 kilometers?

69. If 24 grams of water will yield 4 grams of hydrogen, how many grams of hydrogen will there be in 216 grams of water?

70. The operating instructions for a gasoline chain saw call for a 16 gallons : 1 pint fuel-to-oil mixture. How many *pints* of oil are needed to mix with 88 gallons of fuel?

71. The power-to-weight ratio of a given engine is 5 : 3. What is the weight of the engine if it produces 650 horsepower?

72. If a 20-pound casting costs \$1.50, at this same rate, how much would a 42-pound casting cost?

73. A copper wire 300 feet long has a resistance of 1,024 ohms. What is the resistance of 2,000 feet of copper wire?

74. In a hydraulic press, the force on the output piston is to the force on the input piston as the area of the output piston is to the area of the input piston. That is, $\frac{F_o}{F_i} = \frac{A_o}{A_i}$ or $F_o : F_i = A_o : A_i$. Find the area of the input piston if F_o is 15.2 pounds, F_i is 6.5 pounds, and A_o is 10.4 inches^2.

75. If the ratio of the wins to the losses of the Chicago Cubs in a given season is 6:5, how many games did they lose if they won 90 games?

76. A rectangular picture that is 10 inches long and 8 inches wide is to be enlarged so that the enlargement will be 36 inches wide. What should be the length of the enlargement?

77. The corresponding sides of the triangles in the diagram are in proportion. Find the dimensions of the missing sides, x and y. (*Hint:* The corresponding sides are 6″ and x, 7″ and 5″, 8″ and y.)

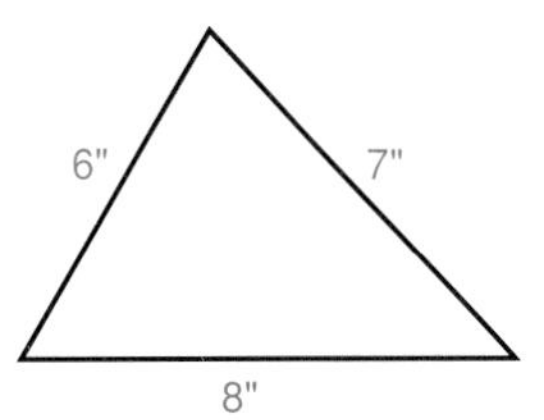

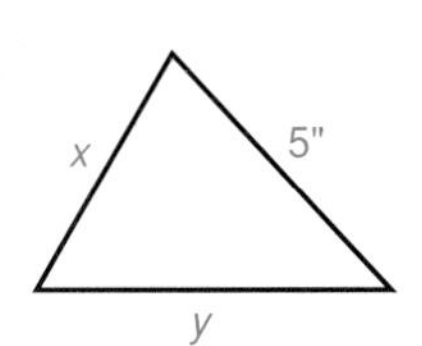

78. Ann is operating a machine that can produce 14 parts in 20 minutes. How long will it take for her to produce 224 parts?

79. A punch machine can make 72 holes in 4 minutes. How many holes can the machine make in 3 hours?

80. A roof rises $6\frac{1}{2}$ feet in a rafter span of 9 feet. At this rate, what would be the rise in a 15-foot span?

81. Nat can type 3 pages of an English paper in 15 minutes. How long would it take him to type 54 pages? (State the answer in hours and minutes.)

82. On a draftsman scale, $\frac{1}{8}$ inch represents 1 foot. What length will a measurement of $2\frac{5}{8}$ inches on the scale represent?

83. An automobile engine uses $\frac{3}{4}$ quart of oil in 900 miles. How much oil will it take in 3,000 miles?

Review exercises

Solve the following equations. See sections 2–4 and 2–7.

1. $6y + 5 = y - 4$ (Find the solution set.)
2. $P = 2\ell + 2w$ for w
3. The product of two consecutive odd integers is 143. Find the integers. See section 4–8.
4. Find the solution set of the inequality $3x - 2 \leq 2(x + 1)$. See section 2–7.

Completely factor each expression. See sections 4–2 and 4–4.

5. $16x^2 - y^2$
6. $x^2 - 16x - 17$
7. $5x^2 - 5x - 10$

Chapter 5 lead-in problem

On a road map, 3 inches represent a distance of 45 miles. If the distance between Detroit and Sault Ste. Marie is 23 inches on the map, how far is it from Detroit to Sault Ste. Marie?

Solution

Let x = the distance from Detroit to Sault Ste. Marie. Then, since 3 inches represents 45 miles on the map, we use the relationship

3 in. is to 45 mi as 23 in. is to x mi

which we write as the proportion

$$\frac{3}{45} = \frac{23}{x}$$

$$3 \cdot x = 45 \cdot 23 \quad \text{Property of proportions}$$

$$x = \frac{45 \cdot 23}{3} \quad \text{Divide each member by 3}$$

$$x = 15 \cdot 23 \quad \text{Reduce by 3}$$

$$x = 345 \quad \text{Multiply in the right member}$$

The distance from Detroit to Sault Ste. Marie is 345 miles.

Chapter 5 summary

1. A **rational expression** can be written in the form $\frac{P}{Q}$, where P and Q are polynomials, $Q \neq 0$.
2. The **domain** of a rational expression in one variable is the set of all replacement values of the variable for which the rational expression is defined.
3. The **fundamental principle of rational expressions** is used to *reduce* rational expressions *to lowest terms* and to obtain equivalent rational expressions having the same denominator for addition and subtraction. It states, $\frac{PR}{QR} = \frac{P}{Q}$, where P, Q, and R are polynomials, Q and R are not equal to zero.
4. To **reduce** a rational expression to lowest terms, we divide the numerator and the denominator by any common factors.
5. To divide a polynomial by a monomial, divide each term of the polynomial by the monomial.
6. To divide a polynomial by a polynomial, the dividend and the divisor must be arranged in descending powers of the same variable with zeros inserted for missing variables.
7. A **ratio** is the comparison of two numbers by division.
8. A **proportion** is a statement of equality of two ratios.
9. If $\frac{a}{b} = \frac{c}{d}$, then $ad = bc$, where ad and bc are called the cross products.

Chapter 5 error analysis

1. Finding the domain of a rational expression
   Example: $\frac{3x+1}{x^2+2x} = \frac{3x+1}{x(x+2)}$
   Domain is all real numbers except -2.
   Correct answer: Domain is all real numbers except 0 and -2.
   What error was made? (*see page 204*)
2. Reducing to lowest terms by "cancelling"
   Example: $\frac{x-3}{x^2-9} = \frac{\overset{1}{\cancel{x}} - \overset{1}{\cancel{3}}}{\underset{x}{\cancel{x^2}} - \underset{3}{\cancel{9}}} = \frac{1-1}{x-3} = \frac{0}{x-3} = 0$
   Correct answer: $\frac{1}{x+3}$
   What error was made? (*see page 208*)

3. Dividing a polynomial by a monomial

Example: $\frac{y - y^2}{y} = \frac{\overset{1}{\cancel{y}} - y^2}{\underset{1}{\cancel{y}}} = 1 - y^2$

Correct answer: $1 - y$

What error was made? (*see page 212*)

4. Dividing a polynomial by a polynomial

Example: $\frac{x^2 - 3x - 4}{x + 1}$

$$\begin{array}{r} x - 2 \\ x + 1\overline{)x^2 - 3x - 4} \\ \underline{x^2 + x} \\ -2x - 4 \\ \underline{-2x - 2} \\ -6 \end{array} = x - 2 - \frac{6}{x + 1}$$

Correct answer: $x - 4$

What error was made? (*see page 213*)

5. Solving proportion problems

Example: On a map, 1 inch represents 10 miles. How many inches represent 24 miles? Let x = inches represented by 24 miles.

$\frac{1}{10} = \frac{24}{x}$

$x = 240$ inches

Correct answer: 2.4 inches

What error was made? (*see page 223*)

6. Order of operations

Example: $10 - 12 \div 2 + 6 = 5$

Correct answer: 10

What error was made? (*see page 57*)

7. Division using zero

Example: $\frac{0}{-3}$ is undefined

Correct answer: $\frac{0}{-3} = 0$

What error was made? (*see page 53*)

8. Combining like terms

Example: $8xy - 4xy = 4$

Correct answer: $4xy$

What error was made? (*see page 80*)

9. Reciprocal of a number

Example: The reciprocal of 1 is -1.

Correct answer: The reciprocal of 1 is itself.

What error was made? (*see page 94*)

10. Properties of exponents

Example: $(3x)^3 = 3x^3$

Correct answer: $27x^3$

What error was made? (*see page 129*)

Chapter 5 critical thinking

If m and n represent 2 integers where $n > m$, how many integers are there from m to n?

Chapter 5 review

[5–1]

Determine the domain of the given rational expression.

1. $\frac{x + 1}{x}$
2. $\frac{y - 3}{y + 7}$
3. $\frac{3x + 1}{x - 9}$
4. $\frac{2z - 5}{3z + 2}$
5. $\frac{x}{5x - 3}$
6. $\frac{x^2 - x + 4}{x^2 + x - 12}$
7. $\frac{x^2 + 3x + 2}{x^2 - 1}$

[5–2]

Reduce the following rational expressions to lowest terms. Assume no denominator is equal to zero.

8. $\frac{18ab^2}{6a^2b}$
9. $\frac{45x^2yz^3}{30xy^3z^2}$
10. $\frac{x^2 - 49}{x^2 + 14x + 49}$
11. $\frac{x^2 - 3x - 18}{x^2 + x - 42}$
12. $\frac{18a - 6b}{15a - 5b}$
13. $\frac{x^2 - y^2}{y - x}$
14. $\frac{3p^2 - 8p + 4}{5p^2 - 9p - 2}$
15. $\frac{2R^2 - 32}{6R^2 + 22R - 8}$
16. $\frac{20 - 9n + n^2}{8 + 2n - n^2}$

[5–3]

Find the indicated quotients.

17. $\dfrac{24x^3}{-3x}$

18. $\dfrac{2a^2 - 3a + 5a^3}{a}$

19. $\dfrac{5x^2y - 3xy^4 + x^2y^2}{xy}$

20. $\dfrac{8a^3b + 12a^2b^2 - 24a^3b^7}{4a^2b}$

21. $\dfrac{8a^2 - 2a - 3}{2a - 1}$

22. $\dfrac{3a^2 - 17a + 11}{a - 5}$

23. $\dfrac{x^2 - 49}{x + 7}$

24. $\dfrac{20x^3 - 19x^2 - 13x + 12}{4x - 3}$

[5–4]

Find the indicated ratios in two different forms reduced to lowest terms.

25. 15 meters to 35 meters

26. 36 pounds to 16 pounds

27. 12 inches to $2\frac{1}{2}$ feet

28. 450 miles to 15 gallons

29. In business, the current ratio compares current assets to current liabilities and represents the measure of the firm's ability to pay off the liabilities over a time period. What is the current ratio, reduced to lowest terms, if the firm's total current assets are $4,386 and total current liabilities are $1,762?

30. The May company has 42 sales representatives who are meeting their sales quota. If another 18 sales representatives have fallen short of their quota, what is the ratio of success to failure?

Find the value of the unknown that makes the statement a proportion.

31. $\dfrac{8}{x} = \dfrac{9}{36}$

32. $\dfrac{5.4}{3.6} = \dfrac{a}{2.4}$

33. $y : 18 = 15 : 25$

34. $\dfrac{5}{6} : \dfrac{1}{2} = \dfrac{2}{3} : p$

35. If a blueprint is drawn to the scale $\frac{1}{8}$ inch = 1 foot, what is the size of the corresponding part of a final product if the blueprint measurement is $4\frac{3}{8}$ inches?

36. An automobile has a 16-quart cooling system. If the ratio of antifreeze to water is 3 to 1, how much of each does the system have? (*Hint:* Let x be the amount of antifreeze. Then $16 - x$ is the amount of water.)

Chapter 5 cumulative test

Perform the indicated operations.

[1–8] 1. $-4[7(12 - 2) - 8^3 + 3]$

[1–7] 2. $\dfrac{(-6)(-8)}{(-2)(0)}$

[1–8] 3. -6^2

Perform the indicated operations and simplify.

[2–3] 4. $(5a - b) - [3a - (4b + 3a)]$

[3–2] 5. $(3x - 2)^2$

[3–2] 6. $(5y - 2)(5y + 2)$

[3–2] 7. $(4a + 3b)(a - 6b)$

[3–2] 8. $(x - y)^3$

[2–2] 9. Given $a = -5$, $b = 3$, $c = 4$, and $d = -6$, evaluate the expression $(2a - 3b) - (5c + d)$.

Find the solution set.

[2–6] 10. $3(x + 2) - 2(x - 4) = 12$

[2–6] 11. $\dfrac{3x}{4} - 5 = 1$

[2–9] 12. $-4 \leq 2x + 5 < 11$

[2–9] 13. $2(3x - 4) > 5(x - 1)$

Write in completely factored form.

[4–1] **14.** $x(m + n) - y(m + n)$

[4–3] **15.** $3a^2 + 7a + 4$

[4–4] **16.** $4x^2 - 20x + 25$

[4–4] **17.** $4z^2 - 9$

[4–4] **18.** $36 - y^2$

[4–2] **19.** $x^2 - 12x - 45$

[4–5] **20.** $2a^3 - 16b^3$

Find the solution set.

[4–7] **21.** $x^2 - 5x - 14 = 0$

[4–7] **22.** $2y^2 + 3y - 9 = 0$

Simplify and leave answers with only positive exponents.

[3–4] **23.** $(4yz^{-1})^{-3}$

[3–4] **24.** $\frac{a^{-7}}{a^{-10}}$

[3–4] **25.** $(4ab^2)(-2a^3b)(-a^{-1}b^{-3})$

[5–4] **26.** Find x when $36 : x = 21{:}14$.

[5–4] **27.** What is the ratio of 52 pounds to 24 pounds?

[5–3] **28.** Divide $(x^2 - 8x + 13) \div (x - 2)$.

Reduce the following expressions to lowest terms. Assume that no denominator is equal to zero.

[1–1] **29.** $\frac{56}{42}$

[3–3] **30.** $\frac{36ab^3}{28a^3b^2}$

[5–2] **31.** $\frac{a^2 - 36}{a^2 - a - 42}$

[5–2] **32.** $\frac{8x - 8y}{5x^2 - 5y^2}$

[5–2] **33.** $\frac{y^2 - y - 20}{y^2 - 25}$

[5–2] **34.** $\frac{2x^2 - 5xy - 3y^2}{6x^2 + 7xy + 2y^2}$

[5–4] **35.** A photograph that is 8 inches by 12 inches is to be enlarged. If the enlargement calls for the longest side to be 32 inches, how many inches should the other side be?

[5–4] **36.** Two lots are proportional in their lengths and widths. If the larger lot is 15 feet wide and 28 feet long and the length of the smaller lot is 24 feet, how wide is the smaller lot?

CHAPTER

6 Operations with Rational Expressions

Marc owns $\frac{5}{8}$ interest in a print shop and his uncle owns $\frac{1}{4}$ interest in the shop. In a given year, they shared earnings of \$140,000. How much did the shop earn that year?

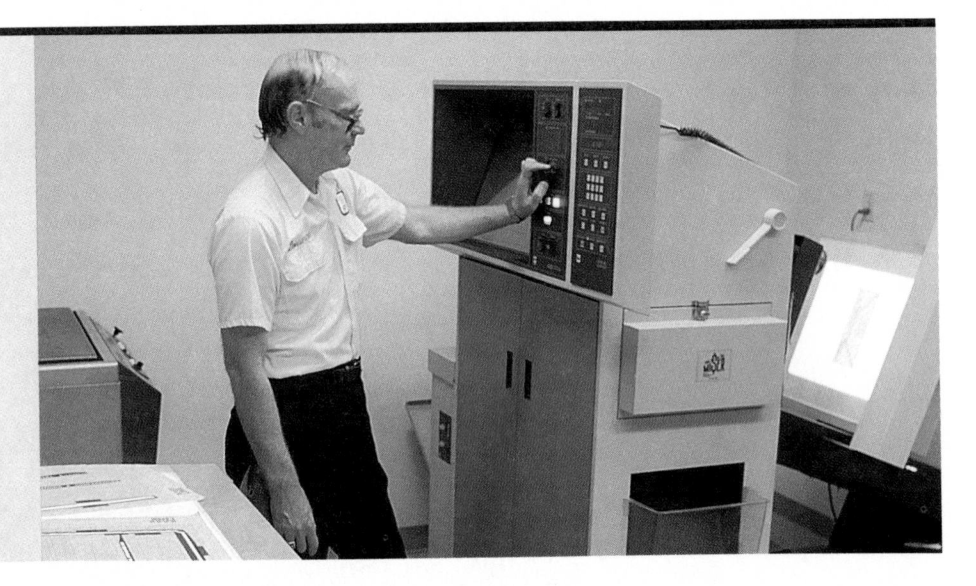

6–1 ■ Multiplication and division of rational expressions

Multiplication of rational expressions

Recall that to multiply two real number fractions we multiply the numerators and multiply the denominators.

Multiplication property of fractions

If a, b, c, and d are real numbers, then

$$\frac{a}{b} \cdot \frac{c}{d} = \frac{a \cdot c}{b \cdot d} \ (b, d \neq 0)$$

Note Any possible reduction is performed *before* the multiplication takes place.

■ Example 6–1 A

Multiply the fractions $\frac{3}{7}$ and $\frac{14}{27}$ and simplify the product.

$\frac{3}{7} \cdot \frac{14}{27} = \frac{3 \cdot 14}{7 \cdot 27}$ Multiply the numerators; Multiply the denominators

$= \frac{3 \cdot 2 \cdot 7}{7 \cdot 3 \cdot 3 \cdot 3}$ Factor the numerator and the denominator

$= \frac{2 \cdot (3 \cdot 7)}{3 \cdot 3 \cdot (3 \cdot 7)}$ Group the common factors $(3 \cdot 7)$

$$= \frac{2}{3 \cdot 3}$$ Divide numerator and denominator by $(3 \cdot 7)$

$$= \frac{2}{9}$$ Multiply the remaining factors ■

This same procedure is followed when we multiply two rational expressions.

Multiplication property of rational expressions

Given rational expressions $\frac{P}{Q}$ and $\frac{R}{S}$, then

$$\frac{P}{Q} \cdot \frac{R}{S} = \frac{P \cdot R}{Q \cdot S} \quad (Q, S \neq 0)$$

Since we want the resulting product to be stated in lowest terms, we apply the *fundamental principle of rational expressions* and divide both the numerator and the denominator by their common factors. That is, we *reduce* by dividing out the common factors.

Multiplication of rational expressions

1. State the numerators and denominators as indicated products. (Do not multiply.)
2. Factor the numerator and the denominator.
3. Divide the numerators and the denominators by the factors that are common.
4. Multiply the remaining factors in the numerator and place this product over the product of the remaining factors in the denominator.

■ Example 6–1 B

Perform the indicated multiplication and simplify your answer. Assume that no denominator equals zero.

1. $\frac{3x}{4} \cdot \frac{5}{2y}$

$$= \frac{3x \cdot 5}{4 \cdot 2y}$$ Multiply numerators and denominators

$$= \frac{15x}{8y}$$ Will not reduce

2. $\frac{x + 1}{x - 3} \cdot \frac{4}{x + 2}$

$$= \frac{(x + 1) \cdot 4}{(x - 3)(x + 2)}$$ Multiply numerators and denominators

$$= \frac{4x + 4}{x^2 - x - 6}$$ Will not reduce

3. $\dfrac{4}{9x} \cdot \dfrac{3x^2}{2}$

$= \dfrac{4 \cdot 3x^2}{9x \cdot 2}$ Multiply numerators / Multiply denominators

$= \dfrac{2 \cdot 2 \cdot 3 \cdot x \cdot x}{3 \cdot 3 \cdot 2 \cdot x}$ Factor numerator and denominator

$= \dfrac{2 \cdot x \cdot (2 \cdot 3 \cdot x)}{3 \cdot (2 \cdot 3 \cdot x)}$ Identify common factors

$= \dfrac{2x}{3}$ Divide numerator and denominator by $(2 \cdot 3 \cdot x)$

4. $\dfrac{x+1}{3-x} \cdot \dfrac{(x-3)^2}{x-2}$

$= \dfrac{(x+1) \cdot (x-3)^2}{(3-x) \cdot (x-2)}$ Multiply numerators / Multiply denominators

$= \dfrac{x-3}{3-x} \cdot \dfrac{(x+1)(x-3)}{x-2}$ Factor opposites

$= \dfrac{-1 \cdot (x+1)(x-3)}{x-2}$ $\dfrac{x-3}{3-x} = -1$

$= \dfrac{-1(x^2-2x-3)}{x-2}$ Multiply as indicated

$= \dfrac{-x^2+2x+3}{x-2}$

5. $\dfrac{x^2-8x+16}{x^2+3x-10} \cdot \dfrac{x^2-4}{x^2-5x+4}$

$= \dfrac{(x^2-8x+16)(x^2-4)}{(x^2+3x-10)(x^2-5x+4)}$ Multiply numerators and denominators

$= \dfrac{(x-4)(x-4)(x-2)(x+2)}{(x-4)(x-2)(x+5)(x-1)}$ Factor numerator and denominator

$= \dfrac{(x-4)(x+2)}{(x+5)(x-1)}$ Reduce by common factors $(x-4)$ and $(x-2)$

$= \dfrac{x^2-2x-8}{x^2+4x-5}$ Multiply remaining factors

► ***Quick check*** Multiply $\dfrac{12}{5y} \cdot \dfrac{15y^2}{4}$ ■

Division of rational expressions

Recall that to divide two fractions $\dfrac{a}{b}$ and $\dfrac{c}{d}$, we multiply $\dfrac{a}{b}$ by the *reciprocal* of $\dfrac{c}{d}$, which is $\dfrac{d}{c}$.

Division property of fractions

If a, b, c, and d are real numbers, then

$$\frac{a}{b} \div \frac{c}{d} = \frac{a}{b} \cdot \frac{d}{c} = \frac{a \cdot d}{b \cdot c} \quad (b, c, d \neq 0)$$

Example 6–1 C

Find the indicated quotient. Reduce your answer to lowest terms.

$\frac{18}{25} \div \frac{9}{5}$

$= \frac{18}{25} \cdot \frac{5}{9}$ — Multiply by the reciprocal of $\frac{9}{5}$

$= \frac{2 \cdot 3 \cdot 3 \cdot 5}{5 \cdot 5 \cdot 3 \cdot 3}$ — Factor in numerator and denominator

$= \frac{2 \cdot (3 \cdot 3 \cdot 5)}{5 \cdot (3 \cdot 3 \cdot 5)}$ — Group the common factors $(3 \cdot 3 \cdot 5)$

$= \frac{2}{5}$ — Reduce by the common factors $(3 \cdot 3 \cdot 5)$ ■

Division of rational expressions is done in the same way.

Division property of rational expressions

If $\frac{P}{Q}$ and $\frac{R}{S}$ are rational expressions, then

$$\frac{P}{Q} \div \frac{R}{S} = \frac{P}{Q} \cdot \frac{S}{R} = \frac{P \cdot S}{Q \cdot R} \quad (Q, R, S \neq 0)$$

Notice that once the operation of division has been changed to multiplication, we proceed exactly as we did with the multiplication of rational expressions.

Division of rational expressions

1. Multiply the first rational expression by the reciprocal of the second.
2. Proceed as in the multiplication of rational expressions.

Example 6–1 D

Find the indicated quotients. Express the answer in reduced form.

1. $\frac{3ab}{5} \div \frac{9abc}{10}$

$= \frac{3ab}{5} \cdot \frac{10}{9abc}$ — Multiply by the reciprocal of $\frac{9abc}{10}$

$= \frac{3ab \cdot 2 \cdot 5}{5 \cdot 3 \cdot 3 \cdot abc}$ — Factor numerator and denominator

$= \frac{2 \cdot (3 \cdot 5 \cdot ab)}{3 \cdot c \cdot (3 \cdot 5 \cdot ab)}$ — Group common factors $(3 \cdot 5 \cdot ab)$

$= \frac{2}{3c}$ — Divide numerator and denominator by common factors $(3 \cdot 5 \cdot ab)$

2. $\dfrac{x^2 - 4}{5} \div \dfrac{x - 2}{15}$

$= \dfrac{x^2 - 4}{5} \cdot \dfrac{15}{x - 2}$ — Multiply by the reciprocal of $\dfrac{x - 2}{15}$

$= \dfrac{(x - 2)(x + 2) \cdot 3 \cdot 5}{5 \cdot (x - 2)}$ — Factor numerator and denominator

$= \dfrac{3 \cdot (x + 2) \cdot 5(x - 2)}{5(x - 2)}$ — Locate common factors $5(x - 2)$

$= \dfrac{3(x + 2)}{1} = 3x + 6$ — Reduce to lowest terms by dividing numerator and denominator by $5(x - 2)$

3. $\dfrac{4x + 2}{x - 1} \div \dfrac{2x + 1}{4 - 4x}$

$= \dfrac{4x + 2}{x - 1} \cdot \dfrac{4 - 4x}{2x + 1}$ — Multiply by the reciprocal of $\dfrac{2x + 1}{4 - 4x}$

$= \dfrac{2(2x + 1)(-4)(x - 1)}{(x - 1)(2x + 1)}$ — Factor $4 - 4x = -4(x - 1)$ and $4x + 2 = 2(2x + 1)$

$= \dfrac{2(-4)(2x + 1)(x - 1)}{(2x + 1)(x - 1)}$ — Reduce by $(x - 1)(2x + 1)$

$= 2(-4)$

$= -8$

4. $\dfrac{x^2 - 9}{2x + 1} \div \dfrac{3 - x}{2x^2 + 7x + 3}$

$= \dfrac{x^2 - 9}{2x + 1} \cdot \dfrac{2x^2 + 7x + 3}{3 - x}$ — Multiply by the reciprocal of $\dfrac{3 - x}{2x^2 + 7x + 3}$

$= \dfrac{(x - 3)(x + 3) \cdot (2x + 1)(x + 3)}{(2x + 1)(3 - x)}$ — Factor numerators

$= \dfrac{x - 3}{3 - x} \cdot \dfrac{(2x + 1)(x + 3)(x + 3)}{2x + 1}$ — Factor opposites

$= -1 \cdot (x + 3)(x + 3)$ — $\dfrac{x - 3}{3 - x} = -1$; reduce by $(2x + 1)$

$= -1(x^2 + 6x + 9)$ — Multiply as indicated

$= -x^2 - 6x - 9$ — Multiply by -1

▶ ***Quick check*** Divide $\dfrac{x^2 - 4}{4x - 1} \div \dfrac{2 - x}{4x^2 + 3x - 1}$ ■

Mastery points

Can you

- Multiply rational expressions?
- Divide rational expressions?

Exercise 6–1

Find the indicated product or quotient. Write your answer in simplest form. Assume all denominators are nonzero. See examples 6–1 A–D.

Example $\frac{12}{5y} \cdot \frac{15y^2}{4}$

Solution

$= \frac{12 \cdot 15y^2}{5y \cdot 4}$ Multiply numerator and denominator

$= \frac{3 \cdot 2 \cdot 2 \cdot 3 \cdot 5 \cdot y \cdot y}{5 \cdot y \cdot 2 \cdot 2}$ Factor numerator and denominator

$= \frac{3 \cdot 3 \cdot y(5 \cdot y \cdot 2 \cdot 2)}{1(5 \cdot y \cdot 2 \cdot 2)}$ Group common factors $(5 \cdot y \cdot 2 \cdot 2)$

$= \frac{9y}{1}$ Divide by common factors $(5 \cdot y \cdot 2 \cdot 2)$

$= 9y$

Example $\frac{x^2 - 4}{4x - 1} \div \frac{2 - x}{4x^2 + 3x - 1}$

Solution

$= \frac{x^2 - 4}{4x - 1} \cdot \frac{4x^2 + 3x - 1}{2 - x}$ Multiply by the reciprocal of $\frac{2 - x}{4x^2 + 3x - 1}$

$= \frac{(x + 2)(x - 2) \cdot (4x - 1)(x + 1)}{(4x - 1)(2 - x)}$ Factor numerator and denominator

$= \frac{x - 2}{2 - x} \cdot \frac{(x + 2)(x + 1)(4x - 1)}{4x - 1}$ Factor opposites

$= -1 \cdot (x + 2)(x + 1)$ $\frac{x - 2}{2 - x} = -1$; reduce by $4x - 1$

$= -x^2 - 3x - 2$ Multiply as indicated

1. $\frac{24}{35} \cdot \frac{7}{8}$
2. $\frac{3}{8} \cdot \frac{5}{9}$
3. $\frac{7}{10} \div \frac{21}{25}$
4. $\frac{56}{39} \div \frac{8}{13}$
5. $\frac{4a}{5} \cdot \frac{5}{2}$
6. $\frac{16b}{7a} \cdot \frac{5}{4}$
7. $\frac{14}{3a} \div \frac{7}{15a}$
8. $\frac{6x}{5y} \div \frac{21x}{15y}$
9. $\frac{5}{6} \cdot \frac{3x}{10y}$
10. $\frac{7a}{12b} \cdot \frac{9b}{28}$
11. $\frac{9x^2}{8} \cdot \frac{4}{6x}$
12. $\frac{36p^2}{7q} \cdot \frac{14q^2}{28p^3}$
13. $\frac{24a}{35x} \div 6a$
14. $\frac{14y}{23x} \div 7y$
15. $6a \div \frac{24a}{35x}$
16. $7y \div \frac{14y}{23}$
17. $\frac{21ab}{16c} \cdot \frac{8c^2}{3ab^2}$
18. $\frac{18x^2y^2}{5ab} \cdot \frac{25a^2b}{12xy}$
19. $\frac{5x^2}{9y^3} \div \frac{20x}{6y}$
20. $\frac{28m}{15n} \div \frac{7m^2}{3n^3}$
21. $\frac{24abc}{7xyz^2} \cdot \frac{14x^2yz}{9a^2}$
22. $\frac{80x^2yz^3}{11mn^2} \cdot \frac{33mn^2}{25xyz}$
23. $\frac{3ab}{8x^2} \div \frac{15b^3}{16x}$
24. $\frac{20mn^3}{9x^2} \div \frac{4mn}{3xy^2}$
25. $\frac{x + y}{3} \cdot \frac{12}{(x + y)^2}$
26. $\frac{5(a - b)}{8} \cdot \frac{12}{10(a - b)}$
27. $\frac{9 - p}{7} \div \frac{4(p - 9)}{21}$
28. $\frac{4x - 2}{15} \div \frac{1 - 2x}{27}$
29. $\frac{3b - 6}{4b + 8} \cdot \frac{5b + 10}{2 - b}$
30. $\frac{8y + 16}{3 - y} \cdot \frac{4y - 12}{3y + 6}$
31. $\frac{4a + 12}{a - 5} \div (a + 3)$
32. $\frac{9 - 3z}{2z + 8} \div (6 - 2z)$

33. $(x^2 - 4x + 4) \cdot \dfrac{18}{x^2 - 4}$

34. $\dfrac{21}{a^2 - 9} \cdot (a^2 + a - 12)$

35. $\dfrac{x^2 - 4}{25y} \cdot \dfrac{24y^2}{x + 2}$

36. $\dfrac{16a^2}{b^2 - 9} \cdot \dfrac{b - 3}{12a^2}$

37. $\dfrac{r^2 - 16}{r + 1} \div \dfrac{r + 4}{r^2 - 1}$

38. $\dfrac{p^2 + 2p + 1}{4p - 1} \div \dfrac{p^2 - 1}{16p^2 - 1}$

39. $\dfrac{9 - x^2}{x + y} \cdot \dfrac{4x + 4y}{x - 3}$

40. $\dfrac{b^2 - a^2}{2a + 4b} \cdot \dfrac{a + b}{a - b}$

41. $\dfrac{a^2 - 5a + 6}{a^2 - 9a + 20} \cdot \dfrac{a^2 - 5a + 4}{a^2 - 3a + 2}$

42. $\dfrac{a^2 - 5a - 14}{a^2 - 9a - 36} \cdot \dfrac{a^2 + 10a + 21}{a^2 + 4a - 77}$

43. $\dfrac{x^2 - 2x - 3}{x^2 + 3x - 4} \div \dfrac{x^2 - x - 6}{x^2 + x - 12}$

44. $\dfrac{y^2 + 3y + 2}{y^2 + 5y + 4} \div \dfrac{y^2 + 5y + 6}{y^2 + 10y + 24}$

45. $\dfrac{2x^2 - 15x + 7}{x^2 - 9x + 8} \cdot \dfrac{x^2 - 2x + 1}{x^2 - 49}$

46. $\dfrac{4x^2 - 4}{3x^2 - 13x - 10} \cdot \dfrac{x^2 - 6x + 5}{4x + 4}$

47. $\dfrac{6r^2 - r - 7}{12r^2 + 16r - 35} \div \dfrac{r^2 - r - 2}{2r^2 + r - 10}$

48. $\dfrac{4x^2 - 9}{x^2 - 9x + 18} \div \dfrac{2x^2 - 5x - 12}{x^2 - 10x + 24}$

49. $(3x^2 - 2x - 8) \div \dfrac{x^2 - 4}{x + 2}$

50. $(8a^2 - 16a) \div \dfrac{a^3 - 16a}{a - 4}$

51. $\dfrac{3x - 4}{2x + 1} \div (6x^2 - 5x - 4)$

52. $\dfrac{m^2 - 3m - 10}{m^2 - 4} \div (2m^2 - 9m - 5)$

53. $\dfrac{10}{a^3 - 27} \cdot \dfrac{a^2 + 3a - 18}{15}$

54. $\dfrac{x^3 - 8}{16} \cdot \dfrac{24}{x^2 + 2x - 8}$

55. $\dfrac{z^2 - 5z - 14}{z - 4} \div \dfrac{5z^3 + 40}{z^2 - z - 12}$

56. $\dfrac{3b^3 + 3}{b - 2} \div \dfrac{b^2 + 2b + 1}{b^2 + 6b - 16}$

57. $\dfrac{y^2 + 8y + 16}{y + 4} \cdot \dfrac{y^2 - 25}{y^2 + 9y + 20} \cdot \dfrac{y^2 + 5y}{y^2 - 5y}$

58. $\dfrac{6m^2 - 7m + 2}{6m^2 + 5m + 1} \cdot \dfrac{2m^2 + m}{4m^2 - 1} \cdot \dfrac{12m^2 - 5m - 3}{12m^2 - 17m + 6}$

Review exercises

Add or subtract the following. See section 1–1.

1. $\dfrac{3}{4} + \dfrac{5}{6}$

2. $\dfrac{7}{8} - \dfrac{5}{12}$

Completely factor the following. See sections 4–2 and 4–3.

3. $2x^2 - 50$

4. $x^2 + 9x - 22$

5. $x^2 + 8x + 16$

Solve the following proportions. See section 5–4.

6. $\dfrac{3}{x} = \dfrac{5}{8}$

7. $\dfrac{5}{9} = \dfrac{y}{27}$

8. Write the number 0.0000789 in scientific notation. See section 3–5.

6–2 ■ Addition and subtraction of rational expressions

Recall that to add or subtract fractions having the same denominator, we add, or subtract, the numerators and place this sum, or difference, over the same denominator.

Addition and subtraction properties for fractions

If a, b, and c are real numbers, $b \neq 0$, then

$$\frac{a}{b} + \frac{c}{b} = \frac{a + c}{b} \quad \text{and} \quad \frac{a}{b} - \frac{c}{b} = \frac{a - c}{b}$$

■ Example 6–2 A

Add or subtract as indicated.

1. $\frac{3}{11} + \frac{4}{11} = \frac{3 + 4}{11}$ — Add numerators

$= \frac{7}{11}$ — $3 + 4 = 7$

2. $\frac{3}{7} - \frac{1}{7} = \frac{3 - 1}{7}$ — Subtract numerators

$= \frac{2}{7}$ — $3 - 1 = 2$ ■

We use the following similar procedure to add or subtract rational expressions.

Addition and subtraction properties for rational expressions

If $\frac{P}{R}$ and $\frac{Q}{R}$ are rational expressions, $R \neq 0$, then

$$\frac{P}{R} + \frac{Q}{R} = \frac{P + Q}{R} \quad \text{and} \quad \frac{P}{R} - \frac{Q}{R} = \frac{P - Q}{R}$$

Note Rational expressions having common denominators are called *like* rational expressions.

Addition and subtraction of like rational expressions

1. Add or subtract the numerators.
2. Place the sum or difference over the common denominator.
3. Reduce the resulting rational expression to lowest terms.

■ Example 6–2 B

Find the indicated sum or difference. Assume all denominators are nonzero.

1. $\frac{3}{x - 2} + \frac{5}{x - 2} = \frac{3 + 5}{x - 2}$ — Add numerators and place over $x - 2$

$= \frac{8}{x - 2}$ — $3 + 5 = 8$

2. $\dfrac{5y}{3y+5} - \dfrac{9y}{3y+5} = \dfrac{5y-9y}{3y+5}$ Subtract numerators and place over $3y + 5$

$= \dfrac{-4y}{3y+5}$ $5y - 9y = -4y$

3. $\dfrac{2x-1}{x^2+5x+6} - \dfrac{4-x}{x^2+5x+6}$

$= \dfrac{(2x-1)-(4-x)}{x^2+5x+6}$ Place numerators in parentheses and subtract

$= \dfrac{2x-1-4+x}{x^2+5x+6}$ Remove parentheses and subtract

$= \dfrac{3x-5}{x^2+5x+6}$ Combine like terms

Note Notice that when we subtracted $4 - x$ from $2x - 1$, we placed parentheses around each polynomial. This step is *important* to avoid the common mistake of failing to change signs in the second expression when subtraction is involved.

4. $\dfrac{2x-1}{x^2+5x+6} + \dfrac{4-x}{x^2+5x+6}$

$= \dfrac{(2x-1)+(4-x)}{x^2+5x+6}$ Place numerators in parentheses and add

$= \dfrac{x+3}{x^2+5x+6}$ Remove parentheses and combine like terms

$= \dfrac{x+3}{(x+3)(x+2)}$ Factor denominator

$= \dfrac{1}{x+2}$ Reduce by common factor $x + 3$

Note In the last step, *always* look for a possible reduction to lowest terms as we did in example 4.

▶ ***Quick check*** $\dfrac{4m-5}{m^2+9} - \dfrac{2m-3}{m^2+9}$ ■

When one denominator is the opposite of the other, as in the indicated sum

$$\frac{2x}{3} + \frac{5}{-3},$$

where 3 and -3 are opposites, we first multiply one of the expressions by $\dfrac{-1}{-1}$ to obtain equivalent expressions with the same denominator.

■ ***Example 6–2 C***

Find the indicated sum or difference. Assume all denominators are not zero.

1. $\dfrac{2x}{3} + \dfrac{5}{-3} = \dfrac{2x}{3} + \dfrac{-1}{-1} \cdot \dfrac{5}{-3}$ Multiply $\dfrac{5}{-3}$ by $\dfrac{-1}{-1}$

$= \dfrac{2x}{3} + \dfrac{-1(5)}{-1(-3)}$ Multiply numerators and denominators

$= \dfrac{2x}{3} + \dfrac{-5}{3}$ Common denominator of 3

$= \dfrac{2x + (-5)}{3}$ Add numerators and place over 3

$= \dfrac{2x - 5}{3}$ Definition of subtraction

2. $\dfrac{5y - 1}{y - 4} + \dfrac{2y + 3}{4 - y}$

$= \dfrac{5y - 1}{y - 4} + \dfrac{-1}{-1} \cdot \dfrac{2y + 3}{4 - y}$ Multiply $\dfrac{2y + 3}{4 - y}$ by $\dfrac{-1}{-1}$

$= \dfrac{5y - 1}{y - 4} + \dfrac{-1(2y + 3)}{-1(4 - y)}$

$= \dfrac{5y - 1}{y - 4} + \dfrac{-2y - 3}{y - 4}$ Same denominator: $-1(4 - y) = y - 4$

$= \dfrac{(5y - 1) + (-2y - 3)}{y - 4}$ Add numerators in parentheses

$= \dfrac{5y - 1 - 2y - 3}{y - 4}$ Remove parentheses

$= \dfrac{3y - 4}{y - 4}$ Combine like terms

Note We could have multiplied $\dfrac{5y - 1}{y - 4}$ by $\dfrac{-1}{-1}$. The resulting denominator would then have been $4 - y$ and the numerator would have been $4 - 3y$. We would then multiply this by $\dfrac{-1}{-1}$ to obtain the same form of the answer.

3. $\dfrac{2x + 1}{x - 5} - \dfrac{x - 4}{5 - x}$

$= \dfrac{2x + 1}{x - 5} - \dfrac{-1}{-1} \cdot \dfrac{x - 4}{5 - x}$ Multiply $\dfrac{x - 4}{5 - x}$ by $\dfrac{-1}{-1}$

$= \dfrac{2x + 1}{x - 5} - \dfrac{-1(x - 4)}{-1(5 - x)}$

$= \dfrac{2x + 1}{x - 5} - \dfrac{4 - x}{x - 5}$ Same denominator: $-1(x - 4) = 4 - x$, $-1(5 - x) = x - 5$

$= \dfrac{(2x + 1) - (4 - x)}{x - 5}$ Place "( )" around numerators and subtract

$= \dfrac{2x + 1 - 4 + x}{x - 5}$ Definition of subtraction

$= \dfrac{3x - 3}{x - 5}$ Combine like terms

▶ ***Quick check*** $\dfrac{x + 7}{x - 1} + \dfrac{3x + 1}{1 - x}$ ■

The least common denominator (LCD)

If the fractions to be added or subtracted do not have the same denominator, we must change at least one of the fractions to an equivalent fraction so the fractions do have a common denominator. There are many such numbers we could use as a common denominator. However, the most convenient denominator to use is the smallest (least) number that is exactly divisible by each of the denominators—called the **least common denominator,** denoted by LCD. For example, the least common denominator (LCD) of the two fractions

$$\frac{5}{6} \text{ and } \frac{2}{9}$$

is 18, since 18 is the smallest (least) number that is exactly divisible by both 6 and 9.

Finding the LCD of a set of denominators

1. Factor each denominator completely. Write each factorization using exponential notation.
2. List each *different* factor that appears in any one of the factorizations in step 1.
3. Raise each factor of step 2 to the *greatest* power that factor has in step 1. Form the product of these factors.

Note The LCD of two or more rational expressions is also called the least common multiple (LCM) of the denominators.

■ ***Example 6–2 D***

Find the LCD of rational expressions having the given denominators.

1. 6 and 9

$$\left.\begin{aligned} 6 &= 2 \cdot 3 \\ 9 &= 3 \cdot 3 = 3^2 \end{aligned}\right\} \quad \text{Factor each denominator}$$

The different factors are 2 and 3. The greatest power of 2 is 2^1 and of 3 is 3^2. The LCD is $2^1 \cdot 3^2 = 2 \cdot 9 = 18$.

2. $16a$ and $8a^3$

$$16a = 2 \cdot 2 \cdot 2 \cdot 2 \cdot a = 2^4 \cdot a$$
$$8a^3 = 2 \cdot 2 \cdot 2 \cdot a \cdot a \cdot a = 2^3 \cdot a^3$$

Since the different factors are 2 and a, the greatest power of 2 is 2^4, and the greatest power of a is a^3, the LCD is $2^4 \cdot a^3 = 16a^3$.

3. $50x^3y^2$ and $20x^2y^3$

$$50x^3y^2 = 2 \cdot 5 \cdot 5 \cdot x^3 \cdot y^2 = 2 \cdot 5^2 \cdot x^3 \cdot y^2$$
$$20x^2y^3 = 2 \cdot 2 \cdot 5 \cdot x^2 \cdot y^3 = 2^2 \cdot 5 \cdot x^2 \cdot y^3$$

The different factors are 2, 5, x, and y. Since the greatest power of 2 is 2^2, of 5 is 5^2, of x is x^3, and of y is y^3, the LCD is $2^2 \cdot 5^2 \cdot x^3 \cdot y^3 = 100x^3y^3$.

4. $x^2 + x - 12$ and $x^2 + 2x - 8$

$x^2 + x - 12 = (x + 4)(x - 3)$
$x^2 + 4x - 8 = (x + 4)(x - 2)$

The different factors are $x + 4$, $x - 3$, and $x - 2$. Each factor is carried to the first power so the LCD is $(x + 4)(x - 3)(x - 2)$.

5. $x^2 - 2x + 1$, $x^2 + 11x - 12$, and $1 - x$

$x^2 - 2x + 1 = (x - 1)^2$
$x^2 + 11x - 12 = (x - 1)(x + 12)$
$1 - x = -1(x - 1)$

The different factors are $x - 1$ and $x + 12$. Since the greatest power of $x - 1$ is $(x - 1)^2$ and of $x + 12$ is $(x + 12)^1$, the LCD is $(x - 1)^2(x + 12)$.

Note Factors $x - 1$ and $1 - x$ are opposites and we will change $1 - x$ to $-1(x - 1)$ when working with the denominators. The factor -1 is then carried along in the problem.

▶ ***Quick check*** Find the LCD for $3x - 6$, $x^2 - 4x + 4$, $x^2 - 2x$. ■

Mastery points

Can you

- Add and subtract like rational expressions?
- Add and subtract rational expressions having denominators that are opposites?
- Find the least common denominator (LCD) of a set of rational expressions?

Exercise 6–2

Combine the given rational expressions and reduce the answer to lowest terms. Assume all denominators are nonzero. See example 6–2 B.

Example $\dfrac{4m - 5}{m^2 + 9} - \dfrac{2m - 3}{m^2 + 9}$

Solution $= \dfrac{(4m - 5) - (2m - 3)}{m^2 + 9}$ Place numerators in parentheses and subtract

$= \dfrac{4m - 5 - 2m + 3}{m^2 + 9}$ Remove parentheses and subtract in numerator

$= \dfrac{2m - 2}{m^2 + 9}$ Combine like terms in numerator

1. $\dfrac{5}{x} + \dfrac{3}{x}$

2. $\dfrac{8}{y^2} + \dfrac{10}{y^2}$

3. $\dfrac{9}{p} - \dfrac{2}{p}$

4. $\dfrac{18}{m^2} - \dfrac{5}{m^2}$

5. $\dfrac{5x}{x+2} + \dfrac{9x}{x+2}$ **6.** $\dfrac{8y}{y-1} + \dfrac{-3y}{y-1}$ **7.** $\dfrac{x-1}{2x} - \dfrac{x+3}{2x}$ **8.** $\dfrac{3y-2}{y^2} - \dfrac{4y-1}{y^2}$

9. $\dfrac{3x+5}{x^2-1} - \dfrac{2x+3}{x^2-1}$ **10.** $\dfrac{b^2+2}{b+3} - \dfrac{b^2+2b-3}{b+3}$

See example 6–2 C.

Example $\dfrac{x+7}{x-1} + \dfrac{3x+1}{1-x}$

Solution

$$= \frac{x+7}{x-1} + \frac{-1}{-1} \cdot \frac{3x+1}{1-x} \qquad \text{Multiply } \frac{3x+1}{1-x} \text{ by } \frac{-1}{-1}$$

$$= \frac{x+7}{x-1} + \frac{-1(3x+1)}{-1(1-x)}$$

$$= \frac{x+7}{x-1} + \frac{-3x-1}{x-1} \qquad -1(1-x) = x-1$$

$$= \frac{x+7-3x-1}{x-1} \qquad \text{Add numerators}$$

$$= \frac{-2x+6}{x-1} \qquad \text{Combine like terms}$$

11. $\dfrac{5}{7} + \dfrac{6}{-7}$ **12.** $\dfrac{9}{10} - \dfrac{3}{-10}$ **13.** $\dfrac{4}{z} - \dfrac{5}{-z}$

14. $\dfrac{6}{y} + \dfrac{9}{-y}$ **15.** $\dfrac{5}{x-2} + \dfrac{12}{2-x}$ **16.** $\dfrac{1}{x-7} - \dfrac{5}{7-x}$

17. $\dfrac{5y}{y-6} - \dfrac{4y}{6-y}$ **18.** $\dfrac{4z}{z-3} + \dfrac{z}{3-z}$ **19.** $\dfrac{x+1}{x-5} + \dfrac{2x-3}{5-x}$

20. $\dfrac{4y+3}{y-9} - \dfrac{2y-7}{9-y}$ **21.** $\dfrac{2y-5}{2y-3} - \dfrac{y+7}{3-2y}$ **22.** $\dfrac{z+5}{4z-3} + \dfrac{4z-1}{3-4z}$

23. $\dfrac{2x+5}{5-2x} + \dfrac{x+9}{2x-5}$ **24.** $\dfrac{5-y}{6-5y} - \dfrac{9y+1}{5y-6}$

Find the least common denominator (LCD) of rational expressions having the following denominators. See example 6–2 D.

Example $3x - 6$, $x^2 - 4x + 4$, and $x^2 - 2x$

Solution

$$\left.\begin{aligned} 3x - 6 &= 3(x-2) \\ x^2 - 4x + 4 &= (x-2)^2 \\ x^2 - 2x &= x(x-2) \end{aligned}\right\} \quad \text{Factor each denominator}$$

LCD is $3x(x-2)^2$

25. $6x$ and $9x$ **26.** $8a$ and $12a$ **27.** $16x^2$ and $24x$

28. $6b^2$ and $14b$ **29.** $28y^2$ and $35y^3$ **30.** $9z^3$ and $7z^4$

[illegible] $32a^2$ and $64a^4$ **32.** $4x^2$, $3x$, and $8x^3$ **33.** $10a^2$, $12a^3$, and $9a$

[illegible] $- 2$ and $2x - 1$ **35.** $x - 4$ and $3x - 12$ **36.** $6x - 12$ and $9x - 18$

37. $18y^3$ and $9y - 36$
38. $32z^2$ and $16z - 32$
39. $a^2 + a$ and $a^2 - 1$
40. $(z - 1)^2$ and $z^2 - 1$
41. $8a + 16$ and $a^2 + 3a + 2$
42. $9p - 18$ and $p^2 - 7p + 10$
43. $a^2 - 5a + 6$ and $a^2 - 4$
44. $y^2 - y - 12$ and $y^2 + 6y + 9$
45. $a^2 - 9$ and $a^2 - 5a + 6$
46. $p^2 - 9$, $p^2 + p - 6$, and $p^2 - 4p + 4$
47. $x^2 - 49$, $7 - x$, and $2x + 14$
48. $5 - y$, $y^2 - 25$, and $y^2 - 10y + 25$

Review exercises

1. The statement $4(y + 3) = 4(3 + y)$ demonstrates what property of real numbers? See section 1–8.

Factor the following expressions. See sections 4–2, 4–3, and 4–4.

2. $5y^2 - 20$
3. $x^2 + 20x + 100$
4. $3y^2 - y - 4$

Find the solution set of the following equations. See sections 2–6 and 4–7.

5. $4(x + 3) = 5(4 - 3x)$
6. $x^2 - 2x = 15$

6–3 ■ Addition and subtraction of rational expressions

Now that we can find the least common denominator (LCD) of a group of rational expressions, let us review the process for changing a fraction (or rational expression) to an equivalent fraction with a new denominator.

■ Example 6–3 A

1. Change $\frac{7}{15}$ to an equivalent fraction having denominator 60.

We want $\frac{7}{15} = \frac{?}{60}$.

Since $60 = 15 \cdot 4$ (factor 4 is found by dividing $60 \div 15 = 4$), we multiply the given fraction by $\frac{4}{4}\left(\frac{4}{4} = 1\right)$.

$$\frac{7}{15} = \frac{7}{15} \cdot \frac{4}{4}$$

$$= \frac{7 \cdot 4}{15 \cdot 4} \quad \text{Multiply numerators, Multiply denominators}$$

$$= \frac{28}{60}$$

Thus, $\frac{7}{15} = \frac{28}{60}$.

2. Change $\frac{x+1}{x-4}$ to an equivalent rational expression having denominator $x^2 - 2x - 8$.

We want $\frac{x+1}{x-4} = \frac{?}{x^2 - 2x - 8}$.

Since $x^2 - 2x - 8 = (x-4)(x+2)$, we multiply the given rational expression by $\frac{x+2}{x+2}$.

$$\begin{aligned} \frac{x+1}{x-4} &= \frac{x+1}{x-4} \cdot \frac{x+2}{x+2} \\ &= \frac{(x+1)(x+2)}{(x-4)(x+2)} && \text{Multiply numerators; Multiply denominators} \\ &= \frac{x^2+3x+2}{x^2-2x-8} && \text{Perform indicated operations} \end{aligned}$$

Thus, $\frac{x+1}{x-4} = \frac{x^2+3x+2}{x^2-2x-8}$. ■

Once equivalent rational expressions are obtained with the LCD as the denominator, we add or subtract as previously learned. Use the following steps to add or subtract rational expressions having unlike denominators.

Addition and subtraction of rational expressions having different denominators

1. Find the LCD of the rational expressions.
2. Write each rational expression as an equivalent rational expression with the LCD as the denominator.
3. Perform the indicated addition or subtraction as before.
4. Reduce the results to lowest terms.

■ *Example 6–3 B*

Add the following rational expressions. Assume the denominators are not equal to zero. Reduce all answers to lowest terms.

1. $\frac{5x}{8} + \frac{7x}{12}$

$$\left.\begin{aligned} 8 &= 2 \cdot 2 \cdot 2 \\ 12 &= 2 \cdot 2 \cdot 3 \end{aligned}\right\} \text{LCD} = 2 \cdot 2 \cdot 2 \cdot 3 = 24 \qquad \text{Find the LCD}$$

Since $\frac{24}{8} = 3$ and $\frac{24}{12} = 2$

$$\begin{aligned} \frac{5x}{8} + \frac{7x}{12} &= \frac{5x}{8} \cdot \frac{3}{3} + \frac{7x}{12} \cdot \frac{2}{2} && \text{Multiply } \frac{5x}{8} \text{ by } \frac{3}{3} \text{ and } \frac{7x}{12} \text{ by } \frac{2}{2} \\ &= \frac{15x}{24} + \frac{14x}{24} && \text{Multiply numerators and denominators} \\ &= \frac{15x + 14x}{24} && \text{Add numerators} \\ &= \frac{29x}{24} && \text{Combine as indicated} \end{aligned}$$

2. $\dfrac{15}{4x^2} + \dfrac{25}{18x}$

$$\left.\begin{aligned} 4x^2 &= 2 \cdot 2 \cdot x^2 \\ 18x &= 2 \cdot 3 \cdot 3 \cdot x \end{aligned}\right\} \text{LCD} = 2 \cdot 2 \cdot 3 \cdot 3 \cdot x^2 = 36x^2$$ Find the LCD

Since $\dfrac{36x^2}{4x^2} = 9$ and $\dfrac{36x^2}{18x} = 2x$

$$\frac{15}{4x^2} + \frac{25}{18x} = \frac{15}{4x^2} \cdot \frac{9}{9} + \frac{25}{18x} \cdot \frac{2x}{2x}$$ Multiply $\dfrac{15}{4x^2}$ by $\dfrac{9}{9}$ and $\dfrac{25}{18x}$ by $\dfrac{2x}{2x}$

$$= \frac{135}{36x^2} + \frac{50x}{36x^2}$$ Multiply numerators and denominators

$$= \frac{135 + 50x}{36x^2}$$ Add numerators

3. $\dfrac{3y + 2}{y^2 - 16} + \dfrac{y - 4}{3y + 12}$

$$\left.\begin{aligned} y^2 - 16 &= (y + 4)(y - 4) \\ 3y + 12 &= 3(y + 4) \end{aligned}\right\} \text{LCD} = 3(y + 4)(y - 4)$$ Find the LCD

Since $\dfrac{3(y + 4)(y - 4)}{(y + 4)(y - 4)} = 3$ and $\dfrac{3(y + 4)(y - 4)}{3(y + 4)} = y - 4$, then

$$\frac{3y + 2}{y^2 - 16} + \frac{y - 4}{3y + 12}$$

$$= \frac{3y + 2}{(y + 4)(y - 4)} \cdot \frac{3}{3} + \frac{y - 4}{3(y + 4)} \cdot \frac{y - 4}{y - 4}$$ Multiply $\dfrac{3y + 2}{(y + 4)(y - 4)}$ by $\dfrac{3}{3}$ and $\dfrac{y - 4}{3(y + 4)}$ by $\dfrac{y - 4}{y - 4}$

$$= \frac{3(3y + 2)}{3(y + 4)(y - 4)} + \frac{(y - 4)(y - 4)}{3(y + 4)(y - 4)}$$ Multiply numerators and denominators

$$= \frac{(9y + 6) + (y^2 - 8y + 16)}{3(y + 4)(y - 4)}$$ Add numerators

$$= \frac{y^2 + y + 22}{3(y + 4)(y - 4)}$$ Remove parentheses and combine

Note When the numerators have two or more terms as in example 3, we place the quantities in parentheses when we add the numerators. This is a good practice to avoid a *most common* mistake when subtracting.

▶ ***Quick check*** Add the rational expressions $\dfrac{20}{3y} + \dfrac{25}{12y^2}$. ■

Example 6–3 C

Subtract the following rational expressions. Assume the denominators are not equal to zero. Reduce to lowest terms.

1. $\dfrac{5y}{2y-1} - \dfrac{y+1}{y+2}$

The LCD of the rational expressions is $(2y-1)(y+2)$.

$$\frac{5y}{2y-1} - \frac{y+1}{y+2}$$

$$= \frac{5y}{2y-1}\cdot\frac{y+2}{y+2} - \frac{y+1}{y+2}\cdot\frac{2y-1}{2y-1}$$ Multiply $\frac{5y}{2y-1}$ by $\frac{y+2}{y+2}$ and $\frac{y+1}{y+2}$ by $\frac{2y-1}{2y-1}$

$$= \frac{5y(y+2)}{(2y-1)(y+2)} - \frac{(y+1)(2y-1)}{(2y-1)(y+2)}$$ Multiply numerators and denominators

$$= \frac{(5y^2+10y)-(2y^2+y-1)}{(2y-1)(y+2)}$$ Subtract numerators

$$= \frac{5y^2+10y-2y^2-y+1}{(2y-1)(y+2)}$$ Remove parentheses and change signs when subtracting

$$= \frac{3y^2+9y+1}{(2y-1)(y+2)}$$ Combine like terms

Don't forget the parentheses

Note The numerator $3y^2 + 9y + 1$ cannot be factored so we are unable to reduce. We should always check this!

2. $\dfrac{5x-4}{x^2-2x+1} - \dfrac{3x}{x^2+4x-5}$

$$\left.\begin{aligned} x^2-2x+1 &= (x-1)^2 \\ x^2+4x-5 &= (x+5)(x-1)\end{aligned}\right\} \text{LCD} = (x-1)^2(x+5)$$ Find the LCD

$$\frac{5x-4}{x^2-2x+1} - \frac{3x}{x^2+4x-5}$$ Factor denominators

$$= \frac{5x-4}{(x-1)^2} - \frac{3x}{(x+5)(x-1)}$$

$$= \frac{5x-4}{(x-1)^2}\cdot\frac{x+5}{x+5} - \frac{3x}{(x+5)(x-1)}\cdot\frac{x-1}{x-1}$$ Multiply $\frac{5x-4}{(x-1)^2}$ by $\frac{x+5}{x+5}$ and $\frac{3x}{(x+5)(x-1)}$ by $\frac{x-1}{x-1}$

$$= \frac{(5x-4)(x+5)}{(x-1)^2(x+5)} - \frac{3x(x-1)}{(x-1)^2(x+5)}$$ Multiply numerators and denominators

$$= \frac{(5x^2+21x-20)-(3x^2-3x)}{(x-1)^2(x+5)}$$ Subtract numerators

$$= \frac{5x^2+21x-20-3x^2+3x}{(x-1)^2(x+5)}$$ Remove parentheses and change signs

$$= \frac{2x^2+24x-20}{(x-1)^2(x+5)}$$ Combine like terms

Don't forget the parentheses

▶ ***Quick check*** Subtract $\dfrac{y+1}{y^2-y-12} - \dfrac{3y+2}{y^2-9y+20}$ ■

Problem solving

■ **Example 6–3 D**

Set up a rational expression for the following word statements.

1. If Dick can mow his lawn in h hours, what part of the lawn can he mow in 2 hours?

   We must find what part of the total time, h hours, is 2 hours. Thus, Dick can mow the fractional part

   $$\frac{2}{h}$$

   of the lawn in 2 hours.

2. If the area (A) of a rectangle is n square inches, what is the expression for the length, ℓ, if the width, w, is 12 inches? ($A = \ell \cdot w$)

   Using $A = \ell \cdot w$, we have $\ell = \dfrac{A}{w}$ and so the length

   $\ell = \dfrac{n}{12}$ inches. Replace A with n and w with 12

▶ ***Quick check*** If the area of a rectangle is p square yards, write an expression for the width, w, if the length is 5 yards. ■

Mastery points

Can you

- Add and subtract rational expressions having unlike denominators?

Exercise 6–3

Perform the indicated addition and reduce the answer to lowest terms. Assume all denominators are not equal to zero. See example 6–3 B.

Example $\dfrac{20}{3y} + \dfrac{25}{12y^2}$

Solution $\left.\begin{aligned} 3y &= 3 \cdot y \\ 12y^2 &= 3 \cdot 2 \cdot 2 \cdot y \cdot y \end{aligned}\right\}$ LCD $= 3 \cdot 2 \cdot 2 \cdot y \cdot y = 12y^2$

$\dfrac{12y^2}{3y} = 4y$ and the denominator of the second expression is the LCD, $12y^2$.

$\dfrac{20}{3y} + \dfrac{25}{12y^2} = \dfrac{20}{3y} \cdot \dfrac{4y}{4y} + \dfrac{25}{12y^2}$ Multiply numerator and denominator of $\dfrac{20}{3y}$ by $4y$

$= \dfrac{80y}{12y^2} + \dfrac{25}{12y^2}$ Perform indicated operations

$= \dfrac{80y + 25}{12y^2}$ Add numerators and place over common denominator

1. $\frac{x}{6} + \frac{3}{4}$

2. $\frac{3z}{10} + \frac{2z}{15}$

3. $\frac{2x - 1}{16} + \frac{x + 2}{24}$

4. $\frac{4}{3x} + \frac{5}{2x}$

5. $\frac{3a + 1}{a} + \frac{2a - 3}{3a}$

6. $\frac{4}{x - 1} + \frac{5}{x + 3}$

7. $\frac{8}{y + 4} + \frac{7}{y - 5}$

8. $\frac{x}{x + 2} + \frac{3x}{4x - 1}$

9. $\frac{15}{5y - 10} + \frac{14}{2y - 4}$

10. $\frac{21}{6x + 12} + \frac{15}{2x + 4}$

11. $\frac{12}{x^2 - 4} + \frac{7}{4x - 8}$

12. $\frac{16}{2y + 6} + \frac{5}{y^2 - 9}$

13. $5 + \frac{4x}{x + 8}$

14. $9 + \frac{y + 9}{y - 1}$

15. $\frac{x}{x - 1} + \frac{3x}{x^2 - 1}$

16. $\frac{2y}{y^2 - 16} + \frac{5y}{2y - 8}$

17. $\frac{4}{x^2 - x - 6} + \frac{5}{x^2 - 9}$

18. $\frac{6}{x^2 - 4x - 12} + \frac{5}{x^2 - 36}$

19. $\frac{2y}{y^2 - 6y + 9} + \frac{5y}{y^2 - 2y - 3}$

20. $\frac{4z}{z^2 + z - 20} + \frac{z}{z^2 - 8z + 16}$

21. $\frac{y - 2}{y^2 - 3y - 10} + \frac{y + 1}{y^2 - y - 6}$

22. $\frac{2x + 1}{x^2 + 6x + 5} + \frac{4x - 3}{x^2 - x - 30}$

Perform the indicated subtraction and reduce to lowest terms. Assume all denominators are not zero. See example 6–3 C.

Example $\frac{y + 1}{y^2 - y - 12} - \frac{3y + 2}{y^2 - 9y + 20}$

Solution $\left.\begin{aligned} y^2 - y - 12 &= (y - 4)(y + 3) \\ y^2 - 9y + 20 &= (y - 4)(y - 5) \end{aligned}\right\}$ LCD is $(y - 4)(y + 3)(y - 5)$.

$$\frac{y + 1}{y^2 - y - 12} - \frac{3y + 2}{y^2 - 9y + 20}$$

$$= \frac{y + 1}{(y - 4)(y + 3)} \cdot \frac{y - 5}{y - 5} - \frac{3y + 2}{(y - 4)(y - 5)} \cdot \frac{y + 3}{y + 3}$$

Multiply numerator and denominator of $\frac{y + 1}{(y - 4)(y + 3)}$ by $y - 5$ and of $\frac{3y + 2}{(y - 4)(y - 5)}$ by $y + 3$

$$= \frac{(y + 1)(y - 5)}{(y - 4)(y + 3)(y - 5)} - \frac{(3y + 2)(y + 3)}{(y - 4)(y + 3)(y - 5)}$$

Multiply numerators and denominators

$$= \frac{(y^2 - 4y - 5) - (3y^2 + 11y + 6)}{(y - 4)(y + 3)(y - 5)}$$

Multiply in each numerator, place parentheses around each product, and subtract (Don't forget the parentheses)

$$= \frac{y^2 - 4y - 5 - 3y^2 - 11y - 6}{(y - 4)(y + 3)(y - 5)}$$

Remove parentheses, change signs, and subtract

$$= \frac{-2y^2 - 15y - 11}{(y - 4)(y + 3)(y - 5)}$$

Combine like terms in numerator

Don't forget the parentheses

Note The denominator is left in factored form in case the answer can be reduced.

23. $\frac{y}{9} - \frac{5}{6}$

24. $\frac{5y}{12} - \frac{y}{8}$

25. $\frac{9}{14y} - \frac{1}{21y}$

26. $\frac{7}{12z} - \frac{10}{9z}$

27. $\frac{5a + 3}{12} - \frac{a - 4}{10}$

28. $\frac{2x + 9}{8} - \frac{x - 7}{20}$

29. $\frac{2x + 5}{6x} - \frac{x - 5}{9x}$

30. $\frac{4y - 1}{3y} - \frac{2y - 3}{15y}$

31. $\frac{7}{2x - 3} - \frac{6}{x - 5}$

32. $\frac{7}{4x - 6} - \frac{12}{3x + 9}$

33. $\frac{12}{3y + 6} - \frac{11}{7y + 14}$

34. $\frac{14}{5x - 15} - \frac{8}{2x - 6}$

35. $9 - \frac{6}{x + 8}$

36. $12 - \frac{7}{z - 12}$

37. $\frac{2x}{3x + 1} - 9$

38. $\frac{4y}{5y - 4} - 10$

39. $\frac{-3}{a^2 - 5a + 6} - \frac{3}{a^2 - 4}$

40. $\frac{8}{x^2 - 25} - \frac{7}{x^2 + 3x - 10}$

41. $\frac{20}{y^2 - 2y - 24} - \frac{8}{y^2 + y - 12}$

42. $\frac{2p}{p^2 - 9p + 20} - \frac{5p - 2}{p - 5}$

43. $\frac{2a - 3}{a^2 - 5a + 6} - \frac{3a}{a - 2}$

Add and subtract as indicated.

44. $\frac{13}{12b} - \frac{2}{9b} + \frac{5}{4b}$

45. $\frac{5}{6z} + \frac{4}{8z} - \frac{1}{4z}$

46. $\frac{3x}{8} - \frac{2x}{5} + \frac{7x}{10}$

47. $\frac{4a}{5} + \frac{7a}{15} - \frac{a}{9}$

48. $\frac{a - 1}{5} - \frac{2a + 3}{15} + \frac{3a - 1}{25}$

49. $\frac{5b + 1}{6} + \frac{3b - 2}{9} - \frac{b + 1}{12}$

50. $\frac{4a}{a^2 + 2a - 15} + \frac{3a}{2a^2 + 11a + 5} - \frac{5a}{2a^2 - 5a - 3}$

51. $\frac{5}{z^2 - 4} - \frac{z}{z^2 - 1} + \frac{4}{z^2 + z - 2}$

Solve the following word problems.

52. For a lens maker to manufacture lenses that will refract light by exactly the right amounts, the following formula is used:

$$\frac{1}{f} = (u - 1)\left(\frac{1}{R_1} + \frac{1}{R_2}\right)$$

Add the expressions containing R_1 and R_2.

53. Women A, B, and C can complete a given job in a, b, and c hours, respectively. Working together they can complete in one hour $\frac{1}{a} + \frac{1}{b} + \frac{1}{c}$ of the job. By combining, obtain a single expression for what they can do together in one hour.

54. In electricity, the total resistance of any parallel circuit may be given by

$$\frac{1}{R_t} = \frac{I_1}{E_1} + \frac{I_2}{E_2} + \frac{I_3}{E_3}$$

Combine the expression in the right member.

See example 6–3 D.

Example If the area of a rectangle is p square yards, write an expression for the width, w, if the length is 5 yards.

Solution Using $A = \ell \cdot w$, we are given that $A = p$ square yards and $\ell = 5$ yards. Substituting, we obtain

$$p = 5 \cdot w$$

and solving for w, we divide each member by 5. Thus $w = \frac{p}{5}$.

55. A faucet when fully open, can fill the sink in m minutes. What part of the sink can it fill in 3 minutes?

56. An inlet pipe to a swimming pool can fill the pool in h hours. What part of the pool can it fill in 9 hours?

57. An outlet pipe can drain a swimming pool in 36 hours. What part of the pool can it drain in h hours?

58. Jane can paint her house in h hours. What part of the house can she paint in 1 hour?

59. The product of two numbers is 48. If one of the numbers is m, what is the other number?

60. The area of a rectangle is 54 square centimeters. What is the length ℓ if the width is w centimeters?

61. The area of a rectangle is A square feet. What is the width if the length is 23 feet?

62. The area of a triangle is 21 square yards. If the triangle has a base length b, what is the altitude h of the triangle? $\left(A = \frac{1}{2}bh\right)$

63. The area of a triangle is A square rods. If the altitude h is 9 rods, what is the length of the base b? (See problem 62.)

64. John drives 25 miles in h hours. At what speed, r, did he travel? [*Hint:* Use distance traveled (d) = rate $(r) \times$ time (t).]

65. Mabel travels d miles at a rate of 55 miles per hour. Write an expression for the time t that she traveled. (See problem 64.)

66. What is the reciprocal of the natural number n?

67. What is the reciprocal of the fraction $\frac{a}{b}$?

Review exercises

Completely factor the following expressions. See sections 4–2, 4–3, and 4–4.

1. $x^2 - 14x + 49$
2. $2x^2 - 11x + 5$
3. $4x^2 - 16$

Multiply the following expressions. See section 3–2.

4. $(x + 9)(x - 9)$
5. $(4x + 3)^2$
6. $(2x + 1)(x - 8)$

Find the LCD of expressions having the following denominators. See section 6–2.

7. 16, 12, 6
8. $4x, 2x^2, 6$
9. $x^2 - 9; (x - 3)^2; x + 3$

Add or subtract the following. See section 6–3.

10. $\frac{5}{x} + \frac{3}{2x}$
11. $\frac{5}{x - 2} - \frac{3}{x - 1}$

6–4 ■ Complex fractions

A complex fraction

A **complex fraction** is a fraction (rational expression) whose numerator, denominator, or both contain fractions (or rational expressions). The expressions

$$\frac{\frac{3}{4}}{\frac{5}{6}}, \quad \frac{\frac{1}{3}+2}{1-\frac{1}{2}}, \quad \text{and} \quad \frac{\frac{x-1}{x-2}}{\frac{x+3}{x}}$$

are examples of complex fractions. Given a complex fraction, we simplify the fraction by *eliminating the fractions within the numerator and/or the denominator* to obtain a simple fraction.

We name the parts of a complex fraction as shown in the following examples.

$$\frac{\frac{3}{x}+\frac{4}{y}}{\frac{1}{x}-\frac{2}{y}}$$

$\frac{3}{x}+\frac{4}{y}$ — Primary numerator

$\frac{1}{x}-\frac{2}{y}$ — Primary denominator

In $\frac{\frac{3}{x}+\frac{4}{y}}{\frac{1}{x}-\frac{2}{y}}$: 3 and 4 — Secondary numerators; x and y — Secondary denominators; 1 and 2 — Secondary numerators; x and y — Secondary denominators

To simplify a complex fraction, we use one of the following methods.

Simplifying a complex fraction

Method 1 Multiply the primary numerator and the primary denominator by the LCD of the secondary denominators and reduce if possible.

Method 2 Form a single fraction in the numerator and in the denominator and divide the primary numerator by the primary denominator.

■ Example 6–4 A

Simplify each complex fraction.

1. $\dfrac{\frac{3}{4}}{\frac{5}{6}}$

Method 1

$$\frac{\frac{3}{4}}{\frac{5}{6}} = \frac{\frac{3}{4}\cdot 12}{\frac{5}{6}\cdot 12}$$

Multiply primary numerator and primary denominator by the LCD of secondary denominators 4 and 6, which is 12

$$= \frac{3\cdot\frac{12}{4}}{5\cdot\frac{12}{6}}$$

Divide 4 into 12

Divide 6 into 12

$$= \frac{3\cdot 3}{5\cdot 2}$$

Multiply in numerator and denominator

$$= \frac{9}{10}$$

Method 2

$$\frac{\frac{3}{4}}{\frac{5}{6}} = \frac{3}{4} \div \frac{5}{6} = \frac{3}{4} \cdot \frac{6}{5}$$ Multiply by the reciprocal of $\frac{5}{6}$

$$= \frac{3 \cdot 6}{4 \cdot 5}$$ Multiply numerators and denominators as indicated

$$= \frac{18}{20}$$ Multiply and reduce to lowest terms

$$= \frac{9}{10}$$ Reduce to lowest terms

2. $\dfrac{\frac{3}{a} - 1}{1 + \frac{4}{b}}$

Method 1

The LCD of the secondary denominators, a and b, is ab. We multiply the primary numerator, $\frac{3}{a} - 1$, and the primary denominator, $1 + \frac{4}{b}$, by ab. Remember to multiply the LCD times *every term* in the numerator and the denominator.

$$\frac{\frac{3}{a} - 1}{1 + \frac{4}{b}} = \frac{\left(\frac{3}{a} - 1\right) \cdot ab}{\left(1 + \frac{4}{b}\right) \cdot ab}$$

$$= \frac{\frac{3}{a} \cdot ab - 1 \cdot ab}{1 \cdot ab + \frac{4}{b} \cdot ab}$$ Multiply each term of numerator and denominator by ab

$$= \frac{3b - ab}{ab + 4a}$$ Perform indicated multiplications

$$= \frac{b(3 - a)}{a(b + 4)}$$ Factor numerator and denominator

Method 2

We first change the primary numerator, $\frac{3}{a} - 1$, and the primary denominator, $1 + \frac{4}{b}$, to single fractions. Thus,

$$\frac{3}{a} - 1 = \frac{3}{a} - \frac{a}{a} = \frac{3 - a}{a}$$ Combine in primary numerator

and

$$1 + \frac{4}{b} = \frac{b}{b} + \frac{4}{b} = \frac{b + 4}{b}$$ Combine in primary denominator

$$\frac{\frac{3}{a} - 1}{1 + \frac{4}{b}} = \frac{\frac{3 - a}{a}}{\frac{b + 4}{b}} = \frac{3 - a}{a} \cdot \frac{b}{b + 4} \qquad \text{Multiply by the reciprocal of } \frac{b + 4}{b}$$

$$= \frac{b(3 - a)}{a(b + 4)}$$

Note Always form a single fraction in the numerator and in the denominator *before* we invert and multiply. That is,

$$\frac{\frac{3}{a} - 1}{1 + \frac{4}{b}} \neq \left(\frac{3}{a} - 1\right) \cdot \left(1 + \frac{b}{4}\right)$$

As you can see, both methods will work. We will use only Method 1 in the remaining examples. Choice of method is left up to the student.

3. $\dfrac{\frac{4}{x} - \frac{3}{y}}{\frac{5}{x} + \frac{7}{y}}$

The LCD of the secondary denominators, x and y, is xy. We multiply the primary numerator, $\frac{4}{x} - \frac{3}{y}$, and the primary denominator, $\frac{5}{x} + \frac{7}{y}$, by xy and get

$$\frac{\frac{4}{x} - \frac{3}{y}}{\frac{5}{x} + \frac{7}{y}} = \frac{\left(\frac{4}{x} - \frac{3}{y}\right) \cdot xy}{\left(\frac{5}{x} + \frac{7}{y}\right) \cdot xy} = \frac{\frac{4}{x} \cdot xy - \frac{3}{y} \cdot xy}{\frac{5}{x} \cdot xy + \frac{7}{y} \cdot xy} \qquad \text{Multiply each term of numerator and denominator by } xy$$

$$= \frac{4y - 3x}{5y + 7x} \qquad \text{Perform indicated multiplications}$$

4. $\dfrac{\frac{a^2 - x^2}{x}}{\frac{a + x}{x^2}}$

The LCD of the secondary denominators, x and x^2, is x^2. We then multiply the primary numerator, $\frac{a^2 - x^2}{x}$, and the primary denominator, $\frac{a + x}{x^2}$, by x^2 and get

$$\frac{\frac{a^2 - x^2}{x}}{\frac{a + x}{x^2}} = \frac{\left(\frac{a^2 - x^2}{x}\right) \cdot x^2}{\left(\frac{a + x}{x^2}\right) \cdot x^2} = \frac{(a^2 - x^2) \cdot x}{a + x}$$

$$= \frac{(a + x)(a - x) \cdot x}{a + x} \qquad \text{Factor the numerator}$$

$$= (a - x) \cdot x \qquad \text{Reduce by the common factor } a + x$$

$$= ax - x^2 \qquad \text{Multiply as indicated}$$

Note In example 4, we proceeded to *reduce before* performing the multiplication in the numerator. *Always do this, if possible.*

▶ ***Quick check*** Simplify $\dfrac{\dfrac{3}{4}+\dfrac{2}{3}}{\dfrac{5}{6}-\dfrac{1}{12}}$ ■

Mastery points

Can you

- Simplify complex fractions?

Exercise 6–4

Simplify each complex fraction to a simple fraction. See example 6–4 A.

Example $\dfrac{\dfrac{3}{4}+\dfrac{2}{3}}{\dfrac{5}{6}-\dfrac{1}{12}}$

Solution *Method 1*

$$\frac{\dfrac{3}{4}+\dfrac{2}{3}}{\dfrac{5}{6}-\dfrac{1}{12}} = \frac{\left(\dfrac{3}{4}+\dfrac{2}{3}\right)12}{\left(\dfrac{5}{6}-\dfrac{1}{12}\right)12} = \frac{\dfrac{3}{4}\cdot 12+\dfrac{2}{3}\cdot 12}{\dfrac{5}{6}\cdot 12-\dfrac{1}{12}\cdot 12}$$ Multiply numerator and denominator by 12

$$= \frac{9+8}{10-1}$$ Multiply as indicated

$$= \frac{17}{9}$$ Combine in the numerator and the denominator

Method 2

$$\frac{\dfrac{3}{4}+\dfrac{2}{3}}{\dfrac{5}{6}-\dfrac{1}{12}} = \frac{\dfrac{9}{12}+\dfrac{8}{12}}{\dfrac{10}{12}-\dfrac{1}{12}} = \frac{\dfrac{17}{12}}{\dfrac{9}{12}}$$ Add in the numerator

Subtract in the denominator

$$= \frac{17}{12} \div \frac{9}{12}$$

$$= \frac{17}{12} \cdot \frac{12}{9}$$ Multiply by the reciprocal of $\frac{9}{12}$

$$= \frac{17}{9}$$

1. $\dfrac{\dfrac{2}{3}}{\dfrac{4}{5}}$

2. $\dfrac{\dfrac{7}{8}}{\dfrac{5}{6}}$

3. $\dfrac{\dfrac{4}{3}}{\dfrac{8}{9}}$

4. $\dfrac{\dfrac{9}{10}}{\dfrac{7}{6}}$

5. $\dfrac{1+\dfrac{3}{5}}{2-\dfrac{1}{5}}$

6. $\dfrac{5 - \dfrac{3}{4}}{1 + \dfrac{5}{8}}$

7. $\dfrac{7}{2 + \dfrac{4}{5}}$

8. $\dfrac{10}{4 - \dfrac{11}{12}}$

9. $\dfrac{4 + \dfrac{3}{5}}{6}$

10. $\dfrac{10 - \dfrac{7}{8}}{3}$

11. $\dfrac{\dfrac{3}{4} + \dfrac{5}{8}}{\dfrac{1}{2} - \dfrac{1}{4}}$

12. $\dfrac{\dfrac{6}{7} - \dfrac{5}{14}}{\dfrac{3}{14} - \dfrac{5}{7}}$

13. $\dfrac{x + \dfrac{1}{4}}{x - \dfrac{3}{4}}$

14. $\dfrac{y - \dfrac{5}{6}}{y + \dfrac{1}{2}}$

15. $\dfrac{\dfrac{1}{a} + 3}{\dfrac{2}{a} - 4}$

16. $\dfrac{5 - \dfrac{3}{b}}{6 + \dfrac{5}{b}}$

17. $\dfrac{\dfrac{3}{a^2} + 4}{5 - \dfrac{3}{a}}$

18. $\dfrac{\dfrac{5}{x} - 5}{6 + \dfrac{4}{x^3}}$

19. $\dfrac{a - \dfrac{3}{b}}{a + \dfrac{4}{b}}$

20. $\dfrac{x + \dfrac{4}{y}}{x - \dfrac{5}{y}}$

21. $\dfrac{\dfrac{1}{x} + \dfrac{1}{y}}{\dfrac{1}{x} - \dfrac{1}{y}}$

22. $\dfrac{\dfrac{3}{x^2} - \dfrac{4}{y}}{\dfrac{5}{x} + \dfrac{2}{y^2}}$

23. $\dfrac{x + y}{\dfrac{1}{x} + \dfrac{1}{y}}$

24. $\dfrac{a - b}{\dfrac{2}{a} + \dfrac{3}{b}}$

25. $\dfrac{\dfrac{4}{a^2} - \dfrac{5}{b}}{a - b}$

26. $\dfrac{\dfrac{6}{x^2} - \dfrac{5}{y^2}}{x + y}$

27. $\dfrac{\dfrac{1}{x + y} - \dfrac{1}{x - y}}{\dfrac{1}{x + y} + \dfrac{1}{x - y}}$

28. $\dfrac{\dfrac{a}{a + b} - \dfrac{b}{a - b}}{\dfrac{b}{a + b} + \dfrac{a}{a - b}}$

29. $\dfrac{\dfrac{x + y}{x - y} + \dfrac{x - y}{x + y}}{x^2 - y^2}$

30. $\dfrac{\dfrac{b}{a + b} - \dfrac{a}{a - b}}{a^2 - b^2}$

31. $\dfrac{\dfrac{2}{ab} + \dfrac{3}{ab^2}}{a^2b^2}$

32. $\dfrac{\dfrac{5}{x^2y^2} - \dfrac{4}{xy}}{xy}$

33. $\dfrac{\dfrac{2}{a + 3} + \dfrac{1}{a - 2}}{\dfrac{3}{a - 2} - \dfrac{4}{a + 3}}$

34. $\dfrac{\dfrac{7}{x - 4} - \dfrac{5}{x + 3}}{\dfrac{5}{x + 3} + \dfrac{9}{x - 4}}$

35. $\dfrac{\dfrac{5}{x^2 - x - 12}}{\dfrac{4}{x + 3} - \dfrac{5}{x - 4}}$

36. $\dfrac{\dfrac{7}{b - 7} + \dfrac{8}{b - 5}}{\dfrac{6}{b^2 - 12b + 35}}$

37. $\dfrac{3 - \dfrac{4}{a + 4}}{\dfrac{5}{a + 4} - \dfrac{6}{a - 1}}$

38. $\dfrac{\dfrac{6}{x - 5} + 7}{\dfrac{8}{x - 5} - \dfrac{9}{x + 3}}$

39. $\dfrac{\dfrac{y^2 - y - 6}{y^2 + 2y + 1}}{\dfrac{y^2 - 2y - 8}{y^2 + 7y + 6}}$

40. $\dfrac{\dfrac{a^2 + 3a - 10}{a^2 - 5a - 14}}{\dfrac{a^2 + 6a + 5}{a^2 - 8a + 7}}$

Solve the following word problems.

41. A refrigeration coefficient-of-performance formula for the ideal refrigerator is given by

$$cp = \frac{1}{\dfrac{T_2}{T_1} - 1}$$

Simplify the right member.

42. In electronics, a formula for coupled inductance in parallel with fields aiding is given by

$$L_t = \frac{1}{\dfrac{1}{L_1 - M} + \dfrac{1}{L_2 + M}}$$

Simplify the right member.

43. In electronics, a formula for self-inductance of circuits in parallel is given by

$$L_t = \frac{1}{\frac{1}{L_1} + \frac{1}{L_2} + \frac{1}{L_3}}$$

Simplify the right member.

44. Coupled inductance with circuits connected in parallel with opposing fields is given by

$$L_t = \frac{1}{\frac{1}{L_1 - M} + \frac{1}{L_2 - M}}$$

Simplify the right member.

Review exercises

Reduce the following expressions to lowest terms. See section 5–2.

1. $\frac{36}{42}$

2. $\frac{x^2 - 5x - 14}{x^2 + 4x + 4}$

3. Subtract $(3x^3 - 2x^2 + x - 12) - (x^3 - 5x^2 + 9)$. See section 2–3.

Solve the following equations. See sections 2–6 and 4–7.

4. $4x + 3x = 21$

5. $8y - 4 = 5y - 10$

6. $x^2 + 2x - 3 = 0$

Determine the domain of the following rational expressions. See section 5–1.

7. $\frac{3}{x + 7}$

8. $\frac{y - 4}{y^2 - 4}$

9. Perform the indicated operations, if possible. See section 1–7.

a. $\frac{4}{0}$ b. $\frac{0}{-3}$

6–5 ■ Rational equations

A rational equation

An algebraic equation that contains *at least one* rational expression is called a **rational equation.**

The basic operations for solving equations that you learned in chapter 2 will apply to rational equations once the denominators in the equation are eliminated. We accomplish this by using the multiplication property of equality. The multiplier is the least common denominator of all denominators in the rational expressions of the equation.

Solving a rational equation

1. Find the LCD of all denominators.
2. Eliminate the denominators by multiplying each term of both members of the equation by the LCD of the denominators in the equation.
3. Use the four steps from section 2–6 to solve the resulting equation.

■ *Example 6–5 A*

Find the solution set of each of the following rational equations.

1. $\dfrac{x-3}{4} = \dfrac{x}{8}$

We can see that the LCD of the denominators 4 and 8 is 8.

$$\frac{8}{1} \cdot \frac{(x-3)}{4} = \frac{8}{1} \cdot \frac{x}{8} \qquad \text{Multiply each member by 8}$$
$$2(x-3) = x \qquad \text{Reduce in each member}$$
$$2x - 6 = x \qquad \text{Multiply as indicated}$$
$$x - 6 = 0 \qquad \text{Subtract } x \text{ from each member}$$
$$x = 6 \qquad \text{Add 6 to each member}$$

The solution set is $\{6\}$. If we wish to check our work, we substitute 6 for x in the original equation to obtain the equivalent equation.

$$\frac{6-3}{4} = \frac{6}{8} \qquad \text{Replace } x \text{ with 6 in the original equation}$$
$$\frac{3}{4} = \frac{3}{4} \qquad \text{(True)}$$

2. $\dfrac{t}{4} - \dfrac{t-4}{5} = \dfrac{7}{10}$

The LCD of the denominators 4, 5, and 10 is 20.

$$\frac{20}{1} \cdot \frac{t}{4} - \frac{20}{1} \cdot \frac{t-4}{5} = \frac{20}{1} \cdot \frac{7}{10} \qquad \text{Multiply each term by the LCD 20}$$
$$5t - 4(t-4) = 2 \cdot 7 \qquad \text{Reduce each term}$$
$$5t - 4t + 16 = 14 \qquad \text{Multiply as indicated}$$
$$t + 16 = 14 \qquad \text{Combine like terms}$$
$$t = -2 \qquad \text{Subtract 16 from each member}$$

The solution set is $\{-2\}$. Check your answer by replacing t with -2 in the original equation.

Note A common error that is made when multiplying $-4(t-4)$ is to get $-4t - 16$. Do not forget that you are using the distributive property to multiply -4 times each term in the group $(t-4)$. The correct result is $-4t + 16$.

3. $\dfrac{5}{3a} + \dfrac{4}{9} = \dfrac{5}{12a}$

We determine that the LCD of the denominators is $36a$.

$$36a \cdot \frac{5}{3a} + 36a \cdot \frac{4}{9} = 36a \cdot \frac{5}{12a} \quad (a \neq 0) \qquad \text{Multiply each term by the LCD } 36a$$
$$12 \cdot 5 + 4a \cdot 4 = 3 \cdot 5 \qquad \text{Reduce each term}$$
$$60 + 16a = 15 \qquad \text{Multiply as indicated}$$
$$16a = -45 \qquad \text{Subtract 60 from each member}$$
$$a = -\frac{45}{16} \qquad \text{Divide each member by 16}$$

The solution set is $\left\{-\dfrac{45}{16}\right\}$. Check your solution.

Note The domain of the variable is every real number *except* 0 since two of the terms are undefined when $a = 0$. This is an important observation to make as shown in example 4.

4. $\dfrac{3}{y} = \dfrac{4}{y^2 - 2y} - \dfrac{2}{y - 2}$

Factor the denominator $y^2 - 2y$ to get $y(y - 2)$. We determine the LCD of the denominators to be $y(y - 2)$.

$$y(y-2) \cdot \frac{3}{y} = y(y-2) \cdot \frac{4}{y(y-2)} - y(y-2) \cdot \frac{2}{y-2} \quad (y \neq 0,2)$$ Multiply each term by the LCD $y(y - 2)$

$$(y - 2) \cdot 3 = 4 - y \cdot 2$$ Reduce each term

$$3y - 6 = 4 - 2y$$ Multiply as indicated

$$3y + 2y = 4 + 6$$ Add 6 and $2y$ to each member

$$5y = 10$$ Combine like terms

$$y = 2$$ Divide each member by 5

Two *is not in the domain* of the variable y, since $y = 2$ makes the denominator $y - 2 = 0$. So 2 cannot be a solution of the equation. Therefore, the solution set is $\emptyset$.

We conclude there is no solution for the equation of example 4. The number 2 in that example is called an *extraneous solution.* An extraneous solution can occur whenever the variable appears in the denominator of one or more of the terms of the equation. Thus, you should *always* check your solution(s) of a rational equation when a variable appears in the denominator.

5. $\dfrac{1}{3}x^2 - \dfrac{5}{2}x + 3 = 0$

Multiply each term of the equation by 6, the LCD of the denominators 2 and 3.

$$6 \cdot \frac{1}{3}x^2 - 6 \cdot \frac{5}{2}x + 6 \cdot 3 = 6 \cdot 0$$

$$2x^2 - 15x + 18 = 0$$ Write the equation in standard form

$$(2x - 3)(x - 6) = 0$$ Factor $2x^2 - 15x + 18 = (2x - 3)(x - 6)$

$$2x - 3 = 0 \quad \text{or} \quad x - 6 = 0$$ Set each factor equal to 0

$$2x = 3 \qquad x = 6$$ Solve each equation

$$x = \frac{3}{2} \qquad x = 6$$

The solution set is $\left\{\dfrac{3}{2}, 6\right\}$. Check your solution.

Note We eliminate all the denominators *only* when solving rational equations. This is *never done* when adding or subtracting rational expressions. To illustrate, given the equation $\dfrac{6}{x} - \dfrac{4}{x^2} = 0$, we multiply each term by x^2, whereas given the subtraction problem $\dfrac{6}{x} - \dfrac{4}{x^2}$, we do *not* multiply each term by x^2.

▶ ***Quick check*** Find the solution set of $\dfrac{6}{4z} + \dfrac{7}{8} = \dfrac{9}{16z}$ ■

Rational equations in more than one variable

In scientific fields, equations and formulas involving rational expressions and *more than one variable* are common. It is often desirable to solve such equations for one variable in terms of the other variables in the equation. The procedures for finding such solutions are identical to those used in solving the preceding equations.

Solving a rational equation in more than one variable

1. Remove the fractions by multiplying each member of the equation by the LCD of the denominators in the equation.
2. Collect all terms containing the variable you are solving for in one member of the equation and all other terms in the other member.
3. Factor out the variable you are solving for if it appears in more than one term.
4. Divide each member by the coefficient of the variable for which you are solving.

■ ***Example 6–5 B***

Solve each rational equation for the indicated variable.

1. Solve $\dfrac{a}{3} + \dfrac{3x}{2} = c$ for x.

Determine that the LCD of 3 and 2 is 6.

$$6 \cdot \frac{a}{3} + 6 \cdot \frac{3x}{2} = 6 \cdot c \quad \text{Multiply each term by 6}$$
$$2 \cdot a + 3 \cdot 3x = 6c \quad \text{Reduce where possible}$$
$$2a + 9x = 6c \quad \text{Multiply in left member}$$
$$2a + 9x - 2a = 6c - 2a \quad \text{Subtract } 2a \text{ from each member}$$
$$9x = 6c - 2a \quad \text{Combine like terms}$$
$$\frac{9x}{9} = \frac{6c - 2a}{9} \quad \text{Divide each member by the coefficient 9}$$
$$x = \frac{6c - 2a}{9}$$

2. Solve $\dfrac{1}{a} = \dfrac{1}{b} + \dfrac{1}{c}$ for c $(a, b, c \neq 0)$.

Determine the LCD of a, b, and c is abc.

$$abc \cdot \frac{1}{a} = abc \cdot \frac{1}{b} + abc \cdot \frac{1}{c} \quad \text{Multiply each term by } abc$$
$$bc = ac + ab \quad \text{Reduce where possible}$$

Note Since we are solving for c, we must get all terms containing c in the same member of the equation.

$$bc - ac = ac + ab - ac$$ Subtract ac from each member
$$bc - ac = ab$$
$$(b - a)c = ab$$ Factor c from each term in the left member
$$\frac{(b - a)c}{(b - a)} = \frac{ab}{(b - a)}$$ Divide each member by the coefficient of c, $(b - a)$
$$c = \frac{ab}{b - a}$$ Reduce in left member

▶ ***Quick check*** Solve $\frac{5}{a} - \frac{3}{b} = 3$ for a. ■

Mastery points

Can you

- Solve equations containing rational expressions?
- Solve rational equations for one variable in terms of the other variables?

Exercise 6–5

Find the solution set of each rational equation. Indicate any restrictions on the domain of the variable. See example 6–5 A.

Example $\frac{6}{4z} + \frac{7}{8} = \frac{9}{16z} \; (z \neq 0)$

Solution The LCD of the denominators is $16z$.

$$16z \cdot \frac{6}{4z} + 16z \cdot \frac{7}{8} = 16z \cdot \frac{9}{16z}$$ Multiply each term by $16z$
$$4 \cdot 6 + 2z \cdot 7 = 9$$ Reduce in each term
$$24 + 14z = 9$$ Multiply in each term
$$14z = -15$$ Subtract 24 from each member
$$z = -\frac{15}{14}$$ Divide each member by 14

The solution set is $\left\{-\frac{15}{14}\right\}$.

1. $\frac{y}{4} = \frac{2}{3}$
2. $\frac{4x}{5} - \frac{2}{3} = 4$
3. $\frac{p}{6} = \frac{7}{9}$
4. $\frac{a}{3} + \frac{5}{2} = 6$
5. $\frac{z}{8} + 3 = \frac{1}{4}$
6. $\frac{3R}{4} - 5 = \frac{5}{6}$
7. $\frac{3a}{6} + \frac{2a}{5} = 1$
8. $\frac{5x}{8} - \frac{x}{12} = 3$
9. $\frac{2x + 1}{7} - \frac{2x - 3}{14} = 1$
10. $\frac{3a + 1}{9} + \frac{1}{12} = \frac{2a - 1}{3}$
11. $\frac{3}{2x} = \frac{4}{5} + \frac{2}{x}$
12. $\frac{4}{b} - \frac{7}{3b} = \frac{2}{5}$

13. $\frac{2}{3R} + \frac{3}{2R} + \frac{1}{R} = 4$

14. $\frac{3}{w} - \frac{6}{5w} + \frac{1}{2w} = 5$

15. $\frac{4}{6y} + 5 = \frac{1}{9y} + 2$

16. $\frac{16}{5a} - 1 = 5 + \frac{3}{4a}$

17. $\frac{5}{3b} - \frac{1}{2} = \frac{7}{6b}$

18. $3 - \frac{5}{9x} = \frac{4}{6x}$

19. $\frac{3p + 2}{7p} - 3 = \frac{p}{14p}$

20. $\frac{5 - x}{8x} = \frac{2x + 5}{6x}$

21. $\frac{a - 4}{3a} = \frac{2a - 1}{4a}$

22. $\frac{R + 2}{10R} + \frac{4R - 1}{4R} = 2$

23. $\frac{4}{x - 4} = \frac{5}{x + 4}$

24. $\frac{9}{3 - x} = \frac{8}{2x + 1}$

25. $\frac{10}{3z + 1} = \frac{3}{5}$

26. $\frac{3}{2a + 3} = \frac{1}{7}$

27. $\frac{y}{y - 2} - 2 = \frac{2}{y - 2}$

28. $\frac{1}{x + 5} = \frac{3}{x - 5} - \frac{10}{x^2 - 25}$

29. $\frac{b - 1}{b^2 - 4} = \frac{6}{b + 2}$

30. $\frac{5}{R + 3} = \frac{4R + 3}{R^2 - 9}$

31. $\frac{5}{a^2 - 25} + \frac{3}{a - 5} = \frac{4}{a + 5}$

32. $\frac{6}{2p - 4} + \frac{4}{p - 2} = \frac{1}{p^2 - 4}$

33. $\frac{5}{x^2 + x - 6} = \frac{2}{x^2 + 3x - 10}$

34. $\frac{8}{a^2 - 6a + 8} = \frac{1}{a^2 - 16}$

35. $\frac{1}{a - 3} + \frac{2}{a + 4} = \frac{6}{a^2 + a - 12}$

36. $\frac{5}{2x - 1} - \frac{7}{3x + 2} = \frac{9}{6x^2 + x - 2}$

37. $3x^2 + 4x + \frac{4}{3} = 0$

38. $3x^2 + \frac{11}{2}x + \frac{3}{2} = 0$

39. $b^2 + \frac{3}{2}b = \frac{9}{2}$

40. $\frac{2}{3}x^2 + x = \frac{20}{3}$

41. $\frac{3}{4}z^2 = 2 - \frac{5}{2}z$

42. $x^2 - \frac{5}{6}x = \frac{2}{3}$

43. $x^2 = \frac{2}{3} - \frac{x}{3}$

44. $x^2 + \frac{4}{3}x = -\frac{4}{9}$

Solve the following equations for the indicated letter. Assume all denominators are nonzero. See example 6–5 B.

Example $\frac{5}{a} - \frac{3}{b} = 3$ for a

Solution

	$ab \cdot \frac{5}{a} - ab \cdot \frac{3}{b} = ab \cdot 3$	Multiply each term by ab
	$5b - 3a = 3ab$	Reduce each term
	$5b = 3ab + 3a$	Add $3a$ to each member
So	$5b = a(3b + 3)$	Factor a in the right member
	$a = \frac{5b}{3b + 3}$	Divide each member by $3b + 3$

45. $\frac{2}{x} + \frac{1}{y} = 3$ for x

46. $\frac{5}{I} - 6 = \frac{8}{E}$ for I

47. $\frac{1}{c} = \frac{1}{c_1} + \frac{1}{c_2}$ for c_1

48. $\frac{1}{R} = \frac{1}{R_1} + \frac{1}{R_2} + \frac{1}{R_3}$ for R

49. $\frac{1}{8} = \frac{1}{a} + \frac{1}{b}$ for a

50. $\frac{1}{x} - \frac{1}{y} - \frac{1}{z} = 6$ for y

51. $\frac{3}{a} - \frac{4}{b} = \frac{5}{ab}$ for a

52. $\frac{5}{m} + 4 = \frac{6a}{2m} + 3b$ for m

53. $\frac{x + 4}{2} + \frac{y - 3}{5} = \frac{2}{10}$ for x

Solve the following word problems.

54. The principal amount of money P in a savings account paying interest rate r, over a given period of time t, which pays interest I is given by $P = \frac{I}{rt}$. Solve the equation for r.

55. The pressure per square inch of steam or water in a pipe, p, is given by $p = \frac{P}{LD}$, where
$P =$ the total pressure on a diametral plane,
$L =$ the length of the pipe in inches, and
$D =$ the diameter of the pipe in inches.
Solve for D.

56. The safe internal unit pressure, p, in a given pipe of given thickness is given by $p = \frac{2st}{D}$, where
$s =$ the unit tensile stress,
$t =$ the thickness of the pipe, and
$D =$ the diameter of the pipe. Solve for s.

57. The pitch diameter of a gear, P, is given by $P = \frac{D \cdot N}{N + 2}$, where $D =$ the outside diameter of the gear and $N =$ the number of teeth in the gear. Solve for N.

58. The rule governing the speeds of two gears, one the driver gear and the other the driven gear, is given by $\frac{T_A}{T_B} = \frac{R_B}{R_A}$, where $T_A =$ the number of teeth in the driven gear, $T_B =$ the number of teeth in the driver gear, $R_B =$ the revolutions per minute of the driven gear, and $R_A =$ the revolutions per minute of the driver gear. Solve for T_B.

59. Given F is the force on the large piston and f is the force on the small piston of a hydraulic press, then $\frac{F}{f} = \frac{A}{a}$, where A is the area of the large piston and a is the area of the small piston. Solve for f.

60. The kinetic energy of a body, KE, is computed by $KE = \frac{Wv^2}{2g}$, where v is the velocity expressed in feet per second, W is the weight in pounds, and g is acceleration due to gravity. Solve for W.

61. Charles' Law is in regard to the relationship between pressure P, volume V, and the absolute temperature T of a gas as it expands when heated. Charles' Law is represented by $\frac{P_1V_1}{T_1} = \frac{P_2V_2}{T_2}$, where P_1, V_1, and T_1 are the initial conditions of the gas and P_2, V_2, and T_2 are the final conditions. Solve for T_2.

62. The coefficient of linear expansion, k, of a solid when heated is given by $k = \frac{L_t - L_0}{L_0t}$, where L_0 is the length at 0° C, L_t is the length at t° C, and t is any given temperature in Celsius. Solve for L_t. Solve for L_0.

63. The coefficient of linear expansion α (alpha) when metal is heated is given by $\alpha = \frac{L_2 - L_1}{L_1(t_2 - t_1)}$. Solve for L_1.

64. Carnot's ideal efficiency of any heat engine operating between the temperature limits T_1 and T_2 is given by $CE = \frac{T_1 - T_2}{T_1}$. Solve for T_1.

65. A formula for resistors in parallel is given by $R = \frac{R_1R_2}{R_1 + R_2}$. Solve for R_1.

66. In electronics, a formula for the unknown resistance R_x in a battery is given by $R_x = R_m\left(\frac{E_1}{E_2} - 1\right)$. Solve for E_2.

Review exercises

Evaluate each expression for the given values. See sections 2–2 and 5–1.

1. $5x - 3y + z$ when $x = 1$, $y = -2$, and $z = 3$
2. $\frac{3a - b}{2a + b}$ when $a = 4$ and $b = -5$
3. $\frac{y_1 - y_2}{x_1 - x_2}$ when $x_1 = 3$, $x_2 = 1$, $y_1 = 3$, and $y_2 = -5$

Solve each equation for y. See section 2–7.

4. $5x + y = 4$ **5.** $2x + 3y = 6$ **6.** $x - 4y = 8$

Simplify the following expressions. Assume all denominators are nonzero. Answer with positive exponents only. See section 3–4.

7. $(3y^{-1})(y^2x^{-2})$ **8.** $\dfrac{x^{-3}y^3}{x^2y^{-1}}$

6–6 ■ Rational expression applications

Rational expressions occurring in rational equations have many applications in the physical and scientific world. We now wish to discuss some of the more common applications.

■ Example 6–6 A

Choose a variable, set up an appropriate equation, and solve the following problems.

1. A water holding tank is fed by two pipes. If it takes the smaller pipe 12 hours to fill the tank and the larger pipe 9 hours to fill the tank, how long would it take to fill the tank if both pipes are open? (This is called a *work* problem.)

Let x = the number of hours it takes the two pipes to fill the tank. Now,

a. the smaller pipe can fill $\dfrac{1}{12}$ of the tank in 1 hour,

b. the larger pipe can fill $\dfrac{1}{9}$ of the tank in 1 hour,

c. the two pipes together can fill $\dfrac{1}{x}$ of the tank in 1 hour.

The amount of the tank filled by the smaller pipe in 1 hour plus the amount of the tank filled by the larger pipe in 1 hour must be equal to the amount of the tank filled by the two pipes together in 1 hour. Thus, the equation is

Amount by larger pipe ↓

Amount by smaller pipe → $\dfrac{1}{12} + \dfrac{1}{9} = \dfrac{1}{x}$ ← Amount together

Multiply each term by the LCM of 12, 9, and x, which is $36x$.

$$36x \cdot \frac{1}{12} + 36x \cdot \frac{1}{9} = 36x \cdot \frac{1}{x}$$

$$3x + 4x = 36 \quad \text{Reduce in each term}$$

$$7x = 36 \quad \text{Combine in left member}$$

$$x = \frac{36}{7} \quad \text{Divide each member by 7}$$

Therefore, together the two pipes can fill the tank in $\dfrac{36}{7}$ hours, or $5\dfrac{1}{7}$ hours.

Note That is, approximately 5 hours and 9 minutes.

2. Jim James drove a distance of 120 miles, part at 50 miles per hour (mph) and part at 60 mph. If he drove the 120 miles in $2\frac{1}{4}$ hours, how many miles did he drive at 50 mph? (This is called a *time–rate–distance* problem.)

Note In a distance (d)–rate (r)–time (t) problem, we use $d = rt$, $t = \frac{d}{r}$, or $r = \frac{d}{t}$.

Let x = the distance he drove at 50 mph. Then $120 - x$ = the distance he drove at 60 mph.

The time traveled at 50 mph plus the time traveled at 60 mph equals the total time traveled, $2\frac{1}{4}$ hr $\left(\text{or } \frac{9}{4} \text{ hr}\right)$. We use the following table for time–rate–distance problems.

distance (d)	rate (r)	time (t)
x	50	$\frac{x}{50}$
$120 - x$	60	$\frac{120 - x}{60}$

$t = \frac{d}{r}$

The equation is then

Time at 60 mph ↓

Time at 50 mph → $\frac{x}{50} + \frac{120 - x}{60} = \frac{9}{4}$ ← Total time

$$300 \cdot \frac{x}{50} + 300 \cdot \frac{120 - x}{60} = 300 \cdot \frac{9}{4}$$ Multiply each term by the LCD 300

$$6x + 5(120 - x) = 675$$ Reduce in each term

$$6x + 600 - 5x = 675$$ Distributive property

$$x + 600 = 675$$ Combine in left member

$$x = 75$$ Subtract 600 from each member

Thus, Jim James drove 75 miles at 50 mph.

Note He drove $120 - x = 120 - 75 = 45$ miles at 60 mph.

3. The denominator of a fraction is 3 more than the numerator. If 4 is added to the numerator and the denominator, the resulting fraction is $\frac{3}{4}$. Find the original fraction.

Let x = the numerator of the original fraction. Then $x + 3$ = the denominator of the original fraction. The equation we get is

$$\frac{x + 4}{(x + 3) + 4} = \frac{3}{4}$$

then

$$\frac{x+4}{x+7} = \frac{3}{4}$$

The LCD of 4 and $x + 7$ is $4(x + 7)$.

$$4(x+7) \cdot \frac{x+4}{x+7} = 4(x+7) \cdot \frac{3}{4} \quad (x \neq -7)$$ Multiply each term by $4(x + 7)$

$$4(x+4) = (x+7) \cdot 3$$ Reduce each term

$$4x + 16 = 3x + 21$$ Multiply as indicated

$$x = 5$$ Subtract $3x$ and 16 from each member

$$\text{and } x + 3 = 8$$

The original fraction is $\frac{5}{8}$.

4. In an electrical circuit, when two resistors are connected in parallel, the total resistance R of the circuit in ohms is given by

$$\frac{1}{R} = \frac{1}{R_1} + \frac{1}{R_2}$$

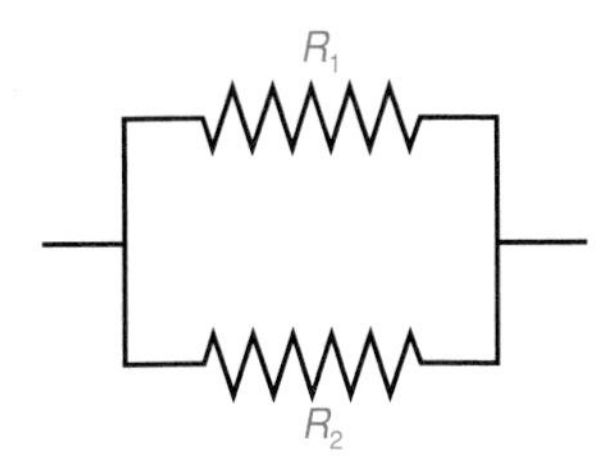

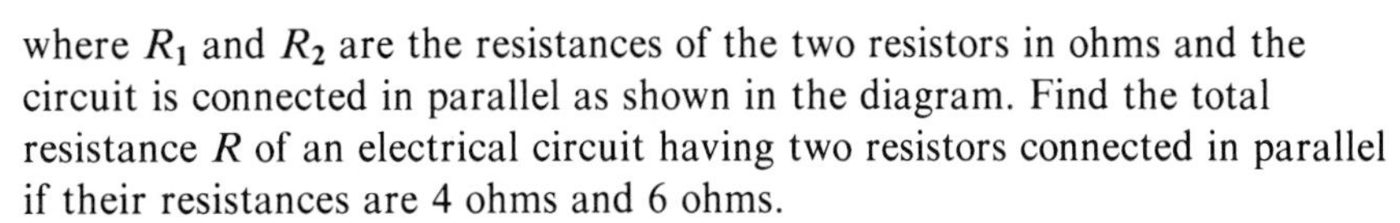

where R_1 and R_2 are the resistances of the two resistors in ohms and the circuit is connected in parallel as shown in the diagram. Find the total resistance R of an electrical circuit having two resistors connected in parallel if their resistances are 4 ohms and 6 ohms.

We want R if $R_1 = 4$ ohms and $R_2 = 6$ ohms.

$$\frac{1}{R} = \frac{1}{4} + \frac{1}{6}$$

Multiply both members by the LCM of 4, 6, and R, which is $12R$.

$$12R \cdot \frac{1}{R} = 12R \cdot \frac{1}{4} + 12R \cdot \frac{1}{6} \quad (R \neq 0)$$

$$12 = 3R + 2R$$

$$12 = 5R$$

$$\frac{12}{5} = R$$

Therefore the total resistance is $\frac{12}{5}$, or $2\frac{2}{5}$, ohms.

▶ ***Quick check*** If the same number is added to the numerator and the denominator of $\frac{3}{4}$, the result is the fraction $\frac{5}{6}$. What is the number? ■

Mastery points

Can you

- Solve work problems?
- Solve distance-rate-time problems?
- Solve number problems?
- Solve resistance in electric circuits problems?

Exercise 6–6

Choose a variable, set up an equation, and solve the following problems.

Example If the same number is added to the numerator and the denominator of $\frac{3}{4}$, the result is the fraction $\frac{5}{6}$. What is the number?

Solution Let x = the number to be added to the numerator and the denominator. The equation is then

$$\frac{3+x}{4+x} = \frac{5}{6}$$

$$6(4+x)\frac{3+x}{4+x} = 6(4+x)\cdot\frac{5}{6} \quad \text{Multiply each member by the LCD } 6(4+x)$$

$$6(3+x) = (4+x)\cdot 5 \quad \text{Reduce in each member}$$

$$18 + 6x = 20 + 5x \quad \text{Multiply as indicated}$$

$$x = 2 \quad \text{Solve for } x$$

The number to be added is 2. $\left(\textit{Check: } \frac{3+2}{4+2} = \frac{5}{6}\right)$

See example 6–6 A–1.

1. Jim can mow his parents' lawn in 50 minutes. His younger brother Kenny can mow the lawn in 70 minutes. How long would it take for the boys to mow the lawn if they mow together?
2. In a factory, worker A can do a certain job in 6 hours while worker B can do the same job in 5 hours. How long would it take workers A and B to do the same job working together?
3. Three different sized pipes feed water into a swimming pool. If the pipes can fill the same pool individually in 6 hours, 8 hours, and 9 hours, respectively, how long would it take to fill the same pool if all three pipes were open?
4. During "clean-up week" in Podunk Junction, Jane, Joan, and Ruth work to clean up a vacant lot. If the girls can do the job individually in 2 hours, 3 hours, and 4 hours, respectively, how long will it take them, working together, to clean the lot?
5. It takes two men—Harry and Dick—working together 4 hours to paint the exterior of a house. If Harry could do the job in 6 hours working alone, how long would it take Dick to paint the house alone?
6. A water tank has two drain pipes. If the larger pipe could empty the tank in 45 minutes and the two pipes together could drain the tank in 30 minutes, how long would it take for the smaller pipe to drain the tank?
7. Three combines—A, B, and C—working together can harvest a field of oats in $1\frac{1}{2}$ hours. If A could do the job alone in 5 hours and B could do it alone in 6 hours, how long would it take for combine C to do the same job alone?
8. In a pizzeria, three women are at work making pizzas. The three could make 50 pizzas in $1\frac{1}{2}$ hours working together. If two of the women could make all the pizzas in 5 hours and 6 hours, respectively, working alone, how long would it take the third woman to make the 50 pizzas working alone?
9. A tank has one inlet pipe and one outlet pipe. If the inlet pipe can fill the tank in $3\frac{1}{3}$ hours and the outlet pipe can empty the tank in 5 hours, how long would it take to fill the tank if both pipes are left open? (*Hint:* Subtract the drainage.)
10. A sink drain, when left open, can empty a sink full of water in 4 minutes. If the cold water and hot water faucets can, when fully open, fill the sink in $2\frac{1}{2}$ minutes and 3 minutes, respectively, how long would it take to fill the sink if all three are open simultaneously?

See example 6–6 A–2.

11. Jack drove 320 miles in $5\frac{1}{2}$ hours. If he drove part of the trip averaging 55 mph and the rest averaging 60 mph, how many miles did he drive at each speed?

12. A. J. Foyt, when driving the Indianapolis 500 Race, averaged 198 mph over part of the race. Due to an accident on the track, he averaged 160 mph over the rest. If the race took 2 hours and 40 minutes to run, how many miles did he drive at an average of 198 mph? (Round the answer to the nearest integer.)

13. Car A travels 120 miles in the same time that car B travels 150 miles. If car B averages 10 mph faster than car A, how fast is each car traveling?

14. A freight train travels 260 kilometers in the same time a passenger train travels 320 kilometers. If the passenger train averages 15 kilometers per hour faster than the freight train, what was the average speed of the freight train?

15. Sheila can row a boat 2 mph in still water. How fast is the current of a river if she takes the same length of time to row 4 miles upstream as she does to row 10 miles downstream? (*Hint:* Subtract current upstream and add downstream.)

16. An airplane flew 1,000 miles with the wind in the same length of time it took to fly 850 miles against the wind. If the wind was blowing at 25 miles per hour, what was the average air speed of the plane?

17. On a trip from Detroit to Columbus, Ohio, Mrs. Smith drove at an average speed of 60 mph. Returning, her average speed was 55 mph. If it took her $\frac{1}{3}$ hour longer on the return trip, how far is it from Detroit to Columbus?

18. A jet plane flew at an average speed of 240 miles per hour going from city A to city B and averaged 300 miles per hour on the return flight. Its return flight took 1 hour and 40 minutes less time. How far is it from city A to city B? (Disregard any wind.)

See example 6–6 A–3.

19. The numerator of a given fraction is 4 less than the denominator. If 5 is added to both the numerator and the denominator, the resulting fraction is $\frac{5}{7}$. What is the original fraction?

20. The denominator of a fraction exceeds the numerator by 7. If 3 is added to the numerator and 1 is subtracted from the denominator, the resulting fraction is $\frac{4}{5}$. Find the original fraction.

21. One number is four times another number. The sum of their reciprocals is $\frac{5}{12}$. What are the numbers?

22. One number is four times another number. The sum of their reciprocals is $\frac{1}{4}$. What are the numbers?

23. If $\frac{1}{2}$ is added to three times the reciprocal of a number, the result is 1. Find the number.

24. If $\frac{1}{2}$ is subtracted from four times the reciprocal of a number, the result is 0. Find the number.

25. When a certain number is added to the numerator and subtracted from the denominator of the fraction $\frac{2}{5}$, the result is 6. What is the number?

26. When a certain number is added to the numerator and subtracted from the denominator of the fraction $\frac{5}{9}$, the result is $\frac{4}{7}$. What is the number?

27. A prescription for an illness calls for a child's dosage to be $\frac{3}{5}$ of the number of pills that an adult takes. Together they use 24 pills. How many pills are taken by each?

28. An apprentice electrician receives $\frac{3}{8}$ of the hourly wage of a journeyman electrician. Together their hourly wage is $29.70. Find the hourly wage of each.

29. Pat and Mike earn a total of $65.50 per week delivering papers. If Pat's earnings are $\frac{5}{7}$ of Mike's, how much does Mike earn each day?

Use the formula $\frac{1}{R} = \frac{1}{R_1} + \frac{1}{R_2}$ to solve the following exercises. See example 6–6 A–5.

30. Two resistors in an electric circuit have resistances of 6 ohms and 8 ohms and are connected in parallel. Find the total resistance of the circuit.

31. Two resistors of an electric circuit are connected in parallel. If one has a resistance of 5 ohms and the other has a resistance of 12 ohms, what is the total resistance in the circuit?

32. Three resistors connected in parallel have resistances of 4 ohms, 6 ohms, and 10 ohms. What is the total resistance in the electric circuit? $\left(\textit{Hint: } \text{Use } \frac{1}{R} = \frac{1}{R_1} + \frac{1}{R_2} + \frac{1}{R_3}.\right)$

33. The total resistance in a parallel wiring circuit is 12 ohms. If the resistance in one branch is 30 ohms, what is the resistance in the other branch?

34. The resistance in one branch of a two-resistor parallel wiring circuit is 10 ohms. If the total resistance in the circuit is 6 ohms, what is the resistance in the other branch?

35. A three-resistor parallel wiring circuit has a total resistance of 10 ohms. If two of the branches of the circuit have resistances of 20 ohms and 30 ohms, what is the resistance in the third branch? (See hint in exercise 32.)

Review exercises

Find the solution set of the following equations. See sections 2–6 and 2–7.

1. $2y + 3 = 4y - 1$

2. $2y - 3x = 6$ for y

Factor the following expressions. See sections 4–2, 4–3, and 4–4.

3. $8y^2 - 32$

4. $x^2 + 20x + 100$

5. $3y^2 - 4y - 4$

Combine the following rational expressions. See section 6–3.

6. $\frac{3x}{x - 1} + \frac{2x}{x + 3}$

7. $\frac{4y}{2y + 1} - \frac{3y}{y - 5}$

Chapter 6 lead-in problem

Marc owns $\frac{5}{8}$ interest in a print shop and his uncle owns $\frac{1}{4}$ interest in the shop. In a given year, they shared earnings of \$140,000. How much did the shop earn that year?

Solution

Let x = the total earnings of the print shop.

	Interest	Shop earnings	Income
Marc	$\frac{5}{8}$	x	$\frac{5}{8}x$
Uncle	$\frac{1}{4}$	x	$\frac{1}{4}x$

Then $\frac{5}{8}x + \frac{1}{4}x = 140{,}000$ — Together they earned \$140,000

$8 \cdot \frac{5}{8}x + 8 \cdot \frac{1}{4}x = 8 \cdot 140{,}000$ — Multiply each term by the LCM of 4 and 8, which is 8

$5x + 2x = 1{,}120{,}000$ — Perform indicated multiplications

$7x = 1{,}120{,}000$ — Combine like terms

$x = 160{,}000$ — Divide each member by 7

Chapter 6 summary

1. To *multiply* two rational expressions, we multiply the numerators and place that product over the product of the denominators.
2. To *divide* two rational expressions, $\frac{P}{Q} \div \frac{R}{S}$, we multiply $\frac{P}{Q}$ by the reciprocal of $\frac{R}{S}$, which is $\frac{S}{R}$.
3. To find the *least common denominator* (LCD) of two or more rational expressions, we
   a. write each denominator in completely factored form;
   b. take each different factor that appears in the factorizations in step a and write them as a product;
   c. raise each factor of step b to the greatest power it has in step a.
4. To obtain an *equivalent rational expression,* we multiply, or divide, the numerator and the denominator of the given rational expression by a *common factor.*
5. To *add* or *subtract* rational expressions, we must
   a. write each rational expression as an equivalent rational expression with the same denominator, preferably the LCD;
   b. add, or subtract, the numerators and place this sum, or difference, over the common denominator.
6. A **complex fraction** is a fraction whose numerator or denominator, or both, contains at least one fraction.
7. In the *complex fraction*

$$\frac{\frac{a}{b}}{\frac{c}{d}}$$

   we call $\frac{a}{b}$ the *primary numerator,* $\frac{c}{d}$ the *primary denominator,* and b and d the *secondary denominators.*
8. To *simplify a complex fraction,* we
   a. divide the primary numerator by the primary denominator after making a single fraction of each; or
   b. multiply the primary numerator and the primary denominator by the LCM of the secondary denominators and reduce the result, where possible.
9. A **rational equation** is an equation that contains at least one rational expression.
10. To *solve a rational equation,* we
   a. multiply each term of the equation by the LCD to eliminate the fractions;
   b. use the basic procedures for solving equations;
   c. check for extraneous solutions.

Chapter 6 error analysis

1. Subtracting rational expressions
   Example: $\frac{3x-1}{x-2} - \frac{x+2}{x-2} = \frac{3x-1-x+2}{x-2} = \frac{2x+1}{x-2}$
   Correct answer: $\frac{2x-3}{x-2}$
   What error was made? (*see page 240*)
2. Solving rational equations
   Example: Find the solution set of $\frac{y}{3} - \frac{y+1}{4} = \frac{2}{12}$

$$12 \cdot \frac{y}{3} - 12 \cdot \frac{y+1}{4} = 12 \cdot \frac{2}{12}$$
$$4y - 3(y+1) = 2$$
$$4y - 3y + 3 = 2$$
$$y = -1 \qquad \{-1\}$$

   Correct answer: $\{5\}$
   What error was made? (*see page 259*)
3. Adding and subtracting rational expressions
   Example: $\frac{3}{x-5} + \frac{5}{5-x} = \frac{3+5}{x-5} = \frac{8}{x-5}$
   Correct answer: $\frac{-2}{x-5}$
   What error was made? (*see page 241*)
4. Adding and subtracting rational expressions
   Example: $\frac{3}{x^2} + \frac{2}{x^2} = x^2 \cdot \frac{3}{x^2} + x^2 \cdot \frac{2}{x^2} = 3 + 2 = 5$
   Correct answer: $\frac{5}{x^2}$
   What error was made? (*see page 239*)
5. Exponents
   Example: $(-4)^2 = -16$
   Correct answer: 16
   What error was made? (*see page 57*)
6. Order of operations
   Example: $3^2 + 12 \cdot 2 - 8 \div 2 = 17$
   Correct answer: 29
   What error was made? (*see page 57*)
7. Combining polynomials
   Example: $(4a^2b - 2ab^2) - (a^2b + ab^2) = 3a^2b - ab^2$
   Correct answer: $3a^2b - 3ab^2$
   What error was made? (*see page 82*)
8. Solving linear inequalities
   Example: If $4 < -2x \le 6$, then $-2 < x \le -3$.
   Correct answer: If $4 < -2x \le 6$, then $-2 > x \ge -3$.
   What error was made? (*see page 116*)

9. Multiplication of like bases
Example: $5 \cdot 5^3 = 5^3$
Correct answer: 5^4
What error was made? (*see page 129*)

10. Squaring a binomial
Example: $(x - 6)^2 = (x)^2 - (6)^2 = x^2 - 36$
Correct answer: $x^2 - 12x + 36$
What error was made? (*see page 135*)

Chapter 6 critical thinking

Which integers can be written as the sum of three consecutive integers?

Example: $6 = 1 + 2 + 3$

Chapter 6 review

[6–1]

Perform the indicated multiplication and reduce the product to lowest terms.

1. $\frac{15}{8a} \cdot \frac{12a}{5}$
2. $\frac{24b}{7a} \cdot \frac{21a^2}{8b^2}$
3. $\frac{x - 3y}{2x + y} \cdot \frac{2x - y}{x - 3y}$
4. $\frac{5x - 10}{x + 3} \cdot \frac{3x + 9}{15}$
5. $\frac{m^2 - n^2}{14x} \cdot \frac{35x^2}{5m + 5n}$
6. $\frac{y}{y^2 - 1} \cdot \frac{y + 1}{y^2 - y}$
7. $\frac{b^2 - 36}{b^2 + 12b + 36} \cdot \frac{2b + 12}{b - 6}$
8. $\frac{5 - x}{6a - 3} \cdot \frac{2a - 1}{x^2 - 25}$
9. $\frac{x^2 + x - 2}{x^2 - 4x - 12} \cdot \frac{x^2 - 7x + 6}{x^2 - 1}$

Find the indicated quotients and state the answer reduced to lowest terms. Assume all denominators are nonzero.

10. $\frac{14a}{9} \div \frac{7}{3}$
11. $\frac{10}{9b} \div \frac{35}{12b^2}$
12. $\frac{24ab}{7} \div \frac{16a^2b^2}{21}$
13. $\frac{2x - 1}{3x + 4} \div \frac{3x}{7x}$
14. $\frac{x + 6}{x^2 - 4} \div \frac{(x + 6)^2}{x + 2}$
15. $\frac{4a - 8}{2a + 1} \div \frac{6a - 12}{10a + 5}$
16. $\frac{x^2 + 16x + 64}{x^2 + 9x + 8} \div \frac{x^2 - 64}{x + 1}$
17. $\frac{1 - b}{(b + 3)^2} \div \frac{b^2 - 2b + 1}{b^2 - 9}$
18. $\frac{9a^2 + 15a + 6}{a^2 + 3a - 4} \div \frac{36a^2 - 16}{3a^2 + 10a - 8}$

[6–2]

Find the least common denominator of rational expressions having the given denominators. Assume all denominators are nonzero.

19. $9x^2$ and $12x$
20. $14ab$ and $21a^2b^2$
21. $x^2 - 9$ and $2x + 6$
22. $y^2 - 2y - 15$ and $y^2 - 25$
23. $4z - 8$, $z^2 - 4$, and $z + 1$
24. $x^2 + x$, $x^2 + 2x + 1$, and $3x^2 - 2x - 5$

Find the indicated sum or difference. Assume all denominators are nonzero.

25. $\frac{4}{x} + \frac{1}{-x}$
26. $\frac{y}{y - 2} - \frac{3y}{2 - y}$
27. $\frac{3x - 2}{3x - 1} - \frac{x + 6}{1 - 3x}$
28. $\frac{x + 7}{2x - 5} + \frac{2x - 6}{5 - 2x}$

[6–3]

Add or subtract as indicated and reduce the answer to lowest terms. Assume all denominators are nonzero.

29. $\frac{4x}{15} + \frac{8x}{21}$
30. $\frac{25}{16a} - \frac{13}{12a}$
31. $\frac{5}{3x + 1} + \frac{9}{4x - 3}$
32. $\frac{4x}{x + 4} - \frac{7x}{x^2 - 16}$
33. $\frac{4}{ab^2} + \frac{12}{5a^2b} - \frac{3}{4ab}$
34. $4a + \frac{6}{a + 1}$

35. $\dfrac{7}{x^2 + 1} - 10$

36. $\dfrac{x + 1}{4x} - \dfrac{3x - 5}{8x^2} + \dfrac{5x + 4}{12x}$

37. $\dfrac{4y}{y^2 - 7y - 18} + \dfrac{9y}{y^2 - 4}$

38. $\dfrac{5a}{x + 3} + \dfrac{2}{x^2 - 9} - \dfrac{7a}{x - 3}$

39. $\dfrac{9}{x} - \dfrac{4}{x - 5} + \dfrac{3}{x + 4}$

40. $\dfrac{5}{x^2 + 2xy + y^2} + \dfrac{5}{x^2 - xy - 2y^2}$

[6–4]

Simplify the given complex fractions and reduce to lowest terms.

41. $\dfrac{\frac{4}{7}}{\frac{9}{4}}$

42. $\dfrac{\frac{3}{4} - \frac{1}{2}}{\frac{1}{4} + \frac{3}{2}}$

43. $\dfrac{\frac{4}{5} + 1}{2 - \frac{1}{5}}$

44. $\dfrac{\frac{1}{x} - \frac{1}{y}}{\frac{1}{x} + \frac{1}{y}}$

45. $\dfrac{\frac{4}{x^2} + \frac{3}{x}}{\frac{2}{x^2} - \frac{5}{x}}$

46. $\dfrac{\frac{a}{b} - \frac{b}{a}}{\frac{a}{b} + \frac{b}{a}}$

47. $\dfrac{\frac{1}{x} - \frac{1}{y}}{\frac{1}{xy}}$

48. $\dfrac{\frac{1}{x} - \frac{1}{y}}{\frac{1}{x^2} - \frac{1}{y^2}}$

49. $\dfrac{\frac{a^2}{b^2} - 2a - 3}{\frac{a - 3b}{ab}}$

[6–5]

Find the solution set of the following rational equations.

50. $\dfrac{x}{8} - 3 = \dfrac{2x}{12} + 1$

51. $\dfrac{4a + 1}{4} + \dfrac{5}{6} = 3a - \dfrac{5}{8}$

52. $\dfrac{12}{4a} - \dfrac{5}{6a} = 4$

53. $\dfrac{7}{a + 1} - \dfrac{1}{a^2 - 1} = \dfrac{2}{a + 1}$

54. $\dfrac{1}{y + 3} - \dfrac{6}{y} = \dfrac{7}{y} + \dfrac{2}{y + 3}$

55. $\dfrac{7}{2c^2} + \dfrac{5}{6c^2} = \dfrac{7}{12c^3}$

56. $\dfrac{x^2}{4} - 3x = 0$

57. $\dfrac{2}{3}y^2 = \dfrac{3}{2}$

58. $\dfrac{x^2}{2} - \dfrac{15}{2} = -x$

59. $\dfrac{a^2}{3} = \dfrac{5a}{2} - 3$

Solve for the indicated variable.

60. $\dfrac{a}{x} + \dfrac{b}{x} = 3$ for x

61. $\dfrac{y}{x + 1} = \dfrac{y^2}{x - 3}$ for x

62. $\dfrac{3}{4 - y} = \dfrac{a}{b}$ for y

63. $\dfrac{4}{y - c} - \dfrac{5}{y + b} = 0$ for y

Solve the following word problems.

64. An equation for tensile and compressional stresses is given by $y = \dfrac{Fl}{Ae}$. Solve for l.

65. The efficiency of a screw jack is calculated by the formula $E = \dfrac{Wp}{2 + LF}$. Solve for L.

66. A formula for the theoretical mechanical advantage, M, of a chain fall is given by $M = \dfrac{2R}{R - r}$. Solve for R.

67. The total reaction force F of air against a plane at the bottom of a vertical loop in centripetal force is given by $F = \dfrac{mv^2}{r} + m \cdot g$. Solve for m.

[6–6]

68. With different equipment, one painter can paint a house three times faster than a second painter. Working together they can do it in 4 hours. How long would it take each of them to paint the house working alone?

69. Paul can row his boat at a rate of 4 miles per hour in still water. It takes him as long to row 20 miles downstream as it takes him to row 8 miles upstream. What is the rate of the current?

70. The sum of three times a number and twice its reciprocal is 5. Find the number.

Chapter 6 cumulative test

Compute and simplify.

[1–5] **1.** $-\frac{1}{3}+\frac{1}{4}+(-6)+\frac{1}{6}$

[1–6] **2.** $-\frac{4}{5}\left(-\frac{15}{16}\right)$

[1–5] **3.** $\frac{5}{6}-\left(-\frac{3}{5}\right)$

[1–7] **4.** $-\frac{24}{35}\div\left(-\frac{3}{7}\right)$

Simplify the following expressions with only positive exponents.

[3–3] **5.** $x^8\cdot x^3\cdot x^0$

[3–3] **6.** $\frac{y^{-5}}{y^4}$

[3–4] **7.** $(-5x^2y^{-3})^2$

[2–3] **8.** $(9x^3-3x^2+4x-8)-(-x^3+x^2-7x+1)$

Find the solution set of the following equations and inequalities.

[2–6] **9.** $-5(2x+5)=-3x+1$

[2–6] **10.** $\frac{1}{3}x-\frac{3}{4}=6$

[4–7] **11.** $2y^2-y=6$

[2–9] **12.** $3x-4\leq 5x+6$

Completely factor each expression.

[4–4] **13.** $2x^2-18$

[4–1] **14.** $6x^5-36x^3+9x^2$

[4–3] **15.** $4x^2+16x+15$

[4–4] **16.** $3-12x^6$

Perform the indicated operations.

[3–2] **17.** $(7x-6)^2$

[3–2] **18.** $\left(5-\frac{1}{2}x\right)\left(5+\frac{1}{2}x\right)$

[3–2] **19.** $(2x-3)(3x^2-5x+11)$

[3–2] **20.** $(3-2x^2)(5-4x^2)$

[5–4] **21.** Find x when $\frac{x}{24}=\frac{5}{7}$.

[5–4] **22.** Power (in foot-pounds per minute) is given by the ratio of work, w, to time, t. Find the power exerted by pushing an 80-pound load 180 feet in 6 minutes. (*Hint:* w = weight · length.)

Perform the indicated operations and reduce the answers to lowest terms.

[6–1] **23.** $\frac{4a}{5b}\cdot\frac{a}{b}$

[6–1] **24.** $\frac{y+3}{y-6}\cdot\frac{y+7}{y-2}$

[6–1] **25.** $\frac{y^2-4}{y+3}\cdot\frac{y^2-9}{y+2}$

[6–1] **26.** $\frac{x^2-y^2}{9}\div\frac{y^2+xy}{9x-9}$

[6–3] **27.** $\frac{2}{x^2y}+\frac{4}{xy^2}$

[6–3] **28.** $\frac{6}{x-y}-\frac{3}{x+y}$

[6–3] **29.** $\frac{a+1}{a-2}+\frac{a-8}{2-a}$

[6–3] **30.** $\frac{7}{x^2-49}+\frac{6}{x^2-5x-14}$

[6–3] **31.** $\frac{2x}{x^2+x-6}-\frac{3x}{x^2-3x-18}$

Simplify the following complex rational expressions.

[6–4] 32. $\dfrac{3 - \dfrac{1}{y}}{4 + \dfrac{5}{y}}$

[6–4] 33. $\dfrac{\dfrac{1}{x} + \dfrac{1}{y}}{\dfrac{3}{y} - \dfrac{4}{x}}$

[6–4] 34. $\dfrac{\dfrac{x - 4}{x + 7}}{\dfrac{x + 5}{x^2 - 49}}$

Find the solution set of the following equations.

[6–5] 35. $\dfrac{4}{x} = \dfrac{6}{2x - 1}$

[6–5] 36. $x^2 - 2 = \dfrac{17x}{3}$

[6–5] 37. Solve the equation $\dfrac{1}{f} = \dfrac{1}{p} + \dfrac{1}{q}$ for q.

Choose a variable, set an appropriate equation, and solve the following problems.

[6–6] 38. John can build a fence in 6 days and Harry can build the same fence in 4 days. How long would it take them to build the fence working together?

[6–6] 39. The denominator of a fraction is two times the numerator. If 2 is added to the numerator and 2 is subtracted from the denominator, the resulting fraction is $\dfrac{5}{7}$. Find the original fraction.

CHAPTER

7 Linear Equations in Two Variables

Tickets to a football game at Smith High School cost $1.25 for students and $3.00 for adults. The total receipts for a game with University High School were $1,020. Write an equation using x for the number of students and y for the number of adults attending the game.

7–1 ■ Ordered pairs and the rectangular coordinate system

Linear equations in two variables

In chapter 2, we developed methods for solving linear equations (first-degree equations) in one variable. All such equations could be stated in the form

$$ax + b = 0$$

where a and b are real number constants and $a \neq 0$. In this chapter, we expand our work to linear equations in *two variables*. The equations

$$3x + 4y = 8 \quad \text{and} \quad 4y - x = 0$$

are examples of linear equations in two variables.

Definition

A **linear equation in two variables** x and y is any equation that can be written in the form

$$ax + by = c$$

where a, b, and c are real numbers and a and b are not both zero.

Our primary concern with all equations is finding their **solution(s).** These are the replacement values for the variable(s) that satisfy the equation. In an equation in two variables, x and y, any *pair* of values for x and y that satisfies the equation is a solution of the equation.

■ Example 7–1 A

Given the linear equation $3x + 2y = 6$, determine if the given values of x and y are solutions of the equation.

1. Let $x = 2$ and $y = 0$

$$3x + 2y = 6$$
$$3(2) + 2(0) = 6 \quad \text{Replace } x \text{ with 2 and } y \text{ with 0 in the equation}$$
$$6 + 0 = 6 \quad \text{Multiply as indicated}$$
$$6 = 6 \quad \text{(True)}$$

We see that the given values satisfy the equation, so $x = 2$ and $y = 0$ is a solution.

2. Let $x = 1$ and $y = \frac{3}{2}$

$$3x + 2y = 6$$
$$3(1) + 2\left(\frac{3}{2}\right) = 6 \quad \text{Replace } x \text{ with 1 and } y \text{ with } \frac{3}{2} \text{ in the equation}$$
$$3 + 3 = 6$$
$$6 = 6 \quad \text{(True)}$$

Again we see that the values $x = 1$ and $y = \frac{3}{2}$ satisfy the equation and thereby form a solution.

3. Let $x = 3$ and $y = 1$

$$3x + 2y = 6$$
$$3(3) + 2(1) = 6 \quad \text{Replace } x \text{ with 3 and } y \text{ with 1 in the equation}$$
$$9 + 2 = 6$$
$$11 = 6 \quad \text{(False)}$$

We conclude that $x = 3$ and $y = 1$ *do not* satisfy the equation and hence *do not form a solution.*

▶ ***Quick check*** Determine if $x = 1$ and $y = -1$ and if $x = 2$ and $y = 0$ are solutions for $3x + y = 2$ ■

Ordered pairs of numbers

The pairs of values for x and y used in example 7–1 A may be written as a pair of numbers. We separate them by a comma and place them inside the parentheses. The value of x is *always* given first. That is, the pair of numbers is written (x,y). In the examples, we used the pairs

$$(2,0),\ \left(1,\frac{3}{2}\right),\ \text{and } (3,1)$$

Pairs of numbers written in this special *order* (with the x value always first) are called **ordered pairs of numbers.** The *first number* of the ordered pair (the value of x) is called the *first component* of the ordered pair. The *second number* (the value of y) is called the *second component* of the ordered pair.

To determine ordered pairs that are solutions of an equation in two variables, we use the following procedure:

Finding ordered pair solutions

1. Choose a value for one of the variables.
2. Replace the variable with this known value and solve the resulting equation for the other variable.

■ Example 7–1 B

Using the given value of one of the variables, find the remaining variable. Write the ordered pair solution of the equation.

1. Given $y = 2x + 1$

a. Let $x = 3$

$y = 2x + 1$	
$y = 2(3) + 1$	Replace x with 3 in the equation
$y = 6 + 1$	Multiply as indicated
$y = 7$	Add in right member

The ordered pair (3,7) is a solution.

b. Let $x = -2$.

$y = 2x + 1$	
$y = 2(-2) + 1$	Replace x with -2 in the equation
$y = -4 + 1$	Solve for y
$y = -3$	

The ordered pair $(-2,-3)$ is a solution.

c. Let $y = 5$

$y = 2x + 1$	
$(5) = 2x + 1$	Replace y with 5 in the equation
$4 = 2x$	Solve for x
$2 = x$	

The ordered pair (2,5) is a solution.

Note We can choose *infinitely many* values for x and get a corresponding value of y for each one. That means there are infinitely many solutions.

2. Given $3x - 2y = 4$

a. Let $x = 2$

$3x - 2y = 4$	
$3(2) - 2y = 4$	Replace x with 2 in the equation
$6 - 2y = 4$	Solve for y
$-2y = -2$	
$y = 1$	

The ordered pair (2,1) is a solution.

b. Let $y = 7$

$$\begin{aligned} 3x - 2y &= 4 \\ 3x - 2(7) &= 4 && \text{Replace } y \text{ with 7 in the equation} \\ 3x - 14 &= 4 && \text{Solve for } x \\ 3x &= 18 \\ x &= 6 \end{aligned}$$

The ordered pair (6,7) is a solution.

3. Given $y = 6$
We can rewrite this equation as

$$y + 0 \cdot x = 6$$

a. Let $x = 3$

$$\begin{aligned} y + 0 \cdot x &= 6 \\ y + 0(3) &= 6 && \text{Replace } x \text{ with 3 in the equation} \\ y + 0 &= 6 && \text{Solve for } y \\ y &= 6 \end{aligned}$$

The ordered pair (3,6) is a solution.

b. Let $x = -7$

$$\begin{aligned} y + 0 \cdot x &= 6 \\ y + 0(-7) &= 6 && \text{Replace } x \text{ with } -7 \text{ in the equation} \\ y + 0 &= 6 && \text{Solve for } y \\ y &= 6 \end{aligned}$$

The ordered pair $(-7,6)$ is a solution.

Note No matter what *value* we choose for *x*, *y* will *always* be 6. Then *every* ordered pair will have a 6 in the second position.

4. Given $x + 3 = 0$
We can rewrite this equation as $x = -3$ and then as $x + 0 \cdot y = -3$.

a. Let $y = 1$

$$\begin{aligned} x + 0 \cdot y &= -3 \\ x + 0(1) &= -3 && \text{Replace } y \text{ with 1 in the equation} \\ x + 0 &= -3 && \text{Solve for } x \\ x &= -3 \end{aligned}$$

The ordered pair $(-3,1)$ is a solution.

b. Let $y = -4$

$$\begin{aligned} x + 0 \cdot y &= -3 \\ x + 0(-4) &= -3 && \text{Replace } y \text{ with } -4 \text{ in the equation} \\ x + 0 &= -3 && \text{Solve for } x \\ x &= -3 \end{aligned}$$

The ordered pair $(-3,-4)$ is a solution.

Note No matter what *value* we choose for *y*, *x* will *always* be -3. Then every ordered pair will have a -3 in the first position.

From examples 3 and 4, we can see that solutions of linear equations in two variables that can be written in the form

$$x = a \quad \text{or} \quad y = b$$

where a and b are constants, have very special characteristics. Given

1. $x = a$, the *first component* in every ordered pair is always the number a.
2. $y = b$, the *second component* in every ordered pair is always the number b.

▶ ***Quick check*** In the following equations, find the missing variables value for each variable given.
a. $2x + y = 1$; $x = 2$, $x = -3$
b. $2y - 3x = 1$; $y = -4$, $y = 0$ ■

The rectangular coordinate plane

In chapter 1, we associated the set of real numbers with points on a straight line and called this the number line. This number line was then used to draw the graph of the solution of an equation or inequality in one variable. Now we associate the solutions of a linear equation *in two variables* with *points* on a flat surface, called a *plane*.

We take two number lines, one horizontal and the other vertical, on the plane that are perpendicular to each other, and call them *axes*. These two number lines are represented in figure 7–1. The horizontal line (x-axis) is associated with values of x, and the vertical line (y-axis) is associated with values of y. Together, the x- and y-axes form the **rectangular coordinate plane** (or Cartesian coordinate plane).

Note The invention of this graphing method (relating the algebraic concept of an ordered number pair with the geometric concept of a point in the plane) is attributed to French mathematician and philosopher René Descartes (1596–1650). This combination of algebra and geometry has become known as *analytic geometry* or *coordinate geometry*.

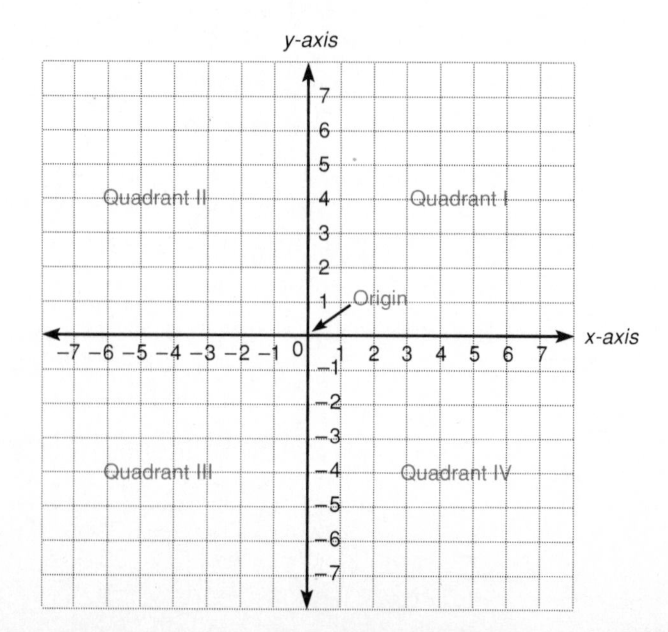

Figure 7–1

The point at which the two axes intersect is their common zero point and is called the *origin*. The origin corresponds to the ordered pair (0,0). The two axes separate the coordinate plane into four regions called *quadrants*. These quadrants are named as shown in figure 7–1. Points that lie on the x-axis or y-axis *do not* lie in any of the quadrants.

On the x-axis, numbers to the right of the origin are positive and those to the left of the origin are negative. On the y-axis, the positive numbers are above the origin, and the negative numbers are below the origin. For each point on the x-axis, $y = 0$, and for each point on the y-axis, $x = 0$.

Note We have chosen to make each unit on the axes equal to 1. Other choices are possible and, in fact, may be necessary in some instances.

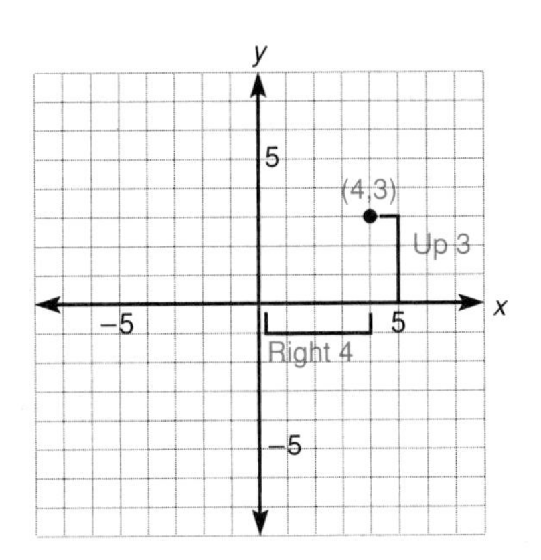

Figure 7–2

Each ordered pair (x,y) corresponds to *exactly* one point, called the *graph* of the ordered pair. To find the location of such a point is called *plotting* the point. We can plot any ordered pair, (x,y), on our coordinate plane if we consider the ordered pair as two instructions to direct us from the origin to the proper location of the point. To plot the point that corresponds to the ordered pair (4,3), we start at the origin. Since the x-value, also called the **abscissa** of the point, is 4, we move four units *to the right* (the positive direction) along the x-axis. From this position, the y-value, also called the **ordinate** of the point, instructs us to move *up* (the positive direction) three units parallel to the y-axis. The abscissa and ordinate are usually called the *coordinates* of a point. We have plotted the point that is the graph of the ordered pair (4,3) as shown in figure 7–2.

Similarly, to plot the graph of $(-3,-5)$, we start at the origin. Since $x = -3$, we move three units *to the left* (the negative direction). Next, $y = -5$ instructs us to move *down* (the negative direction) five units parallel to the y-axis. We have plotted the graph of the ordered pair $(-3,-5)$ as shown in figure 7–3.

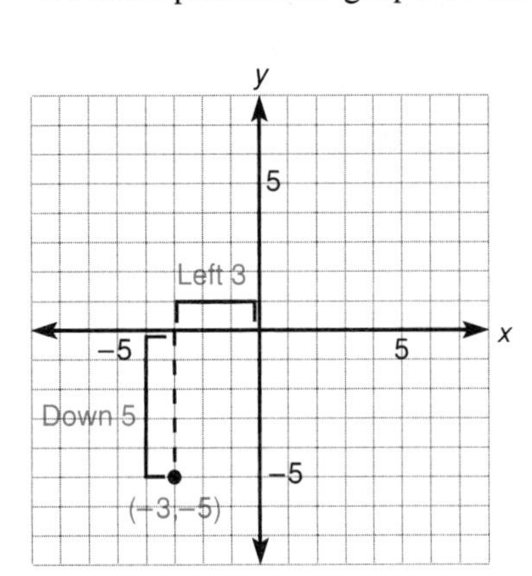

Figure 7–3

Points are always named by capital letters. The notation $A(x,y)$ indicates that the *name* of the point is A and the *coordinates* of the point are (x,y).

■ Example 7–1 C

Plot the following points.

a. $A(2,4)$
b. $B(-1,-3)$
c. $C(-4,3)$
d. $D(5,0)$
e. $E(2,-3)$
f. $F(0,4)$
g. $G(-3,0)$
h. $H\left(2,\frac{3}{2}\right)$

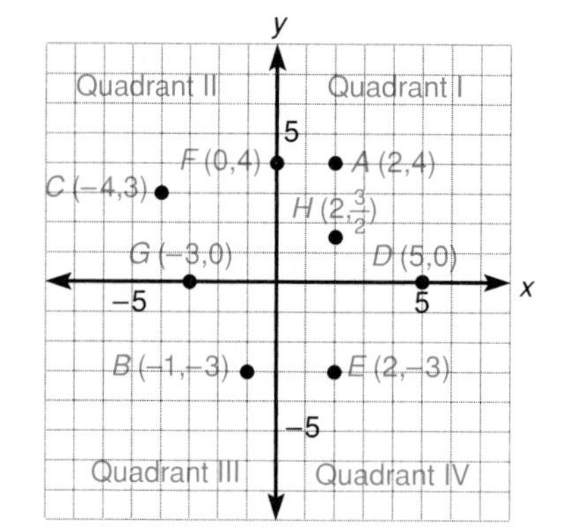

Note Whenever either one of the coordinates of the point is zero, the point is located on an axis. If the first component is zero, the point is on the y-axis. If the second component is zero, the point is on the x-axis. If both components are zero, the point is the origin.

We can see in the diagram that a point is in

1. quadrant I when x and y are both positive,
2. quadrant II when x is negative and y is positive,
3. quadrant III when x and y are both negative,
4. quadrant IV when x is positive and y is negative. ■

Graphs of solutions—linear equations in two variables

Now consider the graphs of some of the ordered pairs that are solutions of the linear equation in two variables $y = 2x + 1$. Suppose we let $x = -3$, $x = 0$, $x = 2$, and $x = 3$. Then

when $x = -3$, $y = 2(-3) + 1 = -6 + 1 = -5$ Replace x with -3
when $x = 0$, $y = 2(0) + 1 = 0 + 1 = 1$ Replace x with 0
when $x = 2$, $y = 2(2) + 1 = 4 + 1 = 5$ and Replace x with 2
when $x = 3$, $y = 2(3) + 1 = 6 + 1 = 7$ Replace x with 3

The ordered pairs $(-3,-5)$, $(0,1)$, $(2,5)$, and $(3,7)$ are solutions of the equation $y = 2x + 1$. These ordered pairs are often shown in an x-y table

x	$y = 2x + 1$	Ordered pair
-3	$2(-3) + 1 = -5$	$(-3,-5)$
0	$2(0) + 1 = 1$	$(0,1)$
2	$2(2) + 1 = 5$	$(2,5)$
3	$2(3) + 1 = 7$	$(3,7)$

We now plot these points in figure 7–4.

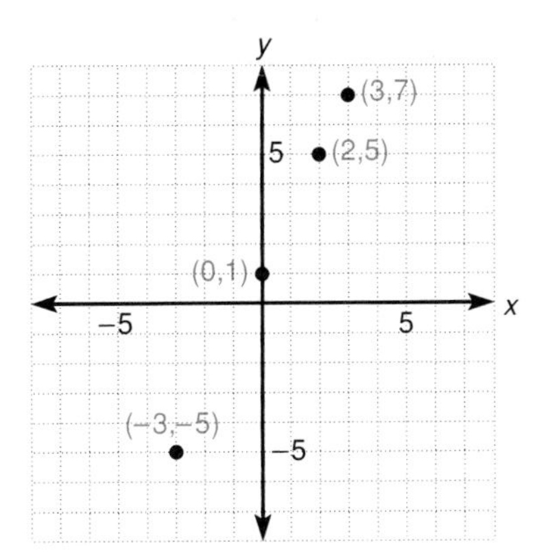

Figure 7–4

> ***Mastery points***
>
> *Can you*
>
> - Determine whether or not an ordered pair is a solution of a given equation?
> - Find the value of one variable, given the value of the other variable?
> - Plot ordered pairs in the rectangular coordinate plane?
> - Plot ordered pair solutions of linear equations?

Exercise 7–1

Determine whether or not the given ordered pairs are solutions of the given equation. See example 7–1 A.

Example Consider $3x + y = 2$; $(1,-1)$, $(2,0)$.

Solution a.

$$\begin{aligned} x + y &= 2 \\ 3(1) + (-1) &= 2 \quad \text{Replace } x \text{ with 1 and } y \text{ with } -1 \\ 3 + (-1) &= 2 \\ 2 &= 2 \quad \text{(True)} \end{aligned}$$

Therefore $(1,-1)$ is a solution.

b.

$$\begin{aligned} 3x + y &= 2 \\ 3(2) + (0) &= 2 \quad \text{Replace } x \text{ with 2 and } y \text{ with 0} \\ 6 + 0 &= 2 \\ 6 &= 2 \quad \text{(False)} \end{aligned}$$

Therefore $(2,0)$ is *not* a solution.

1. $y = 3x - 1$; $(1,2)$, $(-1,-4)$, $(2,3)$

2. $y = 2x + 4$; $(-1,3)$, $(0,4)$, $(2,8)$

3. $x + 2y = 3$; $(1,2)$, $(-1,2)$, $(3,0)$

4. $3x - y = 4$; $(1,-1)$, $\left(\frac{1}{3},2\right)$, $(0,-4)$

5. $3y - 4x = 2$; $(1,2)$, $(-2,1)$, $\left(\frac{1}{2},1\right)$

6. $5x - 2y = 6$; $(2,2)$, $(0,-3)$, $(4,-2)$

7. $3x = 2y$; $(2,3)$, $(3,2)$, $(0,0)$

8. $3y = -4x$; $(4,-3)$, $(-4,3)$, $(8,-6)$

9. $x = -4$; $(-4,1)$, $(4,2)$, $(-4,-4)$

10. $y = 3$; $(2,3)$, $(-5,2)$, $\left(\frac{3}{4},3\right)$

11. $x + 5 = 0$; $(3,-5)$, $(-5,3)$, $(-5,8)$

12. $y - 2 = 0$; $(-2,2)$, $\left(\frac{2}{3},-2\right)$, $(5,2)$

Find the value for y corresponding to the given values for x in each equation. Express the answer as an ordered pair. See Example 7–1 B.

Example Consider $2x + y = 1$; $x = 2$, $x = -3$.

Solution a. Let $x = 2$ then

$2x + y = 1$
$2(2) + y = 1$ Replace x with 2
$4 + y = 1$ Solve for y
$y = -3$

The ordered pair is $(2,-3)$.

b. Let $x = -3$ then

$2x + y = 1$
$2(-3) + y = 1$ Replace x with -3
$-6 + y = 1$ Solve for y
$y = 7$

The ordered pair is $(-3,7)$.

13. $y = 3x + 2$; $x = 1$, $x = -2$, $x = 0$

14. $y = 4x - 3$; $x = -1$, $x = 2$, $x = 0$

15. $y = 3 - 2x$; $x = 3$, $x = -4$, $x = 0$

16. $y = -5 - x$; $x = \frac{1}{5}$, $x = 5$, $x = 0$

17. $3x + y = 4$; $x = 3$, $x = -2$, $x = 0$

18. $y - 4x = 1$; $x = 1$, $x = \frac{5}{4}$, $x = 0$

19. $x - 5y = 3$; $x = -2$, $x = 3$, $x = 0$

20. $x + 4y = 0$; $x = -4$, $x = 8$, $x = 0$

21. $5x + 2y = -3$; $x = 1$, $x = -1$, $x = 0$

22. $2x - 3y = 1$; $x = \frac{1}{4}$, $x = -4$, $x = 0$

23. $y = 5$; $x = 1$, $x = -6$, $x = 0$

24. $y = -4$; $x = -5$, $x = -4$, $x = 0$

25. $y + 1 = 0$; $x = 7$, $x = -\frac{3}{5}$, $x = 0$

26. $y - 4 = 0$; $x = 1$, $x = 2$, $x = 0$

Find the value for x corresponding to the given value for y in each equation. Express the answer as an ordered pair. See example 7–1 B.

Example Consider $2y - 3x = 1$; $y = -4$, $y = 0$.

Solution a. Let $y = -4$, then

$2y - 3x = 1$
$2(-4) - 3x = 1$ Replace y with -4
$-8 - 3x = 1$ Solve for x
$-3x = 9$
$x = -3$

The ordered pair is $(-3,-4)$.

b. Let $y = 0$, then

$2y - 3x = 1$
$2(0) - 3x = 1$ Replace y with 0
$0 - 3x = 1$ Solve for x
$-3x = 1$
$x = -\frac{1}{3}$

The ordered pair is $\left(-\frac{1}{3},0\right)$

27. $x = 2y + 3$; $y = 1$, $y = -2$, $y = 0$

28. $x = -3y + 1$; $y = -1$, $y = 2$, $y = 0$

29. $3y - 2x = 0$; $y = 2$, $y = -4$, $y = 0$

30. $2x + y = 3$; $y = -3$, $y = 5$, $y = 0$

31. $x = 1$; $y = -2$, $y = 7$, $y = 0$

32. $x + 7 = 0$; $y = -1$, $y = 3$, $y = 0$

Solve the following word problems.

33. The total cost c in dollars of producing x units of a certain commodity is given by the equation $c = 2x + 20$. Find the cost of producing (a) 75; (b) 300; (c) 1,000 units of the commodity. Write the answers as ordered pairs (x,c).

34. In exercise 33, find the number of units produced when the total cost is (a) \$430; (b) \$700; (c) \$1,400. Write the answers as ordered pairs (x,c).

35. The cholesterol level in the blood, y, is related to the dosage of a new anticholesterol drug in grams, x, by the equation $y = 240 - 2x$. Find the cholesterol level in the blood when the dosage is (a) 2 grams, (b) 12 grams, (c) 0 grams. Write the answers as ordered pairs (x,y).

36. In exercise 35, determine the number of grams in the dosage when the cholesterol level in the blood is (a) 200, (b) 0, (c) 210. Write the answers as ordered pairs (x,y).

37. Suppose the equation $y = 3x + 20$ represents the number of students present in a mathematics class, where x represents the number of hours of study required and y represents the number of students present. Find the number of students present when the class requires (a) 0 hours of study, (b) 5 hours of study, (c) 3 hours of study. Write the answers as ordered pairs (x,y).

38. In exercise 37, find the number of hours of study required when there are (a) 26 students present, (b) 23 students present, (c) 32 students present. Write the answers as ordered pairs (x,y).

39. If the distance, y, (in miles) that Mary Jane travels in x hours of driving is given by the equation $y = 55x$, how far does Mary Jane travel in (a) 3 hours, (b) 8 hours, (c) 5 hours and 12 minutes $\left(5\frac{1}{5}\text{ hours}\right)$? Write the answers as ordered pairs (x,y).

40. In exercise 39, how many hours does it take Mary Jane to drive (a) 110 miles, (b) 385 miles, (c) 500 miles? Write the answers as ordered pairs (x,y).

Plot the following ordered pairs on a rectangular coordinate plane. See example 7–1 C.

41. $(2,4)$ **42.** $(5,1)$ **43.** $(-1,3)$ **44.** $(-4,4)$ **45.** $(-6,-1)$

46. $(-2,-3)$ **47.** $(0,4)$ **48.** $(0,2)$ **49.** $(5,0)$ **50.** $(-4,0)$

51. $(-7,0)$ **52.** $(0,0)$ **53.** $\left(\frac{1}{2},3\right)$ **54.** $\left(\frac{2}{3},-2\right)$ **55.** $\left(\frac{3}{2},0\right)$

Determine the coordinates, (x,y), of the given points in the diagram.

56. A **57.** B **58.** C

59. D **60.** E **61.** F

62. G **63.** H **64.** I

65. J **66.** K

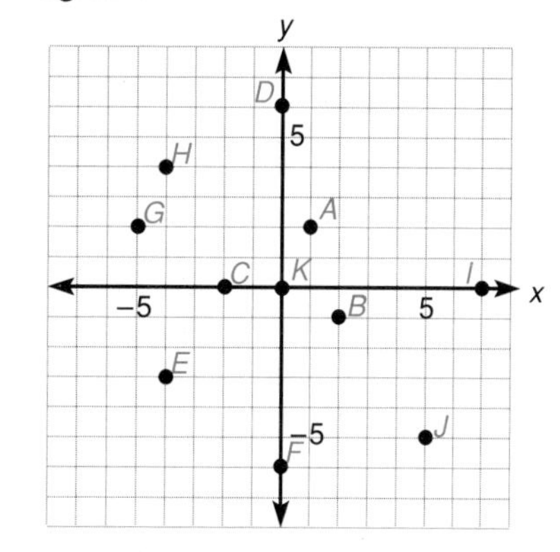

State the quadrant in which each point lies.

Example $(-1,3)$ lies in the second quadrant because the x-component is negative and the y-component is positive.

67. $(-2,-5)$ **68.** $(4,-1)$ **69.** $(5,3)$ **70.** $(-7,-9)$ **71.** $\left(\frac{1}{2},-4\right)$

72. $\left(-\frac{2}{3},7\right)$ **73.** $\left(-\frac{5}{2},-\frac{3}{4}\right)$ **74.** $\left(\frac{7}{8},-\frac{7}{8}\right)$ **75.** $(5,-14)$ **76.** $(-3,-3)$

Solve the following word problems.

77. In what quadrant does a point (x,y) lie if (a) $x > 0$ and $y < 0$, (b) $x < 0$ and $y > 0$, (c) $x < 0$ and $y < 0$, (d) $x > 0$ and $y > 0$?

78. What is the value of x for any point on the y-axis?

79. What is the value of y for any point on the x-axis?

80. Using the diagram for exercises 56–66, what is the *abscissa* of each of the points A, C, E, G, I, and K?

81. Using the diagram for exercises 56–66, what is the *ordinate* of each of the points B, D, F, H, and J?

82. Plot five points whose abscissa is -3. Connect the plotted points. Describe the resulting figure.

83. Using the same axes as in exercise 82, plot five points whose ordinate is 4. Connect the plotted points. Describe the resulting figure.

84. Choose five ordered pairs whose first and second coordinates are the same. Plot these points and connect the points. What kind of figure do we get? In what quadrants does the figure lie?

85. Choose five ordered pairs whose first component is the opposite of the second component. Plot the points and connect them. What kind of figure do we get? In what quadrants does the figure lie?

Plot the points in the following exercises using the indicated scale.

Note Exercises 86 to 90 require the use of different scales on the two axes. This is sometimes necessary in certain problems, especially in applications in the sciences and other fields.

Example 1 square = 5 units (on x-axis)
1 square = 10 units (on y-axis)

Note A "square" is one unit on a grid.

$A(0,-10)$, $B(5,20)$, $C(-15,20)$, $D(-25,-40)$, $E(20,0)$

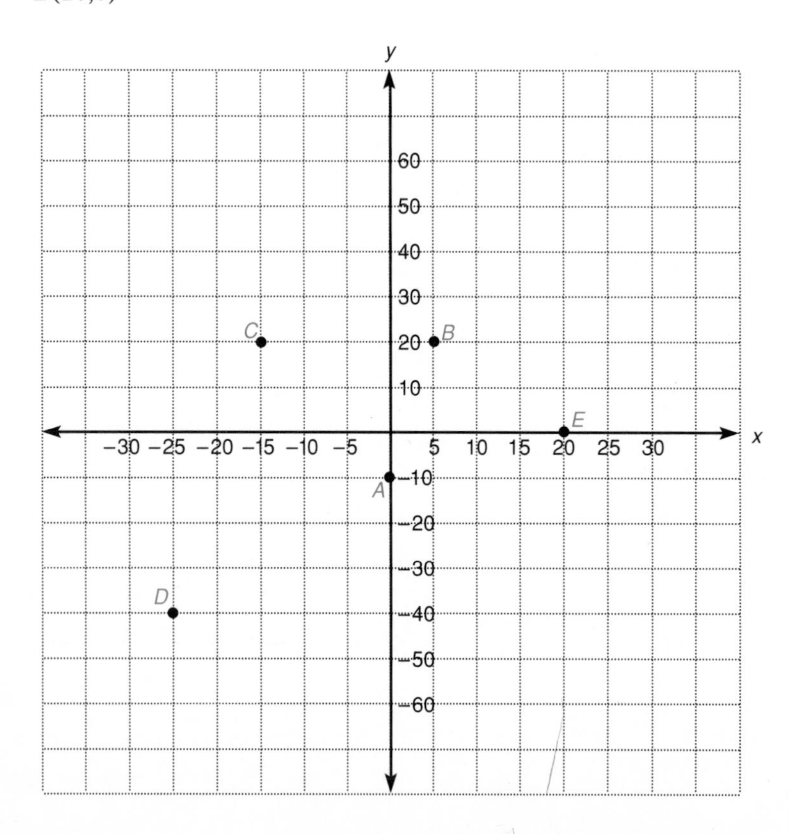

86. 1 square = 10 units
$A(20,20)$, $B(-30,40)$, $C(0,-50)$, $D(40,0)$, $E(50,-70)$

87. 1 square = 1 unit (on x-axis)
1 square = 5 units (on y-axis)
$A(4,10)$, $B(-6,35)$, $C(5,5)$, $D(0,-45)$, $E(-3,-50)$

88. 1 square = 5 units (on x-axis)
1 square = 1 unit (on y-axis)
$A(-20,3)$, $B(-5,5)$, $C(40,-6)$, $D(25,0)$, $E(-20,-4)$

89. 1 square = 10 units
$A(-35,20)$, $B(35,-40)$, $C(25,100)$, $D(0,0)$, $E(-40,90)$

90. 1 square = 25 units (on x-axis)
1 square = 100 units (on y-axis)
$A(75,-600)$, $B(200,-200)$, $C(0,-100)$, $D(-175,500)$, $E(-225,0)$

Example The following set of points represents the percent grades Bruce received on ten algebra quizzes.

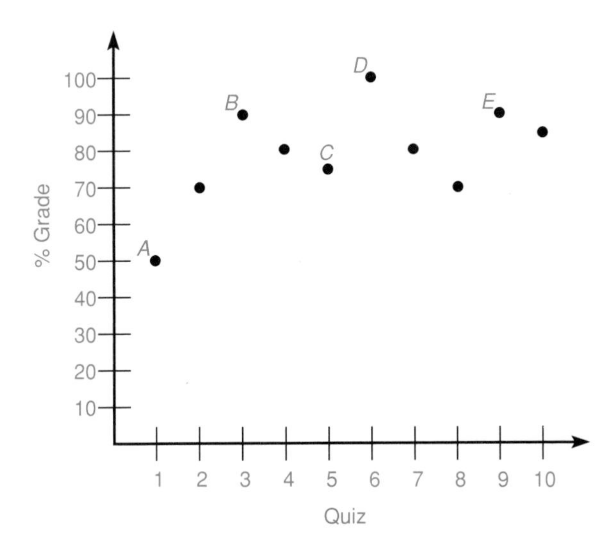

Find the coordinates of points A, B, C, D, and E and interpret the coordinates.

Solution $A(1,50)$: Grade was 50 percent on first quiz
$B(3,90)$: Grade was 90 percent on third quiz
$C(5,75)$: Grade was 75 percent on fifth quiz
$D(6,100)$: Grade was 100 percent on sixth quiz
$E(9,90)$: Grade was 90 percent on ninth quiz

91. The following sets of points represent the average temperature in Detroit during a seven-day period last summer. Find the approximate coordinates of points A, B, C, and D and interpret the coordinates.

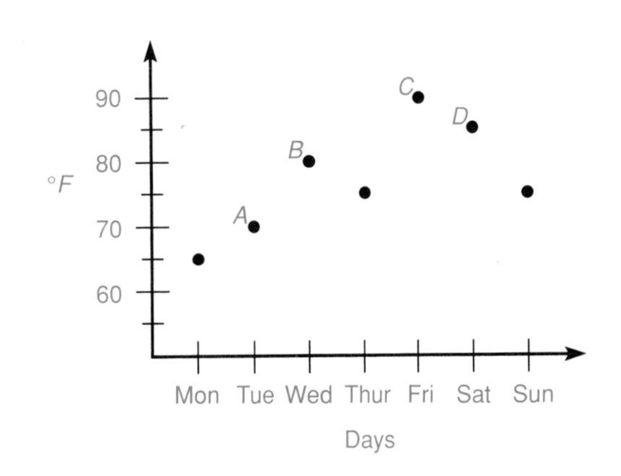

92. The following sets of points represent the approximate number of housing starts in the United States for the years 1973 to 1979. Find the coordinates of points A, B, C, and D and interpret the coordinates.

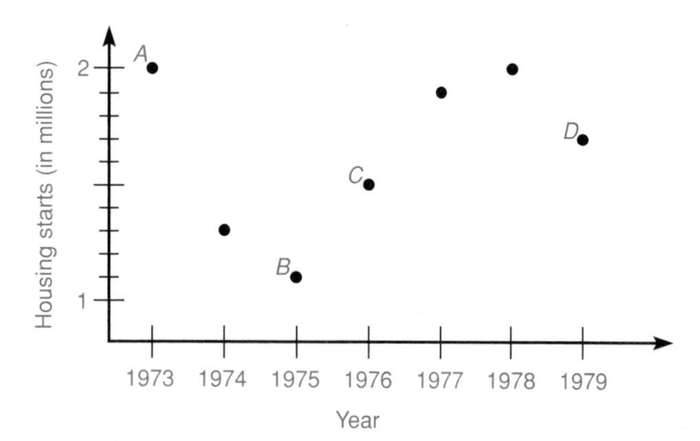

Find the missing component in each ordered pair using the given equation. Then plot the ordered pairs using a separate coordinate system for each problem. (See the diagram.)

Example $y = 4x + 2$; $\quad (0,\ \ ), (-2,\ \ ), (1,\ \ ), (2,\ \ )$

Solution When $x = 0, y = 4(0) + 2 = 0 + 2 = 2; (0,2)$
$x = -2, y = 4(-2) + 2 = -8 + 2 = -6; (-2,-6)$
$x = 1, y = 4(1) + 2 = 4 + 2 = 6; (1,6)$
$x = 2, y = 4(2) + 2 = 8 + 2 = 10; (2,10)$

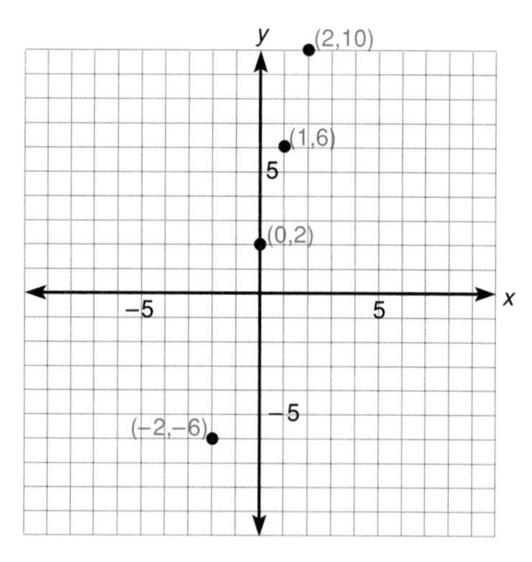

93. $y = x + 4; (0,\ \ ), (-4,\ \ ), \left(\frac{5}{2},\ \ \right), (2,\ \ )$

94. $y = x - 3; (0,\ \ ), (3,\ \ ), (1,\ \ ), \left(-\frac{1}{2},\ \ \right)$

95. $y = 2x + 1; (0,\ \ ), \left(-\frac{1}{2},\ \ \right), (2,\ \ ), (-1,\ \ )$

96. $y = 3x - 4; (0,\ \ ), (-1,\ \ ), \left(-\frac{1}{3},\ \ \right), (2,\ \ )$

97. $y = 2x; (0,\ \ ), \left(\frac{1}{2},\ \ \right), (2,\ \ ), (-1,\ \ )$

98. $y = 3x; (0,\ \ ), (3,\ \ ), (-3,\ \ ), \left(\frac{1}{3},\ \ \right)$

99. $y = -x + 3; (0,\ \ ), (-3,\ \ ), (3,\ \ ), \left(\frac{5}{2},\ \ \right)$

100. $y = -x + 1; (0,\ \ ), (1,\ \ ), (-2,\ \ ), (-4,\ \ )$

101. $y = -2x + 3; (0,\ \ ), (3,\ \ ), (-2,\ \ ), \left(\frac{3}{2},\ \ \right)$

102. $3x + y = 1; (0,\ \ ), \left(\frac{1}{3},\ \ \right), (-2,\ \ ), (1,\ \ )$

103. $y = 4 - x; (0,\ \ ), (-3,\ \ ), (2,\ \ ), (3,\ \ )$

104. $2y - 3x = 2; (0,\ \ ), \left(\frac{2}{3},\ \ \right), (-2,\ \ ), (2,\ \ )$

Review exercises

Perform the indicated operations. See sections 6–1 and 6–2.

1. $\dfrac{x^2 - 4}{3x} \cdot \dfrac{6x^2}{x - 2}$
2. $\dfrac{2x - 1}{x + 3} \div \dfrac{4x^2 - 1}{x^2 + 6x + 9}$
3. $\dfrac{4}{x - 2} - \dfrac{3}{x + 1}$
4. Find $-|-8|$. See section 1–3.
5. Multiply $x^2 \cdot x \cdot x^0$. See section 3–3.
6. Multiply $(-2)(-3)(4)(0)$. See section 1–6.

7–2 ■ Graphs of linear equations

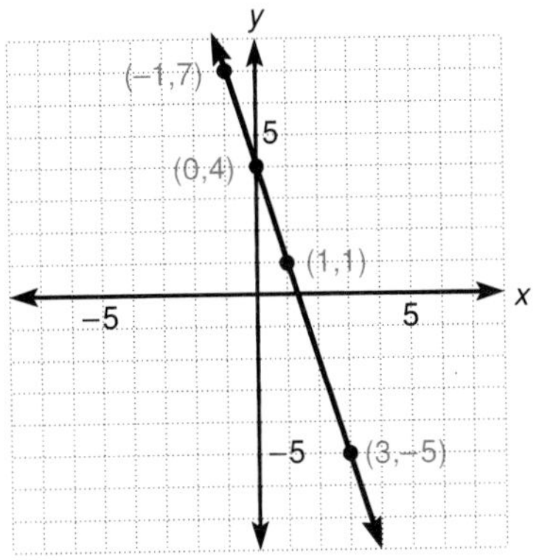

Graph of a linear equation

In section 7–1, we learned that there are infinitely many ordered pairs that will satisfy an equation in two variables. That is, given the linear equation $3x + y = 4$, we can find many ordered pairs (as many as we wish) that are solutions of the equation. To list all of these solutions is impossible. However, these solutions can be represented geometrically by a graph of the ordered pairs that are solutions.

In section 7–1, we plotted the graphs of several of the ordered pairs that satisfied the given equations. To illustrate again, consider the equation $3x + y = 4$. In figure 7–5, we plot the ordered pairs

$$(0,4), (1,1), (-1,7), \text{ and } (3,-5)$$

that are solutions of the equation.

y
(−1,7)
5
(0,4)
(1,1)
x
−5
5
−5
(3,−5)

Figure 7–5

Connecting the points, we find they all lie on the same straight line. We have drawn arrowheads in each direction at each end of the line to indicate that the line goes on indefinitely in each direction. *Any point whose coordinates satisfy the equation* $3x + y = 4$ *will lie on this line,* and *the coordinates of any point on this line will satisfy the equation.* We have a graphical representation of a portion of the solutions of the equation. The straight line in figure 7–5 is called the *graph of the equation* $3x + y = 4$.

Note Any point that *does not* lie on this line has coordinates that will not satisfy the equation $3x + y = 4$ and so *does not* represent a solution of the equation. For example, the point $(-1,1)$ does not lie on the line. If we substitute, we obtain

$$3(-1) + 1 = 4$$
$$-3 + 1 = 4$$
$$-2 = 4 \quad \text{(False)}$$

Straight line

In general, the graph of *any* linear equation in two variables is a *straight line.* A geometric fact we now use is that *through any two given points in the plane we can draw one and only one straight line.* Thus, since we know the graph of a linear equation in two variables is a straight line, we can determine the graph of the equation using only two points. However, it is a good idea to find a third point as a check on our work. (Remember, the word *line* appears in the name *line*ar equation.)

Example 7–2 A

Graph the equation $y = 2x + 4$.

We choose *three* arbitrary values of x and find corresponding values for y by substituting the value of x and solving for y.

$$\begin{aligned} \text{Let } x &= 0, \text{ then } y = 2(0) + 4 = 0 + 4 = 4 \\ x &= -2, \text{ then } y = 2(-2) + 4 = -4 + 4 = 0 \\ x &= 2, \text{ then } y = 2(2) + 4 = 4 + 4 = 8 \end{aligned}$$

Now we graph the points (0,4), (−2,0), and (2,8) and draw a straight line through the points.

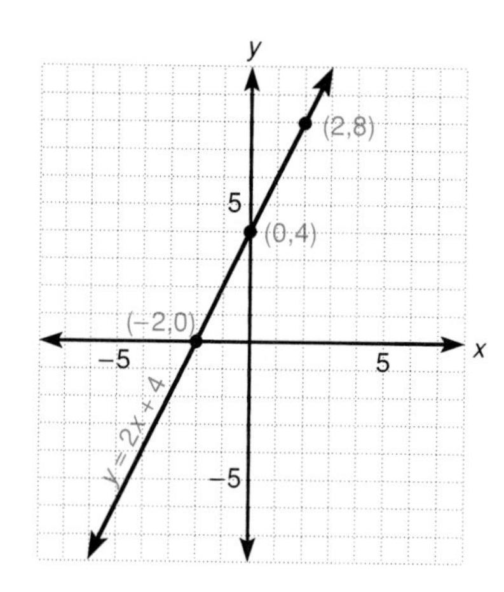

The x- and y-intercepts

Notice the graph of $y = 2x + 4$ crosses the y-axis at (0,4) and the x-axis at (−2,0). The points (0,4) and (−2,0) are called the *y-intercept* and the *x-intercept,* respectively. Since we need only two points to sketch the graph of a linear equation in two variables, in many cases we use the x- and y-intercepts. Observe from the example that when the line crosses the x-axis, the value of y is zero. When the line crosses the y-axis, the value of x is zero.

Finding x- and y-intercepts

1. To find the x-intercept, we let $y = 0$ and find the corresponding value of x. This is the point $(x,0)$.
2. To find the y-intercept, we let $x = 0$ and find the corresponding value for y. This is the point $(0,y)$.

Example 7–2 B

Graph the following linear equations using x- and y-intercepts.

1. $y = -2x + 3$

Let $x = 0$ to find the y-intercept; let $y = 0$ to find the x-intercept.

$y = -2x + 3$		$y = -2x + 3$	
$y = -2(0) + 3$	Replace x with 0	$(0) = -2x + 3$	Replace y with 0
$y = 0 + 3$	Multiply as indicated	$2x = 3$	Add $2x$ to each member
$y = 3$		$x = \dfrac{3}{2}$	

The point (0,3) is the y-intercept.

The point $\left(\dfrac{3}{2},0\right)$ is the x-intercept.

To find another point, choose $x = 2$.

$y = -2x + 3$
$y = -2(2) + 3$ Replace x with 2
$y = -4 + 3$ Multiply as indicated
$y = -1$

The third point is $(2, -1)$.

We now plot the three points $(0,3)$, $\left(\frac{3}{2},0\right)$ and $(2,-1)$ and draw a straight line through them.

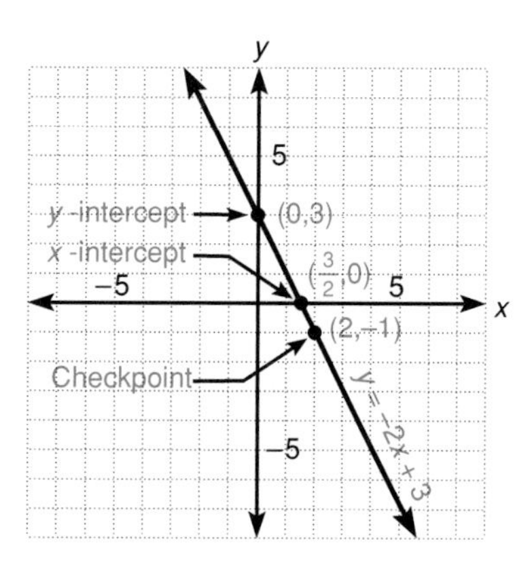

Note The components of these points are often stated in a table of related values as illustrated in the following examples.

2. $3y - 2x = 9$
Let $x = 0$ to find the y-intercept; let $y = 0$ to find the x-intercept.

When $x = 0$

$3y - 2x = 9$
$3y - 2(0) = 9$ Replace x with 0
$3y - 0 = 9$
$3y = 9$
$y = 3$

The point $(0,3)$ is the y-intercept.

When $y = 0$

$3y - 2x = 9$
$3(0) - 2x = 9$ Replace y with 0
$0 - 2x = 9$
$-2x = 9$
$x = -\frac{9}{2}$

The point $\left(-\frac{9}{2},0\right)$ is the x-intercept.

For the third point, let $x = 3$.

$3y - 2x = 9$
$3y - 2(3) = 9$ Replace x with 3
$3y - 6 = 9$
$3y = 15$
$y = 5$

The checkpoint is $(3,5)$.

Plot the points $(0,3)$, $\left(-\frac{9}{2},0\right)$, and $(3,5)$ and draw a straight line through them.

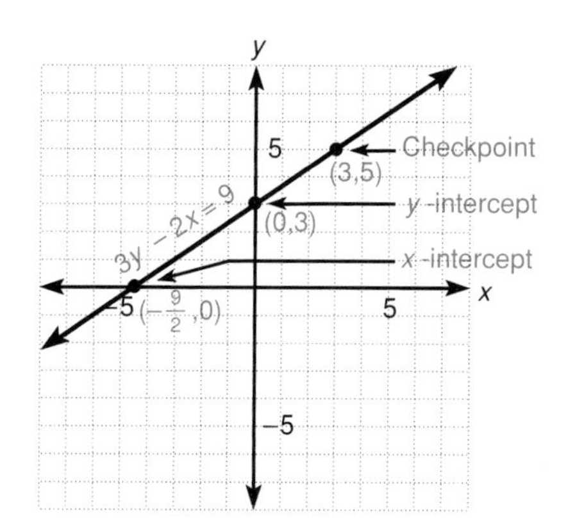

In the previous examples, the x- and y-intercepts were different points. For some equations, the x- and y-intercepts are the same point, as demonstrated in example 3.

3. $y = 3x$
If we let $x = 0$, then $y = 3(0) = 0$, giving the ordered pair (0,0). When $y = 0$, then $0 = 3x$ and $x = 0$, giving the same point (0,0). We must choose two additional values for x or y. Let $x = 1$ and $x = -1$.

x	**y = 3x**	**Ordered pair (x,y)**	
0	$3(0) = 0$	(0,0)	x- and y-intercepts
1	$3(1) = 3$	(1,3)	Arbitrary second point
-1	$3(-1) = -3$	$(-1,-3)$	Checkpoint

Plot the points (0,0), (1,3), and $(-1,-3)$ and draw a straight line through them.

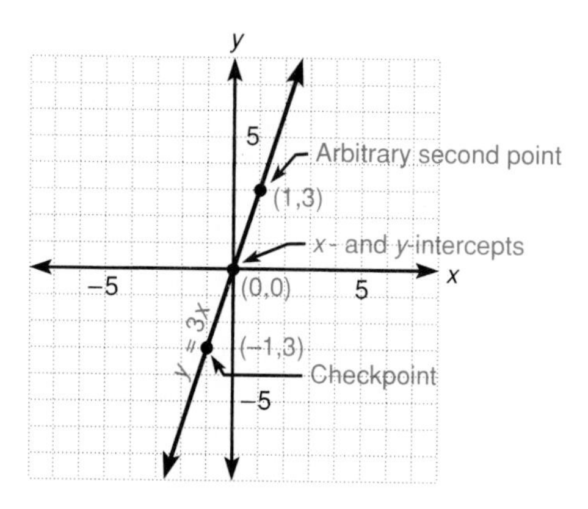

To generalize, any linear equation that can be written in the form

$$y = kx \quad \text{or} \quad x = ky$$

where k is a real number, will pass through the origin (0,0). The next two examples show what happens when one of the variables is missing.

4. $y = -3$
Recall that this equation could be written $y = 0 \cdot x - 3$ and that for *any* value of x we might choose, y is *always* equal to -3. Therefore, we choose any three values for x and obtain $y = -3$.

x	y
-2	-3
0	-3
2	-3

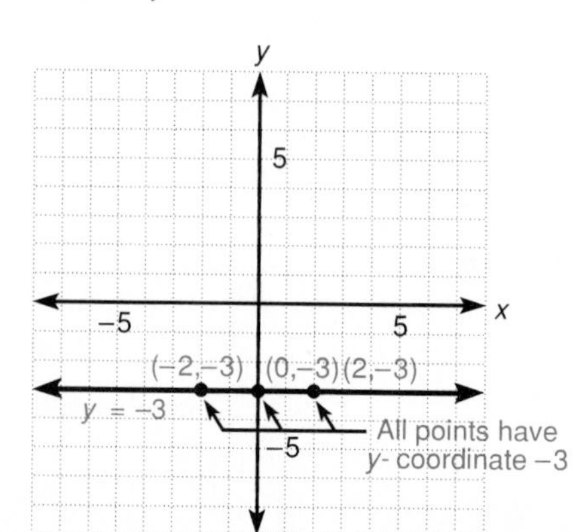

Note The graph has a y-intercept, -3, but no x-intercept. The graph is a horizontal straight line passing through the point $(0,-3)$ (parallel to the x-axis). In fact, the graph of any equation of the form $y = b$ will be a horizontal line passing through the point with coordinates $(0,b)$, the y-intercept.

5. $x = 4$

Recall that this equation can be written $x = 0 \cdot y + 4$ and that the value of x will be 4 for *any* value of y.

x	y
4	-2
4	0
4	2

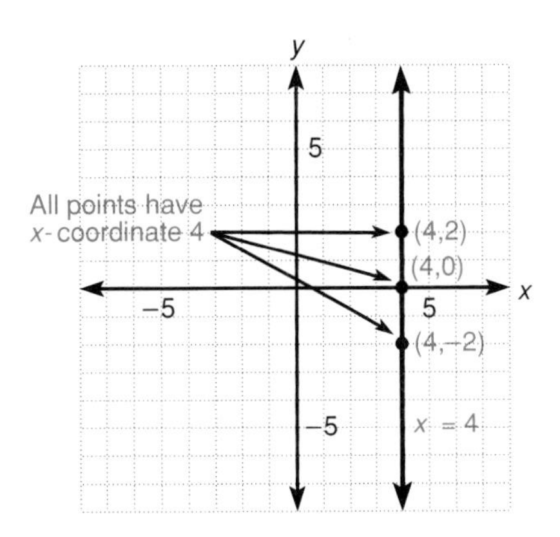

Note The graph has an x-intercept, 4, but no y-intercept. The graph is a *vertical* straight line (parallel to the y-axis) passing through the point with coordinates $(4,0)$. In fact, the graph of any equation of the form $x = a$ will be a vertical line passing through the point with coordinates $(a,0)$, the x-intercept.

▶ ***Quick check*** a. Find the x- and y-intercepts for the graph of $5x + 3y = 15$
b. Graph the linear equation $2y - 3x = 12$ using the x- and y-intercepts. ■

We now summarize the different forms that linear equations might take.

1. $ax + by = c$ Graph by finding the x-intercept (let $y = 0$), the y-intercept (let $x = 0$), and a third checkpoint by choosing any value for x or y not yet used.

Example:
$2x + y = 4$

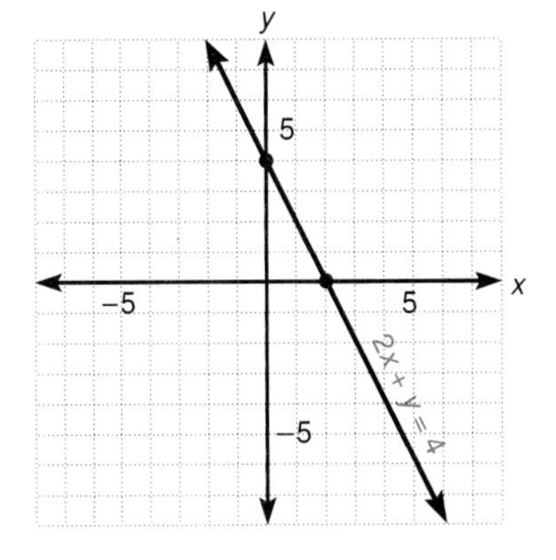

2. $y = kx$ or $x = ky$

Graph goes through the origin (0,0). Find two other points by choosing values for x or y other than 0.

Examples: $y = 2x$ and $x = -3y$

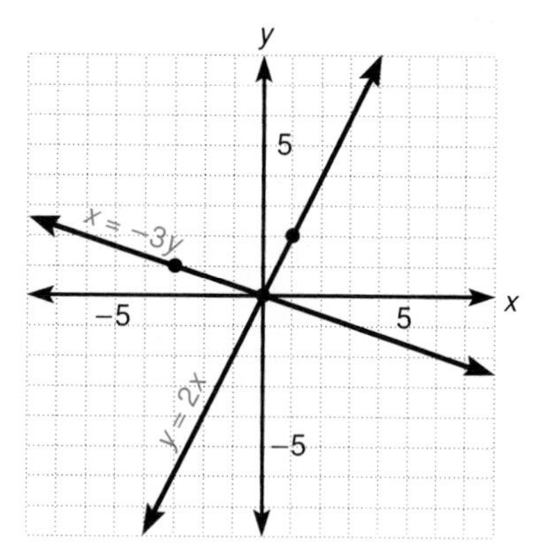

3. $x = a$

Graph is a vertical line through $(a,0)$.

Example: $x = 1$

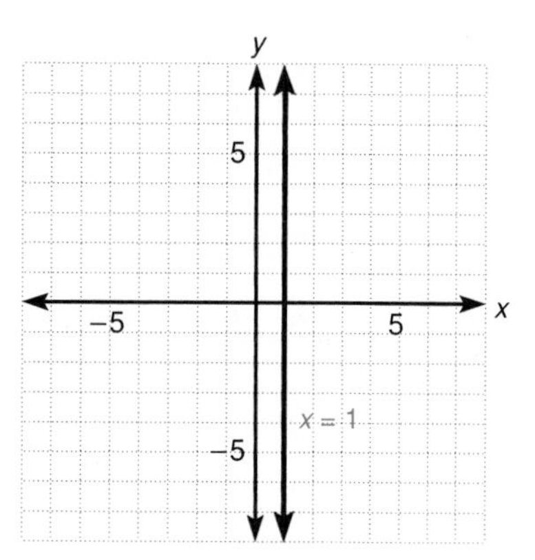

4. $y = b$

Graph is a horizontal line through $(0,b)$.

Example: $y = -6$

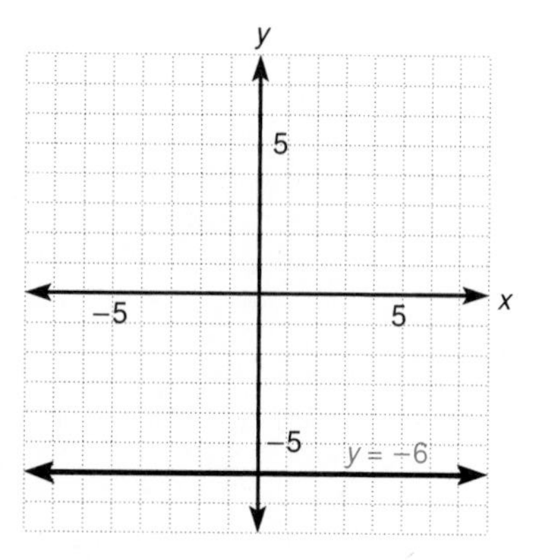

Mastery points

Can you

- Plot the graph of linear equations using ordered pairs?
- Find the x- and y-intercepts of a linear equation?
- Plot the graph of linear equations using the x- and y-intercepts?
- Plot graphs of the equations $x = a$ and $y = b$, where a and b are constants?

Exercise 7–2

Find the x- and y-intercepts. Write ordered pairs representing the points where the line crosses the axes. See example 7–2 B.

Example $5x + 3y = 15$

Solution Let $x = 0$, then $5(0) + 3y = 15$ — Replace x with 0

$$0 + 3y = 15$$
$$3y = 15$$
$$y = 5 \quad \text{y-intercept}$$

Let $y = 0$, then $5x + 3(0) = 15$ — Replace y with 0

$$5x + 0 = 15$$
$$5x = 15$$
$$x = 3 \quad \text{x-intercept}$$

The line crosses the y-axis at (0,5) and the x-axis at (3,0).

1. $y = 2x + 4$ **2.** $y = 5x - 10$ **3.** $y = 3x + 1$ **4.** $y = 2x - 3$

5. $2x + 3y = 6$ **6.** $x + 4y = 12$ **7.** $2x + 5y - 11 = 0$ **8.** $x - y = 4$

9. $y = 5x$ **10.** $y = -2x$ **11.** $3x - 2y = 0$ **12.** $y - 4x = 0$

13. $(1.2)x + (2.4)y = 4.8$ **14.** $(0.3)x - (0.4)y = 0.7$ **15.** $y = \frac{1}{2}x - \frac{3}{2}$ **16.** $y = \frac{2}{3}x - \frac{1}{3}$

Plot the graphs of the given linear equations using the x- and y-intercepts. See example 7–2 B.

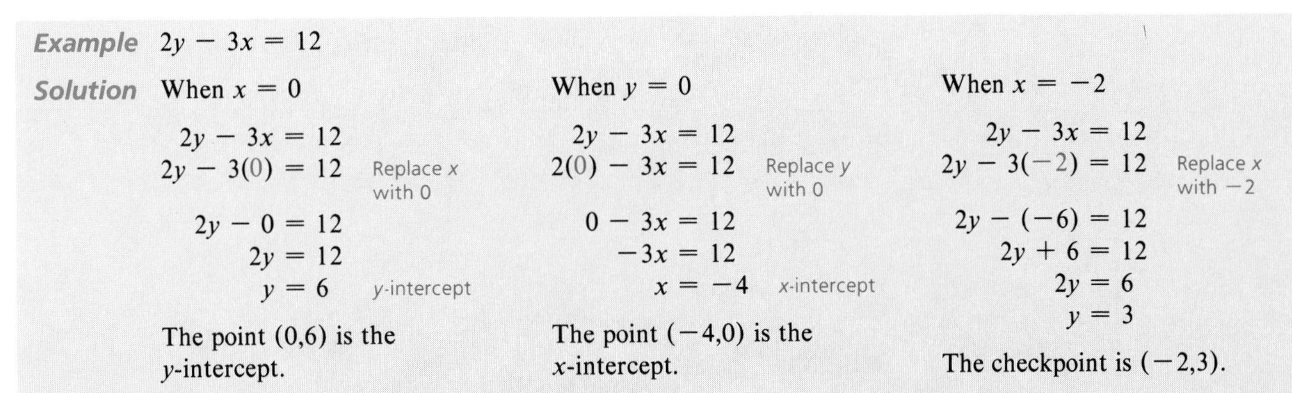

Example $2y - 3x = 12$

Solution

When $x = 0$

$$2y - 3x = 12$$
$$2y - 3(0) = 12 \quad \text{Replace } x \text{ with } 0$$
$$2y - 0 = 12$$
$$2y = 12$$
$$y = 6 \quad \text{y-intercept}$$

The point (0,6) is the y-intercept.

When $y = 0$

$$2y - 3x = 12$$
$$2(0) - 3x = 12 \quad \text{Replace } y \text{ with } 0$$
$$0 - 3x = 12$$
$$-3x = 12$$
$$x = -4 \quad \text{x-intercept}$$

The point $(-4,0)$ is the x-intercept.

When $x = -2$

$$2y - 3x = 12$$
$$2y - 3(-2) = 12 \quad \text{Replace } x \text{ with } -2$$
$$2y - (-6) = 12$$
$$2y + 6 = 12$$
$$2y = 6$$
$$y = 3$$

The checkpoint is $(-2,3)$.

We now plot the points (0,6), $(-4,0)$, and $(-2,3)$ and draw a line through them.

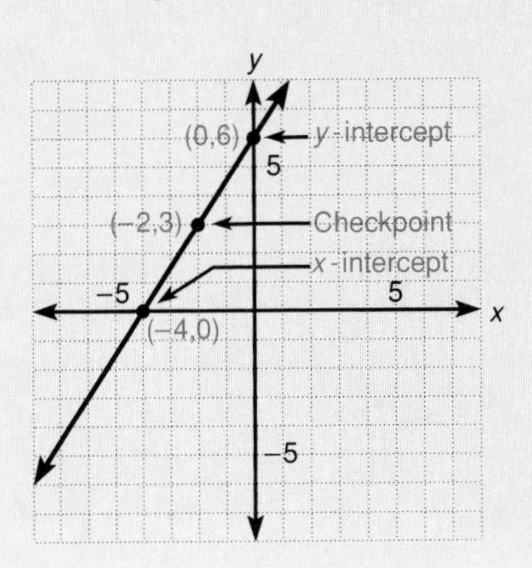

17. $y = 3x + 6$ **18.** $y = x + 5$ **19.** $y = x - 2$ **20.** $y = 2x + 4$
$y = 2x - 8$ **22.** $y = 3x$ **23.** $y = x$ **24.** $y = -3x$
25. $y = -2x$ **26.** $x + y = 0$ **27.** $y - 2x = 0$ **28.** $x - 3y = 0$
29. $4x - 3y = 12$ **30.** $3x - 2y = 6$ **31.** $2y + 5x = 10$ **32.** $2x + 2y = 3$
33. $5x - 6y = 30$ **34.** $3y - 5x = 15$ **35.** $y = 6$ **36.** $y = -2$
37. $x = 5$ **38.** $x = -2$ **39.** $x = 0$ **40.** $y = 0$

Solve the following equations for y in terms of x. Write the equations in the form $y = mx + b$, where m and b are rational numbers. Identify m and b.

Example $3y - 2x = 4$

Solution $3y = 2x + 4$ Add $2x$ to each member

$y = \frac{2}{3}x + \frac{4}{3}$ Divide each member by 3

$m = \frac{2}{3}$ and $b = \frac{4}{3}$

41. $y - 2x + 7 = 0$ **42.** $y - 3x - 4 = 0$ **43.** $3y - 4x = 9$ **44.** $5y - 2x = 7$
45. $7x + 3y = 10$ **46.** $5x + 3y = 4$ **47.** $x - 5y + 7 = 0$ **48.** $x - 3y + 9 = 0$
49. $5y - 8x + 14 = 0$ **50.** $7x + 5y - 11 = 0$

51. Plot the graph of the equation $y = -2x + b$ for (a) $b = 5$, (b) $b = 0$, and (c) $b = -3$ all on the same coordinate system.

52. Plot the graph of the equation $y = mx + 1$ for (a) $m = 1$, (b) $m = \frac{1}{2}$, and (c) $m = -2$ all on the same coordinate system.

Write an equation for each of the statements in exercises 53 to 56 and plot the graph of the equation.

Example The value of y is 6 more than twice the value of x.

Solution The equation is $y = 2x + 6$.

x	y	
0	6	y-intercept
-3	0	x-intercept
-2	2	Checkpoint

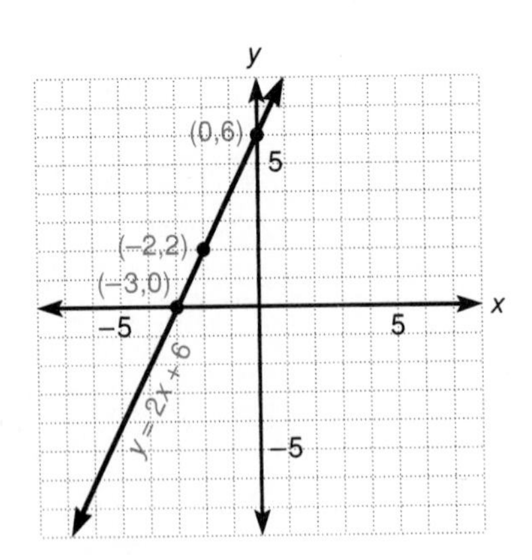

53. The value of y is 3 less than two times the value of x.

54. The value of x is 4 more than the value of y.

55. Two times x taken away from three times y is 6.

56. Five times x less the product of 2 and y gives 20.

Review exercises

Reduce the following expressions to lowest terms. See sections 1–1 and 5–2.

1. $\dfrac{18}{15}$

2. $\dfrac{x - 3}{x^2 + x - 12}$

3. $\dfrac{x^2 - 8x + 16}{x^2 + x - 20}$

4. Given $\dfrac{a - b}{c - d}$, evaluate the expression when $a = 2$, $b = 1$, $c = -3$, and $d = 2$. See section 5–1.

5. Given $y = mx + b$, find y when $m = -3$ and $b = 6$. See section 2–2.

6. If the product of a number and three times that number is 48, find the number. See section 4–8.

7–3 ■ The slope of a line

The slope

Consider the portions of the two roadways denoted by R_1 and R_2 (read "R sub-one" and "R sub-two") shown in figure 7–6.

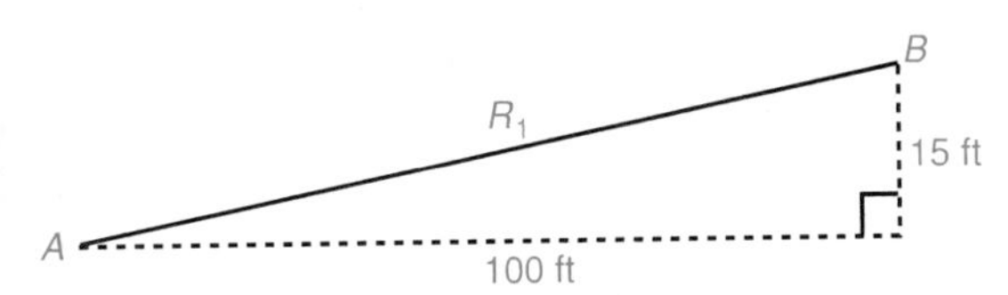

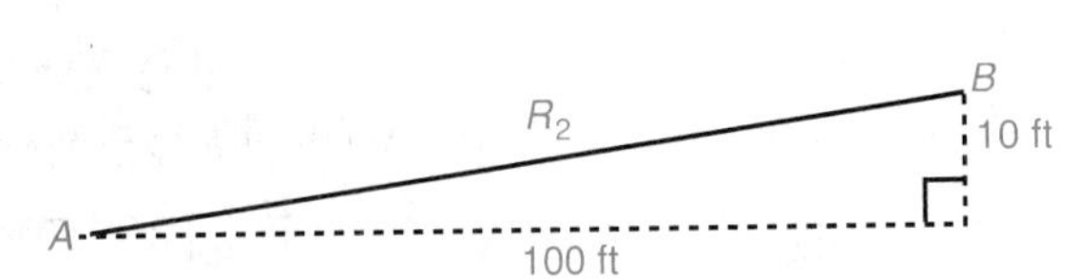

Figure 7–6

We would say roadway R_1 is "steeper" than roadway R_2. In moving from point A to point B on each roadway, a horizontal change in position of 100 feet, the vertical change in position is

$$15 \text{ feet on roadway } R_1$$

and

$$10 \text{ feet on roadway } R_2$$

If we measure this "steepness" by the ratio

$$\frac{\text{vertical change}}{\text{horizontal change}}$$

the roadway

$$R_1 \text{ has "steepness"} = \frac{15 \text{ ft}}{100 \text{ ft}} = \frac{3}{20}$$

and

$$R_2 \text{ has "steepness"} = \frac{10 \text{ ft}}{100 \text{ ft}} = \frac{1}{10}$$

Note $\frac{3}{20}$ is greater than $\frac{1}{10}$, so R_1 is "steeper" than R_2.

When applying this concept to any straight line, "steepness" is called the **slope** of the line. Thus, the slope of any line L is given by

$$\text{slope} = \frac{\text{vertical change}}{\text{horizontal change}}$$

Observe that the *slope is a ratio*. (See figure 7–7.)

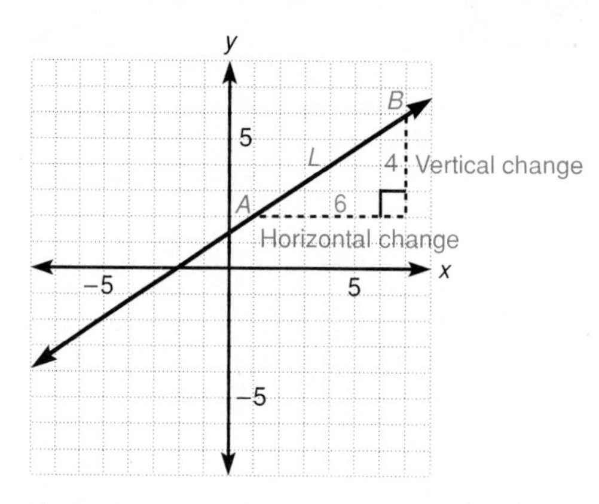

Figure 7–7

In figure 7–7, line L has from point A to point B

$$\text{a vertical change} = 4 \text{ units}$$

and

$$\text{a horizontal change} = 6 \text{ units}$$

Thus, the

$$\text{slope of } L = \frac{\text{vertical change}}{\text{horizontal change}} = \frac{4}{6} = \frac{2}{3}$$

To obtain the slope of a nonvertical straight line, given points P_1 and P_2 (read "P sub-one" and "P sub-two") which have coordinates (x_1,y_1) and (x_2,y_2), respectively, use the following definition.

Definition of the slope of a nonvertical line

The slope m of the nonvertical line through the points $P_1(x_1,y_1)$ and $P_2(x_2,y_2)$ is given by

$$m = \frac{y_2 - y_1}{x_2 - x_1} \qquad (x_1 \neq x_2)$$

Concept

The slope of a nonvertical line is determined by dividing the change in y-values by the change in x-values of any two points on the line.

See figure 7–8.

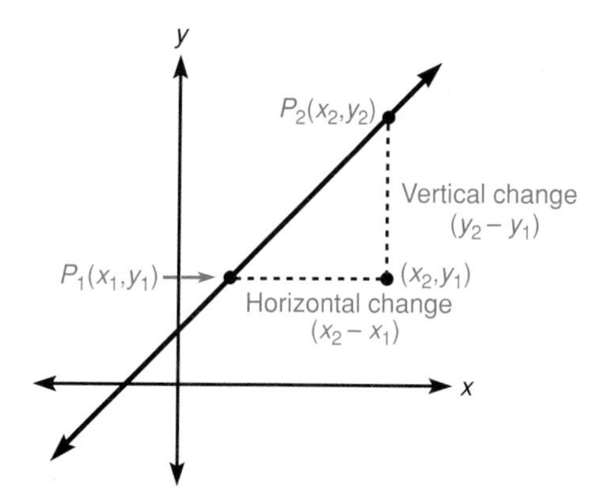

Figure 7–8

Note The vertical change is sometimes called the *rise* and the horizontal change is called the *run.* Thus, slope m can be defined

$$m = \frac{\text{rise}}{\text{run}}$$

■ *Example 7–3 A*

1. Find the slope of the line through the points $P_1(2,3)$ and $P_2(5,9)$.

$$\begin{aligned} \text{slope } m &= \frac{y_2 - y_1}{x_2 - x_1} \\ &= \frac{(9) - (3)}{(5) - (2)} \\ &= \frac{6}{3} \\ &= 2 \end{aligned}$$

Replace y_2 with 9, y_1 with 3, x_2 with 5, and x_1 with 2

This means that for every unit moved to the right, there is a rise of 2 units.

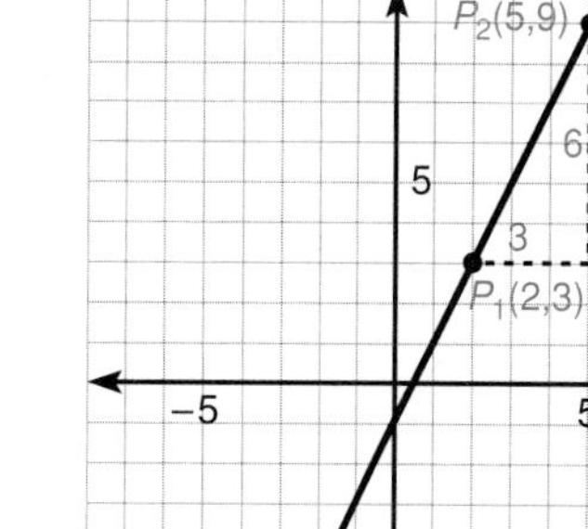

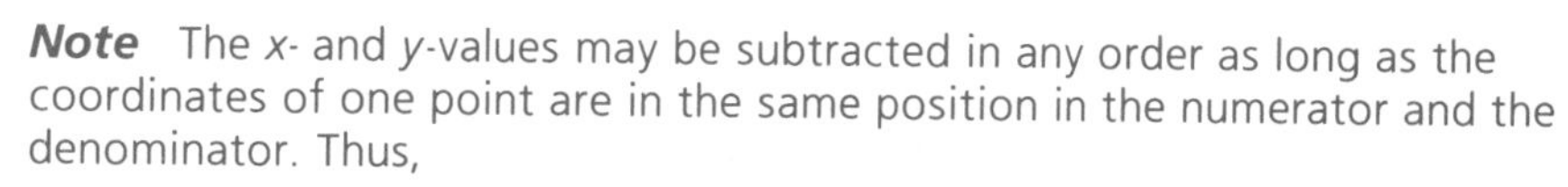

Note The x- and y-values may be subtracted in any order as long as the coordinates of one point are in the same position in the numerator and the denominator. Thus,

$$m = \frac{3 - 9}{2 - 5} = \frac{-6}{-3} = 2$$

in which we used

$$m = \frac{y_1 - y_2}{x_1 - x_2}$$

2. Find the slope of the line through points $P_1(-3,2)$ and $P_2(5,-4)$.

$$\text{slope } m = \frac{y_2 - y_1}{x_2 - x_1} = \frac{(-4) - (2)}{(5) - (-3)} = \frac{-6}{8} = \frac{-3}{4}$$

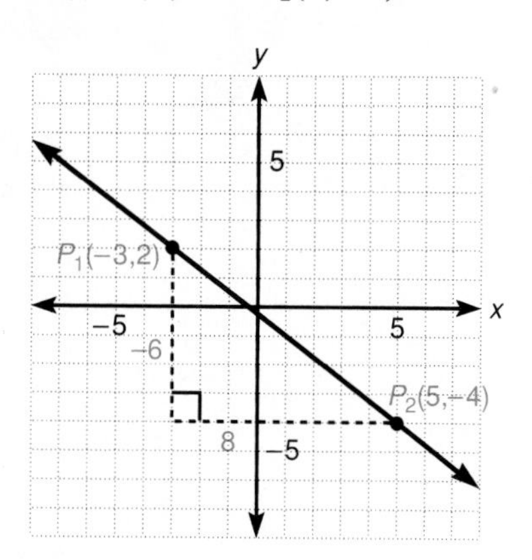

This means for every move of 4 units to the right, there is a "negative rise" (fall) of 3 units.

3. Find the slope of the horizontal line through $P_1(-3,4)$ and $P_2(2,4)$.

$$\text{slope } m = \frac{y_2 - y_1}{x_2 - x_1} = \frac{(4) - (4)}{(2) - (-3)} = \frac{0}{5} = 0$$

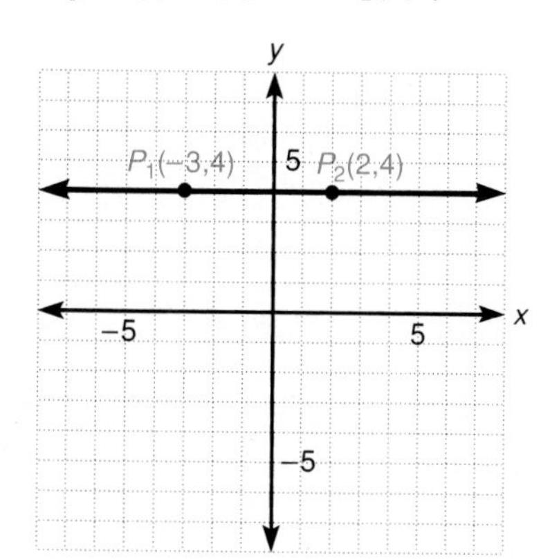

The slope $m = 0$ and the line is a horizontal line. Recall that the equation of any horizontal line is of the form $y = b$, or $y - b = 0$, where b is the y-intercept. In this case, the equation of the horizontal line is $y = 4$. Therefore we can generalize as follows:

Slope of a horizontal line

The slope m of any horizontal line having equation $y = b$ is $m = 0$.

4. Find the slope of the vertical line through $P_1(-4,1)$ and $P_2(-4,-3)$.

$$\text{slope } m = \frac{y_2 - y_1}{x_2 - x_1} = \frac{(-3) - (1)}{(-4) - (-4)} = \frac{-4}{0} \quad \text{(Undefined)}$$

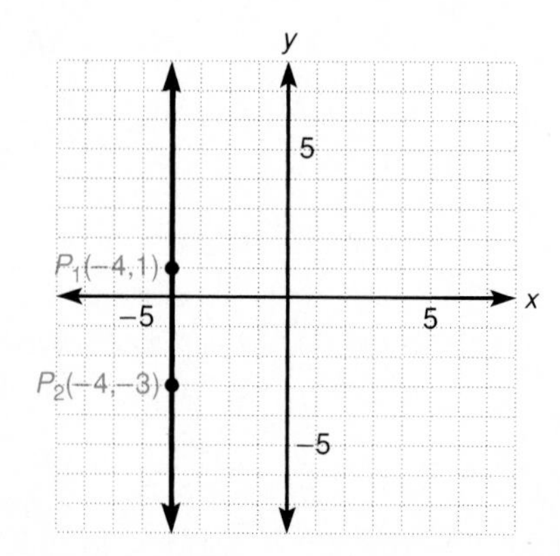

The line is a vertical line and the slope is undefined. Recall that the equation of any vertical line is of the form $x = a$, where a is the x-intercept. In this case, the equation of the vertical line is $x = -4$.

Slope of a vertical line

Any vertical line having equation $x = a$ has undefined slope.

▶ ***Quick check*** Find the slope of the line passing through the points $(-2,1)$ and $(3,4)$. Draw the graph of the line. ■

It is important for us to realize that the *slope of a nonvertical line is the same no matter what two points on the line we use to compute the slope.* Consider figure 7–9.

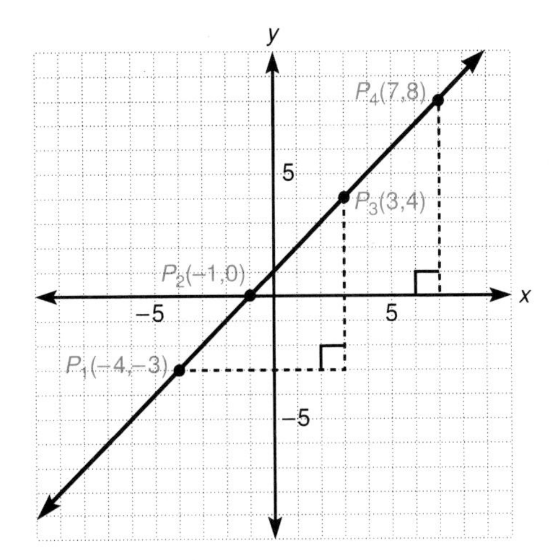

Figure 7–9

$$\text{Using } P_1 \text{ and } P_3,\ m = \frac{(4)-(-3)}{(3)-(-4)} = \frac{7}{7} = 1 \text{ and}$$
$$\text{using } P_2 \text{ and } P_4,\ m = \frac{(8)-(0)}{(7)-(-1)} = \frac{8}{8} = 1$$

We obtain the same slope, $m = 1$

The slope of a line through two plotted points

Suppose we are given the graph of a line and we wish to determine its slope. (See figure 7–10.) We inspect the graph and look for two points on the line that apparently meet the grid lines at points of intersection. In figure 7–10, we can see that the graph goes through the points $(-1,4)$ and $(3,-2)$. Then the slope m is

$$m = \frac{(-2)-(4)}{(3)-(-1)} = \frac{-6}{4} = \frac{-3}{2}$$

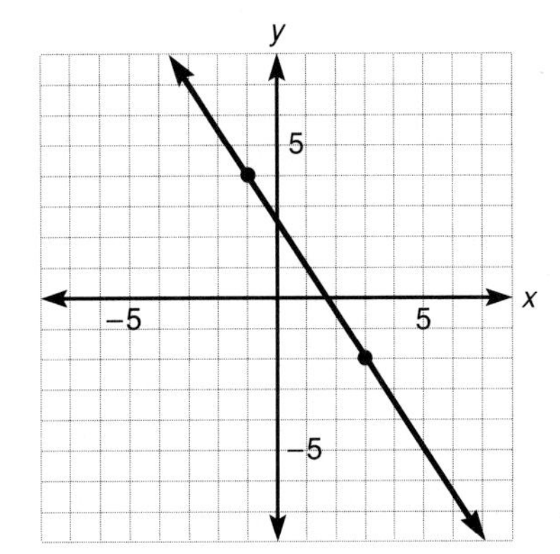

Figure 7–10

■ Example 7–3 B

Find the slope of the line whose graph is plotted.

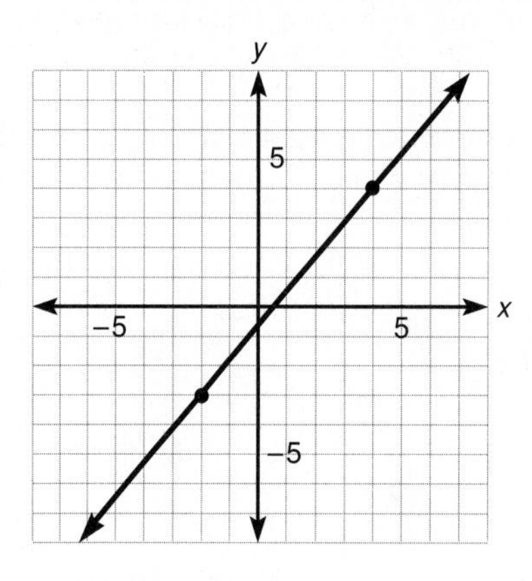

The line passes through points whose coordinates are (4,4) and (−2,−3). Therefore

$$m = \frac{(4) - (-3)}{(4) - (-2)} = \frac{4 + 3}{4 + 2} = \frac{7}{6}$$

▶ ***Quick check*** Find the slope of the line whose graph is plotted.

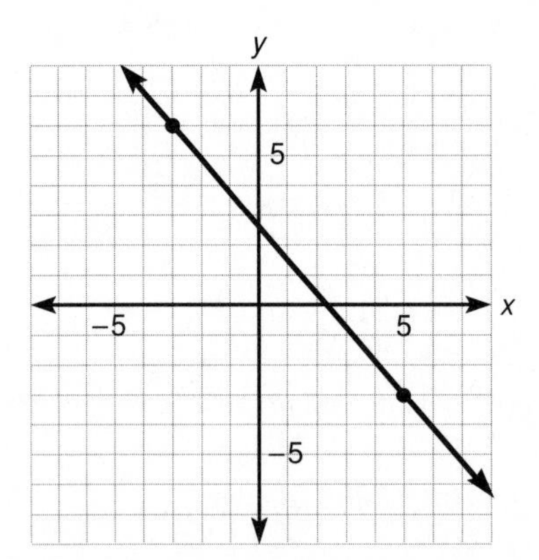

■

Mastery points

Can you

- Find the slope of a line given two points on the line?
- Remember the slope of a horizontal line and of a vertical line?
- Find the slope of a line given the graph of the line?

Exercise 7–3

Find the slope of the line passing through each of the following pairs of points. Draw the graph of the line. See example 7–3 A.

Example $(-2,1)$ and $(3,4)$

Solution $m = \dfrac{(4) - (1)}{(3) - (-2)}$

$= \dfrac{3}{5}$

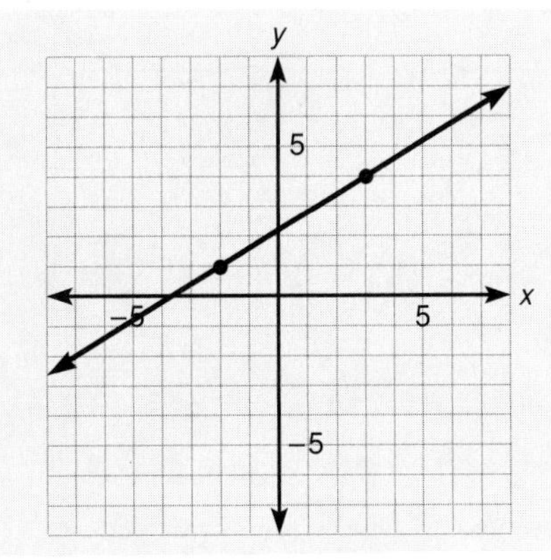

1. $(5,2), (3,3)$
2. $(4,3), (2,2)$
3. $(-4,-1), (2,3)$
4. $(-2,3), (5,5)$
5. $(-4,3), (2,3)$
6. $(4,-4), (2,-4)$
7. $(6,-7), (6,1)$
8. $(-1,4), (-1,-3)$

Find the slope of the line passing through each of the following pairs of points. See example 7–3 A.

9. $(-3,1), (0,3)$
10. $(4,0), (-2,8)$
11. $(-3,9), (-3,-5)$
12. $(5,0), (-3,0)$
13. $(5,6), (3,4)$
14. $(0,0), (4,3)$
15. $(0,7), (0,-8)$
16. $(4,-8), (0,-2)$
17. $(-8,-3), (-1,-2)$
18. $(-6,-5), (-4,-3)$
19. $(-10,4), (2,-5)$
20. $(7,-3), (8,-7)$
21. $(5,7), (-6,-3)$
22. $(9,4), (-1,-1)$

Find the slope of each of the following lines. See example 7–3 B.

Example

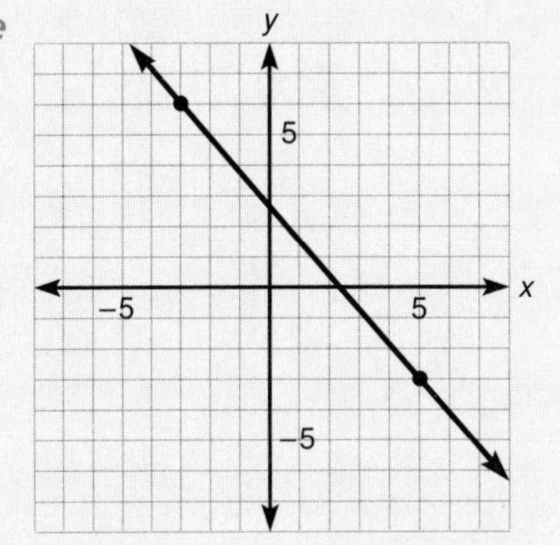

Solution We note that the line passes through the points $(-3,6)$ and $(5,-3)$.

Then slope $m = \dfrac{(6) - (-3)}{(-3) - (5)}$

$= \dfrac{9}{-8}$

$= -\dfrac{9}{8}$

23.

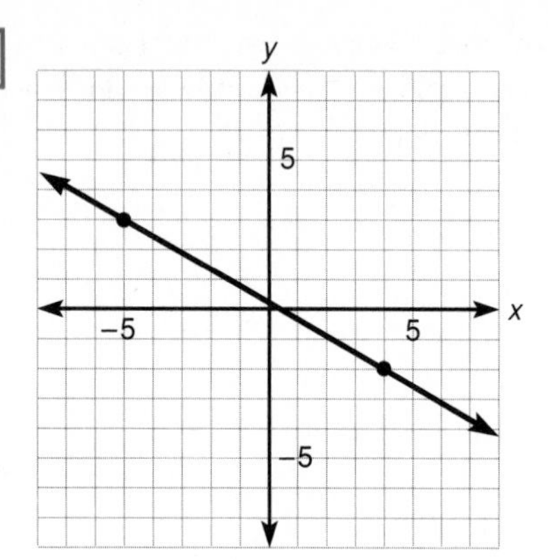

24.

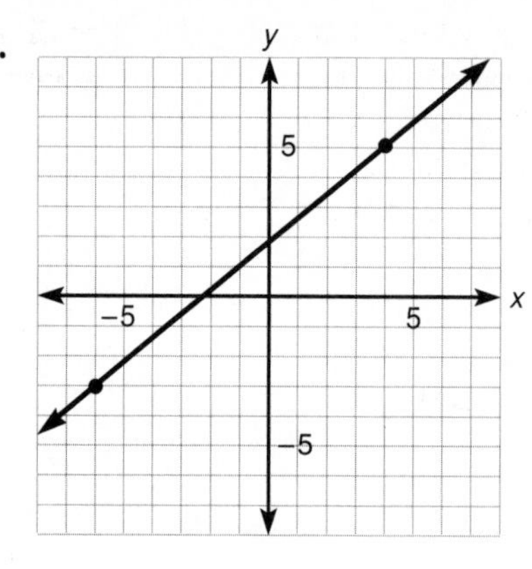

25.

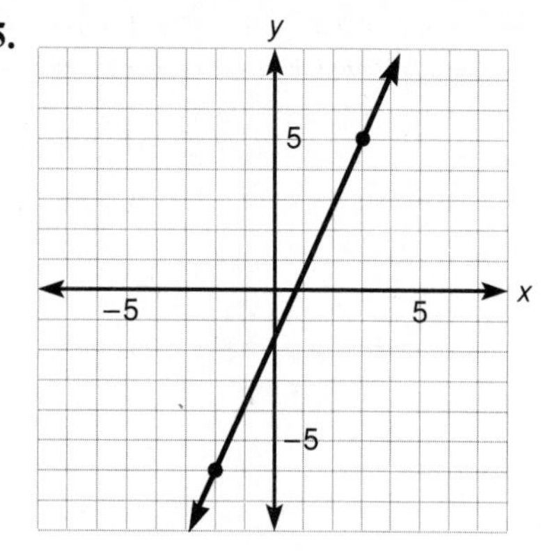

26.

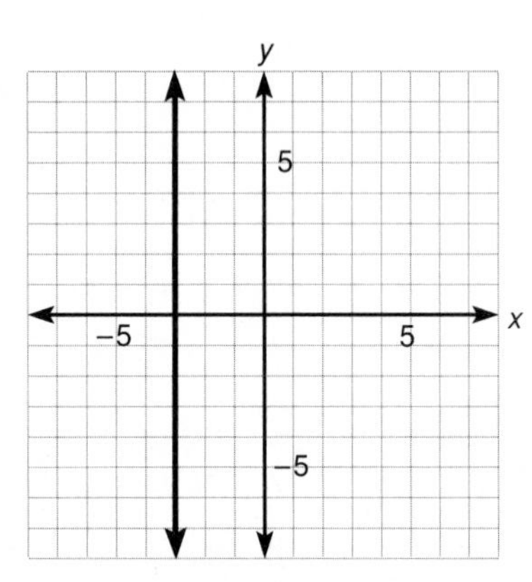

27.

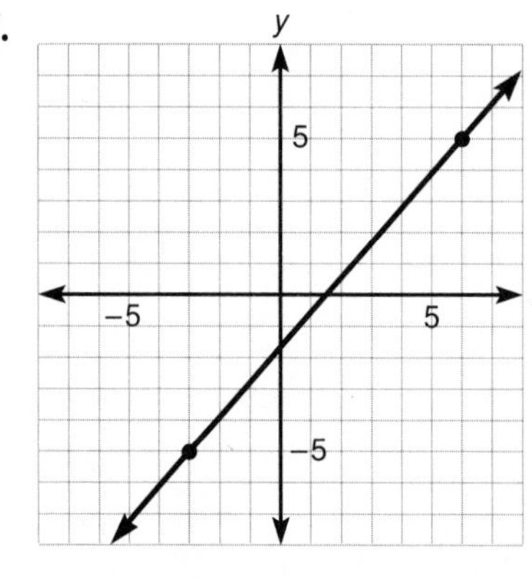

28.

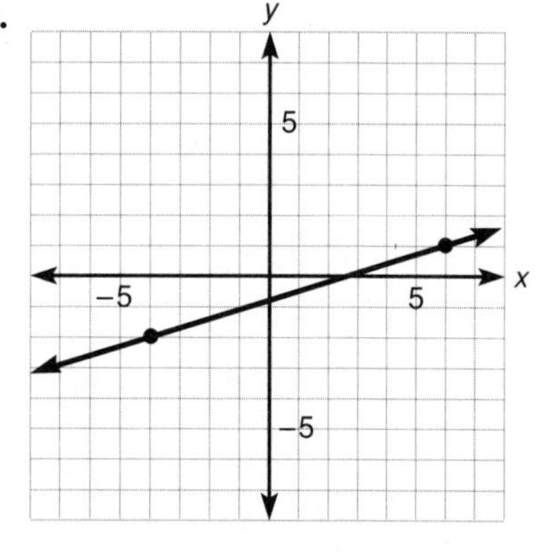

Example A boy is flying his kite. The kite is 35 feet above the ground, and the distance from the boy to a point directly below the kite is 40 feet. If the string from the boy to the kite is a straight line, what is the slope of the string?

Solution Using $m = \dfrac{\text{vertical change}}{\text{horizontal change}}$, the vertical change is 35 feet and the horizontal change is 40 feet, then

$$m = \frac{35 \text{ ft}}{40 \text{ ft}} = \frac{7}{8}$$

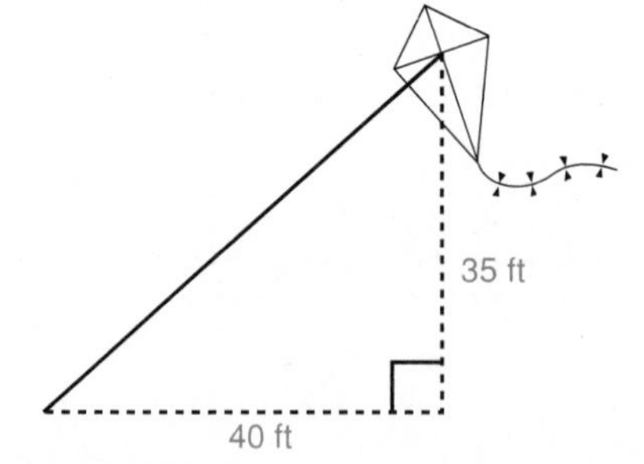

29. The roof of a home rises vertically a distance of 8 feet through a horizontal distance of 12 feet. Find the pitch (slope) of the roof.

30. The roof of a factory rises vertically 6 feet through a horizontal run of 27 feet. What is the pitch of the roof?

31. A ladder leaning against the side of a building touches the building at a point 12 meters from the ground. If the foot of the ladder is 18 meters from the base of the building, what is the slope of the ladder?

32. A guy wire is attached to a telephone pole. If the wire is attached to the ground at a point 15 feet from the base of the pole and to the pole at a point 10 feet up on the pole, what is the slope of the wire?

33. A company's profits (P) are related to the number of items produced (x) by a linear equation. If profits rise by $1,000 for every 250 items produced, what is the slope of the graph of the equation?

34. A company's profits (P) are related to increases in a worker's average pay (x) by a linear equation. If the company's profits drop by $1,500 per month for every increase of $450 per year in the worker's average pay, what is the slope of the graph of the equation?

35. The diagram shows a linear representation of a jogger's heartbeat in beats per minute as his speed is increased (in feet per second). What is the slope of the line?

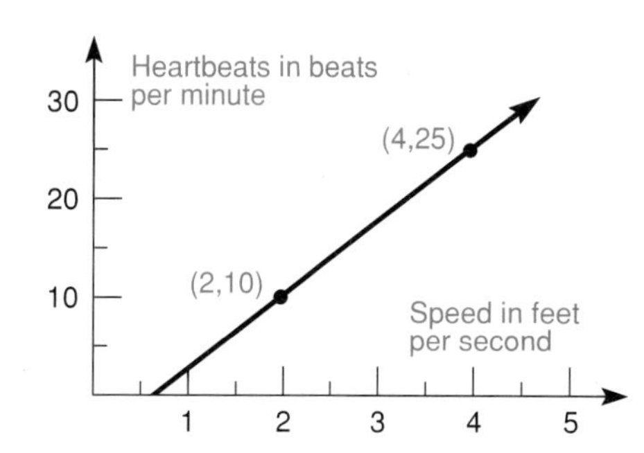

36. The diagram shows a linear representation of the bacteria count in a culture as related to the hours it exists. What is the slope of the line?

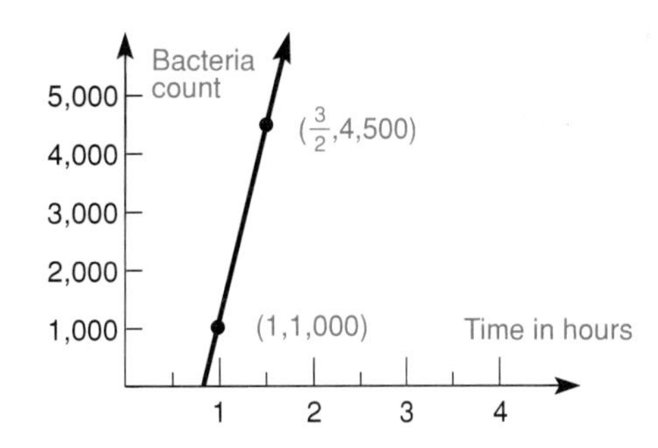

Review exercises

1. Simplify the expression $[3 - 4(5 - 2)]$. See section 1–6.

2. Evaluate $4^2 - 2^3 - 3 \cdot 5 - 2 \cdot 3$ See section 1–6.

Simplify the following. Use only positive exponents. Assume all variables are nonzero. See section 3–4.

3. $(-2)^{-3}$

4. $\dfrac{a^{-2}b^3}{a^3b^{-2}}$

5. $x^{-2} \cdot x^3 \cdot x^0$

6. When four times a number is increased by 12, the result is 64. What is the number? See section 2–8.

Solve the following equations for y. See section 2–6.

7. $3x + y = -1$

8. $3x - 3y = 6$

7–4 ■ The equation of a line

In section 7–2, we discussed the straight line graph of a linear equation in two variables. In this section, we discuss how to determine the equation of a straight line when we know certain facts about the line. There are three forms of the equation of a straight line that are of use to us: the *standard* form, the *point-slope* form, and the *slope-intercept* form.

The standard form

The **standard form** of the equation of a straight line is stated here.

Standard form of the equation of a line

$$ax + by = c$$

where a, b, and c are integers, $a \geq 0$, a and b not both zero.

To illustrate, the equations

$$2x + 3y = 6,\ 4x - 2y = 0, \text{ and } x - 6y = -2$$

are written in standard form. In most situations, equations will be written in this form.

The point-slope form

Consider again the definition for the slope of a line

$$m = \frac{y_2 - y_1}{x_2 - x_1} \qquad (x_1 \neq x_2)$$

where (x_1,y_1) and (x_2,y_2) are coordinates of two *known* points on the line. Suppose we replace the known point (x_2,y_2) with *any other arbitrary point* (x,y) on the line. Then the slope is given by

$$m = \frac{y - y_1}{x - x_1} \qquad (x_1 \neq x)$$

$$m(x - x_1) = y - y_1 \qquad \text{Multiply both members of the equation by } x - x_1$$

$$y - y_1 = m(x - x_1) \qquad \text{Symmetric property}$$

We call this the **point-slope form** of the equation of a line, where (x_1,y_1) is a known point on the line, m is the slope, and (x,y) is *any* other unknown point on the line.

Point-slope form of the equation of a line

$$y - y_1 = m(x - x_1)$$

where (x_1,y_1) is a known point on the line and m is the slope of the line.

We can use this form to find the equation of a line if we know the slope of the line and the coordinates of at least one point on the line.

■ Example 7–4 A

1. Find the standard form of the equation of a line having slope $m = 2$ and passing through the point $(4,-3)$.

   Using the point-slope form, we know $m = 2$ and $(x_1,y_1) = (4,-3)$.

   $$y - y_1 = m(x - x_1)$$

   $$y - (-3) = 2(x - 4) \qquad \text{Replace } y_1 \text{ with } -3,\ x_1 \text{ with 4, and } m \text{ with 2}$$

   $$y + 3 = 2x - 8 \qquad \text{Multiply and subtract as indicated}$$

   $$y = 2x - 11 \qquad \text{Add } -3 \text{ to both members}$$

   $$-2x + y = -11 \qquad \text{Add } -2x \text{ to both members}$$

   $$2x - y = 11 \qquad \text{Multiply each member by } -1$$

Note We wrote our final answer in *standard form*, $ax + by = c$ where $a > 0$.

2. Find the standard form of the equation of the line passing through the points $(-3,2)$ and $(5,1)$.

We use the two points to find the slope and then choose one of the points to find the equation.

$$m = \frac{y_2 - y_1}{x_2 - x_1}$$

$$m = \frac{(1) - (2)}{(5) - (-3)} \quad \text{Replace } y_2 \text{ with 1, } y_1 \text{ with 2, } x_2 \text{ with 5, and } x_1 \text{ with } -3$$

$$m = \frac{-1}{8} = -\frac{1}{8}$$

Choosing *either* of the points $(-3,2)$ or $(5,1)$ together with the slope, we use the point-slope form.

$$y - y_1 = m(x - x_1)$$

$$y - 2 = -\frac{1}{8}[x - (-3)] \quad \text{Use point } (-3,2)\text{. Replace } y_1 \text{ with 2, } x_1 \text{ with } -3, \text{ and } m \text{ with } -\frac{1}{8}$$

$$y - 2 = -\frac{1}{8}(x + 3) \quad x - (-3) = x + 3$$

$$8(y - 2) = -1(x + 3) \quad \text{Multiply each member by 8}$$

$$8y - 16 = -x - 3 \quad \text{Perform indicated multiplications}$$

$$8y = -x + 13 \quad \text{Add 16 to both members}$$

$$x + 8y = 13 \quad \text{Add } x \text{ to each member}$$

Note If we had used the point $(5,1)$, then

$$y - y_1 = m(x - x_1)$$

$$y - 1 = -\frac{1}{8}(x - 5)$$

$$8(y - 1) = -1(x - 5)$$

$$8y - 8 = -x + 5$$

$$8y = -x + 13$$

$$x + 8y = 13 \quad \text{(Produces the same equation.)}$$

▶ ***Quick check*** a. Find the equation of the line passing through $(-1,3)$ and having slope $m = -2$.
b. Find the equation of the line passing through the points $(-1,2)$ and $(3,4)$. ■

The slope-intercept form

Suppose a given line L, having slope m, passes through the point $(0,b)$, the y-intercept of the line. Using the point-slope form of the equation of a line,

$$y - b = m(x - 0) \quad \text{Replace } y_1 \text{ with } b \text{ and } x_1 \text{ with 0}$$

Then $$y - b = mx \quad x - 0 = x$$

and $$y = mx + b \quad \text{Add } b \text{ to each member}$$

Slope → m; y-intercept → b

We call $y = mx + b$ the **slope-intercept** form of the equation of a line.

Slope-intercept form of the equation of a line

When a linear equation is stated in the form $y = mx + b$, m is the slope of the line and b is the y-intercept.

We can use the slope-intercept form of the equation of a line to find the slope and the y-intercept of a line and to graph a linear equation in two variables.

■ Example 7–4 B

1. Find the slope and the y-intercept of the line given by the equation $3y - 5x = 9$.

$$\begin{aligned} 3y - 5x &= 9 && \text{Solve for } y \\ 3y &= 5x + 9 && \text{Add } 5x \text{ to each member} \\ y &= \frac{5x + 9}{3} && \text{Divide each member by 3} \\ y &= \frac{5}{3}x + 3 && \text{Write in slope-intercept form by dividing 3 into 5 and into 9} \end{aligned}$$

Then slope $m = \frac{5}{3}$ and y-intercept $b = 3$. [The y-intercept is the point (0,3).]

▶ ***Quick check*** Write the equation $3x + 4y = -12$ in slope-intercept form and determine the slope m and the y-intercept b.

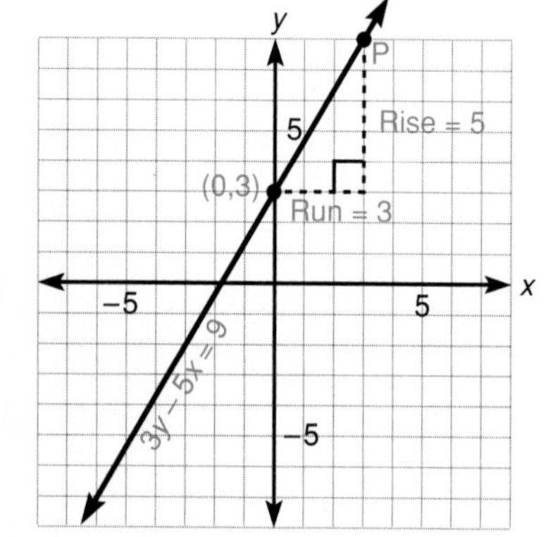

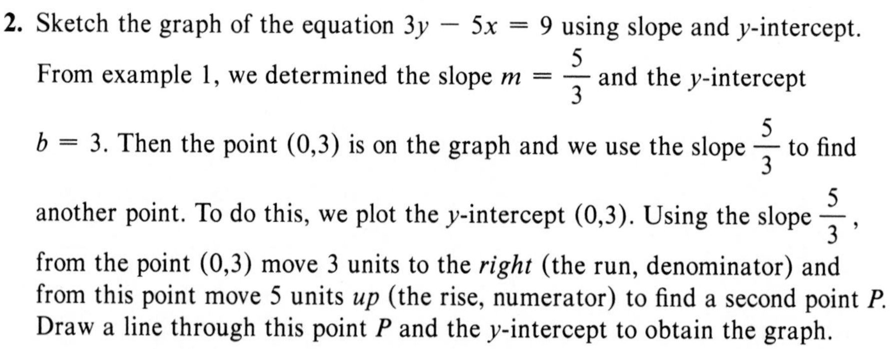

2. Sketch the graph of the equation $3y - 5x = 9$ using slope and y-intercept. From example 1, we determined the slope $m = \frac{5}{3}$ and the y-intercept $b = 3$. Then the point (0,3) is on the graph and we use the slope $\frac{5}{3}$ to find another point. To do this, we plot the y-intercept (0,3). Using the slope $\frac{5}{3}$, from the point (0,3) move 3 units to the *right* (the run, denominator) and from this point move 5 units *up* (the rise, numerator) to find a second point P. Draw a line through this point P and the y-intercept to obtain the graph.

3. Graph the line through $(-3,4)$ having slope $m = -3$. First, plot the point$(-3,4)$. Write the slope

$$m = -3 = \frac{-3}{1} = \frac{\text{negative rise (fall)}}{\text{run}}$$

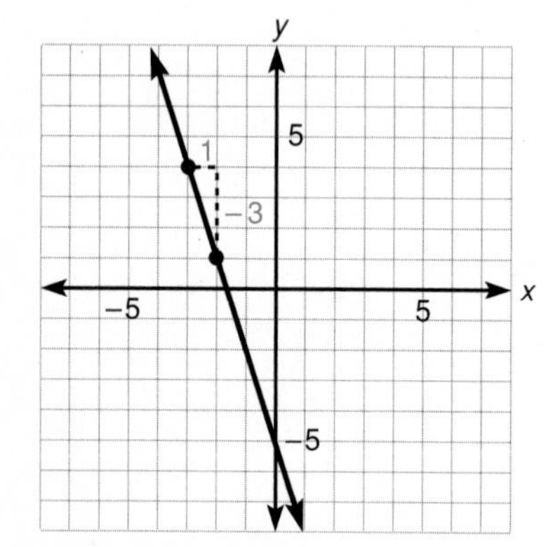

From the point $(-3,4)$, move 1 unit right and then 3 units *down* (because of the negative sign) to find a second point P. Draw a line through point P and the given point $(-3,4)$.

▶ ***Quick check*** a. Graph the equation $3x + y = 2$ using the slope m and the y-intercept b.

b. Graph the line through $(-3,2)$ with slope $m = \frac{1}{3}$ ■

Given the slope of a line and its y-intercept, it is then possible to determine the equation of the line.

■ ***Example 7–4 C***

1. Find the equation of the line having slope $m = -3$ and y-intercept $b = 2$. Leave your answer in slope-intercept form.

   Using $y = mx + b$

   $y = (-3)x + (2)$ Replace m with -3 and b with 2
   $y = -3x + 2$

2. Find the equation of the line having slope $m = \frac{2}{3}$ and passing through the point $(0,-4)$. Leave your answer in slope-intercept form.

   Since the point $(0,-4)$ is on the y-axis, it is the y-intercept, so $b = -4$.

   $y = mx + b$
   $y = \left(\frac{2}{3}\right)x + (-4)$ Replace m with $\frac{2}{3}$ and b with -4
   $y = \frac{2}{3}x - 4$

3. Find the equation of the line with slope $m = 0$ and having y-intercept 3.

   $y = mx + b$
   $y = (0)x + (3)$ Replace m with 0 and b with 3
   $y = 0 + 3$
   $y = 3$

Note The equation is of the form $y = b$ so the graph is a horizontal line.

4. Find the equation of the line with undefined slope and passing through $(3,-4)$.

   Since m is undefined, the line is vertical and the equation is of the form $x = a$. Thus, $x = 3$ is the equation. ■

Parallel and perpendicular lines

Given two distinct straight lines in a plane, they will either be parallel (never meet no matter how far they are extended) or intersect in one point.

For two nonvertical lines to be parallel, they must *have the same slope and different y-intercepts.*

Slopes of parallel lines

Two distinct nonvertical lines having slopes m_1 and m_2 are parallel if and only if $m_1 = m_2$.

Note All vertical lines (whose slopes are undefined) are parallel to one another.

■ Example 7–4 D

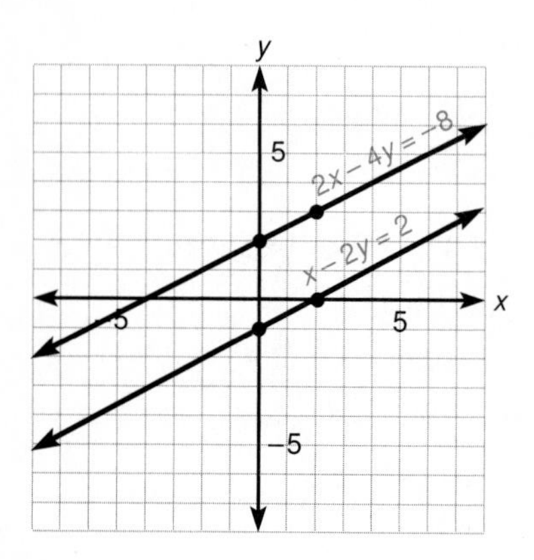

Show that distinct lines $x - 2y = 2$ and $2x - 4y = -8$ are parallel lines.

We solve each equation for y to write them in slope-intercept form $y = mx + b$.

$$\begin{aligned} x - 2y &= 2 \\ -2y &= -x + 2 \\ y &= \frac{-x + 2}{-2} \\ y &= \frac{1}{2}x - 1 \\ m_1 &= \frac{1}{2} \end{aligned} \qquad \begin{aligned} 2x - 4y &= -8 \\ -4y &= -2x - 8 \\ y &= \frac{-2x - 8}{-4} \\ y &= \frac{1}{2}x + 2 \\ m_2 &= \frac{1}{2} \end{aligned}$$

Since $m_1 = \frac{1}{2} = m_2$, the two lines are parallel. ■

Two lines that intersect in a single point can be *perpendicular.* Perpendicular lines make *right angles* with one another. Two nonvertical lines having slopes m_1 and m_2, respectively, are perpendicular if and only if the *product* of their slopes is -1.

Slopes of perpendicular lines

Two nonvertical lines having slopes m_1 and m_2 are perpendicular if and only if $m_1 m_2 = -1$.

To illustrate, lines having slopes $\frac{3}{4}$ and $-\frac{4}{3}$ are perpendicular since

$$\left(\frac{3}{4}\right)\left(-\frac{4}{3}\right) = -1$$

Note The slopes $\frac{3}{4}$ and $-\frac{4}{3}$ are *negative reciprocals* of each other. From this, we can conclude that the slopes of perpendicular lines will be negative reciprocals.

■ Example 7–4 E

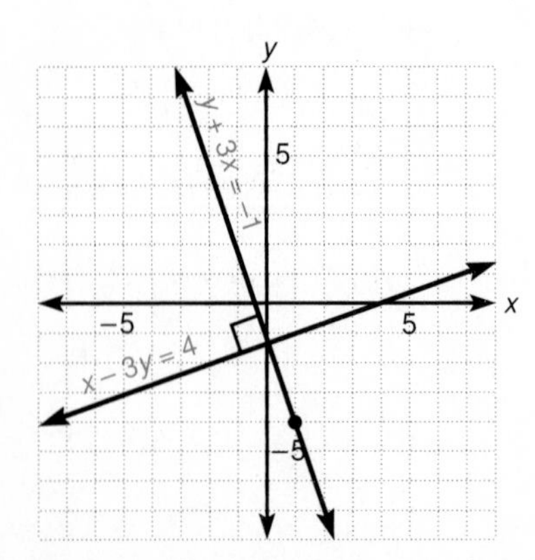

Show that the lines $x - 3y = 4$ and $3x + y = -1$ are perpendicular lines.

We solve for y to write each equation in slope-intercept form $y = mx + b$.

$$\begin{aligned} x - 3y &= 4 \\ -3y &= -x + 4 \\ y &= \frac{1}{3}x - \frac{4}{3} \\ m_1 &= \frac{1}{3} \end{aligned} \qquad \begin{aligned} 3x + y &= -1 \\ y &= -3x - 1 \\ m_2 &= -3 \end{aligned}$$

Since $m_1 m_2 = \frac{1}{3} \cdot (-3) = -1$, the lines are perpendicular.

Note $\frac{1}{3}$ and -3 are *negative reciprocals* of each other. The box where the lines intersect indicates perpendicular lines.

▶ ***Quick check*** Find the slopes of the lines $3x + y = 4$ and $2x - y = 1$ and determine if the lines are parallel, perpendicular, or neither. ■

All vertical lines are perpendicular to all horizontal lines even though the product of their slopes does not exist. For example, the lines $x = -2$ and $y = 4$ are perpendicular.

Mastery points

Can you

- Write the equation of a line in standard form?
- Find the equation of a line knowing the slope and a point or two points on the line?
- Find the slope and y-intercept of a line knowing the equation of the line?
- Graph a linear equation in two variables using the slope and y-intercept?
- Graph a linear equation in two variables using the slope and a point of the line?
- Find the equation of a line given the slope and the y-intercept?
- Determine whether two lines are parallel or perpendicular?

Exercise 7–4

Write the equation of the line passing through the given point and having the given slope. See example 7–4 A–1, C.

Example $(-1,3)$; $m = -2$

Solution Use the *point-slope form* of the equation of a line.

$$y - y_1 = m(x - x_1)$$

$$y - (3) = (-2)[x - (-1)] \quad \text{Replace } y_1 \text{ with 3, } x_1 \text{ with } -1\text{, and } m \text{ with } -2$$

$$y - 3 = -2(x + 1) \quad \text{Definition of subtraction}$$

$$y - 3 = -2x - 2$$

$$y = -2x + 1 \text{ or } 2x + y = 1 \quad \text{Written in standard form } ax + by = c$$

1. $(1,3)$; $m = 4$
2. $(-3,1)$; $m = 2$
3. $(0,-1)$; $m = -2$
4. $(0,4)$; $m = -3$

5. $(2,7)$; $m = \frac{3}{5}$
6. $(-1,5)$; $m = \frac{4}{7}$
7. $(5,0)$; $m = -\frac{7}{6}$
8. $(-3,0)$; $m = -\frac{5}{8}$

9. $(-3,-7)$; $m = \frac{5}{4}$
10. $(-9,-2)$; $m = \frac{3}{2}$
11. $(2,-3)$; $m = 0$
12. $(-7,3)$; $m = 0$

13. $(1,-8)$; slope is undefined.
14. $(0,4)$; slope is undefined.

Find the standard form of the equation of the line passing through each pair of given points. Write the equation in standard form $ax + by = c$, where a, b, and c are integers, $a \geq 0$. See example 7–4 A–2.

Example $(-1,2)$ and $(3,4)$

Solution Find the slope using

$$m = \frac{y_2 - y_1}{x_2 - x_1}$$

$$= \frac{(4) - (2)}{(3) - (-1)}$$ Replace y_2 with 4, y_1 with 2, x_2 with 3, and x_1 with -1

$$= \frac{2}{4}$$ Subtract as indicated

$$= \frac{1}{2}$$

Use the point-slope form of the equation of a line and *one of the given points,* (3,4).

$$y - y_1 = m(x - x_1)$$

$$y - (4) = \left(\frac{1}{2}\right)[x - (3)]$$ Replace y_1 with 4, x_1 with 3, and m with $\frac{1}{2}$

$$2y - 8 = x - 3$$ Mulitiply each member by 2 to clear denominator

$$2y = x + 5$$ Add 8 to each member

$$-x + 2y = 5$$ Subtract x from each member

$$x - 2y = -5$$ Multiply each member by -1

15. $(2,1)$ and $(6,3)$ **16.** $(3,2)$ and $(4,5)$ **17.** $(-3,2)$ and $(5,-1)$ **18.** $(-6,2)$ and $(4,-3)$
19. $(0,4)$ and $(-2,2)$ **20.** $(1,7)$ and $(0,-3)$ **21.** $(-6,0)$ and $(1,-1)$ **22.** $(5,6)$ and $(5,0)$
23. $(0,8)$ and $(-3,0)$ **24.** $(4,0)$ and $(0,-1)$ **25.** $(0,0)$ and $(-5,8)$ **26.** $(7,-1)$ and $(0,0)$

Write the following equations in slope–intercept form and determine the slope m and y-intercept b. See example 7–4 B–1.

Example $3x + 4y = -12$

Solution Solve for y.

$$3x + 4y = -12$$

$$4y = -3x - 12$$ Add $-3x$ to each member

$$y = \frac{-3x - 12}{4}$$ Divide each member by 4

$$y = -\frac{3}{4}x - 3$$ Divide -3 and -12 by 4

Then slope $m = -\frac{3}{4}$ and y-intercept $b = -3$

27. $x + y = 2$ **28.** $y - x = 3$ **29.** $3x + y = -2$ **30.** $y - 4x = 5$
31. $2x + 5y = 10$ **32.** $3x + 2y = 8$ **33.** $7x - 2y = -4$ **34.** $9x - 3y = 3$
35. $2y - 9x = -6$ **36.** $4y - 5x = -12$ **37.** $8x - 9y = 1$ **38.** $-7x + 4y = -5$

Use the slope m and the y-intercept b to graph the following equations. See example 7–4 B–2.

Example $3x + y = 2$

Solution Write the equation in slope-intercept form $y = mx + b$.

$y = -3x + 2$ so $m = -3$ and $b = 2$

Since $-3 = \frac{-3}{1}$, from the point (0,2), move 1 unit to the right and 3 units down to get a second point. Draw a straight line through the two points.

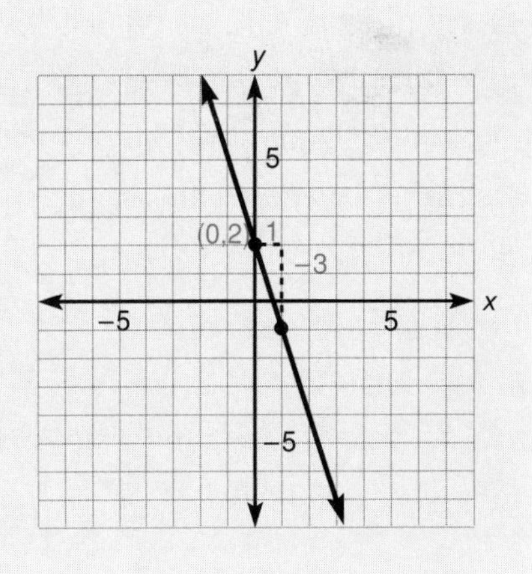

39. $y = 2x - 4$
40. $y = 3x + 1$
41. $y = -5x + 2$
42. $y = -2x - 3$
43. $y = \frac{2}{3}x - 1$
44. $y = \frac{-4}{3}x + 3$
45. $2x + y = -3$
46. $3x + y = -2$
47. $5x - 2y = 6$
48. $3x - 4y = 8$
49. $4x + 3y = -9$
50. $3x + 8y = -16$

Graph the line through the given point having the given slope m. See example 7–4 B–3.

Example $(-3,2)$; $m = \frac{1}{3}$

Solution Plot the point $(-3,2)$. Then move 3 units to the right and 1 unit up to find the second point. Draw a straight line through the two points.

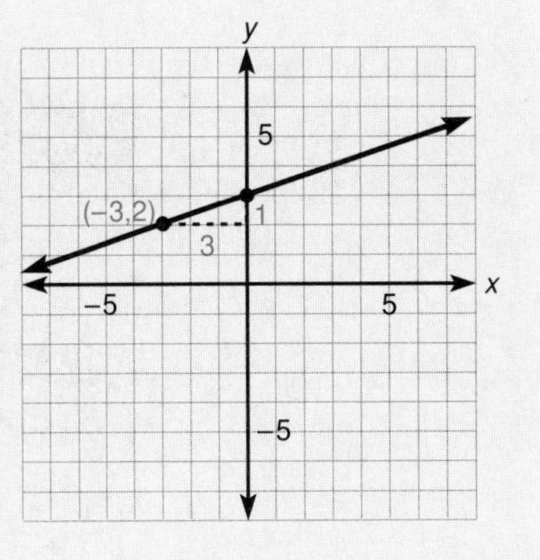

51. $(4,3)$; $m = \frac{3}{4}$
52. $(-2,-2)$; $m = \frac{2}{3}$
53. $(-1,2)$; $m = \frac{-1}{3}$
54. $(2,-1)$; $m = \frac{-3}{2}$
55. $(-4,-5)$; $m = 4$
56. $(2,-3)$; $m = 2$
57. $(0,3)$; $m = -1$
58. $(1,5)$; $m = -3$

(*Hint:* For exercises 59 to 62, recall the types of lines that have slope $m = 0$ or undefined slope.)

59. $(5,6)$; $m = 0$
60. $(-3,1)$; $m = 0$
61. $(-1,-1)$; slope undefined
62. $(4,7)$; slope undefined

Find the equation of the line (in slope-intercept form) having the following characteristics. See Example 7–4 C.

63. Having slope $m = 4$ and y-intercept $b = -1$

64. Having slope $m = -8$ and y-intercept $b = 6$

65. Having slope $m = -\frac{5}{3}$ and passing through (0,2)

66. Having slope $m = \frac{3}{4}$ and passing through (0,5)

67. Having slope $m = 0$ and passing through (0,−1)

68. Having slope $m = 0$ and y-intercept4

Find the slope of each line and determine if each pair of lines is parallel, perpendicular, or neither. See example 7–4 D and E.

Example $3x + y = 4$
$2x - y = 1$

Solution Write each equation in slope-intercept form.

$3x + y = 4$ | $2x - y = 1$
$y = -3x + 4$ | $y = 2x - 1$
so $m_1 = -3$ | so $m_2 = 2$

Since $m_1 \neq m_2$, the lines are *not* parallel; and since $m_1m_2 = -3 \cdot 2 = -6$, the lines are not perpendicular. Therefore the lines are neither.

69. $x + y = 4$
$x + y = -7$

70. $-x + y = 9$
$-x + y = -2$

71. $x + y = 5$
$-x + y = -1$

72. $-x + y = 8$
$x + y = 3$

73. $3x - y = 2$
$6x + 2y = -5$

74. $x + 2y = 5$
$-6x + 3y = -1$

75. $5x + y = 1$
$10x + 3y = 2$

76. $8x - 9y = 1$
$16x + 7y = 3$

77. $4x - y = 5$
$12x - 3y = 1$

78. $4x + 6y = -3$
$6x + 9y = 4$

79. $y = 5$
$x = 2$

80. $x - 3y = -6$
$3x - 9y = -18$

Find the equation for the given line using the slope, one of the points, and the point-slope form $y - y_1 = m(x - x_1)$, where point (x_1,y_1) is a known point on the line.

Example It appears that the line passes through points (−4,3) and (3,−4).

Then $m = \frac{(3) - (-4)}{(-4) - (3)}$
$= \frac{7}{-7} = -1$

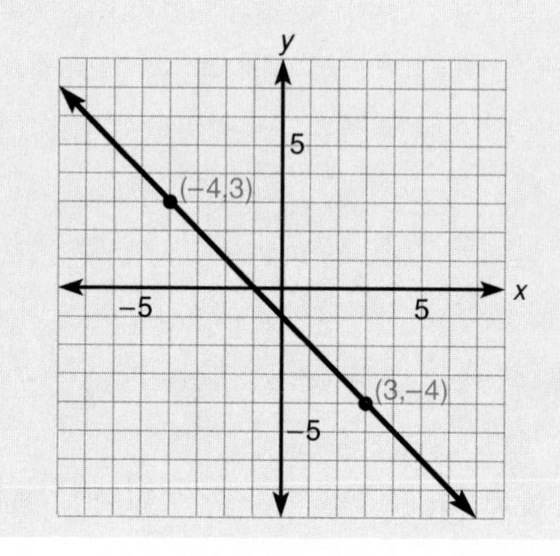

Use $y - y_1 = m(x - x_1)$ and the point (−4,3).

$y - (3) = (-1)[x - (-4)]$
$y - 3 = -1(x + 4)$
$y - 3 = -x - 4$
$y = -x - 1$
$x + y = -1$

81.

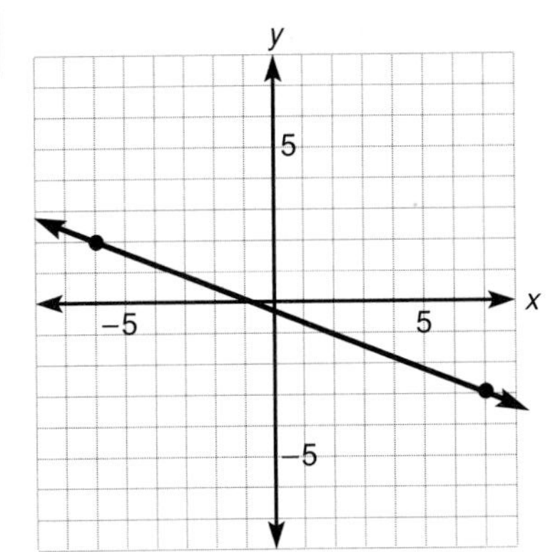

82.

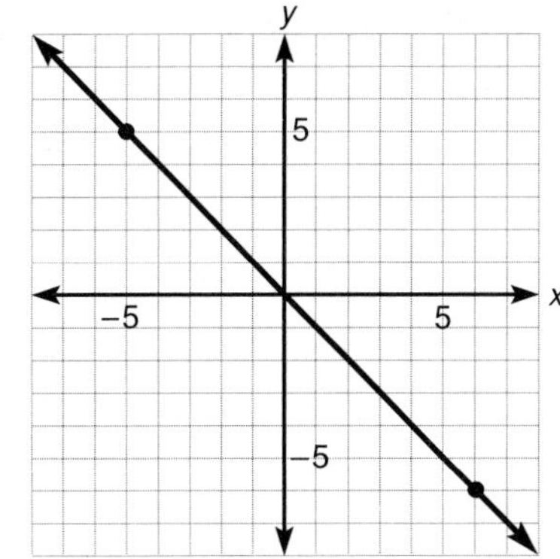

83.

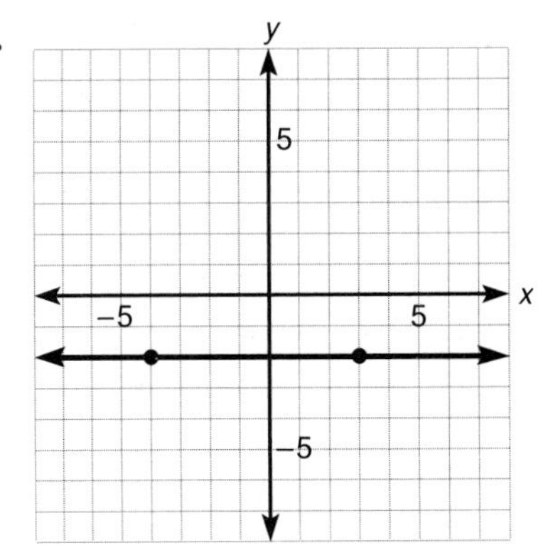

84.

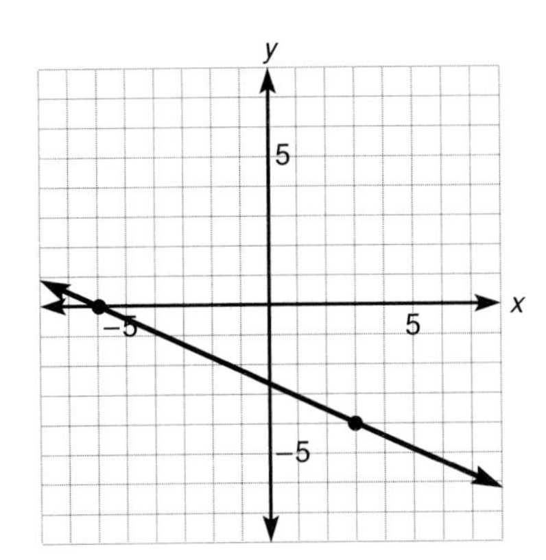

85.

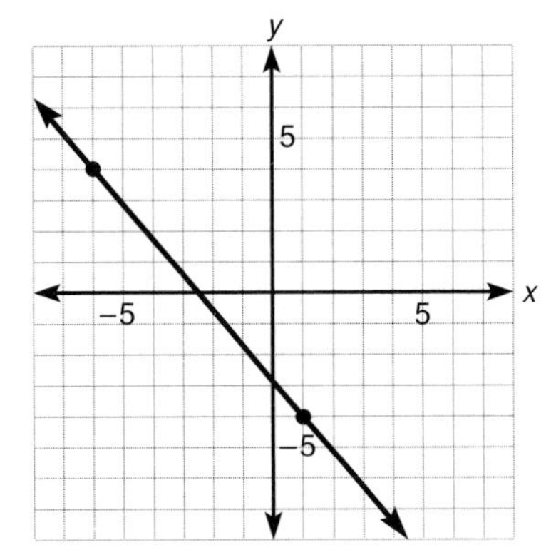

86.

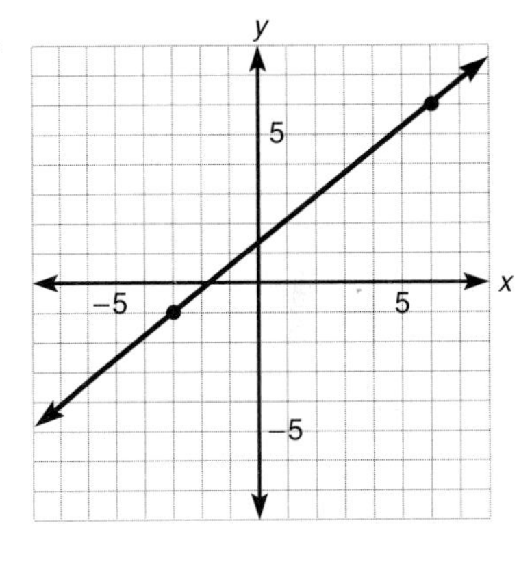

Review exercises

1. A 100-foot extension cord is to be cut into two pieces so that one piece is 17 feet longer than the other. What is the length of each piece? See section 2–8.

2. From the numbers 0, -3, $\frac{3}{4}$, 9, $-\frac{7}{8}$, $\frac{9}{3}$, and $\frac{0}{4}$, choose the integers. See section 1–3.

Find the solution set of the following equations. See section 2–6.

3. $4y + 1 = 2y - 5$

4. $3(2y - 1) = -1(y + 2)$

5. $\frac{3x}{5} - \frac{x}{2} = 1$

Factor the following expressions. See sections 4–2, 4–3, and 4–4. If the expression does not factor, so state.

6. $3x^2 + 4x - 4$

7. $x^2 + 5x + 7$

8. $8y^2 - 32x^2$

7–5 ■ Graphing linear inequalities in two variables

In section 7–2, we graphed linear equations in two variables such as $2x + y = 6$. In this section, we consider the graphs of *linear inequalities* in two variables such as

$$2x + y < 6 \quad \text{and} \quad 2x + y \geq 6$$

where the equals to sign has been replaced by one of the four inequality symbols

$$<, \leq, >, \text{or} \geq$$

The graph of the equation $2x + y = 6$ is a straight line in the plane. This line divides the plane into two regions called *half-planes* and serves as the *boundary line* for each half-plane. See figure 7–11.

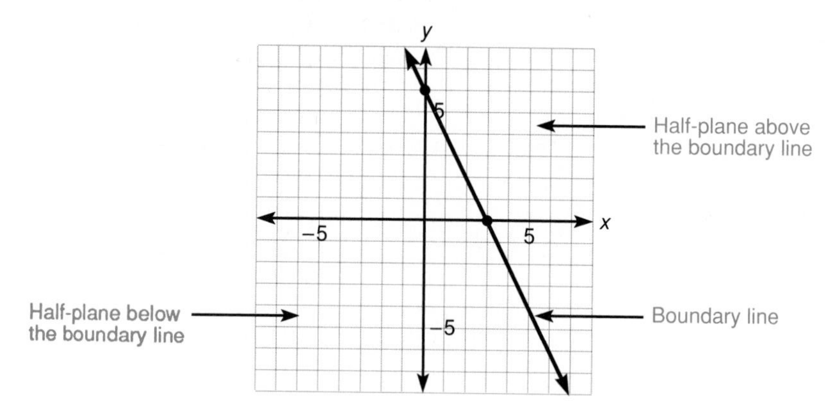

Figure 7–11

Graph of a linear inequality

The graph of any linear inequality in two variables is a half-plane.

The inequality $2x + y \geq 6$ is read "$2x + y$ is greater than or equal to 6"
So,

$$2x + y > 6 \quad \text{or} \quad 2x + y = 6$$

The graph *will consist of the boundary line* and the proper half-plane. The inequality $2x + y > 6$ is read "$2x + y$ is greater than 6" so the boundary line *is not* a part of the graph of the inequality. To indicate this, the boundary line is drawn as a dashed line rather than a solid line.

1. For $2x + y \geq 6$, the boundary line $2x + y = 6$ is a *solid* line, to show that the points on the line are included.
2. For $2x + y > 6$, the boundary line $2x + y = 6$ is a *dashed* line to show that the points on the line are *not* included.

See figure 7–12.

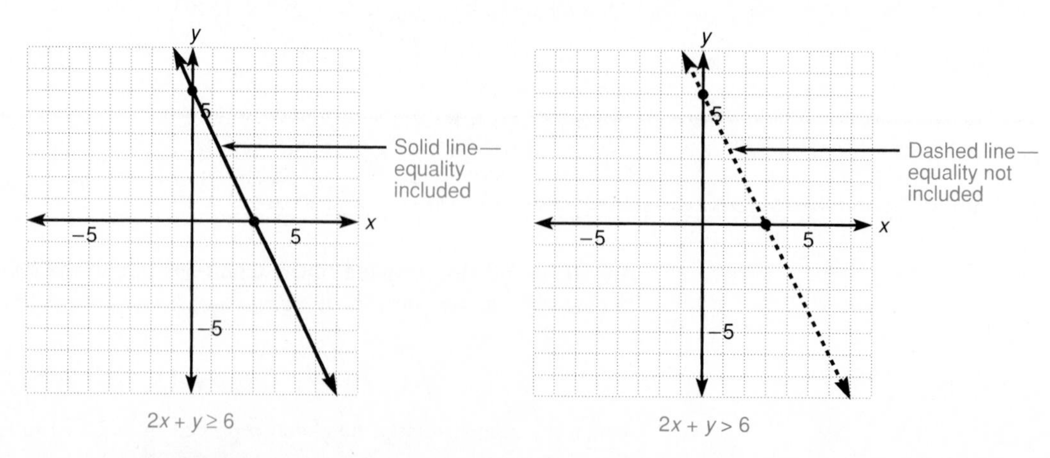

Figure 7–12

To determine the proper half-plane to shade, we choose a *test point* in one of the half-planes [usually the origin, (0,0)] and substitute into the inequality.

1. If the test point satisfies the inequality, shade the half-plane containing that point,
2. If the test point *does not satisfy* the inequality, shade the half-plane that does not contain the point.

To illustrate, consider the inequality $2x + y > 6$ and the test point (0,0).

$$2x + y > 6$$
$$2(0) + (0) > 6 \quad \text{Replace } x \text{ with 0 and } y \text{ with 0}$$
$$0 + 0 > 6$$
$$0 > 6 \quad \text{(False statement)}$$

Shade the half-plane that *does not* contain the origin (0,0). See figure 7–13(a).

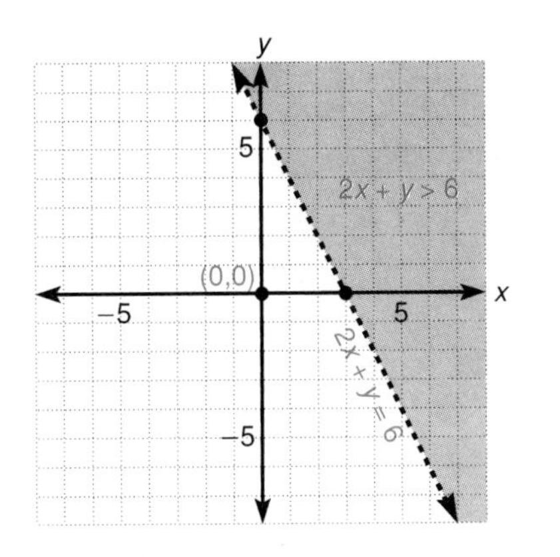

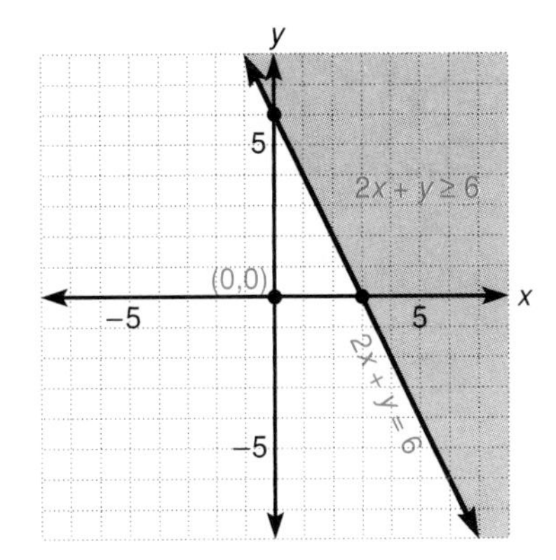

Figure 7–13

Note We will use the origin (0,0) as the test point in all cases except when the boundary line passes through the origin.

Graphing a linear inequality in two variables

1. Replace the inequality symbol with the equality symbol.
2. Graph the resulting equation for the boundary line which will be a
   a. *solid line* if the inequality symbol is $\leq$ or $\geq$,
   b. *dashed line* if the inequality symbol is $<$ or $>$.
3. Choose some test point that is *not* on the line [usually the origin (0,0)] and substitute the coordinates of the point into the inequality.
4. If the resulting statement is
   a. *true,* shade the half-plane containing the test point,
   b. *false,* shade the half-plane that *does not* contain the test point.

■ *Example 7–5 A*

Graph the following linear inequalities.

1. $x + y < 2$

a. Replace $<$ with $=$ to get the equation $x + y = 2$.

b. Graph $x + y = 2$ as a *dashed* line since we have $<$.

c. Since the boundary line does not go through the origin, choose test point (0,0).

$$x + y < 2$$
$$(0) + (0) < 2 \quad \text{Replace } x \text{ with 0 and } y \text{ with 0}$$
$$0 < 2 \quad \text{(True)}$$

d. Shade the half-plane containing the origin (0,0).

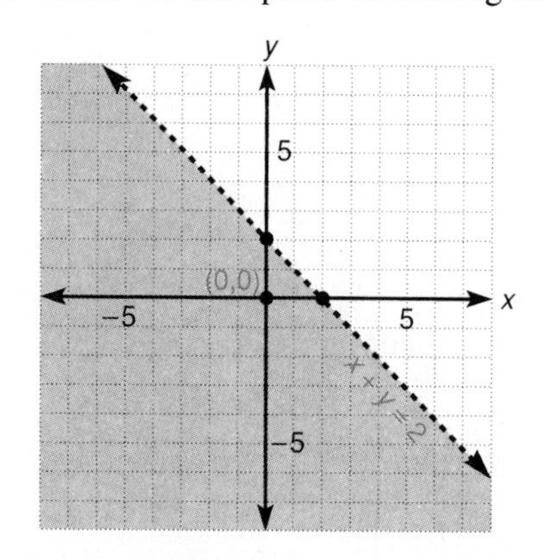

2. $3x - 4y \geq 12$

a. Replace $\geq$ with $=$ to get the equation $3x - 4y = 12$.

b. Graph $3x - 4y = 12$ as a *solid* line since we have $\geq$.

c. Choose test point (0,0).

$$3x - 4y \geq 12$$
$$3(0) - 4(0) \geq 12$$
$$0 \geq 12 \quad \text{(False)}$$

d. Shade the half-plane that *does not* contain the origin (0,0).

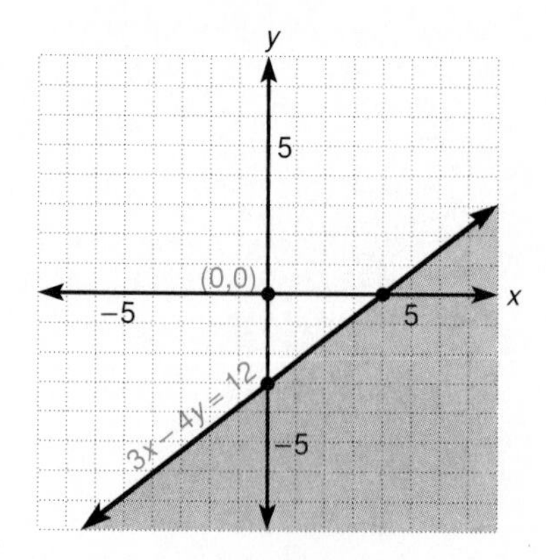

3. $y \leq -2$

a. Replace $\leq$ with $=$ to get the equation $y = -2$.

b. Graph $y = -2$ as a *solid* horizontal line since we have $\leq$.

c. Choose the test point (0,0).

$$y \leq -2$$
$$0 \leq -2 \quad \text{(False)}$$

d. Shade the half-plane that *does not* contain the origin.

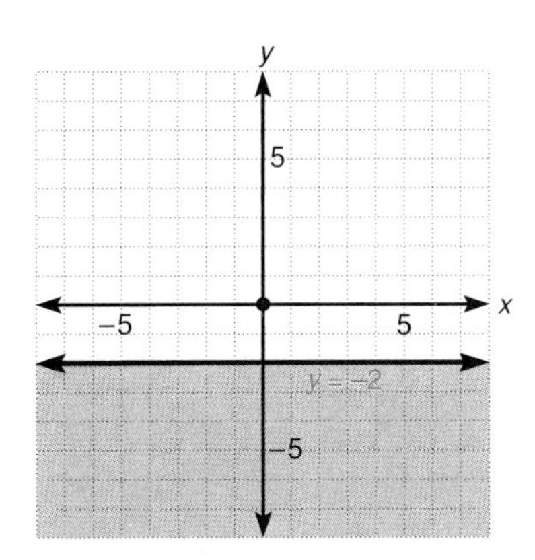

4. $3y - 2x > 0$

a. Replace $>$ with $=$ to get the equation $3y - 2x = 0$.

b. Graph $3y - 2x = 0$ as a *dashed* line since we have the inequality $>$.

c. Since the graph goes through the origin, choose another test point not on the line, say (2,4).

$$3y - 2x > 0$$
$$3(4) - 2(2) > 0 \quad \text{Replace } x \text{ with 2 and } y \text{ with 4}$$
$$12 - 4 > 0$$
$$8 > 0 \quad \text{(True)}$$

d. Shade the half-plane containing the point (2,4).

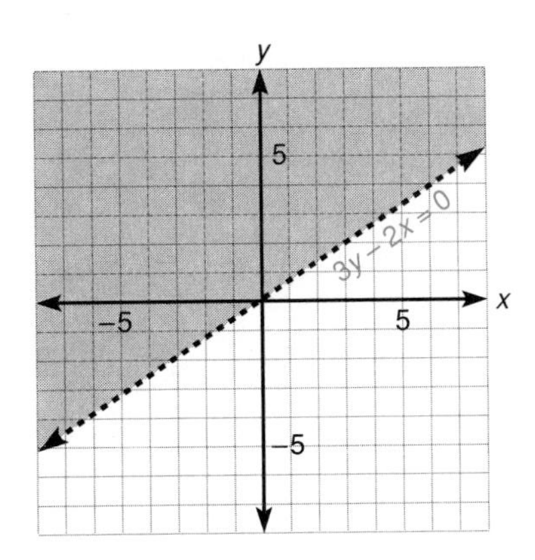

▶ ***Quick check*** Graph the linear inequality. $2y + x < 5$ ■

Mastery points

Can you

- Graph the solution set of a linear inequality in two variables?
- Distinguish when the boundary line is solid and when it is dashed?

Exercise 7–5

Complete the graph of the inequality by shading the correct half-plane. The correct boundary line has been drawn. See example 7–5 A.

Examples a. $x \geq -3$

Solutions Since (0,0) makes $x \geq -3$ true, $0 \geq -3$, shade to the right of $x = -3$.

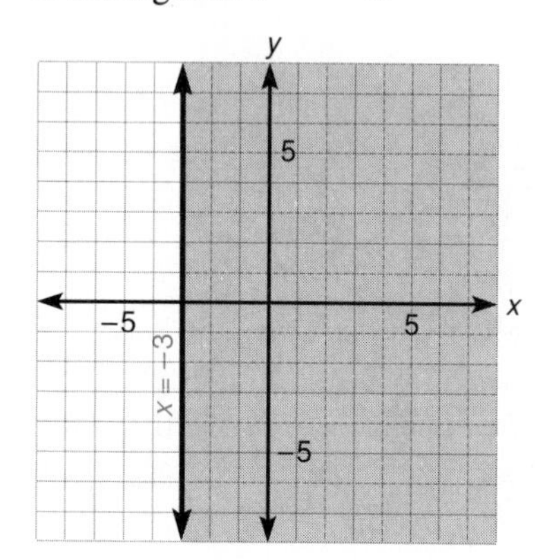

b. $4x - y < 8$

Since (0,0) makes $4x - y < 8$ true, $4(0) - 0 < 8$, shade to the left of $4x - y = 8$.

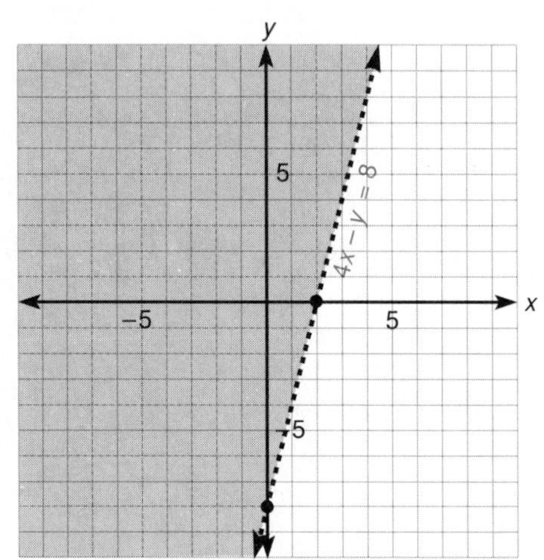

1. $x < 4$

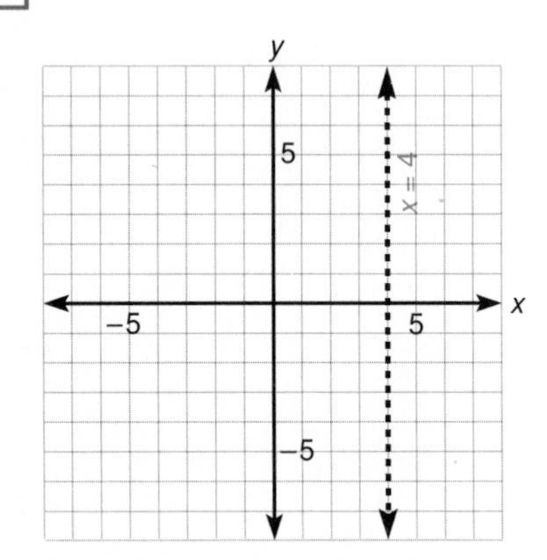

2. $y \geq -5$

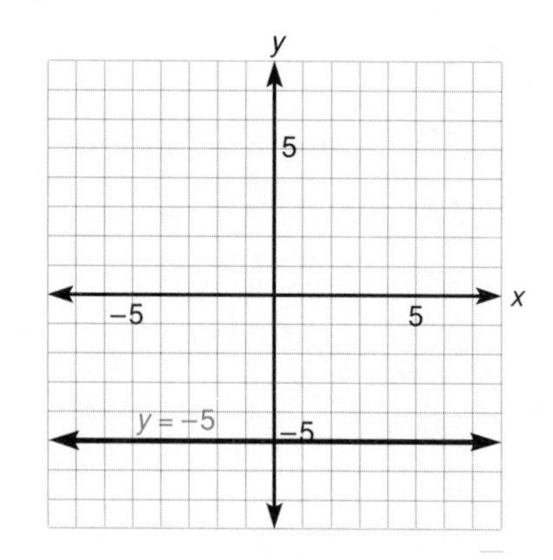

3. $x + y \geq 4$

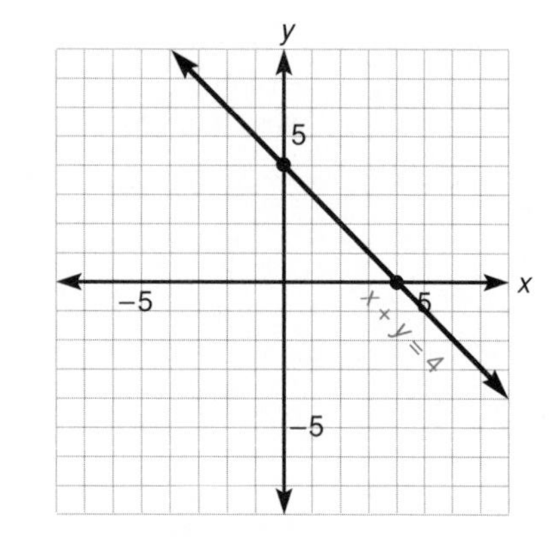

4. $x - y < 1$

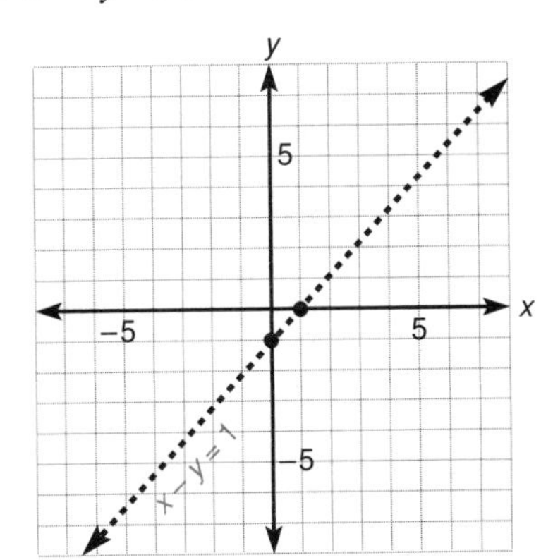

5. $3x - y \leq 4$

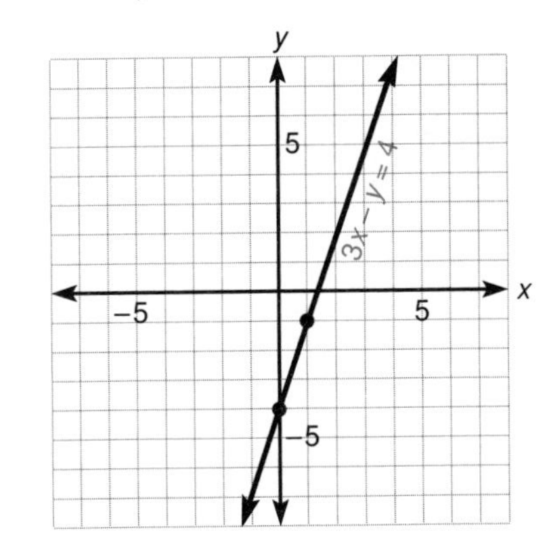

6. $2x - 4y > 8$

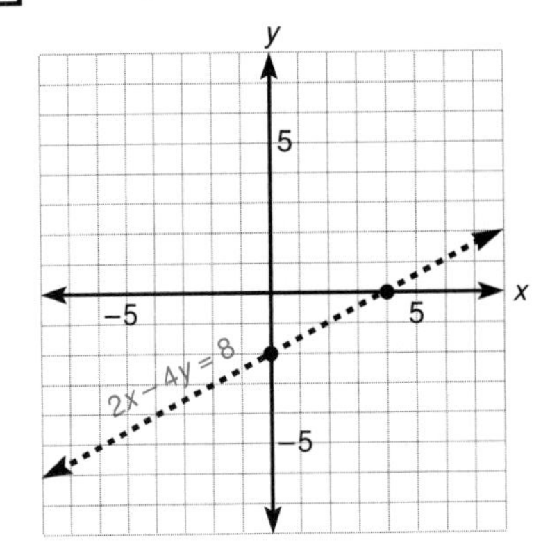

7. $3x + 5y \geq 0$

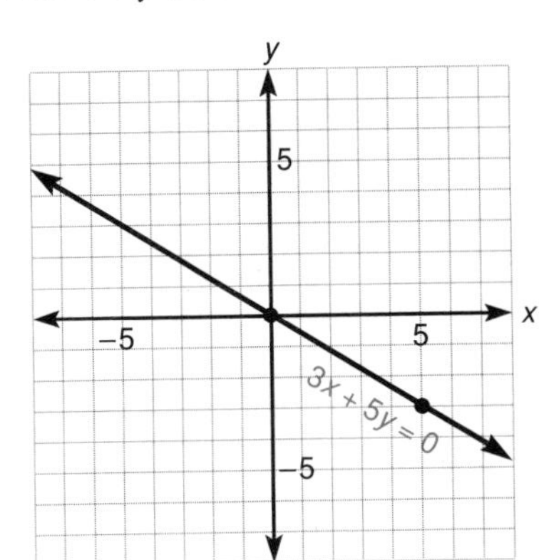

8. $y - 2x < 0$

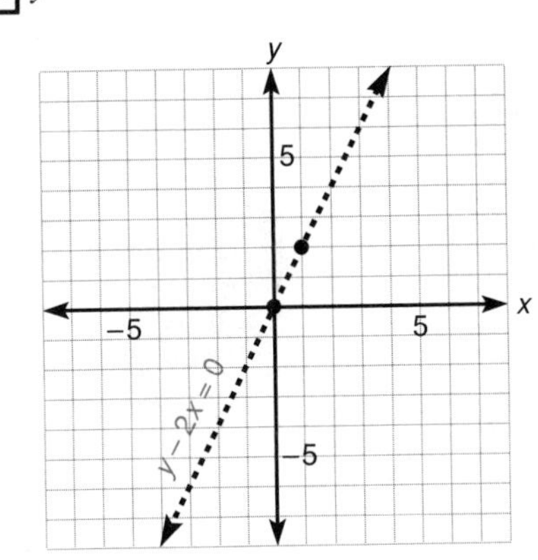

9. $2y \leq 3x$

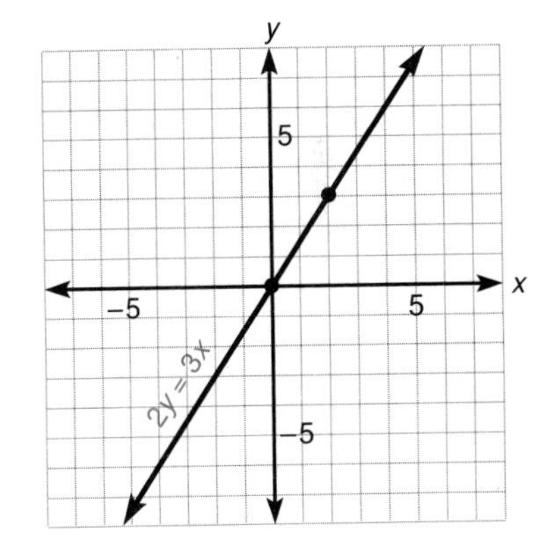

10. $-4x \geq y$

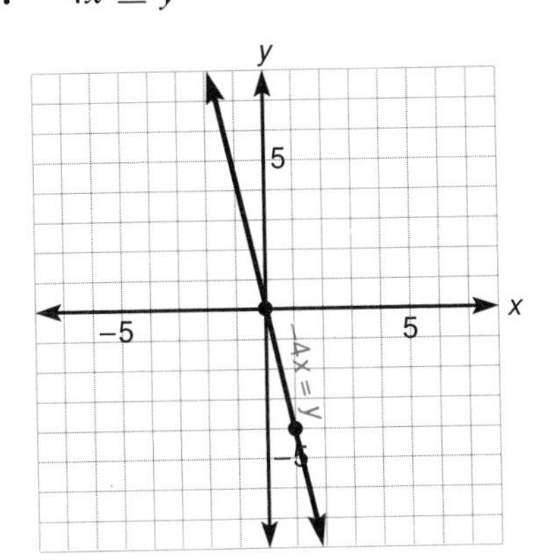

Graph the given linear inequality. See example 7–5 A.

Example $2y + x < 5$

Solution Graph the equation $2y + x = 5$ making the line *dashed*. Substitute test point (0,0) into the inequality.

$$2y + x < 5$$
$$2(0) + (0) < 5 \quad \text{Replace } x \text{ by 0 and } y \text{ by 0}$$
$$0 + 0 < 5$$
$$0 < 5 \quad \text{(True)}$$

Shade the half-plane containing (0,0).

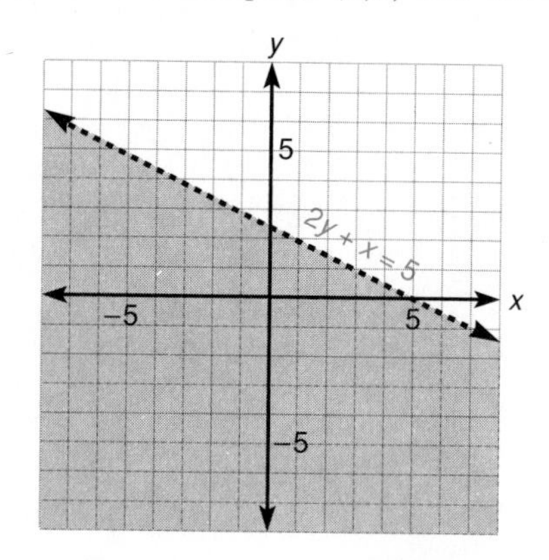

11. $x + y \leq -2$
12. $x + y > 3$
13. $x - y < 6$
14. $x - y \geq -5$
15. $x + 3y > 1$
16. $x + 2y \leq -2$
17. $y \geq 4x + 1$
18. $y < 2x - 3$
19. $3x - 5y < -15$
20. $2x - 7y > 14$
21. $5x - 4y \leq 20$
22. $5x + 2y \leq -10$
23. $x > 3$
24. $x > -4$
25. $x \leq -2$
26. $x < 5$
27. $x + 4 > 0$
28. $x + 5 \leq -2$
29. $y + 3 \leq 5$
30. $y - 2 > 1$
31. $y > x - 2$
32. $3x \geq y$
33. $x < -y$
34. $y \leq 4x$
35. $x \geq 3y$
36. $x + y \geq 0$
37. $8x - 3y \geq 0$
38. $x - y < 0$
39. $3x + 2y < 0$

40. Write an inequality describing all points in the plane that lie (a) below the x-axis, (b) to the right of the y-axis.

41. Write inequalities describing all points that lie in (a) quadrant I, (b) quadrant II, (c) quadrant III, (d) quadrant IV.

Review exercises

Find the domain of the following rational expressions. See section 5–1.

1. $\dfrac{2x + 1}{x - 3}$
2. $\dfrac{y - 4}{y^2 + 3y - 10}$
3. $\dfrac{4 - x}{x^2 + 1}$

Find the standard form of the value of the unknown variable given the value of the other variable. See section 7–1.

4. $4x - 2y = 3$; $x = 2$
5. $4y + x = 1$; $y = -3$

Find the standard form of the equation of the straight line having the following characteristics. See section 7–4.

6. Through points $(-2,3)$ and $(4,1)$
7. Having slope $\dfrac{3}{4}$ and y-intercept -2.
8. Vertical line through $(-3,6)$
9. Horizontal line through $(8,-3)$

7–6 ■ Functions defined by linear equations in two variables

Functions

In section 7–1, we studied equations in two variables such as

$$3x + 2y = 4 \quad \text{or} \quad y = 4 - 3x$$

which related values for x and y that satisfied the given equation. We showed this relationship by ordered pairs (x,y), where, in *most* cases, the value of y was dependent on a chosen value for x.

In all phases of mathematics, from the most elementary to the most sophisticated, the idea of a function is a cornerstone for each mathematical development. Phrases such as "price is a function of cost" or "price is determined by the cost to manufacture," in the business world, and "velocity is a function of time" or "velocity depends on time," of a falling object in science, are common. What is being said in each case is that "the price of an item will change when the cost of producing the item changes" and "the velocity of a falling object will change as the falling time changes."

In mathematics, a function is a special kind of relation, so we first define a **relation.**

Relation

A **relation** is any set of ordered pairs.

Thus, the set of ordered pairs $\{(1,2), (-3,6), (0,-5), (3,4)\}$ is a relation. Every relation has a **domain** and a **range.**

Domain and range of a relation

1. The **domain** of a relation is the set of all first components of the ordered pairs.
2. The **range** of a relation is the set of all second components of the ordered pairs.

Thus, in the relation $\{(1,2), (-3,6), (0,-5), (3,4)\}$,

$$\text{the domain} = \{1,-3,0,3\} \text{ and the range} = \{2,6,-5,4\}$$

The correspondence between the elements of the domain and the elements of the range is shown in figure 7–14.

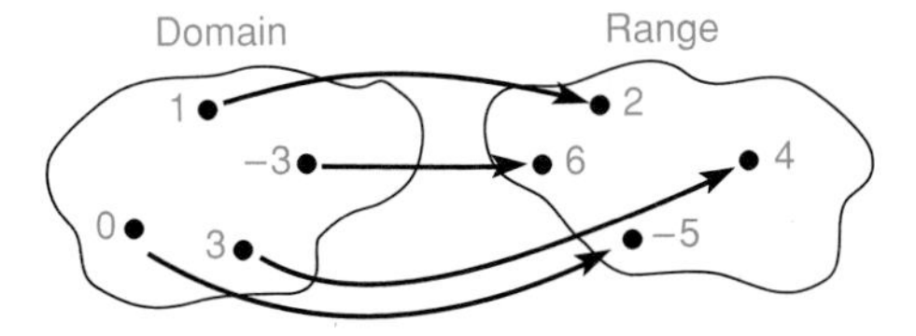

Figure 7–14

Recall, we initially said that a function is a special relation—a set of ordered pairs with a special characteristic.

Function

A **function** is a relation in which no two distinct ordered pairs have the same first component.

From this definition, each member of the domain is assigned to one and only one member of the range.

Example 7–6 A

Determine whether the following relations are functions.

1. $\{(1,2), (-3,6), (0,-5), (3,4)\}$
   No two distinct ordered pairs have the same first component, so the relation is a function.
2. $\{(-3,5), (2,4), (2,-6), (-7,0)\}$
   Since the ordered pairs $(2,4)$ and $(2,-6)$ have the same first component, this *is not* a function.
3. $\{(1,3), (-2,3), (0,3), (4,3)\}$
   Even though the second component of every ordered pair is the same number, this does not violate the definition. This relation is a function. ■

Many useful functions have infinitely many ordered pairs and must be defined by an equation that tells how to get the second component when we are given the first component.

Usually we use x and y to denote the variables in a mathematical function. By our definition, for *y to be a function of x,* the two variables *must be related* so that for each value assigned to x, there is a unique (one and only one) corresponding value assigned to y. This is a very important concept for us to remember. When we use an equation to determine the outcome in a given situation, we want our equation to be of the type that gives us only one result.

Consider the following equations involving the two variables x and y.

Example 7–6 B

1. $y = x + 1$
   This equation defines y as a function of x because, for each chosen value of x, the value of y is one more than the value of x. Thus the ordered pairs $(1,2)$, $(-3,-2)$, $\left(\frac{1}{3},\frac{4}{3}\right)$, and so on, all satisfy this equation. *Only one* value of y has been assigned to each value of x. Therefore, $y = x + 1$ defines a function.
2. $y = 2x - 1$
   This equation defines y as a function of x because, for each chosen value of x, the value of y will be one less than two times x. The ordered pairs $(2,3)$, $(-3,-7)$, $(0,-1)$, $\left(\frac{3}{2},2\right)$, and so on, all satisfy this equation. Only one value of y is associated with each value of x. Therefore, $y = 2x - 1$ defines a function.

3. $x = -4$
This equation *does not* define y as a function of x. Recall that in section 7–2 we found this equation defines a set of ordered pairs whose first components are all -4. Thus while x can only have the value -4, y can take on values 3 and -5 (as well as many other values). The two distinct ordered pairs, $(-4,3)$ and $(-4,-5)$, have the *same first component* and are solutions of this equation. Thus $x = -4$ does not define a function.

4. $y = -3$
This equation *does* define a function since it can be written $y = 0 \cdot x - 3$ and for *any* real value of x, y is determined to be -3. That is, $(2,-3)$, $(-1,-3)$, $(0,-3)$, and so on, are all solutions, and for each value of x, there is only *one* value of y.

Note The function defined by $y = a$ is often called the "constant function." ■

From the definitions, we see that if y is defined to be a function of x, then x represents an element in the *domain* and y represents an element in the *range* of the function. The domain of a function must always be stated, or implied by the nature of the function. In this text, if the domain is not stated, we assume it to be the *set of all real numbers* for which y is defined or makes sense.

■ ***Example 7–6 C***

Determine the domain and range for the function defined by

$2x + y = 1$, where $x \in \{-4,-2,0,2,4\}$.

The domain is restricted to be the set $\{-4,-2,0,2,4\}$, even though x could take on infinitely many values.

Since we are given values of x, we will find it easier to solve for y *once* and then substitute the values of x. Otherwise, we will have to solve for y *each time* there is a new value of x. Solving for y, $y = 1 - 2x$. When

$x = -4$, then $y = 1 - 2(-4) = 1 + 8 = 9$
$x = -2$, then $y = 1 - 2(-2) = 1 + 4 = 5$
$x = 0$, then $y = 1 - 2(0) = 1 - 0 = 1$
$x = 2$, then $y = 1 - 2(2) = 1 - 4 = -3$
$x = 4$, then $y = 1 - 2(4) = 1 - 8 = -7$

The range is the set $\{9,5,1,-3,-7\}$.
The function defined by the equation $2x + y = 1$ and the given restricted domain is represented by the set of ordered pairs $\{(-4,9), (-2,5), (0,1), (2,-3), (4,-7)\}$.

▶ ***Quick check*** State the domain and range of the function defined by the equation $3x = 2y + 1$; $x \in \{1,3,5,7\}$. ■

We use the letters f, g, and h to name functions. For example, the function defined by the equation $y = 5x + 2$ is often written

$$f(x) = 5x + 2$$

Read the symbol "$f(x)$" as "f at x" or "f of x," which means "the value of the function f at x."

Note If we call a function f, then $f(x)$ means y in equations. Since the set of values of y is called its range, all values of $f(x)$ are also in the range.

Example 7–6 D

In the function f defined by $f(x) = 5x + 2$, if $x = 3$, find $f(3)$.

$$\begin{aligned} f(3) &= 5(3) + 2 \quad \text{Replace } x \text{ with 3 in } f(x) = 5x + 2 \\ &= 15 + 2 \\ &= 17 \end{aligned}$$

Thus $f(3) = 17$, which means the value of the function f is 17 when $x = 3$. In like fashion, when

1. $x = -2$, then $f(-2) = 5(-2) + 2$ Replace x with -2
$$\begin{aligned} &= -10 + 2 \\ &= -8 \end{aligned}$$
Thus $f(-2) = -8$.

2. $x = \frac{1}{5}$, then $f\left(\frac{1}{5}\right) = 5\left(\frac{1}{5}\right) + 2$ Replace x with $\frac{1}{5}$
$$\begin{aligned} &= 1 + 2 \\ &= 3 \end{aligned}$$
Thus $f\left(\frac{1}{5}\right) = 3$.

3. $x = 0$, then $f(0) = 5(0) + 2$ Replace x with 0
$$\begin{aligned} &= 0 + 2 \\ &= 2 \end{aligned}$$
Thus $f(0) = 2$.

We have just determined that the function f defined by $f(x) = 5x + 2$ contains the ordered pairs (3,17), (−2,−8), $\left(\frac{1}{5},3\right)$, and (0,2). There *are* many more ordered pairs in function f.

▶ ***Quick check*** a. Given $f(x) = 3x - 2$, find $f(-3)$, $f(0)$, and $f(4)$.
b. Given $f(x) = 5$, find $f(2)$, $f(-3)$, and $f(0)$.
c. Given $f(x) = 2 - 4x$, find $f(-1)$, $f(0)$, and $f(1)$ and state the answers as second components of ordered pairs belonging to the function. ■

When discussing arbitrary functions involving the variables x and y, we often see the statement

$$y = f(x)$$

This states that "y is a function of x."

Functions may be defined by many equations other than linear equations. In chapter 10, we will discuss another type of equation that defines a function.

Mastery points

Can you

- Determine the domain and range of a function?
- Understand what $f(x)$ means and find particular values of $f(x)$?

Exercise 7–6

State the domain and range of each of the functions defined by the equation using the restricted values of x. See example 7–6 C.

Example $3x = 2y + 1;\ x \in \{1,3,5,7\}$

Solution The domain $= \{1,3,5,7\}$.
Solving for y:

$$2y + 1 = 3x$$
$$2y = 3x - 1$$
$$y = \frac{3x - 1}{2}$$

When $x = 1,\ y = \frac{3(1) - 1}{2} = \frac{2}{2} = 1$

$x = 3,\ y = \frac{3(3) - 1}{2} = \frac{8}{2} = 4$

$x = 5,\ y = \frac{3(5) - 1}{2} = \frac{14}{2} = 7$

$x = 7,\ y = \frac{3(7) - 1}{2} = \frac{20}{2} = 10$

The range $= \{1,4,7,10\}$.

1. $y = 3x - 1;\ x \in \{-5,-3,-1,0,1,3,5\}$
2. $y = 2 - x;\ x \in \{-4,-2,0,2,4\}$
3. $2x - 3y = 1;\ x \in \{0,1,2,3,4,5\}$
4. $5x + y = 3;\ x \in \{-4,-3,-2,-1,0\}$
5. $y = -4;\ x \in \{-10,-5,0,5,10\}$
6. $y = 12;\ x \in \{-8,-4,-2,0,2,4,8\}$

Find the indicated value of the given function. See example 7–6 D.

Examples a. $f(x) = 3x - 2$; find $f(-3), f(0), f(4)$.

b. $f(x) = 5$; find $f(2), f(-3), f(0)$.
Rewrite $f(x) = 5$ as $f(x) = 0 \cdot x + 5$

Solutions a. $f(-3) = 3(-3) - 2 = -9 - 2 = -11$
$f(0) = 3(0) - 2 = 0 - 2 = -2$
$f(4) = 3(4) - 2 = 12 - 2 = 10$

$f(-3) = -11;\ f(0) = -2;\ f(4) = 10$

b. $f(2) = 0 \cdot 2 + 5 = 5$
$f(-3) = 0 \cdot -3 + 5 = 5$
$f(0) = 0 \cdot 0 + 5 = 5$

$f(2) = 5;\ f(-3) = 5;\ f(0) = 5$

7. $f(x) = 2x;\ f(2), f(0), f(-1)$
8. $f(x) = 3x;\ f(3), f(0), f(-4.3)$
9. $f(t) = t + 5;\ f(5), f(-5), f(0)$
10. $f(x) = x - 4;\ f(4), f(2), f(5.2)$
11. $f(x) = 3x + 5;\ f(-3), f\left(-\frac{5}{3}\right), f(1)$
12. $f(s) = 5s + 2;\ f(-2), f(3), f\left(-\frac{2}{5}\right)$
13. $f(x) = 3;\ f(-7), f(0), f(0.41)$
14. $f(x) = -5;\ f(-5), f(9), f(0)$

Find the indicated value of the given function. State the answers as second components of ordered pairs belonging to the function. See example 7–6 D.

Example $f(x) = 2 - 4x$; find $f(-1), f(0), f(1)$.

Solution
$f(-1) = 2 - 4(-1) = 2 + 4 = 6$ Answer: $(-1,6)$
$f(0) = 2 - 4(0) = 2 - 0 = 2$ Answer: $(0,2)$
$f(1) = 2 - 4(1) = 2 - 4 = -2$ Answer: $(1,-2)$

15. $f(x) = 5x + 1;\ f(-3), f\left(-\frac{2}{5}\right), f(0), f(4), f\left(\frac{1}{5}\right)$
16. $f(x) = 7 - 2x;\ f(-4), f\left(-\frac{1}{2}\right), f(0), f\left(\frac{1}{2}\right), f(1.7)$
17. $f(x) = \frac{2}{3}x + 1;\ f(6), f(-3), f(0), f\left(\frac{3}{2}\right), f\left(-\frac{3}{2}\right)$
18. $f(x) = 5 - \frac{1}{4}x;\ f(-8), f\left(-\frac{4}{3}\right), f(0), f(4), f(0.13)$

19. $f(x) = 0; f(-1), f(-10), f(0), f\left(\frac{8}{3}\right), f(2)$

20. $f(x) = -8; f(-3), f\left(-\frac{3}{2}\right), f(0), f\left(\frac{5}{7}\right), f(14.7)$

Solve the following word problems.

21. The temperature in degrees Fahrenheit (F) can be expressed as a function of degrees Celsius (C) by $F = f(C)$. If $f(C) = \frac{9}{5}C + 32$, find $f(100), f(0), f(25), f(-10)$.

22. The temperature in degrees Celsius (C) can be expressed as a function of degrees Fahrenheit (F) by $C = g(F)$. If $g(F) = \frac{5}{9}(F - 32)$, find $g(32)$, $g(212)$, $g(-4)$, $g(59)$.

23. The cost c in cents of sending a letter by first-class mail is a function of the weight w in ounces of the letter defined by $c = 29w$. Then $c = h(w) = 29w$. Find $h(2)$, $h(3)$, $h(4)$. What is the domain of function h?

24. The perimeter P of a square can be expressed as a function of its sides s. If $P = 4s$, then $P = f(s) = 4s$. Find $f(4), f\left(\frac{3}{4}\right), f(4.7), f\left(\frac{5}{2}\right)$. What is the domain of f?

25. The resistance R in an electric circuit can be expressed as a function of the voltage E. If $R = 0.07E$, then $R = g(E) = 0.07E$. Find $g(2)$, $g(3)$, $g(1.2)$.

26. When two objects are pressed together, the frictional force F_f between their surfaces is related to the perpendicular (or normal) force N holding the surfaces together by $F_f = \mu N$, where μ is a constant coefficient of friction. For wood on wood, the coefficient of static friction is $\mu = 0.5$. Using h, define F_f as a function of N and find $h(100)$, $h(50)$, $h(25)$.

27. Using exercise 26, for a rubber tire on dry concrete, the coefficient of sliding friction is $\mu = 0.7$. Define a function g and find $g(5)$, $g(10)$, $g(3)$.

28. The circumference C of a circle may be expressed as a function of its radius r. Given $C = 2\pi r$, where 2π is a constant, express C as a function of r using f and find $f(7)$, $f(0.8)$, and $f\left(\frac{5}{2}\right)$. State the domain of function f.

29. The temperature F (in degrees Fahrenheit) can be approximated by the formula

$$F = \frac{n}{4} + 40$$

where n is the number of times a cricket chirps in 1 minute. Express F as a function of n using h to name the function and find $h(50)$, $h(12)$, $h(120)$.

30. Simple interest, I, on a loan for one year at 12% interest is given by $I = (.12)P$, where P is the amount borrowed. Using f to name the function, express I as a function of P and find $f(1{,}000)$, $f(5{,}000)$, $f(12{,}350)$.

31. The gross wages W of an hourly worker are determined by
$W =$ hourly rate (r) $\times$ hours worked (h) or
$W = rh$.
Define W as a function of h when $r = \$10.50$. Calling this function g, find $g(40)$, $g(48)$, $g(28)$.

32. A company's income statement, I, for one quarter is given by $I = R - E$, where R is the revenue in sales and gains and E is the expenses incurred. Define $I = h(R,E) = R - E$ and find $h(1{,}000;350)$, $h(875;490)$, $h(2{,}515;1{,}031)$. (*Hint:* $h(1{,}000;350)$ means that $R = \$1{,}000$ and $E = 350$.)

33. The selling price, S, of some merchandise is given by $S = C + M$, where C is the cost price of the merchandise and M is the markup (expenses incurred in operations). Using f, define S as a function of C and M; that is $S = f(C,M) = C + M$. Find $f(25,7)$, $f(32,11)$, $f(146,27)$.

Review exercises

Solve the following equations for y. See section 2–7.

1. $2x + y = 7$
2. $3x - 4y = 8$
3. Given $x = 4y - 2$ and $3y - 2x = 1$, replace x with $4y - 2$ in the second equation and solve for y. See section 2–2.
4. Graph the equations $3x - 2y = 6$ and $x + y = 2$ on the same set of axes. Determine the coordinates of the point of intersection. See section 7–2.

Find the value of the unknown variable in each of the following equations. See section 7–1.

5. $x + 3y = 4$ when $y = -2$
6. $4y - 5x = 3$ when $x = 1$

Chapter 7 lead-in problem

Tickets to a football game at Smith High School cost \$1.25 for students and \$3.00 for adults. The total receipts for a game with University High School were \$1,020. Write an equation using x for the number of students and y for the number of adults attending the game.

Solution

Let x = the number of students at the game
y = the number of adults at the game

	Cost per ticket	Number	Receipts
Students	1.25	x	$1.25x$
Adults	3.00	y	$3.00y$

We add the individual receipts from the students and the adults to obtain the total receipts, \$1,020.

$$1.25x + 3.00y = 1{,}020$$

Chapter 7 summary

1. A **linear equation in two variables** is an equation that can be written in the form $ax + by = c$, where a, b, and c are real numbers, and a and b are not both zero.
2. The **solutions** of a linear equation in two variables are the ordered pairs of numbers of the form (x,y) that satisfy the equation.
3. Solutions of linear equations in two variables are found by choosing a value for one variable and evaluating to find the value of the other variable.
4. **Ordered pairs of numbers** may be graphed as points in the **rectangular coordinate plane.**
5. In the ordered pair (x,y), we call x the **abscissa** and y the **ordinate** of the point in the plane that is its graph.
6. Graphs of linear equations in two variables are *straight lines.*
7. To find the *x-intercept,* the abscissa of the point where the graph crosses the x-axis, let $y = 0$ and solve for x.
8. To find the *y-intercept,* the ordinate of the point where the graph crosses the y-axis, let $x = 0$ and solve for y.
9. The graph of the equation $y = b$ is a *horizontal line* passing through the point $(0,b)$.
10. The graph of the equation $x = a$ is a *vertical line* passing through the point $(a,0)$.
11. The **slope** m of a line is the "steepness" of the line and is defined by $m = \dfrac{y_2 - y_1}{x_2 - x_1} = \dfrac{y_1 - y_2}{x_1 - x_2}$, where $P_1(x_1,y_1)$ and $P_2(x_2,y_2)$ are any two points lying on the line.
12. The **standard form** of the equation of a line is $ax + by = c$, where a, b, and c are integers, $a \geq 0$, and a and b not both 0.
13. The **point-slope** form of the equation of a line is given by $y - y_1 = m(x - x_1)$, where m is the slope and the point $P_1(x_1,y_1)$ lies on the line.
14. The **slope-intercept** form of the equation of a line is $y = mx + b$, where m is the slope and b is the y-intercept.

15. Two distinct nonvertical lines having slopes m_1 and m_2 are **parallel** if and only if $m_1 = m_2$. (The slopes are the same.)
16. Two nonvertical lines having slopes m_1 and m_2 are **perpendicular** if and only if $m_1m_2 = -1$. (The slopes are negative reciprocals of each other.)
17. Vertical lines and horizontal lines are perpendicular to one another.
18. A **linear inequality in two variables** is an inequality of the form $ax + by < c$, $ax + by \leq c$, $ax + by > c$, or $ax + by \geq c$.
19. The graph of a linear inequality is a **half-plane** that
    a. includes the boundary line when the symbols $\leq$ or $\geq$ are used. The line is solid.
    b. excludes the boundary line when the symbols $<$ or $>$ are used. The line is dashed.
20. A **function** is a relation (set of ordered pairs) in which no two distinct ordered pairs have the same first component.
21. The **domain** of a function is the set of all first components of the ordered pairs in the function.
22. The **range** of a function is the set of all second components of the ordered pairs in the function.
23. Given a function f, $f(x)$ represents the value of f at the number x.

Chapter 7 error analysis

1. Ordered pair solutions of linear equations in two variables
   Example: The ordered pair (1,5) is a solution of $x = 5$.
   Correct answer: The ordered pair (5,1) is a solution of $x = 5$.
   What error was made? (*see page 278*)
2. Coordinates of points on axes
   Example: The point (0,4) lies on the x-axis.
   Correct answer: The point (4,0) lies on the x-axis.
   What error was made? (*see page 282*)
3. Ordered pairs on the graph of an equation
   Example: The ordered pair $(-1,2)$ lies on the graph of $2x - y = 3$.
   Correct answer: $(-1,2)$ is not a solution, the point does not lie on the graph.
   What error was made? (*see page 289*)
4. Finding the slope of a line
   Example: The slope of the line through $(-2,3)$ and $(2,1)$ is
   $$m = \frac{3-1}{2-(-2)} = \frac{2}{2+2} = \frac{2}{4} = \frac{1}{2}.$$
   Correct answer: $-\frac{1}{2}$
   What error was made? (*see page 299*)
5. Zero exponents
   Example: $(a^2b^3c^0)^2 = a^4b^6c^2$
   Correct answer: a^4b^6
   What error was made? (*see page 142*)
6. Negative exponents
   Example: $(-3)^{-3} = \frac{1}{3^3} = \frac{1}{27}$
   Correct answer: $-\frac{1}{27}$
   What error was made? (*see page 141*)
7. Multiplying like bases
   Example: $3^2 \cdot 3^3 = 9^5$
   Correct answer: 3^5
   What error was made? (*see page 128*)
8. Exponents
   Exponents: $-\ (5)^2 = 25$
   Correct answer: -25
   What error was made? (*see page 128*)
9. Graphing linear inequalities
   Example: The graph of $-2 \leq x < 4$ is

   Correct answer:

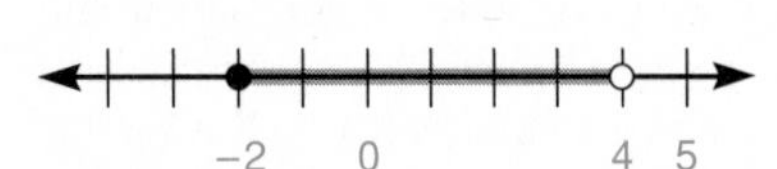

   What error was made? (*see page 118*)
10. Power to a power
    Example: $(-5x^2)^3 = -5x^5$
    Correct answer: $-125x^6$
    What error was made? (*see page 130*)

Chapter 7 critical thinking

The owner of a store has made up a work schedule for Bill and Ted so that at least one of them will be at the store each day (they never will both be off at the same time). Starting today Bill has every fourth day off and starting tomorrow Ted will have every sixth day off. Will the owner's plan work and explain why or why not.

Chapter 7 review

[7–1]

Find the value of y corresponding to the given values for x. Express the answer as an ordered pair.

1. $y = 3x + 4; x = -1, x = 0, x = 4$

2. $2x - 3y = -1; x = -2, x = 0, x = 1$

3. $y + 3 = 0; x = -7, x = 0, x = 5$

4. $5x + y = 0; x = -3, x = 0, x = 3$

Find the value of x corresponding to the given values of y. Express the answer as an ordered pair.

5. $x = -3y + 1; y = -2, y = 0, y = 5$

6. $4x + 2y = 7; y = -1, y = 0, y = 3$

7. $x - 1 = 0; y = -8, y = 0, y = 2$

8. $3x - 2y = 0; y = -3, y = 0, y = 1$

Plot the following ordered pairs on a rectangular coordinate system.

9. $(1,5)$ **10.** $(4,-4)$ **11.** $(-1,-6)$ **12.** $(0,-4)$ **13.** $\left(2,\frac{1}{2}\right)$ **14.** $\left(5,-\frac{2}{3}\right)$

Determine to the nearest integer the coordinates, (x,y), of the given points.

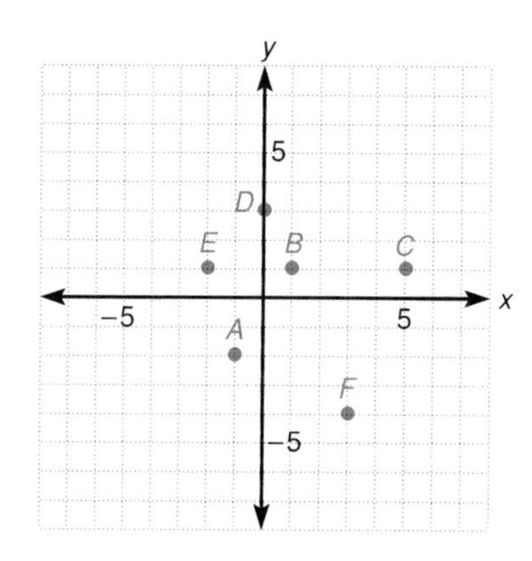

15. A **16.** B **17.** C **18.** D **19.** E **20.** F

Find the missing component in each ordered pair using the given equation. Then plot the ordered pairs using a separate coordinate system for each problem.

21. $y = 3x + 4; (2,\ ), (0,\ ), \left(\frac{4}{3},\ \right), \left(-\frac{2}{3},\ \right)$

22. $y = -2x + 5; (-2,\ ), \left(\frac{5}{2},\ \right), (0,\ ), (3,\ )$

23. $y = x^2 + 2; (1,\ ), (-1,\ ), (0,\ ), (2,\ )$

24. $y = x^2 - 16; (4,\ ), (3,\ ), (0,\ ), (-4,\ )$

[7–2]

Find the x- and y-intercepts. State the answer as an ordered pair.

25. $y = 3x + 5$ **26.** $y = -4x$ **27.** $y + 2 = 0$ **28.** $x - 6 = 0$

29. $2x - 7y = 4$ **30.** $4y - x = 0$

Graph the following linear equations using the x- and y-intercepts, where possible.

31. $y = -x + 7$ **32.** $y = \frac{1}{2}x - 1$ **33.** $y = -x$ **34.** $y = 7$

35. $x = -4$ **36.** $2x - 3y = -6$ **37.** $5x - 3y = 15$

[7–3]

Find the slope of the line passing through the given pairs of points.

38. $(-4,3)$ and $(1,0)$ **39.** $(-5,3)$ and $(-5,-1)$ **40.** $(1,-2)$ and $(5,-2)$ **41.** $(-4,-4)$ and $(1,1)$

[7–4]

Express the following equations in *slope-intercept* form $y = mx + b$ and determine the slope m and y-intercept b.

42. $3x - 4y = 8$ **43.** $4x + 3y = 2$ **44.** $8x - 3y + 1 = 0$ **45.** $y - 5x = -10$

Find the slope of the following lines and find the equation of the line using the point-slope form of the equation of a line.

46.

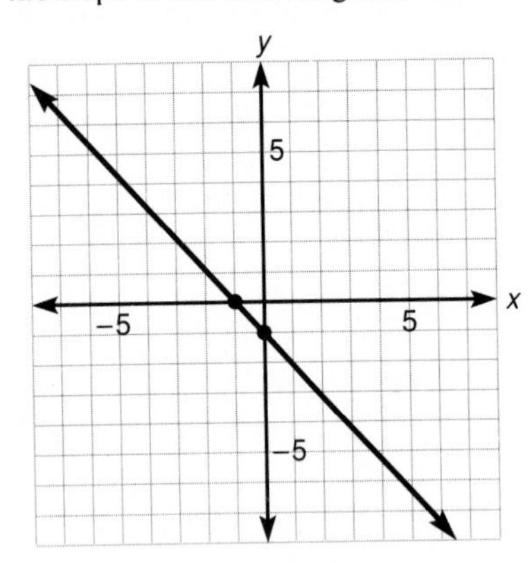

47.

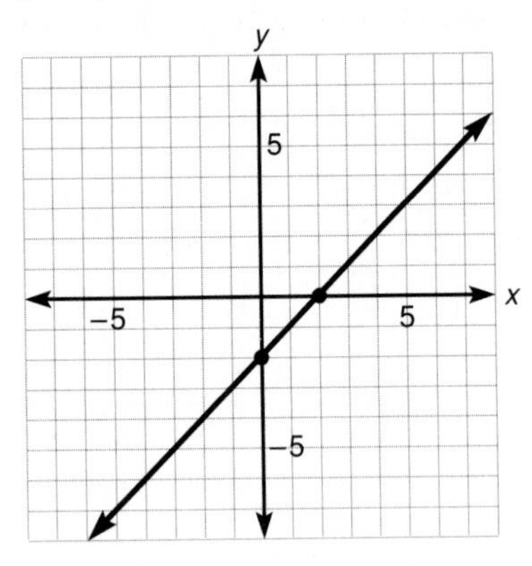

48. A roadway rises vertically 150 feet through a horizontal distance of 1,050 feet. What is the slope of the roadway?

Find the equation of the lines having the given conditions. Write the answer in standard form.

49. Passing through (4,5) having slope $m = -4$

50. Having slope $m = \frac{1}{3}$ and y-intercept 4

51. Passing through the points (5,2) and (−3,1)

52. Passing through the points (0,4) and (−3,0)

Graph the following equations using the slope m and y-intercept b.

53. $y = 3x + 4$ **54.** $2x - y = 5$ **55.** $4y - 3x = 8$ **56.** $3y + 4x = 9$

[7–5]

Graph the following inequalities.

57. $x \leq 2y + 1$ **58.** $y > 3x - 2$ **59.** $2x - y \geq 4$ **60.** $y \leq 5$

61. $x < -1$ **62.** $2y - 5x > 0$ **63.** $3x + 2y \geq 9$ **64.** $5x - 2y < 10$

[7–6]

Determine the domain and range of the function defined by each of the following equations

65. $y = -3x - 1; x \in \{-5,-3,-1,0,1,3,5\}$

66. $x + 4y = 3; x \in \{-4,-1,0,2,3\}$

67. $x - 2y = 0; x \in \{-7,-3,0,2,6\}$

Find the indicated value of each function. State the answers as ordered pairs.

68. $f(x) = 4x; f(-1), f(0), f(4)$

69. $f(x) = 4 - x; f(-2), f(0), f(2)$

70. $f(x) = 5x - 3; f(-4), f(0), f(6)$

71. $f(x) = 5; f(-8), f(0), f(9)$

72. The perimeter P of a regular hexagon (six-sided figure whose sides are the same length and whose angles have the same measure) is given by $P = 6s$, where s is the length of a side. Express P as a function of s using f and find $f(3)$ and $f(14)$. What is the domain of f?

Chapter 7 cumulative test

[2–2] **1.** Evaluate the expression
$xy + (-xy)^2 - x + y$ when $x = 3$ and $y = -2$.

[2–3] **2.** Remove the grouping symbols and simplify the expression.
$3 - [5(6x - 8) - (12x - 7) + 9x]$

Find the solution set of the following equations and inequalities.

[2–6] **3.** $6(x + 2) = 9x + 4$

[6–5] **4.** $\frac{3}{4} = \frac{5}{x} - \frac{2}{3}$

[2–6] **5.** $-4(y + 1) - 8 = -10 - (-3 + y)$

[6–5] **6.** $3 - \frac{4}{x^2} = \frac{4}{x}$

[2–9] **7.** $-7 \leq 5 - 3x < 8$

Completely factor the following expressions.

[4–3] **8.** $7x^2 - 6x - 1$

[4–1] **9.** $y^8 + y^7 - y^6$

[4–6] **10.** $p^5 + p^4 - 4p^2 - 4p$

[4–4] **11.** $16a^2 - 16b^2$

[4–6] **12.** $t^3 - 8t^2 + 7t$

[4–1] **13.** $-35y - 28z$

Perform the indicated operations.

[3–2] **14.** $(4y - 3x)^2$

[3–2] **15.** $(4x - 7)(5x + 8)$

[3–2] **16.** $\left(5 - \frac{3}{4}y\right)\left(5 + \frac{3}{4}y\right)$

[3–2] **17.** $(x - y)^3$

[3–2] **18.** $(3y^2 + 7)(5y^3 + 6y^2 - y + 8)$

[6–2] **19.** $\frac{2x}{x^2 - x - 42} + \frac{3x}{x^2 - 49}$

[6–2] **20.** $\frac{a + 1}{a + 5} - \frac{a - 3}{a^2 + 3a - 10}$

[6–1] **21.** $\frac{2y - 6}{y + 8} \cdot \frac{3y + 24}{y^2 - 9}$

Simplify the following. Express the answer with positive exponents.

[3–3] **22.** $\frac{x^{-6}}{x^3}$

[3–3] **23.** $8^{-5} \cdot 8^{-3}$

[3–3] **24.** $(-5a^{-3})^3$

[3–5] **25.** Write the number 0.00000776 in scientific notation.

[5–3] **26.** Divide. $(3x^3 - 2x^2 + 4x - 5) \div (x + 2)$

[6–4] **27.** Simplify the complex fraction. $\dfrac{\frac{2x}{x - 3} - \frac{3}{x + 1}}{\frac{3x + 4}{x^2 - 2x - 3}}$

Graph the following equations using the x- and y-intercepts if possible.

[7–2] **28.** $2x - y = 7$

[7–2] **29.** $x - 7 = 0$

[7–5] **30.** Graph the linear inequality.
$3x - 5y > 15$

[7–3] **31.** Find the slope of the line through the points (2,3) and (−1,−4).

Find the equation of the line.

[7–4] **32.** Passing through the points (1,4) and (−3,1)

[7–4] **33.** Having slope $-\frac{3}{5}$ and passing through the point (4,3)

Determine if the following lines are parallel, perpendicular, or neither.

[7–4] **34.** $2x - 3y = 4$ and $4x - 6y = -1$

[7–4] **35.** $x + 2y = 6$ and $3x - 6y = 1$

[7–6] **36.** Given the function defined by $f(x) = 3x - 2$, find (a) $f(0)$, (b) $f(-3)$, (c) $f(4)$.

Solve the following word problems.

[2–8] **37.** Three-fourths of what number is 102?

[4–8] **38.** The product of a whole number times the next consecutive even whole number is 168. What is the whole number?

[4–8] **39.** A right triangle has a hypotenuse of 13 inches. If one of the legs is 7 inches longer than the other, find the lengths of the legs. (*Hint:* Use $c^2 = a^2 + b^2$, where c is the hypotenuse and a and b are the legs.)

CHAPTER

8 Systems of Linear Equations

A keypunch operator at a local firm works for \$9 per hour while an entry-level typist works for \$6.50 per hour. The total pay for an 8-hour day is \$476. If there are two more typists than keypunch operators, how many keypunch operators does the firm employ?

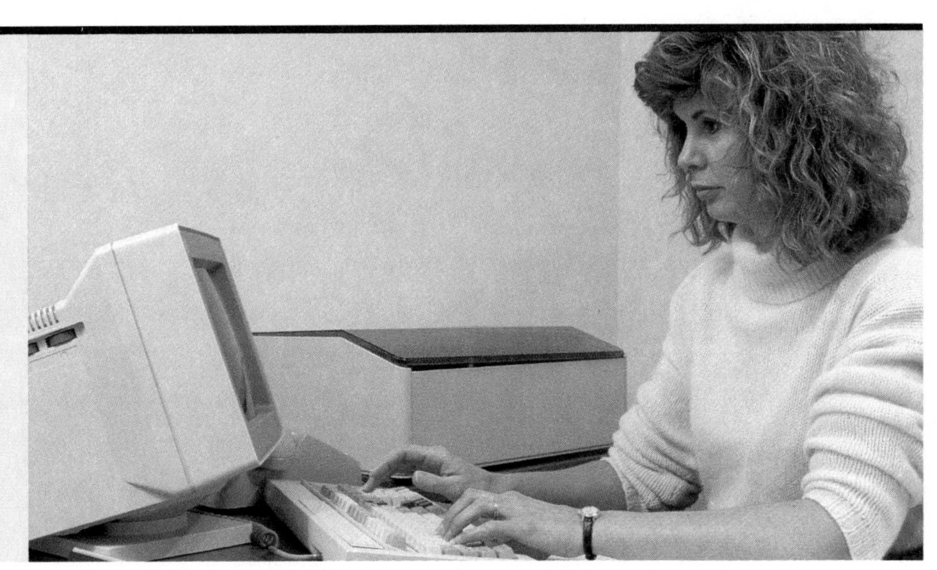

8–1 ■ Solutions of systems of linear equations by graphing

In chapter 7, we showed by graphs the relationship between the variables in a linear equation in two variables. We now consider the relationship that exists when *two* linear equations in two variables are graphed on the same coordinate plane. Two or more linear equations involving the same variables form a **system of linear equations.** Here are examples of systems of linear equations.

I	II	III	IV
$4x - y = 1$	$3x - 7y = 5$	$x = 3y + 1$	$x = 4$
$x + y = 4$	$5x + 2y = 11$	$x + 3y = 4$	$3x - 2y = 5$

To solve any one of these systems of linear equations, we must find the ordered pair(s) of real numbers (x,y) that satisfy both equations at the same time.

Solutions of systems of linear equations

A **solution** of a system of linear equations is an ordered pair of numbers that satisfies (makes true) all equations in the system.

■ Example 8–1 A

Determine whether the given ordered pair is a solution of the system of linear equations.

1. $4x - y = 1$
$x + y = 4$ $(1,3)$

$$4x - y = 1$$
$$4(1) - (3) = 1 \quad \text{Replace } x \text{ with 1 and } y \text{ with 3}$$
$$4 - 3 = 1$$
$$1 = 1 \quad \text{(True)}$$

$$x + y = 4$$
$$(1) + (3) = 4$$
$$4 = 4 \quad \text{(True)}$$

Both resulting statements are true. Therefore, the ordered pair $(1,3)$ satisfies both equations and is a solution.

2. $5x + 3y = -20$
$7x + 2y = -17$ $(-4,0)$

$$5x + 3y = -20$$
$$5(-4) + 3(0) = -20 \quad \text{Replace } x \text{ with } -4 \text{ and } y \text{ with 0}$$
$$-20 + 0 = -20$$
$$-20 = -20 \quad \text{(True)}$$

$$7x + 2y = -17$$
$$7(-4) + 2(0) = -17$$
$$-28 + 0 = -17$$
$$-28 = -17 \quad \text{(False)}$$

The ordered pair $(-4,0)$ is *not* a solution of the system since it does not satisfy the equation $7x + 2y = -17$.

▶ ***Quick check*** Determine if the ordered pair $(1,1)$ is a solution of the system.
$3x - y = 2$
$4x + 3y = 9$ ■

Graphical solutions

The first method that we shall examine is finding the solution of a system of linear equations by **graphing.** We are looking for the coordinates of any point where the graphs intersect. This point of intersection will give a solution of the system. Using the methods for graphing that we learned in chapter 7, we graph each equation on the same coordinate plane. Using the x- and y-intercepts and a checkpoint, we graph each equation.

■ Example 8–1 B

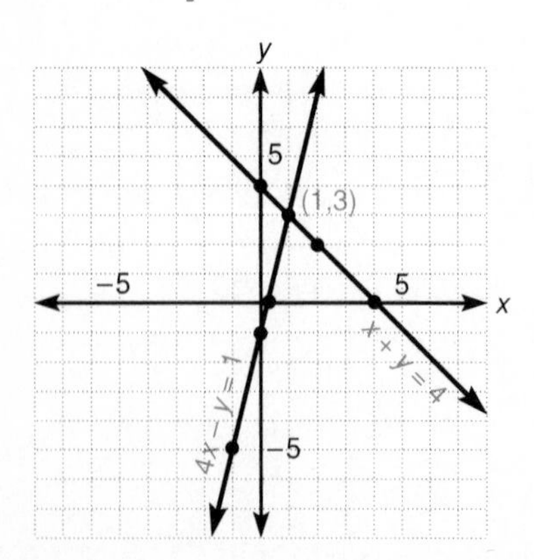

Find the solution of each system by graphing.

1. $4x - y = 1$
$x + y = 4$

$4x - y = 1$

x	y	
0	-1	y-intercept; $x = 0$
$\frac{1}{4}$	0	x-intercept; $y = 0$
-1	-5	Checkpoint

$x + y = 4$

x	y
0	4
4	0
2	2

It appears the point of intersection of the two graphs is the point (1,3). We must check this as the possible solution.

$4x - y = 1$		$x + y = 4$	
$4(1) - (3) = 1$	Replace x with 1 and y with 3	$(1) + (3) = 4$	
$4 - 3 = 1$		$4 = 4$	(True)
$1 = 1$	(True)		

The solution of the system is the ordered pair (1,3).

2. $5x - y = 5$
$x + y = 2$

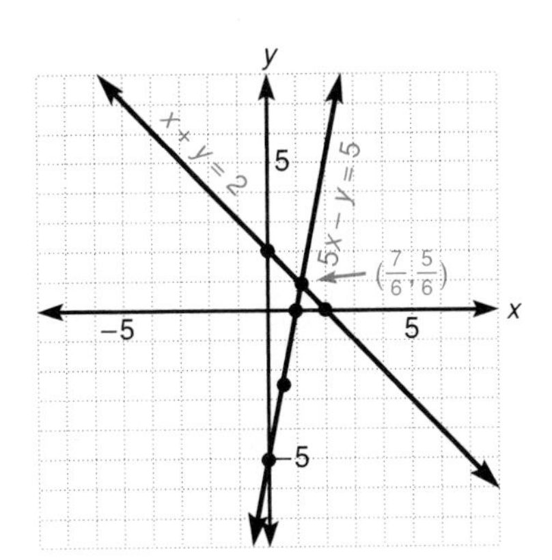

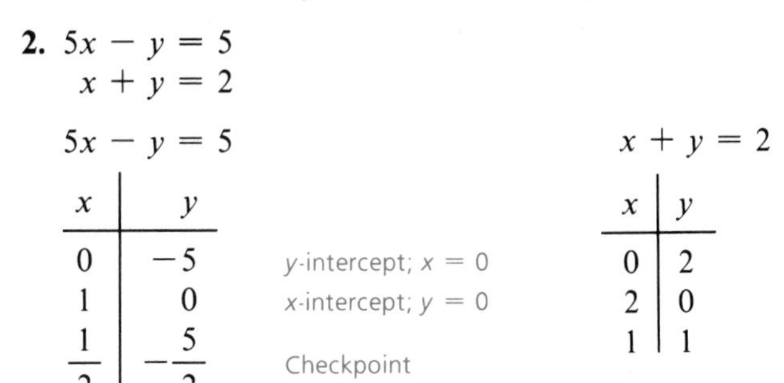

$5x - y = 5$

x	y	
0	-5	y-intercept; $x = 0$
1	0	x-intercept; $y = 0$
$\frac{1}{2}$	$-\frac{5}{2}$	Checkpoint

$x + y = 2$

x	y
0	2
2	0
1	1

An estimate of the point of intersection is (1,1). A check shows this is not a correct solution.

$5x - y = 5$		$x + y = 2$	
$5(1) - (1) = 5$	Replace x with 1 and y with 1	$(1) + (1) = 2$	
$4 = 5$	(False)	$2 = 2$	(True)

We will see in section 8–2 that the actual point of intersection is $\left(\frac{7}{6}, \frac{5}{6}\right)$.

► ***Quick check*** Estimate the solution of the system by graphing.
$3x - y = 6$
$2y + x = 4$ ■

It should be obvious that our graphical method of finding the solution may not be exact because the drawing of the graph depends on measuring devices. Inaccuracies are thereby often produced in the graphing process.

When the graph of a system consists of intersecting lines and thus has a simultaneous solution, the system is said to be **consistent** and **independent.**

Inconsistent and dependent systems

Sometimes the graphs of the two equations in the system do not intersect at all (are parallel lines) or are one and the same line.

■ Example 8–1 C

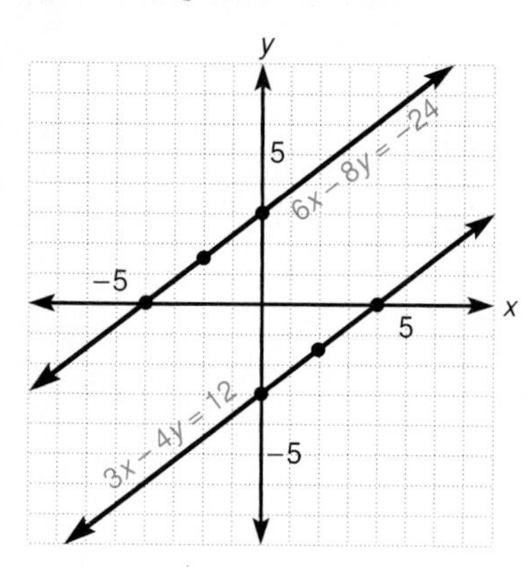

1. $3x - 4y = 12$
   $6x - 8y = -24$

$3x - 4y = 12$

x	y	
0	-3	y-intercept; $x = 0$
4	0	x-intercept; $y = 0$
2	$-\frac{3}{2}$	Checkpoint

$6x - 8y = -24$

x	y
0	3
-4	0
-2	$\frac{3}{2}$

The lines appear to be parallel (they are) and will not intersect. When this condition exists, the system is said to be *inconsistent* and there are no solutions (no points of intersection). We will show later how you can determine by algebra if the lines are or are not parallel.

2. $4x - 2y = 6$
   $2x - y = 3$

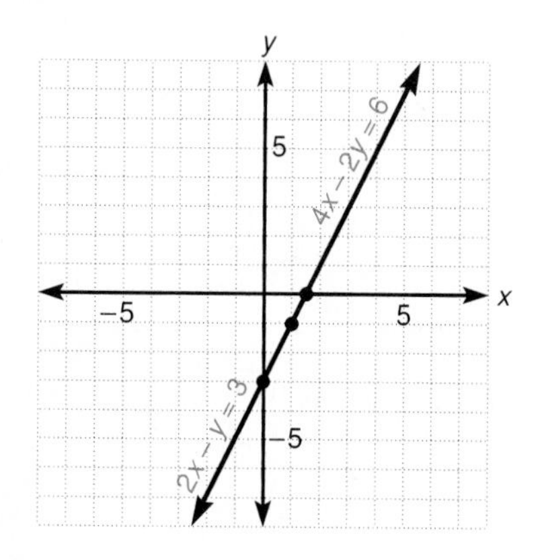

$4x - 2y = 6$

x	y	
0	-3	y-intercept; $x = 0$
$\frac{3}{2}$	0	x-intercept; $y = 0$
1	-1	Checkpoint

$2x - y = 3$

x	y
0	-3
$\frac{3}{2}$	0
1	-1

We observe that the two intercepts and checkpoint $(1, -1)$, are the same for both lines. Therefore, the two lines coincide and the graph is a single line. Such a system is called a **dependent** system of equations. In a dependent system of equations, an unlimited number of solutions are possible because each ordered pair that satisfies the first equation will also satisfy the second. ■

To solve a system of linear equations by graphing

1. Graph each equation on the same set of axes.
2. Their graphs may intersect in
   a. one point—one solution (the system is *consistent* and *independent*).
   b. no points—the lines are parallel and there are no solutions (the system is *inconsistent*).
   c. all points—the lines are one and the same (the system is *dependent*).
3. If the graphs intersect in one point, check the coordinates of the point in each equation.

Mastery points

Can you

- Determine if an ordered pair is a solution of a system of linear equations?
- Solve a system of linear equations by graphing?
- Recognize an inconsistent and a dependent system of linear equations?

Exercise 8–1

Determine if the given ordered pair is a solution to the system of equations. See example 8–1 A.

Example Determine if the ordered pair (1,1) is a simultaneous solution of the system. $3x - y = 2$
$4x + 3y = 9$

Solution Substitute 1 for x and 1 for y in each equation.

$3(1) - (1) = 2$ Replace x with 1 and y with 1
$3 - 1 = 2$
$2 = 2$ (True)

$4(1) + 3(1) = 9$
$4 + 3 = 9$
$7 = 9$ (False)

Since (1,1) satisfies only one equation, it is *not* a solution of the system of equations.

1. $x + y = 2$, $3x - y = 10$ (3,−1)
2. $x + 2y = 4$, $x + 4y = 10$ (−2,3)
3. $x - y = 1$, $x + y = 5$ (3,2)
4. $3x - y = -10$, $3x + y = 4$ (−1,7)
5. $2x + y = 2$, $6x - y = 22$ (3,−4)
6. $x - 3y = -1$, $3x + y = -2$ (−1,0)
7. $3x + 4y = 14$, $2x + y = 6$ (1,4)
8. $3x + y = 12$, $2x + 3y = 15$ (2,6)
9. $x = 6$, $2x - y = 3$ (6,9)
10. $y = 4x - 3$, $y = -1$ (−1,−1)
11. $3x - 2y = 4$, $y = x + 3$ (10,2)
12. $y = 2 - 5x$, $x - y = 6$ (−1,5)

Estimate the solution of each system by graphing. All answers in the back of the book will be given exactly. If the system is inconsistent or dependent, so state. See examples 8–1 B, C.

Example $3x - y = 6$ $2y + x = 4$

Solution $3x - y = 6$

x	y	
0	−6	y-intercept
2	0	x-intercept
1	−3	Checkpoint

$2y + x = 4$

x	y
0	2
4	0
1	$\frac{3}{2}$

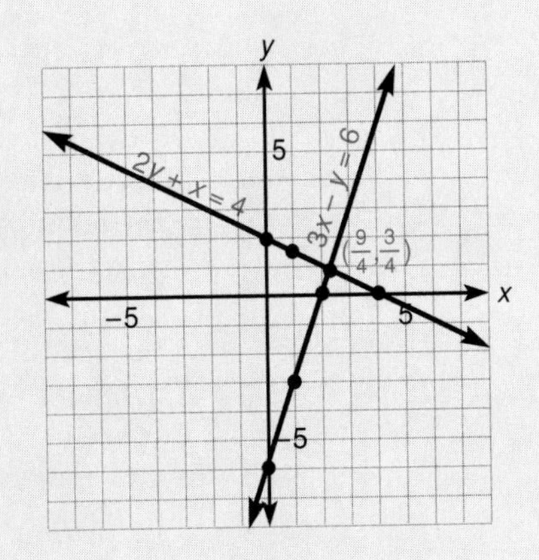

We estimate the solution to be $\left(\frac{9}{4},\frac{3}{4}\right)$.

Note The actual solution is $\left(\frac{16}{7},\frac{6}{7}\right)$.

13. $x - y = 1$
$x + y = 5$

14. $x + 2y = 4$
$x + 4y = 10$

15. $3x - y = 10$
$x + y = 2$

16. $x - y = 3$
$2x + 3y = 11$

17. $2x - y = 3$
$y + 2x = -7$

18. $y = 1$
$3x - 2y = -9$

19. $2x + 3y = 13$
$3x - 2y = 0$

20. $3x + y = 10$
$6x + 2y = 5$

21. $2x + y = 2$
$-6x - 3y = 6$

22. $x = 4$
$3x + 2y = 4$

23. $3x + y = -3$
$x = -1$

24. $y + 2x = 6$
$4y + 3x = 4$

25. $3x + y = 3$
$6x + 2y = 3$

26. $2x - y = 3$
$6x - 3y = 9$

27. $5x + y = -5$
$-10x - 2y = 10$

28. $\frac{3}{2}x - \frac{1}{2}y = -5$
$x + \frac{1}{3}y = \frac{4}{3}$

29. $\frac{1}{3}x + \frac{1}{6}y = 1$
$\frac{1}{4}x + \frac{1}{3}y = 1$

30. $6x - 5y = 14$
$y = 8$

Review exercises

1. Find the solution set of the equation $\frac{3x}{2} - \frac{x}{4} = 1$. See section 6–5.

2. Divide as indicated. $(3x^3 - x^2 + 1) \div (x + 3)$. See section 5–3.

Find the solution of the following inequalities. See section 2–6.

3. $4y - 3 \leq 2y + 7$

4. $-5 < 2x + 1 \leq 7$

Perform the indicated operations. See sections 3–2 and 3–3.

5. $x^{-3} \cdot x^5 \cdot x^2$

6. $x^3(x^2 - 2x + 1)$

7. $(2x - y)^2$

8–2 ■ Solutions of systems of linear equations by elimination

In section 8–1, we found the solution of a system of consistent and independent linear equations by graphing the equations and by observing the coordinates of the point of intersection of the two lines. This technique will often produce only an approximate solution.

An algebraic method for obtaining an *exact* solution to a system involves the operation of *addition,* together with the following property of real numbers.

Addition property of equality

Given a, b, c, and d are real numbers, if $a = b$ and $c = d$,

$$a + c = b + d$$

Concept

Adding equal quantities to each member of an equation will result in equal sums.

This property of real numbers is used to *eliminate* one of the variables in the system. The resulting single equation has only one unknown that we solve to obtain one of the components of the solution. We illustrate this property's use in the following examples.

■ *Example 8–2 A*

Find the solution set of the following systems of linear equations by elimination.

1. $2x - y = 4$
$x + y = 2$
Adding the left members and the right members, we obtain

$$\begin{aligned} 2x - y &= 4 \\ \underline{x + y} &\underline{= 2} \\ 3x \quad &= 6 && \text{Add left and right members} \\ x &= 2 && \text{Divide each member by 3} \end{aligned}$$

We have found the value of x to be 2. To find y, we replace x with 2 in *either* of the equations and solve for y.

$$\begin{aligned} x + y &= 2 && \text{Choose either equation } (x + y = 2) \\ (2) + y &= 2 && \text{Replace } x \text{ with 2} \\ y &= 0 && \text{Subtract 2 from each member} \end{aligned}$$

The solution of our system is (2,0). The solution set is $\{(2,0)\}$.

Check:

$$\begin{aligned} 2x - y &= 4 \\ 2(2) - (0) &= 4 \\ 4 - 0 &= 4 \\ 4 &= 4 \quad \text{(True)} \end{aligned}$$

$$\begin{aligned} x + y &= 2 \\ (2) + (0) &= 2 \\ 2 &= 2 \quad \text{(True)} \end{aligned}$$

2. $x = 4 - 3y$
$2x - 3y = 2$
To solve the system by elimination, like terms must be in a column and the equations must be in standard form.

$$\begin{aligned} x + 3y &= 4 && \text{Add } 3y \text{ to each member of first equation} \\ \underline{2x - 3y} &\underline{= 2} \\ 3x \quad &= 6 && \text{Add left and right members} \\ x &= 2 && \text{Divide each member by 3} \end{aligned}$$

Replace x with 2 in either equation and solve for y.

$$\begin{aligned} x + 3y &= 4 && \text{Choose either equation } (x + 3y = 4) \\ (2) + 3y &= 4 && \text{Replace } x \text{ with 2} \\ 3y &= 2 && \text{Subtract 2 from each member} \\ y &= \frac{2}{3} && \text{Divide each member by 3} \end{aligned}$$

The solution of the system is the ordered pair $\left(2,\frac{2}{3}\right)$. The solution set is $\left\{\left(2,\frac{2}{3}\right)\right\}$. Check the solution in *both original* equations. ■

In the previous examples, we observe that one of the variables was easily eliminated by addition because in the original equations that variable had coefficients that were opposites. When this condition is not present in the equation, we must first use the following procedure to obtain the necessary opposite coefficients.

> Multiply one or both equations by some nonzero constant so that equivalent equations are formed where one of the variables has coefficients that are opposites.

Example 8–2 B

Find the solution set of each system of linear equations by elimination.

1. $3x + 4y = 14$ → $3x + 4y = 14$

$2x + y = 6$ Multiply each member by -4 → $-8x - 4y = -24$

Add the equations: $-5x = -10$

Solve for x: $x = 2$

$3(2) + 4y = 14$ Replace x with 2 in $3x + 4y = 14$

$6 + 4y = 14$ Solve for y

$4y = 8$

$y = 2$

Solution: (2,2). The solution set is $\{(2,2)\}$. Check the solution.

2. $7x + 5y = -9$ Multiply each member by 2 → $14x + 10y = -18$

$3x - 2y = -8$ Multiply each member by 5 → $15x - 10y = -40$

Add the equations: $29x = -58$

Solve for x: $x = -2$

Using $7x + 5y = -9$, we substitute and solve for y.

$7(-2) + 5y = -9$ Replace x with -2 in $7x + 5y = -9$

$-14 + 5y = -9$ Solve for y

$5y = 5$

$y = 1$

Solution: $(-2,1)$. The solution set is $\{(-2,1)\}$. Check the solution.

Note In the previous example, both equations had to be multiplied by a constant to form opposite coefficients of y. We could have eliminated x by multiplying the first equation by 3 and the second equation by -7, forming opposite coefficients of 21 and -21. It is an arbitrary choice to decide which variable to eliminate, but for ease in solving the system, it is best to eliminate the variable whose opposite coefficients will be least in value.

3. $\frac{1}{2}x + \frac{2}{3}y = \frac{1}{6}$

$\frac{3}{4}x - \frac{1}{4}y = \frac{1}{2}$

The first step in this problem will be to clear the fractions.

$\frac{1}{2}x + \frac{2}{3}y = \frac{1}{6}$ Multiply each member by the LCD, 6 $6\left(\frac{1}{2}x + \frac{2}{3}y\right) = 6 \cdot \frac{1}{6}$

$3x + 4y = 1$

$\frac{3}{4}x - \frac{1}{4}y = \frac{1}{2}$ Multiply each member by the LCD, 4 $4\left(\frac{3}{4}x - \frac{1}{4}y\right) = 4 \cdot \frac{1}{2}$

$3x - y = 2$

We now solve the system.

$3x + 4y = 1$
$3x - y = 2$ Multiply each member by $(-1) \rightarrow$

$3x + 4y = 1$
$-3x + y = -2$

Add the equations $5y = -1$

Solve for y $y = -\frac{1}{5}$

$3x - \left(-\frac{1}{5}\right) = 2$ Replace y with $-\frac{1}{5}$ in $3x - y = 2$

$3x + \frac{1}{5} = 2$ Definition of subtraction

$15x + 1 = 10$ Multiply each term by the LCD, 5

$15x = 9$ Subtract 1 from each member

$x = \frac{3}{5}$ Divide each term by 15 and reduce

Solution: $\left(\frac{3}{5}, -\frac{1}{5}\right)$. The solution set is $\left\{\left(\frac{3}{5}, -\frac{1}{5}\right)\right\}$.

4. $4x - 2y = 6$
$2x - y = 3$ Multiply each member by $(-2) \rightarrow$

$4x - 2y = 6$
$-4x + 2y = -6$

Add the equations $0x + 0y = 0$

$0 = 0$ (True)

At this point, both variables have been eliminated, and we have formed the true statement $0 = 0$. This indicates that any ordered pair that satisfies either equation is a solution and there are an unlimited number of solutions. The system is *dependent* and the solution set is all ordered pairs satisfying equation $2x - y = 3$ or $4x - 2y = 6$. The same line is graphed.

5. $3x - 4y = 12$ Multiply each member by $(-2) \rightarrow$ $-6x + 8y = -24$
$6x - 8y = -24$ $6x - 8y = -24$

Add the equations $0x + 0y = -48$

$0 = -48$ (False)

The resulting statement, $0 = -48$, is false, which indicates there is no ordered pair that can satisfy this system. The system is *inconsistent* and the solution set is $\emptyset$ (the null set). Parallel lines are graphed.

▶ ***Quick check*** Find the solution set of the system of equations by elimination.
$5x - 2y = 0$
$y - 3x = 4$

■

To solve a system of linear equations by elimination

1. If needed multiply the members of one or both equations by a constant that will make the coefficients of one of the variables opposites of each other.
2. Add the corresponding members of the equations.
3. Solve this new equation for the remaining variable.
4. Substitute the value found in step 3 into either of the original equations and solve this equation for the other variable. The solution is the ordered pair of numbers obtained in step 3 and this step.
5. If, when adding in step 2,
   a. both variables are eliminated and a *false statement* is obtained, there are *no solutions* and the system is *inconsistent* (parallel lines).
   b. both variables are eliminated and a *true statement* such as $0 = 0$ is obtained, there are infinitely *many* solutions and the system is *dependent* (same line).
6. If the system has a solution found in step 4, check the solution in both equations.

Mastery points

Can you

- Solve a system of linear equations by elimination?
- Recognize a dependent or an inconsistent system of linear equations while solving by elimination?

Exercise 8–2

Find the solution set of the following systems of linear equations by elimination. If a system is inconsistent or dependent, so state. See examples 8–2 A and B.

Example $5x - 2y = 0$
$y - 3x = 4$

Solution Rewrite the second equation to get the system.

$$\begin{aligned} 5x - 2y &= 0 \\ -3x + y &= 4 \end{aligned}$$

Multiply each term by 2 →

$$\begin{aligned} 5x - 2y &= 0 \\ -6x + 2y &= 8 \end{aligned}$$

Add equations

$$-x = 8$$

Multiply by -1

$$x = -8$$

Replace x with -8 in $5x - 2y = 0$

$$5(-8) - 2y = 0$$

Solve for y

$$\begin{aligned} -40 - 2y &= 0 \\ -2y &= 40 \\ y &= -20 \end{aligned}$$

Solution: $(-8,-20)$. The solution set is $\{(-8,-20)\}$.

1. $x - y = 1$
$x + y = 5$

2. $x + y = 2$
$3x - y = 10$

3. $3x - y = 10$
$6x + y = 5$

4. $2x + y = 2$
$-6x - y = 4$

5. $x - 2y = 4$
$-x + 2y = 4$

6. $-x + y = 6$
$x + 4y = 4$

7. $-3x + y = -3$
$3x - 3y = -1$

8. $2x + y = 1$
$-2x + y = 1$

9. $x + 2y = 4$
 $x + 4y = 10$

10. $x - y = 3$
 $2x + 3y = 11$

11. $3x - 2y = -9$
 $2x + y = 1$

12. $5x - y = 4$
 $x + 3y = 2$

13. $4x + y = 11$
 $2x + 3y = 4$

14. $x - 2y = 1$
 $3x - 6y = 3$

15. $8x - 4y = 12$
 $2x - y = 3$

16. $-2x + 3y = 6$
 $4x + y = 1$

17. $10x + y = 5$
 $2x + 5y = -23$

18. $10x - 5y = 7$
 $2x - y = 4$

19. $-6x + 3y = 9$
 $2x - y = -3$

20. $3x - 2y = 6$
 $4x - 3y = 7$

21. $5x + 2y = 3$
 $3x - 3y = 4$

22. $4x + 7y = 11$
 $8x - 3y = -4$

23. $5x - 2y = 4$
 $3x + 3y = 5$

24. $5x - 3y = 34$
 $4x + 5y = 5$

25. $3x - 2y = 19$
 $5x + 3y = 19$

26. $4x + 5y = -2$
 $3x - 2y = -36$

27. $\frac{1}{2}x + \frac{1}{3}y = 1$
 $\frac{2}{3}x - \frac{1}{4}y = \frac{1}{12}$

28. $\frac{2}{3}x - \frac{1}{4}y = 4$
 $\frac{1}{3}x + y = 2$

29. $\frac{1}{2}x + \frac{2}{5}y = \frac{7}{10}$
 $\frac{3}{2}x + \frac{6}{5}y = \frac{3}{10}$

30. $\frac{6}{7}x - \frac{3}{5}y = \frac{9}{10}$
 $\frac{2}{7}x - \frac{1}{5}y = \frac{5}{10}$

31. $\frac{9}{2}x + \frac{1}{2}y = -10$
 $\frac{1}{3}x + \frac{1}{4}y = \frac{3}{4}$

32. $\frac{7}{2}x - y = -\frac{17}{2}$
 $\frac{2}{3}x + \frac{1}{3}y = 1$

33. $(0.4)x - (0.7)y = 0.7$
 $(0.6)x - (0.5)y = 2.7$

34. $x + (0.4)y = 3.4$
 $(0.6)x - (1.4)y = 0.4$

35. Let x represent the width of a rectangle and y represent the length of the rectangle. Write an equation stating that the length is five times the width.

36. The perimeter of the rectangle in exercise 35 is 60 inches. If the perimeter of a rectangle equals twice the width plus twice the length, write an equation for the perimeter of the rectangle.

37. Let x represent the speed of automobile A and y represent the speed of automobile B. Write an equation stating that automobile B travels 20 mph faster than automobile A.

38. If the two automobiles in exercise 37 travel in opposite directions, write an equation that states the two automobiles are 500 miles apart after 3 hours. (*Note:* distance = speed times the time traveled)

39. Let x represent the amount of money Jane invests at 8% and y represent the amount she invests at 7%. Write an equation stating that Jane invests a total of $14,000.

40. In exercise 39, Jane receives a total of $1,000 from her investments. If income equals the amount invested times the rate of interest, write an equation relating Jane's total income to her two investments.

Review exercises

1. Given $2x + y = 3$, solve for x when $y = x + 2$. See sections 1–8 and 2–6.

Find the equation in standard form of each line having the given properties. See section 7–4.

2. Through $(-1,6)$, slope $\frac{2}{3}$

3. Through the points $(0,5)$ and $(-7,1)$

Factor the following expressions. See sections 4–3 and 4–4.

4. $16x^2 - 4y^2$

5. $9x^2 - 12x + 4$

6. $5y^2 - 6y - 8$

7. Find the solution set of the equation $4(3x - 2) + 2x = 6$. See section 2–6.

8–3 ■ Solutions of systems of linear equations by substitution

A basic property of equality is that if two expressions are equal (represent the same quantity), one expression may replace the other expression in any equation to form an equivalent equation. We use this property to solve systems of equations by a third method, called the **substitution** method. This method is useful when

1. one of the equations is solved for one variable in terms of the other; for example, $y = 3x - 4$, or
2. one of the equations can *easily* be solved for one variable in terms of the other (usually when the coefficient of the variable is 1 or -1), for example, $x - 2y = 3$.

To illustrate, consider the system of linear equations

$$\begin{aligned} 3x + 2y &= 9 \\ y &= x + 2 \end{aligned}$$

Since the second equation is already solved for y in terms of x, we *substitute* $x + 2$ for y in the first equation.

$$\begin{aligned} 3x + 2y &= 9 \\ 3x + 2(x + 2) &= 9 && \text{Replace } y \text{ with } x + 2 \\ 3x + 2x + 4 &= 9 && \text{Solve for } x \\ 5x + 4 &= 9 \\ 5x &= 5 \\ x &= 1 \end{aligned}$$

Now substitute 1 for x in the second equation.

$$\begin{aligned} y &= x + 2 \\ y &= (1) + 2 && \text{Replace } x \text{ with } 1 \\ y &= 3 \end{aligned}$$

The solution is $(1,3)$. The solution set is $\{(1,3)\}$.

Note We can check our solution by substituting 1 for x and 3 for y in *both* original equations.

$$\begin{aligned} 3x + 2y &= 9 \\ 3(1) + 2(3) &= 9 \\ 3 + 6 &= 9 \\ 9 &= 9 \end{aligned} \qquad \begin{aligned} y &= x + 2 \\ (3) &= (1) + 2 \\ 3 &= 3 \end{aligned}$$

■ Example 8–3 A

Find the solution set of the following systems of linear equations by substitution.

1. $3x + 2y = 14$
$y = 2x - 7$

$$\begin{aligned} 3x + 2y &= 14 \\ 3x + 2(2x - 7) &= 14 && \text{Replace } y \text{ with } 2x - 7 \\ 3x + 4x - 14 &= 14 && \text{Multiply as indicated} \\ 7x - 14 &= 14 && \text{Combine like terms} \\ 7x &= 28 && \text{Add 14 to each member} \\ x &= 4 && \text{Divide each member by 7} \end{aligned}$$

Using the equation

$y = 2x - 7$
$y = 2(4) - 7$ — Replace x with 4
$y = 8 - 7$
$y = 1$

The solution of the given system is (4,1). Check the solution. The solution set is $\{(4,1)\}$.

2. $2x - 4y = 7$
$x = 3y - 2$

$2x - 4y = 7$
$2(3y - 2) - 4y = 7$ — Replace x with $3y - 2$
$6y - 4 - 4y = 7$ — Multiply as indicated
$2y = 11$ — Combine like terms; add 4 to each member
$y = \frac{11}{2}$ — Divide each member by 2

Using the equation $x = 3y - 2$,

$x = 3\left(\frac{11}{2}\right) - 2$ — Replace y with $\frac{11}{2}$
$x = \frac{33}{2} - 2$
$x = \frac{29}{2}$

The system has the solution $\left(\frac{29}{2},\frac{11}{2}\right)$. Check the solution. The solution set is $\left\{\left(\frac{29}{2},\frac{11}{2}\right)\right\}$.

3. $3x + y = 12$
$2x + 5y = 8$
To use substitution, we must first solve the first equation for y in terms of x.

$3x + y = 12$
$y = 12 - 3x$ — Subtract $3x$ from each member

Using the equation $2x + 5y = 8$,

$2x + 5(12 - 3x) = 8$ — Replace y with $12 - 3x$
$2x + 60 - 15x = 8$ — Solve for x
$60 - 13x = 8$
$-13x = -52$
$x = 4$

Using the equation $y = 12 - 3x$,

$y = 12 - 3x$
$y = 12 - 3(4)$ — Replace x with 4
$y = 12 - 12$ — Multiply in right member
$y = 0$

The solution is the ordered pair (4,0). Check the solution. The solution set is $\{(4,0)\}$.

4. $8x + 2y = 6$
$y = -4x + 3$
Using the equation $8x + 2y = 6$,

$$\begin{aligned} 8x + 2(-4x + 3) &= 6 \quad \text{Replace } y \text{ with } -4x + 3 \\ 8x - 8x + 6 &= 6 \\ 6 &= 6 \quad \text{(True)} \end{aligned}$$

Since $6 = 6$ is a true statement, the system is dependent and there are infinitely many solutions that satisfy the equations $8x + 2y = 6$ and $y = -4x + 3$.

5. $-4x + 2y = 7$
$y = 2x + 9$
Since we have $y = 2x + 9$, we substitute $2x + 9$ for y in the equation $-4x + 2y = 7$.

$$\begin{aligned} -4x + 2y &= 7 \\ -4x + 2(2x + 9) &= 7 \quad \text{Replace } y \text{ with } 2x + 9 \\ -4x + 4x + 18 &= 7 \\ 18 &= 7 \quad \text{(False)} \end{aligned}$$

Since $18 = 7$ is a *false* statement, there are no solutions, and the system is inconsistent. The solution set is $\emptyset$ (the null set). Lines are parallel.

▶ ***Quick check*** Find the solution set of the system of linear equations using substitution. $3x - y = 7$
$4x - 5y = 2$ ■

To solve a system of linear equations by substitution

1. Solve one of the equations for one of the variables in terms of the other (if this is not already done).
2. Substitute the expression obtained in step 1 into the other equation and solve.
3. Substitute the value obtained in step 2 into either equation and solve for the other variable.
4. The solution is the ordered pair obtained from steps 2 and 3.
5. If step 2 results in
   a. the variables being eliminated and a true statement is obtained, the system is dependent and there are infinitely many solutions (same line).
   b. the variables being eliminated and a false statement is obtained, the system is inconsistent and there are no simultaneous solutions (parallel lines).

As a final note, when the method of solution for a system of linear equations is not specified, the following are good criteria by which a method is selected.

1. The process of elimination is generally the easiest.
2. Substitution is most useful when the coefficient of one of the variables is 1 or -1 or an equation is solved for one variable in terms of the other variable.

3. Graphical solutions are approximations and can be used when exact answers are not necessary or to verify the results that we find algebraically.

Mastery points

Can you

- Solve systems of linear equations by substitution?
- Choose an appropriate method for solving a system of linear equations?

Exercise 8–3

Find the solution set of the following systems of linear equations by *substitution*. If a system is inconsistent or dependent, so state. See example 8–3 A.

Example $3x - y = 7$
$4x - 5y = 2$

Solution Solve $3x - y = 7$ for y. Then $y = 3x - 7$.

$$4x - 5y = 2$$
$$4x - 5(3x - 7) = 2 \quad \text{Replace } y \text{ with } 3x - 7$$
$$4x - 15x + 35 = 2 \quad \text{Solve for } x$$
$$-11x + 35 = 2$$
$$-11x = -33$$
$$x = 3$$

Using the equation

$$3x - y = 7$$
$$3(3) - y = 7 \quad \text{Replace } x \text{ with } 3$$
$$9 - y = 7$$
$$-y = -2$$
$$y = 2$$

The solution is (3,2) and the solution set is $\{(3,2)\}$.

1. $3x - y = 10$, $y = -x + 2$
2. $2x + 3y = 9$, $x = y - 3$
3. $-2x + 5y = 17$, $y = 2x + 5$
4. $5x + y = 10$, $x = 2 - 3y$
5. $2y - x = 3$, $x = 4y - 1$
6. $3x + 4y = -1$, $y = 2 - 3x$
7. $4y - x = 3$, $x = 2y + 1$
8. $3y + 4x = -4$, $x = 4 - y$
9. $5x - 2y = 3$, $y = x - 4$
10. $3x + y = 10$, $2x + y = 5$
11. $2x + y = 13$, $3x + y = 17$
12. $3y + x = -10$, $3y - x = 4$
13. $2x - y = 2$, $6x - y = 22$
14. $x - y = 3$, $2x + 3y = 11$
15. $2x + y = 3$, $3x - y = 4$
16. $x + 5y = 7$, $2x + 3y = 5$
17. $5x - 3y = 4$, $x + 2y = -2$
18. $x + y = 4$, $3x - 5y = 7$
19. $x - y = 3$, $2x - 2y = 11$
20. $x + y = 5$, $3x + 3y = 3$
21. $3x - y = 7$, $6x - 2y = 14$
22. $-x + 3y = -3$, $9y - 3x = -9$
23. $3x + y = -3$, $x - 3y = -1$
24. $2x + y = 1$, $-3x + y = 1$

25. $2x + y = -3$
$5x - 4y = 1$

26. $5x + 2y = 11$
$7x - y = 4$

27. $2x + 3y = 2$
$6x - y = 5$

28. $x - 2y = 4$
$3x + 2y = 4$

29. $2x + y = 6$
$3x + 4y = 4$

30. $-x - 4y = 3$
$2x + 8y = -6$

31. $x + 4y = -3$
$-2x - 8y = 6$

35. $-3x - 2y = 6$
$6x + 4y = 5$

32. $x - y = 7$
$3x + 3y = 4$

33. $5x - 3y = 11$
$x + y = 5$

34. $-x - 2y = 6$
$6x + 4y = 5$

36. $4x + 3y = 1$
$y = -4$

37. $3x - 5y = -6$
$x = 5$

38. $-5x + 2y = 11$
$x = -3$

39. $3x - 7y = 14$
$y = 2$

40. $y = 2x + 1$
$y = -x + 3$

41. $y = 5x + 3$
$y = 2x - 4$

42. $y = 3x - 4$
$y = 3x + 5$

Find the solution set of each system of linear equations by either elimination or substitution. Try to choose the most suitable method.

43. $3x - 2y = -1$
$2x + 2y = 1$

44. $x - y = 2$
$-5x + 2y = -2$

45. $x - 3 = 0$
$3x + 2y = 1$

46. $y + 2x = 4$
$x = -6y + 1$

47. $\frac{1}{2}x - y = 3$
$2x + \frac{1}{3}y = 1$

48. $\frac{2x}{3} - \frac{y}{2} = 1$
$3x + \frac{y}{4} = 2$

49. $\frac{y}{3} - \frac{x}{4} = 3$
$\frac{x}{2} + \frac{y}{5} = -2$

50. $4x - 3y = -10$
$y + 2 = 0$

51. $4x - y = 8$
$5x + 3y = -4$

52. $7x - 2y = 3$
$4x + 3y = -2$

53. Let x represent one current in an electrical circuit and y represent a second current. If the first current has twice as many amperes as the second, write an equation stating this.

54. If the sum of the two currents in exercise 53 is 56 amperes, write an equation stating this.

55. A clothier has two kinds of suits. Let x represent the number of suits selling for one price and y represent the number of suits selling for another price. If he has a total of 80 suits, write an equation stating this.

56. In exercise 55, if the cost of the first kind of suit (x) is \$190 per suit and the cost per suit of the second kind (y) is \$250, write an equation stating that the total income from the suits was \$8,400.

Review exercises

1. A 42-foot piece of wood is cut into two pieces. If one piece is 6 feet less than twice the length of the other piece, how long are the pieces of wood? See section 2–8.

2. The length of a rectangle is 1 yard longer than twice its width. If the area of the rectangle is 36 square yards, find the length and width of the rectangle. See section 4–8. (*Hint:* Area = length × width.)

3. Solve the equation $A = \frac{1}{2}h(b + c)$ for b. See section 2–7.

4. Find the slope of the line through the points $(-4,4)$ and $(5,6)$. See section 7–3.

5. Solve the rational equation $\frac{4}{x} - \frac{3}{2x} = 4$. See section 6–5.

6. Solve the inequality $-4 \leq 1 - 5x < 6$. See section 2–6.

8–4 ■ Applications of systems of linear equations

In chapter 2, we learned how to take a verbal statement and translate it into an algebraic equation in one unknown. Many practical problems that can be solved using single equations in one unknown can more easily be solved by translating into *two* equations involving *two* unknowns.

While there is no standard procedure *for solving applied problems,* the following guidelines should be useful.

Solve a word problem using systems of linear equations

1. Read the problem carefully, noting the information given and what you are asked to find.
2. Determine any prior knowledge that may be useful in setting up the equations: formulas, distance-rate-time, interest, and so on. Draw a sketch, if appropriate.
3. Choose two variables to represent the unknown quantities.
4. Translate separate statements from the verbal statement of the problem into individual equations.
5. Solve the resulting system of linear equations by one of the methods we have studied.
6. Write the answer—usually a sentence containing the numbers from the simultaneous solution—using the correct units of measure. Check your results in the original statement of the problem.

■ **Example 8–4 A**

1. The length of a rectangle is 3 times as long as its width. The perimeter of the rectangle is 40 inches. What are the dimensions of the rectangle?

Note We need the prior knowledge that the perimeter, P, of a rectangle is given by the formula $P = 2\ell + 2w$, where ℓ is the length and w is the width of the rectangle.

Let ℓ represent the length of the rectangle and w represent the width of the rectangle.

length	is	3 times the width
ℓ	$=$	$3w$

w

$\ell = 3w$

We use the formula for the perimeter, $P = 2\ell + 2w$, and since the perimeter is 40 inches, we have $40 = 2\ell + 2w$. Therefore, we have the system of linear equations

$$\ell = 3w$$
$$2\ell + 2w = 40$$

Using substitution and the equation $2\ell + 2w = 40$,

$$2(3w) + 2w = 40 \quad \text{Replace } \ell \text{ with } 3w$$
$$8w = 40 \quad \text{Solve for } w$$
$$w = 5$$

Substituting 5 for w in the equation $\ell = 3w$, we get

$\ell = 3(5) = 15$

The rectangle is 15 inches long and 5 inches wide.

Check:

a. The perimeter is 40 inches.

$$2(15) + 2(5) = 40$$
$$30 + 10 = 40$$
$$40 = 40 \quad \text{(True)}$$

b. The length is 3 times the width.

$$15 = 3(5)$$
$$15 = 15 \quad \text{(True)}$$

2. A woman has \$10,000, part of which she invests at 11% and the rest at 8% annual interest. If her total income for one year from the two investments is \$980, how much does she invest at each rate?

Note The prior knowledge needed for this problem is that Simple interest = Principal × Rate × Time. Time in this example is 1 year. When using this formula, time should be stated in years.

Let x = amount invested at 11% and y = amount invested at 8%.
Then, $(0.11)x$ = interest from the 11% investment for one year
$(0.08)y$ = interest from the 8% investment for one year

11% investment	plus	8% investment	is	\$980
$(0.11)x$	$+$	$(0.08)y$	$=$	980

From "A woman has \$10,000 . . . ,"

$x + y = 10{,}000$

The system of linear equations is

$$x + y = 10{,}000$$
$$(0.11)x + (0.08)y = 980$$

Multiply the second equation by 100 to clear the decimal fractions.

$$100(0.11)x + 100(0.08)y = 100(980)$$
$$11x + 8y = 98{,}000$$

We then have the system

$x + y = 10{,}000$ Multiply each term by (-8) → $-8x - 8y = -80{,}000$

$11x + 8y = 98{,}000$ → $11x + 8y = 98{,}000$

$3x = 18{,}000$ Add

$x = 6{,}000$ Divide by 3

Using $x + y = 10{,}000$,

$(6{,}000) + y = 10{,}000$ Replace x with 6,000

$y = 4{,}000$

The woman invested \$6,000 at 11% and \$4,000 at 8% interest.

Check:

a. The total yearly income is \$980.

$$\begin{aligned}(0.11)(6{,}000) + (0.08)(4{,}000) &= 980\\ 660 + 320 &= 980\\ 980 &= 980 \quad \text{(True)}\end{aligned}$$

b. The woman has \$10,000.

$$\begin{aligned}6{,}000 + 4{,}000 &= 10{,}000\\ 10{,}000 &= 10{,}000 \quad \text{(True)}\end{aligned}$$

3. The sum of two lengths of wire is 11 meters. If four times the first length is added to the second length, the result is 25 meters. Find the two lengths of wire.

Let x = the first length of wire and y = the second length of wire.

the sum of the two lengths	is	11 meters
$x + y$	$=$	11

four times the first length	added to	the second length	is	25 meters
$4x$	$+$	y	$=$	25

Therefore, we have the system of linear equations

$$\begin{aligned}x + y &= 11 \quad \text{Multiply each term by } (-1) \quad & -x - y &= -11\\ 4x + y &= 25 & 4x + y &= 25\\ & & 3x &= 14 \quad \text{Add}\\ & & x &= \frac{14}{3} = 4\frac{2}{3}\end{aligned}$$

Using the equation $x + y = 11$, substitute $\frac{14}{3}$ for x and solve for y.

$$\left(\frac{14}{3}\right) + y = 11 \quad \text{Replace } x \text{ with } \frac{14}{3}$$

$$y = 11 - \frac{14}{3} \quad \text{Subtract } \frac{14}{3} \text{ from each member}$$

$$y = \frac{33}{3} - \frac{14}{3} = \frac{19}{3} \quad \text{Change to common denominator 3 and subtract}$$

$$y = 6\frac{1}{3}$$

Thus the two wires have length $4\frac{2}{3}$ meters and $6\frac{1}{3}$ meters.

Check:

a. The sum of the two lengths is 11.

$$\begin{aligned}\frac{14}{3} + \frac{19}{3} &= 11\\ \frac{33}{3} &= 11\\ 11 &= 11 \quad \text{(True)}\end{aligned}$$

b. Four times the first length added to the second length is 25.

$$4\left(\frac{14}{3}\right) + \frac{19}{3} = 25$$
$$\frac{56}{3} + \frac{19}{3} = 25$$
$$\frac{75}{3} = 25$$
$$25 = 25 \quad \text{(True)}$$

4. What quantities of silver that are 65% pure and 45% pure must be mixed together to get 100 grams of silver that is 50% pure?

Note 65% pure silver means the metal is 65% silver and 35% of some other metal(s).

Let x = the number of grams of 65% pure silver, and y = the number of grams of 45% pure silver.

We can use the following table to summarize the given information.

Solution	Amount of solution	Percent of silver	Amount of silver in solution
First	x	65%	$0.65x$
Second	y	45%	$0.45y$
Final	100	50%	$0.50(100) = 50$

$$x + y = 100 \qquad\qquad 0.65x + 0.45y = 50$$

We must now solve the system of linear equations.

$$x + y = 100$$
$$0.65x + 0.45y = 50$$

Multiply the second equation by 100 to clear decimals.

$$-65x - 65y = -6{,}500 \quad \text{Multiply } x + y = 100 \text{ by } -65$$
$$65x + 45y = 5{,}000$$
$$-20y = -1{,}500$$
$$y = 75$$

Using the equation,

$$x + y = 100$$
$$x + (75) = 100 \quad \text{Replace } y \text{ with } 75$$
$$x = 25$$

Then 25 grams of 65% pure silver must be mixed with 75 grams of 45% pure silver to get 100 grams of 50% pure silver.

Check: 1. $25 + 75 = 100$ Total of 100 grams

2. $0.65(25) + 0.45(75) = 50$ Sum of mixes yields 50 grams

$$16.25 + 33.75 = 50$$
$$50 = 50$$

5. Two automobiles start at town A and go in opposite directions. After traveling for three hours, they are 351 miles apart. If one automobile is averaging 13 miles per hour faster than the other, what is the average speed of each automobile in miles per hour?

Use the formula that relates distance, rate, and time, $d = rt$. Time $t = 3$ since each automobile travels 3 hours.

Let $x =$ the average speed of the slower automobile and $y =$ the average speed of the faster automobile.

We can use the following table to organize the information.

	r	*t*	*d*
Slower automobile	x	3	$3x$
Faster automobile	y	3	$3y$

↓

$3x + 3y = 351$

From "If one automobile is averaging 13 miles per hour faster than the other," we get the equation

$y = x + 13$

Thus we determine the system to be

$$\begin{aligned} y &= x + 13 \\ 3x + 3y &= 351 \end{aligned}$$

Substitute $x + 13$ for y in the second equation.

$$\begin{aligned} 3x + 3(x + 13) &= 351 && \text{Replace } y \text{ with } x + 13 \\ 3x + 3x + 39 &= 351 \\ 6x + 39 &= 351 \\ 6x &= 312 \\ x &= 52 \end{aligned}$$

Replace x by 52 in the equation.

$$\begin{aligned} y &= x + 13 \\ y &= (52) + 13 = 65 && \text{Replace } x \text{ with } 52 \end{aligned}$$

The faster automobile is traveling at 65 mph and the slower automobile is traveling at 52 mph.

Check:

$$\begin{aligned} 3x + 3y &= 351 \\ 3(52) + 3(65) &= 351 \\ 156 + 195 &= 351 \\ 351 &= 351 \end{aligned} \qquad \begin{aligned} y &= x + 13 \\ (65) &= (52) + 13 \\ 65 &= 65 \end{aligned}$$

▶ ***Quick check*** Karen and her mother are in cities 400 miles apart. They meet at a location on a line between the two cities. If Karen drives 20 miles per hour faster than her mother, and they meet after 4 hours, how fast was each person driving? ■

Mastery points

Can you

- Set up and solve a system of linear equations in two variables given a word problem?

Exercise 8–4

Solve the following problems by setting up a system of two linear equations with two unknowns.

Example Karen and her mother are in cities 400 miles apart. They meet at a location on a line between the two cities. If Karen drives 20 miles per hour faster than her mother, and they meet after 4 hours, how fast was each person driving?

Solution Let x = the speed of Karen's automobile and y = the speed of her mother's automobile.

We use the formula distance (d) = rate (r) · time (t). Time t for each automobile is 4 since each drives 4 hours. The following chart compiles this information using $r \cdot t = d$.

	rate (r)	time (t)	distance (d)
Karen	x	4	$4x$
Mother	y	4	$4y$

Then, since they are 400 miles apart,

$$\begin{aligned} \text{total distance} &= 400 \text{ miles} \\ 4x + 4y &= 400 \\ x + y &= 100 \quad \text{Divide each term by 4} \end{aligned}$$

Since Karen drives 20 miles per hour faster than her mother,

$$x = y + 20$$

We solve the system of linear equations.

$$\begin{aligned} x + y &= 100 \\ x &= y + 20 \end{aligned}$$

Using substitution,

$$\begin{aligned} (y + 20) + y &= 100 \quad \text{Replace } x \text{ with } y + 20 \text{ in } x + y = 100 \\ 2y + 20 &= 100 \\ 2y &= 80 \\ y &= 40 \end{aligned}$$

$$\begin{aligned} \text{Since } x &= y + 20 \\ &= (40) + 20 \quad \text{Replace } y \text{ with } 40 \\ &= 60 \end{aligned}$$

Karen was driving 60 miles per hour and her mother was driving 40 miles per hour.

See example 8–4 A–1.

1. The perimeter of a rectangle is 36 meters and the length is 2 meters more than three times the width. Find the dimensions.

2. The perimeter of a room is 40 feet, and twice the length increased by three times the width is 48 feet. What are the dimensions?

3. The perimeter of a rectangle is 100 meters. The rectangle is 22 meters longer than it is wide. Find the dimensions.

4. A rectangular field has a length that is 21 yards more than twice the width. The perimeter is 620 yards. What are the dimensions of the field?

5. The length of a room is 2 feet less than twice its width. If its perimeter is 62 feet, what are the dimensions?

See example 8–4 A–2.

6. Phil has $20,000, part of which he invests at 8% interest and the rest at 6%. If his total income for one year from the two investments was $1,460, how much did he invest at each rate?

7. Nancy has $18,000. She invests part of her money at $7\frac{1}{2}$% interest and the rest at 9%. If her yearly income from the two investments was $1,560, how much did she invest at each rate?

8. Rich has $18,000, part of which he invests at 10% interest and the rest at 8%. If his yearly income from each investment was the same, what did he invest at each rate?

9. Dick has $30,000, part of which he invests at 9% interest and the rest at 7%. If his yearly income from the 7% investment was $820 more than that from the 9% investment, how much was invested at each rate?

10. Lynne made two investments totaling $25,000. On one investment she made an 18% profit, but on the other investment she took an 11% loss. If her yearly net gain was $2,180, how much was in each investment?

11. Anne made two investments totaling $21,000. One investment made her a 13% profit, but on the other investment she took a 9% loss. If her net yearly loss was $196, how much was in each investment?

See example 8–4 A–3.

12. A piece of pipe is 19 feet long. The pipe must be cut so that one piece is 5 feet longer than the other piece. What are the lengths of the two pieces of pipe?

13. A 12-foot board is cut into two pieces so that one piece is 4 feet longer than the other. How long is each piece?

14. A 24-foot rope is cut into two pieces so that one piece is twice as long as the other. How long is each piece?

15. The sum of two currents is 80 amperes. If the greater current is 24 amperes more than the lesser current, find their values.

16. A 50-foot extension cord is cut into two pieces so that one piece is 12 feet longer than the other piece. How long is each piece?

17. The sum of the number of teeth on two gears is 64 and their difference is 12. How many teeth are on each gear?

18. Two gears have a total of 83 teeth. One gear has 15 less teeth than the other. How many teeth are on each gear?

19. Two electrical voltages have a total of 126 volts. If one voltage is 32 volts more than the other, find the voltages.

20. The sum of two voltages is 85 and their difference is 32. Find the voltages.

21. The sum of the resistances of two resistors is 24 ohms and their difference is 14 ohms. How many ohms are in each resistor?

22. The resistance of one resistor exceeds that of another resistor by 25 ohms and their sum is 67 ohms. How many ohms are in each resistor?

23. A clothing store sells suits at $125 and $185 each. The store owners observe that they sold 40 suits for a total of $5,720. How many suits of each type did they sell?

24. Three times the number of bolts in a piece of machinery is 3 less than twice the number of spot welds, while seven times the number of bolts is 5 more than four times the number of spot welds. How many bolts and spot welds are there?

25. When two batteries are connected in series, we add the voltages together to get a total voltage of 27 V. When the batteries are connected in opposition, the resulting voltage, 5 V, is their difference. Find the voltages of the two batteries.

26. When two batteries are connected in series, the total voltage is 53 V. When the batteries are connected in opposition, the total voltage is 7 V. Find the voltage of the two batteries. (Refer to exercise 25.)

27. Three times the number of teeth on a first gear is 1 more than the number of teeth on a second gear. Also, five times the number of teeth on the first gear is 4 less than twice the number of teeth on the second gear. Find the number of teeth on each gear.

28. The specific gravity of an object is defined as its weight in air divided by the difference between its weight in air and its weight when submerged in water. If an object has a specific gravity of 3 and the sum of its weight in air and in water is 30 pounds, find its weight in air and its weight in water.

See example 8–4 A–4.

29. An auto mechanic has two bottles of battery acid. One contains a 10% solution and the other a 4% solution. How many cubic centimeters (cc) of each solution must be used to make 120 cubic centimeters of a solution that is 6% acid?

30. A metallurgist wishes to form 2,000 kilograms (kg) of an alloy that is 80% copper. This alloy is to be obtained by fusing some alloy that is 68% copper and some that is 83% copper. How many kilograms of each alloy must be used?

31. If a jeweler wishes to form 12 ounces of 75% pure gold from sources that are 60% and 80% pure gold, how much of each substance must be mixed together to produce this?

32. A chemist wishes to make 1,000 liters of a 3.5% acid solution by mixing a 2.5% solution with a 4% solution. How many liters of each solution are necessary?

33. A drum contains a mixture of antifreeze and water. If 6 liters of antifreeze are added, the mixture will be 90% antifreeze, but if 6 liters of water are added, the mixture will be 70% antifreeze. What is the percentage of antifreeze in the mixture presently in the drum?

34. A pharmacist is able to fill 200 3-grain and 2-grain capsules using 500 grains of a certain drug. How many capsules of each kind does he fill?

35. A solution that is 38% silver nitrate is to be mixed with a solution that is 3% silver nitrate to obtain 100 centiliters (cl) of solution that is 5% silver nitrate. How many centiliters of each solution should be used in the mixture?

36. A druggist has two solutions, one 60% hydrogen peroxide and the other 30% hydrogen peroxide. How many liters of each should be mixed to obtain 30 liters of a solution that is 40% hydrogen peroxide?

See example 8–4 A–5.

37. Two cars are 100 miles apart. If they drive toward each other they will meet in 1 hour. If they drive in the same direction they will meet in 2 hours. Find their speeds.

38. A cyclist and a pedestrian are 20 miles apart. If they travel toward each other, they will meet in 75 minutes, but if they travel in the same direction, the cyclist will overtake the pedestrian in 150 minutes. What are their speeds?

39. Jane and Jim leave from a drugstore at the same time, walking in opposite directions. After 1 hour they are 9,680 yards apart. If Jim walked at a rate of 0.5 mph faster than Jane, how fast was each walking in miles per hour?
(*Hint:* 1 mile = 1,760 yards.)

40. A boat can travel 24 miles downstream in 2 hours and 16 miles upstream in the same amount of time. What is the speed of the boat in still water and what is the speed of the current? (*Hint:* Add the speed of the current when going downstream and subtract the speed of the current when going upstream.)

41. An airplane can fly at 460 miles per hour with the wind and 322 miles per hour against the wind. What is the speed of the wind and what is the speed of the airplane in still air?

42. If a boat takes $1\frac{1}{2}$ hours to go 12 miles downstream and 6 hours to return, what is the speed of the current and of the boat in still water?

Example A line whose equation is of the form $y = mx + b$ passes through a point (x,y) if and only if the coordinates of the point satisfy the equation. Find the values of m and b so that the line contains the points $(-1,-7)$ and $(2,5)$. Write the equation of the line.

Solution The ordered pairs $(-1,-7)$ and $(2,5)$ are solutions of the equation. We then substitute these values for x and y into $y = mx + b$ to obtain two equations in the variables m and b.

1. When $(x,y) = (-1,-7)$

$$y = mx + b$$

$$(-7) = m(-1) + b \quad \text{Replace } y \text{ with } -7 \text{ and } x \text{ with } -1$$

$$-7 = -m + b$$

$$-m + b = -7$$

$$m - b = 7 \quad \text{Write in standard form}$$

2. When $(x,y) = (2,5)$

$$y = mx + b$$

$$(5) = m(2) + b \quad \text{Replace } y \text{ with 5 and } x \text{ with 2}$$

$$5 = 2m + b$$

$$2m + b = 5 \quad \text{Write in standard form}$$

Then we have the system of linear equations.

$$m - b = 7$$

$$\underline{2m + b = 5}$$

$$3m = 12 \quad \text{Add the equations}$$

$$m = 4$$

Substitute 4 for m in $2m + b = 5$.

$$2(4) + b = 5 \quad \text{Replace } m \text{ with 4}$$

$$8 + b = 5$$

$$b = -3$$

Replace b with -3 and m with 4 in $y = mx + b$ and the equation of the line containing the two points is $y = 4x - 3$.

43. The line with equation $y = mx + b$ passes through a point (x,y) if and only if the coordinates of the point satisfy the equation. Find the values of m and b so that the line will contain the points $(1,-1)$ and $(2,2)$. Write the equation of the line.

44. Find the values of m and b for the line that passes through the points $(2,1)$ and $(-1,7)$. Write the equation of the line. (Refer to exercise 43.)

45. Find the values of m and b for the line that passes through the points $(-2,3)$ and $(3,-7)$. Write the equation of the line. (Refer to exercise 43.)

46. Find the values of m and b for the line that passes through the points $(-1,7)$ and $(2,-2)$. Write the equation of the line. (Refer to exercise 43.)

47. Find the values of m and b and write the equation of the line that passes through $(0,-4)$ and $(5,0)$. (Refer to exercise 43.)

48. Find the values of m and b and write the equation of the line that passes through $(0,0)$ and $(-4,5)$. (Refer to exercise 43.)

Review exercises

Perform the indicated operations. Express your answer with positive exponents. Assume all variables are nonzero. See sections 3–3 and 3–4.

1. $(-4x^2y^3)$

2. $\dfrac{x^{-2}y^3}{xy^2}$

3. $4^{-1} + 3^{-1}$

See sections 2–3 and 3–2.

4. $(2x^2 - x + 3) - (x^2 + 4x - 1)$

5. $4x^2(3x^2 + 2x - 3)$

6. $(3y + 2)(3y - 2)$

7. $(4x - 5y)^2$

8. Find the solution set of the equation $5(x - 2) = -3(2 - 3x)$. See section 2–6.

9. Find the solution set of the quadratic equation $x^2 - 3x - 10 = 0$. See section 4–7.

8–5 ■ Solving systems of linear inequalities by graphing

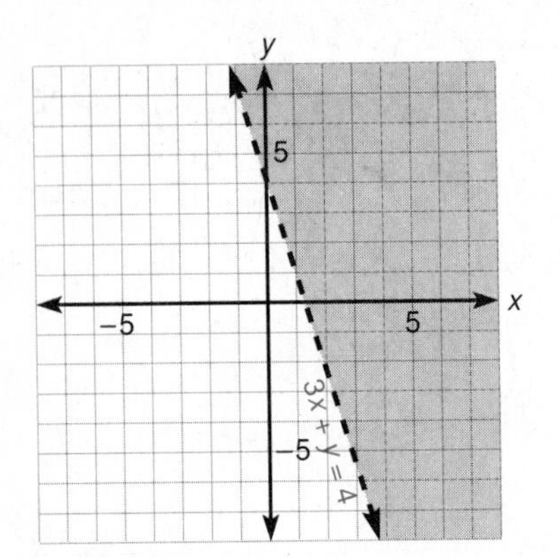

Figure 8–1

In section 7–5, we learned how to graph a linear inequality in two variables. To review this method, let us consider the inequality $3x + y > 4$. We first replace the inequality symbol, $>$, with the equal symbol, $=$, and graph the resulting equation $3x + y = 4$. Since we have the symbol $>$, the coordinates of the points on the line do not satisfy the inequality and the line should be dashed. We then choose a point on either side of the line, usually (0,0), and substitute its coordinates into the inequality.

$$\begin{aligned} 3x + y &> 4 \\ 3(0) + (0) &> 4 \\ 0 &> 4 \quad \text{(False)} \end{aligned}$$

We then shade the half-plane that does *not* contain (0,0) to represent the graph of the inequality. (See figure 8–1.)

In this section, we will consider the graphical method of obtaining the solution of systems of linear inequalities such as

$$\begin{aligned} 2x + y &> 3 \\ x - y &> 1 \end{aligned}$$

A system of linear inequalities, like a system of linear equations, has as its solution set the set of all ordered pairs of real numbers that are solutions of the inequalities of the system. The graph of a system of linear inequalities is the *intersection* of the graphs of the individual inequalities that make up the system.

■ Example 8–5 A

Graph the solution of the following systems of linear inequalities.

1. $2x + y > 3$
   $x - y > 1$
   First, graph each inequality on the same axes, shading each graph. (See the diagram.) The solution of the system is the overlap of the regions of the two graphs—the double-shaded region.

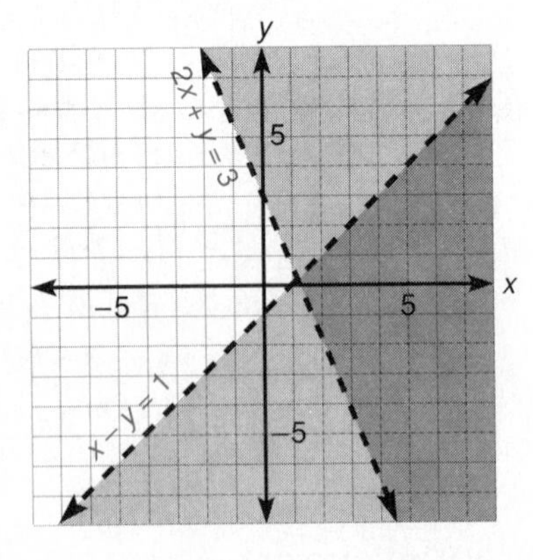

Note The dashed lines that show the ordered pairs of numbers for the points on the lines *are not solutions.*

2. $2x + 3y \le 4$
$x - 4y \le 1$
Graph $2x + 3y \le 4$ and $x - 4y \le 1$ on the same axes. The graph of the solution of the system is the overlapping (double-shaded) region of the diagram.

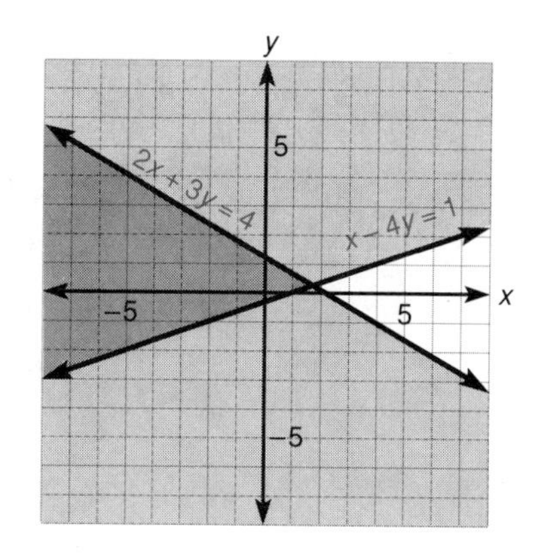

Note The symbol, $\le$, involved in each inequality indicates that the points of that portion of the line bordering the double-shaded region are included as solutions. Therefore when graphing the equations, make the lines *solid.*

The solutions are in the shaded region and *include both* half-lines that form the boundary of this region.

3. $\frac{1}{2}x - \frac{2}{3}y > 1$
$y \le 3$

Graph the inequalities $\frac{1}{2}x - \frac{2}{3}y > 1$ and $y \le 3$ on the same axes. First, clear the denominators in the first inequality by multiplying by the LCD, 6.

$$6 \cdot \frac{1}{2}x - 6 \cdot \frac{2}{3}y > 6 \cdot 1$$
$$3x - 4y > 6$$

Recall, the graph of $y = 3$ is a horizontal line passing through the point (0,3). The graph of the solutions is the double shading in the diagram, together with that portion of the line $y = 3$ that borders the double-shaded region.

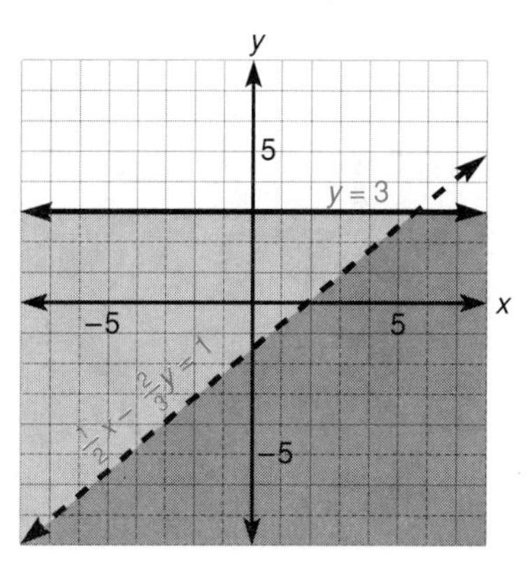

Note The point of intersection (6,3), of the dashed line and the solid line is *not* in the solution.

▶ ***Quick check*** Graph the solution of the system of linear inequalities.
$2x + y \ge 1$
$2y < x + 2$ ■

Mastery points

Can you

- Solve systems of linear inequalities by graphing?

Exercise 8–5

Graph the solution set of the following systems of linear inequalities. See example 8–5 A.

Example $2x + y \geq 1$
$2y < x + 2$

Solution Graph $2x + y \geq 1$ and $2y < x + 2$ on the same axes. A solid line is used in the graph of the first inequality and a dashed line is used in the graph of the second inequality. The graph is the double-shaded portion of the diagram, together with that portion of the line $2x + y = 1$ bordering the double-shaded region.

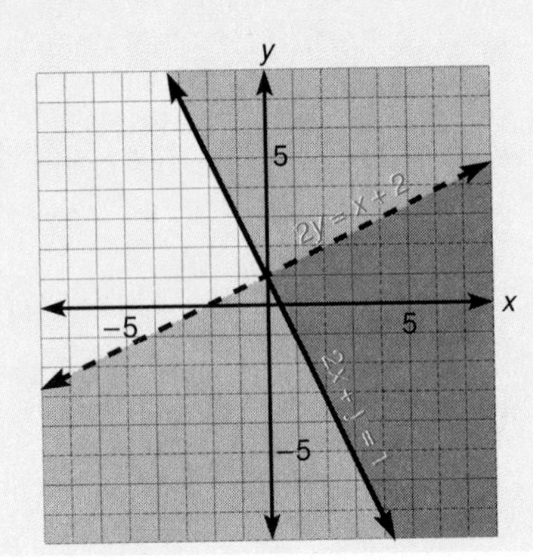

1. $x + y < 2$
$x - y < 4$

2. $x + y < 3$
$x - y > 1$

3. $2x + y > 1$
$x - 2y > 2$

4. $x + 3y > 4$
$2x - y < 0$

5. $-4x + y < 5$
$2x - y < 0$

6. $5x - y < 5$
$2x + y \geq 1$

7. $2x + 3y > 5$
$3x - 2y < 4$

8. $4x + 3y > 12$
$5x - 3y > 15$

9. $2x + 3y < 5$
$8x + 10y > 26$

10. $5x - 2y > 15$
$2x - 3y < 6$

11. $3x - 2y \geq 1$
$2x \geq 5y$

12. $x - 4y \geq 3$
$3y \leq 2x$

13. $y \leq 2x - 7$
$2y + x \geq 5$

14. $4x + 5y \geq 20$
$y \geq x + 1$

15. $5y + 2x \geq 3$
$x \leq 2y - 1$

16. $x + y \geq 5$
$x \geq 3y + 2$

17. $2x - 3y \geq 1$
$-4x + 6y > -3$

18. $-4x + 10y \leq 5$
$-2x + 5y < -1$

19. $\frac{1}{3}x - y \geq 2$
$\frac{1}{3}x + 2y \leq 4$

20. $\frac{3}{4}x - \frac{2}{3}y \leq 1$
$2x + \frac{1}{2}y \geq 3$

21. $\frac{1}{5}x \geq 3y + 1$
$x - \frac{2}{3}y \leq 2$

22. $\frac{3}{2}y - 4x \geq 2$
$\frac{1}{3}y > 1 - x$

23. $3x - 5y > 2$
$x \leq 3$

24. $x - 4y \geq 6$
$x > -2$

25. $5x - 2y \leq 0$
$y > -1$

26. $2y - x > 1$
$y \leq 3$

27. $x \geq -2$
$y < 4$

28. $x \leq 5$
$y > -3$

29. $x \geq 0$
$y \geq 0$

30. $x \leq 0$
$y \geq 0$

Note Exercises 29 and 30 define quadrants I and II, respectively, together with a portion of each axis.

31. State systems of linear inequalities that define (a) quadrant III, (b) quadrant IV.

Review exercises

Perform the indicated operations. See sections 2–3, 3–2, and 3–4.

1. $(3x^2 - 2x + 1) - (x^2 - 5x - 6)$

2. $(4x + 1)(x - 3)$

3. $(5y - 2)^2$

4. $(7z + 1)(7z - 1)$

5. $a^3a^2a^{-1}$

6. $\frac{x^2y^3}{xy}$

7. $(-y^3)^3$

Factor the following expressions. See sections 4–1 and 4–3.

8. $5x^2 - 3x$

9. $4x^2 + 8x - 5$

10. $9y^2 - 30y + 25$

Chapter 8 lead-in problem

A keypunch operator at a local firm works for \$9 per hour while an entry-level typist works for \$6.50 per hour. The total pay for an 8-hour day is \$476. If there are two more typists than keypunch operators, how many keypunch operators does the firm employ?

Solution

Let x = the number of keypunch operators and y = the number of entry-level typists.
From "there are two more typists than keypunch operators,"

$$y = x + 2$$

In an 8-hour day, the pay for each

keypunch operator is $8 \cdot 9 = \$72$
typist is $8 \cdot \$6.50 = \$52$

The table shows the pay for all the workers.

	Number	Pay per worker	Total pay
Keypunch operators	x	72	$72x$
Typists	y	52	$52y$

From "The total pay for an 8-hour day is \$476,"

$$72x + 52y = 476$$

We must solve the system of linear equations.

$$\begin{aligned} y &= x + 2 \\ 72x + 52y &= 476 \end{aligned}$$

Using substitution, substitute $x + 2$ for y in the second equation.

$$\begin{aligned} 72x + 52y &= 476 \\ 72x + 52(x + 2) &= 476 && \text{Replace } y \text{ with } x + 2 \\ 72x + 52x + 104 &= 476 && \text{Distribute in left member} \\ 124x + 104 &= 476 && \text{Combine in left member} \\ 124x &= 372 && \text{Subtract 104 from each member} \\ x &= 3 && \text{Divide each member by 124} \end{aligned}$$

There are 3 keypunch operators working for the firm.

Chapter 8 summary

1. A **system of linear equations** consists of two or more linear equations in the same variables. We have limited the number of variables to two.
2. A **simultaneous solution** of a system of two linear equations is the ordered pair (x,y) that is a solution of both of the equations.
3. A **graphical solution** to a system of two linear equations is found by estimating the coordinates of the point of intersection.
4. An **inconsistent** system of equations has no solution.
5. A **dependent** system of equations has an unlimited number of solutions.
6. A system of linear equations can be solved by **elimination** or **substitution** when an exact answer is required.
7. **Systems of linear inequalities** can be solved by graphing. The graph will consist of all points in a region of the plane bounded by the two lines.

Chapter 8 error analysis

1. Solving systems of linear equations
   Example: Solve the system.
   $$\begin{array}{lll} 2x - 3y = -1 & & 2x - 3y = -1 \\ x + 4y = -2 & \text{Multiply by } 2 \rightarrow & \underline{2x + 8y = -4} \\ & \text{Add} & 4x + 5y = -5 \end{array}$$
   Correct answer:
   $$\begin{array}{lll} 2x - 3y = -1 & & 2x - 3y = -1 \\ x + 4y = -2 & \text{Multiply by } -2 \rightarrow & \underline{-2x - 8y = 4} \\ & \text{Add} & -11y = 3 \\ & & y = -\dfrac{3}{11} \end{array}$$
   What error was made? (*see page 342*)
2. Solutions of systems of equations
   Example: $(2,3)$ is the solution of the system
   $4x - 3y = -1$
   $2x + y = 6$
   Check: $4(2) - 3(3) = -1$
   $8 - 9 = -1$
   $-1 = -1$
   $(2,3)$ is the solution.
   Correct answer: $(2,3)$ is not a solution of $2x + y = 6$ so it is not a solution of the system.
   What error was made? (*see page 336*)
3. Subtracting real numbers
   Example: $(-10) - (-5) = -15$
   Correct answer: -5
   What error was made? (*see page 41*)
4. Division of real numbers
   Example: $\dfrac{-16}{-4} = -4$
   Correct answer: $\dfrac{-16}{-4} = 4$
   What error was made? (*see page 52*)
5. Product of a monomial and a multinomial
   Example: $-4y(y^2 + 3y - 4) = 4y^3 + 12y^2 - 16y$
   Correct answer: $-4y^3 - 12y^2 + 16y$
   What error was made? (*see page 134*)
6. Dividing unlike bases
   Example: $\dfrac{x^3}{y} = x^2$
   Correct answer: $\dfrac{x^3}{y} = \dfrac{x^3}{y}$
   What error was made? (*see page 140*)
7. Zero exponent
   Example: $4y^0 = 1$
   Correct answer: 4
   What error was made? (*see page 142*)
8. Scientific notation
   Example: $-4.21 \times 10^2 = 421$
   Correct answer: -421
   What error was made? (*see page 149*)
9. Factoring the sum of two squares
   Example: $25a^2 + 36b^2 = (5a + 6b)(5a - 6b)$
   Correct answer: $25a^2 + 36b^2$ cannot be factored.
   What error was made? (*see page 176*)
10. Completely factoring the difference of two squares
    Example: $x^4 - y^4 = (x^2 + y^2)(x^2 - y^2)$
    Correct answer: $(x^2 + y^2)(x + y)(x - y)$
    What error was made? (*see page 177*)

Chapter 8 critical thinking

Pick any integer from 20 to 29 and subtract the sum of its digits from it. The answer is always 18. Why is this true?

Chapter 8 review

[8–1]

Find the solution set of the following systems of linear equations by graphing. All answers in the back of the book will be given exactly. If the system is inconsistent or dependent, so state.

1. $x + y = 4$
$x - y = 2$

2. $3x - y = 2$
$x + y = 6$

3. $x - 2y = 0$
$2x - y = 6$

4. $x + y = 4$
$-x + y = 8$

5. $x - \frac{1}{3}y = 1$
$2x - \frac{2}{3}y = -2$

6. $x - y = 4$
$-3x + 3y = -12$

[8–2]

Find the solution set of the following systems of linear equations by elimination. If the system is inconsistent or dependent, so state.

7. $x - 2y = 3$
$x + y = 3$

8. $x - y = 5$
$3x + 2y = 25$

9. $2x + 3y = 1$
$-x + y = 2$

10. $5x + y = 4$
$5x - 3y = 8$

11. $x + 3y = 6$
$2x - 3y = -6$

12. $\frac{1}{2}x + \frac{1}{4}y = \frac{7}{4}$
$\frac{2}{3}x - \frac{1}{3}y = \frac{13}{3}$

13. $(0.3)x - (0.2)y = 1.1$
$(0.3)x + (0.2)y = 1.9$

14. $x - 2y = 5$
$-3x + 6y = 4$

15. $2x - y = 3$
$3x + y = 7$

16. $3x - y = 2$
$9x - 3y = 6$

[8–3]

Find the solution set of the following systems of linear equations by substitution. If the system is inconsistent or dependent, so state.

17. $x - y = 2$
$2x - 3y = 1$

18. $2x + y = 7$
$2x - y = 13$

19. $3x - 2y = 4$
$6x - 4y = 5$

20. $2x - 3y = 3$
$3x + y = 7$

21. $x + y = 4$
$x - y = 2$

22. $x - y = 5$
$\frac{1}{2}x + \frac{1}{3}y = \frac{25}{6}$

23. $12x - 9y = 15$
$4x - 3y = 5$

24. $5x + 3y = 4$
$2x - y = 5$

[8–4]

Solve the following problems by setting up a system of two linear equations with two unknowns.

25. The perimeter of a rectangle is 64 feet. Find the dimensions if three times the width is 4 less than the length.

26. Bruce made two investments totaling $18,000. On one investment he made a 14% profit, but on the other investment he took a 23% loss. If his net yearly loss was $70, how much was in each investment?

27. Four times the number of teeth on a first gear is 6 less than the number of teeth on a second gear. If the total number of teeth on both gears is 66, how many teeth are on each gear?

28. A chemist wishes to make 120 milliliters (ml) of a 75% acid solution by mixing a 60% solution with an 80% solution. How many milliliters of each solution are necessary?

[8–5]

Solve the following systems of linear inequalities by graphing.

29. $x + y \geq 5$
$2x - y < 3$

30. $4x - 3y < 1$
$2x + 5y \leq 2$

31. $4y - 3x \geq 2$
$y \leq 1 - 2x$

32. $2y \leq 4 - x$
$5y - x > 2$

33. $x > 4y + 3$
$y \leq 4$

34. $x > 1$
$5y - x \leq 0$

Chapter 8 cumulative test

Evaluate each of the following.

[3–3] 1. -4^{-3}

[3–3] 2. $\left(\frac{3}{4}\right)^{-1}$

[3–3] 3. $\left(\frac{3}{5} + \frac{1}{8}\right)^0$

[3–3] 4. $(-5)^{-2}$

[3–3] 5. $\frac{1}{2^{-4}}$

[2–2] 6. Evaluate the expression $\frac{ab - bc}{a}$, when $a = \frac{1}{2}$, $b = \frac{2}{3}$, and $c = \frac{3}{4}$.

Completely factor the following expressions.

[4–1] 7. $3a^3 + 15a^2 + 27a$

[4–4] 8. $x^2 - 121$

[4–1] 9. $ax + ay - 5x - 5y$

[4–3] 10. $3a^2 - 5ab - 2b^2$

[4–4] 11. $7x^3 - 7x$

[4–3] 12. $5b^2 - 22b + 8$

Perform the indicated operations and simplify the results.

[6–1] 13. $\frac{4x}{5y} \div \frac{24x^2}{15y^3}$

[6–3] 14. $\frac{5}{x^2 - 9x} - \frac{3}{x}$

[3–4] 15. $(-9x^2y^3z^3)^2$

[5–3] 16. $\frac{36x^3y^2 - 56xy^3}{4xy}$

[6–1] 17. $\frac{a^2 - 8a}{a^2 - a - 56} \cdot \frac{a^2 - 49}{7a}$

[3–2] 18. $(5x - 3y)^2$

[5–3] 19. $(8y^2 + 10y - 42) \div (4y - 7)$

Find the solution set of the following equations.

[6–5] 20. $\frac{4a - 1}{4} = \frac{5a + 2}{7}$

[4–7] 21. $x^2 = -9x$

[6–5] 22. $\frac{3}{4x} - 2 = \frac{5}{3x} + 1$

[4–7] 23. $7x^2 + 4x = 3$

Evaluate the following.

[1–3] 24. $|-6|$

[1–3] 25. $\left|\frac{4}{5}\right|$

[1–3] 26. $-\left|-\frac{1}{3}\right|$

Solve the following systems of linear equations by the appropriate method. If the system is inconsistent or dependent, so state.

[8–3] 27. $3y - x = 2$
$8y + x = 20$

[8–3] 28. $4x - y = 6$
$3x + 2y = -1$

[8–3] 29. $5y - 2z = 4$
$y = -3z + 1$

[8–3] 30. $x - 2y = -6$
$3x - 6y = 1$

[8–3] 31. $8x - 3y = 2$
$y - 1 = 0$

[8–3] 32. $y + 3x = 8$
$6x + 3y = 24$

[8–5] 33. Solve the system of linear inequalities by graphing.
$3x + 2y \leq 6$
$x - 5y > 5$

[8–4] 34. The sum of two numbers is 25. If one number is 1 more than three times the other, what are the numbers? (Solve by system of linear equations.)

[8–4] 35. The sum of two numbers is 79. The difference between the two numbers is 5. Find the numbers. (Solve by system of linear equations.)

[4–8] 36. The sum of the squares of two consecutive integers is 61. What are the numbers?

[4–8] 37. The square of a number is equal to five times that number. Find the number.

[6–6] 38. The sum of a number and twice its reciprocal is 3. Find the number.

[6–6] 39. A tank can be filled in four hours by pipe A and in five hours by pipe B. How long will it take to fill the tank if both pipes are open?

[2–8] 40. The length of a rectangle is 1 inch less than twice its width. If the perimeter is 40 inches, what are the dimensions of the rectangle?

CHAPTER

9 Roots and Radicals

Roger has a ladder that will extend to a length of 21 feet. For the ladder to be safe to climb on, it must be placed 7 feet away from the house. The roof is 20 feet above the ground. Will the ladder be able to reach the roof safely? If not, how far up will the ladder reach? (Leave your answer rounded to one decimal place.)

9–1 ■ Principal roots

Square root

In chapter 4, quadratic equations were solved by factoring, but many quadratic equations such as

$$x^2 - 7 = 0 \quad \text{or} \quad x^2 + 3x + 1 = 0$$

will not factor over the set of rational numbers. We need to be able to solve equations that involve a squared variable. Therefore, we want a process that is the inverse of squaring a number.

In chapter 1, we discussed how to square a number. Consider the following examples:

$$\text{If } x = 3, \quad \text{then } x^2 = (3)^2 = (3)(3) = 9$$
$$\text{If } x = -3, \text{ then } x^2 = (-3)^2 = (-3)(-3) = 9$$

Reversing the process, we ask:

$$\text{If } x^2 = 9, \text{ then what number is } x \text{ equal to?}$$

This inverse operation is called *finding the square root of a number.*

Definition

For every pair of real numbers a and b, if $a^2 = b$, then a is called a square root of b.

Concept

A square root of a number is one of two equal factors of the number.

From this discussion and the definition of square root, we can see that the answer to the question we asked,

If $x^2 = 9$, then what number is x equal to?

is 3 or -3 since $(3)^2 = 9$ and $(-3)^2 = 9$. To distinguish between the two square roots, we define the *principal square root* of a positive number to be positive. The $\sqrt{\ }$ symbol denotes the principal square root.
Thus, if $x = \sqrt{9}$, then $x = 3$ and we say 3 is the principal square root of 9.

$$\sqrt{9} = 3 \text{ (principal square root)}$$

The parts of the principal square root are

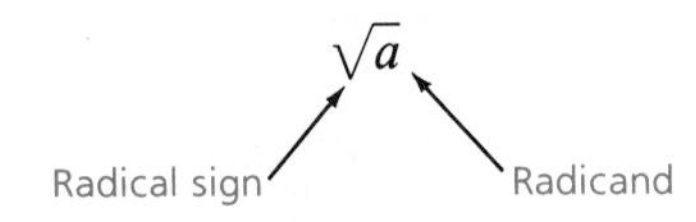

The entire expression is called a *radical* and is read "the principal square root of a."

Example 9–1 A

Find the principal square root.

1. $\sqrt{16} = 4$, since $4 \cdot 4 = 4^2 = 16$.
2. $\sqrt{49} = 7$, since $7 \cdot 7 = 7^2 = 49$.
3. $\sqrt{25} = 5$, since $5 \cdot 5 = 5^2 = 25$.
4. $\sqrt{36} = 6$, since $6 \cdot 6 = 6^2 = 36$.
5. $\sqrt{0} = 0$, since $0 \cdot 0 = 0^2 = 0$.

Note In the examples, 0, 16, 49, 25, and 36 are called **perfect-square integers** because their square roots are integers. ■

Whenever we wish to express the negative value of the square root of a number, we use the symbol $-\sqrt{\ }$. For example, $-\sqrt{9}$ would indicate that we want the negative square root value. That is, $-\sqrt{9} = -3$.

Example 9–1 B

Find the indicated root.

1. $-\sqrt{4} = -2$
2. $-\sqrt{49} = -7$
3. $-\sqrt{25} = -5$
4. $-\sqrt{36} = -6$

▶ ***Quick check*** Find the square root. $-\sqrt{16}$ ■

Our first examples of finding the square root of a number have dealt only with perfect-square integers. We shall now try to find the $\sqrt{2}$. We could use 1.414 but when we square 1.414, we do not get 2 as an answer. We can show that no matter how many decimal places the answer is carried to, when the result is squared, it will be close to, but not equal, 2. The $\sqrt{2}$ is called an **irrational number** because it has the property that it cannot be expressed as a terminating or a repeating decimal number. Another number that is irrational is π, which is used in geometric formulas involving circles.

Whenever you work with irrational numbers in a problem, you may have to approximate the number to as many decimal places as are needed in the problem by using a calculator or a table of values.

■ Example 9–1 C

For the following irrational numbers, find the decimal approximation to three decimal places by using a calculator.

1. $\sqrt{3} \approx 1.732$

Note "≈" is read "is approximately equal to" and is used when our answer is not exact. Square roots of integers that are not perfect squares will be irrational.

2. $-\sqrt{41} \approx -6.403$ **3.** $\sqrt{50} \approx 7.071$

▶ ***Quick check*** Find the decimal approximation to three decimal places by using a calculator. $\sqrt{48}$ and $-\sqrt{56}$ ■

Not all real numbers have a rational or an irrational square root. Consider the following:

$$\sqrt{-4} = \text{what?}$$

We know that all real numbers are either positive, negative, or zero. If we square a real number, the product is never negative. Hence, there is no real number that when squared produces a negative answer. *The square root of a negative number does not exist in the set of real numbers.*

■ Example 9–1 D

Suppose an automotive engineer wishes to determine the diameter of the cylinder bore (D) required to produce H horsepower from N cylinders of an engine that is turning 1,000 rpm. The engineer will use the formula

$$D = \sqrt{\frac{H}{(0.4)\,N}}$$

What would be the bore diameter (in inches) required to produce 40 horsepower at 1,000 rpm from a 4-cylinder engine?

We substitute 40 for H and 4 for N.

$$D = \sqrt{\frac{(40)}{(0.4)(4)}} = \sqrt{\frac{40}{1.6}} = \sqrt{25} = 5$$

Therefore, we need a bore diameter of 5 inches. ■

Pythagorean Theorem

The following is an important property of right triangles called the **Pythagorean Theorem.**

In a right triangle, the square of the length of the hypotenuse (the side opposite the right angle) is equal to the sum of the squares of the lengths of the two legs (the sides that form the right angle). If c is the length of the hypotenuse and a and b are the lengths of the legs, this property can be stated as:

$$c^2 = a^2 + b^2 \text{ or } c = \sqrt{a^2 + b^2}$$

also as $$a^2 = c^2 - b^2 \text{ or } a = \sqrt{c^2 - b^2}$$

and as $$b^2 = c^2 - a^2 \text{ or } b = \sqrt{c^2 - a^2}$$

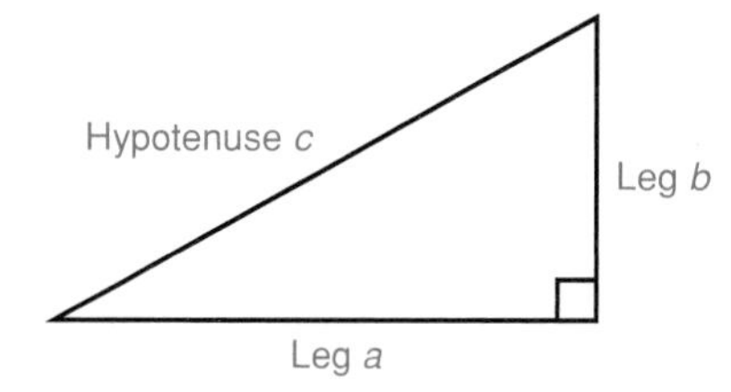

■ Example 9–1 E

Find the length of the hypotenuse of a right triangle whose legs are 6 centimeters and 8 centimeters.

We want to find c when $a = 6$ cm and $b = 8$ cm.
By the Pythagorean Theorem,

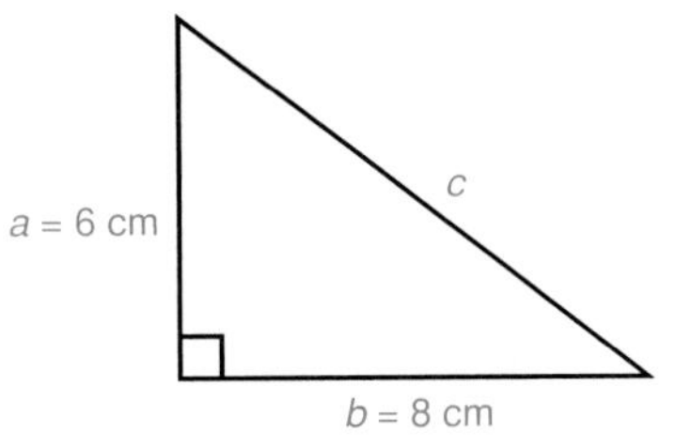

$$
\begin{aligned}
c &= \sqrt{a^2 + b^2} \\
&= \sqrt{6^2 + 8^2} && \text{Substitute} \\
&= \sqrt{36 + 64} && \text{Square the values} \\
&= \sqrt{100} = 10
\end{aligned}
$$

Hence, $c = 10$ cm.

▶ ***Quick check*** Find the second leg of a right triangle whose hypotenuse has length 13 inches and whose first leg is 5 inches long. ■

*n*th roots

The concept of square root can be extended to find cube roots (third root of a number), fourth roots, fifth roots, and so on. A cube root is one of three equal factors of a number. The symbol that is used to express the principal cube root is $\sqrt[3]{}$. The 3 is called the **index** of the radical expression. The index denotes what root we are looking for. The principal fourth root would be indicated by $\sqrt[4]{}$. General notation for the principal *n*th root would be $\sqrt[n]{}$, where n is a natural number greater than 1.

The principal *n*th root

The principal *n*th root of number *a*, denoted by

$$\sqrt[n]{a}$$

is one of n equal factors such that $\sqrt[n]{a} = b$ and

$$\overbrace{b \cdot b \cdot b \cdots b}^{n \text{ factors}} = b^n = a$$

where n is a natural number greater than 1.

The parts of the principal *n*th root are

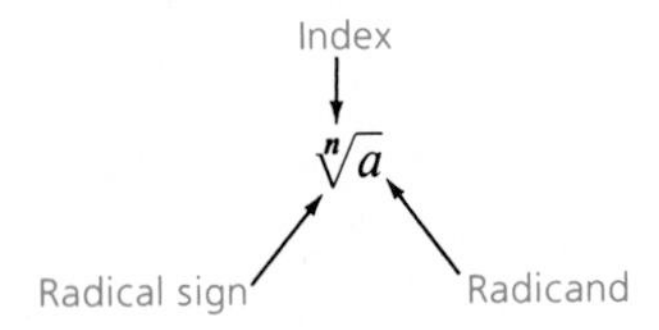

Note If there is no index associated with a radical symbol, it is understood to be 2.

If we exclude even roots of negative numbers, which do not exist in the set of real numbers, we can extend our idea of principal square root to all other roots by saying: **the principal *n*th root** *of a number, denoted by* $\sqrt[n]{}$, *has the same sign as the number itself.*

■ Example 9–1 F

Find the indicated root.

1. $\sqrt[4]{16} = 2$, since $2 \cdot 2 \cdot 2 \cdot 2 = 2^4 = 16$.
2. $\sqrt[3]{-27} = -3$, since $(-3)(-3)(-3) = (-3)^3 = -27$.
3. $\sqrt[3]{-125} = -5$, since $(-5)(-5)(-5) = (-5)^3 = -125$.
4. $\sqrt{-16}$ Does not exist in the set of real numbers.

▶ ***Quick check*** Find the cube root. $\sqrt[3]{-64}$ ■

Mastery points

Can you

- Find the principal square root of a perfect-square integer?
- Find the principal root of a number?
- Find the decimal approximation with a calculator for a root that is an irrational number?

Exercise 9–1

Find the indicated square root. See examples 9–1 A and B.

Example $-\sqrt{16}$

Solution $= -4$ Since $4 \cdot 4 = 16$, and we want the negative value of the square root

1. $\sqrt{100}$ **2.** $\sqrt{36}$ 3. $\sqrt{4}$ **4.** $\sqrt{64}$ 5. $-\sqrt{144}$ **6.** $-\sqrt{81}$

7. $-\sqrt{121}$ **8.** $-\sqrt{4}$ **9.** $\sqrt{121}$ **10.** $\sqrt{81}$ **11.** $-\sqrt{16}$ **12.** $-\sqrt{64}$

Find the decimal approximation to three decimal places of the indicated square root by using a calculator. See example 9–1 C.

Examples $\sqrt{48}$ $-\sqrt{56}$

Solutions ≈ 6.928 ≈ -7.483

13. $\sqrt{18}$ **14.** $\sqrt{24}$ **15.** $\sqrt{41}$ **16.** $\sqrt{47}$ 17. $-\sqrt{52}$ **18.** $-\sqrt{10}$

Solve the following problems. See example 9–1 D.

19. The current I (amperes) in a circuit is found by the formula $I = \sqrt{\frac{\text{watts}}{\text{ohms}}}$. What is the current of a circuit that has 3 ohms resistance and uses 1,728 watts?

20. What is the current of a circuit that has 2 ohms resistance and uses 450 watts? (Refer to exercise 19.)

21. The slant height, S, of a right circular cone is found by the formula $S = \sqrt{r^2 + h^2}$, where r is the radius of the base and h is the height of the cone. What is the slant height of a right circular cone whose base radius is 5 units and whose height is 12 units?

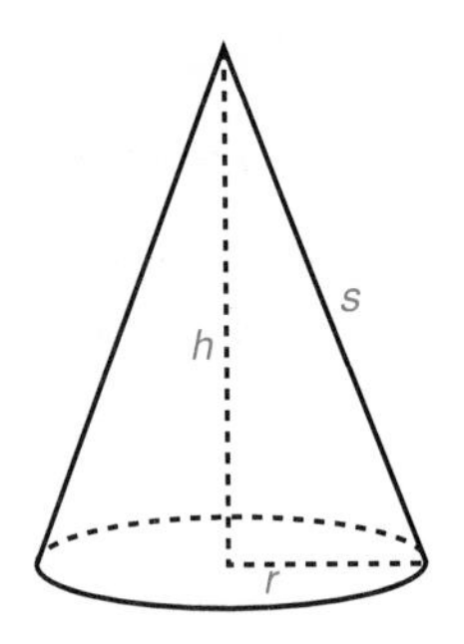

22. What is the slant height of a right circular cone whose base radius is 3 units and whose height is 4 units? (Refer to exercise 21.)

In the following right triangles, find the length of the unknown side. See example 9–1 E.

Example Find the second leg of a right triangle whose hypotenuse has length 13 inches and whose first leg is 5 inches long.

Solution We want to find b given that $c = 13$ in. and $a = 5$ in. Using one of the forms of the theorem,

$$\begin{aligned} b &= \sqrt{c^2 - a^2} \\ &= \sqrt{(13)^2 - (5)^2} && \text{Substitute} \\ &= \sqrt{169 - 25} && \text{Square the values} \\ &= \sqrt{144} = 12 \end{aligned}$$

Hence $b = 12$ in.

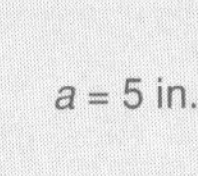

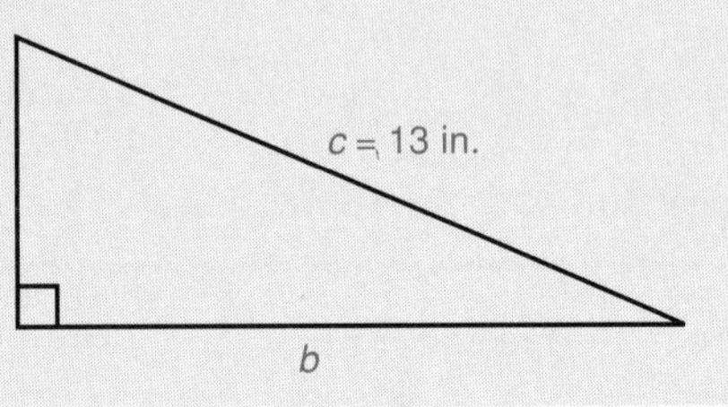

23. $a = 3$ m, $b = 4$ m **24.** $a = 8$ ft, $c = 10$ ft 25. $a = 12$ in., $b = 5$ in.

26. $a = 15$ cm, $b = 8$ cm 27. $a = 6$ yd, $c = 10$ yd **28.** $b = 16$ m, $c = 20$ m

29. $a = 12$ mm, $b = 16$ mm **30.** $a = 10$ in., $b = 24$ in.

Find the indicated root. See example 9–1 F.

Example $\sqrt[3]{-64}$

Solution $= -4$ Since $(-4)^3 = -64$

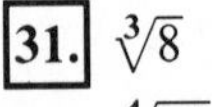

31. $\sqrt[3]{8}$ **32.** $\sqrt[3]{27}$ 33. $\sqrt[3]{125}$ **34.** $\sqrt[3]{64}$ 35. $\sqrt[3]{-8}$ **36.** $\sqrt[3]{-64}$

37. $\sqrt[4]{81}$ **38.** $\sqrt[4]{16}$ 39. $-\sqrt[4]{81}$ **40.** $-\sqrt[4]{625}$ 41. $\sqrt[5]{32}$ **42.** $\sqrt[6]{64}$

43. $-\sqrt[5]{243}$ **44.** $\sqrt[5]{-243}$ 45. $\sqrt[10]{1}$ **46.** $\sqrt[14]{1}$ 47. $\sqrt[9]{-1}$ **48.** $\sqrt[15]{-1}$

Solve the following problems. See example 9–1 D.

49. The formula for finding the length of an edge, e, of a cube when the volume, v, is known is $e = \sqrt[3]{v}$. What is the length of the edge of a cube whose volume is 729 cubic units?

50. What is the length of the edge of a cube whose volume is 216 cubic units? (Refer to exercise 49.)

Review exercises

Rewrite the following numbers in prime factor form. See section 1–1.

1. 9 **2.** 12 **3.** 8 **4.** 40 **5.** 50 **6.** 81 **7.** 64 **8.** 16

9–2 ■ Product property for radicals

Multiplying square roots

In this section, we are going to develop properties for simplifying radicals. Consider the example

$$\sqrt{4} \cdot \sqrt{25} = 2 \cdot 5 = 10$$

We also observe that

$$\sqrt{4 \cdot 25} = \sqrt{100} = 10$$

From our example, we can conclude that

$$\sqrt{4} \cdot \sqrt{25} = \sqrt{4 \cdot 25}$$

We now can generalize the product property for square roots.

Product property for square roots

For all nonnegative real numbers a and b,

$$\sqrt{a} \cdot \sqrt{b} = \sqrt{ab}$$

Concept

The product of two square roots is equal to the square root of their product.

■ **Example 9–2 A**

Perform the indicated operations. Assume that all variables represent nonnegative real numbers.

1. $\sqrt{3}\sqrt{5} = \sqrt{3 \cdot 5} = \sqrt{15}$
2. $\sqrt{6}\sqrt{7} = \sqrt{6 \cdot 7} = \sqrt{42}$
3. $\sqrt{3}\sqrt{a} = \sqrt{3a}$
4. $\sqrt{x}\sqrt{y}\sqrt{z} = \sqrt{xyz}$ ■

Simplifying principal square roots

An important use of our product property is in simplifying radicals. Consider the following example:

Since 12 can be factored into $4 \cdot 3$, by our product property, we can write

$$\sqrt{12} = \sqrt{4 \cdot 3} = \sqrt{4}\sqrt{3} = 2\sqrt{3}$$

We are able to simplify the radical because the radicand contains a perfect-square integer factor, 4. In our example, $2\sqrt{3}$ is called the *simplified form* of $\sqrt{12}$.

Simplifying the principal square root

1. If the radicand is a perfect-square integer, write the corresponding square root.
2. If possible, factor the radicand so that at least one factor is a perfect-square integer. Write the corresponding square root as a coefficient of the radical.
3. The square root is in simplest form when the radicand has no perfect-square integer factors other than 1.

The following property is used when changing radicals involving variables in the radicand to simplest form.

$\sqrt{a^2}$ property

If a is any nonnegative real number, then

$$\sqrt{a^2} = a$$

■ Example 9–2 B

Simplify the following expressions. Assume that all variables represent nonnegative real numbers.

1. $\sqrt{50} = \sqrt{25 \cdot 2}$ Factor having a perfect-square integer
$= \sqrt{25}\sqrt{2}$ Product property
$= 5\sqrt{2}$ $\sqrt{25} = 5$
2. $\sqrt{28} = \sqrt{4 \cdot 7} = \sqrt{4}\sqrt{7} = 2\sqrt{7}$
3. $\sqrt{9a} = \sqrt{9 \cdot a} = \sqrt{9}\sqrt{a} = 3\sqrt{a}$
4. $\sqrt{a^3} = \sqrt{a^2 \cdot a} = \sqrt{a^2}\sqrt{a} = a\sqrt{a}$
5. $\sqrt{a^3b^4} = \sqrt{a^2 \cdot a \cdot b^2 \cdot b^2} = \sqrt{a^2}\sqrt{a}\sqrt{b^2}\sqrt{b^2} = a\sqrt{a}bb = ab^2\sqrt{a}$
6. $\sqrt{a^2 + b^2}$ will not simplify because we are not able to factor the radicand. *The radicand must always be in a factored form before we can simplify.*

Note $\sqrt{a^2 + b^2} \neq \sqrt{a^2} + \sqrt{b^2}$. For example, $\sqrt{3^2 + 4^2} = \sqrt{9 + 16} \neq \sqrt{9} + \sqrt{16}$. Since $\sqrt{9 + 16} = \sqrt{25} = 5$, whereas $\sqrt{9} + \sqrt{16} = 3 + 4 = 7$.

7. $\sqrt{2}\sqrt{6} = \sqrt{2 \cdot 6} = \sqrt{12} = \sqrt{4 \cdot 3} = \sqrt{4} \cdot \sqrt{3} = 2\sqrt{3}$
8. $\sqrt{10}\sqrt{5} = \sqrt{10 \cdot 5} = \sqrt{50} = \sqrt{25 \cdot 2} = \sqrt{25}\sqrt{2} = 5\sqrt{2}$
9. $\sqrt{14x}\sqrt{2x} = \sqrt{14x \cdot 2x} = \sqrt{28x^2} = \sqrt{4 \cdot 7 \cdot x^2} = \sqrt{4}\sqrt{7}\sqrt{x^2}$
$= 2\sqrt{7}x = 2x\sqrt{7}$

▶ ***Quick check*** Simplify. $\sqrt{8a^4}$ and $\sqrt{3a}\sqrt{6a}$ ■

Multiplying nth roots

In section 9–1, we observed that even roots of negative numbers do not exist in the set of real numbers. Therefore, **we will consider all variables to be representing nonnegative real numbers whenever the index of the radical is even.**

Our product property for square roots can be extended to radicals with any index.

Product property for radicals

$$\sqrt[n]{a}\sqrt[n]{b} = \sqrt[n]{ab}$$

Concept

When we multiply two radicals *having the same index,* we multiply the radicands and put the product under a radical symbol with the common index.

■ **Example 9–2 C**

Perform the indicated operations.

1. $\sqrt[3]{3}\sqrt[3]{2} = \sqrt[3]{3 \cdot 2} = \sqrt[3]{6}$
2. $\sqrt[5]{7}\sqrt[5]{9} = \sqrt[5]{7 \cdot 9} = \sqrt[5]{63}$
3. $\sqrt[3]{3}\sqrt[4]{5}$ These radicals cannot be multiplied together in this form since they do not have the same index. ■

Simplifying principal nth roots

We simplify nth roots, where n is greater than 2, as we did square roots. As long as the radicand can be factored so that one or more factors is a

1. perfect cube when the index is 3,
2. perfect fourth-power when the index is 4,
3. perfect fifth-power when the index is 5, and so on,

the radical can be simplified. To do this, we use the property

$$\sqrt[n]{a^n} = a$$

where a is a nonnegative real number.

■ **Example 9–2 D**

Simplify the following radical expressions. Assume that all variables represent nonnegative real numbers.

1. $\sqrt[3]{81} = \sqrt[3]{27 \cdot 3}$ Factor having a perfect-cube integer
   $= \sqrt[3]{27}\sqrt[3]{3}$ Product property
   $= 3\sqrt[3]{3}$ $\sqrt[3]{27} = 3$
2. $\sqrt[4]{32} = \sqrt[4]{16 \cdot 2} = \sqrt[4]{16}\,\sqrt[4]{2} = 2\sqrt[4]{2}$
3. $\sqrt[3]{x^5} = \sqrt[3]{x^3 \cdot x^2} = \sqrt[3]{x^3}\,\sqrt[3]{x^2} = x\sqrt[3]{x^2}$
4. $\sqrt[5]{y^{10}} = \sqrt[5]{y^5 \cdot y^5} = \sqrt[5]{y^5}\,\sqrt[5]{y^5} = y \cdot y = y^2$

Note In example 4, the exponent 10 is evenly divisible by the index 5, and the radical is eliminated. When the exponent of a factor is evenly divisible by the index, that factor will no longer remain under the radical symbol.

5. $\sqrt[4]{x^7y^4} = \sqrt[4]{x^4 \cdot x^3 \cdot y^4} = \sqrt[4]{x^4}\,\sqrt[4]{x^3}\,\sqrt[4]{y^4} = x\sqrt[4]{x^3}y = xy\sqrt[4]{x^3}$
6. $\sqrt[3]{a^7b^2} = \sqrt[3]{a^3a^3ab^2} = \sqrt[3]{a^3}\,\sqrt[3]{a^3}\,\sqrt[3]{ab^2} = a \cdot a\sqrt[3]{ab^2} = a^2\sqrt[3]{ab^2}$

Note No simplification relative to b is possible because the exponent of b is less than the value of the index.

7. $\sqrt[3]{54x^3y^5} = \sqrt[3]{27 \cdot 2 \cdot x^3 \cdot y^3 \cdot y^2} = \sqrt[3]{27}\,\sqrt[3]{x^3}\,\sqrt[3]{y^3}\,\sqrt[3]{2y^2} = 3xy\sqrt[3]{2y^2}$

▶ ***Quick check*** Simplify. $\sqrt[3]{8a^4}$ ■

Observe from the preceding examples that **we can simplify a radical if the radicand has a factor(s) whose exponent is equal to or greater than the index.**

■ Example 9–2 E

Perform the indicated multiplication and simplify. Assume that all variables represent nonnegative real numbers.

1. $\sqrt[3]{a^2}\sqrt[3]{a} = \sqrt[3]{a^2 \cdot a}$ Product property
$= \sqrt[3]{a^3}$ Multiply like bases
$= a$ Perfect cube

2. $\sqrt[4]{8a^2}\sqrt[4]{4a^3} = \sqrt[4]{8a^2 \cdot 4a^3} = \sqrt[4]{32a^5} = \sqrt[4]{16 \cdot 2 \cdot a^4 \cdot a}$
$= \sqrt[4]{16}\sqrt[4]{a^4}\sqrt[4]{2a} = 2a\sqrt[4]{2a}$

▶ ***Quick check*** Simplify. $\sqrt[3]{9a^2}\sqrt[3]{3a^2}$ ■

Note A very common error in problems involving radicals is to forget to carry along the correct index for the radical symbol.

Mastery points

Can you

- Multiply radicals having the same index?
- Simplify radicals?

Exercise 9–2

Perform any indicated operations and simplify. Assume that all variables represent nonnegative real numbers. See examples 9–2 A and B.

Examples $\sqrt{8a^4}$

Solutions
$= \sqrt{4 \cdot 2a^2a^2}$ Factor having perfect squares
$= \sqrt{4}\sqrt{a^2}\sqrt{a^2}\sqrt{2}$ Product property
$= 2aa\sqrt{2}$ $\sqrt{4} = 2$, $\sqrt{a^2} = a$
$= 2a^2\sqrt{2}$ Multiply

$\sqrt{3a}\sqrt{6a}$
$= \sqrt{3a \cdot 6a}$ Product property
$= \sqrt{18a^2}$ Multiply
$= \sqrt{9 \cdot 2 \cdot a^2}$ Factor having perfect squares
$= \sqrt{9}\sqrt{2}\sqrt{a^2}$ Product property
$= 3\sqrt{2}\,a$ $\sqrt{9} = 3$, $\sqrt{a^2} = a$
$= 3a\sqrt{2}$ Multiply

1. $\sqrt{16}$
2. $\sqrt{63}$
3. $\sqrt{20}$
4. $\sqrt{75}$
5. $\sqrt{45}$
6. $\sqrt{48}$
7. $\sqrt{32}$
8. $\sqrt{27}$
9. $\sqrt{80}$
10. $\sqrt{54}$
11. $\sqrt{98}$
12. $\sqrt{96}$
13. $\sqrt{a^7}$
14. $\sqrt{a^5}$
15. $\sqrt{4a^2b^3}$
16. $\sqrt{9ab^4c^3}$
17. $\sqrt{27a^3b^5}$
18. $\sqrt{24x^5yz^3}$
19. $\sqrt{6}\sqrt{3}$
20. $\sqrt{27}\sqrt{6}$
21. $\sqrt{15}\sqrt{15}$
22. $\sqrt{11}\sqrt{11}$
23. $\sqrt{6}\sqrt{10}$
24. $\sqrt{18}\sqrt{24}$
25. $\sqrt{25}\sqrt{15}$
26. $\sqrt{20}\sqrt{20}$
27. $\sqrt{5}\sqrt{15}$
28. $\sqrt{2a}\sqrt{3a}$
29. $\sqrt{5x}\sqrt{15x}$
30. $\sqrt{6x}\sqrt{14xy}$
31. $\sqrt{2a}\sqrt{24b^2}$

32. A square-shaped television picture tube has a surface area of 121 square inches. What is the length of the side of the tube? (*Hint:* Area of a square is found by squaring the length of a side. $A = s^2$.)

33. A room in the shape of a square is 169 square feet. What is the length of a side? (See exercise 32.)

34. The formula for approximating the velocity V in miles per hour of a car based on the length of its skid marks S (in feet) on wet pavement is given by $V = 2\sqrt{3S}$. If the skid marks are 75 feet long, what was the velocity?

35. The formula for approximating the velocity V in miles per hour of a car based on the length of its skid marks S (in feet) on dry pavement is given by $V = 2\sqrt{6S}$. If the skid marks are 24 feet long, what was the velocity?

Perform any indicated operations and simplify. Assume that all variables represent positive real numbers. See examples 9–2 D and E.

Examples $\sqrt[3]{8a^4}$

Solutions
$= \sqrt[3]{8 \cdot a^3 \cdot a}$ Factor having perfect cubes
$= \sqrt[3]{8}\sqrt[3]{a^3}\sqrt[3]{a}$ Product property
$= 2a\sqrt[3]{a}$ $\sqrt[3]{8} = 2$, $\sqrt[3]{a^3} = a$

$\sqrt[3]{9a^2}\sqrt[3]{3a^2}$
$= \sqrt[3]{9a^2 \cdot 3a^2}$ Product property
$= \sqrt[3]{27a^4}$ Multiply radicands
$= \sqrt[3]{27 \cdot a^3 \cdot a}$ Factor having perfect cubes
$= \sqrt[3]{27}\sqrt[3]{a^3}\sqrt[3]{a}$ Product property
$= 3a\sqrt[3]{a}$ $\sqrt[3]{27} = 3$, $\sqrt[3]{a^3} = a$

36. $\sqrt[3]{48}$ **37.** $\sqrt[5]{64}$ **38.** $\sqrt[4]{32}$ 39. $\sqrt[3]{24}$ **40.** $\sqrt[5]{a^7}$
41. $\sqrt[3]{b^8}$ **42.** $\sqrt[3]{x^9}$ **43.** $\sqrt[5]{y^{15}}$ **44.** $\sqrt[3]{a^{12}}$ **45.** $\sqrt[3]{4a^2b^3}$
46. $\sqrt[3]{8r^2s^8}$ 47. $\sqrt[3]{16a^4b^5}$ **48.** $\sqrt[5]{64x^{10}y^{14}}$ **49.** $\sqrt[3]{81a^5b^{11}}$ **50.** $\sqrt[5]{a^2}\sqrt[5]{a}$
51. $\sqrt[3]{b^2}\sqrt[3]{b^2}$ **52.** $\sqrt[5]{b^4}\sqrt[5]{b^3}$ 53. $\sqrt[5]{a}\sqrt[5]{a^4}$ **54.** $\sqrt[3]{5a^2b}\sqrt[3]{75a^2b^2}$
55. $\sqrt[3]{3ab^2}\sqrt[3]{18a^2b^2}$ **56.** $\sqrt[4]{8a^3b}\sqrt[4]{4a^2b^2}$ 57. $\sqrt[4]{27a^2b^3}\sqrt[4]{9ab}$ **58.** $\sqrt[3]{25x^5y^7}\sqrt[3]{15xy^3}$
59. $\sqrt[3]{16a^{11}b^4}\sqrt[3]{12a^4b^6}$ **60.** $\sqrt[4]{8xy}\sqrt[4]{4x^3y^3}$

61. The moment of inertia for a rectangle is given by the formula $I = \dfrac{bh^3}{12}$. If we know the values of I and b, we can solve for h as follows: $h = \sqrt[3]{\dfrac{12I}{b}}$. Find h if $I = 2$ in.^{4} and $b = 3$ in.

62. Use exercise 61 to find h if $I = 27$ in.^{4} and $b = 4$ in.

63. The moment of inertia for a circle is given by the formula $I = \dfrac{\pi r^4}{4}$. If we know the value of I, we can solve for r as follows: $r = \sqrt[4]{\dfrac{4I}{\pi}}$. Find r if $I = 12.56$ in.^{4} and we use 3.14 for π.

64. Use exercise 63 to find r if $I = 63.585$ in.^{4}

Review exercises

Reduce the following fractions and rational expressions to lowest terms. Assume that no variable is equal to zero. See sections 1–1, 3–3, and 5–2.

1. $\dfrac{49}{56}$ **2.** $\dfrac{16x^3y^2}{-4xy}$ **3.** $\dfrac{2y^2 - 50}{y^2 - 4y - 5}$

Perform the indicated operations and simplify. Assume that all variables represent nonnegative real numbers. See sections 9–1 and 9–2.

4. $\sqrt{5}\sqrt{5}$ **5.** $\sqrt{3}\sqrt{3}$ **6.** $\sqrt{2}\sqrt{8}$ **7.** $\sqrt{12}\sqrt{3}$ **8.** $\sqrt{x}\sqrt{x}$

9–3 ■ Quotient property for radicals

The square root of a fraction

The following example will help us develop a property for division involving radicals.

$$\sqrt{\frac{4}{9}} = \sqrt{\left(\frac{2}{3}\right)^2} = \frac{2}{3}$$

We also observe that

$$\frac{\sqrt{4}}{\sqrt{9}} = \frac{2}{3}$$

From our example, we can conclude that

$$\sqrt{\frac{4}{9}} = \frac{\sqrt{4}}{\sqrt{9}}$$

We can now generalize this idea.

Quotient property for square roots

For any nonnegative real numbers a and b, where $b \neq 0$,

$$\sqrt{\frac{a}{b}} = \frac{\sqrt{a}}{\sqrt{b}}$$

Concept

The square root of a fraction can be written as the square root of the numerator divided by the square root of the denominator.

■ ***Example 9–3 A***

Simplify the following expressions. Assume that all variables represent positive real numbers.

1. $\sqrt{\frac{16}{25}} = \frac{\sqrt{16}}{\sqrt{25}} = \frac{4}{5}$ Rewrite as the square root of the numerator over the square root of the denominator and simplify

2. $\sqrt{\frac{36}{49}} = \frac{\sqrt{36}}{\sqrt{49}} = \frac{6}{7}$

3. $\sqrt{\frac{81}{100}} = \frac{\sqrt{81}}{\sqrt{100}} = \frac{9}{10}$

4. $\sqrt{\frac{x^4}{64}} = \frac{\sqrt{x^4}}{\sqrt{64}} = \frac{x^2}{8}$

5. $\sqrt{\frac{x^3}{y^4}} = \frac{\sqrt{x^3}}{\sqrt{y^4}} = \frac{x\sqrt{x}}{y^2}$

▶ ***Quick check*** Simplify. $\sqrt{\frac{16}{49}}$ and $\sqrt{\frac{a^2}{b}}$ ■

Rationalizing the denominator

When simplifying and evaluating radical expressions containing a radical in the denominator, it is easier if we can eliminate the radical in the denominator. For example,

$$\sqrt{\frac{4}{5}} = \frac{\sqrt{4}}{\sqrt{5}} = \frac{2}{\sqrt{5}}$$

Since $\sqrt{5} \cdot \sqrt{5} = 5$, we can eliminate the radical $\sqrt{5}$ in the denominator by multiplying the numerator and the denominator of the fraction by $\sqrt{5}$.

$$\frac{2}{\sqrt{5}} \cdot \frac{\sqrt{5}}{\sqrt{5}} = \frac{2\sqrt{5}}{\sqrt{25}} = \frac{2\sqrt{5}}{5}$$

The process of changing the denominator from a radical to a rational number is called **rationalizing the denominator.**

Rationalizing the denominator

1. Multiply the numerator and the denominator by the square root that is in the denominator. The radicand in the denominator will be a perfect-square integer.
2. Simplify the radical expressions in the numerator and the denominator.
3. Reduce the resulting fraction if possible.

■ ***Example 9–3 B***

Simplify the following expressions. Leave no radicals in the denominator. Assume that all variables represent positive real numbers.

1. $\dfrac{5}{\sqrt{7}} = \dfrac{5}{\sqrt{7}} \cdot \dfrac{\sqrt{7}}{\sqrt{7}}$ Multiply numerator and denominator by $\sqrt{7}$

$= \dfrac{5\sqrt{7}}{\sqrt{49}}$ Multiply in numerator and denominator

$= \dfrac{5\sqrt{7}}{7}$ $\sqrt{49} = 7$

2. $\dfrac{4}{\sqrt{6}} = \dfrac{4}{\sqrt{6}} \cdot \dfrac{\sqrt{6}}{\sqrt{6}} = \dfrac{4\sqrt{6}}{\sqrt{36}} = \dfrac{4\sqrt{6}}{6} = \dfrac{2\sqrt{6}}{3}$

Note In example 2, we were able to reduce the fraction as a final step. Always check to see that the answer is in reduced form.

3. $\dfrac{a}{\sqrt{a}} = \dfrac{a}{\sqrt{a}} \cdot \dfrac{\sqrt{a}}{\sqrt{a}} = \dfrac{a\sqrt{a}}{\sqrt{a^2}} = \dfrac{a\sqrt{a}}{a} = \sqrt{a}$

4. $\sqrt{\dfrac{a^3}{b}} = \dfrac{\sqrt{a^3}}{\sqrt{b}} = \dfrac{a\sqrt{a}}{\sqrt{b}} \cdot \dfrac{\sqrt{b}}{\sqrt{b}} = \dfrac{a\sqrt{ab}}{\sqrt{b^2}} = \dfrac{a\sqrt{ab}}{b}$

▶ ***Quick check*** Simplify. $\dfrac{6}{\sqrt{6}}$ ■

The following is a summary of the conditions necessary for a radical expression to be in **simplest form,** also called **standard form.**

1. The radicand contains no factors that can be written with an exponent greater than or equal to the index. ($\sqrt{a^3}$ violates this.)
2. The radicand contains no fractions. $\left(\sqrt{\dfrac{a}{b}} \text{ violates this.}\right)$
3. No radicals appear in the denominator. $\left(\dfrac{1}{\sqrt{a}} \text{ violates this.}\right)$

The nth root of a fraction

Our quotient property for square roots can be extended to radicals with any index.

Quotient property for radicals

$$\sqrt[n]{\frac{a}{b}} = \frac{\sqrt[n]{a}}{\sqrt[n]{b}} \qquad (b \neq 0)$$

Concept

The nth root of a fraction can be written as the nth root of the numerator divided by the nth root of the denominator.

■ ***Example 9–3 C***

Simplify the following expressions. Assume that all variables represent positive real numbers.

1. $\sqrt[3]{\frac{8}{27}} = \frac{\sqrt[3]{8}}{\sqrt[3]{27}}$ Rewrite as the cube root of the numerator over the cube root of the denominator and simplify

$= \frac{2}{3}$

2. $\sqrt[5]{\frac{32}{a^5}} = \frac{\sqrt[5]{32}}{\sqrt[5]{a^5}} = \frac{2}{a}$

3. $\sqrt[3]{\frac{x^5}{y^6}} = \frac{\sqrt[3]{x^5}}{\sqrt[3]{y^6}} = \frac{x\sqrt[3]{x^2}}{y^2}$

▶ ***Quick check*** Simplify. $\sqrt[3]{\frac{1}{27}}$ and $\sqrt[3]{\frac{8a^5}{b^3}}$ ■

Rationalizing the denominator (nth root)

The following example will help us develop a general rule for rationalizing a denominator that has a single term.

$$\sqrt[3]{\frac{1}{a}} = \frac{\sqrt[3]{1}}{\sqrt[3]{a}} = \frac{1}{\sqrt[3]{a}}$$

At this point, a radical still remains in the denominator. We must now determine what we can do to the fraction to remove the radical from the denominator.

Observations:

1. We can multiply the numerator and the denominator by the same number and form equivalent fractions.
2. If we multiply by a radical, the indices must be the same to carry out the multiplication.
3. To bring a factor out from under the radical symbol and not leave any of the factor behind, the index must divide evenly into the exponent.

With these observations in mind, we rationalize the fraction as follows:

$= \frac{1}{\sqrt[3]{a}} \cdot \frac{\sqrt[3]{\ }}{\sqrt[3]{\ }}$ Indices are the same

$= \frac{1}{\sqrt[3]{a}} \cdot \frac{\sqrt[3]{a^2}}{\sqrt[3]{a^2}}$ Multiply numerator and denominator by the same number

$$= \frac{\sqrt[3]{a^2}}{\sqrt[3]{a^3}}$$ The sum of the exponents of a in the denominator is equal to the index

$$= \frac{\sqrt[3]{a^2}}{a}$$ The index divides evenly into the exponent, the radical is eliminated

To rationalize an nth root denominator

1. Multiply the numerator and the denominator by a radical with the same index as the radical that we wish to eliminate from the denominator.
2. The exponent of each factor under the radical must be such that when we add it to the original exponent of the factor under the radical in the denominator, the sum will be equal to or divisible by the index of the radical.
3. Carry out the multiplication and reduce the fraction if possible.

■ *Example 9–3 D*

Simplify the following expressions. Leave no radicals in the denominator. Assume that all variables represent positive real numbers.

1. $\dfrac{1}{\sqrt[3]{7}} = \dfrac{1}{\sqrt[3]{7}} \cdot \dfrac{\sqrt[3]{7^2}}{\sqrt[3]{7^2}}$ Multiply numerator and denominator by $\sqrt[3]{7^2}$

$= \dfrac{\sqrt[3]{7^2}}{\sqrt[3]{7^3}}$ Multiply in numerator and denominator ($\sqrt[3]{7}\sqrt[3]{7^2} = \sqrt[3]{7^3}$)

$= \dfrac{\sqrt[3]{7^2}}{7}$ $\sqrt[3]{7^3} = 7$

$= \dfrac{\sqrt[3]{49}}{7}$ $7^2 = 49$

2. $$\frac{a}{\sqrt[5]{b^2}} = \frac{a}{\sqrt[5]{b^2}} \cdot \frac{\sqrt[5]{b^3}}{\sqrt[5]{b^3}} = \frac{a\sqrt[5]{b^3}}{\sqrt[5]{b^5}} = \frac{a\sqrt[5]{b^3}}{b}$$

3. $$\frac{x}{\sqrt[4]{x}} = \frac{x}{\sqrt[4]{x}} \cdot \frac{\sqrt[4]{x^3}}{\sqrt[4]{x^3}} = \frac{x\sqrt[4]{x^3}}{\sqrt[4]{x^4}} = \frac{x\sqrt[4]{x^3}}{x} = \sqrt[4]{x^3}$$

4. $$\frac{1}{\sqrt[5]{a^2b}} = \frac{1}{\sqrt[5]{a^2b}} \cdot \frac{\sqrt[5]{a^3b^4}}{\sqrt[5]{a^3b^4}} = \frac{\sqrt[5]{a^3b^4}}{\sqrt[5]{a^5b^5}} = \frac{\sqrt[5]{a^3b^4}}{ab}$$

► ***Quick check*** Simplify. $\dfrac{1}{\sqrt[5]{b^2}}$ ■

Mastery points

Can you

- Simplify radicals containing fractions?
- Rationalize denominators?

Exercise 9–3

Simplify the following expressions. Leave no radicals in the denominator. Assume that all variables represent positive real numbers. See examples 9–3 A and B.

Examples $\sqrt{\frac{16}{49}}$ $\qquad\sqrt{\frac{a^2}{b}}$ $\qquad\frac{6}{\sqrt{6}}$

Solutions $= \frac{\sqrt{16}}{\sqrt{49}} = \frac{4}{7}$

Rewrite as the square root of the numerator over the square root of the denominator and simplify

$= \frac{\sqrt{a^2}}{\sqrt{b}} = \frac{a}{\sqrt{b}} \cdot \frac{\sqrt{b}}{\sqrt{b}} = \frac{a\sqrt{b}}{b}$

$= \frac{6}{\sqrt{6}} \cdot \frac{\sqrt{6}}{\sqrt{6}} = \frac{6\sqrt{6}}{6} = \sqrt{6}$

Multiply the numerator and the denominator by the square root in the denominator and simplify

1. $\sqrt{\frac{9}{25}}$ **2.** $\sqrt{\frac{25}{36}}$ **3.** $\sqrt{\frac{25}{49}}$ **4.** $\sqrt{\frac{81}{100}}$ **5.** $\sqrt{\frac{3}{4}}$

6. $\sqrt{\frac{5}{9}}$ **7.** $\sqrt{\frac{64}{a^2}}$ **8.** $\sqrt{\frac{y^4}{16}}$ **9.** $\sqrt{\frac{1}{2}}$ **10.** $\sqrt{\frac{1}{3}}$

11. $\sqrt{\frac{4}{7}}$ **12.** $\sqrt{\frac{9}{11}}$ **13.** $\sqrt{\frac{1}{15}}$ **14.** $\sqrt{\frac{1}{14}}$ **15.** $\sqrt{\frac{4}{75}}$

16. $\sqrt{\frac{5}{12}}$ **17.** $\frac{2}{\sqrt{2}}$ **18.** $\frac{6}{\sqrt{3}}$ **19.** $\frac{10}{\sqrt{8}}$ **20.** $\frac{15}{\sqrt{27}}$

21. $\sqrt{\frac{x^2}{y}}$ **22.** $\sqrt{\frac{1}{a}}$ **23.** $\sqrt{\frac{1}{x}}$ **24.** $\sqrt{\frac{a^2}{b^3}}$ **25.** $\frac{\sqrt{a^5}}{\sqrt{a}}$

26. Find the width of a rectangle whose diagonal is 17 feet and length is 8 feet.

27. Find the diagonal of a rectangle whose length is 5 meters and whose width is 4 meters.

28. A 13-foot ladder is placed against the wall of a house. If the bottom of the ladder is 5 feet from the house, how far from the ground is the top of the ladder?

29. At an altitude of h feet above the sea or level ground, the distance d in miles that a person can see an object is found by using the equation $d = \sqrt{\frac{3h}{2}}$. How far can someone see who is in a tower 96 feet above the ground?

Simplify the following expressions. Leave no radicals in the denominator. Assume that all variables represent positive real numbers. See examples 9–3 C and D.

Examples $\sqrt[3]{\frac{1}{27}}$ $\qquad\sqrt[3]{\frac{8a^5}{b^3}}$ $\qquad\frac{1}{\sqrt[5]{b^2}}$

Solutions $= \frac{\sqrt[3]{1}}{\sqrt[3]{27}} = \frac{1}{3}$

$= \frac{\sqrt[3]{8a^5}}{\sqrt[3]{b^3}} = \frac{\sqrt[3]{2^3a^5}}{b} = \frac{2a\sqrt[3]{a^2}}{b}$

Rewrite as the cube root of the numerator over the cube root of the denominator and simplify

$= \frac{1}{\sqrt[5]{b^2}} \cdot \frac{\sqrt[5]{b^3}}{\sqrt[5]{b^3}} = \frac{\sqrt[5]{b^3}}{\sqrt[5]{b^5}} = \frac{\sqrt[5]{b^3}}{b}$

Multiply the numerator and the denominator by $\sqrt[5]{b^3}$

Perfect 5th root

$\sqrt[5]{b^5} = b$

30. $\sqrt[3]{\frac{8}{27}}$

31. $\sqrt[3]{\frac{1}{8}}$

32. $\sqrt[4]{\frac{16}{81}}$

33. $\sqrt[3]{\frac{27}{125}}$

34. $\sqrt[3]{\frac{a^2}{b^2}}$

35. $\sqrt[3]{\frac{3a^6}{b^3}}$

36. $\sqrt[3]{\frac{x}{y^{12}}}$

37. $\sqrt[5]{\frac{a^4}{b^{10}}}$

38. $\sqrt[5]{\frac{32x^4}{y^5}}$

39. $\sqrt[4]{\frac{a^4b^9}{c^{11}}}$

40. $\sqrt[4]{\frac{a^9b^{13}}{c^8}}$

41. $\sqrt[5]{\frac{x^3y^2}{z^{15}}}$

42. $\sqrt[3]{\frac{8}{9}}$

43. $\sqrt[3]{\frac{4}{25}}$

44. $\sqrt[3]{\frac{27}{16}}$

45. $\sqrt[4]{\frac{16}{125}}$

46. $\sqrt[4]{\frac{3}{4}}$

47. $\sqrt[3]{\frac{x^3}{y^2}}$

48. $\sqrt[3]{\frac{x^6}{y}}$

49. $\frac{ab}{\sqrt[3]{a^2}}$

50. $\frac{xy}{\sqrt[5]{y^3}}$

51. $\sqrt[3]{\frac{a^3}{b^2c}}$

52. $\sqrt[3]{\frac{8}{xy^2}}$

53. $\sqrt[3]{\frac{a^2}{b^2c}}$

54. $\frac{a}{\sqrt[5]{a^2b^4}}$

55. $\frac{ab}{\sqrt[3]{ab^2}}$

56. $\frac{xy^2}{\sqrt[5]{x^4y}}$

57. If we wish to construct a sphere of specific volume, we can find the length of radius necessary by the formula $r = \sqrt[3]{\frac{3V}{4\pi}}$. Find the radius necessary for a sphere to have a volume of 113.04 cubic units. (Use 3.14 for π.)

58. Use exercise 57 to find r if $V = 904.32$ cubic units. (Use 3.14 for π.)

Review exercises

Combine in the following. See section 2–3.

1. $4x + 2x$
2. $9y - 5y$
3. $5ab + 3ab$
4. $xy + 4xy$

Multiply the following. See section 3–2.

5. $(x + 3)(x - 3)$
6. $(x + y)(x - y)$

9–4 ■ Sums and differences of radicals

Like radicals

We have learned that in addition and subtraction of algebraic expressions, we can only combine like terms. This same idea applies when we are dealing with radicals. **We can add or subtract only like radicals.** Like radicals are radicals having the same index and the same radicand. For example, $3\sqrt{5x}$ and $-2\sqrt{5x}$ are like radicals, but $5\sqrt{7x}$ and $7\sqrt{5x}$ are not, because the radicands are not the same.

Addition and subtraction involving square roots

Addition and subtraction of radicals follow the same procedure as addition and subtraction of algebraic expressions. That is, *once we have determined that we have like radicals, the operations of addition and subtraction are performed only with the numerical coefficients.*

■ Example 9–4 A

Perform the indicated operations and simplify. Assume that all variables represent nonnegative real numbers.

1. $5\sqrt{2} + 3\sqrt{2} = (5 + 3)\sqrt{2} = 8\sqrt{2}$ — Apply the distributive property
2. $12\sqrt{3} - \sqrt{3} = (12 - 1)\sqrt{3} = 11\sqrt{3}$
3. $2\sqrt{a} + 3\sqrt{a} = 5\sqrt{a}$

▶ ***Quick check*** Simplify. $3\sqrt{6} + 2\sqrt{6} - \sqrt{6}$ ■

Consider the example

$$\sqrt{27} + 4\sqrt{3}$$

It appears that the indicated addition cannot be performed since we do not have like radicals. However we should have observed that the $\sqrt{27}$ can be simplified as

$$\sqrt{27} = \sqrt{9 \cdot 3} = 3\sqrt{3}$$

Our problem then becomes

$$\sqrt{27} + 4\sqrt{3} = 3\sqrt{3} + 4\sqrt{3} = 7\sqrt{3}$$

and we are able to add the like radicals. Therefore, *whenever we are working with radicals, we must be certain that all radicals are in simplest form.*

To combine like radicals

1. Perform any simplification within the terms.
2. Use the distributive property to combine terms that have like radicals.

■ Example 9–4 B

Perform the indicated operations. Assume that all variables represent nonnegative real numbers.

1. $\sqrt{45} + \sqrt{20} = \sqrt{9 \cdot 5} + \sqrt{4 \cdot 5}$ — Factor $45 = 9 \cdot 5$ and $20 = 4 \cdot 5$
   $= 3\sqrt{5} + 2\sqrt{5}$ — $\sqrt{4} = 2$; $\sqrt{9} = 3$
   $= 5\sqrt{5}$ — Add coefficients
2. $\sqrt{32} + 5\sqrt{8} = \sqrt{16 \cdot 2} + 5\sqrt{4 \cdot 2}$ — Factor $32 = 16 \cdot 2$; $8 = 4 \cdot 2$
   $= 4\sqrt{2} + 5 \cdot 2\sqrt{2}$ — $\sqrt{16} = 4$ and $\sqrt{4} = 2$
   $= 4\sqrt{2} + 10\sqrt{2}$ — Multiply $5 \cdot 2 = 10$
   $= 14\sqrt{2}$ — Add coefficients
3. $3\sqrt{3a} - \sqrt{12a} + 5\sqrt{48a}$
   $= 3\sqrt{3a} - \sqrt{4 \cdot 3a} + 5\sqrt{16 \cdot 3a}$ — Factor $12 = 4 \cdot 3$; $48 = 16 \cdot 3$
   $= 3\sqrt{3a} - 2\sqrt{3a} + 5 \cdot 4\sqrt{3a}$ — $\sqrt{4} = 2$; $\sqrt{16} = 4$
   $= 3\sqrt{3a} - 2\sqrt{3a} + 20\sqrt{3a}$ — $5 \cdot 4 = 20$
   $= 21\sqrt{3a}$ — Combine coefficients

▶ ***Quick check*** Simplify. $5\sqrt{2} + \sqrt{18}$ ■

Addition and subtraction involving nth roots

Addition and subtraction of radicals other than square roots follow the same procedure as addition and subtraction of expressions containing square roots. That is, *once we have determined that we have like radicals, the operations of addition and subtraction are performed only with the numerical coefficients.*

■ ***Example 9–4 C***

Perform the indicated operations and simplify. Assume that all variables represent nonnegative real numbers.

1. $\sqrt[3]{5} + 6\sqrt[3]{5} = (1 + 6)\sqrt[3]{5} = 7\sqrt[3]{5}$ — Combine coefficients

2.
$$\begin{aligned} 4\sqrt[3]{81} - \sqrt[3]{24} &= 4\sqrt[3]{27 \cdot 3} - \sqrt[3]{8 \cdot 3} && \text{Factor } 81 = 27 \cdot 3;\ 24 = 8 \cdot 3 \\ &= 4 \cdot 3\sqrt[3]{3} - 2\sqrt[3]{3} && \sqrt[3]{27} = 3;\ \sqrt[3]{8} = 2 \\ &= 12\sqrt[3]{3} - 2\sqrt[3]{3} && 4 \cdot 3 = 12 \\ &= 10\sqrt[3]{3} && \text{Subtract coefficients} \end{aligned}$$

3.
$$\begin{aligned} \sqrt[3]{16x^2y} + \sqrt[3]{54x^2y} &= \sqrt[3]{8 \cdot 2x^2y} + \sqrt[3]{27 \cdot 2x^2y} && \text{Factor } 16 = 8 \cdot 2;\ 54 = 27 \cdot 2 \\ &= 2\sqrt[3]{2x^2y} + 3\sqrt[3]{2x^2y} && \sqrt[3]{8} = 2;\ \sqrt[3]{27} = 3 \\ &= 5\sqrt[3]{2x^2y} && \text{Add coefficients} \end{aligned}$$ ■

Mastery points

Can you

- Identify like radicals?
- Add and subtract like radicals?

Exercise 9–4

Perform the indicated operations and simplify. Assume that all variables represent nonnegative real numbers. See examples 9–4 A and B.

Examples $3\sqrt{6} + 2\sqrt{6} - \sqrt{6}$

Solutions
$$\begin{aligned} &= (3 + 2 - 1)\sqrt{6} && \text{Distributive property} \\ &= 4\sqrt{6} && \text{Combine coefficients} \end{aligned}$$

$5\sqrt{2} + \sqrt{18}$
$$\begin{aligned} &= 5\sqrt{2} + \sqrt{9 \cdot 2} && \text{Factor } 18 = 9 \cdot 2 \\ &= 5\sqrt{2} + 3\sqrt{2} && \sqrt{9} = 3 \\ &= (5 + 3)\sqrt{2} && \text{Distributive property} \\ &= 8\sqrt{2} && \text{Add coefficients} \end{aligned}$$

1. $5\sqrt{3} + 4\sqrt{3}$
2. $8\sqrt{7} - 2\sqrt{7}$
3. $6\sqrt{5} + 4\sqrt{5}$
4. $9\sqrt{6} - 6\sqrt{6}$
5. $2\sqrt{3} + 7\sqrt{3} - 3\sqrt{3}$
6. $5\sqrt{5} - 4\sqrt{5} + 6\sqrt{5}$
7. $\sqrt{7} + 5\sqrt{7} - 3\sqrt{7}$
8. $2\sqrt{10} + 11\sqrt{10} - 9\sqrt{10}$
9. $\sqrt{a} + 2\sqrt{a}$
10. $3\sqrt{x} + 4\sqrt{x}$
11. $5\sqrt{a} - 4\sqrt{a} + 7\sqrt{a}$
12. $6\sqrt{y} - \sqrt{y} + 4\sqrt{y}$
13. $5\sqrt{xy} + 2\sqrt{xy}$
14. $3\sqrt{x} + 2\sqrt{y} - \sqrt{x}$
15. $5\sqrt{a} + 2\sqrt{ab} + 3\sqrt{a}$
16. $\sqrt{ab} + 2\sqrt{ab} + 3\sqrt{a}$
17. $5\sqrt{xy} - \sqrt{xy} + 3\sqrt{y}$
18. $\sqrt{20} + 3\sqrt{5}$
19. $\sqrt{8} + 5\sqrt{2}$
20. $\sqrt{12} + \sqrt{75}$
21. $\sqrt{48} - \sqrt{27}$
22. $2\sqrt{3} + 3\sqrt{12}$
23. $5\sqrt{7} + 4\sqrt{63}$
24. $5\sqrt{3} + \sqrt{27} - \sqrt{12}$
25. $4\sqrt{2} - \sqrt{8} + \sqrt{50}$
26. $\sqrt{75} - 4\sqrt{3} + 2\sqrt{27}$
27. $\sqrt{12} + \sqrt{18} + \sqrt{50}$

28. $\sqrt{63} - \sqrt{28} + \sqrt{24}$ **29.** $\sqrt{50a} + \sqrt{8a}$ **30.** $\sqrt{32a} - \sqrt{18a}$

31. $3\sqrt{9x} - 5\sqrt{4x}$ **32.** $2\sqrt{4x^2y} + 3\sqrt{25x^2y}$ **33.** $2\sqrt{8a} + 4\sqrt{50a} - 7\sqrt{2a}$

34. $3\sqrt{48b} - 2\sqrt{12b} + \sqrt{3b}$ **35.** $\sqrt{50a} + 3\sqrt{12a} - \sqrt{18a}$ **36.** $4\sqrt{25x^2y} + 3\sqrt{81x^2y} - 2\sqrt{2y}$

37. We can find the height, h, of the given figure by finding b from the formula $b = \sqrt{c^2 - s^2}$. If $c = 10$ units and $s = 6$ units, find h.

38. Use exercise 37 to find the height of the figure if $c = 13$ feet and $s = 5$ feet.

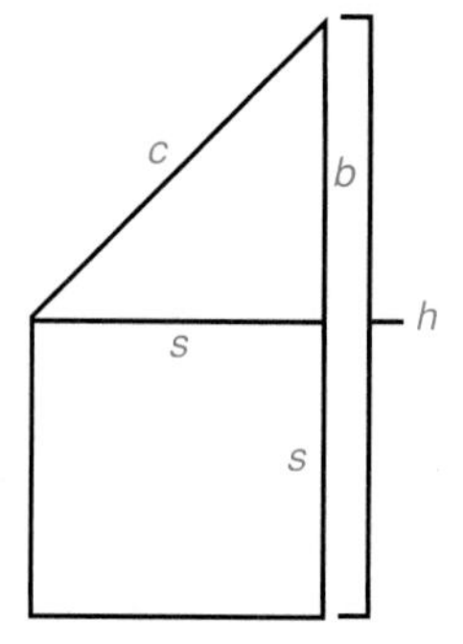

39. The figure is made up of 9 equal squares in which each square has an area of 7.29 square units. What are the dimensions of the figure?

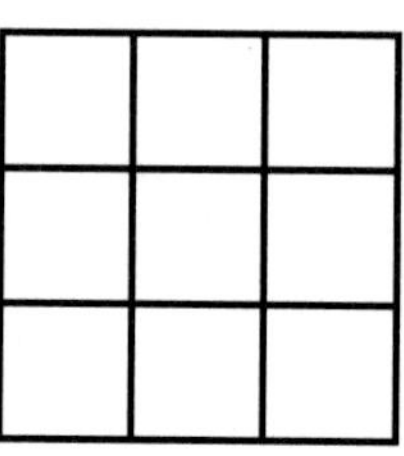

Perform the indicated operations and simplify. Assume that all variables represent nonnegative real numbers. See example 9–4 C.

40. $3\sqrt[3]{4} + 5\sqrt[3]{4}$ **41.** $7\sqrt[5]{2} - 4\sqrt[5]{2} + 3\sqrt[5]{2}$ **42.** $9\sqrt[4]{3} + 6\sqrt[4]{3} + 2\sqrt[4]{3}$

43. $\sqrt[3]{16} + \sqrt[3]{54}$ **44.** $\sqrt[3]{24} - \sqrt[3]{81}$ **45.** $\sqrt[3]{81} + 2\sqrt[3]{250}$

46. $\sqrt[3]{8a^2} + \sqrt[3]{27a^2}$ **47.** $\sqrt[4]{16x^3} + \sqrt[4]{81x^3}$ **48.** $\sqrt[4]{625a} - \sqrt[4]{81a}$

49. $\sqrt[3]{64x^2y} - \sqrt[3]{27x^2y}$ **50.** $\sqrt[3]{x^6y} + 2x^2\sqrt[3]{y}$ **51.** $3a\sqrt[3]{b^2} - \sqrt[3]{a^3b^2}$

52. $\sqrt[3]{16a^2b} + \sqrt[3]{54a^2b}$

Review exercises

Multiply the following expressions. See section 3–2.

1. $3x(2x - y)$ **2.** $2a^2(a^2 - b^2)$ **3.** $(x - 1)(x - 1)$ **4.** $(y - 1)(y + 1)$

5. $(2x + 1)(2x - 1)$ **6.** $(x + 3y)(x + 2y)$ **7.** $(x - y)^2$ **8.** $(a + 2b)^2$

9–5 ■ Further operations with radicals

Multiplication of radical expressions

In section 9–2, we learned the procedure for multiplying two radicals. We now combine those ideas along with the *distributive property*, $a(b + c) = ab + ac$, to perform multiplication of radical expressions containing more than one term.

Example 9–5 A

Perform the indicated operations and simplify. Assume that all variables represent nonnegative real numbers.

1. $\sqrt{3}(3+\sqrt{3}) = \sqrt{3}\cdot 3 + \sqrt{3}\sqrt{3}$ — Distributive property
$= 3\sqrt{3} + \sqrt{9}$ — $\sqrt{3}\sqrt{3} = \sqrt{9}$
$= 3\sqrt{3} + 3$ — $\sqrt{9} = 3$

2. $\sqrt{3}(\sqrt{6}+\sqrt{21}) = \sqrt{3}\sqrt{6} + \sqrt{3}\sqrt{21}$ — Distributive property
$= \sqrt{18} + \sqrt{63}$ — Multiply radicands
$= \sqrt{9\cdot 2} + \sqrt{9\cdot 7}$ — Factor $18 = 9\cdot 2$; $63 = 9\cdot 7$
$= 3\sqrt{2} + 3\sqrt{7}$ — Simplify radicals

3. $(\sqrt{2}+\sqrt{3})(\sqrt{2}+5\sqrt{3})$

Note In this example, we are multiplying two binomials. Therefore, as we did in chapter 3, we will *multiply each term in the first parentheses by each term in the second parentheses.*

$= \sqrt{2}\sqrt{2} + \sqrt{2}\cdot 5\sqrt{3} + \sqrt{3}\sqrt{2} + \sqrt{3}\cdot 5\sqrt{3}$ — Distributive property
$= \sqrt{4} + 5\sqrt{6} + \sqrt{6} + 5\cdot\sqrt{9}$ — Multiply radicands
$= 2 + 5\sqrt{6} + \sqrt{6} + 5\cdot 3$ — $\sqrt{4} = 2$, $\sqrt{9} = 3$
$= 2 + 5\sqrt{6} + \sqrt{6} + 15$
$= 17 + 6\sqrt{6}$ — Combine like terms

4. $(3-\sqrt{2})(3+\sqrt{2}) = 9 + 3\sqrt{2} - 3\sqrt{2} - \sqrt{4}$ — Distributive property
$= 9 + 3\sqrt{2} - 3\sqrt{2} - 2$ — Simplify radicals
$= 9 - 2$ — Combine like terms
$= 7$ — Subtract

We observe that when we simplified, there were no longer any radicals in the answer.

5. $(\sqrt{a}-\sqrt{b})(\sqrt{a}+\sqrt{b})$
$= \sqrt{a}\sqrt{a} + \sqrt{a}\sqrt{b} - \sqrt{b}\sqrt{a} - \sqrt{b}\sqrt{b}$ — Distributive property
$= \sqrt{a^2} + \sqrt{ab} - \sqrt{ab} - \sqrt{b^2}$ — Multiply radicands
$= a + \sqrt{ab} - \sqrt{ab} - b$ — $\sqrt{a^2} = a$ and $\sqrt{b^2} = b$
$= a - b$ — Combine like terms

6. $(\sqrt{3}+2\sqrt{2})^2$
$= (\sqrt{3}+2\sqrt{2})(\sqrt{3}+2\sqrt{2})$
$= \sqrt{3}\sqrt{3} + \sqrt{3}\cdot 2\sqrt{2} + \sqrt{3}\cdot 2\sqrt{2} + 2\sqrt{2}\cdot 2\sqrt{2}$ — Distributive property
$= \sqrt{9} + 2\sqrt{6} + 2\sqrt{6} + 4\sqrt{4}$ — Multiply radicands
$= 3 + 2\sqrt{6} + 2\sqrt{6} + 4\cdot 2$ — Simplify radicals
$= 3 + 2\sqrt{6} + 2\sqrt{6} + 8$ — Multiply
$= 11 + 4\sqrt{6}$ — Combine like terms

▶ ***Quick check*** Simplify. $2(\sqrt{3}+\sqrt{5})$; $\sqrt{2}(\sqrt{14}+\sqrt{6})$; and $(\sqrt{2}+\sqrt{3})(\sqrt{2}-2\sqrt{3})$ ■

Conjugate factors

The type of factors that we are multiplying in examples 4 and 5 are called **conjugate factors.** The conjugate is used to rationalize the denominator of a fraction when the denominator contains two terms where one or both terms contain a square root. The idea of conjugate factors is derived from the factorization of the difference of two squares. When multiplying conjugate factors, we can simply write our answer as the square of the second term subtracted from the square of the first term.

In examples 4 and 5, we could have performed the multiplication as follows:

4. $(3 - \sqrt{2})(3 + \sqrt{2}) = (3)^2 - (\sqrt{2})^2 = 9 - 2 = 7$;
5. $(\sqrt{a} - \sqrt{b})(\sqrt{a} + \sqrt{b}) = (\sqrt{a})^2 - (\sqrt{b})^2 = a - b$.

To determine what the conjugate of a given factor is, we write the original factor and change the sign of the second term.

■ ***Example 9–5 B***

Form the conjugates of the given expressions.

1. $\sqrt{7} + 2$ — The conjugate is $\sqrt{7} - 2$.

2. $\sqrt{11} - \sqrt{6}$ — The conjugate is $\sqrt{11} + \sqrt{6}$.

3. $-5 - 2\sqrt{3}$ — The conjugate is $-5 + 2\sqrt{3}$.

4. $\sqrt{a} + \sqrt{b}$ — The conjugate is $\sqrt{a} - \sqrt{b}$.

▶ ***Quick check*** Form the conjugate. $6 - 3\sqrt{2}$ ■

Rationalizing the denominator

If we wish to rationalize the denominator of the fraction

$$\frac{1}{3 - \sqrt{2}}$$

we recall from example 9–5 A–4 that when we multiplied $3 - \sqrt{2}$ by $3 + \sqrt{2}$, there were no radicals left in our product. This result is precisely what we want to occur in our denominator. Therefore, to rationalize this fraction, we apply the fundamental principle of fractions and multiply the numerator and the denominator by $3 + \sqrt{2}$, the conjugate of the denominator.

$$\frac{1}{3 - \sqrt{2}} = \frac{1}{3 - \sqrt{2}} \cdot \frac{3 + \sqrt{2}}{3 + \sqrt{2}}$$ $3 + \sqrt{2}$ is the conjugate of the denominator

$$= \frac{1(3 + \sqrt{2})}{(3)^2 - (\sqrt{2})^2}$$ (first term)^2 − (second term)^2

$$= \frac{3 + \sqrt{2}}{9 - 2}$$ No radicals remain in the denominator

$$= \frac{3 + \sqrt{2}}{7}$$ Denominator is rationalized

■ ***Example 9–5 C***

Rationalize the denominators.

1. $\dfrac{2}{\sqrt{7}+2} = \dfrac{2}{\sqrt{7}+2}\cdot\dfrac{\sqrt{7}-2}{\sqrt{7}-2}$ Multiply by the conjugate of the denominator

$= \dfrac{2(\sqrt{7}-2)}{(\sqrt{7})^2-(2)^2}$ $(x+y)(x-y)=x^2-y^2$

$= \dfrac{2\sqrt{7}-4}{7-4}$ Simplify in numerator and denominator

$= \dfrac{2\sqrt{7}-4}{3}$ Subtract in denominator

2. $\dfrac{5}{\sqrt{11}-\sqrt{6}} = \dfrac{5}{\sqrt{11}-\sqrt{6}}\cdot\dfrac{\sqrt{11}+\sqrt{6}}{\sqrt{11}+\sqrt{6}}$ Multiply by the conjugate of the denominator

$= \dfrac{5(\sqrt{11}+\sqrt{6})}{(\sqrt{11})^2-(\sqrt{6})^2}$ $(x+y)(x-y)=x^2-y^2$

$= \dfrac{5(\sqrt{11}+\sqrt{6})}{11-6}$ Simplify radicals

$= \dfrac{5(\sqrt{11}+\sqrt{6})}{5}$ Subtract in denominator

$= \sqrt{11}+\sqrt{6}$ Reduce (by 5) to lowest terms

3. $\dfrac{\sqrt{3}}{5-2\sqrt{3}} = \dfrac{\sqrt{3}}{5-2\sqrt{3}}\cdot\dfrac{5+2\sqrt{3}}{5+2\sqrt{3}}$ Multiply by the conjugate of the denominator

$= \dfrac{\sqrt{3}(5+2\sqrt{3})}{(5)^2-(2\sqrt{3})^2}$ $(x+y)(x-y)=x^2-y^2$

$= \dfrac{5\sqrt{3}+2\sqrt{9}}{5^2-2^2(\sqrt{3})^2}$ Simplify radicals

$= \dfrac{5\sqrt{3}+2\cdot 3}{25-4\cdot 3}$ Simplify radicals

$= \dfrac{5\sqrt{3}+6}{25-12}$ Perform operations

$= \dfrac{5\sqrt{3}+6}{13}$ Subtract in denominator

▶ ***Quick check*** Rationalize the denominators. $\dfrac{3}{7+\sqrt{2}}$ and $\dfrac{2}{\sqrt{11}-3}$ ■

Mastery points

Can you

- Multiply radical expressions containing more than one term?
- Form conjugate factors?
- Multiply conjugate factors?
- Rationalize a denominator that has two terms in which one or both terms contain a square root?

Exercise 9–5

Perform the indicated operations and simplify. Assume that all variables represent positive real numbers. See example 9–5 A.

Examples $2(\sqrt{3} + \sqrt{5})$

Solutions $= 2\sqrt{3} + 2\sqrt{5}$ Distributive property

$\sqrt{2}(\sqrt{14} + \sqrt{6})$

$= \sqrt{2}\sqrt{14} + \sqrt{2}\sqrt{6}$ Distributive property

$= \sqrt{28} + \sqrt{12}$ Product property

$= \sqrt{4 \cdot 7} + \sqrt{4 \cdot 3}$ Factor $28 = 4 \cdot 7$; $12 = 4 \cdot 3$

$= 2\sqrt{7} + 2\sqrt{3}$ $\sqrt{4} = 2$

Example $(\sqrt{2} + \sqrt{3})(\sqrt{2} - 2\sqrt{3})$

Solution $= \sqrt{2}\sqrt{2} - \sqrt{2} \cdot 2\sqrt{3} + \sqrt{3}\sqrt{2} - \sqrt{3} \cdot 2\sqrt{3}$ Distributive property

$= 2 - 2\sqrt{6} + \sqrt{6} - 2 \cdot 3$ $\sqrt{2}\sqrt{2} = 2$; $\sqrt{3}\sqrt{3} = 3$

$= 2 - \sqrt{6} - 6$ Combine like radicals

$= -4 - \sqrt{6}$ $2 - 6 = -4$

1. $3(\sqrt{2} + \sqrt{3})$
2. $5(2\sqrt{6} + \sqrt{2})$
3. $\sqrt{2}(\sqrt{3} + \sqrt{7})$
4. $\sqrt{5}(\sqrt{7} - \sqrt{3})$
5. $3\sqrt{2}(2\sqrt{3} - \sqrt{11})$
6. $\sqrt{6}(\sqrt{2} + \sqrt{3})$
7. $\sqrt{5}(\sqrt{15} - \sqrt{10})$
8. $\sqrt{14}(\sqrt{21} + \sqrt{10})$
9. $2\sqrt{7}(\sqrt{35} - 3\sqrt{14})$
10. $\sqrt{a}(\sqrt{ab} + \sqrt{b})$
11. $\sqrt{a}(3\sqrt{a} + \sqrt{b})$
12. $(5 + \sqrt{3})(4 - \sqrt{3})$
13. $(3 + \sqrt{2})(4 + \sqrt{2})$
14. $(5 - \sqrt{5})(5 - \sqrt{5})$
15. $(3 - 4\sqrt{a})(4 - 3\sqrt{a})$
16. $(7 + 2\sqrt{y})(6 + 5\sqrt{y})$
17. $(\sqrt{3} + \sqrt{2})(\sqrt{3} - \sqrt{2})$
18. $(\sqrt{7} + \sqrt{5})(\sqrt{7} - \sqrt{5})$
19. $(2 + \sqrt{6})(2 - \sqrt{6})$
20. $(5 - \sqrt{3})(5 + \sqrt{3})$
21. $(2 + \sqrt{5})^2$
22. $(3 - \sqrt{7})^2$
23. $(\sqrt{x} + \sqrt{y})^2$
24. $(\sqrt{a} - \sqrt{b})^2$
25. $(\sqrt{x} - \sqrt{y})(\sqrt{x} + \sqrt{y})$
26. $(2\sqrt{a} - \sqrt{b})(2\sqrt{a} + \sqrt{b})$
27. $(x\sqrt{y} + \sqrt{z})(x\sqrt{y} - \sqrt{z})$
28. $(a\sqrt{b} + c)(a\sqrt{b} - c)$
29. $(2\sqrt{x} + y)^2$
30. $(3\sqrt{a} + \sqrt{b})^2$

Form the conjugate of the given expressions. See example 9–5 B.

Example $6 - 3\sqrt{2}$

Solution $6 + 3\sqrt{2}$ First term remains the same, change the sign of the second term

31. $11 - \sqrt{3}$
32. $-5\sqrt{7} - \sqrt{2}$
33. $\sqrt{a} + 3\sqrt{b}$
34. $a\sqrt{b} - \sqrt{c}$

Rationalize the denominators. Assume that all variables represent positive real numbers and that no denominator is equal to zero. See example 9–5 C.

Simplify the following expressions, leaving all denominators rationalized. Assume that all variables represent positive real numbers and that no denominator is equal to zero. See example 9–5 C.

Examples $\frac{3}{7+\sqrt{2}}$

Solutions

$= \frac{3}{7+\sqrt{2}} \cdot \frac{7-\sqrt{2}}{7-\sqrt{2}}$ Multiply by the conjugate

$= \frac{3(7-\sqrt{2})}{(7)^2-(\sqrt{2})^2}$ $(x+y)(x-y)=x^2-y^2$

$= \frac{3(7-\sqrt{2})}{49-2}$ Simplify denominator

$= \frac{3(7-\sqrt{2})}{47}$ Simplify denominator

$= \frac{21-3\sqrt{2}}{47}$ Multiply in numerator

Examples $\frac{2}{\sqrt{11}-3}$

Solutions

$= \frac{2}{\sqrt{11}-3} \cdot \frac{\sqrt{11}+3}{\sqrt{11}+3}$ Multiply by the conjugate

$= \frac{2(\sqrt{11}+3)}{(\sqrt{11})^2-(3)^2}$ $(x+y)(x-y)=x^2-y^2$

$= \frac{2(\sqrt{11}+3)}{11-9}$ Simplify denominator

$= \frac{2(\sqrt{11}+3)}{2}$ Simplify denominator

$= \sqrt{11}+3$ Reduce fraction

35. $\frac{1}{\sqrt{2}+3}$ **36.** $\frac{1}{\sqrt{3}-2}$ 37. $\frac{7}{2+\sqrt{7}}$ **38.** $\frac{6}{3-\sqrt{6}}$

39. $\frac{3}{\sqrt{6}-\sqrt{3}}$ **40.** $\frac{1}{\sqrt{a}+b}$ **41.** $\frac{3}{2\sqrt{3}-\sqrt{5}}$ **42.** $\frac{4}{2\sqrt{3}-\sqrt{6}}$

43. $\frac{1+\sqrt{5}}{1-\sqrt{5}}$ **44.** $\frac{\sqrt{3}-\sqrt{7}}{\sqrt{3}+\sqrt{7}}$ **45.** $\frac{\sqrt{a}+b}{\sqrt{a}-b}$

Review exercises

Perform the indicated operations and leave your answer with only positive exponents. Assume that no variable is equal to zero. See sections 3–1 and 3–3.

1. $(2^2)^3$ **2.** $2^2 \cdot 2^3$ **3.** $3^{-2} \cdot 3^4$ **4.** $\frac{3^2}{3^5}$

5. $(x^{-2})^3$ **6.** $(2a^2b)^3$ **7.** $\frac{x^2y^5}{x^3y^2}$ **8.** $x^{-4} \cdot x^{-5}$

9–6 ■ Fractional exponents

Fractional exponents

In this section, we are going to develop the idea of a fraction used as an exponent. Consider the example

$$(a^{1/2})^2 = a^{1/2 \cdot 2} = a^1$$

When we raise a power to a power, we multiply the exponents. In the previous section, we observed that if a represented only nonnegative real numbers, then

$$(\sqrt{a})^2 = a$$

Therefore, for our properties of exponents and our procedures for radicals to be consistent, the following statement must be true:

$$a^{1/2} = \sqrt{a}$$

We generalize as follows:

Definition of $a^{1/n}$

$$a^{1/n} = \sqrt[n]{a}$$

where n is a natural number greater than 1. Whenever n is even, a represents only nonnegative real numbers.

Concept

The expression $a^{1/n}$ represents the principal nth root of a.

Example 9–6 A

Rewrite the following in radical notation and simplify where possible.

1. $5^{1/2} = \sqrt{5}$
2. $(64)^{1/2} = \sqrt{64} = 8$
3. $a^{1/3} = \sqrt[3]{a}$
4. $(-8)^{1/3} = \sqrt[3]{(-8)} = -2$ ■

Once we become acquainted with fractional exponents, the process of changing the fractional exponent to radical form for simplification will become unnecessary.

Consider the following problem:

$$(\sqrt[3]{a})^2 = (a^{1/3})^2 = a^{2/3}$$

We observe that *when a number is raised to a fractional exponent, the numerator of the fractional exponent indicates the power to which the base is to be raised, and the denominator indicates the root to be taken.*

Definition of $a^{m/n}$

If a is any real number, m is any integer, and n is any positive integer, then if $\dfrac{m}{n}$ is reduced to lowest terms,

$$a^{m/n} = (\sqrt[n]{a})^m = \sqrt[n]{a^m}$$

provided that $\sqrt[n]{a}$ is a real number.

Concept

The expression $a^{m/n}$ represents the principal nth root of a raised to the mth power.

We calculate $a^{m/n}$ *by first finding the principal* nth *root of* a *and then raising the resulting number to the* mth *power.*

■ ***Example 9–6 B***

Simplify. Assume that all variables represent nonnegative real numbers.

Numerator is the power — Denominator is the index

1. $(16)^{3/4} = (\sqrt[4]{16})^3$
   $= 2^3$ 4th root of 16 is 2
   $= 8$ Standard form
2. $(32)^{2/5} = (\sqrt[5]{32})^2 = 2^2 = 4$
3. $(-27)^{2/3} = (\sqrt[3]{-27})^2 = (-3)^2 = 9$
4. $(x^6)^{1/3} = \sqrt[3]{x^6} = x^2$

▶ ***Quick check*** Simplify. $(49)^{1/2}$ and $(8)^{2/3}$ ■

Operations with fractional exponents

We can extend the properties and definitions involving integer exponents to expressions that involve fractional exponents.

■ ***Example 9–6 C***

Perform the indicated operations and simplify. Assume that all variables represent positive real numbers.

1. $5^{1/2} \cdot 5^{1/3} = 5^{1/2 + 1/3}$ Multiply like bases, add exponents
   $= 5^{3/6 + 2/6}$ Least common denominator is 6
   $= 5^{5/6}$ Add numerators
2. $\dfrac{2^{1/2}}{2^{1/4}} = 2^{1/2 - 1/4} = 2^{2/4 - 1/4} = 2^{1/4}$
3. $(-27)^{-2/3} = \dfrac{1}{(-27)^{2/3}} = \dfrac{1}{(\sqrt[3]{-27})^2} = \dfrac{1}{(-3)^2} = \dfrac{1}{9}$
4. $(2^3x^9y^{15})^{1/3} = (2^3)^{1/3}(x^9)^{1/3}(y^{15})^{1/3} = 2^{3 \cdot 1/3} \cdot x^{9 \cdot 1/3} \cdot y^{15 \cdot 1/3} = 2^1x^3y^5$
   $= 2x^3y^5$
5. $x^{2/3} \cdot x^{3/4} = x^{2/3 + 3/4} = x^{8/12 + 9/12} = x^{17/12}$
6. $(y^{1/2})^{4/3} = y^{1/2 \cdot 4/3} = y^{2/3}$
7. $\dfrac{z^{1/2}}{z^{2/3}} = z^{1/2 - 2/3} = z^{3/6 - 4/6} = z^{-1/6} = \dfrac{1}{z^{1/6}}$

Note In example 7, $z^{1/6}$ in the denominator is simply another form of $\sqrt[6]{z}$. Therefore, if we want our answer in a rationalized form, we would proceed as follows:

$$\frac{1}{z^{1/6}} = \frac{1}{\sqrt[6]{z}} \cdot \frac{\sqrt[6]{z^5}}{\sqrt[6]{z^5}} = \frac{\sqrt[6]{z^5}}{z} = \frac{z^{5/6}}{z}$$

▶ ***Quick check*** Simplify. $y^{2/3} \cdot y^{3/2}$; $\dfrac{a^{3/4}b^{5/3}}{a^{1/4}b}$; and $(36)^{-1/2}$ ■

Mastery points

Can you

- Express fractional exponents in radical form?
- Express radicals in fractional exponent form?
- Apply the properties and definitions involving integer exponents to fractional exponents?

Exercise 9–6

Simplify the given expressions. See examples 9–6 A and B.

Examples $(49)^{1/2}$

Solutions $= \sqrt{49}$ Square root
$= 7$

$8^{2/3}$
$= (\sqrt[3]{8})^2$ Numerator is the power; Denominator is the index
$= (2)^2$ Cube root of 8 is 2
$= 4$ Standard form

$(36)^{-1/2}$
$= \dfrac{1}{(36)^{1/2}}$ $a^{-n} = \dfrac{1}{a^n}$
$= \dfrac{1}{\sqrt{36}}$ Square root
$= \dfrac{1}{6}$ Simplify

1. $(36)^{1/2}$ 2. $(25)^{1/2}$ 3. $(a^6)^{1/3}$ 4. $(b^{12})^{1/3}$ 5. $(8)^{1/3}$ 6. $(32)^{1/5}$
7. $(-27)^{1/3}$ 8. $(-8)^{1/3}$ 9. $(27)^{2/3}$ 10. $(16)^{3/4}$ 11. $(9)^{3/2}$ 12. $(16)^{3/2}$

Perform the indicated operations and simplify. Assume that all variables represent positive real numbers. See example 9–6 C.

Examples $y^{2/3} \cdot y^{3/2}$

Solutions $= y^{2/3 + 3/2}$ Multiply like bases, add exponents
$= y^{4/6 + 9/6}$ Least common denominator is 6
$= y^{13/6}$ Add numerators

$\dfrac{a^{3/4}b^{5/3}}{a^{1/4}b}$
$= a^{3/4 - 1/4}b^{5/3 - 1}$ Divide like bases, subtract exponents
$= a^{2/4}b^{5/3 - 3/3}$ Common denominators
$= a^{1/2}b^{2/3}$ Subtract numerators

13. $(25)^{-1/2}$ 14. $(9)^{-1/2}$ 15. $(16)^{-3/4}$ 16. $(27)^{-2/3}$ 17. $(-8)^{-1/3}$ 18. $(-27)^{-1/3}$
19. $2^{1/2} \cdot 2^{3/2}$ 20. $3^{1/3} \cdot 3^{2/3}$ 21. $2^{1/3} \cdot 2^{1/2}$ 22. $5^{1/5} \cdot 5^{1/2}$ 23. $x^{1/4} \cdot x^{3/4}$ 24. $b^{1/3} \cdot b^{2/3}$
25. $c^{1/2} \cdot c^{1/4}$ 26. $x^{1/4} \cdot x^{1/3}$ 27. $\dfrac{2^{3/2}}{2^{1/2}}$ 28. $\dfrac{3^{4/3}}{3^{1/3}}$ 29. $\dfrac{2^{1/2}}{2^{1/3}}$ 30. $\dfrac{7^{3/4}}{7^{2/3}}$
31. $\dfrac{a^{4/5}}{a^{1/5}}$ 32. $\dfrac{x^{3/4}}{x^{1/2}}$ 33. $\dfrac{y^{2/3}}{y^{1/2}}$ 34. $\dfrac{x^{5/6}}{x^{2/3}}$ 35. $(a^{2/3})^{1/2}$ 36. $(c^{1/2})^{1/2}$
37. $(x^{1/2})^{4/3}$ 38. $(y^{2/3})^{3/4}$ 39. $(c^{-1/4})^{-2/3}$ 40. $(y^{-1/2})^{-2/5}$ 41. $(a^{-2/3})^{-1/2}$ 42. $(b^{-1/2})^{-1/2}$
43. $(x^{1/4})^{-2/3}$ 44. $(x^{1/3})^{-3/4}$ 45. $(y^{-3/4})^{1/3}$ 46. $(c^{-2/5})^{1/2}$ 47. $(16a^4)^{3/4}$ 48. $(x^3y^{12})^{1/3}$
49. $(8a^6b^3)^{2/3}$ 50. $(27x^3y^{12})^{2/3}$ 51. $\dfrac{b^{3/4}c^{1/2}}{b^{1/4}c^{1/4}}$ 52. $\dfrac{xy^{3/4}}{x^{1/2}y^{1/4}}$ 53. $\dfrac{ab}{a^{1/2}b^{1/3}}$ 54. $\dfrac{xy^{3/4}}{x^{2/5}y^{1/2}}$

Solve the following word problems.

55. Find the number whose principal fourth root is 3.

56. Find the number whose principal cube root is -2.

57. The formula for approximating the velocity V in miles per hour of a car based on the length of its skid marks S (in feet) on dry pavement is given by

$V = 4.9\ S^{1/2}$

If the skid marks are 100 feet long, what was the approximate velocity?

58. The formula for approximating the velocity V in miles per hour of a car based on the length of its skid marks S (in feet) on wet pavement is given by

$V = 3.46\ S^{1/2}$

If the skid marks are 81 feet long, what was the approximate velocity of the car?

59. At an altitude of h feet above the sea or level ground, the distance d in miles that a person can see an object is found by using the equation

$d = 1.2\ h^{1/2}$

How far can someone see who is in a tower 400 feet above the ground?

60. How can you find the principal fourth root of a number on a calculator using only the square root key?

61. How can you find the principal eighth root of a number on a calculator using only the square root key?

Review exercises

Perform the indicated operations. Assume that all radicands are nonnegative. See sections 9–1 and 9–2.

1. $(\sqrt{7})^2$ **2.** $(\sqrt{x})^2$ **3.** $(\sqrt{x+1})^2$

Perform the indicated operations. See section 3–2.

4. $(x + 1)^2$ **5.** $(x - 2)^2$

Find the solution set for the following equations. See section 4–7.

6. $x + 6 = x^2$ **7.** $x + 2 = x^2 - 9x + 18$ **8.** $x + 1 = x^2 + 2x + 1$

9–7 ■ Equations involving radicals

Radical equations

An equation in which the unknown quantity appears under a radical symbol is called a **radical equation.** Examples of radical equations are

$$\sqrt{x} = 5; \quad \sqrt{x+2} = 7; \quad 4 + \sqrt{x+2} = x$$

In this section, we will consider radical equations containing only square roots.

Solving radical equations involves squaring both members of an equation to eliminate the square roots. When we square both members of an equation, we use the squaring property of equality.

Squaring property of equality

If P and Q are algebraic expressions and if

$$P = Q$$

then all solutions of $P = Q$ are also solutions of the equation

$$P^2 = Q^2$$

Concept

If each member of an equation is squared, the solution(s) of the original equation are solution(s) of the resulting equation.

Extraneous solutions

This property implies that there *may be* solutions of the equation $P^2 = Q^2$ that are not solutions of the original equation $P = Q$. If such solutions exist, they are called **extraneous solutions** (roots). Thus, all possible solutions must be checked in the original equation.

To solve a radical equation

1. Rewrite the equation (if necessary) so that a radical is by itself in one member of the equation.
2. Square each member (side) of the equation and combine like terms.
3. Repeat steps 1 and 2 if a radical remains in the equation.
4. Solve the resulting equation.
5. Check all possible solutions in the original equation.

■ *Example 9–7 A*

Find the solution set.

1.

$\sqrt{x} = 5$	
$(\sqrt{x})^2 = (5)^2$	Square both members
$x = 25$	Squaring a square root gives the radicand
Check:	
$\sqrt{x} = 5$	Original equation
$\sqrt{(25)} = 5$	Substitute into original equation
$5 = 5$	True

The solution set is $\{25\}$.

2.

$\sqrt{x + 2} = 7$	
$(\sqrt{x + 2})^2 = (7)^2$	Square both members
$x + 2 = 49$	Squaring a square root gives the radicand
$x = 47$	Subtract 2 from both members
Check:	
$\sqrt{x + 2} = 7$	Original equation
$\sqrt{(47) + 2} = 7$	Substitute into original equation
$\sqrt{49} = 7$	Simplify the radicand
$7 = 7$	True

The solution set is $\{47\}$.

3.
$$4 + \sqrt{x + 2} = x$$
$$\sqrt{x + 2} = x - 4 \quad \text{Isolate the radical by subtracting 4}$$
$$(\sqrt{x + 2})^2 = (x - 4)^2 \quad \text{Square both members}$$
$$x + 2 = (x - 4)(x - 4) \quad \text{Simplify}$$
$$x + 2 = x^2 - 4x - 4x + 16 \quad \text{Multiply}$$
$$x + 2 = x^2 - 8x + 16 \quad \text{Subtract } x \text{ and } 2$$
$$0 = x^2 - 9x + 14 \quad \text{Solve the resulting quadratic equation}$$
$$0 = (x - 7)(x - 2) \quad \text{Factor}$$
$$x - 7 = 0 \text{ or } x - 2 = 0 \quad \text{Set each factor equal to 0 and solve}$$
$$x = 7 \text{ or } x = 2 \quad \text{Check for extraneous roots}$$

Check:

for 7
$$4 + \sqrt{x + 2} = x$$
$$4 + \sqrt{(7) + 2} = (7)$$
$$4 + \sqrt{9} = 7$$
$$4 + 3 = 7$$
$$7 = 7 \quad \text{True}$$

for 2
$$4 + \sqrt{x + 2} = x \quad \text{Original equation}$$
$$4 + \sqrt{(2) + 2} = (2) \quad \text{Substitute}$$
$$4 + \sqrt{4} = 2 \quad \text{Simplify}$$
$$4 + 2 = 2 \quad \text{Simplify}$$
$$6 = 2 \quad \text{False}$$

Therefore, 7 is the only solution to the equation. 2 is an *extraneous root* and the solution set is $\{7\}$.

4.
$$\sqrt{x + 1} = x + 1$$
$$(\sqrt{x + 1})^2 = (x + 1)^2 \quad \text{Square both members}$$
$$x + 1 = (x + 1)(x + 1) \quad \text{Simplify}$$
$$x + 1 = x^2 + x + x + 1 \quad \text{Multiply}$$
$$x + 1 = x^2 + 2x + 1 \quad \text{Combine like terms}$$
$$0 = x^2 + x \quad \text{Solve the resulting quadratic equation}$$
$$0 = x(x + 1) \quad \text{Factor}$$
$$x = 0 \text{ or } x + 1 = 0 \quad \text{Set each factor equal to 0 and solve}$$

Therefore, $x = 0$ or $x = -1$.

Check:

for 0
$$\sqrt{x + 1} = x + 1$$
$$\sqrt{(0) + 1} = (0) + 1$$
$$\sqrt{1} = 1$$
$$1 = 1 \quad \text{True}$$

for -1
$$\sqrt{x + 1} = x + 1 \quad \text{Original equation}$$
$$\sqrt{(-1) + 1} = (-1) + 1 \quad \text{Substitute}$$
$$\sqrt{0} = 0 \quad \text{Simplify}$$
$$0 = 0 \quad \text{True}$$

Therefore, 0 and -1 are solutions of the equation. The solution set is $\{0, -1\}$.

5. $\sqrt{3x + 4} = \sqrt{x + 14}$

Note There are two square roots in this problem, but we can eliminate both radical symbols by squaring both members.

$$(\sqrt{3x + 4})^2 = (\sqrt{x + 14})^2 \quad \text{Square both members}$$
$$3x + 4 = x + 14 \quad \text{Solve for } x$$
$$2x + 4 = 14 \quad \text{Subtract } x \text{ from both members}$$
$$2x = 10 \quad \text{Subtract 4 from both members}$$
$$x = 5 \quad \text{Divide both members by 2}$$

Check:
$$\sqrt{3x + 4} = \sqrt{x + 14} \quad \text{Original equation}$$
$$\sqrt{3(5) + 4} = \sqrt{(5) + 14} \quad \text{Substitute}$$
$$\sqrt{15 + 4} = \sqrt{19} \quad \text{Simplify}$$
$$\sqrt{19} = \sqrt{19} \quad \text{True}$$

The solution set is $\{5\}$.

6. $\sqrt{x+3} = -6$

$(\sqrt{x+3})^2 = (-6)^2$ — Square both members

$x + 3 = 36$ — Solve for x

$x = 33$ — Subtract 3 from both members

Check:

$\sqrt{x+3} = -6$ — Original equation

$\sqrt{(33)+3} = -6$ — Substitute

$\sqrt{36} = -6$ — Simplify

$6 = -6$ — False

$x = 33$ does not check because $\sqrt{36} = 6$, not -6. We conclude that there is *no solution* to this equation, and the solution set is $\emptyset$.

▶ ***Quick check*** Find the solution set. $\sqrt{x+5} = 6$ and $\sqrt{2x+5} = \sqrt{x+8}$ ■

Mastery points

Can you

- Solve equations containing radicals?

Exercise 9–7

Find the solution set. See example 9–7 A.

Examples $\sqrt{x+5} = 6$

Solutions

$(\sqrt{x+5})^2 = (6)^2$

$x + 5 = 36$

$x = 31$

Check:

$\sqrt{(31)+5} = 6$

$\sqrt{36} = 6$

$6 = 6$ — True

The solution set is $\{31\}$.

$\sqrt{2x+5} = \sqrt{x+8}$

$(\sqrt{2x+5})^2 = (\sqrt{x+8})^2$ — Square both members

$2x + 5 = x + 8$ — Simplify

$x + 5 = 8$ — Solve for x

$x = 3$

Check:

$\sqrt{2(3)+5} = \sqrt{(3)+8}$ — Substitute into original equation

$\sqrt{6+5} = \sqrt{11}$ — Simplify

$\sqrt{11} = \sqrt{11}$ — True

The solution set is $\{3\}$.

1. $\sqrt{x} = 4$ **2.** $\sqrt{x} = 5$ **3.** $\sqrt{x} = 9$ **4.** $\sqrt{x} = 7$

5. $\sqrt{x+5} = 4$ **6.** $\sqrt{x-3} = 5$ **7.** $\sqrt{x-7} = 6$ **8.** $\sqrt{x+3} = 7$

9. $\sqrt{2x+1} = 5$ **10.** $\sqrt{2x+6} = 6$ **11.** $\sqrt{3x+1} = 4$ **12.** $\sqrt{5x-4} = 6$

13. $\sqrt{x} + 4 = 7$ **14.** $\sqrt{x} + 2 = 9$ **15.** $\sqrt{x} - 5 = 1$ **16.** $\sqrt{x} - 4 = 2$

17. $\sqrt{x} + 7 = 5$ **18.** $\sqrt{x} + 8 = 4$ **19.** $\sqrt{x} + 6 = 3$ **20.** $\sqrt{x} + 10 = 5$

21. $\sqrt{2x+1} = \sqrt{x+5}$ **22.** $\sqrt{2x+4} = \sqrt{3x-2}$ **23.** $\sqrt{5x-3} = \sqrt{2x+9}$

24. $\sqrt{7x-4} = \sqrt{3x+20}$ **25.** $\sqrt{2x+7} = \sqrt{4x+1}$ **26.** $\sqrt{6x-3} = \sqrt{4x+5}$

27. $\sqrt{3x+5} = \sqrt{5x+1}$ **28.** $\sqrt{4x-4} = \sqrt{x+5}$ **29.** $\sqrt{5x-7} = \sqrt{2x+8}$

30. $\sqrt{4x-15} = \sqrt{7x-9}$ **31.** $\sqrt{5+2x} = \sqrt{2+3x}$ **32.** $\sqrt{9-5x} = \sqrt{15-7x}$

33. $\sqrt{x}\sqrt{x - 15} = 4$

34. $\sqrt{x}\sqrt{x - 3} = 2$

35. $\sqrt{x}\sqrt{x - 8} = 3$

36. $\sqrt{x}\sqrt{x - 6} = 4$

37. $\sqrt{x}\sqrt{x + 6} = 4$

38. $\sqrt{x + 3}\sqrt{x} = 2$

39. $\sqrt{x^2 + 1} = x + 2$

40. $\sqrt{x^2 + 3x} = x + 1$

41. $\sqrt{x^2 + 3x} = x - 3$

42. $\sqrt{x^2 + 12} = x + 2$

43. $\sqrt{x + 6} = x$

44. $\sqrt{5x + 6} = x$

45. $\sqrt{2x + 8} = x$

46. $\sqrt{4x + 12} = x$

47. $\sqrt{x - 4} = x - 6$

48. $\sqrt{x + 2} = x + 2$

49. $\sqrt{x + 4} + 8 = x$

50. $\sqrt{x} + 6 = x$

51. $\sqrt{x + 7} = 2x - 1$

52. $\sqrt{2x - 1} + 2x = 7$

Find the unknown number in problems 53–60.

53. The square root of the sum of a number and 6 is 5. Find the number.

54. The square root of the sum of a number and 9 is 7. Find the number.

55. The square root of the product of a number and 6 is 12. Find the number.

56. The square root of the product of a number and 9 is 6. Find the number.

57. A certain number is equal to the square root of the sum of that number and 12.

58. The square root of the product of a number and 12 is equal to the number increased by 3.

59. The square root of the sum of a number and 11 is 1 less than the number.

60. The square root of the product of a number and 4 is 3 less than the number.

61. At an altitude of h ft above the sea or level ground, the distance d in miles that a person can see an object is given by

$$d = \sqrt{\frac{3h}{2}}$$

How high must a person be to see an object 6 miles away?

62. The formula for approximating the velocity V in miles per hour of a car based on the length of its skid marks S (in feet) on dry pavement is given by

$$V = 2\sqrt{6S}$$

If the velocity is 36 mph, how long will the skid marks be?

63. On wet pavement, the formula in exercise 62 is given by

$$V = 2\sqrt{3S}$$

How long will the skid marks be if the car is traveling at 36 mph on wet pavement?

64. Find the number whose principal square root is 3. (*Hint:* Let $\sqrt{x} = 3$)

65. Find the number whose principal square root is 10.

66. Find the number whose principal square root is 11.

Review exercises

Completely factor the following expressions. See sections 4–2 and 4–4.

1. $x^2 - 4$

2. $x^2 + 9x + 18$

3. $x^2 - 3x - 10$

4. $x^2 - 6x + 9$

Simplify. See section 9–1.

5. $\sqrt{81}$

6. $\sqrt{49}$

7. $\sqrt{121}$

8. Find the solution set for the equation $x^2 = 64$. See section 4–7.

Chapter 9 lead-in problem

Roger has a ladder that will extend to a length of 21 feet. For the ladder to be safe to climb on, it must be placed 7 feet away from the house. The roof is 20 feet above the ground. Will the ladder be able to reach the roof safely? If not, how far up will the ladder reach? (Leave your answer rounded to one decimal place.)

Solution

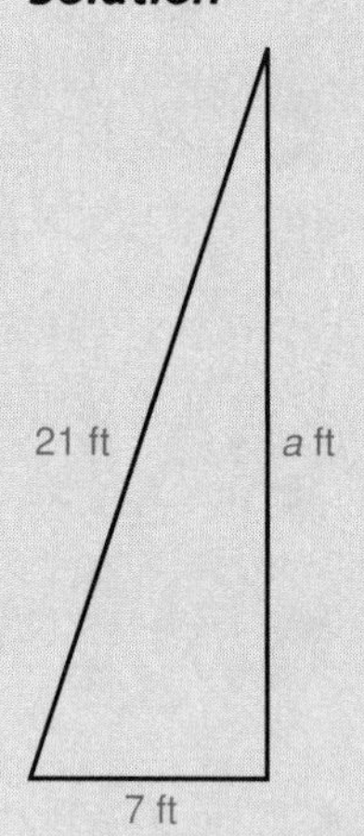

$a^2 + b^2 = c^2$	Pythagorean Theorem
$a^2 + (7)^2 = (21)^2$	Substitute 21 for c and 7 for b
$a^2 + 49 = 441$	Simplify
$a^2 = 392$	Isolate a^2
$a \approx 19.8$	Round $\sqrt{392}$ to one decimal place

The ladder will not be able to reach the roof safely. The ladder's maximum safe reach is approximately 19.8 feet.

Chapter 9 summary

1. $\sqrt[n]{a} = b$ if $\overbrace{b \cdot b \cdot b \cdots b}^{n \text{ factors}} = b^n = a$, where n is a natural number greater than 1.
2. If we exclude even roots of negative numbers, which do not exist in the set of real numbers, the **principal *n*th root** of a number, denoted by $\sqrt[n]{\ }$, has the same sign as the number itself.
3. When we multiply two radicals *having the same index,* we multiply the radicands and put the product under a radical symbol with the common index. $\sqrt{a}\sqrt{b} = \sqrt{ab}$ and $\sqrt[n]{a}\sqrt[n]{b} = \sqrt[n]{ab}$, **product property for radicals.**
4. We can simplify a radical if the radicand has a factor(s) whose exponent is equal to or greater than the value of the index.
5. The *n*th root of a *fraction* can be written as the *n*th root of the numerator over the *n*th root of the denominator.
$\sqrt{\dfrac{a}{b}} = \dfrac{\sqrt{a}}{\sqrt{b}}$ and
$\sqrt[n]{\dfrac{a}{b}} = \dfrac{\sqrt[n]{a}}{\sqrt[n]{b}}$, where $b \neq 0$, **quotient property for radicals.**
6. We eliminate radicals from the denominator of a fraction by **rationalizing** the denominator.
7. We can only add or subtract **like radicals.**
8. **Conjugate factors** are used to rationalize a denominator when the denominator has two terms where one or both terms contain a square root.
9. When a number is raised to a **fractional exponent,** the numerator of the fractional exponent indicates the power to which the base is to be raised, and the denominator indicates the root to be taken.
10. If a is any real number, m is any integer, and n is any positive integer, then if $\dfrac{m}{n}$ is reduced to lowest terms, $a^{m/n} = (\sqrt[n]{a})^m = \sqrt[n]{a^m}$ provided that $\sqrt[n]{a}$ is a real number.
11. The **squaring property of equality** states that if both members of an equation are squared, the solution(s) of the original equation are solution(s) of the resulting equation.
12. When solving equations where the unknown is under a radical symbol, we must check for **extraneous roots.**

Chapter 9 error analysis

1. Principal square root
   Example: $\sqrt{36} = -6$
   Correct answer: 6
   What error was made? (*see page 368*)
2. Rationalizing the denominator of an nth root
   Example: $\dfrac{x}{\sqrt[3]{x}} = \dfrac{x}{\sqrt[3]{x}} \cdot \dfrac{\sqrt[3]{x}}{\sqrt[3]{x}} = \dfrac{x\sqrt[3]{x}}{x} = \sqrt[3]{x}$
   Correct answer: $\sqrt[3]{x^2}$
   What error was made? (*see page 380*)
3. Addition in the radicand of a radical expression
   Example: $\sqrt{9+4} = \sqrt{9} + \sqrt{4} = 3 + 2 = 5$
   Correct answer: $\sqrt{13}$
   What error was made? (*see page 374*)
4. Subtraction of radical expressions
   Example: $\sqrt{3} - \sqrt{2} = \sqrt{3-2} = \sqrt{1} = 1$
   Correct answer: $\sqrt{3} - \sqrt{2}$
   What error was made? (*see page 374*)
5. Sums and differences of radical numbers
   Example:
   $\sqrt{8} + 3\sqrt{2} - \sqrt{18} = 2\sqrt{2} + 3\sqrt{2} - \sqrt{18}$
   $= 5\sqrt{2} - \sqrt{18}$
   Correct answer: $2\sqrt{2}$
   What error was made? (*see page 384*)
6. Multiplication of radical expressions
   Example: $\sqrt{3}(\sqrt{3} + \sqrt{2}) = \sqrt{3} \cdot \sqrt{3} + \sqrt{2} = 3 + \sqrt{2}$
   Correct answer: $3 + \sqrt{6}$
   What error was made? (*see page 387*)
7. Squaring a radical binomial
   Example: $(\sqrt{2} - \sqrt{3})^2 = (\sqrt{2})^2 - (\sqrt{3})^2 = 2 - 3$
   $= -1$
   Correct answer: $5 - 2\sqrt{6}$
   What error was made? (*see page 387*)
8. Fractional exponents to radical expressions
   Example: $x^{3/2} = \sqrt[3]{x^2}$
   Correct answer: $x\sqrt{x}$
   What error was made? (*see page 392*)
9. Extraneous solutions of radical equations
   Example: Find the solution set of $\sqrt{x+2} = x - 4$
   $(\sqrt{x+2})^2 = (x-4)^2$
   $x + 2 = x^2 - 8x + 16$
   $x^2 - 9x + 14 = 0$
   $(x-7)(x-2) = 0$
   $x - 7 = 0$ or $x - 2 = 0$
   $x = 7$ or $x = 2$ $\{2,7\}$
   Correct answer: $\{7\}$
   What error was made? (*see page 396*)
10. Negative exponents
    Example: $6^{-2} = -36$
    Correct answer: $\dfrac{1}{36}$
    What error was made? (*see page 141*)

Chapter 9 critical thinking

Given the numbers 33 and 27, determine a method by which you can multiply these numbers mentally.

Chapter 9 review

Assume that all variables in problems 1–42 represent positive real numbers and that no denominator is equal to zero.

[9–1]

Find the indicated root.

1. $\sqrt{81}$ **2.** $\sqrt{25}$ **3.** $-\sqrt{9}$ **4.** $-\sqrt{49}$

[9–2]

Perform any indicated operations and simplify.

5. $\sqrt{40}$ **6.** $\sqrt{18a^2b^3}$ **7.** $\sqrt{2}\sqrt{14}$ **8.** $\sqrt{18}\sqrt{10}$

[9–3]

Express the given radicals in simplest form with all denominators rationalized.

9. $\sqrt{\dfrac{16}{17}}$ **10.** $\sqrt{\dfrac{7}{18}}$ **11.** $\sqrt{\dfrac{a}{b}}$ **12.** $\sqrt{\dfrac{x}{y^3}}$ **13.** $\dfrac{a}{\sqrt{ab}}$ **14.** $\dfrac{2x}{\sqrt{xy}}$

[9–4]

Perform the indicated operations and simplify.

15. $3\sqrt{7} + 4\sqrt{7}$
16. $\sqrt{18} + 5\sqrt{2}$
17. $3\sqrt{20} - \sqrt{45}$
18. $2\sqrt{75} - \sqrt{3} + 5\sqrt{27}$
19. $\sqrt{50a} - 2\sqrt{8a}$
20. $7\sqrt{9x} - 2\sqrt{4x}$

[9–5]

Perform the indicated operations and simplify.

21. $\sqrt{3}(\sqrt{5} - \sqrt{7})$
22. $\sqrt{10}(\sqrt{14} + \sqrt{6})$
23. $(5 + \sqrt{7})(3 - \sqrt{7})$
24. $(6 - \sqrt{3})^2$
25. $(\sqrt{3} + \sqrt{5})^2$
26. $(2\sqrt{a} - \sqrt{b})(2\sqrt{a} + \sqrt{b})$

Express the given radicals in simplest form with all denominators rationalized.

27. $\dfrac{1}{\sqrt{3} - 2}$
28. $\dfrac{2}{\sqrt{6} + 4}$
29. $\dfrac{1}{\sqrt{a} + b}$
30. $\dfrac{x}{\sqrt{xy} + x}$
31. $\dfrac{a}{a + \sqrt{b}}$
32. $\dfrac{\sqrt{2} + 3}{\sqrt{2} - 3}$

[9–6]

Simplify the given expressions.

33. $(36)^{1/2}$
34. $(8)^{2/3}$
35. $(-8)^{1/3}$
36. $(32)^{-2/5}$

Perform the indicated operations and simplify.

37. $a^{2/5} \cdot a^{3/5}$
38. $b^{1/3} \cdot b^{3/4}$
39. $\dfrac{a^{1/2}}{a^{1/4}}$
40. $(a^{3/2})^{3/4}$
41. $(16a^4b^8)^{3/4}$
42. $\dfrac{a^2b^2}{a^{1/2}b^{3/2}}$

[9–7]

Find the solution set.

43. $\sqrt{x} = 8$
44. $\sqrt{x - 4} = 7$
45. $\sqrt{5x - 3} = \sqrt{3x + 5}$
46. $\sqrt{x}\sqrt{x + 8} = 3$
47. $\sqrt{x}\sqrt{x + 6} = 4$
48. $\sqrt{x^2 + 16} = x + 2$
49. $\sqrt{x + 6} = x + 4$
50. $\sqrt{x - 3} = x - 3$

Chapter 9 cumulative test

Perform the indicated operations and simplify. Assume that all variables represent positive real numbers and that no denominator is equal to zero.

[1–8] 1. $3[4(6 - 2) + (-5 + 4)]$

[3–1] 2. $x^3 \cdot x^2 \cdot x^2$

[3–3] 3. $x^{12} \div x^6$

[2–3] 4. $(3x^2 - 2x + 5) - (x^2 - 4x - 8)$

[3–2] 5. $4a^2b^3(2a^2 - 3ab + 4b^2)$

[3–3] 6. $\dfrac{8a^{-2}b^3c^0}{4a^{-5}b}$

[1–8] 7. -6^2

[3–2] 8. $(3a - b)^2$

[6–1] 9. $\dfrac{a^2 - 9}{3a + 12} \cdot \dfrac{a + 4}{a^2 + 5a + 6}$

[9–4] 10. $5\sqrt{12} + 2\sqrt{27}$

[9–1] 11. $\sqrt[3]{-8}$

[6–2] 12. $\dfrac{2x}{x - 1} - \dfrac{x - 3}{x - 1}$

[9–5] 13. $\dfrac{\sqrt{x}}{x - \sqrt{y}}$

[9–2] 14. $\sqrt[3]{81x^4y^6z}$

[3–2] 15. $(5x - y)(5x + y)$

[2–3] 16. $3x - [2x - (x - y) + 3y]$

Factor completely.

[4–1] **17.** $6a^3b^4 - 2a^3b^5 + 8a^5b^3$

[4–4] **18.** $25c^2 - d^2$

[4–3] **19.** $2x^2 + 7x - 4$

[4–4] **20.** $y^4 - 4z^2$

[4–3] **21.** $6x^2 + 11x + 4$

[4–2] **22.** $x^2 + 3x - 28$

Find the solution set for problems 23–28 and solve problems 29 and 30.

[2–6] **23.** $2(4x - 3) = 5x - 7$

[4–7] **24.** $x^2 = 9$

[2–6] **25.** $3(x + 4) - 2(x - 3) = 12$

[2–6] **26.** $\frac{1}{4}x + 2 = \frac{1}{2}x - 1$

[6–5] **27.** $\frac{3x + 1}{9} + \frac{1}{12} = \frac{2x - 1}{3}$

[4–7] **28.** $2x^2 + 3x + 1 = 0$

[2–9] **29.** $-3 < 2x - 5 < 11$

[2–9] **30.** $3x + 5 > x + 12$

[7–3] **31.** Find the slope of the line passing through points (4,2) and (3,5).

[7–4] **32.** Write $8x - 2y = 4$ in slope-intercept form and determine the slope and y-intercept.

[8–2] **33.** Solve the system of equations.

$$5x - y = 4$$
$$x + 3y = 2$$

[2–8] **34.** One number is four times a second number and their sum is 70. Find the numbers.

[5–4] **35.** A punch machine can make 18 holes in 4 minutes. How many holes can the machine make in 5 hours?

[4–8] **36.** The product of two consecutive positive even integers is 288. Find the integers.

[2–8] **37.** The width of a rectangle is 6 feet less than its length. The perimeter of the rectangle is 96 feet. Find the dimensions.

CHAPTER

10 Solutions of Quadratic Equations

A rock is dropped from the top of the Washington Monument. If the monument is 555 feet tall, how long will it take the rock to strike the ground?

10–1 ■ Solutions of quadratic equations by extracting the roots

In section 4–7, we solved quadratic equations of the form

$$ax^2 + bx + c = 0, a \neq 0$$

by factoring. It was necessary that the quadratic expression $ax^2 + bx + c$ be factorable to use the method discussed. Let us review the procedures used in solving quadratic equations by factoring.

To solve a quadratic equation by factoring

1. Write the equation in standard form

$$ax^2 + bx + c = 0, a \neq 0$$

   if the equation is not written in this form.
2. Factor the expression $ax^2 + bx + c$.
3. Set each of the resulting factors involving the variable equal to 0 and solve each linear equation for the variable.
4. Check your solutions in the original equation.

Example 10–1 A

Find the solution set of each quadratic equation by factoring.

1. $x^2 - x - 12 = 0$

$(x - 4)(x + 3) = 0$ — Factor the left member

$x - 4 = 0$ or $x + 3 = 0$ — Set each factor equal to 0

$x = 4$ $\qquad x = -3$ — Solve each equation for x

The solution set is $\{-3,4\}$.

2. $3y^2 = 7y + 6$

$3y^2 - 7y - 6 = 0$ — Write the equation in standard form

$(3y + 2)(y - 3) = 0$ — Factor the left member

$3y + 2 = 0$ or $y - 3 = 0$ — Set each factor equal to 0

$3y = -2$ $\qquad y = 3$ — Solve each equation for y

$y = -\dfrac{2}{3}$ $\qquad y = 3$

The solution set is $\left\{-\dfrac{2}{3},3\right\}$. ■

Extracting the roots

Given the quadratic equation $x^2 - 9 = 0$, factoring the left member and solving the resulting equations, we get

$$(x - 3)(x + 3) = 0$$
$$x - 3 = 0 \quad \text{or} \quad x + 3 = 0$$
$$x = 3 \quad \text{or} \quad x = -3$$

The solutions of the equation are 3 or -3.

We can obtain the same result if we write the equation in the form

$$x^2 = 9$$

Since 9 is positive, we can take the square root of each member of the equation. Then

$$x = \sqrt{9} = 3 \quad \text{or} \quad x = -\sqrt{9} = -3$$

and we obtain the same result. This development justifies the following method of solving a quadratic equation by **extracting the roots** using the **square root property**.

Square root property

If k is a nonnegative number and $x^2 = k$, then

$$x = \sqrt{k} \quad \text{or} \quad x = -\sqrt{k}$$

Example 10–1 B

Find the solution set of the following quadratic equations by extracting the roots.

1. $x^2 = 25$

$x = \sqrt{25}$ or $x = -\sqrt{25}$ — Extract the roots

$x = 5$ $\qquad x = -5$ — $\sqrt{25} = 5$

The solution set is $\{-5,5\}$.

2. $y^2 = 18$

$$y = \sqrt{18} \quad \text{or} \quad y = -\sqrt{18}$$
$$y = 3\sqrt{2} \qquad\quad y = -3\sqrt{2}$$

Extract the roots
$\sqrt{18} = 3\sqrt{2}$

The solution set is $\{-3\sqrt{2},3\sqrt{2}\}$.

3. $x^2 - 12 = 0$

$$x^2 = 12$$
$$x = \sqrt{12} \quad \text{or} \quad x = -\sqrt{12}$$
$$x = 2\sqrt{3} \qquad\quad x = -2\sqrt{3}$$

Add 12 to each member
Extract the roots
$\sqrt{12} = 2\sqrt{3}$

The solution set is $\{-2\sqrt{3},2\sqrt{3}\}$.

4. $z^2 = -9$

Since -9 is negative and the property requires that k is a nonnegative number, we are not able to solve this equation in the set of real numbers. The equation has no solution so the solution set is $\emptyset$.

5. $2x^2 = 98$

To extract the roots, the squared term must have coefficient 1.

$$2x^2 = 98$$
$$x^2 = 49$$
$$x = \sqrt{49} \quad \text{or} \quad x = -\sqrt{49}$$
$$x = 7 \qquad\quad x = -7$$

Divide each term by 2
Extract the roots
$\sqrt{49} = 7$

The solution set is $\{-7,7\}$.

▶ ***Quick check*** Find the solution set of the equation $3x^2 = 24$ by extracting the roots. ■

Any equation that is written in the form

$$(x + q)^2 = k \text{ or } (px + q)^2 = k$$

can be solved by extracting the roots. Consider the following examples.

■ ***Example 10–1 C***

Find the solution set of the following quadratic equations by extracting the roots.

1. $(x - 2)^2 = 4$

$$x - 2 = \sqrt{4} \quad \text{or} \quad x - 2 = -\sqrt{4}$$
$$x - 2 = 2 \qquad\quad x - 2 = -2$$
$$x = 2 + 2 = 4 \qquad\quad x = 2 - 2 = 0$$

Extract the roots
$\sqrt{4} = 2$ or -2
Add 2 to each member

The solution set is $\{0,4\}$.

2. $(2y - 1)^2 = 24$

$$2y - 1 = \sqrt{24} \quad \text{or} \quad 2y - 1 = -\sqrt{24}$$
$$2y - 1 = 2\sqrt{6} \qquad\quad 2y - 1 = -2\sqrt{6}$$
$$2y = 1 + 2\sqrt{6} \qquad\quad 2y = 1 - 2\sqrt{6}$$
$$y = \frac{1 + 2\sqrt{6}}{2} \qquad\quad y = \frac{1 - 2\sqrt{6}}{2}$$

Extract the roots
$\sqrt{24} = 2\sqrt{6}$
Add 1 to each member
Divide each member by 2

The solution set is $\left\{\frac{1 - 2\sqrt{6}}{2}, \frac{1 + 2\sqrt{6}}{2}\right\}$.

▶ ***Quick check*** Find the solution set of the equation $(x + 4)^2 = 9$ by extracting roots. ■

Mastery points

Can you

- Solve a quadratic equation by factoring?
- Solve quadratic equations of the form $x^2 = k$ and $(px + q)^2 = k$ by extracting the roots?

Exercise 10–1

Find the solution set of each quadratic equation by extracting the roots or by factoring. Express radicals in simplest form. All variables represent nonnegative numbers. See examples 10–1 A and B.

Example $3x^2 = 24$

Solution $x^2 = 8$ — Divide each member by 3

$x = \sqrt{8}$ or $x = -\sqrt{8}$ — Extract the roots

$x = 2\sqrt{2}$ $\quad x = -2\sqrt{2}$ — $\sqrt{8} = \sqrt{4 \cdot 2} = 2\sqrt{2}$

The solution set is $\{2\sqrt{2}, -2\sqrt{2}\}$.

1. $x^2 + 2x - 15 = 0$
2. $x^2 - 4x + 3 = 0$
3. $2y^2 - y - 6 = 0$
4. $5z^2 - 16z + 3 = 0$
5. $x^2 = 4$
6. $x^2 = 49$
7. $x^2 = 64$
8. $x^2 = 81$
9. $y^2 = 11$
10. $x^2 = 5$
11. $a^2 = 20$
12. $x^2 = 28$
13. $x^2 - 3 = 0$
14. $x^2 - 13 = 0$
15. $p^2 - 32 = 0$
16. $x^2 - 40 = 0$
17. $4x^2 = 36$
18. $5x^2 = 75$
19. $3z^2 = 18$
20. $5x^2 = 15$
21. $2x^2 - 100 = 0$
22. $4x^2 - 64 = 0$
23. $7x^2 - 56 = 0$
24. $9a^2 - 162 = 0$
25. $\frac{3}{4}x^2 - 6 = 0$
26. $\frac{1}{5}x^2 - \frac{3}{5} = 0$
27. $\frac{1}{3}x^2 = \frac{2}{3}$
28. $\frac{2}{3}x^2 = 8$
29. $\frac{5}{2}x^2 - \frac{3}{5} = 0$
30. $\frac{4x^2}{3} - 3 = 0$
31. $\frac{1}{2}y^2 - \frac{3}{2} = 4$
32. $\frac{z^2}{4} - 6 = \frac{3}{4}$

See example 10–1 C.

Example $(x + 4)^2 = 9$

Solution $x + 4 = 3$ or $x + 4 = -3$ — Extract the roots

$x = -4 + 3$ $\quad x = -4 - 3$ — Add -4 to each member

$x = -1$ $\quad x = -7$

The solution set is $\{-1, -7\}$.

33. $(x + 2)^2 = 4$
34. $(x + 6)^2 = 16$
35. $(x - 4)^2 = 25$
36. $(x - 3)^2 = 49$
37. $(x + 3)^2 = 6$
38. $(x - 1)^2 = 7$
39. $(x - 9)^2 = 18$
40. $(x + 8)^2 = 8$
41. $(x + 5)^2 = 32$
42. $(x - 10)^2 = 27$
43. $(x + a)^2 = 36$
44. $(x - a)^2 = 50$
45. $(x - 6)^2 = a^2$
46. $(x + 7)^2 = m^2$
47. $(x - p)^2 = q^2$
48. $(x + 5)^2 = t^2$
49. $(2x - 3)^2 = 16$
50. $(3x + 2)^2 = 24$

Solve by setting up a quadratic equation and extracting the roots.

Example A square has an area of 16 square inches. Find the length of each side.

Solution Use the formula $A = s^2$, where A is the area and s is the length of a side. Then $16 = s^2$ or $s^2 = 16$, and

$$s = \sqrt{16} \quad \text{or} \quad s = -\sqrt{16} \qquad \text{Exract the roots}$$
$$s = 4 \qquad\qquad s = -4 \qquad \sqrt{16} = 4$$

Since a square cannot have a side that is -4 inches long, then $s = 4$ inches. The length of each side of the square is 4 inches.

51. Find the length of each side of a square whose area is 25 square meters.

52. Given a square whose area is 45 square centimeters, how long is each side of the square?

53. A circle has an area of approximately 12.56 square feet. Find the approximate length of the radius r of the circle if $A \approx 3.14r^2$.

54. Find the approximate length of the radius of a circle whose area is approximately 50.24 square yards. (Refer to exercise 53 for the formula.)

55. The square of a number less 81 is equal to zero. Find the number.

56. Four times the square of a number is 100. Find the number.

57. The square of a number is equal to nine times the number. Find the number.

58. If you subtract eight times a number from two times the square of the number, you get zero. Find the number.

59. The sum of the areas of two squares is 245 square inches. If the length of the side of the larger square is twice the length of the side of the smaller square, find the lengths of the sides of the two squares.

60. The length of the side of one square is three times the length of the side of a second square. If the difference in their areas is 128 square centimeters, find the lengths of the sides of the two squares.

61. The width of a rectangle is one-fourth of the length. If the area is 144 square meters, find the length of the rectangle. (*Hint:* $A = \ell w$.)

62. The length of a rectangle is three times the width. If the area of the rectangle is 147 square feet, find the dimensions of the rectangle.

63. The sum of the areas of two circles is 80π. Find the length of the radius of each circle if one radius is twice as long as the other.

Solve by using the relationship that exists for a square: The sum of the squares of two sides is equal to the square of the diagonal of the square.

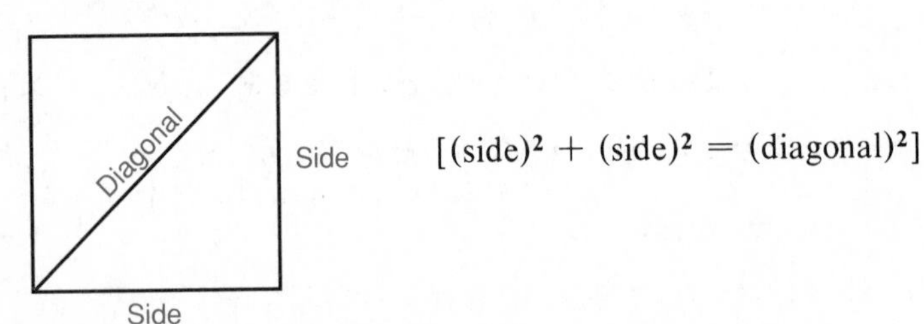

64. Find the length of the side of a square whose diagonal is 16 inches long.

65. Find the length of the side of a square whose diagonal is 10 centimeters long.

66. Find the length of the side of a square whose diagonal is 24 feet long. Simplify the radical answer.

Review exercises

Multiply the following. See section 3–2.

1. $(x - 2)^2$

2. $(3z + 2)^2$

Completely factor the following. See sections 4–2 and 4–3.

3. $x^2 + 18x + 81$

4. $9y^2 + 30y + 25$

Perform the indicated operations. See sections 6–1 and 6–2.

5. $\dfrac{3x}{x + 2} - \dfrac{x}{x^2 - 4}$

6. $\dfrac{x - 3}{x^2 - x - 2} \div \dfrac{x^2 - 9}{x + 1}$

10–2 ■ Solutions of quadratic equations by completing the square

Building perfect square trinomials

The methods we have used to solve quadratic equations thus far have applied to special cases of the quadratic equation. The method that we call **completing the square** involves transforming the quadratic equation

$$ax^2 + bx + c = 0$$

into the form

$$(x + q)^2 = k,\ k \geq 0$$

where q and k are constants. This latter equation can then be solved by extracting the roots, as we did in section 10–1.

Consider the following perfect square trinomials and their equivalent binomial squares.

$$\begin{aligned} x^2 + 2x + 1 &= (x + 1)(x + 1) = (x + 1)^2 \\ x^2 - 10x + 25 &= (x - 5)(x - 5) = (x - 5)^2 \\ x^2 - 14x + 49 &= (x - 7)(x - 7) = (x - 7)^2 \end{aligned}$$

In each of the perfect square trinomials in the left member,

a. the coefficient of x^2 is 1.
b. the third term, the constant, is the square of one-half of the coefficient of the variable x in the middle term.

We further observe that in the square of the binomial in the right member, the constant term in the binomial is one-half of the coefficient of the variable x in the middle term. That is,

1. In the trinomial $x^2 + 2x + 1$, the constant term, 1, is the square of one-half the coefficient of the middle term, 2. Thus,

$$\left[\frac{1}{2}(2)\right]^2 = (1)^2 = 1$$

Constant term of the binomial

2. In the trinomial $x^2 - 10x + 25$, the constant term, 25, is the square of one-half of -10. Thus,

$$\left[\frac{1}{2}(-10)\right]^2 = (-5)^2 = 25$$

Constant term of the binomial (points to -5)

3. In the trinomial $x^2 - 14x + 49$, the constant term, 49, is the square of one-half of -14. Thus,

$$\left[\frac{1}{2}(-14)\right]^2 = (-7)^2 = 49$$

Constant term of the binomial (points to -7)

Now we can use these observations to "build" perfect square trinomials by **completing the square** and obtain the equivalent square of a binomial.

■ *Example 10–2 A*

Complete the square in each of the following expressions. Write the resulting expression as the square of a binomial.

1. $x^2 + 6x$
Since the coefficient of x is 6, the constant term is the square of one-half of 6.

$$\left[\frac{1}{2}(6)\right]^2 = (3)^2 = 9$$

The trinomial becomes

$$x^2 + 6x + 9$$

which factors into

$$(x + 3)(x + 3) = (x + 3)^2$$

2. $x^2 - 8x$
Since the coefficient of x is -8, the constant term is the square of one-half of -8.

$$\left[\frac{1}{2}(-8)\right]^2 = (-4)^2 = 16$$

The trinomial becomes

$$x^2 - 8x + 16$$

which factors into

$$(x - 4)(x - 4) = (x - 4)^2$$

3. $y^2 - 3y$
Since the coefficient of y is -3, the constant term is the square of one-half of -3.

$$\left[\frac{1}{2}(-3)\right]^2 = \left(-\frac{3}{2}\right)^2 = \frac{9}{4}$$

The trinomial becomes

$$y^2 - 3y + \frac{9}{4}$$

which factors into

$$\left(y-\frac{3}{2}\right)\left(y-\frac{3}{2}\right)=\left(y-\frac{3}{2}\right)^2$$

▶ ***Quick check*** Complete the square in the expression $x^2+\frac{1}{3}x$. Write as the square of a binomial. ■

Solutions by completing the square

The following examples show how we use the previous procedure to find the solution set of a quadratic equation by **completing the square.**

■ Example 10–2 B

Find the solution set by completing the square.

1. $x^2-2x-8=0$

We first isolate the terms containing the variable in the left member.

$x^2-2x=8$ — Add 8 to each member

Complete the square in the left member.

$\left[\frac{1}{2}(-2)\right]^2=(-1)^2=1$ — Square one-half of the coefficient of x

$x^2-2x+1=8+1$ — Add 1 to each member

$(x-1)^2=9$ — Factor the left member and combine like terms in the right member

$x-1=\sqrt{9}$ or $x-1=-\sqrt{9}$ — Extract the roots

$x-1=3$ $x-1=-3$ — $\sqrt{9}=3$

$x=1+3$ $x=1-3$ — Add 1 to each member

$x=4$ $x=-2$ — Combine in right member

The solution set is $\{-2,4\}$.

Check:

1. When $x=4$

$x^2-2x-8=0$

$(4)^2-2(4)-8=0$ — Replace x with 4

$16-8-8=0$ — Order of operations

$0=0$ (True)

2. When $x=-2$

$x^2-2x-8=0$

$(-2)^2-2(-2)-8=0$ — Replace x with -2

$4+4-8=0$ — Order of operations

$0=0$ (True)

In future examples, we will not show a check but you should always do this.

2. $x^2 - 6x + 2 = 0$
Isolate the terms containing the variable in the left member.

$x^2 - 6x = -2$ — Add -2 to each member

Complete the square in the left member.

$\left[\frac{1}{2}(-6)\right]^2 = (-3)^2 = 9$ — Square one-half of the coefficient of x

$x^2 - 6x + 9 = -2 + 9$ — Add 9 to each member

$(x - 3)^2 = 7$ — Factor the left member and combine like terms in the right member

$x - 3 = \sqrt{7}$ or $x - 3 = -\sqrt{7}$ — Extract the roots

$x = 3 + \sqrt{7}$ $\quad x = 3 - \sqrt{7}$ — Add 3 to each member

The solution set is $\{3 - \sqrt{7}, 3 + \sqrt{7}\}$.

3. $4x^2 + 4x = 3$
To complete the square using the method we have described, it is necessary for the coefficient of x^2 to be 1. To get this, we divide each term of the equation by 4.

$\frac{4x^2}{4} + \frac{4x}{4} = \frac{3}{4}$ — Divide each term by 4

$x^2 + x = \frac{3}{4}$ — Reduce where possible

Complete the square in the left member.

$\left[\frac{1}{2}(1)\right]^2 = \left(\frac{1}{2}\right)^2 = \frac{1}{4}$ — Square one-half of the coefficient of x

$x^2 + x + \frac{1}{4} = \frac{3}{4} + \frac{1}{4}$ — Add $\frac{1}{4}$ to each member

$\left(x + \frac{1}{2}\right)^2 = 1$ — Factor the left member and combine like terms in the right member

$x + \frac{1}{2} = \sqrt{1}$ or $x + \frac{1}{2} = -\sqrt{1}$ — Extract the roots

$x + \frac{1}{2} = 1$ $\quad x + \frac{1}{2} = -1$ — $\sqrt{1} = 1$

$x = -\frac{1}{2} + 1$ $\quad x = -\frac{1}{2} - 1$ — Add $-\frac{1}{2}$ to each member

$x = -\frac{1}{2} + \frac{2}{2} = \frac{1}{2}$ $\quad x = -\frac{1}{2} - \frac{2}{2} = -\frac{3}{2}$ — Combine like terms

The solution set is $\left\{\frac{1}{2}, -\frac{3}{2}\right\}$.

▶ ***Quick check*** Find the solution set of $x^2 + 7x - 2 = 0$ by completing the square. ■

To find the solution set of the quadratic equation $ax^2 + bx + c = 0$, $a \neq 0$, by completing the square, we proceed as follows:

1. If $a = 1$, proceed to step 2. If $a \neq 1$, divide each term of the equation by a and simplify.
2. Write the equation with the variable terms in the left member and the constant term in the right member.
3. Add to each member of the equation the square of one-half of the numerical coefficient of the middle term of the original equation.
4. Write the left member as a trinomial and factor it as the square of a binomial. Combine like terms in the right member.
5. Extract the roots and solve the resulting linear equations.
6. Check the solutions in the original equation.

Mastery points

Can you

- Complete the square of an expression in the form $x^2 + bx$?
- Find the solution set of a quadratic equation by completing the square?

Exercise 10–2

Complete the square of each of the following and factor as the square of a binomial. See example 10–2 A.

Example $x^2 + \frac{1}{3}x$

Solution $\left[\frac{1}{2}\left(\frac{1}{3}\right)\right]^2 = \left(\frac{1}{6}\right)^2 = \frac{1}{36}$ Square one-half of the coefficient of x

$$x^2 + \frac{1}{3}x + \frac{1}{36} = \left(x + \frac{1}{6}\right)^2$$

1. $x^2 + 10x$
2. $x^2 + 4x$
3. $a^2 - 12a$
4. $y^2 - 18y$
5. $x^2 + 24x$
6. $b^2 + 16b$
7. $y^2 - 20y$
8. $x^2 - 22x$
9. $x^2 + x$
10. $x^2 + 5x$
11. $x^2 - 7x$
12. $y^2 - 9y$
13. $x^2 + \frac{1}{2}x$
14. $b^2 + \frac{1}{4}b$
15. $s^2 - \frac{1}{5}s$
16. $x^2 - \frac{3}{8}x$
17. $y^2 + \frac{2}{3}y$
18. $x^2 + \frac{3}{4}x$
19. $m^2 - \frac{2}{5}m$
20. $x^2 - \frac{1}{8}x$
21. $a^2 - \frac{3}{2}a$
22. $b^2 + \frac{5}{2}b$

Find the solution set by completing the square. See example 10–2 B.

Example $x^2 + 7x - 2 = 0$

Solution $x^2 + 7x = 2$ — Add 2 to each member

Complete the square in the left member.

$$\left[\frac{1}{2}(7)\right]^2 = \left(\frac{7}{2}\right)^2 = \frac{49}{4}$$ Square one-half of the coefficient of x

$$x^2 + 7x + \frac{49}{4} = 2 + \frac{49}{4}$$ Add $\frac{49}{4}$ to each member

$$\left(x + \frac{7}{2}\right)^2 = \frac{57}{4}$$ Factor the left member and combine in the right member

$$x + \frac{7}{2} = \sqrt{\frac{57}{4}} \quad \text{or} \quad x + \frac{7}{2} = -\sqrt{\frac{57}{4}}$$ Extract the roots

$$x + \frac{7}{2} = \frac{\sqrt{57}}{2} \qquad x + \frac{7}{2} = -\frac{\sqrt{57}}{2}$$ $\sqrt{\frac{57}{4}} = \frac{\sqrt{57}}{2}$

$$x = -\frac{7}{2} + \frac{\sqrt{57}}{2} \qquad x = -\frac{7}{2} - \frac{\sqrt{57}}{2}$$ Add $-\frac{7}{2}$ to each member

$$x = \frac{-7 + \sqrt{57}}{2} \qquad x = \frac{-7 - \sqrt{57}}{2}$$ Combine like terms

The solution set is $\left\{\frac{-7 - \sqrt{57}}{2}, \frac{-7 + \sqrt{57}}{2}\right\}$.

23. $x^2 + 8x + 7 = 0$
24. $x^2 + 12x + 11 = 0$
25. $a^2 - 4a - 12 = 0$
26. $y^2 - 10y + 9 = 0$
27. $x^2 - 4x = -3$
28. $x^2 + 14x = -13$
29. $u^2 - u - 1 = 0$
30. $a^2 + 3a - 1 = 0$
31. $x^2 - 5x + 2 = 0$
32. $n^2 - 3n - 2 = 0$
33. $y^2 - 4y = 81$
34. $b^2 - 12b = -25$
35. $h^2 + 21h + 10 = 0$
36. $3x^2 + 6x = 3$
37. $2x^2 + x - 3 = 0$
38. $3x^2 - 10x = -3$
39. $2y^2 + 7y + 3 = 0$
40. $3a^2 + 8a - 4 = 0$
41. $4x^2 - 4x = 3$
42. $2b^2 - b - 15 = 0$
43. $6a^2 - 13a = -6$
44. $6n^2 + n = 1$
45. $y^2 = 3 - y$
46. $x^2 + 1 = -3x$
47. $2 - a = 6a^2$
48. $4 - x^2 = 2x$
49. $-n^2 - 4 = -6n$
50. $1 - n^2 = 3n$
51. $2x^2 - 3 = x + 4$
52. $3a^2 + 5 = 4a + 8$
53. $(x + 3)(x - 2) = 1$
54. $(x - 5)(x - 3) = 4$
55. $3x(2x + 5) = -3$
56. $4x(x - 2) = 1$

Solve the following problems by setting up a quadratic equation and completing the square.

Example A piece of lumber is divided into two pieces so that one piece is 5 inches longer than the other. If the product of their lengths is 104 square inches, what is the length of each piece?

Solution Let x = the length of the shorter piece. Then $x + 5$ = the length of the longer piece.

the product of the lengths	is	104 square inches
$x(x + 5)$	$=$	104

The equation is $x(x + 5) = 104$.

$$x^2 + 5x = 104 \qquad \text{Multiply in the left member}$$

$$\left[\frac{1}{2}(5)\right]^2 = \left(\frac{5}{2}\right)^2 = \frac{25}{4} \qquad \text{Square one-half of the coefficient of } x$$

$$x^2 + 5x + \frac{25}{4} = 104 + \frac{25}{4} \qquad \text{Add } \frac{25}{4} \text{ to each member}$$

$$\left(x + \frac{5}{2}\right)^2 = \frac{416}{4} + \frac{25}{4} \qquad \text{Factor left member}$$

$$\left(x + \frac{5}{2}\right)^2 = \frac{441}{4} \qquad \text{Combine in the right member}$$

$$x + \frac{5}{2} = \sqrt{\frac{441}{4}} \quad \text{or} \quad x + \frac{5}{2} = -\sqrt{\frac{441}{4}} \qquad \text{Extract the roots}$$

$$x = -\frac{5}{2} + \frac{21}{2} \qquad x = -\frac{5}{2} - \frac{21}{2} \qquad \sqrt{441} = 21$$

$$x = \frac{-5 + 21}{2} = \frac{16}{2} = 8 \qquad x = \frac{-5 - 21}{2} = \frac{-26}{2} = -13 \qquad \text{Combine like terms}$$

The solution set of the equation is $\{8, -13\}$. Since we want the length of lumber, -13 is not an appropriate answer. Therefore, $x = 8$ and $x + 5 = 13$. The two pieces have lengths 8 inches and 13 inches.

57. A metal bar is to be divided into two pieces so that one piece is 4 inches shorter than the other. If the sum of the squares of the two lengths is 208 square inches, find the two lengths.

58. To find the total surface area of an automobile cylinder, we use the formula $A = 2\pi r^2 + 2\pi rh$, where π is approximately equal to the constant $\frac{22}{7}$.
If the area A of the cylinder is approximately 88 square inches and the height h is 7 inches, find the approximate value of radius r.

59. One surface of a rectangular solid has a width w that is 8 millimeters shorter than its length ℓ. If the area A of the surface is 105 square millimeters, what are its dimensions? (*Hint:* $A = \ell w$.)

60. The length of a rectangular-shaped piece of paper is 7 inches longer than its width. What are the dimensions of the paper if it has an area of 78 square inches?

61. The perimeter of a rectangle is 52 inches and its area is 153 square inches. What are its dimensions? The formula for the perimeter of a rectangle is $P = 2\ell + 2w$. If we substitute 52 in place of P, we have $52 = 2\ell + 2w$. Divide each member of the equation by 2. Then $26 = \ell + w$. We can use this fact to establish the unknowns.
Width of rectangle: w
Length of rectangle: $26 - w$
Equation: $153 = w(26 - w)$

62. The perimeter of a rectangle is 38 centimeters and its area A is 88 square centimeters. What are its dimensions? (See exercise 61.)

63. The perimeter of a rectangle is 18 meters and its area is $19\frac{1}{4}$ square meters. What are its dimensions?

64. The area of a rectangular piece of sheet metal is 117 square inches. If the sum of the length ℓ and the width w is 22 inches, what are the dimensions of the metal plate?

65. A rectangular lot has an area of 84 square rods. If the sum of the length ℓ and the width w is 20 rods, what are the dimensions?

66. If two metal bars are the same length and if the length of one is increased by 3 centimeters and the second is decreased by 3 centimeters, the product of these two lengths is 27 square centimeters. Find the original lengths.

67. Two rectangular metal surfaces have the same width. If the width of one is increased by 6 inches and the other is increased by 8 inches, the product of the two widths is 99 square inches. Find the original widths.

68. If P dollars is invested at r percent compounded annually, at the end of two years it will grow to an amount $A = P(1 + r)^2$. At what rate will \$200 grow to \$224.72 in two years?

Review exercises

Evaluate the following expressions for the given values. See sections 1–8 and 9–1.

1. $\sqrt{a+b}$; $a = 2$ and $b = 7$
2. $\sqrt{b^2 - 4ac}$; $a = -1$, $b = 5$, and $c = 5$
3. Find the solution set of the system of equations
   $2x - y = 4$
   $3x + 2y = 6$
   by any method.
   See section 8–3.
4. Graph the equation $4x - 3y = -12$.
   See section 7–1.
5. If 2 dozen oranges cost $2.48, how many dozens of oranges can you buy for $11.16? Set up a proportion. See section 5–4.

10–3 ■ Solutions of quadratic equations by the quadratic formula

Identifying a, b, and c in a quadratic equation

We have found solution sets of quadratic equations by factoring, extracting the roots, and by completing the square. Even though the solution set of *any* such quadratic equation can be found by completing the square, a general formula, which is called the **quadratic formula,** can be derived that will enable us to find the solution set in an easier fashion.

To use the quadratic formula, the equation must be written in standard form,

$$ax^2 + bx + c = 0, \; a > 0$$

and we must be able to identify the coefficients a, b, and c. In identifying a, b, and c, we note that

1. a is the coefficient of x^2
2. b is the coefficient of x
3. c is the constant term

■ Example 10–3 A

Write each quadratic equation in standard form and identify the values of a, b, and c.

1. $3x^2 - 2x + 1 = 0$
   The equation is in standard form.

   $3x^2 - 2x + 1 = 0$

   Constant term, $c = 1$
   Coefficient of x, $b = -2$
   Coefficient of x^2, $a = 3$

2. $3x^2 - 4 = x$
   The equation must be written in standard form.

   $3x^2 - x - 4 = 0$ Add $-x$ to each member

   $a = 3$, $b = -1$, $c = -4$

3. $4x(x - 3) = 2x - 1$

The equation must be written in standard form.

$$4x^2 - 12x = 2x - 1 \quad \text{Multiply in left member}$$
$$4x^2 - 14x + 1 = 0 \quad \text{Add } -2x \text{ and 1 to each member}$$
$$a = 4,\ b = -14,\ c = 1$$

▶ ***Quick check*** Write the quadratic equation $2x^2 = 4 - 5x$ in standard quadratic form and identify the values of a, b, and c. ■

Solving quadratic equations using the quadratic formula

To derive the quadratic formula, we solve the equation $ax^2 + bx + c = 0$ by completing the square.

$$ax^2 + bx + c = 0,\ a > 0$$

$$x^2 + \frac{b}{a}x + \frac{c}{a} = 0 \quad \text{Divide each term of the equation by } a$$

$$x^2 + \frac{b}{a}x = -\frac{c}{a} \quad \text{Subtract } \frac{c}{a} \text{ from each member}$$

$$\left[\frac{1}{2}\left(\frac{b}{a}\right)\right]^2 = \left(\frac{b}{2a}\right)^2 = \frac{b^2}{4a^2} \quad \text{Square one-half of the coefficient of } x$$

$$x^2 + \frac{b}{a}x + \frac{b^2}{4a^2} = -\frac{c}{a} + \frac{b^2}{4a^2} \quad \text{Add } \frac{b^2}{4a^2} \text{ to each member}$$

$$\left(x + \frac{b}{2a}\right)^2 = \frac{b^2}{4a^2} - \frac{c}{a} \quad \text{Write left member as the square of a binomial and change the order of terms in the right member}$$

$$\left(x + \frac{b}{2a}\right)^2 = \frac{b^2}{4a^2} - \frac{4ac}{4a^2} \quad \text{Subtract fractions in the right member}$$

$$\left(x + \frac{b}{2a}\right)^2 = \frac{b^2 - 4ac}{4a^2}$$

$$x + \frac{b}{2a} = \sqrt{\frac{b^2 - 4ac}{4a^2}} \quad \text{or} \quad x + \frac{b}{2a} = -\sqrt{\frac{b^2 - 4ac}{4a^2}} \quad \text{Extract the roots}$$

$$x + \frac{b}{2a} = \frac{\sqrt{b^2 - 4ac}}{\sqrt{4a^2}} \quad \text{or} \quad x + \frac{b}{2a} = -\frac{\sqrt{b^2 - 4ac}}{\sqrt{4a^2}}$$

$$x + \frac{b}{2a} = \frac{\sqrt{b^2 - 4ac}}{2a} \quad \text{or} \quad x + \frac{b}{2a} = -\frac{\sqrt{b^2 - 4ac}}{2a} \quad \sqrt{4a^2} = 2a, \text{ since } a > 0$$

$$x = \frac{-b}{2a} + \frac{\sqrt{b^2 - 4ac}}{2a} \quad \text{or} \quad x = \frac{-b}{2a} - \frac{\sqrt{b^2 - 4ac}}{2a} \quad \text{Subtract } \frac{b}{2a} \text{ from each member}$$

$$x = \frac{-b + \sqrt{b^2 - 4ac}}{2a} \quad \text{or} \quad x = \frac{-b - \sqrt{b^2 - 4ac}}{2a}$$

By combining the expressions, these results can be summarized by the **quadratic formula.**

Quadratic formula

$$x = \frac{-b \pm \sqrt{b^2 - 4ac}}{2a}$$

Note We read $\pm$ "plus or minus," which allows us to write the two solutions

$$x = \frac{-b + \sqrt{b^2 - 4ac}}{2a} \quad \text{or} \quad x = \frac{-b - \sqrt{b^2 - 4ac}}{2a}$$

as a single statement.

Note When writing the quadratic formula, be sure that the fraction bar extends all the way beneath the numerator.

$$\frac{-b \pm \sqrt{b^2 - 4ac}}{2a}$$

Fraction bar

A common mistake is to write this as

$$-b \pm \frac{\sqrt{b^2 - 4ac}}{2a}$$

To solve a quadratic equation by the quadratic formula

1. Write the equation in standard form, if it is not already in this form. ($a > 0$).
2. Identify a (coefficient of x^2), b (coefficient of x), and c (the constant).
3. Substitute the values of a, b, and c into the quadratic formula.

$$x = \frac{-b \pm \sqrt{b^2 - 4ac}}{2a}$$

4. Simplify the resulting expressions.

■ *Example 10–3 B*

Find the solution set using the quadratic formula.

1. $x^2 - 2x - 8 = 0$

The equation is already in standard form where $a = 1$, $b = -2$, and $c = -8$.

$$x = \frac{-(-2) \pm \sqrt{(-2)^2 - 4(1)(-8)}}{2(1)}$$

Replace a with 1, b with -2, and c with -8 in the quadratic formula

$$x = \frac{2 \pm \sqrt{4 - (-32)}}{2}$$

Simplify by performing indicated operations

$$x = \frac{2 \pm \sqrt{36}}{2}$$

$$x = \frac{2 \pm 6}{2}$$

$\sqrt{36} = 6$

$$x = \frac{2 + 6}{2} = \frac{8}{2} = 4 \quad \text{or} \quad x = \frac{2 - 6}{2} = \frac{-4}{2} = -2$$

The solution set is $\{-2,4\}$.

2. $x^2 = 4 - x$

Write the equation in standard form: $x^2 + x - 4 = 0$. Then $a = 1$, $b = 1$, and $c = -4$.

Substitute these values into the quadratic formula.

$$x = \frac{-1 \pm \sqrt{(1)^2 - 4(1)(-4)}}{2(1)}$$ Replace a with 1, b with 1, and c with -4

$$x = \frac{-1 \pm \sqrt{1 + 16}}{2}$$

$$x = \frac{-1 \pm \sqrt{17}}{2}$$

Then $x = \dfrac{-1 + \sqrt{17}}{2}$ or $x = \dfrac{-1 - \sqrt{17}}{2}$.

The solution set is $\left\{\dfrac{-1 + \sqrt{17}}{2}, \dfrac{-1 - \sqrt{17}}{2}\right\}$.

3. $x^2 - 7 = 0$

The equation can be written $x^2 + 0x - 7 = 0$, so $a = 1$, $b = 0$, and $c = -7$.

Substitute these values.

$$x = \frac{-0 \pm \sqrt{(0)^2 - 4(1)(-7)}}{2(1)}$$ Replace a with 1, b with 0, and c with -7

$$x = \frac{\pm\sqrt{28}}{2}$$

$$x = \pm\frac{2\sqrt{7}}{2}$$

$$x = \pm\sqrt{7}$$ Simplify by reducing

Thus $x = \sqrt{7}$ or $x = -\sqrt{7}$.

The solution set is $\{\sqrt{7}, -\sqrt{7}\}$.

4. $4x^2 - 3x = 0$

The equation could be written $4x^2 - 3x + 0 = 0$, so $a = 4$, $b = -3$, and $c = 0$.

Substitute these values.

$$x = \frac{-(-3) \pm \sqrt{(-3)^2 - 4(4)(0)}}{2(4)}$$ Replace a with 4, b with -3, and c with 0

$$x = \frac{3 \pm \sqrt{9 - 0}}{8}$$

$$x = \frac{3 \pm \sqrt{9}}{8}$$

$$x = \frac{3 \pm 3}{8}$$

$$x = \frac{3 + 3}{8} = \frac{3}{4} \quad \text{or} \quad x = \frac{3 - 3}{8} = \frac{0}{8} = 0$$

The solution set is $\left\{\dfrac{3}{4}, 0\right\}$.

▶ ***Quick check*** Find the solution set.

a. $5x^2 - 2 = 3x$ b. $3 - \dfrac{2}{3}x^2 - \dfrac{7}{3}x = 0$ ■

Problem solving

Many useful formulas in the physical world have a second-degree term and are solved using the methods for quadratic equations. The following example illustrates this and some more application problems that require a quadratic equation to solve.

Example 10–3 C

1. The position of a particle moving on a straight line at time t in seconds is given by

$s = 3t^2 - 5t (t > 0)$

where s is the distance from the starting point in feet. How many seconds will it take to move the particle 8 feet in a positive direction?

We want t when $s = 8$ feet.

$(8) = 3t^2 - 5t$ — Replace s with 8

$3t^2 - 5t - 8 = 0$ — Write equation in standard form

$(3t - 8)(t + 1) = 0$ — Factor left member

$3t - 8 = 0$ or $t + 1 = 0$ — Set each factor equal to 0

$t = \frac{8}{3}$ $\quad t = -1$ — Solve each equation

Since we do not want a negative answer, the particle will move 8 feet in $\frac{8}{3}\left(\text{or } 2\frac{2}{3}\right)$ seconds.

Set up quadratic equation and solve the following problems.

2. Find two consecutive whole numbers whose product is 156.

Let n = the lesser whole number. Then $n + 1$ = the next consecutive whole number.

product of consecutive whole numbers	is	156
$n \cdot (n + 1)$	$=$	156

$n(n + 1) = 156$

$n^2 + n = 156$ — Multiply in the left member

$n^2 + n - 156 = 0$ — Add -156 to each member

We will use the quadratic formula.

$$n = \frac{-1 \pm \sqrt{(1)^2 - 4(1)(-156)}}{2(1)}$$ Replace a with 1, b with 1, and c with -156

$$n = \frac{-1 \pm \sqrt{1 + 624}}{2}$$

$$n = \frac{-1 \pm \sqrt{625}}{2}$$

$$n = \frac{-1 \pm 25}{2}$$ $\sqrt{625} = 25$

$$n = \frac{-1 + 25}{2} = 12 \text{ or } n = \frac{-1 - 25}{2} = -13$$

We reject the negative number since n is a whole number. So $n = 12$ and $n + 1 = 13$. The consecutive whole numbers are 12 and 13.

3. The sum of a number and its reciprocal is $\frac{25}{12}$. Find the number and its reciprocal.

Let n = the number. Then $\frac{1}{n}$ = the reciprocal of the number.

the sum of a number and its reciprocal	is	$\frac{25}{12}$
$n + \frac{1}{n}$	$=$	$\frac{25}{12}$

$$n + \frac{1}{n} = \frac{25}{12}$$

$$12n \cdot n + 12n \cdot \frac{1}{n} = 12n \cdot \frac{25}{12}$$ Multiply each member by $12n$ (the LCD) to clear the fractions

$$12n^2 + 12 = 25n$$ Reduce in each term

$$12n^2 - 25n + 12 = 0$$ Write in standard form

$$(3n - 4)(4n - 3) = 0$$ Factor left member

$$3n - 4 = 0 \text{ or } 4n - 3 = 0$$ Set each factor equal to 0 and solve

$$n = \frac{4}{3} \qquad n = \frac{3}{4}$$ Solve each equation

The number is $\frac{3}{4}$ and its reciprocal is $\frac{4}{3}$ or the number is $\frac{4}{3}$ and its reciprocal is $\frac{3}{4}$. ■

Mastery points

Can you

- Identify the values of a, b, and c in any quadratic equation?
- Use the quadratic formula

$$x = \frac{-b \pm \sqrt{b^2 - 4ac}}{2a}$$

to solve any quadratic equation?

Exercise 10–3

Write each quadratic equation in standard form and identify the values of a, b, and c, $a > 0$. See example 10–3 A.

Example $2x^2 = 4 - 5x$

Solution Add $-4 + 5x$ to each member to get the equation in standard form

$$2x^2 + 5x - 4 = 0$$

Then $a = 2$, $b = 5$, and $c = -4$.

1. $5x^2 - 3x + 8 = 0$
2. $4x^2 + x - 2 = 0$
3. $-6z^2 - 2z + 1 = 0$
4. $-3x^2 + x + 9 = 0$
5. $4x^2 = 2x - 1$
6. $y^2 = 5y + 3$
7. $x^2 = -3x$
8. $4x - 3x^2 = 0$
9. $5x^2 = 2$

10. $-8b^2 = -3$

11. $p(p + 3) = 4$

12. $2x(x - 9) = 1$

13. $(x + 3)(x - 1) = 6$

14. $(z - 4)(2z + 1) = -6$

15. $8m^2 - (m + 3) = 2m - 1$

16. $3x^2 - (2x - 5) = x - 6$

Find the solution set, using the quadratic formula. See example 10–3 B.

Example $5x^2 - 2 = 3x$

Solution Write the equation in standard form by adding $-3x$ to each member.

$5x^2 - 3x - 2 = 0$

Then $a = 5$, $b = -3$, and $c = -2$.

$$x = \frac{-(-3) \pm \sqrt{(-3)^2 - 4(5)(-2)}}{2(5)}$$ Replace a with 5, b with -3, and c with -2

$$x = \frac{3 \pm \sqrt{9 + 40}}{10}$$

$$x = \frac{3 \pm \sqrt{49}}{10}$$

$$x = \frac{3 \pm 7}{10}.$$

$$x = \frac{3 + 7}{10} = 1 \text{ or } x = \frac{3 - 7}{10} = -\frac{4}{10} = -\frac{2}{5}$$

The solution set is $\left\{1, -\frac{2}{5}\right\}$.

17. $x^2 - 3x + 2 = 0$

18. $y^2 + 6y + 9 = 0$

19. $a^2 - 2a + 1 = 0$

20. $x^2 + 10x + 24 = 0$

21. $x^2 - 25 = 0$

22. $2x^2 - 8 = 0$

23. $5x^2 - 10 = 0$

24. $3x^2 - 9 = 0$

25. $-x^2 = -3x$

26. $x^2 = 4x$

27. $5x^2 - 9x = 0$

28. $2x^2 = 7x$

29. $x^2 - 9x + 4 = 0$

30. $a^2 - 5a = 6$

31. $x^2 + 2x - 6 = 0$

32. $y^2 + y - 1 = 0$

33. $a^2 + 1 = 8a$

34. $2x^2 = 7x - 6$

35. $3y^2 = 5y + 6$

36. $4t^2 = 8t - 3$

37. $3a^2 = -9a - 2$

38. $x^2 - 9x = 6$

39. $3r^2 = r + 10$

40. $3a^2 + 5a = 4$

41. $x^2 + 8x + 16 = 0$

42. $2v^2 + 5v = -2$

43. $4x^2 + 25 = 20x$

44. $4x^2 - 7 = 12x$

45. $4x^2 + 12x + 9 = 0$

46. $4t^2 = 9t + 6$

47. $3a^2 - 2a - 7 = 0$

48. $4x^2 = 8 - 2x$

49. $3x^2 = 18 - 6x$

50. $9x^2 + 4 = 12x$

51. $3r^2 - 3r = 8$

Example $3 - \frac{2}{3}x^2 - \frac{7}{3}x = 0$

Solution $9 - 2x^2 - 7x = 0$ Multiply each member by the LCD, 3

Write the equation in standard form.

$2x^2 + 7x - 9 = 0$

Then $a = 2$, $b = 7$, $c = -9$.

$$x = \frac{-(7) \pm \sqrt{(7)^2 - 4(2)(-9)}}{2(2)}$$ Replace a with 2, b with 7, and c with -9

$$x = \frac{-7 \pm \sqrt{49 + 72}}{4}$$

$$x = \frac{-7 \pm \sqrt{121}}{4}$$

$$x = \frac{-7 \pm 11}{4}$$

$$x = \frac{-7 + 11}{4} = \frac{4}{4} = 1 \text{ or } x = \frac{-7 - 11}{4} = -\frac{18}{4} = -\frac{9}{2}$$

The solution set is $\left\{-\frac{9}{2}, 1\right\}$.

52. $a^2 + a = \frac{15}{4}$

53. $y^2 - y = \frac{3}{5}$

54. $2x^2 - \frac{7}{2} + \frac{x}{2} = 0$

55. $\frac{2}{3}x^2 - x = \frac{4}{3}$

56. $\frac{3}{4}x^2 - \frac{1}{2}x - 4 = 0$

57. $\frac{1}{3}x^2 - \frac{3}{2} = \frac{1}{2}x$

58. $\frac{2}{3}y^2 - \frac{4}{9}y = \frac{1}{3}$

59. $\frac{2a}{3} = \frac{2}{9} - a^2$

Solve the following problems using methods for solving quadratic equations. See example 10–3 C–1.

60. Use the formula $s = vt + \frac{1}{2}at^2$, where s is the distance traveled, v is the velocity, t is the time, and a is the acceleration of an object. Find t when (a) $s = 8$, $v = 3$, $a = 4$; (b) $s = 5$, $v = 4$, $a = 2$.

61. The distance s through which an object will fall in t seconds is $s = \frac{1}{2}gt^2$ feet, where $g = 32$ ft/sec^2. Find t (correct to tenth of a second) when (a) $s = 64$, (b) $s = 96$, (c) $s = 120$.

62. If a certain projectile is fired vertically into the air, the distance in feet above the ground in t seconds is given by $s = 160t - 16t^2$. Find t when (a) $s = 0$; (b) $s = 1{,}600$; (c) $s = 160$.

63. In a certain electric circuit, the relationship between i (in amperes), E (in volts), and R (in ohms) is given by $i^2R + iE = 8{,}000$. Find i ($i > 0$) when (a) $R = 2$ and $E = 80$, (b) $R = 4$ and $E = 60$.

64. A triangular-shaped plate has an altitude that is 5 inches longer than its base. If the area of the plate is 52 square inches, what is the length of the base b and the altitude h if the area of a triangle, A, is given by $A = \frac{1}{2}bh$?

65. The area of a triangle is 135 square inches. If the altitude is one-third the base, what are the lengths of the altitude and base? (Area $= \frac{1}{2}$ times base times altitude.)

66. A metal strip is shaped into a right triangle. In any right triangle, $c^2 = a^2 + b^2$, where c is the longest side, or hypotenuse, and a and b are the lengths of the other two sides, called legs. Find x when $a = x$, $b = x + 14$, and $c = x + 16$. (*Hint:* Substitute for a, b, and c in the above relationship and solve for x.)

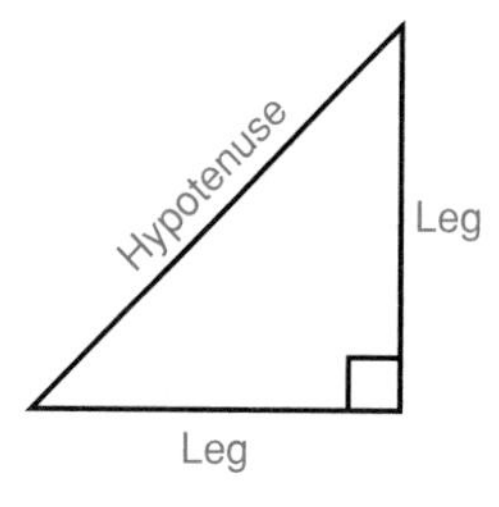

67. The hypotenuse of a right triangle is 10 millimeters long. One leg is 2 millimeters longer than the other. What are the lengths of the two legs? (Refer to exercise 66 for information about the hypotenuse and legs of a right triangle.)

68. The lengths of the legs of a right triangle are consecutive integers. If the hypotenuse is 5 centimeters long, what are the lengths of the legs of the triangle?

69. One leg of a right triangle is 2 feet longer than the other leg. If the hypotenuse is 4 inches long, what are the lengths of the legs of the triangle?

70. A 15-foot ladder is leaning against a building. If the base of the ladder is 6 feet from the base of the building, how high up the building does the ladder reach?

71. Joe leans a 50-foot ladder against his house. If the top of the ladder is 45 feet above the ground, how far out is the foot of the ladder from the house?

See example 10–3 C–2 and 3.

72. Find two consecutive whole numbers whose product is 210.

73. Find two consecutive negative even integers whose product is 224. (*Hint:* Use n and $n + 2$.)

74. Find two consecutive odd positive integers whose product is 143.

75. The sum of a number and its reciprocal is $\frac{50}{7}$. What is the number?

76. The sum of a number and its reciprocal is $\frac{61}{30}$. What is the number?

Review exercises

Perform the indicated operations. See sections 2–3 and 3–2.

1. $(4x^2 + 2x - 1) - (x^2 + x - 6)$

2. $(5y - 2)(y + 7)$

3. $(4z - 3)(4z + 3)$

4. $(3x - 5)^2$

Find the solution set of the following quadratic equations. See sections 10–1 and 10–3.

5. $3y^2 = 12$

6. $2x^2 - 7x - 4 = 0$

7. $x^2 - x = 10$

Subtract as indicated. See section 6–2.

8. $\frac{x+2}{x^2-4} - \frac{x-1}{x^2-x-6}$

10–4 ■ Complex solutions to quadratic equations

Complex numbers

Recall given the equation $x^2 = a$, we placed the restriction that $a \geq 0$. Suppose that $a < 0$ as in the equation $x^2 = -9$. Extracting the roots as we did in section 10–1, we obtain

$$x = \sqrt{-9} \quad \text{or} \quad x = -\sqrt{-9}$$

But $\sqrt{-9}$ is not a real number since there is no real number whose square is -9. Thus, in the set of real numbers, the equation $x^2 = -9$ does not have a solution. We introduce a new set of numbers called the *complex numbers*. To define this set, we need the definition of i.

Definition of *i*

The number i is defined by

$$i = \sqrt{-1}$$

We can now write

1. $\sqrt{-9} = \sqrt{9 \cdot -1} = \sqrt{9} \cdot \sqrt{-1} = 3i$
2. $\sqrt{-49} = \sqrt{49 \cdot -1} = \sqrt{49} \cdot \sqrt{-1} = 7i$

Observe two facts about the number i.

1. i is *not* a real number.
2. $i^2 = -1$ since $i = \sqrt{-1}$ and $i^2 = (\sqrt{-1})^2 = -1$.

We can now define a complex number.

Definition of a complex number

A complex number is any number that can be written in the form

$$a + bi \quad \text{or} \quad a - bi$$

where a and b are real numbers and $i = \sqrt{-1}$.

We call $a + bi$ and $a - bi$ the standard forms of a complex number.

Example 10–4 A

The following are complex numbers.

1. $2 + 3i$, where $a = 2$ and $b = 3$

2. $4 - 2i$, where $a = 4$ and $b = -2$

3. $5i$ since $5i = 0 + 5i$, where $a = 0$ and $b = 5$

4. $\sqrt{-4}$ since $\sqrt{-4} = \sqrt{4 \cdot -1} = \sqrt{4} \cdot \sqrt{-1} = 2i = 0 + 2i$ where $a = 0$ and $b = 2$

5. $2 + \sqrt{-5}$ since $2 + \sqrt{-5} = 2 + \sqrt{5 \cdot -1} = 2 + \sqrt{5} \cdot i = 2 + i\sqrt{5}$, where $a = 2$ and $b = \sqrt{5}$

6. 7 since $7 = 7 + 0i$ where $a = 7$ and $b = 0$

▶ ***Quick check*** Write $3 + \sqrt{-49}$ as a complex number $a + bi$. ■

From the last example, we can see that all real numbers are complex numbers, so the set of real numbers is a subset of the set of complex numbers.

Addition and subtraction of complex numbers

We add and subtract complex numbers in the same manner that we add and subtract polynomials. That is, we combine the similar terms.

Example 10–4 B

Combine the following complex numbers. Write the answer in standard form $a + bi$ or $a - bi$.

1. $(2 + 3i) + (4 - 5i)$

$= (2 + 4) + (3i - 5i)$ — Commutative and associative properties

$= 6 + (-2i)$ — Combine like terms.

$= 6 - 2i$

2. $(3 + i) - (4 - 3i)$
$= 3 + i - 4 + 3i$ — Definition of subtraction
$= (3 - 4) + (i + 3i)$ — Commutative and associative properties
$= -1 + 4i$ — Combine like terms

▶ ***Quick check*** Subtract $(4 - 2i) - (6 + 5i)$ ■

To multiply complex numbers, we apply the distributive property as we did when multiplying polynomials.

■ *Example 10–4 C*

Multiply the following complex numbers. Write the answer in standard form $a + bi$ or $a - bi$.

1. $2i(3 + 4i)$
$= 2i(3) + 2i(4i)$ — Distributive property
$= 6i + 8i^2$ — Multiply
$= 6i + 8(-1)$ — $i^2 = -1$
$= -8 + 6i$ — Write in form $a + bi$

2. $(1 - 4i)(2 + 5i)$
$= 1(2 + 5i) - 4i(2 + 5i)$ — Distributive property
$= 1(2) + 1(5i) + (-4i)(2) + (-4i)(5i)$
$= 2 + 5i - 8i - 20i^2$ — Multiply
$= 2 + 5i - 8i - 20(-1)$ — $i^2 = -1$
$= 2 + 5i - 8i + 20$
$= (2 + 20) + (5i - 8i)$ — Combine like terms
$= 22 + (-3i)$
$= 22 - 3i$

Note In this example, we should be reminded once again of the FOIL method for multiplying binomials. Remember, whenever i^2 appears, it must be replaced with -1.

3. $(5 + 2i)(5 - 2i)$
$= 5(5 - 2i) + 2i(5 - 2i)$ — Distributive property
$= 25 - 10i + 10i - 4i^2$
$= 25 - 4(-1)$ — Combine like terms
$= 25 + 4$
$= 29$

▶ ***Quick check*** Multiply $(2 - 3i)^2$ ■

Given polynomials $a + b$ and $a - b$, we call these *conjugates* of one another and know that $(a + b)(a - b) = a^2 - b^2$. In like fashion, $5 + 2i$ and $5 - 2i$ are called *complex conjugates* of one another. We found that

$$(5 + 2i)(5 - 2i) = 5^2 + 2^2 = 25 + 4 = 29$$

In general,

$$(a + bi)(a - bi) = a^2 + b^2$$

Thus, we find that the product of a complex number and its conjugate yields a *real number*. We use this property of complex numbers to rationalize the denominator of an indicated division of two complex numbers.

Example 10–4 D

Rationalize the denominator. Write the answer in standard form $a + bi$ or $a - bi$.

1. $\dfrac{4i}{2 + i}$

$= \dfrac{4i}{2 + i} \cdot \dfrac{2 - i}{2 - i}$ Multiply the numerator and the denominator by the conjugate of $2 + i$, which is $2 - i$

$= \dfrac{4i(2 - i)}{(2 + i)(2 - i)}$

$= \dfrac{8i - 4i^2}{2^2 + 1^2}$ Distributive property

$= \dfrac{8i - 4(-1)}{4 + 1}$ $i^2 = -1$

$= \dfrac{4 + 8i}{5}$

$= \dfrac{4}{5} + \dfrac{8}{5}i$ Divide each term of the numerator by 5

2. $\dfrac{4 + 3i}{3 - 5i}$

$= \dfrac{4 + 3i}{3 - 5i} \cdot \dfrac{3 + 5i}{3 + 5i}$ Multiply the numerator and the denominator by the conjugate of $3 - 5i$, which is $3 + 5i$

$= \dfrac{(4 + 3i)(3 + 5i)}{(3 - 5i)(3 + 5i)}$

$= \dfrac{12 + 20i + 9i + 15i^2}{3^2 + 5^2}$

$= \dfrac{12 + 29i + 15(-1)}{9 + 25}$

$= \dfrac{-3 + 29i}{34}$

$= -\dfrac{3}{34} + \dfrac{29}{34}i$ Divide each term of the numerator by 34

▶ ***Quick check*** Rationalize the denominator. $\dfrac{3 - 2i}{1 + i}$. Write the answer in standard form $a + bi$ or $a - bi$. ■

Quadratic equations with complex solutions

When solving equations of the form $ax^2 + bx + c = 0$, the solutions can be determined by the quadratic formula

$$x = \frac{-b \pm \sqrt{b^2 - 4ac}}{2a}$$

In our work thus far, we obtained rational or irrational solutions and the radicand, $b^2 - 4ac$, was always positive or zero. When $b^2 - 4ac < 0$ (negative), we obtain complex solutions of the quadratic equations.

■ Example 10–4 E

Find the solution set of the following quadratic equations.

1. $(x - 3)^2 = -4$

This is in the form $(x + q)^2 = k$.

$x - 3 = \pm\sqrt{-4}$ — Extract the roots

$x - 3 = \pm 2i$ — $\sqrt{-4} = 2i$

$x = 3 \pm 2i$ — Add 3 to each member

The solution set is $\{3 + 2i, 3 - 2i\}$.

Note We could have solved the problem by expanding the left member, writing the equation in standard form, and using the quadratic formula.

2. $x^2 - 3x = -7$

Write the equation in standard form and use the quadratic formula.

$x^2 - 3x + 7 = 0$ — Add 7 to each member

Now $a = 1$, $b = -3$, and $c = 7$.

$$x = \frac{-(-3) \pm \sqrt{(-3)^2 - 4(1)(7)}}{2(1)}$$

Replace a with 1, b with -3, and c with 7

$$= \frac{3 \pm \sqrt{9 - 28}}{2}$$

Perform indicated operations

$$= \frac{3 \pm \sqrt{-19}}{2}$$

$$= \frac{3 \pm i\sqrt{19}}{2}$$

$\sqrt{-19} = \sqrt{-1 \cdot 19} = \sqrt{-1} \cdot \sqrt{19} = i\sqrt{19}$

The solution set is $\left\{\frac{3 + i\sqrt{19}}{2}, \frac{3 - i\sqrt{19}}{2}\right\}$.

3. $(x + 1)(2x - 3) = -8$

$2x^2 - x - 3 = -8$ — Multiply in the left member

$2x^2 - x + 5 = 0$ — Write in standard form

Then $a = 2$, $b = -1$, and $c = 5$. — Use the quadratic formula

$$x = \frac{-(-1) \pm \sqrt{(-1)^2 - 4(2)(5)}}{2(2)}$$

Replace a with 2, b with -1, and c with 5

$$= \frac{1 \pm \sqrt{1 - 40}}{4}$$

Perform indicated operations

$$= \frac{1 \pm \sqrt{-39}}{4}$$

$$= \frac{1 \pm i\sqrt{39}}{4}$$

$\sqrt{-39} = i\sqrt{39}$

The solution set is $\left\{\frac{1 + i\sqrt{39}}{4}, \frac{1 - i\sqrt{39}}{4}\right\}$.

► ***Quick check*** Find the solution set. $2y^2 - y = -5$ ■

Because the expression $b^2 - 4ac$ determines the type of solutions a quadratic equation will have, we call $b^2 - 4ac$ the *discriminant*.

Thus when

1. $b^2 - 4ac > 0$, the equation has *two* distinct rational solutions if $b^2 - 4ac$ is a perfect square or two distinct irrational solutions if $b^2 - 4ac$ is *not* a perfect square.
2. $b^2 - 4ac = 0$, the equation has *one* rational solution.
3. $b^2 - 4ac < 0$, the equation has *two* complex solutions.

■ Example 10–4 F

Determine the type of solution(s) that the following quadratic equations yield by using the discriminant.

1. $x^2 - x - 6 = 0$

Since $a = 1$, $b = -1$, and $c = -6$, then

$$\begin{aligned} b^2 - 4ac &= (-1)^2 - 4(1)(-6) \quad \text{Replace } a \text{ with 1, } b \text{ with } -1\text{, and } c \text{ with } -6 \\ &= 1 + 24 \\ &= 25 \end{aligned}$$

Then $b^2 - 4ac > 0$ and, since 25 is a perfect square, we would obtain *two distinct rational* solutions.

2. $3y^2 + 2y - 2 = 0$

Since $a = 3$, $b = 2$, and $c = -2$, then

$$\begin{aligned} b^2 - 4ac &= (2)^2 - 4(3)(-2) \quad \text{Replace } a \text{ with 3, } b \text{ with 2, and } c \text{ with } -2 \\ &= 4 + 24 \\ &= 28 \end{aligned}$$

Then $b^2 - 4ac > 0$ and, since 28 is *not* a perfect square, we would obtain *two distinct irrational* solutions.

3. $x^2 - 10x + 25 = 0$

Since $a = 1$, $b = -10$, and $c = 25$, then

$$\begin{aligned} b^2 - 4ac &= (-10)^2 - 4(1)(25) \\ &= 100 - 100 \\ &= 0 \end{aligned}$$

Then $b^2 - 4ac = 0$ and we would obtain *one rational* solution.

4. $2y^2 + 2y + 7 = 0$

Since $a = 2$, $b = 2$, and $c = 7$, then

$$\begin{aligned} b^2 - 4ac &= (2)^2 - 4(2)(7) \\ &= 4 - 56 \\ &= -52 \end{aligned}$$

Then $b^2 - 4ac < 0$ and we would obtain *two complex* solutions.

▶ ***Quick check*** Determine the type of solutions $3x^2 + 2x = 4$ would yield by using the discriminant. ■

Mastery points

Can you

- Write a complex number in the standard form, $a + bi$?
- Add, subtract, and multiply complex numbers?
- Rationalize the denominator of an indicated quotient of two complex numbers?
- Find the complex solutions of a quadratic equation?
- Determine the type of solutions of any quadratic equation?

Exercise 10–4

Write the following complex numbers in standard form, $a + bi$ or $a - bi$. See example 10–4 A.

Example $3 + \sqrt{-49}$

Solution
$= 3 + \sqrt{49 \cdot -1}$
$= 3 + \sqrt{49} \cdot \sqrt{-1}$ $\quad$ $\sqrt{ab} = \sqrt{a} \cdot \sqrt{b}$
$= 3 + 7i$ $\quad$ $\sqrt{49} = 7;\ \sqrt{-1} = i$

1. 9 **2.** -5 **3.** $4i$ **4.** $10i$
5. $\sqrt{-25}$ **6.** $\sqrt{-29}$ **7.** $4 + 2\sqrt{-4}$ **8.** $-3 - \sqrt{-10}$

Add or subtract as indicated. Write the answer in standard form $a + bi$ or $a - bi$. See example 10–4 B.

Example $(4 - 2i) - (6 + 5i)$

Solution
$= 4 - 2i - 6 - 5i$ $\quad$ Definition of subtraction
$= (4 - 6) + (-2i - 5i)$ $\quad$ Combine like terms
$= -2 + (-7i)$
$= -2 - 7i$ $\quad$ Definition of subtraction

9. $(1 + 2i) + (3 - i)$ **10.** $(5 + 4i) + (3 + 5i)$ **11.** $(5 - i) - (4 + 3i)$
12. $(1 - 5i) - (2 - i)$ **13.** $(3 + \sqrt{-1}) + (2 - 3\sqrt{-1})$ **14.** $(1 - \sqrt{-4}) - (3 + \sqrt{-9})$
15. $(-5 - \sqrt{-7}) - (4 + \sqrt{-7})$ **16.** $(2 + \sqrt{-11}) + (3 - \sqrt{-11})$

Find the indicated products. Write the answer in standard form. See example 10–4 C.

Example $(2 - 3i)^2$

Solution
$= (2 - 3i)(2 - 3i)$
$= 2(2 - 3i) - 3i(2 - 3i)$ $\quad$ Distributive property
$= 4 - 6i - 6i + 9i^2$
$= 4 - 12i + 9(-1)$ $\quad$ Combine similar terms
$= 4 - 12i - 9$ $\quad$ $i^2 = -1$
$= -5 - 12i$

17. $3i(2 + 4i)$ **18.** $4i(5 - 2i)$ **19.** $(3 + 2i)(4 + i)$ **20.** $(5 - i)(3 + 4i)$
21. $(5 - 4i)(5 + 4i)$ **22.** $(5 - 5i)(5 + 5i)$ **23.** $(4 + 7i)^2$ **24.** $(3 - 2i)^2$

Rationalize the denominator. Write the answer in standard form $a + bi$ or $a - bi$. See example 10–4 D.

Example $\dfrac{3 - 2i}{1 + i}$

Solution $= \dfrac{(3 - 2i)(1 - i)}{(1 + i)(1 - i)}$ Multiply numerator and denominator by conjugate of $1 + i$

$= \dfrac{3 - 5i + 2i^2}{1^2 + 1^2}$ Multiply as indicated

$= \dfrac{3 - 5i - 2}{2}$ $2i^2 = 2(-1) = -2$

$= \dfrac{1 - 5i}{2}$

$= \dfrac{1}{2} - \dfrac{5}{2}i$ Divide each term of numerator by 2

25. $\dfrac{5i}{2 + 3i}$ **26.** $\dfrac{6i}{6 - 7i}$ 27. $\dfrac{4 + 2i}{3 - 5i}$ **28.** $\dfrac{1 + i}{2 - i}$ **29.** $\dfrac{5 - i}{4 + 3i}$ **30.** $\dfrac{4 - 7i}{5 + 2i}$

Find the solution set of the following quadratic equations. See example 10–4 E.

Example $2y^2 - y = -5$

Solution $2y^2 - y + 5 = 0$ Write in standard form

$a = 2, b = -1, c = 5$

Using quadratic formula,

$y = \dfrac{-(-1) \pm \sqrt{(-1)^2 - 4(2)(5)}}{2(1)}$ Replace a with 2, b with -1, and c with 5

$= \dfrac{1 \pm \sqrt{1 - 40}}{2}$

$= \dfrac{1 \pm \sqrt{-39}}{2}$

$= \dfrac{1 \pm i\sqrt{39}}{2}$ $\sqrt{-39} = \sqrt{39} \cdot \sqrt{-1} = \sqrt{39} \cdot i = i\sqrt{39}$

The solution set is $\left\{\dfrac{1 - i\sqrt{39}}{2}, \dfrac{1 + i\sqrt{39}}{2}\right\}$.

31. $(x + 2)^2 = -16$ **32.** $(x - 5)^2 = -3$ **33.** $x^2 + x + 2 = 0$

34. $x^2 - 3x + 7 = 0$ **35.** $x^2 - 3x = -5$ **36.** $x^2 + 5x = -9$

37. $2y^2 + y + 4 = 0$ **38.** $3y^2 - 2y + 3 = 0$ **39.** $(x + 3)(x - 2) = -11$

40. $(2y - 1)(3y - 2) = -3$

Determine the type of solution(s) that the following quadratic equations yield, using the discriminant. See example 10–4 F.

Example $3x^2 + 2x = 4$

Solution $3x^2 + 2x - 4 = 0$ Write in standard form

$a = 3, b = 2, c = -4$

$$\begin{aligned} b^2 - 4ac &= (2)^2 - 4(3)(-4) \quad \text{Replace } a \text{ with 3, } b \text{ with 2, and } c \text{ with } -4 \\ &= 4 + 48 \\ &= 52 \end{aligned}$$

The two solutions are distinct and irrational since $52 > 0$ and 52 is *not* a perfect square.

41. $y^2 + 2y - 5 = 0$
42. $2x^2 + x + 3 = 0$
43. $4x^2 - 12x + 9 = 0$
44. $3y^2 + y - 1 = 0$
45. $9x^2 - 3x = 0$
46. $3y^2 + 5y + 2 = 0$
47. $(x + 4)(x + 3) = 1$
48. $(2x + 3)(x + 5) = -3$

Review exercises

Graph the following equations by finding *three* ordered pair solutions and then graphing the points. See section 7–2.

1. $y = 4x - 3$
2. $y = 2 - 3x$
3. Graph the system of equations
   $2x - y = 1$
   $x + y = 3$
   and find the solution. See section 8–1.
4. Find the equation of the line through the points $(1,-3)$ and $(4,5)$. Write the answer in standard form. See section 7–4.

Perform the indicated operation. See section 6–1.

5. $\dfrac{3x}{x - 2} \cdot \dfrac{x^2 - x - 2}{4x^2}$

10–5 ■ The graphs of quadratic equations in two variables—quadratic functions

In chapter 7, we graphed *linear* equations in two variables such as

$$2x - 3y = 12 \text{ and } y = 4x - 5$$

The graphs of such equations were straight lines. In this section, we graph **quadratic equations in two variables** of the form

$$y = ax^2 + bx + c \quad (a \neq 0)$$

For example,

$$y = x^2 + 2x - 1$$

Values of x are arbitrarily chosen and corresponding values of y are found by substituting the value for x into the equation. Thus, when

a. $x = 1, y = (1)^2 + 2(1) - 1 = 1 + 2 - 1 = 2; (1,2)$
b. $x = -3, y = (-3)^2 + 2(-3) - 1 = 9 - 6 - 1 = 2; (-3,2)$
c. $x = 0, y = (0)^2 + 2(0) - 1 = -1; (0,-1)$

Thus, the ordered pairs $(1,2)$, $(-3,2)$, and $(0,-1)$ are solutions of the equation $y = x^2 + 2x - 1$.

Inspection will show us that for any chosen value of x, we get a unique (only one) value of y. Thus, *any quadratic equation in two variables of the form* $y = ax^2 + bx + c$ *does define a function.* Recall that we used the symbol $f(x)$ to replace y when defining a function. Thus, the **quadratic function** may be defined by

$$f(x) = ax^2 + bx + c, a \neq 0$$

Note $y = ax^2 + bx + c$ and $f(x) = ax^2 + bx + c$ define *the same function.* The latter gives the function a specific name, in this case f.

For example, the quadratic equations

$$y = x^2 + 2x + 1, \quad y = x^2 - 4x - 12, \quad \text{and} \quad y = 4x^2 - 3$$

can be used to define the quadratic functions f, g, and h such that

$$f(x) = x^2 + 2x + 1, \quad g(x) = x^2 - 4x - 12, \quad \text{and} \quad h(x) = 4x^2 - 3$$

As functions, the quadratic functions f, g, and h each have a domain and a range. *The domain of every quadratic function of the form* $f(x) = ax^2 + bx + c$ *is the set of real numbers* because we can evaluate the function for any real number we choose.

■ ***Example 10–5 A***

1. Given $f(x) = x^2 + 2x + 1$, find $f(3)$.

$$\begin{aligned} f(3) &= (3)^2 + 2(3) + 1 && \text{Replace } x \text{ with } 3 \\ &= 9 + 6 + 1 \\ &= 16 \end{aligned}$$

Therefore, $f(3) = 16$ and the ordered pair $(3,16)$ is an element of the function f.

2. Given $g(x) = x^2 - 4x - 12$, find $g(6)$.

$$\begin{aligned} g(6) &= (6)^2 - 4(6) - 12 && \text{Replace } x \text{ with } 6 \\ &= 36 - 24 - 12 \\ &= 0 \end{aligned}$$

Therefore, $g(6) = 0$ and the ordered pair $(6,0)$ is an element of the function g.

3. Given $h(x) = 4x^2 - 3$, find $h\left(\frac{1}{2}\right)$.

$$\begin{aligned} h\left(\frac{1}{2}\right) &= 4\left(\frac{1}{2}\right)^2 - 3 && \text{Replace } x \text{ with } \frac{1}{2} \\ &= 1 - 3 \\ &= -2 \end{aligned}$$

Therefore, $h\left(\frac{1}{2}\right) = -2$ and the ordered pair $\left(\frac{1}{2}, -2\right)$ is an element of function h.

Remember a function is *always* a set of ordered pairs.

▶ ***Quick check*** Given $f(x) = 2x^2 - 3x + 1$, find $f(-3)$, $f(0)$, and $f(1)$. State the answers as ordered pairs. ■

The parabola

Recall that the graph of a linear (first-degree) equation in two variables, or linear function, is a straight line. The graph of a quadratic (second-degree) equation in two variables, or quadratic function, is *not* a straight line. For this reason, we will require a number of points to plot the graph. The same procedure we used to graph linear equations can be applied to graph quadratic equations. The graph of any equation of the form $f(x) = y = ax^2 + bx + c$, $a \neq 0$ is, in fact, a special *curve,* called a **parabola.**

Consider the quadratic equation given by $f(x) = y = x^2 + 2x - 8$. Since we do not know what the graph looks like, we will need to choose a sufficient number of points.

x	$f(x)$ or y	(x,y)
-5	$y = (-5)^2 + 2(-5) - 8 = 25 + (-10) - 8 = 7$	$(-5,7)$
-4	$y = (-4)^2 + 2(-4) - 8 = 16 + (-8) - 8 = 0$	$(-4,0)$
-3	$y = (-3)^2 + 2(-3) - 8 = 9 + (-6) - 8 = -5$	$(-3,-5)$
-2	$y = (-2)^2 + 2(-2) - 8 = 4 + (-4) - 8 = -8$	$(-2,-8)$
-1	$y = (-1)^2 + 2(-1) - 8 = 1 + (-2) - 8 = -9$	$(-1,-9)$
0	$y = (0)^2 + 2(0) - 8 = 0 + 0 - 8 = -8$	$(0,-8)$
1	$y = (1)^2 + 2(1) - 8 = 1 + 2 - 8 = -5$	$(1,-5)$
2	$y = (2)^2 + 2(2) - 8 = 4 + 4 - 8 = 0$	$(2,0)$
3	$y = (3)^2 + 2(3) - 8 = 9 + 6 - 8 = 7$	$(3,7)$

Plotting all the points and passing a *smooth curve* through them, we have the graph of the quadratic equation $f(x) = y = x^2 + 2x - 8$ (figure 10–1).

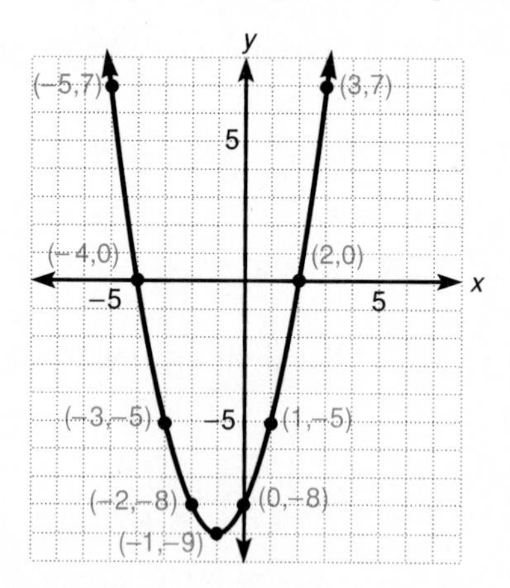

Figure 10–1

Note We *do not* connect the points with straight lines.

The x- and y-intercepts of a parabola

No matter how many points we plot, we cannot be sure that connecting all the points with a smooth curve will produce the correct graph and reveal all the important features of a curve. There are certain features of a parabola that we always wish to include in our graph, namely the x- and y-intercepts and the extreme (lowest or highest) point of the graph, called the *vertex*.

We observe from our example that when the curve crosses the x-axis, the value of y is zero, and when the curve crosses the y-axis, the value of x is zero. This is the same observation that we made with the linear equation, and we can generalize finding the x- and y-intercepts for any graph as follows:

1. To find the x-intercept(s), if there are any, we let $y = 0$ in the equation and solve for x.
2. To find the y-intercept, we let $x = 0$ in the equation and solve for y.

■ *Example 10–5 B*

Find the x- and y-intercepts.

1. $y = x^2 - 4$

a. Let $y = 0$

$(0) = x^2 - 4$ — Replace y with 0

$0 = (x - 2)(x + 2)$ — Factor right member, set each factor equal to 0 and solve

$x = 2$ or -2

The x-intercepts are 2 and -2. [The points are $(2,0)$ and $(-2,0)$.]

b. Let $x = 0$

$y = (0)^2 - 4$ — Replace x with 0

$y = -4$

Therefore, the y-intercept is -4. [The point is $(0,-4)$.]

2. $y = x^2 - 6x + 9$

a. Let $y = 0$

$(0) = x^2 - 6x + 9$ — Replace y with 0

$0 = (x - 3)(x - 3)$ — Factor, set each factor equal to 0 and solve

$x = 3$

Therefore, the x-intercept is 3. [The point is $(3,0)$.]

b. Let $x = 0$

$y = (0)^2 - 6(0) + 9$ — Replace x with 0

$y = 9$

Hence, the y-intercept is 9. [The point is $(0,9)$.]

3. $y = -x^2 + 2x + 3$

a. Let $y = 0$

$(0) = -x^2 + 2x + 3$

$0 = x^2 - 2x - 3$ — Multiply each member by -1

$0 = (x - 3)(x + 1)$ — Factor right member and set each factor equal to 0

$x = 3$ or $x = -1$ — Solve each equation

The x-intercepts are 3 and -1. [The points are $(3,0)$ and $(-1,0)$.]

b. Let $x = 0$

$$y = -(0)^2 + 2(0) + 3$$
$$= 3$$

The y-intercept is 3. [The point is (0,3).]

4. $y = x^2 + 1$

a. Let $y = 0$

$$(0) = x^2 + 1$$

Then $x^2 = -1$ and $x = \pm\sqrt{-1}$. Since $\sqrt{-1}$ is not a real number, there are no real solutions for x. Hence, the graph has no x-intercepts.

b. Let $x = 0$

$$y = (0)^2 + 1$$
$$= 1$$

The y-intercept is 1. [The point is (0,1).]

Note From these examples, we see that the y-intercept is always the constant c. If the quadratic equation is in standard form, the y-intercept will be the point $(0,c)$.

▶ ***Quick check*** Find the x- and y-intercepts. $y = 2x^2 - 3x + 1$ ■

The vertex and the axis of symmetry of a parabola

We wish to find one remaining point of interest on the graph—the vertex. The vertex is the extreme point on the graph, that is, either the maximum or minimum value that the graph will attain. If our equation is in standard form, $y = ax^2 + bx + c$, we can show, but will not do so here, the x-component of the vertex is given by $x = \frac{-b}{2a}$. Once we have determined the x-component, we replace x with this value in our original function and generate the corresponding y-value. Recall our original example: $y = x^2 + 2x - 8$. The value of a is 1 and b is 2. Therefore

$$x = \frac{-b}{2a} = \frac{-2}{2(1)} = \frac{-2}{2} = -1$$

We then substitute this value for x in our original equation and we obtain

$$y = (-1)^2 + 2(-1) - 8 = 1 + (-2) - 8 = -9$$

Hence, in this case, the vertex is the point with coordinates $(-1,-9)$. This means no matter what value x takes, y, or $f(x)$, is *never* less than -9.

Note When the value of a, the coefficient of the squared term, is positive (as in this case), the parabola opens *upward* and the vertex is the *lowest* point of the graph. When a is negative, the parabola opens *downward* and the vertex is the *highest* point of the graph.

■ Example 10–5 C

Find the coordinates of the vertex of each parabola.

1. $y = x^2 - 4$

$y = x^2 + 0x - 4$ Write equation in standard form

Since $a = 1$ and $b = 0$, the vertex has coordinates

$$x = -\frac{b}{2a} = -\frac{(0)}{2(1)} = 0$$ Replace a with 1 and b with 0

$$y = (0)^2 - 4 = -4$$ Replace x with 0 in $y = x^2 - 4$

The vertex is the point $(0, -4)$.

2. $y = x^2 - 6x + 9$

Since $a = 1$ and $b = -6$, the vertex has coordinates

$$x = \frac{-b}{2a} = -\frac{(-6)}{2(1)} = \frac{6}{2} = 3$$

and $y = (3)^2 - 6(3) + 9$
$= 9 - 18 + 9$
$= 0$

The vertex is the point $(3,0)$.

3. $y = -x^2 + 2x + 3$

Since $a = -1$ and $b = 2$, so the vertex has coordinates

$$x = -\frac{b}{2a} = -\frac{(2)}{2(-1)} = \frac{2}{2} = 1$$

and $y = -(1)^2 + 2(1) + 3$
$= -1 + 2 + 3$
$= 4$

The vertex is the point $(1,4)$.

4. $y = x^2 + 1$

Since $a = 1$ and $b = 0$, so the vertex has coordinates

$$x = -\frac{b}{2a} = -\frac{(0)}{2(1)} = -\frac{0}{2} = 0$$

and $y = (0)^2 + 1$
$= 1$

The vertex is the point $(0,1)$.

▶ ***Quick check*** Find the vertex. $y = 2x^2 - 3x + 1$ ■

If the vertex is the point, (h,k), the vertical line, $x = h$, which passes through the vertex, is called the *axis of symmetry* in the graph of the parabola. The parabola is a symmetric curve and if we fold the graph along the axis of symmetry, the left half of the curve will coincide with the right half of the curve. For this reason, we choose two values of x that are greater than h and two values of x that are less than h when finding our arbitrary points to graph.

The graph of a quadratic equation

To draw a reasonably accurate graph of any quadratic equation in two variables, we take the following steps.

Graphing a quadratic equation in two variables

1. Find the coordinates of the x- and y-intercepts.
   a. Let $x = 0$, solve for the y-intercept.
   b. Let $y = 0$, solve for the x-intercept(s).
2. Find the coordinates of the vertex.
   a. $x = -\frac{b}{2a}$
   b. Replace x with the value of $-\frac{b}{2a}$ in the original equation and solve for y.
3. Find the coordinates of four arbitrarily chosen points. Choose values of x such that, if the point (h,k) is the vertex of the parabola,
   a. two values are less than h and
   b. two values are greater than h.
4. Draw a smooth curve through the resulting points.

■ *Example 10–5 D*

Graph the following quadratic equations. Determine the equation of the axis of symmetry.

1. $y = x^2 - 4$

In our previous examples, we found the x- and y-intercepts and the vertex. We need only determine four more points and we will be ready to graph the function.

x	y	
2	0	x-intercepts
−2	0	
0	−4	y-intercept and vertex
−1	−3	Arbitrary points
1	−3	
−3	5	
3	5	

The axis of symmetry is the line $x = 0$ (y-axis).

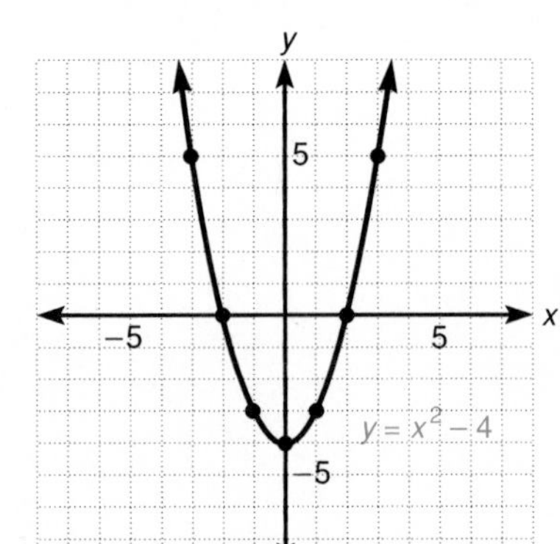

2. $y = x^2 - 6x + 9$

x	y	
3	0	x-intercept and vertex
0	9	y-intercept
1	4	Arbitrary points
2	1	
4	1	
5	4	
6	9	

The axis of symmetry is the line $x = 3$.

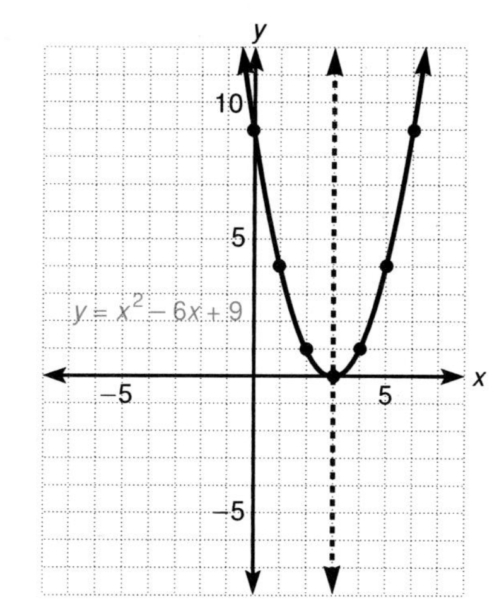

3. $y = -x^2 + 2x + 3$

x	y	
3	0	x-intercepts
−1	0	
0	3	y-intercept
1	4	vertex
−2	−5	Arbitrary points
2	3	
4	−5	

The axis of symmetry is the line $x = 1$.

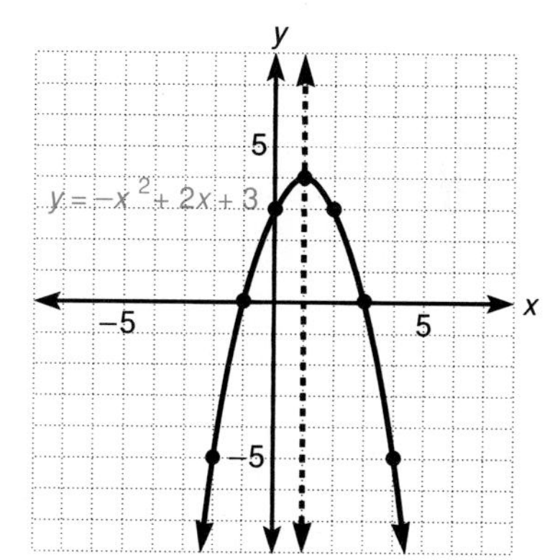

4. $y = x^2 + 1$

x	y	
0	1	y-intercept and vertex
−3	10	Arbitrary points
−2	5	
−1	2	
1	2	
2	5	
3	10	

The axis of symmetry is the line $x = 0$ (y-axis).

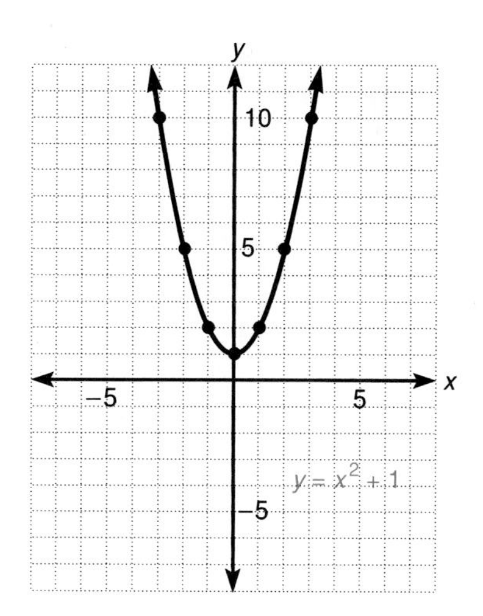

▶ ***Quick check*** Graph the equation and determine the equation of the axis of symmetry. $y = 2x^2 - 3x + 1$ ■

Quadratic equations are used in many physical situations. For example, if an object is thrown into the air, the graph of the distance the object travels versus the time it travels is a parabola.

Example 10–5 E

A projectile is fired vertically into the air. Its distance s in feet above the ground in t seconds is given by $s = 160t - 16t^2$.

Note s is defined as a quadratic function of t. That is, the distance the object travels *changes with* time.

Find the highest point of the projectile (the vertex of the parabola) and the moment when the projectile will strike the ground. Graph the equation.

a. The vertex is the highest point since the parabola opens downward.

$$s = -16t^2 + 160t$$

where $a = -16$, $b = 160$. The t value of the vertex is

$$t = -\frac{b}{2a} = -\frac{(160)}{2(-16)} = -\frac{160}{-32} = 5 \quad \text{Replace } a \text{ with } -16 \text{ and } b \text{ with } 160$$

The height will be

$$s = -16(5)^2 + 160(5) = -400 + 800 = 400 \text{ feet}$$

The maximum height, $s = 400$ feet, is attained when $t = 5$ seconds.

b. The projectile will strike the ground when $s = 0$. Therefore, we set $s = 0$ and solve for t.

$$\begin{aligned} 0 &= -16t^2 + 160t && \text{Replace } s \text{ with } 0 \\ &= -16t(t - 10) && \text{Set each factor equal to 0 and solve} \\ t &= 0 \text{ or } 10 \end{aligned}$$

The value $t = 0$ seconds represents when the projectile was fired. Hence, the value $t = 10$ seconds represents the time when the projectile will strike the ground.

c. To graph the equation, we plot time, t, along the horizontal axis and distance, s, along the vertical axis.

t	s	
5	400	vertex
0	0	t- and s-intercept
10	0	t-intercept
2	256	Arbitrary points
4	384	
6	384	
8	256	

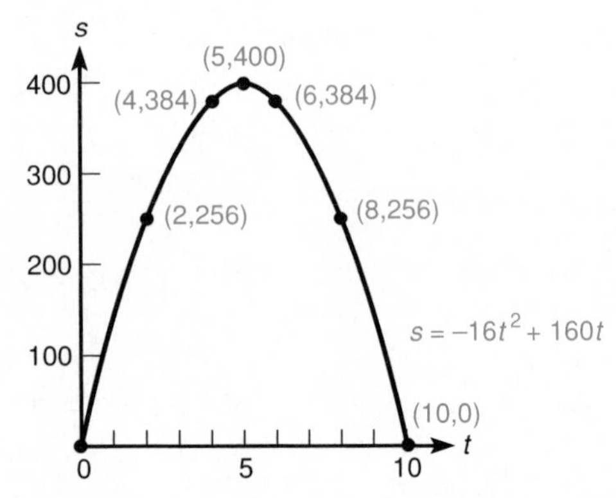

Note We have used different scales on the t- and s-axes, and we do not represent values on the graph for negative time or distance, since they do not have any meaning in this example. ■

Mastery points

Can you

- Evaluate a quadratic function?
- Find the x- and y-intercepts of the graph of a quadratic equation?
- Find the coordinates of the vertex of the graph of a quadratic equation?
- Graph a quadratic function in two variables?
- Determine the equation of the axis of symmetry?

Exercise 10–5

Find the value of the following quadratic functions at the given values of x. State the answer as ordered pairs. See example 10–5 A.

Example Given $f(x) = 2x^2 - 3x + 1$, find $f(-3), f(0), f(1)$.

Solution $f(-3) = 2(-3)^2 - 3(-3) + 1 = 18 + 9 + 1 = 28;$ $(-3,28)$
$f(0) = 2(0)^2 - 3(0) + 1 = 0 - 0 + 1 = 1;$ $(0,1)$
$f(1) = 2(1)^2 - 3(1) + 1 = 2 - 3 + 1 = 0;$ $(1,0)$

1. $f(x) = x^2 + 3x - 4; f(-1), f(0), f(3)$

2. $g(x) = x^2 - x - 1; g(-2), g(0), g(1)$

3. $h(x) = 4x^2 + x + 5; h(-3), h(0), h(2)$

4. $f(x) = 3x^2 - 4x + 7; f(-4), f(0), f\left(\frac{1}{3}\right)$

5. $f(x) = 5x^2 - 3x; f(-6), f(0), f(6)$

6. $g(x) = 8x - 2x^2; g(-2), g(0), g\left(\frac{1}{2}\right)$

7. $h(x) = 5x^2 - 1; h(-3), h(0), h(5)$

8. $f(x) = 4x^2 + 2; f\left(\frac{-1}{2}\right), f(0), f(8)$

9. $g(x) = 4 - 2x - 3x^2; g(-2), g(0), g\left(\frac{3}{4}\right)$

10. $h(x) = 10 + 7x - 2x^2; h(-4), h(0), h(1.2)$

Find the x- and y-intercepts. If they do not exist, so state. See example 10–5 B.

Example $y = 2x^2 - 3x + 1$

Solution a. Let $y = 0$

$(0) = 2x^2 - 3x + 1$ Replace y with 0
$0 = (2x - 1)(x - 1)$ Factor right member
$2x - 1 = 0$ or $x - 1 = 0$ Set each factor equal to 0 and solve
$x = \frac{1}{2}$ $x = 1$

The x-intercepts are $\frac{1}{2}$ and 1. $\left[\text{The points are } \left(\frac{1}{2},0\right) \text{ and } (1,0).\right]$

b. Let $x = 0$
Then $y = 1$. The y-intercept is 1. [The point is $(0,1)$.]

11. $y = x^2 - 16$

12. $y = x^2 - 9$

13. $y = x^2 - 6x + 8$

14. $y = x^2 + 11x - 12$

15. $y = x^2 + 8x + 12$

16. $y = x^2 - 4x - 12$

17. $y = 5 - x^2$	**18.** $y = 7 - x^2$	**19.** $y = x^2 + 6x + 9$
20. $y = x^2 - 4x + 4$	**21.** $y = x^2 + 5$	**22.** $y = x^2 + 6$
23. $y = x^2 + 4x + 6$	**24.** $y = x^2 + 2x + 5$	**25.** $y = -x^2 + 8x - 16$
26. $y = 25 - x^2$	**27.** $y = 2x^2 + 3x + 1$	**28.** $y = 2x^2 - 7x + 6$
29. $y = -2x^2 - x + 6$	**30.** $y = -3x^2 + 7x + 6$	

Find the vertex. Find the equation of the axis of symmetry. See example 10–5 C.

Example $y = 2x^2 - 3x + 1$

Solution Now $a = 2$, $b = -3$, and $c = 1$

so $x = -\dfrac{b}{2a} = -\dfrac{(-3)}{2(2)} = \dfrac{3}{4}$ Replace a with 2 and b with -3

then $y = 2\left(\dfrac{3}{4}\right)^2 - 3\left(\dfrac{3}{4}\right) + 1$ Replace x with $\dfrac{3}{4}$

$$= 2\left(\frac{9}{16}\right) - \frac{9}{4} + 1$$

$$= \frac{9}{8} - \frac{9}{4} + 1$$

$$= \frac{9}{8} - \frac{18}{8} + \frac{8}{8}$$

$$= -\frac{1}{8}$$

The vertex is the point $\left(\dfrac{3}{4}, -\dfrac{1}{8}\right)$. The axis of symmetry is $x = \dfrac{3}{4}$.

31. $y = x^2 - 16$	**32.** $y = x^2 - 9$	**33.** $y = x^2 - 6x + 8$
34. $y = x^2 + 11x - 12$	**35.** $y = x^2 + 8x + 12$	**36.** $y = x^2 - 4x - 12$
37. $y = 5 - x^2$	**38.** $y = 7 - x^2$	**39.** $y = x^2 + 6x + 9$
40. $y = x^2 - 4x + 4$	**41.** $y = x^2 + 5$	**42.** $y = x^2 + 6$
43. $y = x^2 + 4x + 6$	**44.** $y = x^2 + 2x + 5$	**45.** $y = -x^2 + 8x - 16$
46. $y = 25 - x^2$	**47.** $y = 2x^2 + 3x + 1$	**48.** $y = 2x^2 - 7x + 6$
49. $y = -2x^2 - x + 6$	**50.** $y = -3x^2 + 7x + 6$	

Graph the following equations. Include the vertex and the points at which the graph crosses each axis. See example 10–5 D.

Example $y = 2x^2 - 3x + 1$

Note $a = 2$, which is positive so the parabola opens up.

Solution The vertex is at $\left(\frac{3}{4}, -\frac{1}{8}\right)$.

Let $x = 0$. Then

$y = 2(0) - 3(0) + 1 = 1$

so the y-intercept is at $(0,1)$.

Let $y = 0$. Then

$$(0) = 2x^2 - 3x + 1$$
$$= (2x - 1)(x - 1)$$

so $2x - 1 = 0$ or $x - 1 = 0$

$2x = 1$ $\quad x = 1$

$x = \frac{1}{2}$ $\quad x = 1$

The x-intercepts are at $\left(\frac{1}{2}, 0\right)$ and $(1,0)$.

x	y	
0	1	y-intercept
1	0	x-intercepts
$\frac{1}{2}$	0	
$\frac{3}{4}$	$-\frac{1}{8}$	Vertex
2	3	Arbitrary points
−1	6	
3	10	
−2	15	

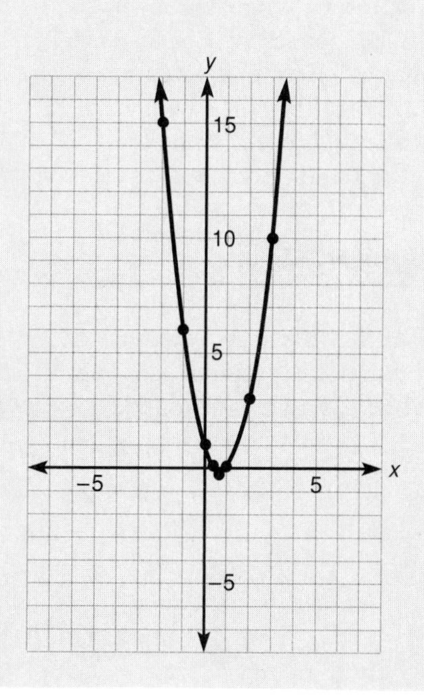

51. $y = x^2 - 16$
52. $y = x^2 - 9$
53. $y = x^2 - 6x + 8$
54. $y = x^2 + 11x - 12$
55. $y = x^2 + 8x + 12$
56. $y = x^2 - 4x - 12$
57. $y = 5 - x^2$
58. $y = 7 - x^2$
59. $y = x^2 - 4x + 4$
60. $y = x^2 + 6x + 9$
61. $y = x^2 + 6$
62. $y = x^2 + 5$
63. $y = x^2 + x - 6$
64. $y = x^2 + 2x + 5$
65. $y = -x^2 + 4x - 3$
66. $y = 25 - x^2$
67. $y = 2x^2 + 3x + 1$
68. $y = 2x^2 - 7x + 6$
69. $y = -2x^2 - x + 6$
70. $y = -3x^2 + 7x + 6$

Graph each equation by plotting the variable for which the equation is solved along the vertical axis and by plotting the other variable along the horizontal axis. Graph the equation only in the regions for which the equation would have meaning. See example 10–5 E.

71. When a ball rolls down an inclined plane, it travels a distance $d = 6t + \frac{t^2}{2}$ feet in t seconds. Plot the graph showing how d depends on t. How long will it take the ball to travel 14 feet?

72. The output power P of a 100-volt electric generator is defined by $P = 100I - 5I^2$, where I is in amperes. Plot the graph showing how P depends on I.

73. The current in a circuit flows according to the equation $i = 12 - 12t^2$, where i is the current and t is the time in seconds. Plot the graph of the relation given by the equation labeling the horizontal axis the t-axis.

74. If a projectile is fired vertically into the air with an initial velocity of 80 feet per second, the distance in feet above the ground in t seconds is given by $s = 80t - 16t^2$. Find the projectile's maximum height and when the projectile will strike the ground. Plot the graph showing how s depends on t.

75. Referring to exercise 74, if the initial velocity is 96 ft/sec, the equation is $s = 96t - 16t^2$. Find the maximum height and when the projectile will strike the ground. Plot the graph of this equation using the t-axis as the horizontal axis.

76. The distance s through which an object will fall in t seconds is $s = 16t^2$. Plot the graph showing the relation between s and t for the first 5 seconds.

77. An object is dropped from the top of the Empire State Building (1,250 feet tall), and the distance that the object is from the ground is given by the equation $s = 1{,}250 - 16t^2$. Plot the graph showing how s depends on t and determine when the object will strike the ground. (t is time in seconds.)

Chapter 10 lead-in problem

A rock is dropped from the top of the Washington Monument. If the monument is 555 feet tall, how long will it take the rock to strike the ground?

Solution

We use $s = 16t^2$, where s is the distance the rock fell and t is the time in seconds.

$555 = 16t^2$ — Replace s with 555

$t^2 = \frac{555}{16}$ — Divide each member by 16

$t = \sqrt{\frac{555}{16}} = \frac{\sqrt{555}}{4} \approx 5.9$ — Extract the roots

or $t = -\sqrt{\frac{555}{16}} = -\frac{\sqrt{555}}{4} \approx -5.9$

Reject the negative value since time t must be positive. Thus, the rock will strike the ground in approximately 5.9 seconds.

Chapter 10 summary

1. We can solve equations of the form $x^2 = k$ and $(x + q)^2$ or $(px + q)^2 = k$, $k \geq 0$, by **extracting the roots.**
2. If $x^2 = k$, then $x = \pm\sqrt{k}$, and if $(x + q)^2 = k$, then $x + q = \pm\sqrt{k}$ and $x = -q \pm \sqrt{k}$.
3. Any quadratic equation can be solved by **completing the square.**
4. Given the quadratic equation $ax^2 + bx + c = 0$, the **quadratic formula** states $x = \dfrac{-b \pm \sqrt{b^2 - 4ac}}{2a}$.
5. A **quadratic function** is defined by $f(x) = ax^2 + bx + c$, where a, b, and c are constants, $a \neq 0$.
6. A **complex number** is any number that can be written in the form $a + bi$, where a and b are real numbers and i equals $\sqrt{-1}$.
7. The graph of a quadratic function is a **parabola.**
8. The coordinates of the vertex of the graph of a quadratic function are found by $x = -\dfrac{b}{2a}$ and replacing x with $-\dfrac{b}{2a}$ to find y.
9. In the quadratic equation in two variables, $y = ax^2 + bx + c$, the graph of the parabola
   a. opens up if $a > 0$
   b. opens down if $a < 0$
10. The equation of the *axis of symmetry* is $x = h$, where the vertex is the point (h,k).

Chapter 10 error analysis

1. Solving quadratic equations by completing the square
   Example: Find the solution set of $x^2 + x - 8 = 0$ by completing the square.
   $$x^2 + x = 8$$
   $$x^2 + x + \frac{1}{4} = 8$$
   $$\left(x + \frac{1}{2}\right)^2 = 8$$
   $$x + \frac{1}{2} = \pm\sqrt{8} = \pm 2\sqrt{2}$$
   $$x = -\frac{1}{2} \pm 2\sqrt{2} = \frac{-1 \pm 4\sqrt{2}}{2} \quad \left\{\frac{-1 - 4\sqrt{2}}{2}, \frac{-1 + 4\sqrt{2}}{2}\right\}$$
   Correct answer: $\left\{\dfrac{-1 - \sqrt{33}}{2}, \dfrac{-1 + \sqrt{33}}{2}\right\}$
   What error was made? (*see page 411*)
2. Solving quadratic equations by quadratic formula
   Example: Find the solution set of $x^2 + 2x - 5 = 0$ by quadratic formula.
   $a = 1, b = 2, c = -5$
   $$x = -2 \pm \frac{\sqrt{(2)^2 - 4(1)(-5)}}{2(1)}$$
   $$= -2 \pm \frac{\sqrt{24}}{2} = -2 \pm \frac{2\sqrt{6}}{2}$$
   $$= -2 \pm \sqrt{6} \quad \{-2 - \sqrt{6}, -2 + \sqrt{6}\}$$
   Correct answer:
   $\{-1 - \sqrt{6}, -1 + \sqrt{6}\}$
   What error was made? (*see page 418*)
3. Finding the solution set of a quadratic equation by quadratic formula
   Example: Find the solution set of $2x^2 - x = 4$ by quadratic formula.
   $a = 2, b = -1, c = 4$
   Correct answer: $a = 2, b = -1, c = -4$
   What error was made? (*see page 418*)
4. Extracting the roots
   Example: Find the solution set of $x^2 = 25$ by extracting roots.
   $x^2 = 25$
   $x = \sqrt{25} = 5 \quad \{5\}$
   Correct answer: $\{-5,5\}$
   What error was made? (*see page 405*)
5. Product of complex numbers
   Example: $(2 + 3i)(1 - i) = (2)(1) - 2i + 3i - 3i^2$
   $= 2 - 2i + 3i - 3$
   $= -1 + i$
   Correct answer: $5 + i$
   What error was made? (*see page 426*)
6. Finding the y-intercept of a parabola
   Example: Given the quadratic equation $y = 2x^2 - 2x - 3$, so the y-intercept is $(0,3)$.
   Correct answer: The y-intercept is $(0,-3)$.
   What error was made? (*see page 435*)

7. Graph of a parabola
Example: The parabola that is the graph of $y = 2x^2 + x - 3$ opens *downward.*
Correct answer: opens *upward*
What error was made? (*see page 436*)

8. Squaring a radical binomial
Example: $(\sqrt{6} + \sqrt{5})^2 = (\sqrt{6})^2 + (\sqrt{5})^2 = 6 + 5 = 11$
Correct answer: $11 + 2\sqrt{30}$
What error was made? (*see page 387*)

9. Multiplying square roots
Example: $\sqrt{-6} \cdot \sqrt{-6} = \sqrt{-6 \cdot -6} = \sqrt{36} = 6$
Correct answer: -6
What error was made? (*see page 426*)

10. Rationalizing the denominator of an nth root
Example: $\dfrac{3}{\sqrt[4]{xy^2}} = \dfrac{3}{\sqrt[4]{xy^2}} \cdot \dfrac{\sqrt[4]{xy^2}}{\sqrt[4]{xy^2}} = \dfrac{3\sqrt[4]{xy^2}}{xy^2}$
Correct answer: $\dfrac{3\sqrt[4]{x^3y^2}}{xy}$
What error was made? (*see page 381*)

Chapter 10 critical thinking

Given three consecutive integers, the product of the first and third is always one less than the square of the middle one.

Chapter 10 review

[10–1]

Find the solution set of the following quadratic equations by extracting the roots or by factoring.

1. $x^2 = 100$
2. $x^2 - 25 = 0$
3. $z^2 = 2$
4. $y^2 - 6 = 0$
5. $6x^2 = 24$
6. $8x^2 - 96 = 0$
7. $\frac{3}{4}x^2 = 12$
8. $\frac{2}{3}x^2 - 8 = 0$
9. $\frac{x^2}{3} - 8 = \frac{1}{4}$
10. $x^2 - x - 42 = 0$
11. $3y^2 - 7y + 4 = 0$

[10–2]

Find the solution set by completing the square.

12. $x^2 - 6x + 4 = 0$
13. $z^2 - 10z + 4 = 0$
14. $2a^2 - 8a = -1$
15. $3y^2 - 6y - 5 = 0$
16. $4 - x^2 = 5x$
17. $5 = 11y - y^2$
18. $x(4x - 1) = 3$
19. $(x - 2)(x + 1) = 1$
20. $4x^2 - 3 = 3x - 1$
21. $\frac{3}{5}x^2 + \frac{1}{5}x = 2$
22. The length of a rectangle is 2 meters more than three times the width. Its area is 16 square meters. What are its dimensions? (Solve by completing the square.)

[10–3]

Find the solution set using the quadratic formula.

23. $x^2 - 2x - 5 = 0$
24. $x^2 - 8 = -4x$
25. $2y^2 - 3y = 5$
26. $3a^2 = 8 - 7a$
27. $2x^2 - 9 = 0$
28. $4x^2 = -7x$
29. $x^2 - \frac{2}{3}x = \frac{4}{3}$
30. $2x + \frac{3}{4} = \frac{3}{2}x^2$
31. A metal bar is to be divided into two pieces so that one piece is 3 inches longer than the other. If the sum of the squares of the two lengths is 117 square inches, find the two lengths. (Use the quadratic formula.)

[10–4]

Perform the indicated operations on the given complex numbers. Write the answer in standard form $a + bi$ or $a - bi$.

32. $(3 - 5i) + (2 + 4i)$
33. $(7 + i) - (3 - 8i)$
34. $(6 + i)(2 - 3i)$
35. $(-3 + 9i)(2 - i)$
36. $(6 - 4i)(6 + 4i)$
37. $(5 - 3i)^2$

Rationalize the denominator. Write the answer in standard form $a + bi$ or $a - bi$.

38. $\dfrac{3i}{1 + i}$

39. $\dfrac{-4i}{2 - i}$

40. $\dfrac{2 - i}{3 - i}$

41. $\dfrac{5 + 2i}{4 + 3i}$

Find the solution set of the following quadratic equations.

42. $x^2 + 4x + 7 = 0$

43. $4y^2 - y = -5$

44. $(x + 1)(x - 1) = -8$

45. $(2x + 3)^2 = -4$

Determine the type of solutions the following quadratic equations will yield, using the discriminant.

46. $y^2 - 16y + 64 = 0$

47. $3x^2 - x - 2 = 0$

48. $5y^2 - 2y + 3 = 0$

49. $3x^2 + x - 3 = 0$

[10–5]

Evaluate each quadratic function at the given values of x.

50. $f(x) = x^2 - 3x - 5$; $f(-5), f(0), f(1)$

51. $g(x) = 4 + 5x - 2x^2$; $g(-1), g(0), g(3)$

52. $h(x) = 4x^2 + 2x$; $h(-3), h(0), h(4)$

53. $f(x) = 12 - 3x^2$; $f(-4), f(0), f(2)$

Find the x- and y-intercepts and the vertex in the graph of each quadratic equation. Find the equation of the axis of symmetry. Sketch the graph.

54. $y = x^2 - 4x - 12$

55. $y = 5x^2 - 6x + 1$

56. $y = 8 - 2x - x^2$

57. $y = 2 + x - 3x^2$

58. $y = 5x^2 - 2x$

59. $y = x - 3x^2$

60. $y = 4x^2 - 8$

61. $y = 9 - x^2$

62. $y = x^2 + 2$

63. $y = x^2 + 2x + 3$

[10–2]

64. A particular projectile is distance d in feet from its starting point after t seconds of time has elapsed according to the formula $d = 2t^2 - 7t + 3$. How many seconds will it take to travel 12 feet?

65. The area of a tennis court is 2,800 sq. ft. Find the length of the court if the length is $3\frac{1}{9}$ times the width.

Final examination

[1–3] 1. Insert the proper inequality symbol, $<$ or $>$, to make the statement $|-5| \quad |4|$ true.

Perform the indicated operations and simplify the expression.

[1–8] 2. $38 - 10 \div 5 + 3 \cdot 4 - 2^3 + \sqrt{4}$

[1–8] 3. $-4[9 - 3(9 - 4) + 6]$

[2–2] 4. Evaluate the expression $a - b(2c - d)$ when $a = -4$, $b = 3$, $c = 5$, and $d = -6$.

Simplify the following and leave the answers with only positive exponents.

[3–4] 5. $x^5 \cdot x^{-3} \cdot x^4$

[3–4] 6. $\dfrac{2x^{-3}}{4x^2}$

[3–4] 7. $(3x^2y^3)(-4xy^2)$

[3–4] 8. $(3xy^{-2})^{-3}$

[3–4] 9. $(7x^3z^2)^0$

Remove the grouping symbols and combine.

[2–3] 10. $(4x^2 - y^2) - (2x^2 + y^2) + (5x^2 - 6y^2)$

[2–3] 11. $5x - (x - y) - 2x + y - (2x + 5y)$

Perform the indicated operations and simplify.

[3–2] 12. $(y + 9)(y - 9)$

[3–2] 13. $(7z - 3w)^2$

[3–2] 14. $(x + 4)(5x^2 - 3x + 1)$

[5–3] 15. $\dfrac{5xy^2 - 3x^4y + x^2y^2}{xy}$

[5–3] 16. $(8y^2 - 2y - 3) \div (2y - 1)$

Find the solution set of the following equations.

[2–6] **17.** $2(x + 1) - 3(x - 3) = 4$

[4–7] **18.** $x^2 - 11x - 12 = 0$

[6–5] **19.** $3 + \frac{2}{x^2} - \frac{7}{x} = 0$

[6–5] **20.** $\frac{3x}{6} - 2 = \frac{5x}{4}$

[4–7] **21.** $8x^2 = 12x$

Completely factor the following expressions.

[4–1] **22.** $3x^2 - 6xy + 9x$

[4–2] **23.** $a^2 - 4a - 21$

[4–3] **24.** $4x^2 - 12x + 5$

[4–4] **25.** $9a^2 - 64$

[4–6] **26.** $6ax - 2ay + 3bx - by$

[4–4] **27.** $x^2 - 10x + 25$

[4–8] **28.** The product of two consecutive integers is 132. Find the integers.

Perform the indicated operations and reduce to lowest terms.

[6–1] **29.** $\frac{x^2 + 7x + 6}{x^2 - 4} \cdot \frac{x - 2}{x + 6}$

[6–1] **30.** $\frac{3x}{4x - 8} \div \frac{9x}{x^2 - 4x + 4}$

[6–2] **31.** $\frac{9}{x - 6} - \frac{5}{6 - x}$

[6–2] **32.** $\frac{x - 2}{x + 5} + \frac{x + 4}{x^2 - 25}$

[6–4] **33.** Simplify the complex fraction $\dfrac{5 + \frac{4}{y}}{4 - \frac{6}{y}}$.

[5–4] **34.** Find the value of x if $15 : 6 = 8 : x$.

[5–4] **35.** What is the ratio of 42 oz to 5 lb?

[7–4] **36.** Find the equation of the line passing through points $(-2,5)$ and $(1,-1)$. Write the answer in standard form.

[7–4] **37.** Given the equation $2x - 3y = 9$, find the slope m and the y-intercept b of the line.

[10–5] **38.** Given $f(x) = 3x^2 + 3x - 1$, find (a) $f(2)$, (b) $f(0)$, (c) $f(-1)$. Write the answers as ordered pairs.

[8–3] **39.** Find the solution set of the system of equations
$x - 2y = 3$
$2x - 3y = -5$.

[2–8] **40.** The perimeter of a rectangle is 34 feet. If the length is 2 more than twice the width, what are the dimensions of the rectangle?

Simplify the following expressions by performing the indicated operations. Rationalize all denominators.

[9–4] **41.** $\sqrt{27} - \sqrt{48}$

[9–5] **42.** $\sqrt{3}(\sqrt{2} + \sqrt{3})$

[9–5] **43.** $(4 + \sqrt{3})(4 - \sqrt{3})$

[9–5] **44.** $(2 - \sqrt{7})^2$

[9–3] **45.** $\frac{3}{3 - \sqrt{5}}$

[9–1] **46.** $\sqrt[3]{-27}$

[9–6] **47.** $(16)^{3/4}$

[9–7] **48.** Find the solution set of the equation $\sqrt{x + 1} - 1 = x$.

[7–4] **49.** Sketch the graph of $3x + 2y = 12$ using the x- and y-intercepts.

[10–5] **50.** Sketch the graph of $y = x^2 - 5x - 6$ using the vertex, x- and y-intercepts, and four arbitrary points.

[10–3] **51.** Find the solution set of the quadratic equation $4x^2 - 7x = 3$ by any method.

Perform the indicated operations on the following complex numbers.

[10–4] **52.** $(3 - 9i) - (2 + 10i)$

[10–4] **53.** $(5 + 7i)(2 - i)$

[10–4] **54.** $(8 - i)(8 + i)$

[10–4] **55.** $(5 + 7i)^2$

[10–4] **56.** $(3 - \sqrt{-4})(2 + \sqrt{-4})$

[10–4] **57.** $(1 - \sqrt{-5})(1 + \sqrt{-5})$

Rationalize the denominator of the expression.

[10–4] **58.** $\frac{5 - 3i}{6 + i}$

59. $\frac{3 + 4i}{2 - 3i}$

APPENDIX

Answers and Solutions

Chapter 1

Exercise 1–1

Answers to odd-numbered problems

1. $\frac{1}{2}$ **3.** $\frac{5}{6}$ **5.** $\frac{8}{9}$ **7.** $\frac{7}{9}$ **9.** 2 **11.** $\frac{17}{20}$ **13.** $\frac{1}{2}$

15. $\frac{49}{96}$ **17.** $\frac{7}{12}$ **19.** $\frac{15}{28}$ **21.** $\frac{2}{7}$ **23.** $\frac{25}{17}$ **25.** $\frac{51}{7}$ or $7\frac{2}{7}$

27. $\frac{132}{7}$ or $18\frac{6}{7}$ **29.** 12 **31.** 12 **33.** $\frac{21}{32}$ **35.** $\frac{1}{5}$

37. $\frac{32}{21}$ or $1\frac{11}{21}$ **39.** $\frac{20}{7}$ or $2\frac{6}{7}$ **41. a.** $187\frac{17}{48}$ in.^{3} **b.** $31\frac{39}{64}$ in.^{3}

43. 120 **45.** 126 **47.** 144 **49.** 385 **51.** 120 **53.** 60

55. $\frac{2}{3}$ **57.** $\frac{7}{12}$ **59.** $\frac{3}{5}$ **61.** $\frac{13}{8}$ or $1\frac{5}{8}$ **63.** $\frac{17}{5}$ or $3\frac{2}{5}$

65. $\frac{16}{15}$ or $1\frac{1}{15}$ **67.** $\frac{7}{24}$ **69.** $\frac{149}{270}$ **71.** $\frac{11}{20}$ **73.** $\frac{57}{16}$ or $3\frac{9}{16}$

75. $\frac{5}{3}$ or $1\frac{2}{3}$ **77.** $86\frac{1}{2}$ ft **79.** $10\frac{11}{12}$ lb

Solutions to trial exercise problems

7. $\frac{28}{36} = \frac{4 \cdot 7}{4 \cdot 9} = \frac{7}{9}$ **21.** $\frac{6}{7} \div 3 = \frac{6}{7} \cdot \frac{1}{3} = \frac{6 \cdot 1}{7 \cdot 3} = \frac{2 \cdot 3}{7 \cdot 3} = \frac{2}{7}$

27. $7\frac{1}{3} \cdot 2\frac{4}{7} = \frac{22}{3} \cdot \frac{18}{7} = \frac{22 \cdot 3 \cdot 3 \cdot 2}{3 \cdot 7} = \frac{22 \cdot 3 \cdot 2}{7} = \frac{132}{7}$ or $18\frac{6}{7}$

33. $\dfrac{\frac{7}{8}}{\frac{4}{3}} = \frac{7}{8} \div \frac{4}{3} = \frac{7}{8} \cdot \frac{3}{4} = \frac{7 \cdot 3}{8 \cdot 4} = \frac{21}{32}$

35. $\frac{4}{5} \cdot \frac{2}{3} \cdot \frac{3}{8} = \frac{4 \cdot 2 \cdot 3}{5 \cdot 3 \cdot 8} = \frac{1 \cdot (4 \cdot 2 \cdot 3)}{5 \cdot (4 \cdot 2 \cdot 3)} = \frac{1}{5}$

42. $61\frac{1}{2} \div 14 = \frac{123}{2} \cdot \frac{1}{14} = \frac{123}{28} = 4\frac{11}{28}$ in.

45. $6 = 2 \cdot 3$
$14 = 2 \cdot 7$ LCD is $2 \cdot 3 \cdot 3 \cdot 7 = 126$.
$18 = 2 \cdot 3 \cdot 3$

61. $1 + \frac{5}{8} = \frac{8}{8} + \frac{5}{8} = \frac{8 + 5}{8} = \frac{13}{8}$ or $1\frac{5}{8}$

71. $\frac{7}{15} + \frac{5}{6} - \frac{3}{4} = \frac{7}{15} \cdot \frac{4}{4} + \frac{5}{6} \cdot \frac{10}{10} - \frac{3}{4} \cdot \frac{15}{15} = \frac{28}{60} + \frac{50}{60} - \frac{45}{60}$

(LCD is 60) $= \frac{28 + 50 - 45}{60} = \frac{33}{60} = \frac{11}{20}$

77. $P = 2\ell + 2w = 2 \cdot 24\frac{1}{2} + 2 \cdot 18\frac{3}{4} = 2 \cdot \frac{49}{2} + 2 \cdot \frac{75}{4}$

$= 49 + \frac{75}{2} = \frac{98}{2} + \frac{75}{2} = \frac{98 + 75}{2} = \frac{173}{2}$

$= 86\frac{1}{2}$ ft

Exercise 1–2

Answers to odd-numbered problems

1. $\frac{2}{5}$ **3.** $\frac{3}{20}$ **5.** $\frac{1}{8}$ **7.** $\frac{7}{8}$ **9.** 19.019 **11.** 540.2927

13. 13.5585 **15.** 156.9876 **17.** 1.06964 **19.** 9.52816

21. 0.428412 **23.** 0.9100081809 **25.** 1.2 **27.** 40 **29.** 102

31. 2,500 **33.** 0.15 **35.** 0.65 **37.** $0.\overline{2}$ **39.** $12.91

41. 13 cardinals **43.** 0.57 sec **45.** 122.28 gal

47. 1097.222 yd^2 **49.** $0.05 = \frac{1}{20}$ **51.** $0.12 = \frac{3}{25}$

53. $1.35 = \frac{27}{20}$ or $1\frac{7}{20}$ **55.** $3.25 = 3\frac{1}{4}$ or $\frac{13}{4}$ **57.** $\frac{4}{5}$, 80%

59. $\frac{27}{50}$, 54% **61.** $\frac{23}{20}$, 115% **63.** 0.75, 75% **65.** 0.375, 37.5%

67. 2 **69.** 33.8 **71.** 550 **73.** $256.50 **75.** $23

77. $8.50 discount, $25.50 discount price **79.** 0.96 oz

Solutions to trial exercise problems

3. $0.15 = \frac{15}{100} = \frac{3 \cdot 5}{20 \cdot 5} = \frac{3}{20}$

13. $10.03 + 3.113 + 0.3342 + 0.0763 + 0.005 =$

```
 10.0300
  3.1130
  0.3342
  0.0763
  0.0050
 -------
 13.5585
```

19. $(7.006)(1.36) =$

```
   7.006
    1.36
 -------
   42036
  21018
  7006
 -------
 9.52816
```

28. $21.681 \div 8.03 = 2.7$

```
             2.7
 8.03 ) 21.681
        16 06
         5 621
         5 621
         -----
             0
```

33. $\frac{3}{20} = 20\overline{)3.000} = 0.15$

$$\begin{array}{r} .15 \\ 20\overline{)3.000} \\ \underline{20} \\ 100 \\ \underline{100} \end{array}$$

37. $\frac{2}{9} = 9\overline{)2.000} = 0.\overline{2}$ (repeating)

$$\begin{array}{r} .222... \\ 9\overline{)2.000} \\ \underline{18} \\ 20 \\ \underline{18} \\ 20 \end{array}$$

39.

$$\begin{array}{r} 14.36 \\ \underline{0.899} \\ 12924 \\ 12924 \\ \underline{11488} \\ 12.90964 \end{array}$$

(89.9¢ = \$0.899)

12.90964 ≈ \$12.91

51. 12% = 12.% = 0.12

$12\% = \frac{12}{100} = \frac{3 \cdot 4}{25 \cdot 4} = \frac{3}{25}$

59. 0.54 = 0.54. = 54%

$= \frac{54}{100} = \frac{27}{50}$

62. $2.40 = 2\frac{2}{5} = \frac{12}{5} = 240\%$

69. 26% of 130 = 0.26 × 130 = 130

$$\begin{array}{r} 130 \\ \underline{0.26} \\ 780 \\ \underline{260} \\ 33.80 \end{array}$$

= 33.8

77. discount = 25% of 34

= 0.25 × 34 = \$8.50

price = 34.00 − 8.50 = \$25.50

Exercise 1–3

Answers to odd-numbered problems

1. {Sunday,Monday,Tuesday,Wednesday,Thursday,Friday,Saturday} **3.** {January,February,March} **5.** {January,March,May,July, August,October,December} **7.** {a,l,g,e,b,r} **9.** {i,n,t,e,r,m,d,a} **11.** {3,5,7,9} **13.** {Sunday,Saturday} **15.** −\$10, \$150 **17.** −10 yards, 16 yards **19.** −14 points, 8 points

21.

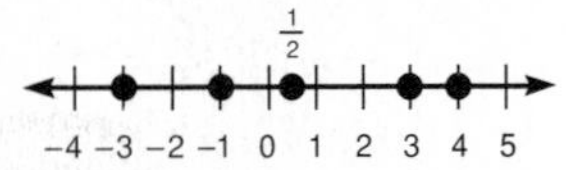

23.

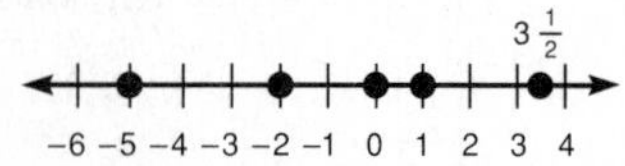

25.

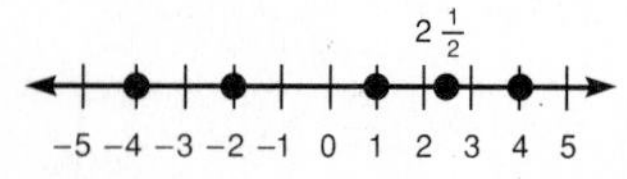

27.

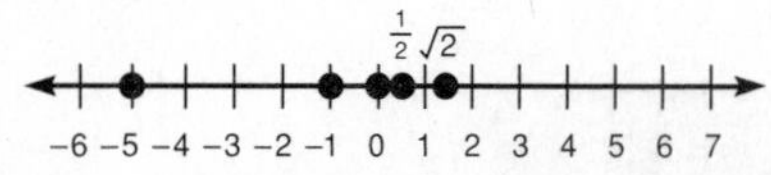

29. −6,−4,0,3,6 **31.** −11,−7,−4,−1,2 **33.** −2,3,4,5,11 **35.** $-1\frac{3}{4},-\frac{1}{2},1\frac{1}{2},3,4$ **37.** 4 < 8 **39.** 9 > 2 **41.** −3 > −8 **43.** −10 < −5 **45.** 0 < 4 **47.** 0 > −6 **49.** 2 **51.** 5 **53.** 4 **55.** $\frac{1}{2}$ **57.** $1\frac{1}{2}$ **59.** $-\frac{5}{8}$ **61.** −6 **63.** |5|<|−7| **65.** |0|<|−2| **67.** |−8|>|−5| **69.** |−6|>|−2| **71.** 7 >|−2| **73.** |−4|< 6 **75.** |−27| **77.** |−9|

Solutions to trial exercise problems

27.

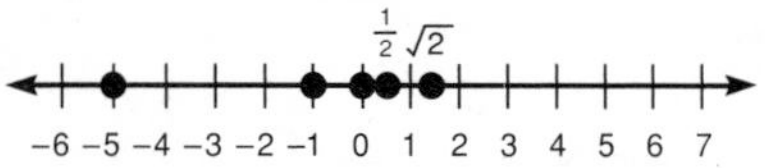

For the location of $\sqrt{2}$, we use the approximation 1.414 from a calculator.

35. $-1\frac{3}{4},-\frac{1}{2},1\frac{1}{2},3,4$ The values represent an approximation of the coordinates. **40.** −2 > −4, since −2 lies to the right of −4 on the number line. **46.** −3 < 0, since −3 lies to the left of 0 on the number line.

Exercise 1–4

Answers to odd-numbered problems

1. −13 **3.** 5 **5.** −5 **7.** −5 **9.** 0 **11.** −13.6 **13.** −11.1 **15.** $-\frac{1}{2}$ **17.** $\frac{1}{10}$ **19.** $-\frac{3}{4}$ **21.** 3 **23.** 10 **25.** −44 **27.** −22 **29.** −10 **31.** 0 **33.** 11 **35.** 15 **37.** 7° *C* **39.** \$32 **41.** 2 mb **43.** 99 **45.** −41° *F* **47.** \$64

Solutions to trial exercise problems

11. (−8.7) + (−4.9) = −13.6 The signs are the same so we add their absolute values 8.7 + 4.9 = 13.6 and prefix this sum by their common sign. **15.** $\left(-\frac{1}{6}\right) + \left(-\frac{1}{3}\right) = -\frac{1}{2}$ The signs are the same so we add their absolute values $\frac{1}{6} + \frac{1}{3} = \frac{1}{6} + \frac{2}{6} = \frac{1+2}{6} = \frac{3}{6} = \frac{1}{2}$ and prefix this sum by their common sign.

21. 10 + (−5) + (−2) = 5 + (−2) = 3 The numbers were added left to right. **23.** (−12) + (−10) + (+8) + (+24) = (−22) + (+8) + (+24) = (−14) + (+24) = 10

34. The sum of 15 and −18 increased by 10

15 + (−18) + 10

15 + (−18) + 10 = −3 + 10 = 7

44. Let t = the temperature at 1 P.M. To find the new temperature, we must *add* the rise in temperature to the original temperature.

temperature at 1 P.M. is temperature at 8 A.M. rose 39°

t = −13 + 39

t = −13 + 39

t = 26

The temperature at 1 P.M. was 26° *F*.

Exercise 1–5

Answers to odd-numbered problems

1. -1 **3.** 6 **5.** -12 **7.** -5 **9.** -4 **11.** 14 **13.** -6 **15.** 7 **17.** $-\frac{1}{4}$ **19.** $2\frac{5}{8}$ **21.** -9.4 **23.** -312 **25.** 301.8 **27.** -24 **29.** 11 **31.** -53 **33.** 16 **35.** -3 **37.** 10 **39.** -8 **41.** -7 **43.** $-38°\ C$ **45.** $-\$372$ **47.** 6 feet **49.** -22 **51.** -9 **53.** 60 **55.** 19 **57.** -13 **59.** -5 **61.** $8 **63.** 38° **65.** 14 years old **67.** $245 **69.** 20,602 feet **71.** $19

Solutions to trial exercise problems

13. $(-6) + 0 = -6$ The sum of zero and a number is that number. **15.** $7 - 0 = 7$ A number minus zero is that number. **26.** $(-12) - (-10) - (8) = (-12) + (10) + (-8)$ $= (-2) + (-8) = -10$ **38.** $12 + 3 - 16 - 10 - (12 + 5)$ $= 12 + 3 - 16 - 10 - (17) = 15 - 16 - 10 - (17)$ $= -1 - 10 - (17) = -11 - (17) = -28$

65. Let a = the age that Erin will be in the year 2000. We must find the *difference* between 2000 and 1986.

age in the year 2000	is	the difference between 2000 and 1986		
		2000	and	1986
a	=	2000	$-$	1986

$a = 2000 - 1986$
$a = 14$
Erin will be 14 years old in the year 2000.

Exercise 1–6

Answers to odd-numbered problems

1. 15 **3.** -28 **5.** -60 **7.** -36 **9.** 20 **11.** -105 **13.** -4.32 **15.** -13.769 **17.** -2.16 **19.** $\frac{9}{16}$ **21.** $\frac{1}{4}$ **23.** -120 **25.** 144 **27.** 0 **29.** 0 **31.** $-5,6$ **33.** $-5,-4$ **35.** $-7,0$ **37.** $3,-4$ **39.** 7,5 **41.** $8,-1$ **43.** $4,-3$ **45.** $6,-3$ **47.** $76 **49.** 1,050 people **51.** $975 **53.** $-\$92$ **55.** 525¢ = $5.25 **57.** 700 gallons

Solutions to trial exercise problems

11. $7 \cdot (-1)(-3)(-5) = (-7)(-3)(-5) = (21)(-5) = -105$; negative answer because there were an odd number of negative factors **27.** $(-2)(0)(3)(-4) = 0$ When zero is one of the factors, zero will be the answer. **30.** $-16, 0$ Since $(-4)(4) = (-16)$ and $(-4) + (4) = 0$, then -4 and 4 are the integers. **31.** $-30, 1$ Since $(-5)(6) = (-30)$ and $(-5) + (6) = (1)$, then -5 and 6 are the integers. **48.** Assets $(5)(-6)$, and his assets would change by (-30) dollars. **57.** Let g = the number of gallons of milk sold in 4 weeks. Since there are 28 days in 4 weeks, we must *multiply* 28 by 25 to determine the gallons of milk sold.

total gallons of milk sold	is	28 days	at	25 gallons per day
g	=	28	$\cdot$	25

$g = 28 \cdot 25$
$g = 700$
The grocer sold 700 gallons of milk.

Exercise 1–7

Answers to odd-numbered problems

1. 2 **3.** -8 **5.** 2 **7.** -8 **9.** undefined **11.** 0 **13.** indeterminate **15.** -7 **17.** -5 **19.** -2 **21.** -4 **23.** 0 **25.** -3 **27.** undefined **29.** indeterminate **31.** $-1°\ C$ **33.** 6 hours **35.** 9 seconds **37.** $4 **39.** 25 miles **41.** 4° **43.** 32 books **45.** 12 minutes

Solutions to trial exercise problems

11. $\frac{0}{-9} = 0$, since $(-9) \cdot 0 = 0$ **19.** $\frac{(-4)(-3)}{-6} = \frac{12}{-6}$ $= -2$; odd number of negative factors **22.** $\frac{(-4)(0)}{-8} = \frac{0}{-8}$ $= 0$, since $(-8) \cdot 0 = 0$ **26.** $\frac{(-2)(-4)}{(0)(4)} = \frac{8}{0}$ is undefined **29.** $\frac{(-6)(0)}{(-3)(0)} = \frac{0}{0}$ = indeterminate **33.** number of hours $= \frac{\text{number of miles}}{\text{rate of travel in miles per hour}}$. Hence $\frac{282}{47} = 6$; 6 hours.

45. Let m = the number of minutes it took Alice to run 1 mile. Since there are 60 minutes in 1 hour, the race took 300 minutes + 12 minutes, which is 312 minutes. We must divide 312 minutes by 26 miles to determine the number of minutes per mile.

minutes per mile	is	number of minutes	divided by	number of miles
m	=	312	$\div$	26

$m = 312 \div 26 = 12$
Alice ran 1 mile in 12 minutes.

Exercise 1–8

Answers to odd-numbered problems

1. 16 **3.** -27 **5.** -36 **7.** -1 **9.** 1 **11.** 4 **13.** 26 **15.** 0 **17.** 3 **19.** 6 **21.** $\frac{12}{7}$ or $1\frac{5}{7}$ **23.** 11 **25.** 46 **27.** 2 **29.** 121 **31.** $\frac{5}{24}$ **33.** 96 **35.** -48 **37.** 78 **39.** 3 **41.** 38 **43.** 4 **45.** 12 **47.** -15.99 **49.** 38.47 **51.** 29 **53.** 33 **55.** $\frac{19}{288}$ **57.** $23\frac{1}{3}°\ C$ **59.** $\frac{110}{7}$ or $15\frac{5}{7}$ square inches **61.** $\frac{288}{41}$ or $7\frac{1}{41}$ inches **63.** $374.50 **65.** 8 **67.** 2,728

Solutions to trial exercise problems

2. $(-5)^4 = (-5)(-5)(-5)(-5) = (25)(-5)(-5)$ $= (-125)(-5) = 625$; positive since we have a negative number to an even power **3.** $(-3)^3 = (-3)(-3)(-3) = (9)(-3)$ $= -27$; negative since we have a negative number to an odd power **8.** $-2^2 = -(2 \cdot 2) = -(4) = -4$ **17.** $0(5 + 2) + 3$ $= 0(7) + 3 = 0 + 3 = 3$ **41.** $4(2 - 5)^2 - 2(3 - 4)$ $= 4(-3)^2 - 2(3 - 4) = 4(-3)^2 - 2(-1) = 4(9) - 2(-1)$ $= 36 - 2(-1) = 36 - (-2) = 38$ **43.** $\frac{5(3 - 5)}{2} - \frac{27}{-3}$ $= \frac{5(-2)}{2} - \frac{27}{-3} = \frac{-10}{2} - \frac{27}{-3} = (-5) - \frac{27}{-3} = (-5) - (-9)$ $= 4$ **50.** $5[10 - 2(4 - 3) + 1] = 5[10 - 2(1) + 1]$ $= 5[10 - 2 + 1] = 5[8 + 1] = 5[9] = 45$

54. $\left(\frac{6-3}{7-4}\right)\left(\frac{14+2\cdot 3}{5}\right) = \left(\frac{3}{3}\right)\left(\frac{14+6}{5}\right) = \left[\frac{3}{3}\right]\left[\frac{20}{5}\right]$
$= 1\cdot 4 = 4$ **59.** $\frac{22}{7}\cdot 3^2 - \frac{22}{7}\cdot 2^2 = \frac{22}{7}\cdot 9 - \frac{22}{7}\cdot 4$
$= \frac{198}{7} - \frac{88}{7} = \frac{198-88}{7} = \frac{110}{7}$ or $15\frac{5}{7}$ square inches

65. Let p = the total number of pieces of lumber. *Dividing* the 16-foot board by 4 and the 12-foot board by 3 will give us the number of pieces of lumber. If we *add* the number of pieces from the 16-foot board to the number of pieces from the 12-foot board, we will have the total number of pieces.

total number of pieces	is	number of pieces from the 16-foot board	combined with	number of pieces from the 12-foot board
p	$=$	$16 \div 4$	$+$	$12 \div 3$

$p = 16 \div 4 + 12 \div 3$
$= 4 + 4$ Priority 3
$= 8$ Priority 4

There will be 8 pieces of lumber.

Chapter 1 review

1. $\frac{5}{7}$ **2.** $\frac{3}{4}$ **3.** $\frac{2}{3}$ **4.** $\frac{10}{7}$ or $1\frac{3}{7}$ **5.** $\frac{3}{5}$ **6.** $\frac{21}{20}$ or $1\frac{1}{20}$
7. $\frac{7}{8}$ **8.** $\frac{25}{8}$ or $3\frac{1}{8}$ **9.** $\frac{25}{3}$ or $8\frac{1}{3}$ **10.** $\frac{5}{8}$ acre **11.** $\frac{2}{5}$ cup
12. $\frac{8}{7}$ or $1\frac{1}{7}$ **13.** $\frac{19}{24}$ **14.** $\frac{5}{6}$ **15.** $\frac{2}{9}$ **16.** $6\frac{17}{20}$ **17.** $\frac{2}{15}$
18. $\frac{7}{12}$ **19.** $\frac{7}{8}$ acre **20.** 263.51 **21.** 31.795 **22.** 1,355.09
23. 14.3 **24.** \$565.49 **25.** 7.86 acres **26.** $\approx$ 12.42 mpg
27. 10 **28.** 68.4 **29.** 25 **30.** 78.72 **31.** {50,51,52,53,54,55}
32. {1,2,3,4} **33.** {0} **34.** {−3,−2,−1,0,1,2,3}

35. $-\frac{1}{2}$
−3 −2 −1 0 1 2 3 4

36. $-\frac{3}{4}$ $\frac{3}{2}$ $\frac{5}{2}$
−1 0 1 2 3

37. $\sqrt{2}$
−5 −4 −3 −2 −1 0 1 2 3 4 5

38. $\frac{1}{2}$ π
−5 −4 −3 −2 −1 0 1 2 3 4 5 6

39. $<$ **40.** $<$ **41.** $>$ **42.** $<$ **43.** $>$ **44.** $>$ **45.** −4
46. 3 **47.** −6 **48.** 1 **49.** −6 **50.** 15 **51.** −4
52. −9 **53.** 3 **54.** 15 **55.** −21 **56.** 12 **57.** 24
58. −144 **59.** 0 **60.** −7 **61.** 2 **62.** −6 **63.** undefined
64. 0 **65.** indeterminate **66.** −1 **67.** −9,−4
68. a. 52,000 + (−3,000) + (−2,560) + (−3,300) **b.** \$43,140
69. 40° **70. a.** +9, +8, −5, −6 **b.** >, +6 **c.** 58°, 69 + (−11) = 58 **71.** 25 **72.** −64 **73.** −16 **74.** −27
75. 98 **76.** −3 **77.** 20 **78.** 49 **79.** −9 **80.** 20
81. −14 **82.** 27

Chapter 2

Exercise 2–1

Answers to odd-numbered problems

1. 2 terms **3.** 3 terms **5.** 1 term **7.** 3 terms **9.** 2 terms **11.** 1 term **13.** 2 terms **15.** 5 is the coefficient of x^2, 1 is understood to be the coefficient of x, −4 is the coefficient of z **17.** 1 is understood to be the coefficient of x, −1 is understood to be the coefficient of y, −3 is the coefficient of z **19.** −2 is the coefficient of a, −1 is understood to be the coefficient of b, 1 is understood to be the coefficient of c **21.** polynomial, binomial **23.** not a polynomial because a variable is in the denominator **25.** not a polynomial because a variable is in the denominator **27.** polynomial, trinomial **29.** $b - 3a$ **31.** $y + 5$ **33.** $x(y + z)$ **35.** $a - b$ **37.** (let x = the number) $x - 12$ **39.** (let x = the number) $3x + 1$ **41.** (let x = the number) $2(x + 4)$

Solutions to trial exercise problems

8. $\frac{15x^2 + y}{8}$ has one term because the fraction bar is a grouping symbol. **24.** $\frac{a+b}{5} - c$ is a polynomial. A constant can appear in the denominator, a variable cannot. **36.** $\frac{1}{2}$ of x, decreased by 2 times x would be $\frac{1}{2}x - 2x$. **39.** 3 times a number, increased by 1: If we let x represent the number, then we would have $3x + 1$.

Review exercises

1. −25 **2.** 64 **3.** −2 **4.** 15 **5.** 22 **6.** 23

Exercise 2–2

Answers to odd-numbered problems

1. 9 **3.** 5 **5.** 48 **7.** 5 **9.** 62 **11.** 288 **13.** 61 **15.** 0 **17.** −1 **19.** 1 **21.** −44 **23.** 20 **25.** 0 **27.** 43 **29.** $\frac{20}{3}$ **31.** 160 **33.** 288 **35.** 54 **37.** 2,140 **39.** 114 **41.** 256 **43.** 6 **45.** $\frac{15{,}000}{857}$ **47.** $\frac{540}{13}$ **49.** $\frac{400}{33}$ **51.** $85m$ **53.** $\frac{y}{10}$ **55.** $5n + 10d$ **57. a.** $p + 12$ **b.** $p - 5$ **59.** $258 - n + m$ **61.** $\frac{c}{50}$ **63.** $y + 2$ **65.** $12f + t$ **67.** $25{,}000n - 2{,}000$ **69.** $(9.95)p + (12.99)q$

Solutions to trial exercise problems

5. $(3a + 2b)(a - c) = [3(\ \ ) + 2(\ \ )][(\ \ ) - (\ \ )]$ $= [3(2) + 2(3)][(2) - (-2)] = [6 + 6][4] = [12][4] = 48$ **14.** $(4a + b) - (3a - b)(c + 2d) = [4(\ \ ) + (\ \ )]$ $- [3(\ \ ) - (\ \ )][(\ \ ) + 2(\ \ )] = [4(2) + (3)]$ $- [3(2) - (3)][(-2) + 2(-3)] = [8 + 3]$ $- [6 - 3][(-2) + (-6)] = [11] - [3][-8] = [11] - [-24]$ $= 35$ **31.** $I = prt$; $I = (\ \ )(\ \ )(\ \ ) = (1{,}000)(0.08)(2) = (80)(2)$ $= 160$ **39.** $A = \frac{I^2R - 120E^2}{R}$; $A = \frac{(\ \ )^2(\ \ ) - 120(\ \ )^2}{(\ \ )}$ $= \frac{(12)^2(100) - 120(5)^2}{(100)} = \frac{(144)(100) - 120(25)}{100}$ $= \frac{14{,}400 - 3{,}000}{100} = \frac{11{,}400}{100} = 114$ **47.** $V = \frac{vn}{N}$; $V = \frac{(\ \ )(\ \ )}{(\ \ )}$

$= \dfrac{(90)(30)}{(65)} = \dfrac{2{,}700}{65} = \dfrac{540}{13}$ **55.** n nickels is represented by $5n$ because there are 5 cents in each nickel. Therefore d dimes would be represented by $10d$. The total is represented by adding the cents from the nickels to the cents from the dimes, $5n + 10d$. **64.** If we use 11 as an example of an odd integer, the next greater odd integer would be 13. To get from 11 to 13, we must add 2. Therefore if z is an odd integer, then $z + 2$ is the next greater odd integer.

Review exercises

1. -6 **2.** -4 **3.** -8 **4.** 18 **5.** 0 **6.** -21

Exercise 2–3

Answers to odd-numbered problems

1. like **3.** like **5.** unlike **7.** $9x$ **9.** $13a + 2b$ **11.** $14x$ **13.** $-3ab$ **15.** $7x^2 + 4x$ **17.** $a^2b + 2a^3 - b^3 - 6ab^2$ **19.** $13a - 5c - 2x^2$ **21.** $-3a - 9b$ **23.** $3x^2 + 11y^2$ **25.** $3x^2 - 3x$ **27.** $x^2y^2 - 2x^2y + 5xy$ **29.** $8x^2 + 2x + 1$ **31.** $3x + 8y$ **33.** $2x + 3y$ **35.** $2a + b + 5c$ **37.** $2x + 4y$ **39.** $-3x^2y + 15xy$ **41.** $4a^3 + a^2b + 6ab^2 - 5b^3$ **43.** $70a + 9b$ **45.** $-5xy + 9y^2z + 14yz$ **47.** $-8b + 10$ **49.** $3a - 9b$ **51.** $3x^2 + 3z$ **53.** $2x$ **55.** $-4a - 3b + 4x + 2y$ **57.** $5a + 5b$ **59.** $-a - 2b$ **61.** a

Solutions to trial exercise problems

19. $3a + b + 2a - 5c - b - 2x^2 + 8a = (3a + 2a + 8a) + (b - b) - 5c - 2x^2 = (3 + 2 + 8)a + (0) - 5c - 2x^2 = 13a - 5c - 2x^2$ **45.** $(8xy + 9y^2z) - (13xy - 14yz) = 8xy + 9y^2z - 13xy + 14yz = (8xy - 13xy) + 9y^2z + 14yz = (8 - 13)xy + 9y^2z + 14yz = -5xy + 9y^2z + 14yz$
52. $2x - [3x - (5x - 3)] = 2x - [3x - 5x + 3] = 2x - [-2x + 3] = 2x + 2x - 3 = 4x - 3$
59. $-[4a + 7b - (3a + 5b)] = -[4a + 7b - 3a - 5b] = -[a + 2b] = -a - 2b$

Review exercises

1. $3x$ **2.** $6(a + 7)$ **3.** $(y - 2) \div 4$ or $\dfrac{y-2}{4}$

4. (let x = the number) $5x$ **5.** (let x = the number) $x - 12$

6. (let x = the number) $\dfrac{x}{8} - 9$

Exercise 2–4

Answers to odd-numbered problems

1. true **3.** true **5.** true **7.** false **9.** true **11.** false **13.** $\{16\}$ **15.** $\{-3\}$ **17.** $\{-2\}$ **19.** $\{-5\}$ **21.** $\{-6\}$ **23.** $\{-7\}$ **25.** $\{14\}$ **27.** $\{-1\}$ **29.** $\{4\}$ **31.** $\{-5\}$ **33.** $\{5\}$ **35.** $\{6\}$ **37.** $\{6\}$ **39.** $\{8\}$ **41.** $\{-6\}$ **43.** $\{5\}$ **45.** $\{-7\}$ **47.** $\{12\}$ **49.** $\{16\}$ **51.** $\{9\}$ **53.** 26 **55.** 33 **57.** \$735 **59.** \$365

Solutions to trial exercise problems

8. $3x + 2 = 5x - 1; \left\{\dfrac{3}{2}\right\}$

$$3\left(\frac{3}{2}\right) + 2 = 5\left(\frac{3}{2}\right) - 1$$
$$\frac{9}{2} + 2 = \frac{15}{2} - 1$$
$$\frac{9}{2} + \frac{4}{2} = \frac{15}{2} - \frac{2}{2}$$
$$\frac{13}{2} = \frac{13}{2} \text{ (true)}$$

23.
$$b + 7 = 0$$
$$b + 7 - 7 = 0 - 7$$
$$b = -7$$
$$\{-7\}$$
$$Check: (-7) + 7 = 0$$
$$0 = 0 \text{ (true)}$$

28.
$$-y - 6 = -2y + 1$$
$$-y + 2y - 6 = -2y + 2y + 1$$
$$y - 6 = 1$$
$$y - 6 + 6 = 1 + 6$$
$$y = 7$$
$$\{7\}$$
$$Check: -(7) - 6 = -2(7) + 1$$
$$-13 = -14 + 1$$
$$-13 = -13 \text{ (true)}$$

38.
$$5(x + 2) = 4(x - 1)$$
$$5x + 10 = 4x - 4$$
$$5x - 4x + 10 = 4x - 4x - 4$$
$$x + 10 = -4$$
$$x + 10 - 10 = -4 - 10$$
$$x = -14$$
$$\{-14\}$$

45.
$$3(z + 7) - (8 + 2z) = 6$$
$$3z + 21 - 8 - 2z = 6$$
$$z + 13 = 6$$
$$z + 13 - 13 = 6 - 13$$
$$z = -7$$
$$\{-7\}$$

58. Let b = the original balance.

original balance	makes a deposit	of $42.50	equals	new balance
b	+	42.50	=	125.30

$$b + 42.50 = 125.30$$
$$b + 42.50 - 42.50 = 125.30 - 42.50$$ Subtract 42.50 from both members.
$$b = 82.80$$

The original balance was \$82.80.

Review exercises

1. 16 **2.** -12 **3.** 1 **4.** 1 **5.** 1 **6.** 1

Exercise 2–5

Answers to odd-numbered problems

1. $\{4\}$ **3.** $\{6\}$ **5.** $\{16\}$ **7.** $\{35\}$ **9.** $\{12\}$ **11.** $\{-3\}$ **13.** $\{-4\}$ **15.** $\{7\}$ **17.** $\{-4\}$ **19.** $\left\{\dfrac{7}{3}\right\}$ **21.** $\left\{\dfrac{3}{2}\right\}$ **23.** $\{0\}$ **25.** $\{0\}$ **27.** $\{15\}$ **29.** $\{-14\}$ **31.** $\{4\}$ **33.** $\{-7\}$ **35.** $\{11\}$ **37.** $\{-26\}$ **39.** $\left\{\dfrac{112}{3}\right\}$ **41.** 9 **43.** -63 **45.** \$4.50 **47.** \$130 **49.** 64

Solutions to trial exercise problems

7.
$$\frac{1}{7}x = 5$$
$$7 \cdot \frac{1}{7}x = 7 \cdot 5$$
$$x = 35$$
$$\{35\}$$
$$Check: \frac{1}{7}(35) = 5$$
$$5 = 5 \text{ (true)}$$

15.
$$-4x = -28$$
$$\frac{-4x}{-4} = \frac{-28}{-4}$$
$$x = 7$$
$$\{7\}$$
$$Check: -4(7) = -28$$
$$-28 = -28 \text{ (true)}$$

23. $5x = 0$

$\frac{5x}{5} = \frac{0}{5}$

$x = 0$

$\{0\}$

Check: $5(0) = 0$

$0 = 0$ (true)

27. $\frac{x}{3} = 5$

$3 \cdot \frac{x}{3} = 3 \cdot 5$

$x = 15$

$\{15\}$

Check: $\frac{(15)}{3} = 5$

$5 = 5$ (true)

31. $2.6x = 10.4$

$\frac{2.6x}{2.6} = \frac{10.4}{2.6}$

$x = 4$

$\{4\}$

Check: $2.6(4) = 10.4$

$10.4 = 10.4$ (true)

38. $\frac{5}{7}x = 8$

$\frac{7}{5} \cdot \frac{5}{7}x = \frac{7}{5} \cdot 8$

$x = \frac{56}{5}$

$\left\{\frac{56}{5}\right\}$

Check: $\frac{5}{7}\left(\frac{56}{5}\right) = 8$

$8 = 8$ (true)

45. Let $w =$ Nancy's hourly wage.

30 hours	at	hourly wage	is	$135.00
30	$\cdot$	w	$=$	135

$30w = 135$

$\frac{30w}{30} = \frac{135}{30}$ Divide both members by 30.

$w = 4.5$

Nancy's hourly wage is $4.50.

Review exercises

1. $5x - 2$ **2.** $2x + 1$ **3.** $12x + 2$
4. $3x - 5$ **5.** $10x - 1$ **6.** $5x + 1$

Exercise 2–6

Answers to odd-numbered problems

1. $\{2\}$ **3.** $\{-2\}$ **5.** $\{36\}$ **7.** $\left\{\frac{16}{3}\right\}$ **9.** $\{4\}$ **11.** $\{0\}$ **13.** $\{3\}$ **15.** $\{0\}$ **17.** $\{1\}$ **19.** $\left\{\frac{8}{5}\right\}$ **21.** $\{-10\}$ **23.** $\left\{-\frac{9}{2}\right\}$ **25.** $\{18\}$ **27.** $\left\{-\frac{51}{8}\right\}$ **29.** $\{12\}$ **31.** $\{3\}$ **33.** $\left\{\frac{10}{7}\right\}$ **35.** $\left\{\frac{16}{11}\right\}$ **37.** $\left\{\frac{37}{10}\right\}$ **39.** $\left\{\frac{2}{5}\right\}$ **41.** $\{3\}$ **43.** $\{1\}$ **45. a.** $-\frac{70}{9}°\ C$ or $-7\frac{7}{9}°\ C$ **b.** $-\frac{295}{9}°\ C$ or $-32\frac{7}{9}°\ C$ **c.** $-\frac{50}{3}°\ C$ or $-16\frac{2}{3}°\ C$ **47.** $\frac{7}{3}$

Solutions to trial exercise problems

31. $3(2x - 1) = 4x + 3$

$6x - 3 = 4x + 3$

$6x - 4x - 3 = 4x - 4x + 3$

$2x - 3 = 3$

$2x - 3 + 3 = 3 + 3$

$2x = 6$

$\frac{2x}{2} = \frac{6}{2}$

$x = 3$

$\{3\}$

35. $8 - 2(3x + 4) = 5x - 16$

$8 - 6x - 8 = 5x - 16$

$-6x = 5x - 16$

$-6x + 6x = 5x + 6x - 16$

$0 = 11x - 16$

$0 + 16 = 11x - 16 + 16$

$16 = 11x$

$\frac{16}{11} = \frac{11x}{11}$

$\frac{16}{11} = x$

$\left\{\frac{16}{11}\right\}$

46b. $W = 243,\ T = -3$

$W = KT^4$

$(243) = K(-3)^4$

$243 = K \cdot 81$

$\frac{243}{81} = \frac{K \cdot 81}{81}$

$3 = K$

Hence the value of K is 3.

Review exercises

1. 108 **2.** 144 **3.** 88 **4.** 360 **5.** 280 **6.** 92

Exercise 2–7

Answers to odd-numbered problems

1. $w = \frac{V}{\ell h}$ **3.** $P = \frac{I}{rt}$ **5.** $m = \frac{F}{a}$ **7.** $V = \frac{K}{P}$ **9.** $R = \frac{W}{I^2}$ **11.** $w = \frac{P - 2\ell}{2}$ **13.** $a = P - b - c$ **15.** $a = \frac{by + c + 3}{y}$ **17.** $k = V - gt$ **19.** $b = \frac{2A - ch}{h}$ **21.** $a = \ell - dn + d$ **23.** $P = \frac{A}{1 + r}$ **25.** $f = \frac{T - g}{2}$ **27.** $q = \frac{D - R}{d}$ **29.** $c = \frac{W - b^2 - R}{2b}$ **31.** $r = \frac{A - P}{Pt}$ **33.** $a = \frac{V + br^2}{r^2}$ **35.** $x = -6y$ **37.** $g = \frac{2vt - 2S}{t^2}$ **39.** $g = \frac{2s - 2vt}{t^2}$ **41.** $S = \frac{P + Cn + e}{n}$ **43.** $e = nS - Cn - P$

Solutions to trial exercise problems

18. $V = k + gt$, for t

$V = k + gt$

$V - k = gt$

$\frac{V - k}{g} = t$

22. $\ell = a + (n - 1)d$, for d

$\ell = a + (n - 1)d$

$\ell - a = (n - 1)d$

$\frac{\ell - a}{n - 1} = d$

29. $R = W - b(2c + b)$, for c

$R = W - 2bc - b^2$

$R + 2bc = W - b^2$

$2bc = W - b^2 - R$

$c = \frac{W - b^2 - R}{2b}$

Review exercises

1. -25 **2.** 25 **3.** -81 **4.** -27 **5.** x^4
6. (let x = the number) x^2 **7.** ab **8.** xy

Exercise 2–8

Answers to odd-numbered problems

1. 22, 40 **3.** 35, 52 **5.** 28 **7.** 7 **9.** 54 **11.** 21 **13.** 15 **15.** 29, 34 **17.** 12, 108 **19.** 15, 17, 19 **21.** 12, 84 **23.** 20, 22, 24 **25.** 30, 31, 32 **27.** 14, 7, 42 **29.** 5, 9 **31.** 24 **33.** 9, 10, 11 **35.** 34, 36, 38 **37.** 19 feet, 16 feet **39.** 9 inches, 26 inches **41.** 18 feet, 23 feet **43.** 4 feet, 8 feet **45.** 19 feet, 31 feet **47.** 22 teeth, 37 teeth **49.** $28\frac{1}{2}$ volts, $60\frac{1}{2}$ volts
51. \$13,000 at 8%; \$7,000 at 6% **53.** \$11,000 at 8%; \$4,000 at 6% **55.** \$8,000 at 10%; \$10,000 at 8% **57.** \$8,000 at 9%; \$22,000 at 7% **59.** \$8,000 at 11%; \$17,000 at 18% **61.** \$7,700 at 13%; \$13,300 at 9% **63.** \$5,000 **65.** \$14,000

Solutions to trial exercise problems

5. number — equation

$x \qquad \frac{x}{4} + 6 = 13$

Solution:
$$\frac{x}{4} + 6 = 13$$
$$\frac{x}{4} = 7$$
$$x = 28$$

26. first — second — third — equation

$x \qquad 3 \cdot x \qquad x - 6 \qquad (x) + (3 \cdot x) + (x - 6) = 44$

Solution:
$$(x) + (3x) + (x - 6) = 44$$
$$x + 3x + x - 6 = 44$$
$$5x - 6 = 44$$
$$5x = 50$$
$$x = 10$$

First is 10, second $(3x)$ is $3(10) = 30$, and the third $(x - 6)$ is $(10) - 6 = 4$.

43. shorter piece — longer piece — equation

$x \qquad x + 4 \qquad x + (x + 4) = 12$

Solution:
$$x + (x + 4) = 12$$
$$x + x + 4 = 12$$
$$2x + 4 = 12$$
$$2x = 8$$
$$x = 4$$

Shorter piece is 4 feet and the longer piece $(x + 4)$ is $(4) + 4 = 8$, 8 feet.

59. number of dollars at 11% loss — number of dollars at 18% profit — equation

$x \qquad 25{,}000 - x \qquad (25{,}000 - x)(0.18) - x(0.11) = 2{,}180$

Solution:
$$(25{,}000 - x)(0.18) - x(0.11) = 2{,}180$$
$$(25{,}000)(0.18) - (0.18)x - (0.11)x = 2{,}180$$
$$4{,}500 - (0.29)x = 2{,}180$$
$$4{,}500 = (0.29)x + 2{,}180$$
$$2{,}320 = (0.29)x$$
$$8{,}000 = x$$

Therefore \$8,000 was invested at the 11% loss and \$25,000 − 8,000 = \$17,000 was invested at 18% profit.

Review exercises

1. 100 **2.** 84 **3.** 204 **4.** 108 **5.** 3,180 **6.** 16 **7.** 256
8. 40

Exercise 2–9

Answers to odd-numbered problems

1.
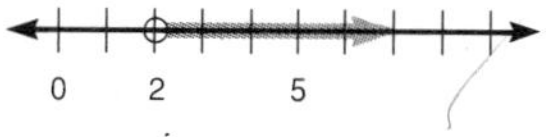

3.

5.
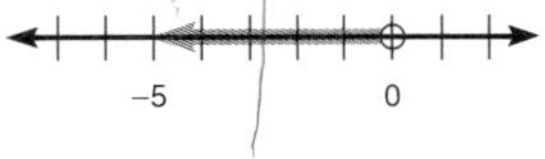

7.
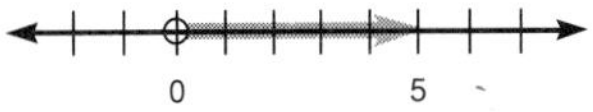

9.
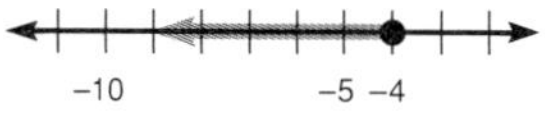

11.
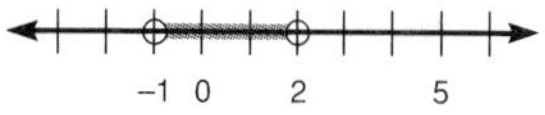

13.

15.
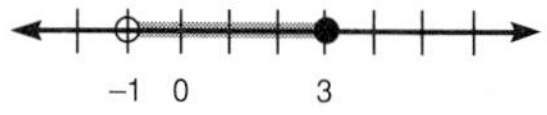

17. $x \le \frac{5}{2}$;
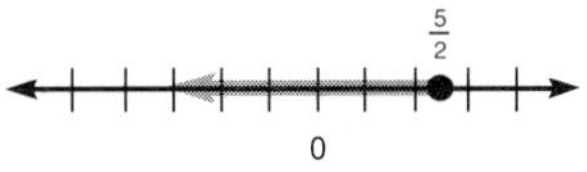

19. $x < 6$;
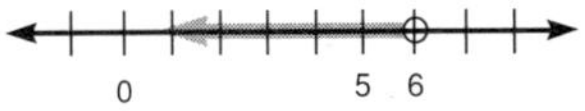

21. $x \ge 18$;
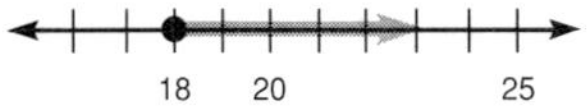

23. $x \ge$
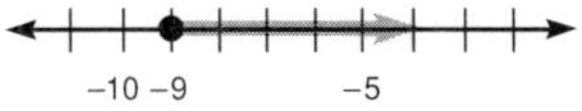

25. $x > -5$;

27. $x >$
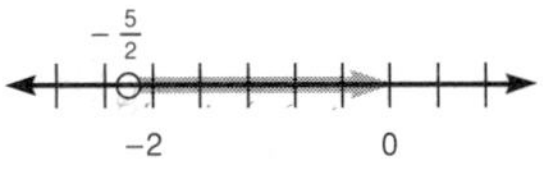

29. $x > \frac{16}{3}$;
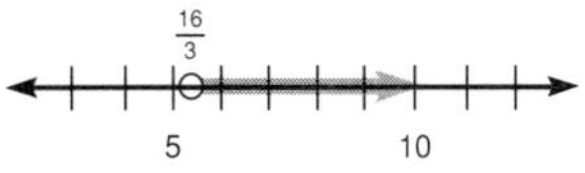

31. $x \le 2$;
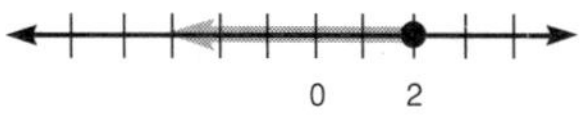

33. $x < 11$;

35. $x > 1$;
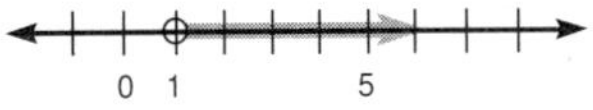

37. $x \geq 3$;

39. $x > \dfrac{10}{7}$;

41. $x > -10$;

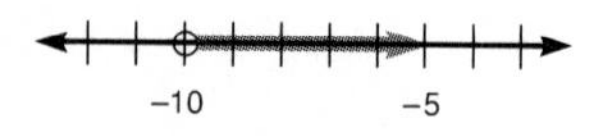

43. $x < \dfrac{16}{11}$;

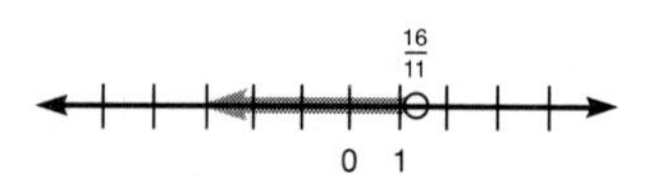

45. $x \leq \dfrac{37}{10}$;

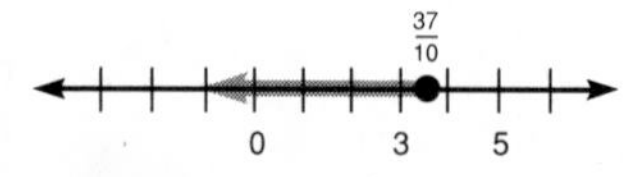

47. $x \geq \dfrac{5}{2}$;

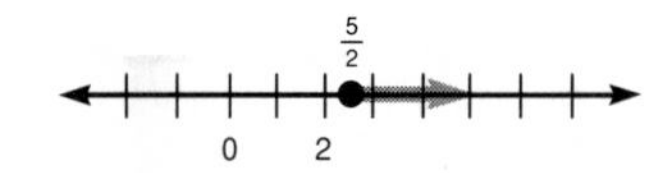

49. $x \geq \dfrac{2}{5}$;

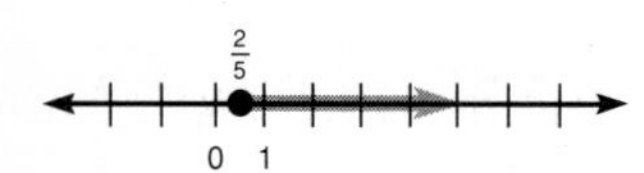

51. $-2 < x < \dfrac{1}{2}$;

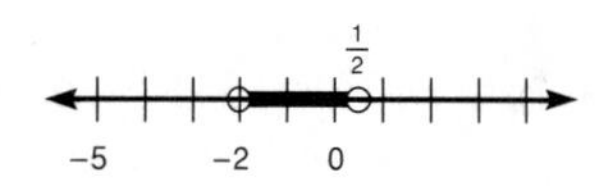

53. $-\dfrac{4}{5} \leq x \leq \dfrac{1}{5}$;

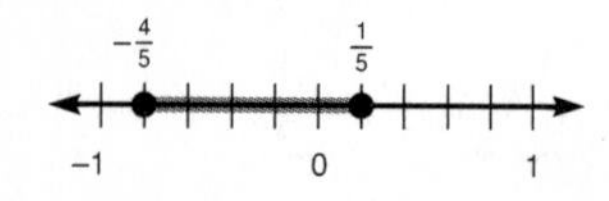

55. $-2 < x \leq \dfrac{5}{4}$;

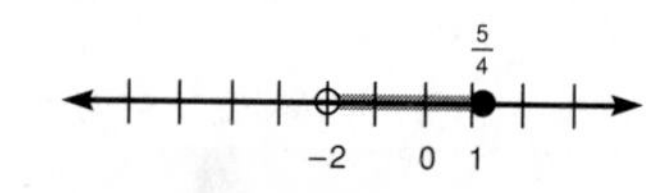

57. $-4 < x \leq 1$;

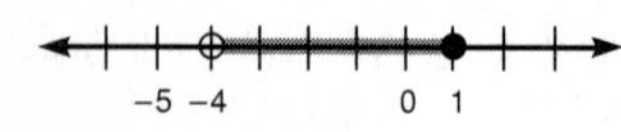

59. $-1 \leq x < 7$;

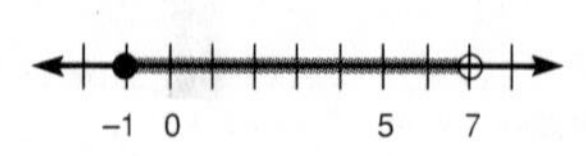

61. $-2 < x \leq \dfrac{1}{3}$;

63. (x = student's score) $x \geq 72$ **65.** (x = number of new lift trucks) $x \geq 8$ **67.** (P = selling price, C = cost) $P \geq 2C$
69. $x < 6$ **71.** $x \geq 12$ **73.** $x \geq 7$ **75.** $1 < x < 5$
77. $5 < x < 27$ **79.** $x > 18$

Solutions to trial exercise problems

22.
$$-4x < 12$$
$$\frac{-4x}{-4} > \frac{12}{-4}$$
$$x > -3$$

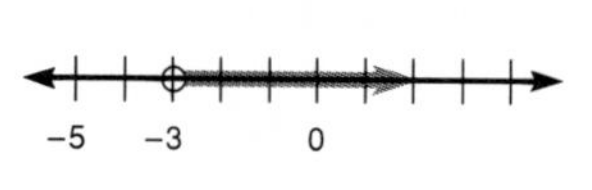

43.

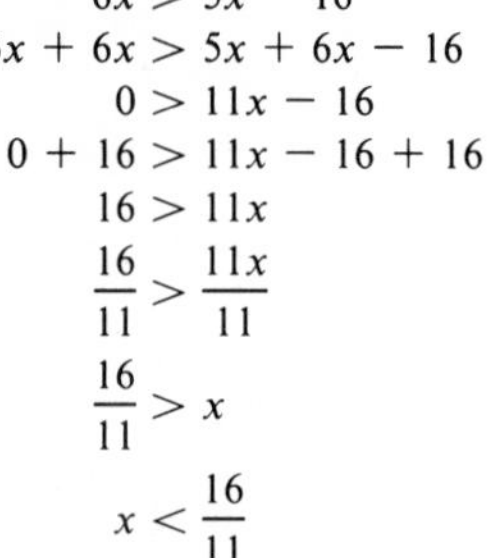

$$8 - 2(3x + 4) > 5x - 16$$
$$8 - 6x - 8 > 5x - 16$$
$$-6x > 5x - 16$$
$$-6x + 6x > 5x + 6x - 16$$
$$0 > 11x - 16$$
$$0 + 16 > 11x - 16 + 16$$
$$16 > 11x$$
$$\frac{16}{11} > \frac{11x}{11}$$
$$\frac{16}{11} > x$$
$$x < \frac{16}{11}$$

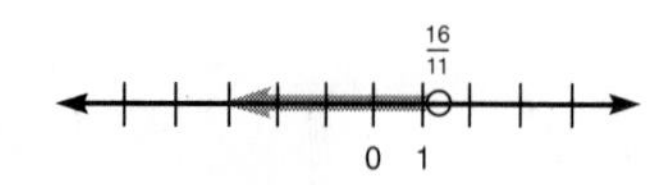

56.
$$-2 < -x \leq 3$$
$$-2 < -1 \cdot x \leq 3$$
$$\frac{-2}{-1} > \frac{-1 \cdot x}{-1} \geq \frac{3}{-1}$$
$$2 > x \geq -3$$
$$-3 \leq x < 2$$

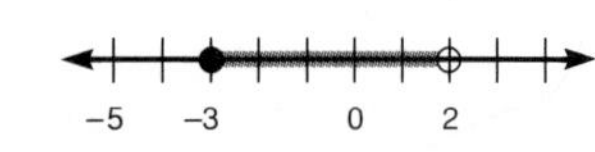

66. A company will hire at least 2 new employees, but not more than 7. Let x be the number of new employees. Then the inequality would be $2 \leq x \leq 7$.

75. Two times a number plus 4 is greater than 6 but less than 14. Let x be the number. Then the inequality would be $6 < 2x + 4 < 14$.
Solving: $6 - 4 < 2x + 4 - 4 < 14 - 4$
$$2 < 2x < 10$$
$$\frac{2}{2} < \frac{2x}{2} < \frac{10}{2}$$
$$1 < x < 5$$

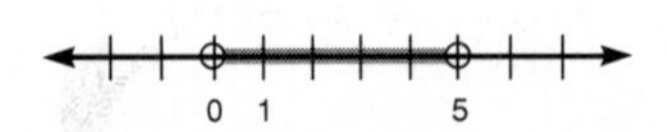

78. The perimeter of a rectangle must be less than 100 feet. If the length is known to be 30 feet, find all numbers that the width could be. Let x represent the width of the rectangle. Then the inequality would be $60 < 2 \cdot 30 + 2 \cdot x < 100$. Since the width of a real rectangle must be greater than zero and we already know the length to be 30, then the smallest value for the perimeter must be greater than 60 (two times the known length).
Solving:
$$60 < 2 \cdot 30 + 2x < 100$$
$$60 < 60 + 2x < 100$$
$$60 - 60 < 60 - 60 + 2x < 100 - 60$$
$$0 < 2x < 40$$
$$\frac{0}{2} < \frac{2x}{2} < \frac{40}{2}$$
$$0 < x < 20$$
The width of the rectangle must be some real number greater than zero feet but less than 20 feet.

Review exercises

1. -16 **2.** 16 **3.** -16 **4.** 16 **5.** x^5 **6.** (let x = the number) x^3 **7.** (let x = the number) x^2 **8.** xy

Chapter 2 review

1. 3 **2.** 1 **3.** 2 **4.** 2 **5.** polynomial **6.** polynomial **7.** polynomial **8.** not a polynomial because a variable is in the denominator **9.** $5x$ **10.** $y - 7$ **11.** $z + 4$ **12.** (let x = the number) $2x + 6$ **13.** 1 **14.** -1 **15.** 72 **16.** -4 **17.** 4 **18.** 7 **19. a.** 3 **b.** $\frac{189}{4}$ **20.** 1,040 **21.** $4x^2 - 3x + 3$ **22.** $-a^2 + a + 11$ **23.** $-6a^2$ **24.** $5x^3 - 7xy^2 - 2y^3 - 4x^2$ **25.** $-11ab + 7b^2c + 11bc$ **26.** $-6y + 1$ **27.** $5ab + 3ac - 4bc$ **28.** -5 **29.** $7x + y$ **30.** $2x + 5y$ **31.** $-6x + 4y$ **32.** $9a - 10b$ **33.** true **34.** false **35.** false **36.** true **37.** $\{7\}$ **38.** $\{21\}$ **39.** $\{-11\}$ **40.** $\{-6\}$ **41.** $\{4\}$ **42.** $\{19\}$ **43.** $\{-2\}$ **44.** $\{9\}$ **45.** $\{3\}$ **46.** $\{3\}$ **47.** $\{-7\}$ **48.** $\{-7\}$ **49.** $\{12\}$ **50.** $\{14\}$ **51.** $\{15\}$ **52.** $\{21\}$ **53.** $\left\{\frac{21}{4}\right\}$ **54.** $\{-18\}$ **55.** $\left\{-\frac{9}{2}\right\}$ **56.** $\{35\}$ **57.** $\{0\}$ **58.** $\{4\}$ **59.** $\{6\}$ **60.** $\{-8\}$ **61.** $\{3\}$ **62.** $\left\{\frac{14}{3}\right\}$ **63.** $\left\{\frac{7}{3}\right\}$ **64.** $\{3\}$ **65.** $\left\{\frac{1}{2}\right\}$ **66.** $\left\{-\frac{22}{3}\right\}$ **67.** $\left\{-\frac{1}{4}\right\}$ **68.** $\left\{\frac{7}{13}\right\}$ **69.** $\left\{-\frac{5}{2}\right\}$ **70.** $\{-14\}$ **71.** $\{-24\}$ **72.** $\left\{\frac{7}{4}\right\}$ **73.** $\{0\}$ **74.** $\left\{\frac{1}{5}\right\}$ **75.** $\left\{-\frac{7}{4}\right\}$ **76.** $\{2\}$ **77.** $a = \frac{F}{m}$ **78.** $I = \frac{E}{R}$ **79.** $P = \frac{k}{V}$ **80.** $g = V - k - t$ **81.** $c = \frac{2A - bh}{h}$ **82.** $x = \frac{4y}{3}$ **83.** 64 and 41 **84.** 36 **85.** 45 **86.** \$11,000 at 8%; \$9,000 at 7% **87.** \$12,000 at 12%; \$13,000 at 19%

88. $x > 4$;

89. $x \le 3$;

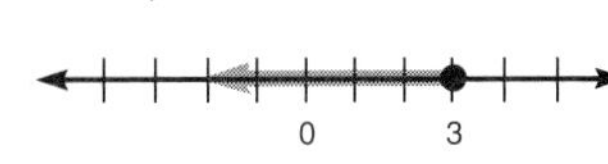

90. $x > -7$;

91. $x < -4$;

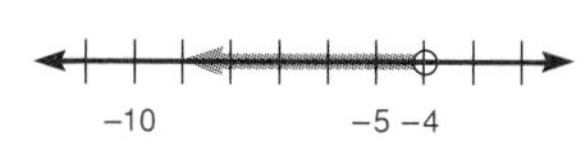

92. $x < 2$;

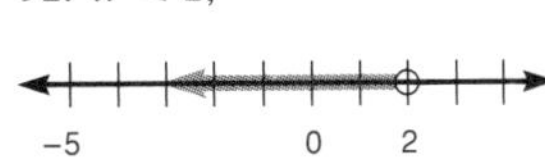

93. $x > \frac{15}{7}$;

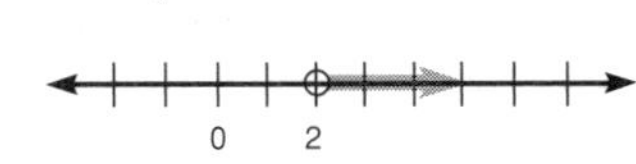

94. $x > \frac{9}{2}$;

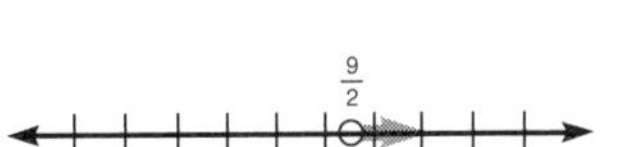

95. $x \ge -\frac{6}{5}$;

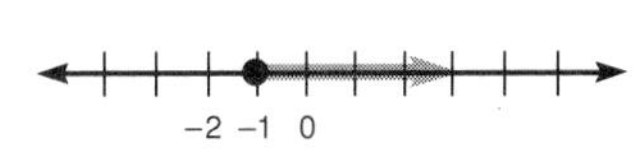

96. $x \le \frac{2}{9}$;

97. $-\frac{11}{5} < x < \frac{3}{5}$;

98. $-1 < x \le \frac{1}{5}$;

99. $\frac{1}{2} < x < \frac{9}{4}$;

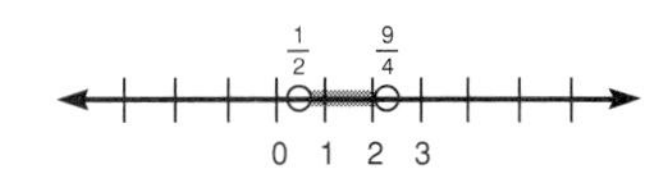

100. $-\frac{13}{3} \le x \le -\frac{1}{3}$;

Chapter 2 cumulative test

1. -12 **2.** 4 **3.** 3 **4.** undefined **5.** -25 **6.** $\frac{1}{2}$ **7.** $-\frac{1}{6}$ **8.** 0 **9.** 19.78 **10.** 41 **11.** 38 **12.** -24 **13.** 26 **14.** 6 **15.** 20 **16.** $4x$ **17.** $2x^2y^2 + 3xy$ **18.** $a + 3b$ **19.** $4x^2y - 7xy^2$ **20.** $a^3 + 2a^2 + a - 1$ **21.** $-3x + 4y$ **22.** $4a + 2b$ **23.** 64 **24.** -25 **25.** 234 **26.** 2 **27.** 4 **28.** 42 **29.** 51 mph **30.** $x - y$ **31.** (let x = the number) $x + 6$ **32.** $10d + 5n + c$ **33.** $\left\{\frac{5}{3}\right\}$ **34.** $\{0\}$ **35.** $\left\{-\frac{19}{2}\right\}$ **36.** $\left\{\frac{17}{16}\right\}$ **37.** $\left\{\frac{30}{11}\right\}$ **38.** $x \le -6$ **39.** $x < -7$ **40.** $x > \frac{8}{5}$ **41.** $-2 < x < 4$ **42.** $-1 \le x \le 6$ **43.** $b = P - a - c$ **44.** $y = \frac{x - az}{a}$ **45.** \$6,000 at 6%; \$4,000 at 5% **46.** 14,16,18 **47.** 12 **48.** \$6,500 at 12% profit; \$10,500 at 19% loss **49.** $x \le 6$

Chapter 3

Exercise 3–1

Answers to odd-numbered problems

1. a^5 **3.** $(-2)^4$ **5.** x^6 **7.** $(xy)^4$ **9.** $(x - y)^3$ **11.** $xxxx$ **13.** $(-2)(-2)(-2)$ **15.** $5 \cdot 5 \cdot 5$ **17.** $(4y)(4y)(4y)(4y)$ **19.** $(x - y)(x - y)$ **21.** x^{11} **23.** R^3 **25.** a^9 **27.** $5^5 = 3{,}125$ **29.** $4^7 = 16{,}384$ **31.** $(x - 2y)^{10}$ **33.** $(a - b)^{11}$ **35.** x^4y^4 **37.** $64x^3y^3z^3$ **39.** x^{15} **41.** b^{25} **43.** $6x^4y^3$ **45.** a^7b^5 **47.** $30x^5$ **49.** $12a^7b^7c$ **51.** $12a^5b^3$ **53.** $-6a^3b^5$ **55. a.** $V = 5^3$, 125 cubic units **b.** $V = 4^3$, 64 cubic units **c.** $V = 6^3$, 216 cubic units **57.** $s = \frac{1}{2}gt^2$ **59.** $V = \frac{4}{3}\pi r^3$ **61.** $m^4 - 8$ **63.** $2x^2 - y^3$ **65.** $\frac{p^3}{q^2}$

Solutions to trial exercise problems

23. $R^2 \cdot R = R^2 \cdot R^1 = R^{2+1} = R^3$ **27.** $5^2 \cdot 5^3 = 5^{2+3} = 5^5 = 3{,}125$ **43.** $(2xy^2)(3x^3y) = 2 \cdot 3 \cdot xx^3y^2y = 6x^{1+3}y^{2+1} = 6x^4y^3$ **55a.** $V = e^3$, then $V = (5)^3 = 5 \cdot 5 \cdot 5 = 25 \cdot 5 = 125$ cubic units **60.** The cube of Johnny's age is given as n^3. Since his mother is 6 years more than the cube of Johnny's age, we add 6, giving $n^3 + 6$.

Review exercises

1. $9a$ **2.** $8x$ **3.** $6ab$ **4.** $7xy$ **5.** $7a^2 + 5a$ **6.** $5x^2 + 5x$ **7.** $x^2y + 7xy^2$ **8.** $3ab^2 + 2a^2b$

Exercise 3–2

Answers to odd-numbered problems

1. $2a^3b - 2ab^2c + 2abc^2$ **3.** $15ab^2 - 21ac^2$ **5.** $-15a^3b^2 + 5a^2b^3 - 20ab^4$ **7.** $3a^3b - 6a^2b^2 - 3ab^3$ **9.** $12a^2b^2 - 6ab^3$ **11.** $x^2 + 7x + 12$ **13.** $y^2 - 13y + 36$ **15.** $a^2 + 2a + 1$ **17.** $R^2 - 6R + 9$ **19.** $a^2 - 9$ **21.** $6a^2 - 31a + 35$ **23.** $4x^2 - 49$ **25.** $3k^2 - 17kw - 6w^2$ **27.** $a^2 + 12ab + 36b^2$ **29.** $4a^2 - 9b^2$ **31.** $a^3 + 2a^2b - 7ab^2 + 4b^3$ **33.** $6x^3 + 21x^2 - 5x + 28$ **35.** $x^4 - x^3 - x^2 - 11x - 12$ **37.** $a^3 - 7a^2 + 4a + 12$ **39.** $2a^3 - 3a^2b - 2ab^2 + 3b^3$ **41.** $a^3 - 3a^2b + 3ab^2 - b^3$ **43.** $a^3 - 6a^2b + 12ab^2 - 8b^3$ **45.** $cx^2 - 4c^2x + 4c^3$

Solutions to trial exercise problems

8. $(2x)(x - y + 5)(5y) = [(2x)(x - y + 5)](5y) = [2x \cdot x - 2x \cdot y + 2x \cdot 5](5y) = [2x^2 - 2xy + 10x](5y) = 2x^2 \cdot 5y - 2xy \cdot 5y + 10x \cdot 5y = 10x^2y - 10xy^2 + 50xy$ **17.** $(R - 3)^2$ is a special product. $(R - 3)^2 = (R)^2 + [2 \cdot R \cdot (-3)] + (-3)^2 = R^2 - 6R + 9$ **18.** $(R + 2)(R - 2)$ is a special product. $(R + 2)(R - 2) = (R)^2 - (2)^2 = R^2 - 4$ **31.** $(a + 4b)(a^2 - 2ab + b^2) = a^3 - 2a^2b + ab^2 + 4a^2b - 8ab^2 + 4b^3 = a^3 + 2a^2b - 7ab^2 + 4b^3$ **37.** $(a - 6)(a - 2)(a + 1) = [(a - 6)(a - 2)](a + 1) = [a^2 - 2a - 6a + 12](a + 1) = [a^2 - 8a + 12](a + 1) = a^3 + a^2 - 8a^2 - 8a + 12a + 12 = a^3 - 7a^2 + 4a + 12$ **40.** $(a + b)^3 = (a + b)(a + b)(a + b) = [(a + b)(a + b)](a + b) = [a^2 + ab + ab + b^2](a + b) = [a^2 + 2ab + b^2](a + b) = a^3 + a^2b + 2a^2b + 2ab^2 + ab^2 + b^3 = a^3 + 3a^2b + 3ab^2 + b^3$

Review exercises

1. -5 **2.** 12 **3.** 3 **4.** -9 **5.** x^{12} **6.** a^{15} **7.** $27a^3b^3$ **8.** $8x^3$

Exercise 3–3

Answers to odd-numbered problems

1. 1 **3.** 5 **5.** 1 **7.** $\frac{1}{R^5}$ **9.** $\frac{1}{9P^2}$ **11.** $\frac{9}{C^4}$ **13.** $\frac{y^3}{2}$ **15.** $\frac{2y^2}{x^4}$ **17.** $\frac{t^5}{r^2}$ **19.** $\frac{a^6}{b^6}$ **21.** $\frac{8}{27}$ **23.** $\frac{16x^4}{y^4}$ **25.** $\frac{27a^3}{b^3}$ **27.** y^2 **29.** b^2 **31.** $\frac{1}{R^4}$ **33.** 4 **35.** $\frac{1}{36}$ **37.** y^4 **39.** a **41.** $\frac{1}{y^6}$ **43.** a^3b^3 **45.** $4x^2y^2$ **47.** $\frac{1}{27a^3}$ **49.** x^3 **51.** $\frac{1}{a^6}$ **53.** x^2 **55.** $\frac{1}{a^2}$ **57.** $-\frac{1}{125}$ **59.** 27 **61.** $\frac{1}{a^6}$ **63.** $\frac{1}{x^{10}}$ **65.** a^6 **67.** 1 **69.** 1 **71.** $\frac{R^5}{S^9}$ **73.** $\frac{ab^4c^2}{8}$

Solutions to trial exercise problems

3. $5a^0 = 5 \cdot 1 = 5$ **10.** $4z^{-2} = 4 \cdot \frac{1}{z^2} = \frac{4}{z^2}$ **15.** $2x^{-4}y^2 = 2 \cdot \frac{1}{x^4} \cdot y^2 = \frac{2y^2}{x^4}$ **32.** $\frac{3^4}{3^2} = 3^{4-2} = 3^2 = 9$ **36.** $\frac{x^4x^3}{x^2} = \frac{x^{4+3}}{x^2} = \frac{x^7}{x^2} = x^{7-2} = x^5$ **48.** $\frac{5^2a^3b}{5^3a^7b^3} = 5^{2-3}a^{3-7}b^{1-3} = 5^{-1}a^{-4}b^{-2} = \frac{1}{5^1} \cdot \frac{1}{a^4} \cdot \frac{1}{b^2} = \frac{1}{5a^4b^2}$ **49.** $x^{-4}x^7 = x^{-4+7} = x^3$ *Alternate:* $x^{-4}x^7 = \frac{1}{x^4} \cdot x^7 = \frac{x^7}{x^4} = x^{7-4} = x^3$ **57.** $(-5)^{-3} = \frac{1}{(-5)^3} = -\frac{1}{125}$ (*Note:* The sign of the base, -5, is unchanged.) **73.** $\frac{4^{-1}a^{-2}b^3c^0}{2a^{-3}b^{-1}c^{-2}} = \frac{\frac{1}{4^1} \cdot \frac{1}{a^2} \cdot b^3 \cdot 1}{2 \cdot \frac{1}{a^3} \cdot \frac{1}{b^1} \cdot \frac{1}{c^2}} = \frac{\frac{b^3}{4a^2}}{\frac{2}{a^3bc^2}} = \frac{b^3}{4a^2} \cdot \frac{a^3bc^2}{2} = \frac{a^3b^{3+1}c^2}{2 \cdot 4a^2} = \frac{a^3b^4c^2}{8a^2} = \frac{b^4c^2}{8} \cdot a^{3-2} = \frac{b^4c^2}{8} \cdot a^1 = \frac{ab^4c^2}{8}$

Review exercises

1. -10 **2.** 21 **3.** 4 **4.** -16 **5.** a^8 **6.** x^{12} **7.** x^6 **8.** $4a^2b^2$

Exercise 3–4

Answers to odd-numbered problems

1. $8a^6$ **3.** $8x^6y^3$ **5.** $x^{16}y^{12}z^4$ **7.** $125a^{15}b^6c^3$ **9.** $\frac{1}{4a^4}$ **11.** $\frac{16}{x^4}$ **13.** $\frac{y^{12}}{27x^3}$ **15.** $\frac{y^{10}}{x^4z^6}$ **17.** $6x^7y^3$ **19.** $-6x^{10}y^4$ **21.** $6x^6y^3z^7$ **23.** $\frac{8x^3}{y^6}$ **25.** $\frac{9a^4}{b^6}$ **27.** $\frac{8x^6}{y^9}$ **29.** $\frac{4x^6y^6}{z^{10}}$ **31.** $\frac{b^8}{a^5}$ **33.** $\frac{R^2}{3S^4}$ **35.** $\frac{4}{ab^9}$ **37.** $\frac{3S^3}{R^4}$ **39.** $a^{10}b^{17}$ **41.** $32a^{12}$ **43.** $32x^{14}y^{13}$ **45.** $\frac{y^2}{xz^4}$ **47.** $\frac{a^6b^{10}}{4}$ **49.** $\frac{b^6}{a^3c^3}$

Solutions to trial exercise problems

9. $(2a^2)^{-2} = \frac{1}{(2a^2)^2} = \frac{1}{2^2(a^2)^2} = \frac{1}{2^2a^4} = \frac{1}{4a^4}$ **31.** $\frac{a^{-2}b^3}{a^3b^{-5}} = \frac{\frac{1}{a^2} \cdot b^3}{a^3 \cdot \frac{1}{b^5}} = \frac{\frac{b^3}{a^2}}{\frac{a^3}{b^5}} = \frac{b^3}{a^2} \cdot \frac{b^5}{a^3} = \frac{b^{3+5}}{a^{2+3}} = \frac{b^8}{a^5}$ **39.** $(a^2b^3)^3(ab^2)^4 = (a^2)^3(b^3)^3 \cdot a^4(b^2)^4 = a^6b^9a^4b^8 = a^{6+4}b^{9+8} = a^{10}b^{17}$

Review exercises

1. 35.34 **2.** 10.36 **3.** 16.72 **4.** 28.98 **5.** 18.81 **6.** 49.5

Exercise 3–5

Answers to odd-numbered problems

1. 2.55×10^2 **3.** 1.2345×10^4 **5.** 1.55×10^5 **7.** 8.55076×10^2 **9.** 1.0076×10^6 **11.** 1.2×10^{-4} **13.** 8.1×10^{-6} **15.** 7×10^{-4} **17.** 9.4×10^{-11} **19.** -4.5×10^3 **21.** -5.85×10^6 **23.** -4.578×10 **25.** -2.985×10^{-8} **27.** 49,900,000 **29.** 7.23 **31.** 0.0042 **33.** 0.00000147 **35.** 0.000789 **37.** -0.00000000482 **39.** -0.00000492 **41.** 1×10^{-9} **43.** 2×10^{12} **45.** 35,600,000 **47.** 0.000 000 000 000 000 000 000 093 **49.** 140,000 **51.** 1.22304×10^{14} **53.** 3.63226×10^{-1} **55.** 1.76979×10^{-7} **57.** 4.84481×10^8 **59.** 1.4×10^3 **61.** 4.6×10^3

Solution to trial exercise problem

58. $(177{,}000) \div (0.15) = \dfrac{1.77 \times 10^5}{1.5 \times 10^{-1}} = \dfrac{1.77}{1.5} \times 10^{5-(-1)} = \dfrac{1.77}{1.5} \times 10^6 = 1.18 \times 10^6$

Review exercises

1. $5x$ **2.** $3a^2$ **3.** $12ab$ **4.** $x^3 + 2x^2$ **5.** $6a^2 - 15a$ **6.** $3x^3y + 2x^2y^2 - 7x^2y$

Chapter 3 review

1. a^{12} **2.** a^{14} **3.** $4^5 = 1{,}024$ **4.** x^4y^4 **5.** a^{15} **6.** $20a^4b^5$ **7.** $6x^3y^7$ **8.** $-15x^5$ **9.** $6a^3b^5$ **10.** $10x^5y^5$ **11.** $-6a^6b^{10}$ **12.** $15x^2 - 10xy$ **13.** $-6a^4b + 9a^3b^2 - 12a^2b^3$ **14.** $30x^2y^2 - 10xy^3$ **15.** $x^2 - x - 12$ **16.** $x^2 + 10x + 25$ **17.** $a^2 - 49$ **18.** $15x^2 + 7xy - 2y^2$ **19.** $x^3 + x^2y - 5xy^2 - 2y^3$ **20.** $3a^3 - 4a^2b - 5ab^2 + 2b^3$ **21.** $8a^3 + 12a^2b + 6ab^2 + b^3$ **22.** $\dfrac{1}{b^2}$ **23.** $\dfrac{5}{a^2}$ **24.** a^4 **25.** $\dfrac{1}{x^3}$ **26.** 1 **27.** $\dfrac{5}{x^3y^2}$ **28.** a^3 **29.** $\dfrac{a^5}{b^5}$ **30.** $\dfrac{4y^2z^2}{x^2}$ **31.** $\dfrac{b^3}{4a^3}$ **32.** $\dfrac{a^9}{b^3}$ **33.** $8a^6b^9$ **34.** $3^{12}x^{16}y^{20} = 531{,}441x^{16}y^{20}$ **35.** $\dfrac{y^6}{4x^2}$ **36.** $\dfrac{xy^3z^3}{2}$ **37.** $\dfrac{2a^2c^3}{b^6}$ **38.** $8x^{26}y^{25}$ **39.** $72a^{14}b^3$ **40.** $\dfrac{a^9b^3}{c^{12}}$ **41.** $\dfrac{9a^6b^4}{c^{10}}$ **42.** $\dfrac{32x^5y^{20}}{z^{30}}$ **43.** 1.84×10^3 **44.** 1.57×10^{-3} **45.** 1.07×10^8 **46.** 8.49×10^{11} **47.** -3.75×10 **48.** -5.43×10^{-3} **49.** 504,000 **50.** 0.00639 **51.** -596 **52.** -0.00886 **53.** 0.000000735 **54.** 812,000,000 **55.** 2.67672×10^5 **56.** 1.54818×10^{-8} **57.** 7.2×10^{-3} **58.** 1.25×10^{-7}

Chapter 3 cumulative test

1. false **2.** false **3.** true **4.** -2 **5.** undefined **6.** 8 **7.** 0 **8.** 16 **9.** 38 **10.** $x^2y + 2xy^2 - 3x^2y^2$ **11.** -40 **12.** 0 **13.** $9x^2 - 6xy + y^2$ **14.** 0 **15.** $4a^4b^{10}$ **16.** -25 **17.** 0 **18.** $9x^2 - 4y^2$ **19.** $6x^3y^5$ **20.** 26 **21.** undefined **22.** $4a + 2b$ **23.** $x^3 - 2x - 1$ **24.** x^2 **25.** a^4 **26.** $\{7\}$ **27.** $\left\{\dfrac{5}{3}\right\}$ **28.** $\left\{-\dfrac{19}{4}\right\}$ **29.** $x > -\dfrac{1}{3}$ **30.** $x > \dfrac{1}{6}$ **31.** $\left\{-\dfrac{3}{2}\right\}$ **32.** $-8 \le x \le -1$ **33.** 97 **34.** 72 **35.** \$18,500 at 8%; \$11,500 at 7%

Chapter 4

Exercise 4–1

Answers to odd-numbered problems

1. $2(y + 3)$ **3.** $4(x^2 + 2y)$ **5.** $3(x^2y + 5z)$ **7.** $7(a - 2b + 3c)$ **9.** $3(5xy - 6z + x^2)$ **11.** $7(6xy - 3y^2 + 1)$ **13.** $2(4x - 5y + 6z - 9w)$ **15.** $5ab(4a - 12 + 9b)$ **17.** $3xy(x + 2)$ **19.** $2R^2(R^2 - 3)$ **21.** $x(2x^2 - x + 1)$ **23.** $3ab(5 + 6b - a)$ **25.** $xy(y + z + 1)$ **27.** $2L(L^2 - 9 + L)$ **29.** $5p(p + 2 + 3p^2)$ **31.** $-3(2x + 3)$ **33.** $2(3x - 4z - 6w)$ **35.** $-3(4L - 5W + 2H)$ **37.** $-x(1 - x + x^2)$ **39.** $-xyz(1 - x + y - z)$ **41.** $-5ab(2ab - 3 + 4a^2b^2)$ **43.** $(a + b)(x + y)$ **45.** $5(2a + b)(3x + 2y)$ **47.** $3(a + 4b)(x + 2y)$ **49.** $(b + 6)(8a - 1)$ **51.** $(a + b)(c + d)$ **53.** $(2a + b)(3x - 2y)$ **55.** $(2x + y)(2a - b)$ **57.** $(5x - 3y)(4x + z)$ **59.** $(a + 3b)(4x - 3y)$ **61.** $(2c - y)(a + 3b)$ **63.** $(c + 4y)(2a + 3b)$ **65.** $(3x + y)(2a + b)$ **67.** $(x - 2d)(3a + b)$ **69.** $(a + 5)(2a^2 + 3)$ **71.** $(4a^2 + 3)(2a - 1)$ **73.** $\pi r(s + r)$ **75.** $\dfrac{wx}{48EI}(2x^3 - 3\ell x^2 - \ell^3)$

Solutions to trial exercise problems

11. $42xy - 21y^2 + 7 = 7 \cdot 6xy - 7 \cdot 3y^2 + 7 \cdot 1 = 7(6xy - 3y^2 + 1)$ **25.** $xy^2 + xyz + xy = xy \cdot y + xy \cdot z + xy \cdot 1 = xy(y + z + 1)$ **33.** $6x - 8z - 12w = 2(\qquad)$; $2 \cdot 3x - 2 \cdot 4z - 2 \cdot 6w = 2(3x - 4z - 6w)$ **34.** $-4a^3 - 36ab + 16ab^2 - 24b^3 = -4(\qquad)$; $(-4)a^3 + (-4)(9ab) + (-4)(-4ab^2) + (-4)(6b^3) = -4(a^3 + 9ab - 4ab^2 + 6b^3)$ **36.** $-3a + a^3b = -a(\qquad)$; $(-a) \cdot 3 + (-a)(-a^2b) = -a(3 - a^2b)$ **45.** $15x(2a + b) + 10y(2a + b) = 5(2a + b) \cdot 3x + 5(2a + b) \cdot 2y = 5(2a + b)(3x + 2y)$ **53.** $6ax - 2by + 3bx - 4ay = 6ax + 3bx - 4ay - 2by = (6ax + 3bx) + (-4ay - 2by) = 3x(2a + b) - 2y(2a + b) = (2a + b)(3x - 2y)$ **75.** $Y = \dfrac{2wx^4}{48EI} - \dfrac{3\ell wx^3}{48EI} - \dfrac{\ell^3wx}{48EI} = \dfrac{wx}{48EI} \cdot 2x^3 - \dfrac{wx}{48EI} - 3\ell x^2 - \dfrac{wx}{48EI} \cdot \ell^3 = \dfrac{wx}{48EI}(2x^3 - 3\ell x^2 - \ell^3)$

Review exercises

1. 4, 5 **2.** 3, 4 **3.** -8, 2 **4.** 8, -2 **5.** 8, 2 **6.** -8, -2 **7.** 6, 6 **8.** 11, 1

Exercise 4–2

Answers to odd-numbered problems

1. $(a + 6)(a + 3)$ **3.** $(x + 12)(x - 1)$ **5.** $(y + 15)(y - 2)$ **7.** $(x - 12)(x - 2)$ **9.** $(a + 8)(a - 3)$ **11.** $(x + 6)(x + 2)$ **13.** $(a - 6)(a + 4)$ **15.** $2(x + 5)(x - 2)$ **17.** $3(x - 8)(x + 2)$ **19.** will not factor, prime polynomial **21.** $(y + 15)(y + 2)$ **23.** $4(x + 2)(x - 3)$ **25.** $5(a - 5)(a + 2)$ **27.** $(xy - 6)(xy + 3)$ **29.** $(xy + 12)(xy + 1)$ **31.** $3(xy + 3)(xy - 4)$ **33.** $(x + 2y)(x + y)$ **35.** $(a - 3b)(a + b)$ **37.** $(a - 3b)(a + 2b)$ **39.** $(x + 3y)(x - 5y)$

Solutions to trial exercise problems

17. $3x^2 - 18x - 48 = 3(x^2 - 6x - 16)$
m and n are -8 and 2. $= 3(x - 8)(x + 2)$
26. $x^2y^2 - 4xy - 21 = (xy)^2 - 4(xy) - 21$
m and n are -7 and 3. $= (xy - 7)(xy + 3)$
33. We need to find m and n that add to $3y$ and multiply to $2y^2$. The values are $2y$ and y. The factorization is $x^2 + 3xy + 2y^2 = (x + 2y)(x + y)$.

Review exercises

1. $x^2(a + b + c)$ **2.** $3x(x^2 + 4x - 2)$ **3.** $(3x + 5)(2x + 1)$
4. $(2x + 3)(3x - 2)$ **5.** $(4x + 1)(5x + 1)$
6. $(6x - 1)(2x + 3)$ **7.** $(x - 2)(3x - 5)$
8. $(7x - 3)(x - 9)$

Exercise 4–3

Answers to odd-numbered problems

1. $(2x - 3)(x + 2)$ **3.** $(2x + 1)(x + 1)$ **5.** $(2R - 3)(R - 2)$
7. $(5x + 3)(x - 2)$ **9.** $(3x - 1)(3x - 1) = (3x - 1)^2$
11. will not factor, prime polynomial **13.** $(2x + 3)(3x + 2)$
15. $(2x + 7)(2x + 3)$ **17.** will not factor, prime polynomial
19. $(3y - 8)(3y + 1)$ **21.** $(5x + 6)(2x - 1)$
23. $(2x - 5)(x - 2)$ **25.** $2(2x + 3)(x + 2)$
27. $2(2x + 3)(x + 1)$ **29.** $(2x + 3)(3x - 1)$
31. $2(x + 5)(x - 2)$ **33.** $(3x - 2)(2x + 3)$
35. $(2x + 3)(2x + 3) = (2x + 3)^2$ **37.** $(7x - 1)(x - 5)$
39. $(3P + 1)(5P - 1)$ **41.** $2x(x - 5)(x + 2)$
43. $a(2a + 1)(a + 7)$ **45.** $(2x - 5)(4x + 3)$
47. $-16(t - 1)(t - 1) = -16(t - 1)^2$

Solutions to trial exercise problems

25. $4x^2 + 14x + 12 = 2(2x^2 + 7x + 6)$
m and n are 3 and 4. $= 2[(2x^2 + 3x) + (4x + 6)]$
$= 2[x(2x + 3) + 2(2x + 3)]$
$= 2(2x + 3)(x + 2)$
41. $2x^3 - 6x^2 - 20x = 2x(x^2 - 3x - 10)$
m and n are -5 and 2. $= 2x(x - 5)(x + 2)$

Review exercises

1. $x^2 - y^2$ **2.** $9a^2 - 4b^2$ **3.** $x^2 - 2xy + y^2$ **4.** $25a^2 - 16b^2$
5. $4a^2 + 4ab + b^2$ **6.** $16x^2 - 8xy + y^2$ **7.** $x^4 - 1$
8. $a^4 - 16$

Exercise 4–4

Answers to odd-numbered problems

1. $(6)^2$ **3.** $(c)^2$ **5.** $(4x)^2$ **7.** $(2z^2)^2$ **9.** $(x + 1)(x - 1)$
11. $(a + 2)(a - 2)$ **13.** $(3 + E)(3 - E)$ **15.** $(1 + k)(1 - k)$
17. $(3b + 4)(3b - 4)$ **19.** $(b + 6c)(b - 6c)$
21. $(2a + 5b)(2a - 5b)$ **23.** $(5p + 9)(5p - 9)$
25. $8(x + 2y)(x - 2y)$ **27.** $5(r + 5s)(r - 5s)$
29. $2(5 + x)(5 - x)$ **31.** $(rs + 5t)(rs - 5t)$
33. $(x^2 - 3)(x^2 + 3)$ **35.** $(r^2 + 9)(r + 3)(r - 3)$
37. $(7x + 8y^2)(7x - 8y^2)$ **39.** $2(7xy + 5pc)(7xy - 5pc)$
41. $(c - 7)^2$ **43.** $(a + 3)^2$ **45.** $(y - 3)^2$ **47.** $(2a - 3b)^2$
49. $(3c - 2d)^2$ **51.** $\frac{V}{8l}(h + 2v_1)(h - 2v_1)$

Solutions to trial exercise problems

17. $9b^2 - 16 = (3b)^2 - (4)^2 = (3b + 4)(3b - 4)$
25. $8x^2 - 32y^2 = 8(x^2 - 4y^2) = 8[(x)^2 - (2y)^2]$
$= 8(x + 2y)(x - 2y)$ **34.** $x^4 - 1 = (x^2)^2 - (1)^2$
$= (x^2 + 1)(x^2 - 1) = (x^2 + 1)[(x)^2 - (1)^2]$
$= (x^2 + 1)(x + 1)(x - 1)$ **41.** $c^2 - 14c + 49$
$= (c)^2 - 2(c)(7) + (7)^2 = (c - 7)^2$

Review exercises

1. $(x + 6)(x + 2)$ **2.** $(7a + 9)(7a - 9)$ **3.** $(x - 4y)(3a + b)$
4. $2x(x + 3)(x + 4)$ **5.** $(2a + 3)(5a + 3)$ **6.** $(2a - 5)^2$
7. $(xy + 5)(xy + 3)$ **8.** $(x + 2y)^2$

Exercise 4–5

Answers to odd-numbered problems

1. 4^3 **3.** 5^3 **5.** $(3x)^3$ **7.** $(a^2)^3$ **9.** $(2b^5)^3$
11. $(r + s)(r^2 - rs + s^2)$ **13.** $(2x + y)(4x^2 - 2xy + y^2)$
15. $(h - k)(h^2 + hk + k^2)$ **17.** $(a - 2)(a^2 + 2a + 4)$
19. $(x - 2y)(x^2 + 2xy + 4y^2)$ **21.** $(4x - y)(16x^2 + 4xy + y^2)$
23. $(3x - 2y)(9x^2 + 6xy + 4y^2)$
25. $(2a + 3b)(4a^2 - 6ab + 9b^2)$
27. $2(a + 2)(a^2 - 2a + 4)$ **29.** $2(x - 2)(x^2 + 2x + 4)$
31. $x^2(x + 3y)(x^2 - 3xy + 9y^2)$ **33.** $(x^2 + y)(x^4 - x^2y + y^2)$
35. $(a^3 - b)(a^6 + a^3b + b^2)$ **37.** $(x^4 - 3)(x^8 + 3x^4 + 9)$
39. $a^2(2b - a)(4b^2 + 2ab + a^2)$ **41.** $2(3r + s)(9r^2 - 3rs + s^2)$
43. $(xy - z)(x^2y^2 + xyz + z^2)$
45. $(a^5b^2 - 2c^3)(a^{10}b^4 + 2a^5b^2c^3 + 4c^6)$
47. $(ab + 2)(a^2b^2 - 2ab + 4)$
49. $(x^3y^4 + z^5)(x^6y^8 - x^3y^4z^5 + z^{10})$

Solutions to trial exercise problems

27. $2a^3 + 16 = 2(a^3 + 8)$
$= 2[(a)^3 + (2)^3]$
Then $2(\ \ +\ \ )[(\ \ )^2 - (\ \ )(\ \ ) + (\ \ )^2]$
and $= 2(a + 2)[(a)^2 - (a)(2) + (2)^2]$
$= 2(a + 2)(a^2 - 2a + 4)$
44. $x^3y^9 - 1 = (xy^3)^3 - (1)^3$
Then $(\ \ -\ \ )[(\ \ )^2 + (\ \ )(\ \ ) + (\ \ )^2]$
and $(xy^3 - 1)[(xy^3)^2 + (xy^3)(1) + (1)^2]$
$= (xy^3 - 1)(x^2y^6 + xy^3 + 1)$

Review exercises

1. $(a - 2)(a - 5)$ **2.** $(3a + b)(2x - y)$ **3.** $(x + 2y)^2$
4. $(3a + b)(2a - b)$ **5.** $5a(a - 3)(a - 5)$
6. $(2x + 3)(3x - 4)$

Exercise 4–6

Answers to odd-numbered problems

1. $(n + 7)(n - 7)$ **3.** $(7b + 1)(b + 5)$ **5.** $(xy + 4)(xy - 2)$
7. $(6 + y)(6 - y)$ **9.** $10(a - b)^2$ **11.** $4(a + 2b)(a - 2b)$
13. $(3a - b)(x + 2y)$ **15.** $(3x + 5)(2x - 1)$
17. $(2m - n)(3a + 2b)$ **19.** $(7b - 5)(b + 3)$
21. $(4x - 3)(x + 5)$ **23.** $6(x^2 - 4xy - 8y^2)$
25. $3xy(x + 5y)(m - 4n)$ **27.** $(3a - 5b)^2$
29. $3a(a^2 + 4)(a + 2)(a - 2)$ **31.** $3ab^3(a + b)^2$
33. $(3b + 7)(b - 13)$ **35.** $(3x + 2y)(a + 2b)$
37. $(6x - 1)(x + 2)$ **39.** $3x^2(x + 4)(x - 4)$
41. $(y + 3z)(y^2 - 3yz + 9z^2)$ **43.** $(x - y^3)(x^2 + xy^3 + y^6)$

Solutions to trial exercise problems

25. $3x^2y(m-4n)+15xy^2(m-4n)=3xy(m-4n)(x+5y)$ common factor of $3xy(m-4n)$ **27.** $9a^2-30ab+25b^2$ $=(3a)^2-2(3a)(5b)+(5b)^2=(3a-5b)^2$ **29.** $3a^5-48a$ $=3a(a^4-16)=3a[(a^2)^2-(4)^2]=3a(a^2+4)(a^2-4)$ $=3a(a^2+4)(a+2)(a-2)$

Review exercises

1. $\{-3\}$ **2.** $\{3\}$ **3.** $\{6\}$ **4.** $\left\{-\frac{3}{5}\right\}$

5. $\left\{-\frac{2}{3}\right\}$ **6.** $\left\{-\frac{1}{3}\right\}$ **7.** $\left\{\frac{1}{4}\right\}$ **8.** $\{0\}$

Exercise 4–7

Answers to odd-numbered problems

1. $\{-5,5\}$ **3.** $\{0,-6\}$ **5.** $\{0,7\}$ **7.** $\left\{3,-\frac{3}{2}\right\}$ **9.** $\left\{-\frac{1}{2},\frac{2}{3}\right\}$

11. $\left\{-\frac{1}{5},\frac{1}{5}\right\}$ **13.** $\{7,8\}$ **15.** $\left\{\frac{4}{3},\frac{8}{5}\right\}$ **17.** $\{-5,5\}$

19. $\left\{-1,0,\frac{3}{2}\right\}$ **21.** $\left\{-\frac{3}{5},\frac{1}{4},10\right\}$ **23.** $\{0,-4\}$ **25.** $\left\{0,\frac{5}{3}\right\}$

27. $\{0,-3\}$ **29.** $\{-3,3\}$ **31.** $\left\{0,-\frac{3}{2}\right\}$ **33.** $\{-5,5\}$

35. $\{-3,3\}$ **37.** $\{-4,4\}$ **39.** $\{-2,2\}$ **41.** $\left\{-\frac{3}{2},\frac{3}{2}\right\}$

43. $\{-8,2\}$ **45.** $\{-7\}$ **47.** $\{-7,2\}$ **49.** $\{-1,-2\}$

51. $\{-1,12\}$ **53.** $\{-4,8\}$ **55.** $\left\{-1,\frac{9}{2}\right\}$ **57.** $\left\{\frac{1}{3},\frac{1}{2}\right\}$

59. $\left\{\frac{4}{3},-\frac{3}{2}\right\}$ **61.** $\left\{-\frac{1}{2},\frac{3}{2}\right\}$ **63.** $\left\{-\frac{1}{9},1\right\}$

65. $\left\{-\frac{4}{3},\frac{5}{2}\right\}$ **67.** $\{-1\}$ **69.** $\left\{-\frac{5}{3}\right\}$ **71.** $\left\{-\frac{7}{3},4\right\}$

73. $\left\{\frac{3}{2},-3\right\}$ **75.** $\{-2,-1\}$ **77.** $\{-2,-1\}$ **79.** $\{-2,9\}$

81. $\left\{\frac{2}{3}\right\}$ **83.** $\{-1,5\}$

Solutions to trial exercise problems

1. $(x+5)(x-5)=0$
$x+5=0$ or $x-5=0$
$x=-5 \qquad x=5$
$\{-5,5\}$

7. $(3x-9)(2x+3)=0$
$3x-9=0$ or $2x+3=0$
$3x=9 \qquad 2x=-3$
$x=3 \qquad x=-\frac{3}{2}$
$\left\{3,-\frac{3}{2}\right\}$

15. $(4-3u)(8-5u)=0$
$4-3u=0$ or $8-5u=0$
$4=3u \qquad 8=5u$
$\frac{4}{3}=u \qquad \frac{8}{5}=u$
$\left\{\frac{4}{3},\frac{8}{5}\right\}$

31. $10a^2=-15a$
$10a^2+15a=0$
$5a(2a+3)=0$
$5a=0$ or $2a+3=0$
$a=0 \qquad 2a=-3$
$a=-\frac{3}{2}$
$\left\{0,-\frac{3}{2}\right\}$

40. $5y^2-45=0$
$5(y^2-9)=0$
$5(y+3)(y-3)=0$
$y+3=0$ or $y-3=0$
$y=-3 \qquad y=3$
$\{3,-3\}$

47. $b^2+5b-14=0$
$(b+7)(b-2)=0$
$b+7=0$ or $b-2=0$
$b=-7 \qquad b=2$
$\{-7,2\}$

52. $x^2-14x=15$
$x^2-14x-15=0$
$(x-15)(x+1)=0$
$x-15=0$ or $x+1=0$
$x=15 \qquad x=-1$
$\{15,-1\}$

59. $6x^2+x-12=0$
$(3x-4)(2x+3)=0$
$3x-4=0$ or $2x+3=0$
$3x=4 \qquad 2x=-3$
$x=\frac{4}{3} \qquad x=\frac{-3}{2}$
$\left\{\frac{4}{3},\frac{-3}{2}\right\}$

64. $-6x=-3x^2-3$
$3x^2-6x+3=0$
$3(x^2-2x+1)=0$
$3(x-1)^2=0$
$x-1=0$
$x=1$
$\{1\}$

71. $3x^2-4x-28=x$
$3x^2-5x-28=0$
$(3x+7)(x-4)=0$
$3x+7=0$ or $x-4=0$
$3x=-7 \qquad x=4$
$x=\frac{-7}{3} \qquad x=4$
$\left\{\frac{-7}{3},4\right\}$

76. $3x(3x+2)=24$
$9x^2+6x=24$
$9x^2+6x-24=0$
$3(3x^2+2x-8)=0$
$3(3x-4)(x+2)=0$
$3x-4=0$ or $x+2=0$
$3x=4 \qquad x=-2$
$x=\frac{4}{3} \qquad x=-2$
$\left\{\frac{4}{3},-2\right\}$

84.
$$x(x+7) = 36 - 2x$$
$$x^2 + 7x = 36 - 2x$$
$$x^2 + 9x - 36 = 0$$
$$(x+12)(x-3) = 0$$
$$x + 12 = 0 \text{ or } x - 3 = 0$$
$$x = -12 \qquad x = 3$$
$$\{-12, 3\}$$

Review exercises

1. (let x = the number) $x + 7$ **2.** (let x = the number) $x - 11$

3. $6(x^2 + x)$ **4.** $\frac{x^2 + 2x}{8}$ **5.** 13 **6.** 6, 32 **7.** 22, 24, 26

Exercise 4–8

Answers to odd-numbered problems

1. 11 and 13; -13 and -11 **3.** 16 and 18; -18 and -16
5. 7 and 13; -13 and -7 **7.** 4 and 6; -2 and 0
9. -4 and -9 **11.** 2 and 7; -6 and $-\frac{7}{3}$ **13.** $w = 4$ meters; $\ell = 6$ meters **15.** $w = 2$ feet; $\ell = 5$ feet **17.** $w = 1$ inch or $w = 6$ inches **19. a.** $t = 2$ **b.** $t = 1$ **21. a.** $I = 12$ or $I = 10$ **b.** $I = 12$ or $I = 20$ **23. a.** $t = 2$ **b.** $t = 4$
25. $t = 1$ second (on way up); $t = 5$ seconds (on way down)
27. $t = 3$ seconds **29.** $b_2 = 8$ inches; $h = 7$ inches
31. $b_2 = 4$ inches; $b_1 = 8$ inches; $h = 4$ inches **33.** $w = 7$ feet; $\ell = 8$ feet **35.** $w = 6$ feet; $\ell = 18$ feet

Solutions to trial exercise problems

7. Let x = the lesser integer.
Let $x + 2$ = the next consecutive even integer.
Then $x(x+2) = 2[x + (x+2)] + 4$
$$x^2 + 2x = 2(2x+2) + 4$$
$$x^2 + 2x = 4x + 4 + 4$$
$$x^2 + 2x = 4x + 8$$
$$x^2 - 2x - 8 = 0$$
$$(x-4)(x+2) = 0$$
$$x - 4 = 0 \text{ or } x + 2 = 0$$
$$x = 4 \qquad x = -2$$
$$x + 2 = 6 \qquad x + 2 = 0$$
The consecutive even integers are 4 and 6 or -2 and 0.

10. Let x = one integer.
Then $-3 - x$ = the other integer.
$$x(-3 - x) = -70$$
$$-3x - x^2 = -70$$
$$0 = x^2 + 3x - 70 = (x+10)(x-7)$$
$$x + 10 = 0 \text{ or } x - 7 = 0$$
$$x = -10 \qquad x = 7$$
$$-3 - x = -3 - (-10) = 7 \text{ or } -3 - x = -3 - 7 = -10$$
The integers are -10 and 7.

15. Let w = the width. Let $\ell = w + 3$. From "the area of a rectangle is numerically equal to twice the length," we get
$$w(w+3) = 2(w+3)$$
$$w^2 + 3w = 2w + 6$$
$$w^2 + w - 6 = 0$$
$$(w+3)(w-2) = 0$$
$$w + 3 = 0 \text{ or } w - 2 = 0$$
$$w = -3 \qquad w = 2$$
(*Note:* A geometric figure cannot have a negative width, so we ignore $w = -3$.)
The width is 2 feet and the length is $w + 3 = 5$ feet.

22a. $P = 100I - 5I^2$, when $P = 480$.
$$480 = 100I - 5I^2$$
$$5I^2 - 100I + 480 = 0$$
$$5(I^2 - 20I + 96) = 0$$
$$5(I-8)(I-12) = 0$$
$$I - 8 = 0 \text{ or } I - 12 = 0$$
$$I = 8 \qquad I = 12$$
Therefore $I = 8$ amperes or $I = 12$ amperes.

25. $s = v_0t - 16t^2$, given $s = 80$ and $v_0 = 96$
$$80 = 96t - 16t^2$$
$$16t^2 - 96t + 80 = 0$$
$$16(t^2 - 6t + 5) = 0$$
$$16(t-5)(t-1) = 0$$
$$t - 5 = 0 \text{ or } t = 1$$
$$t = 5 \qquad t = 1$$
So $t = 1$ second on the way up,
and $t = 5$ seconds on the way down.

30. Given $A = \frac{1}{2}h(b_1 + b_2)$, $A = 21$ square feet, $b_2 = 5$ feet, and using "b_1 is 6 feet longer than the altitude h," we have $b_1 = h + 6$.
Substitute $21 = \frac{1}{2}h[(h+6) + 5]$
$$21 = \frac{1}{2}h(h+11)$$
Multiply both members by 2 to get $42 = h(h+11)$
Then $42 = h^2 + 11h$
Add -42 to both members to get $0 = h^2 + 11h - 42$
So $h^2 + 11h - 42 = 0$
Factor in the left member: $(h+14)(h-3) = 0$
Then $h + 14 = 0$ or $h - 3 = 0$
So $h = -14$ or $h = 3$.
A trapezoid cannot have a negative altitude, so we ignore $h = -14$. The altitude of the trapezoid is 3 feet.

33. Given $V = \ell wh$, $V = 224$ cubic feet and $h = 4$ feet, from "the length is 1 foot longer than the width," $\ell = w + 1$.
Substituting, $224 = w(w+1) \cdot 4$
Multiplying in the right member, $224 = 4w^2 + 4w$
Add -224 to both members and interchange members.
$$0 = 4w^2 + 4w - 224$$
$$4w^2 + 4w - 224 = 0$$
Factor 4 from each term to get $4(w^2 + w - 56) = 0$
Then $4(w+8)(w-7) = 0$ and $w + 8 = 0$ or $w - 7 = 0$
So $w = -8$ or $w = 7$.
The width of a box cannot be negative, so we discard $w = -8$.
The width is then 7 feet and the length is $w + 1 = 8$ feet.

Review exercises

1. $\left\{\frac{2}{3}\right\}$ **2.** $\left\{\frac{5}{4}\right\}$ **3.** $\{-3, 3\}$ **4.** $\{-4, -2\}$ **5.** 18
6. -27 **7.** 0 **8.** undefined

Chapter 4 review

1. $3(x+3)$ **2.** $9x(x-2)$ **3.** $-4y^2(y-2)$
4. $7(a + 2b - 4c)$ **5.** $-5a(a + 3 - 6a^2)$ **6.** $3(a^2 - ab + b)$
7. $xy(x + z + yz)$ **8.** $a^3b(1+b)$ **9.** $R^2(3R - 1 + 5R^2)$
10. $4y(y + 2 + 3y^2)$ **11.** $x^2(x^2 + 3x + 9)$
12. $4R^2S^2(4R - 3R^2S + 6)$ **13.** $5a^2b^2(2a^2b + 3 - 4a)$
14. $(a+b)(2+x)$ **15.** $(x - 3z)(y + 4)$

16. $(3R + 1)(a + b)$ **17.** $(x - 3y)(2a - 3b)$
18. $(3a - b)(2x - y)$ **19.** $(x + 2y)(4a + 3b)$
20. $(a + 3b)(x - 4)$ **21.** $(x^2 + 4)(a - 2b)$
22. $(x - 7)(x - 2)$ **23.** $2a(a - 5)(a + 1)$
24. $(a + 12)(a + 2)$ **25.** $(x - 8)(x + 4)$
26. $(a - 18)(a + 2)$ **27.** $3(x - 5)(x + 2)$
28. $x(x - 3)(x + 2)$ **29.** $x(x - 7)(x + 3)$
30. $(ab + 3)(ab - 2)$ **31.** $(ab + 6)(ab + 4)$
32. $(ab - 6)(ab - 3)$ **33.** $(ab - 10)(ab + 2)$
34. $(2x + 1)(2x + 1)$ or $(2x + 1)^2$ **35.** $9(r - 2)(r - 2)$ or $9(r - 2)^2$ **36.** $(4x - 1)(x - 1)$ **37.** $(3a + 5)(3a - 2)$
38. $(4a - 3)(2a + 1)$ **39.** $(6x + 1)(4x + 3)$
40. $(4a - 3)(2a - 3)$ **41.** $(2a + 3)(a + 6)$
42. $(2a + 3)(2a - 3)$ **43.** $(6b + c)(6b - c)$
44. $(5 + a)(5 - a)$ **45.** $4(2x + y)(2x - y)$
46. $(3x + y^2)(3x - y^2)$ **47.** $(x^2 + 4)(x + 2)(x - 2)$
48. $(y^2 + 9)(y + 3)(y - 3)$ **49.** $(b + 6)^2$ **50.** $(c - 5)^2$
51. $(2x - 3)^2$ **52.** $(3x - 2)^2$ **53.** $(R + 2S)(R^2 - 2RS + 4S^2)$
54. $2(2x - 3)(4x^2 + 6x + 9)$ **55.** $(3a + 5b)(9a^2 - 15ab + 25b^2)$
56. $(xy - 1)(x^2y^2 + xy + 1)$ **57.** $2(x^3 + 5)(x^6 - 5x^3 + 25)$
58. $(4x^4 - y^5)(16x^8 + 4x^4y^5 + y^{10})$
59. $(ab^2 + c^3)(a^2b^4 - ab^2c^3 + c^6)$ **60.** $3x^3(4x - 1)$
61. $(a - 5)(a + 2)$ **62.** $(4a - 1)(a - 5)$
63. $(3y + 2)(3y - 2)$ **64.** $(2a + 3b)(3x - 2)$
65. $(b - 5)(b + 4)$ **66.** $(3x + 2)(3x + 5)$ **67.** $(a + 7)^2$
68. $3x^3(2x + 1)(2x - 1)$ **69.** $c(c + 4)(c + 5)$ **70.** $(4a - 1)^2$
71. $(b^2 + 1)(b + 1)(b - 1)$ **72.** $\{1,-3\}$ **73.** $\{0,8\}$
74. $\left\{-\frac{1}{5},\frac{7}{3}\right\}$ **75.** $\left\{\frac{1}{7},\frac{8}{5}\right\}$ **76.** $\left\{\frac{4}{3},9\right\}$ **77.** $\{0,-9,-4\}$
78. $\left\{\frac{4}{5},-\frac{4}{5}\right\}$ **79.** $\{-1,3,-2\}$ **80.** $\left\{0,\frac{9}{4}\right\}$ **81.** $\{-1,1\}$
82. $\{0,64\}$ **83.** $\{-5,5\}$ **84.** $\{6,-5\}$ **85.** $\{1\}$ **86.** $\left\{-\frac{1}{4},-3\right\}$
87. $\left\{-\frac{2}{5},2\right\}$ **88.** $\{4\}$ **89.** $\{1,3\}$ **90.** $\left\{-1,\frac{3}{4}\right\}$ **91.** $\{2,-3\}$
92. 9,10 **93.** 8 feet and 13 feet **94.** 20 cattle **95.** 6 seconds

Chapter 4 cumulative test

1. 33 **2.** $8a^6b^3$ **3.** $a^2 + 4ab + 4b^2$ **4.** a^6 **5.** 12 **6.** x^5y^4
7. $5x - 7y$ **8.** $9x^2 - 4y^2$ **9.** $\frac{8a^6}{b^3}$ **10.** $\frac{1}{x^2}$ **11.** $4x + 4y$
12. $\frac{x^6}{9y^4}$ **13. a.** 0 **b.** 0 **c.** -6 **14.** $\left\{\frac{9}{5}\right\}$ **15.** $x < 12$
16. $\{7\}$ **17.** $x \geq 3$ **18.** $\{-3,3\}$ **19.** $\{2,5\}$ **20.** $x < \frac{5}{2}$
21. $2 \leq x \leq \frac{14}{3}$ **22.** $x = 3y$ **23.** $x = \frac{5y + 2}{3}$
24. $2ab(1 - 2ab - 4a^2b^4)$ **25.** $(2a + 3)^2$
26. $(5c + 3d)(5c - 3d)$ **27.** $(2a + 3)(2a - 5)$
28. $(x + 3)(x + 6)$ **29.** 14,39 **30.** 11,13
31. \$10,000 at 8%; \$5,000 at 6% **32.** 6 meters by 11 meters

Chapter 5

Exercise 5–1

Answers to odd-numbered problems

1. $\frac{1}{3}$ **3.** undefined **5.** 4 **7.** $\frac{13}{23}$ **9.** $\frac{16}{17}$ **11.** 3
13. undefined **15.** $\frac{15}{2}$ **17.** all real numbers except 0
19. all real numbers except 5 **21.** all real numbers except -3
23. all real numbers except $\frac{3}{4}$ **25.** all real numbers except $\frac{8}{3}$
27. all real numbers except 1 and 6 **29.** all real numbers except -2 and $\frac{4}{3}$ **31.** all real numbers except $-\frac{2}{3}$ and $\frac{2}{3}$
33. all real numbers except -3 and 3 **35.** all real numbers except $\frac{7}{3}$ **37.** all real numbers except -1 and 1 **39.** all real numbers **41.** $L = 0$ **43.** $T_1 \neq 0, T_2 \neq 0$

Solutions to trial exercise problems

6. $\frac{-5b^3}{5 - 2b}$; $b = -2$

$$\frac{-5b^3}{5 - 2b} = \frac{-5(-2)^3}{5 - 2(-2)} = \frac{-5(-8)}{5 + 4} = \frac{40}{9}$$

9. $\frac{(-2x)^2}{x^2 + 3x + 7}$; $x = 2$

$$\frac{(-2x)^2}{x^2 + 3x + 7} = \frac{[(-2)(2)]^2}{(2)^2 + 3(2) + 7} = \frac{(-4)^2}{4 + 6 + 7} = \frac{16}{17}$$

22. $\frac{x + 1}{2x - 1}$ Set $2x - 1 = 0$, then $2x = 1$ and $x = \frac{1}{2}$.

Domain is all real numbers except $\frac{1}{2}$.

25. $\frac{y + 4}{8 - 3y}$ Set $8 - 3y = 0$, then $3y = 8$ and $y = \frac{8}{3}$.

Domain is all real numbers except $\frac{8}{3}$.

28. $\frac{5s^2 + 7}{2s^2 - s - 3}$ Set $2s^2 - s - 3 = 0$ and factor. We have

$(2s - 3)(s + 1) = 0$. Then
$2s - 3 = 0$ or $s + 1 = 0$
$2s = 3$ $\quad s = -1$
$s = \frac{3}{2}$ $\quad s = -1$

Domain is all real numbers except -1 and $\frac{3}{2}$.

32. $\frac{a - 2}{4a^2 - 16}$ Set $4a^2 - 16 = 0$ and factor. We have

$4(a - 2)(a + 2) = 0$.
This is true if and only if $a - 2 = 0$, $a = 2$ or $a + 2 = 0$, $a = -2$.
Domain is all real numbers except -2 and 2.

37. $\frac{17q}{3q^2 - 3}$ Set $3q^2 - 3 = 0$ and factor to get

$3(q^2 - 1) = 0$
$3(q + 1)(q - 1) = 0$
$q + 1 = 0$ or $q - 1 = 0$
Then $q = -1$ or $q = 1$
Domain is all real numbers except -1 and 1.

Review exercises

1. distributive property **2.** -50 **3.** 5 **4.** $\frac{5}{9}$

5. $(2x + 1)(x - 5)$ **6.** $4(y - 5)^2$

Exercise 5-2

Answers to odd-numbered problems

1. $\frac{3}{4}$ **3.** $\frac{2x}{5}$ **5.** $\frac{4x}{3}$ **7.** $-\frac{4}{3x^2}$ **9.** $\frac{4a}{5b}$ **11.** -5 **13.** $\frac{5}{4}$

15. $\frac{6}{x + 3}$ **17.** $\frac{1}{a - b}$ **19.** $\frac{3}{5}$ **21.** $\frac{x - 1}{2(x + 1)}$ **23.** $\frac{x - 3}{x + 3}$

25. $\frac{x - 5}{x - 3}$ **27.** $\frac{2y + 3}{4y - 1}$ **29.** $\frac{1}{x^2 + 3x + 9}$ **31.** $\frac{a - b}{a^2 - ab + b^2}$

33. -4 **35.** $-\frac{2}{3}$ **37.** $-2(x + y)$ **39.** $\frac{y - x}{x + y}$ **41.** $\frac{n - m}{n + m}$

43. $-\frac{4}{x + y}$ **45.** $-\frac{1}{x + 4}$

Solutions to trial exercise problems

10. $\frac{15a^2x^3}{35ax^2} = \frac{3 \cdot 5 \cdot a \cdot a \cdot x \cdot x \cdot x}{7 \cdot 5 \cdot a \cdot x \cdot x} = \frac{3 \cdot a \cdot x \cdot (5 \cdot a \cdot x \cdot x)}{7 \cdot (5 \cdot a \cdot x \cdot x)} = \frac{3ax}{7}$

17. $\frac{a + b}{a^2 - b^2} = \frac{a + b}{(a + b)(a - b)} = \frac{1}{a - b}$

25. $\frac{x^2 - 3x - 10}{x^2 - x - 6} = \frac{(x - 5)(x + 2)}{(x - 3)(x + 2)} = \frac{x - 5}{x - 3}$

30. $\frac{x + 2}{x^3 + 8} = \frac{x + 2}{(x + 2)(x^2 - 2x + 4)} = \frac{1}{x^2 - 2x + 4}$

34. $\frac{8b - 8a}{a - b} = \frac{-8(a - b)}{a - b} = -8$

45. $\frac{x - 3}{12 - x - x^2} = \frac{x - 3}{(3 - x)(4 + x)} = \frac{-1(3 - x)}{(3 - x)(4 + x)} = \frac{-1}{x + 4}$

Review exercises

1. 3.14×10^{-4} **2.** $-\frac{11}{7}$ **3.** 5 ft and 11 ft **4.** 1 **5.** $27y^3$

6. $-8x^9y^6$

Exercise 5-3

Answers to odd-numbered problems

1. $4x^2$ **3.** $-5x^3yz$ **5.** $3(a - b)$ **7.** $2a(b - c)$

9. $-4a^2b(x + y)$ **11.** $-4a^2 + 2a$ **13.** $a^2 - 3a + 2$

15. $5a^2 - 3a + 4 - \frac{2}{a}$ **17.** $-x + y + 2y^2$

19. $10xy^2 + 7 - 6y^2$ **21.** $a - c$ **23.** $a + 9 + \frac{28}{a - 2}$

25. $a + 2 + \frac{4}{a + 3}$ **27.** $a + 3 + \frac{1}{a + 3}$

29. $3a - 4 - \frac{4}{3a - 4}$ **31.** $x^2 + 2x + 4$ **33.** $x^2 + 2x + 3$

35. $b^2 + 7b + 14 + \frac{6}{b - 1}$ **37.** $3a + 8$

39. $x^2 - 1 + \frac{6x - 13}{x^2 + 3x - 5}$ **41.** $y^2 - 2y + 9 + \frac{-26y + 42}{y^2 + 2y - 5}$

43. $4x + 3$ **45.** $-6x^4 + 15x^3 + 4x^2 - 22x + 30$

Solutions to trial exercise problems

5. $\frac{3(a - b)^2}{a - b} = \frac{3(a - b)^2}{(a - b)} = 3(a - b)^{2-1} = 3(a - b)^1 = 3(a - b)$

or $3a - 3b$ (*Note:* A quantity is treated as just one term.)

15. $\frac{15a^3 - 9a^2 + 12a - 6}{3a} = \frac{15a^3}{3a} - \frac{9a^2}{3a} + \frac{12a}{3a} - \frac{6}{3a}$

$= 5a^2 - 3a + 4 - \frac{2}{a}$

17. $\frac{x^2y - xy^2 - 2xy^3}{-xy} = \frac{x^2y}{-xy} - \frac{xy^2}{-xy} - \frac{2xy^3}{-xy} = -x + y + 2y^2$

21. $\frac{a(b - 1) - c(b - 1)}{b - 1} = \frac{a(b - 1)}{(b - 1)} - \frac{c(b - 1)}{(b - 1)} = a - c$

30. $(27a^3 - 1) \div (3a - 1)$

(*Note:* Insert zeros to hold positions where terms are missing.)

$$\begin{array}{r} 9a^2 + 3a + 1 \\ 3a - 1 \overline{\smash{)}\, 27a^3 + 0a^2 + 0a - 1} \\ \underline{27a^3 - 9a^2} \qquad\qquad \\ 9a^2 + 0a \qquad \\ \underline{9a^2 - 3a} \qquad \\ 3a - 1 \\ \underline{3a - 1} \\ 0 \end{array}$$

Answer: $9a^2 + 3a + 1$

42. The length times the width is $x^2 + 6x + 8$. If we know the length to be $x + 4$, then the width is found by

$$\begin{array}{r} x + 2 \\ x + 4 \overline{\smash{)}\, x^2 + 6x + 8} \\ \underline{x^2 + 4x} \qquad \\ 2x + 8 \\ \underline{2x + 8} \\ 0 \end{array}$$

Width $= x + 2$

Review exercises

1. $\left\{-2, -\frac{1}{4}\right\}$ **2.** 6 ft and 7 ft **3.** $16x^2 - 24x + 9$

4. $x^3 + 3x^2 + x - 2$ **5.** $25y^2 - 1$ **6.** $\frac{3y + 1}{2y + 3}$

Exercise 5-4

Answers to odd-numbered problems

1. $\frac{12}{7}$; 12:7 **3.** $\frac{1}{6}$; 1:6 **5.** $\frac{8}{3}$; 8:3 **7.** $\frac{3}{1}$; 3:1

9. $\frac{32}{3}$; 32:3 **11.** $\frac{1}{4}$; 1:4 **13.** $\frac{5}{4}$; 5:4 **15.** $\frac{39}{32}$; 39:32

17. $\frac{2}{1}$; 2:1 **19.** $\frac{3}{7}$; 3:7 **21.** $\frac{3}{7}$; 3:7 **23.** $\frac{5}{2}$; 5:2

25. $\frac{3}{2}$; 3:2 **27.** $\frac{5}{12}$; 5:12 **29.** $\frac{60}{7}$; 60:7 **31.** $\frac{1}{26}$; 1:26

33. $\frac{2 \text{ lb}}{1 \text{ ft}^3} = 2 \text{ lb}:1 \text{ ft}^3$ **35.** $\frac{8}{1}\left(\frac{\text{g}}{\text{cm}^3}\right)$; 8 g:1 cm^3

37. $\frac{60}{1}\left(\frac{\text{miles}}{\text{hr}}\right)$; 60 miles:1 hr **39.** $\frac{15}{17}$ **41.** $\frac{5}{12}$; $\frac{12}{5}$

43. $\frac{8}{3}$ **45.** $\frac{13}{14}$ **47.** $\frac{21}{2}$ **49.** $\frac{31}{2}$ **51.** $\frac{7}{50}$

53. 7 ft-lb/sec **55.** $x = \frac{5}{4}$ **57.** $p = \frac{100}{9}$ **59.** $x = \frac{16}{5}$

61. $a = \frac{12}{19}$ **63.** $x = 3$ **65.** $a = \frac{50}{13}$ **67.** 7 weeks

69. 36 grams of hydrogen **71.** 390 **73.** $6{,}826\frac{2}{3}$ ohms

75. 75 losses **77.** $x = \frac{30}{7}$ inches, $y = \frac{40}{7}$ inches

79. 3,240 holes **81.** 270 minutes $= 4\frac{1}{2}$ hours **83.** $2\frac{1}{2}$ quarts

Solutions to trial exercise problems

9. 8 to $\frac{3}{4}$ is written $\frac{8}{\frac{3}{4}}$ or $8:\frac{3}{4}$

but $\frac{8}{\frac{3}{4}} = 8 \cdot \frac{4}{3} = \frac{32}{3}$

Therefore we have the ratio $\frac{32}{3}$ or 32:3

13. $\frac{5}{6}$ to $\frac{2}{3}$ is written $\frac{\frac{5}{6}}{\frac{2}{3}} = \frac{5}{6} \cdot \frac{3}{2} = \frac{5}{4}$

Therefore we have the ratio $\frac{5}{4}$ or 5:4

17. 4.2 to 2.1 is written $\frac{4.2}{2.1} = \frac{42}{21} = \frac{2}{1}$

so we have the ratio $\frac{2}{1}$ or 2:1

27. Since there are $3 \cdot 12 = 36$ inches in 3 feet, we have

15 inches to 36 inches written $\frac{15}{36} = \frac{5}{12}$ or 5:12

36. 300 miles to 10 gallons is written $\frac{300 \text{ miles}}{10 \text{ gallons}} = \frac{30 \text{ miles}}{1 \text{ gallon}}$,

which we state as 30 miles per gallon.

39. $ME = \frac{\text{output}}{\text{input}} = \frac{375}{425} = \frac{15}{17}$

Therefore the mechanical efficiency is $\frac{15}{17}$

44. $\frac{\text{Cutting speed of tool steel}}{\text{Cutting speed of cast iron}} = \frac{20 \text{ ft/min}}{45 \text{ ft/min}} = \frac{4}{9}$

53. $\frac{F}{t} = \frac{42 \text{ ft-lb}}{6 \text{ sec}} = 7$ ft-lb per sec

55. $\frac{9}{x} = \frac{36}{5}$, then $36 \cdot x = 9 \cdot 5$

$$36x = 45$$

$$x = \frac{45}{36} = \frac{5}{4}$$

Therefore $\frac{9}{\frac{5}{4}} = \frac{36}{5}$

59. $6:15 = x:8$, then $15 \cdot x = 6 \cdot 8$

$$15x = 48$$

$$x = \frac{48}{15} = \frac{16}{5}$$

Therefore $6:15 = \frac{16}{5}:8$

62. $\frac{3}{4}:4 = \frac{1}{2}:b$, then $\frac{3}{4} \cdot b = 4 \cdot \frac{1}{2}$

$$\frac{3}{4}b = 2$$

$$b = 2 \cdot \frac{4}{3} = \frac{8}{3}$$

so $\frac{3}{4}:4 = \frac{1}{2}:\frac{8}{3}$

68. Let $x =$ the number of liters of gasoline to travel 1,428 kilometers.

Then $\frac{8 \text{ liters}}{84 \text{ km}} = \frac{x \text{ liters}}{1{,}428 \text{ km}}$ and

$$x \cdot 84 = 8 \cdot 1{,}428$$

$$84x = 11{,}424$$

$$x = \frac{11{,}424}{84} = 136$$

Therefore at the same rate of gasoline consumption, it would take 136 liters to travel 1,428 kilometers.

76. Let $\ell =$ the length of the enlargement.

Then $\frac{10 \text{ in.}}{8 \text{ in.}} = \frac{\ell \text{ in.}}{36 \text{ in.}}$ and $8 \cdot \ell = 10 \cdot 36$

$$8\ell = 360$$

$$\ell = \frac{360}{8} = 45$$

Therefore the enlargement will be 45 inches long.

82. Let $x =$ the number of feet represented by $2\frac{5}{8}$ inches.

Then $\frac{\frac{1}{8} \text{ in.}}{1 \text{ ft}} = \frac{2\frac{5}{8} \text{ in.}}{x \text{ ft}}$ and $\frac{1}{8} \cdot x = 1 \cdot 2\frac{5}{8}$

$$\frac{1}{8}x = 2\frac{5}{8}$$

$$x = \frac{21}{8} \cdot 8 = 21$$

Therefore $2\frac{5}{8}$ inches represents 21 feet.

Review exercises

1. $\left\{-\frac{9}{5}\right\}$ **2.** $w = \frac{P - 2\ell}{2}$ **3.** 11 and 13; -13 and -11
4. $x \le 4$ **5.** $(4x + y)(4x - y)$ **6.** $(x - 17)(x + 1)$
7. $5(x - 2)(x + 1)$

Chapter 5 review

1. All real numbers except 0 **2.** All real numbers except -7
3. All real numbers except 9 **4.** All real numbers except $-\frac{2}{3}$
5. All real numbers except $\frac{3}{5}$ **6.** All real numbers except -4 and 3 **7.** All real numbers except -1 and 1 **8.** $\frac{3b}{a}$ **9.** $\frac{3xz}{2y^2}$
10. $\frac{x - 7}{x + 7}$ **11.** $\frac{x + 3}{x + 7}$ **12.** $\frac{6}{5}$ **13.** $-(x + y)$ **14.** $\frac{3p - 2}{5p + 1}$

15. $\frac{R-4}{3R-1}$ **16.** $\frac{5-n}{2+n}$ **17.** $-8x^2$ **18.** $2a-3+5a^2$
19. $5x-3y^3+xy$ **20.** $2a+3b-6ab^6$
21. $4a+1+\frac{-2}{2a-1}$ **22.** $3a-2+\frac{1}{a-5}$ **23.** $x-7$
24. $5x^2-x-4$ **25.** $\frac{3}{7}$ or 3:7 **26.** $\frac{9}{4}$ or 9:4 **27.** $\frac{2}{5}$ or 2:5
28. $30\frac{\text{miles}}{\text{gallon}}$ or 30 miles per gallon **29.** $\frac{2{,}193}{881}$ or 2,193:881
30. $\frac{7}{3}$ or 7:3 **31.** $x=32$ **32.** $a=3.6$ **33.** $y=\frac{54}{5}$
34. $p=\frac{2}{5}$ **35.** 35 feet **36.** 12 quarts antifreeze, 4 quarts water

Chapter 5 cumulative test

1. 1,756 **2.** undefined **3.** -36 **4.** $5a+3b$
5. $9x^2-12x+4$ **6.** $25y^2-4$ **7.** $4a^2-21ab-18b^2$
8. $x^3-3x^2y+3xy^2-y^3$ **9.** -33 **10.** $\{-2\}$ **11.** $\{8\}$
12. $-\frac{9}{2}\le x<3$ **13.** $x>3$ **14.** $(m+n)(x-y)$
15. $(3a+4)(a+1)$ **16.** $(2x-5)^2$ **17.** $(2z+3)(2z-3)$
18. $(6-y)(6+y)$ **19.** $(x-15)(x+3)$
20. $2(a-2b)(a^2+2ab+4b^2)$ **21.** $\{7,-2\}$ **22.** $\left\{\frac{3}{2},-3\right\}$
23. $\frac{z^3}{64y^3}$ **24.** a^3 **25.** $8a^3$ **26.** $x=24$ **27.** $\frac{13}{6}$ or 13:6
28. $x-6+\frac{1}{x-2}$ **29.** $\frac{4}{3}$ **30.** $\frac{9b}{7a^2}$ **31.** $\frac{a-6}{a-7}$
32. $\frac{8}{5(x+y)}$ **33.** $\frac{y+4}{y+5}$ **34.** $\frac{x-3y}{3x+2y}$ **35.** $21\frac{1}{3}$ inches
36. $12\frac{6}{7}$ feet

Chapter 6

Exercise 6–1

Answers to odd-numbered problems

1. $\frac{3}{5}$ **3.** $\frac{5}{6}$ **5.** $2a$ **7.** 10 **9.** $\frac{x}{4y}$ **11.** $\frac{3x}{4}$ **13.** $\frac{4}{35x}$
15. $\frac{35x}{4}$ **17.** $\frac{7c}{2b}$ **19.** $\frac{x}{6y^2}$ **21.** $\frac{16bcx}{3az}$ **23.** $\frac{2a}{5b^2x}$
25. $\frac{4}{x+y}$ **27.** $-\frac{3}{4}$ **29.** $-\frac{15}{4}$ **31.** $\frac{4}{a-5}$ **33.** $\frac{18(x-2)}{x+2}$
35. $\frac{24y(x-2)}{25}$ **37.** $(r-4)(r-1)$ **39.** $-4(x+3)$
41. $\frac{a-3}{a-5}$ **43.** $\frac{(x-3)(x+1)}{(x-1)(x+2)}$ **45.** $\frac{(2x-1)(x-1)}{(x-8)(x+7)}$
47. 1 **49.** $3x+4$ **51.** $\frac{1}{(2x+1)^2}$ **53.** $\frac{2a+12}{3a^2+9a+27}$
55. $\frac{(z-7)(z+3)}{5(z^2-2z+4)}$ **57.** $y+5$

Solutions to trial exercise problems

10. $\frac{7a}{12b}\cdot\frac{9b}{28}=\frac{7a\cdot 9b}{12b\cdot 28}=\frac{7\cdot a\cdot 3\cdot 3\cdot b}{2\cdot 2\cdot 3\cdot b\cdot 2\cdot 2\cdot 7}$
$=\frac{3\cdot a\cdot(7\cdot 3\cdot b)}{2\cdot 2\cdot 2\cdot 2\cdot(7\cdot 3\cdot b)}=\frac{3a}{16}$

20. $\frac{28m}{15n}\div\frac{7m^2}{3n^3}=\frac{28m\cdot 3n^3}{15n\cdot 7m^2}=\frac{4n^2}{5m}$

21. $\frac{24abc}{7xyz^2}\cdot\frac{14x^2yz}{9a^2}=\frac{3\cdot 8\cdot a\cdot b\cdot c\cdot 2\cdot 7\cdot x^2\cdot y\cdot z}{7\cdot x\cdot y\cdot z^2\cdot 3\cdot 3\cdot a^2}$
$=\frac{8\cdot b\cdot c\cdot 2\cdot x(3\cdot 7\cdot a\cdot x\cdot y\cdot z)}{z\cdot 3\cdot a(3\cdot 7\cdot a\cdot x\cdot y\cdot z)}=\frac{8\cdot b\cdot c\cdot 2\cdot x}{z\cdot 3\cdot a}=\frac{16bcx}{3az}$

25. $\frac{x+y}{3}\cdot\frac{12}{(x+y)^2}=\frac{2\cdot 2\cdot 3(x+y)}{3(x+y)^2}=\frac{4}{x+y}$

30. $\frac{8y+16}{3-y}\cdot\frac{4y-12}{3y+6}=\frac{8(y+2)\cdot 4(y-3)}{-(y-3)\cdot 3(y+2)}=\frac{8\cdot 4}{-3}=-\frac{32}{3}$

37. $\frac{r^2-16}{r+1}\div\frac{r+4}{r^2-1}=\frac{(r^2-16)(r^2-1)}{(r+1)(r+4)}$
$=\frac{(r-4)(r+4)(r+1)(r-1)}{(r+1)(r+4)}=\frac{(r-4)(r-1)}{1}$
$=r^2-5r+4$

39. $\frac{9-x^2}{x+y}\cdot\frac{4x+4y}{x-3}=\frac{(3-x)(3+x)\cdot 4(x+y)}{(x+y)(x-3)}$
$=\frac{-1(x-3)(x+3)\cdot 4(x+y)}{(x+y)(x-3)}=\frac{-1(x+3)\cdot 4}{1}$
$=-4(x+3)=-4x-12$

41. $\frac{a^2-5a+6}{a^2-9a+20}\cdot\frac{a^2-5a+4}{a^2-3a+2}$
$=\frac{(a-3)(a-2)\cdot(a-4)(a-1)}{(a-4)(a-5)\cdot(a-2)(a-1)}=\frac{a-3}{a-5}$

47. $\frac{6r^2-r-7}{12r^2+16r-35}\div\frac{r^2-r-2}{2r^2+r-10}$
$=\frac{(6r^2-r-7)(2r^2+r-10)}{(12r^2+16r-35)(r^2-r-2)}$
$=\frac{(6r-7)(r+1)(2r+5)(r-2)}{(6r-7)(2r+5)(r-2)(r+1)}=1$

49. $(3x^2-2x-8)\div\frac{x^2-4}{x+2}=\frac{(3x+4)(x-2)}{1}$
$\cdot\frac{x+2}{(x+2)(x-2)}=3x+4$

53. $\frac{10}{a^3-27}\cdot\frac{a^2+3a-18}{15}=\frac{2\cdot 5(a+6)(a-3)}{3\cdot 5(a-3)(a^2+3a+9)}$
$=\frac{2(a+6)}{3(a^2+3a+9)}=\frac{2a+12}{3a^2+9a+27}$

Review exercises

1. $\frac{19}{12}$ **2.** $\frac{11}{24}$ **3.** $2(x+5)(x-5)$ **4.** $(x+11)(x-2)$
5. $(x+4)^2$ **6.** $x=\frac{24}{5}$ **7.** $y=15$ **8.** 7.89×10^{-5}

Exercise 6–2

Answers to odd-numbered problems

1. $\frac{8}{x}$ **3.** $\frac{7}{p}$ **5.** $\frac{14x}{x+2}$ **7.** $-\frac{2}{x}$ **9.** $\frac{x+2}{x^2-1}$

11. $-\frac{1}{7}$ **13.** $\frac{9}{z}$ **15.** $\frac{-7}{x-2}$ **17.** $\frac{9y}{y-6}$

19. $\frac{-x+4}{x-5}$ **21.** $\frac{3y+2}{2y-3}$ **23.** $\frac{-x+4}{2x-5}$ **25.** $18x$

27. $48x^2$ **29.** $140y^3$ **31.** $64a^4$ **33.** $180a^3$ **35.** $3(x-4)$

37. $18y^3(y-4)$ **39.** $a(a+1)(a-1)$ **41.** $8(a+2)(a+1)$

43. $(a+2)(a-2)(a-3)$ **45.** $(a+3)(a-3)(a-2)$

47. $2(x-7)(x+7)$

Solutions to trial exercise problems

5. $\frac{5x}{x+2}+\frac{9x}{x+2}=\frac{5x+9x}{x+2}=\frac{14x}{x+2}$

8. $\frac{3y-2}{y^2}-\frac{4y-1}{y^2}=\frac{(3y-2)-(4y-1)}{y^2}=\frac{3y-2-4y+1}{y^2}=\frac{-y-1}{y^2}$

16. $\frac{1}{x-7}-\frac{5}{7-x}=\frac{1}{x-7}+\frac{5}{x-7}=\frac{6}{x-7}$

21. $\frac{2y-5}{2y-3}-\frac{y+7}{3-2y}=\frac{2y-5}{2y-3}+\frac{y+7}{2y-3}=\frac{3y+2}{2y-3}$

32. $4x^2=2^2\cdot x^2$

$3x=3\cdot x$

$8x^3=2^3\cdot x^3$

$\text{LCD}=2^3\cdot 3\cdot x^3=24x^3$

44. $y^2-y-12=(y-4)(y+3)$

$y^2+6y+9=(y+3)^2$

$\text{LCD}=(y+3)^2(y-4)$

47. $x^2-49=(x+7)(x-7)$

$7-x=-1(x-7)$

$2x+14=2(x+7)$

$\text{LCD}=2(x+7)(x-7)$

Review exercises

1. commutative property of addition **2.** $5(y+2)(y-2)$

3. $(x+10)^2$ **4.** $(3y-4)(y+1)$ **5.** $\left\{\frac{8}{19}\right\}$ **6.** $\{-3,5\}$

Exercise 6–3

Answers to odd-numbered problems

1. $\frac{2x+9}{12}$ **3.** $\frac{8x+1}{48}$ **5.** $\frac{11}{3}$ **7.** $\frac{15y-12}{(y+4)(y-5)}$ **9.** $\frac{10}{y-2}$

11. $\frac{7x+62}{4(x+2)(x-2)}$ **13.** $\frac{9x+40}{x+8}$ **15.** $\frac{x(x+4)}{(x+1)(x-1)}$

17. $\frac{9x+22}{(x+3)(x-3)(x+2)}$ **19.** $\frac{7y^2-13y}{(y-3)^2(y+1)}$

21. $\frac{2y^2-9y+1}{(y-5)(y-3)(y+2)}$ **23.** $\frac{2y-15}{18}$ **25.** $\frac{25}{42y}$

27. $\frac{19a+39}{60}$ **29.** $\frac{4x+25}{18x}$ **31.** $\frac{-5x-17}{(2x-3)(x-5)}$

33. $\frac{17}{7(y+2)}$ **35.** $\frac{9x+66}{x+8}$ **37.** $\frac{-25x-9}{3x+1}$

39. $\frac{-6a+3}{(a-2)(a+2)(a-3)}$ **41.** $\frac{12y-12}{(y-6)(y+4)(y-3)}$

43. $\frac{-3a^2+11a-3}{(a-3)(a-2)}$ **45.** $\frac{13}{12z}$ **47.** $\frac{52a}{45}$ **49.** $\frac{39b-5}{36}$

51. $\frac{-z^3+9z^2-13}{(z-2)(z+2)(z-1)(z+1)}$ **53.** $\frac{bc+ac+ab}{abc}$ **55.** $\frac{3}{m}$

57. $\frac{h}{36}$ **59.** $\frac{48}{m}$ **61.** $w=\frac{A}{23}$ **63.** $b=\frac{2A}{9}$ **65.** $t=\frac{d}{55}$

67. $\frac{b}{a}$

Solutions to trial exercise problems

6. $\frac{4}{x-1}+\frac{5}{x+3}=\frac{4(x+3)+5(x-1)}{(x-1)(x+3)}$

$=\frac{4x+12+5x-5}{(x-1)(x+3)}=\frac{9x+7}{(x-1)(x+3)}$

11. $\frac{12}{x^2-4}+\frac{7}{4x-8}=\frac{12}{(x+2)(x-2)}+\frac{7}{4(x-2)}$

$=\frac{12\cdot 4}{4(x+2)(x-2)}+\frac{7(x+2)}{4(x+2)(x-2)}=\frac{48+(7x+14)}{4(x+2)(x-2)}$

$=\frac{7x+62}{4(x+2)(x-2)}$

13. $5+\frac{4x}{x+8}=\frac{5(x+8)}{x+8}+\frac{4x}{x+8}$

$=\frac{5x+40+4x}{x+8}=\frac{9x+40}{x+8}$

25. $\frac{9}{14y}-\frac{1}{21y}=\frac{9\cdot 3}{2\cdot 3\cdot 7\cdot y}$

$-\frac{1\cdot 2}{2\cdot 3\cdot 7\cdot y}=\frac{27}{42y}-\frac{2}{42y}=\frac{27-2}{42y}=\frac{25}{42y}$

27. $\frac{5a+3}{12}-\frac{a-4}{10}=\frac{5(5a+3)}{60}-\frac{6(a-4)}{60}$

$=\frac{(25a+15)-(6a-24)}{60}=\frac{25a+15-6a+24}{60}=\frac{19a+39}{60}$

34. $\frac{14}{5x-15}-\frac{8}{2x-6}=\frac{14}{5(x-3)}-\frac{8}{2(x-3)}=\frac{14\cdot 2}{10(x-3)}$

$-\frac{8\cdot 5}{10(x-3)}=\frac{28-40}{10(x-3)}=\frac{-12}{10(x-3)}=\frac{-6}{5(x-3)}$

42. $\frac{2p}{p^2-9p+20}-\frac{5p-2}{p-5}=\frac{2p}{(p-5)(p-4)}-\frac{5p-2}{p-5}$

$=\frac{2p}{(p-5)(p-4)}-\frac{(5p-2)(p-4)}{(p-5)(p-4)}=\frac{2p-(5p^2-22p+8)}{(p-5)(p-4)}$

$=\frac{2p-5p^2+22p-8}{(p-5)(p-4)}=\frac{-5p^2+24p-8}{(p-5)(p-4)}$

52. $\frac{1}{f}=(u-1)\left(\frac{1}{R_1}+\frac{1}{R_2}\right)=(u-1)\left(\frac{R_2}{R_1R_2}+\frac{R_1}{R_1R_2}\right)$

$=(u-1)\left(\frac{R_2+R_1}{R_1R_2}\right)$

59. Since n is one of the numbers, let x be the other number. Then $x\cdot n=48$ and x (the other number) $=\frac{48}{n}$.

62. Using $A=\frac{1}{2}bh$, then $2A=bh$ (multiply each member by 2).

Then $h=\frac{2A}{b}=\frac{2(21)}{b}=\frac{42}{b}$.

Review exercises

1. $(x-7)^2$ **2.** $(2x-1)(x-5)$ **3.** $4(x+2)(x-2)$

4. x^2-81 **5.** $16x^2+24x+9$ **6.** $2x^2-15x-8$

7. 48 **8.** $12x^2$ **9.** $(x-3)^2(x+3)$

10. $\frac{13}{2x}$ **11.** $\frac{2x+1}{(x-2)(x-1)}$

Exercise 6–4

Answers to odd-numbered problems

1. $\frac{5}{6}$ **3.** $\frac{3}{2}$ **5.** $\frac{8}{9}$ **7.** $\frac{5}{2}$ **9.** $\frac{23}{30}$ **11.** $\frac{11}{2}$

13. $\frac{4x+1}{4x-3}$ **15.** $\frac{3a+1}{2-4a}$ **17.** $\frac{4a^2+3}{5a^2-3a}$ **19.** $\frac{ab-3}{ab+4}$

21. $\frac{y+x}{y-x}$ **23.** xy **25.** $\frac{4b-5a^2}{a^3b-a^2b^2}$ **27.** $-\frac{y}{x}$

29. $\frac{2(x^2+y^2)}{(x^2-y^2)^2}$ **31.** $\frac{2b+3}{a^3b^4}$ **33.** $\frac{3a-1}{17-a}$ **35.** $\frac{-5}{x+31}$

37. $-\frac{3a^2+5a-8}{a+29}$ **39.** $\frac{(y-3)(y+6)}{(y+1)(y-4)}$ **41.** $\frac{T_1}{T_2-T_1}$

43. $\frac{L_1L_2L_3}{L_2L_3+L_1L_3+L_1L_2}$

Solutions to trial exercise problems

7. $$\frac{7}{2+\frac{4}{5}} = \frac{7\cdot 5}{\left(2+\frac{4}{5}\right)\cdot 5} = \frac{7\cdot 5}{2\cdot 5+\frac{4}{5}\cdot 5} = \frac{35}{10+4} = \frac{35}{14} = \frac{5}{2}$$

12. $$\frac{\frac{6}{7}-\frac{5}{14}}{\frac{3}{14}-\frac{5}{7}} = \frac{\left(\frac{6}{7}-\frac{5}{14}\right)\cdot 14}{\left(\frac{3}{14}-\frac{5}{7}\right)\cdot 14} = \frac{\frac{6}{7}\cdot 14-\frac{5}{14}\cdot 14}{\frac{3}{14}\cdot 14-\frac{5}{7}\cdot 14} = \frac{12-5}{3-10}$$
$$= \frac{7}{-7} = -1$$

15. $$\frac{\frac{1}{a}+3}{\frac{2}{a}-4} = \frac{\left(\frac{1}{a}+3\right)\cdot a}{\left(\frac{2}{a}-4\right)\cdot a} = \frac{\frac{1}{a}\cdot a+3\cdot a}{\frac{2}{a}\cdot a-4\cdot a} = \frac{1+3a}{2-4a} = \frac{3a+1}{2-4a}$$

22. $$\frac{\frac{3}{x^2}-\frac{4}{y}}{\frac{5}{x}+\frac{2}{y^2}} = \frac{\left(\frac{3}{x^2}-\frac{4}{y}\right)\cdot x^2y^2}{\left(\frac{5}{x}+\frac{2}{y^2}\right)\cdot x^2y^2} = \frac{\frac{3}{x^2}\cdot x^2y^2-\frac{4}{y}\cdot x^2y^2}{\frac{5}{x}\cdot x^2y^2+\frac{2}{y^2}\cdot x^2y^2}$$
$$= \frac{3y^2-4x^2y}{5xy^2+2x^2}$$

27. $$\frac{\frac{1}{x+y}-\frac{1}{x-y}}{\frac{1}{x+y}+\frac{1}{x-y}} = \frac{\left(\frac{1}{x+y}-\frac{1}{x-y}\right)\cdot(x+y)(x-y)}{\left(\frac{1}{x+y}+\frac{1}{x-y}\right)\cdot(x+y)(x-y)}$$
$$= \frac{\frac{1}{x+y}\cdot(x+y)(x-y)-\frac{1}{x-y}\cdot(x+y)(x-y)}{\frac{1}{x+y}\cdot(x+y)(x-y)+\frac{1}{x-y}\cdot(x+y)(x-y)}$$
$$= \frac{(x-y)-(x+y)}{(x-y)+(x+y)} = \frac{x-y-x-y}{x-y+x+y} = \frac{-2y}{2x} = \frac{-y}{x}$$

36. $$\frac{\frac{7}{b-7}+\frac{8}{b-5}}{\frac{6}{b^2-12b+35}} = \frac{\left(\frac{7}{b-7}+\frac{8}{b-5}\right)\cdot(b-5)(b-7)}{\frac{6}{(b-5)(b-7)}\cdot(b-5)(b-7)}$$

The LCD is $(b-5)(b-7)$.

$$= \frac{\frac{7}{b-7}\cdot(b-5)(b-7)+\frac{8}{b-5}\cdot(b-5)(b-7)}{6}$$
$$= \frac{7(b-5)+8(b-7)}{6} = \frac{7b-35+8b-56}{6} = \frac{15b-91}{6}$$

Review exercises

1. $\frac{6}{7}$ **2.** $\frac{x-7}{x+2}$ **3.** $2x^3+3x^2+x-21$ **4.** $\{3\}$ **5.** $\{-2\}$ **6.** $\{-3,1\}$ **7.** Domain is all real numbers except -7. **8.** Domain is all real numbers except -2 and 2. **9. a.** undefined **b.** 0

Exercise 6–5

Answers to odd-numbered problems

1. $\left\{\frac{8}{3}\right\}$ **3.** $\left\{\frac{14}{3}\right\}$ **5.** $\{-22\}$ **7.** $\left\{\frac{10}{9}\right\}$ **9.** $\left\{\frac{9}{2}\right\}$

11. $\left\{-\frac{5}{8}\right\}$ $x \neq 0$ **13.** $\left\{\frac{19}{24}\right\}$ $R \neq 0$ **15.** $\left\{-\frac{5}{27}\right\}$, $y \neq 0$

17. $\{1\}$ $b \neq 0$ **19.** $\left\{\frac{4}{37}\right\}$ $p \neq 0$ **21.** $\left\{-\frac{13}{2}\right\}$ $a \neq 0$

23. $\{36\}$ $x \neq 4,-4$ **25.** $\left\{\frac{47}{9}\right\}$ $z \neq \frac{-1}{3}$ **27.** $\emptyset$; $y \neq 2$

29. $\left\{\frac{11}{5}\right\}$ $b \neq 2,-2$ **31.** $\{40\}$ $a \neq 5,-5$

33. $\left\{-\frac{19}{3}\right\}$ $x \neq 2,-3,-5$ **35.** $\left\{\frac{8}{3}\right\}$ $a \neq 3,-4$ **37.** $\left\{-\frac{2}{3}\right\}$

39. $\left\{\frac{3}{2},-3\right\}$ **41.** $\left\{\frac{2}{3},-4\right\}$ **43.** $\left\{\frac{2}{3},-1\right\}$

45. $x = \frac{2y}{3y-1}$ **47.** $c_1 = \frac{cc_2}{c_2-c}$ **49.** $a = \frac{8b}{b-8}$

51. $a = \frac{3b-5}{4}$ **53.** $x = \frac{-2(y+6)}{5}$ **55.** $D = \frac{P}{pL}$

57. $N = \frac{2P}{D-P}$ **59.** $f = \frac{Fa}{A}$ **61.** $T_2 = \frac{T_1P_2V_2}{P_1V_1}$

63. $L_1 = \frac{L_2}{\alpha(t_2-t_1)+1}$ **65.** $R_1 = \frac{RR_2}{R_2-R}$

Solutions to trial exercise problems

2. $\frac{4x}{5}-\frac{2}{3} = 4$ Multiply by 15 to get

The LCD is 15.
$$\frac{4x}{5}\cdot 15-\frac{2}{3}\cdot 15 = 4\cdot 15$$
$$4x\cdot 3-2\cdot 5 = 60$$
$$12x-10 = 60$$
$$12x = 70$$
$$x = \frac{70}{12} = \frac{35}{6}$$

The solution set is $\left\{\frac{35}{6}\right\}$.

6. $\frac{3R}{4}-5 = \frac{5}{6}$ Multiply by 12 to get

The LCD is 12.
$$\frac{3R}{4}\cdot 12-5\cdot 12 = \frac{5}{6}\cdot 12$$
$$3R\cdot 3-60 = 5\cdot 2$$
$$9R-60 = 10$$
$$9R = 70$$
$$R = \frac{70}{9}$$

The solution set is $\left\{\frac{70}{9}\right\}$.

15. $\frac{4}{6y} + 5 = \frac{1}{9y} + 2$ Multiply by $18y$ to get

The LCD is $18y$.

$$\begin{aligned} \frac{4}{6y} \cdot 18y + 5 \cdot 18y &= \frac{1}{9y} \cdot 18y + 2 \cdot 18y \\ 4 \cdot 3 + 90y &= 1 \cdot 2 + 36y \\ 12 + 90y &= 2 + 36y \\ 90y &= -10 + 36y \\ 54y &= -10 \\ y &= \frac{-10}{54} = \frac{-5}{27} \end{aligned}$$

The solution set is $\left\{-\frac{5}{27}\right\} (y \neq 0)$.

22. $\frac{R + 2}{10R} + \frac{4R - 1}{4R} = 2$ Multiply by $20R$ to get

The LCD is $20R$.

$$\begin{aligned} \frac{R + 2}{10R} \cdot 20R + \frac{4R - 1}{4R} \cdot 20R &= 2 \cdot 20R \\ (R + 2) \cdot 2 + (4R - 1) \cdot 5 &= 40R \\ 2R + 4 + 20R - 5 &= 40R \\ 22R - 1 &= 40R \\ -1 &= 18R \\ R &= \frac{-1}{18} \end{aligned}$$

The solution set is $\left\{-\frac{1}{18}\right\} (R \neq 0)$.

34. $\frac{8}{a^2 - 6a + 8} = \frac{1}{a^2 - 16}$ Multiply both members by $(a - 2)(a + 4)(a - 4)$.

$$\begin{aligned} \frac{8}{(a - 4)(a - 2)} \cdot (a - 2)(a + 4)(a - 4) &= \frac{1}{(a + 4)(a - 4)} \cdot (a - 2)(a + 4)(a - 4) \\ 8(a + 4) &= 1(a - 2) \\ 8a + 32 &= a - 2 \\ 8a &= a - 34 \\ 7a &= -34 \\ a &= \frac{-34}{7} \end{aligned}$$

The solution set is $\left\{-\frac{34}{7}\right\} (a \neq 4, 2, -4)$.

37. $3x^2 + 4x + \frac{4}{3} = 0$ Multiply by 3 to obtain

$9x^2 + 12x + 4 = 0$ Factor the left member.

$$\begin{aligned} (3x + 2)^2 &= 0 \\ 3x + 2 &= 0 \\ 3x &= -2 \\ x &= -\frac{2}{3} \end{aligned}$$ So

The solution set is $\left\{-\frac{2}{3}\right\}$.

42. $x^2 - \frac{5}{6}x = \frac{2}{3}$ Multiply by the LCD, 6, to obtain

$$\begin{aligned} 6x^2 - 5x &= 4 \\ 6x^2 - 5x - 4 &= 0 \\ (2x + 1)(3x - 4) &= 0 \\ 2x + 1 &= 0 \text{ or } 3x - 4 = 0 \\ x &= -\frac{1}{2} \text{ or } x = \frac{4}{3} \end{aligned}$$

The solution set is $\left\{-\frac{1}{2}, \frac{4}{3}\right\}$.

47. $\frac{1}{c} = \frac{1}{c_1} + \frac{1}{c_2}$ Multiply by cc_1c_2.

$$\frac{1}{c} \cdot cc_1c_2 = \frac{1}{c_1} \cdot cc_1c_2 + \frac{1}{c_2} \cdot cc_1c_2$$

$c_1c_2 = cc_2 + cc_1$ Add $-cc_1$ to both members.

$c_1c_2 - cc_1 = cc_2$ Factor c_1 in the left member.

$(c_2 - c)c_1 = cc_2$ Divide both members by $c_2 - c$.

$$c_1 = \frac{cc_2}{c_2 - c}$$

52. $\frac{5}{m} + 4 = \frac{6a}{2m} + 3b$ Multiply by the LCD $2m$.

$$\frac{5}{m} \cdot 2m + 4 \cdot 2m = \frac{6a}{2m} \cdot 2m + 3b \cdot 2m$$

$10 + 8m = 6a + 6bm$ Add -10 to both members.

$8m = 6a - 10 + 6bm$ Add $-6bm$ to both members.

$8m - 6bm = 6a - 10$ Factor m in the left member.

$(8 - 6b)m = 6a - 10$ Divide both members by $8 - 6b$.

$$m = \frac{6a - 10}{8 - 6b} = \frac{2(3a - 5)}{2(4 - 3b)} = \frac{3a - 5}{4 - 3b}$$

58. $\frac{T_A}{T_B} = \frac{R_B}{R_A}$ Multiply by the LCD T_BR_A.

$$\frac{T_A}{T_B} \cdot T_BR_A = \frac{R_B}{R_A} \cdot T_BR_A$$

$$T_AR_A = R_BT_B$$

$$T_B = \frac{T_AR_A}{R_B}$$

62. Given $k = \frac{L_t - L_0}{L_0t}$. Multiply both members by L_0t to get $kL_0t = L_t - L_0$.

a. To find L_t, add L_0 to both members to get $L_t = kL_0t + L_0$.

b. To find L_0, use $L_t = kL_0t + L_0$, factor L_0 in the right member to get $L_t = (kt + 1)L_0$. Divide both members by $kt + 1$.

Then $\frac{L_t}{kt + 1} = L_0$ or $L_0 = \frac{L_t}{kt + 1}$.

Review exercises

1. 14 **2.** $\frac{17}{3}$ **3.** 4

4. $y = -5x + 4$ **5.** $y = \frac{-2x + 6}{3}$

6. $y = \frac{x - 8}{4}$ **7.** $\frac{3y}{x^2}$ **8.** $\frac{y^4}{x^5}$

Exercise 6–6

Answers to odd-numbered problems

1. $29\frac{1}{6}$ minutes **3.** $2\frac{14}{29}$ hours **5.** 12 hours **7.** $3\frac{1}{3}$ hours

9. 10 hours **11.** 110 miles at 55 mph, 210 miles at 60 mph

13. car *A*, 40 mph; car *B*, 50 mph **15.** $\frac{6}{7}$ mph **17.** 220 miles

19. $\frac{5}{9}$ **21.** 3 and 12 **23.** 6 **25.** 4 **27.** adult, 15 pills; child, 9 pills **29.** $38.21 **31.** $\frac{60}{17}$ ohms or $3\frac{9}{17}$ ohms **33.** 20 ohms

35. 60 ohms

Solutions to trial exercise problems

1. Let x = the number of minutes required for both boys to mow the lawn. Then,

$$\frac{1}{50} + \frac{1}{70} = \frac{1}{x}$$

Multiply by the LCM of 50, 70, and x, that is, $350x$.

$$350x \cdot \frac{1}{50} + 350x \cdot \frac{1}{70} = 350x \cdot \frac{1}{x}$$

$$7x + 5x = 350$$

$$12x = 350$$

$$x = \frac{350}{12} = 29\frac{1}{6}$$

Together Jim and Kenny could mow the lawn in $29\frac{1}{6}$ minutes.

5. Let x = the time required for Dick to paint the house alone.

Then $\frac{1}{6} + \frac{1}{x} = \frac{1}{4}$.

Multiply both members by the LCM of 4, 6, and x, that is, $12x$.

$$12x \cdot \frac{1}{6} + 12x \cdot \frac{1}{x} = 12x \cdot \frac{1}{4}$$

$$2x + 12 = 3x$$

$$12 = x$$

Therefore Dick would take 12 hours to paint the house alone.

13. Let x = the average speed of car A.

Then $x + 10$ = the average speed of car B.

Using $t = \frac{d}{r}$, let t_A = time of car A, t_B = time of car B, then $t_A = t_B$.

Now $t_A = \frac{120}{x}$ and $t_B = \frac{150}{x + 10}$, so

$$\frac{120}{x} = \frac{150}{x + 10}$$

Multiply both members by the LCM of x and $x + 10$, that is, $x(x + 10)$.

$$x(x + 10) \cdot \frac{120}{x} = x(x + 10) \cdot \frac{150}{x + 10}$$

$$(x + 10)120 = 150x$$

$$120x + 1{,}200 = 150x$$

$$1{,}200 = 30x$$

$$40 = x$$

So $x + 10 = 50$. Therefore car B averaged 50 mph and car A averaged 40 mph.

20. Let x = the numerator of the fraction. Then $x + 7$ = the denominator of the fraction.

Then $\frac{x + 3}{(x + 7) - 1} = \frac{4}{5}$. Simplifying, we have the equation

$$\frac{x + 3}{x + 6} = \frac{4}{5}.$$

Multiply both members by the LCM of 5 and $x + 6$, that is, $5(x + 6)$.

$$5(x + 6) \cdot \frac{x + 3}{x + 6} = 5(x + 6) \cdot \frac{4}{5}$$

$$5(x + 3) = 4(x + 6)$$

$$5x + 15 = 4x + 24$$

$$x + 15 = 24$$

$$x = 9$$

$$x + 7 = 16$$

The original fraction is then $\frac{9}{16}$.

21. Let x = the lesser of the two numbers then $4x$ = the greater of the two numbers.

Their reciprocals are then $\frac{1}{x}$ and $\frac{1}{4x}$ and we get the equation

$$\frac{1}{x} + \frac{1}{4x} = \frac{5}{12}.$$

Multiply both members by the LCM of x, $4x$, and 12, that is, $12x$.

$$12x \cdot \frac{1}{x} + 12x \cdot \frac{1}{4x} = 12x \cdot \frac{5}{12}$$

$$12 + 3 = 5x$$

$$15 = 5x$$

then

$$x = 3$$

$$4x = 12$$

Therefore the two numbers are 3 and 12.

(*Note:* To check, show $\frac{1}{3} + \frac{1}{12} = \frac{5}{12}$

$$\frac{4}{12} + \frac{1}{12} = \frac{5}{12}$$

$$\frac{5}{12} = \frac{5}{12})$$

28. Let x = hourly wage of a journeyman electrician, then $\frac{3}{8}x$ = the hourly wage of an apprentice electrician.

$$x + \frac{3}{8}x = 29.70$$

$$8 \cdot x + 8 \cdot \frac{3}{8}x = 8 \cdot 29.70$$

$$8x + 3x = 237.60$$

$$11x = 237.60$$

$$x = \$21.60$$

$$\frac{3}{8}x = \$8.10$$

The journeyman earns $21.60 per hour and the apprentice earns $8.10 per hour.

30. Let R = the total resistance of the circuit. Then

$$\frac{1}{6} + \frac{1}{8} = \frac{1}{R}.$$

Multiply by the LCM of 6, 8, and R, that is, $24R$.

$$24R \cdot \frac{1}{6} + 24R \cdot \frac{1}{8} = 24R \cdot \frac{1}{R}$$

$$4R + 3R = 24$$

$$7R = 24$$

$$R = \frac{24}{7}$$

Therefore the total resistance in the parallel circuit is $\frac{24}{7}$ or $3\frac{3}{7}$ ohms.

33. Let R_1 = the resistance in the unknown branch. Then

$$\frac{1}{R_1} + \frac{1}{30} = \frac{1}{12}.$$

Multiply by the LCM of R_1, 12, and 30, that is, $60R_1$.

$$60R_1 \cdot \frac{1}{R_1} + 60R_1 \cdot \frac{1}{30} = 60R_1 \cdot \frac{1}{12}$$

$$60 + 2R_1 = 5R_1$$

$$60 = 3R_1$$

$$20 = R_1$$

Therefore the other branch has resistance of 20 ohms.

Review exercises

1. $\{2\}$ **2.** $y = \frac{3x+6}{2}$ **3.** $8(y+2)(y-2)$

4. $(x+10)^2$ **5.** $(3y+2)(y-2)$ **6.** $\frac{5x^2+7x}{(x-1)(x+3)}$

7. $\frac{-2y^2-23y}{(2y+1)(y-5)}$

Chapter 6 review

1. $\frac{9}{2}$ **2.** $\frac{9a}{b}$ **3.** $\frac{2x-y}{2x+y}$ **4.** $x-2$ **5.** $\frac{x(m-n)}{2}$

6. $\frac{1}{y^2-2y+1}$ **7.** 2 **8.** $-\frac{1}{3x+15}$ **9.** $\frac{x-1}{x+1}$

10. $\frac{2a}{3}$ **11.** $\frac{8b}{21}$ **12.** $\frac{9}{2ab}$ **13.** $\frac{14x-7}{9x+12}$

14. $\frac{1}{x^2+4x-12}$ **15.** $\frac{10}{3}$ **16.** $\frac{1}{x-8}$ **17.** $\frac{3-b}{b^2+2b-3}$

18. $\frac{3a+3}{4a-4}$ **19.** $36x^2$ **20.** $42a^2b^2$ **21.** $2(x+3)(x-3)$

22. $(y+5)(y-5)(y+3)$ **23.** $4(z+2)(z-2)(z+1)$

24. $x(x+1)^2(3x-5)$ **25.** $\frac{3}{x}$ **26.** $\frac{4y}{y-2}$ **27.** $\frac{4x+4}{3x-1}$

28. $\frac{-x+13}{2x-5}$ **29.** $\frac{68x}{105}$ **30.** $\frac{23}{48a}$

31. $\frac{47x-6}{(3x+1)(4x-3)}$ **32.** $\frac{4x^2-23x}{(x+4)(x-4)}$

33. $\frac{80a+48b-15ab}{20a^2b^2}$ **34.** $\frac{4a^2+4a+6}{a+1}$ **35.** $\frac{-10x^2-3}{x^2+1}$

36. $\frac{16x^2+5x+15}{24x^2}$ **37.** $\frac{13y^2-89y}{(y-9)(y+2)(y-2)}$

38. $\frac{2-2ax-36a}{(x+3)(x-3)}$ **39.** $\frac{8x^2-40x-180}{x(x-5)(x+4)}$

40. $\frac{10x-5y}{(x+y)^2(x-2y)}$ **41.** $\frac{16}{63}$ **42.** $\frac{1}{7}$ **43.** 1 **44.** $\frac{y-x}{y+x}$

45. $\frac{4+3x}{2-5x}$ **46.** $\frac{a^2-b^2}{a^2+b^2}$ **47.** $y-x$ **48.** $\frac{xy}{x+y}$

49. $\frac{a^3-2a^2b^2-3ab^2}{ab-3b^2}$ **50.** $\{-96\}$ **51.** $\left\{\frac{41}{48}\right\}$ **52.** $\left\{\frac{13}{24}\right\}$

53. $\left\{\frac{6}{5}\right\}$ **54.** $\left\{-\frac{39}{14}\right\}$ **55.** $\left\{\frac{7}{52}\right\}$ **56.** $\{0,12\}$

57. $\left\{-\frac{3}{2},\frac{3}{2}\right\}$ **58.** $\{-5,3\}$ **59.** $\left\{6,\frac{3}{2}\right\}$ **60.** $x = \frac{a+b}{3}$

61. $x = \frac{y+3}{1-y}$ **62.** $y = \frac{4a-3b}{a}$ **63.** $y = 4b+5c$

64. $l = \frac{Aey}{F}$ **65.** $L = \frac{Wp-2E}{EF}$ **66.** $R = \frac{Mr}{M-2}$

67. $m = \frac{Fr}{v^2+gr}$ **68.** $5\frac{1}{3}$ hr, 16 hr **69.** $1\frac{5}{7}$ mph **70.** $\frac{2}{3}$ or 1

Chapter 6 cumulative test

1. $-\frac{71}{12}$ **2.** $\frac{3}{4}$ **3.** $\frac{43}{30}$ **4.** $\frac{8}{5}$ **5.** x^{11} **6.** $\frac{1}{y^9}$ **7.** $\frac{25x^4}{y^6}$

8. $10x^3-4x^2+11x-9$ **9.** $\left\{-\frac{26}{7}\right\}$ **10.** $\left\{\frac{81}{4}\right\}$

11. $\left\{-\frac{3}{2},2\right\}$ **12.** $x \geq -5$ **13.** $2(x+3)(x-3)$

14. $3x^2(2x^3-12x+3)$ **15.** $(2x+5)(2x+3)$

16. $3(1+2x^3)(1-2x^3)$ **17.** $49x^2-84x+36$

18. $25-\frac{1}{4}x^2$ **19.** $6x^3-19x^2+37x-33$

20. $15-22x^2+8x^4$ **21.** $x = \frac{120}{7}$

22. 2,400 foot-pounds/min **23.** $\frac{4a^2cf\ell}{5b^2}$ **24.** $\frac{y^2+10y+21}{y^2-8y+12}$

25. y^2-5y+6 **26.** $\frac{(x-y)(x-1)}{y}$ **27.** $\frac{2y+4x}{x^2y^2}$

28. $\frac{3x+9y}{(x-y)(x+y)}$ **29.** $\frac{9}{a-2}$ **30.** $\frac{13x+56}{(x-7)(x+7)(x+2)}$

31. $\frac{-x^2-6x}{(x+3)(x-2)(x-6)}$ **32.** $\frac{3y-1}{4y+5}$ **33.** $\frac{y+x}{3x-4y}$

34. $\frac{x^2-11x+28}{x-5}$ **35.** $\{2\}$ **36.** $\left\{-\frac{1}{3},6\right\}$ **37.** $q = \frac{fp}{p-f}$

38. $2\frac{2}{5}$ days **39.** $\frac{8}{16}$

Chapter 7

Exercise 7–1

Answers to odd-numbered problems

1. (1,2) and (−1,−4) are solutions. **3.** (−1,2), (3,0) are solutions. **5.** (1,2) is a solution. **7.** (2,3), (0,0) are solutions. **9.** (−4,1) and (−4,−4) are solutions. **11.** (−5,3), (−5,8) are solutions. **13.** (1,5), (−2,−4), (0,2) **15.** (3,−3), (−4,11), (0,3)

17. (3,−5), (−2,10), (0,4) **19.** (−2,−1), (3,0), $\left(0,-\frac{3}{5}\right)$

21. (1,−4), (−1,1), $\left(0,-\frac{3}{2}\right)$ **23.** (1,5), (−6,5), (0,5)

25. (7,−1), $\left(-\frac{3}{5},-1\right)$, (0,−1) **27.** (5,1), (−1,−2), (3,0)

29. (3,2), (−6,−4), (0,0) **31.** (1,−2), (1,7), (1,0)

33. a. (75,$170) **b.** (300,$620) **c.** (1,000,$2,020) **35. a.** (2,236) **b.** (12,216) **c.** (0,240) **37. a.** (0,20) **b.** (5,35) **c.** (3,29)

39. a. (3,165) **b.** (8,440) **c.** $\left(\frac{26}{5},286\right)$

41. (2,4) **43.** (−1,3) **45.** (−6,−1) **47.** (0,4) **49.** (5,0)
51. (−7,0) **53.** $\left(\frac{1}{2},3\right)$ **55.** $\left(\frac{3}{2},0\right)$

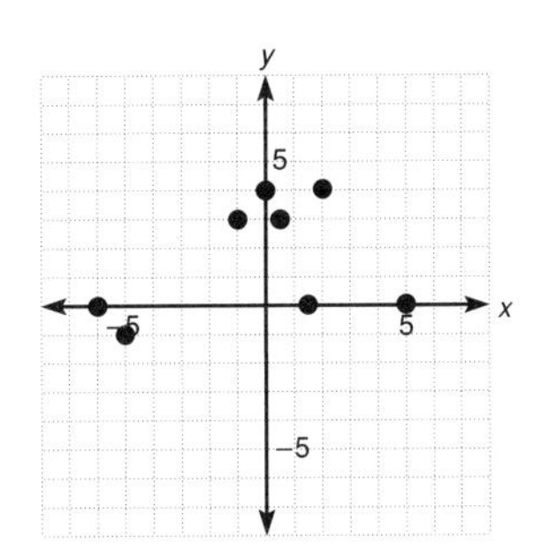

57. $B(2,-1)$ **59.** $D(0,6)$ **61.** $F(0,-6)$ **63.** $H(-4,4)$
65. $J(5,-5)$ **67.** III **69.** I **71.** IV **73.** III **75.** IV
77. a. IV **b.** II **c.** III **d.** I **79.** 0 **81.** B, −1; D, 6; F, −6; H, 4; J, −5 **83.** Figure is a horizontal line parallel to the x-axis having equation $y = 4$.

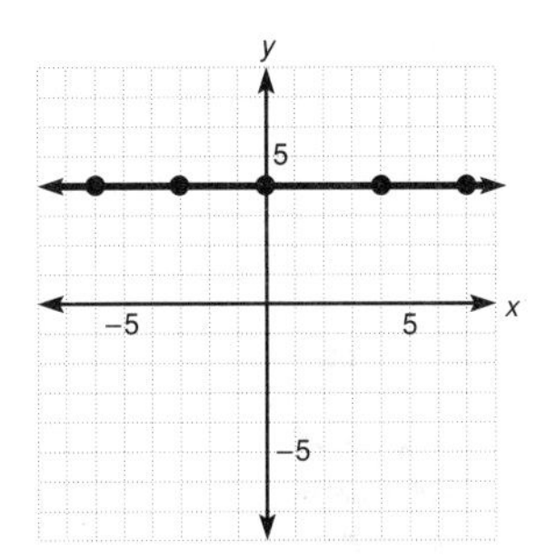

85. (−5,5), (3,−3),(−2,2), (5,−5), (1,−1) The resulting figure is a line through the origin that lies in quadrants II and IV.

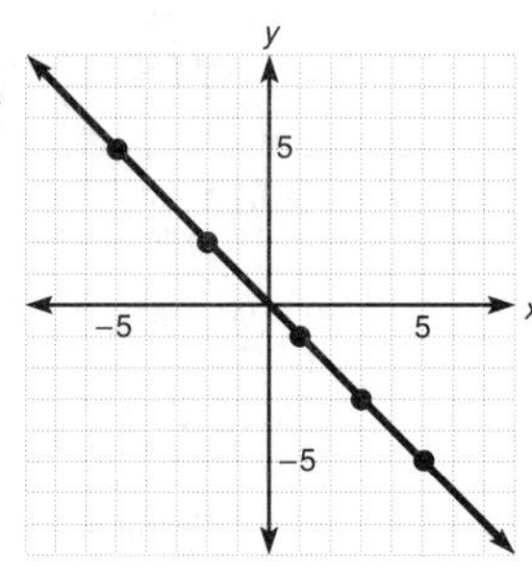

87.

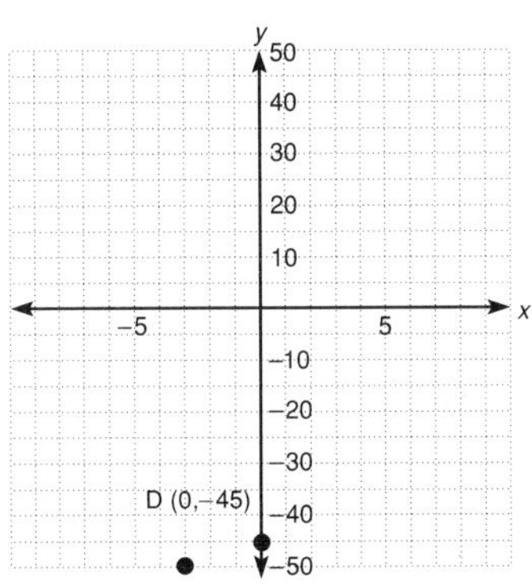

89.

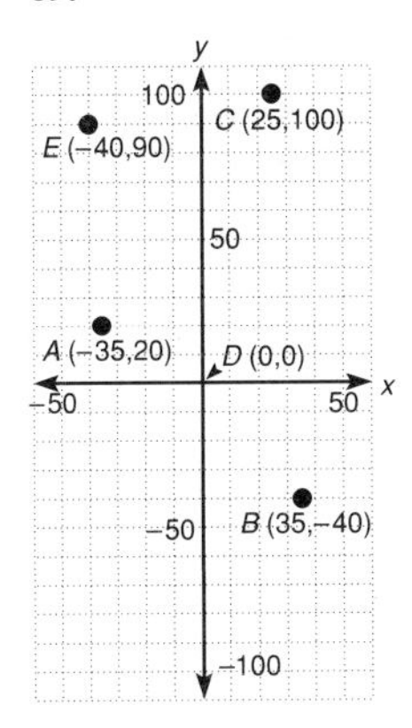

91. A (Tues., 70), the temperature was 70° on Tuesday. B (Wed., 80), the temperature was 80° on Wednesday. C (Fri., 90), the temperature was 90° on Friday. D (Sat., 85), the temperature was 85° on Saturday.

93. (0,4), (−4,0), $\left(\frac{5}{2},\frac{13}{2}\right)$, (2,6)

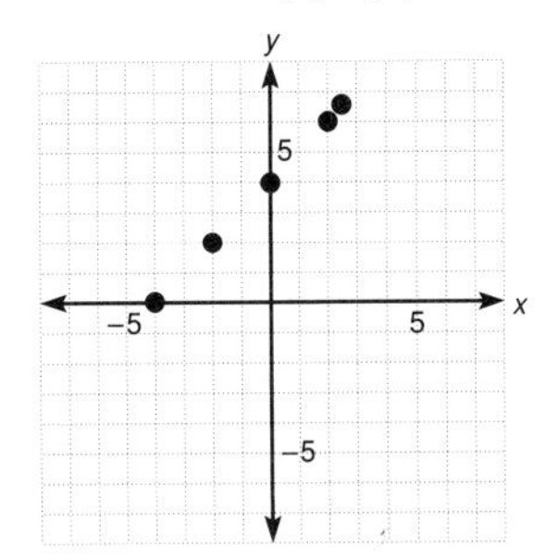

95. (0,1), $\left(-\frac{1}{2},0\right)$, (2,5), (−1,−1)

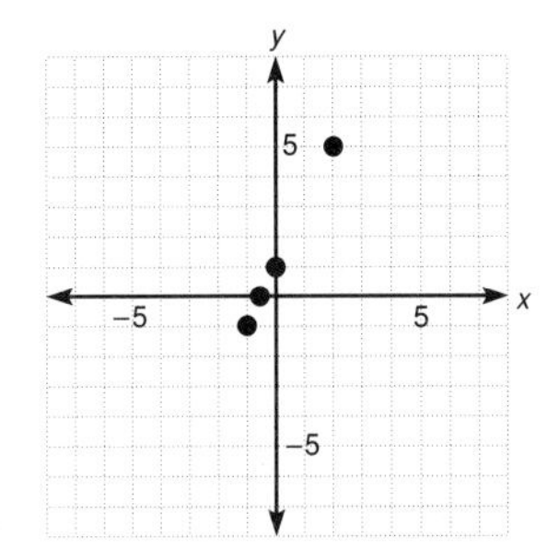

97. (0,0), $\left(\frac{1}{2},1\right)$, (2,4), (−1,−2)

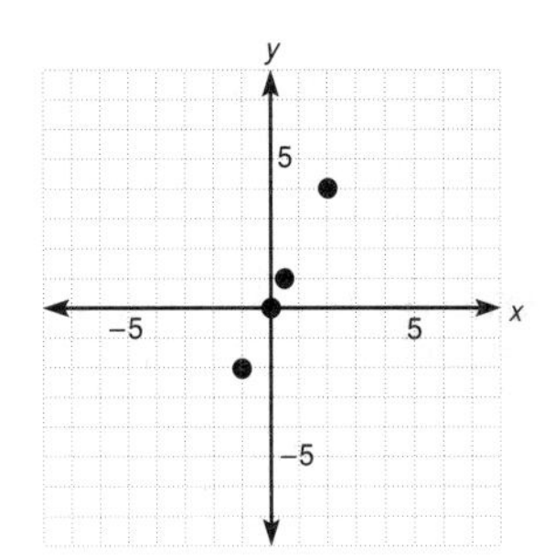

99. (0,3), (−3,6), (3,0), $\left(\frac{5}{2},\frac{1}{2}\right)$

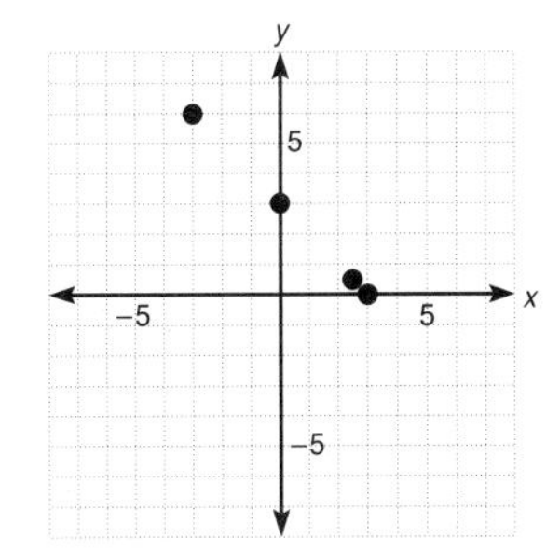

101. (0,3), (3,−3), (−2,7), $\left(\frac{3}{2},0\right)$

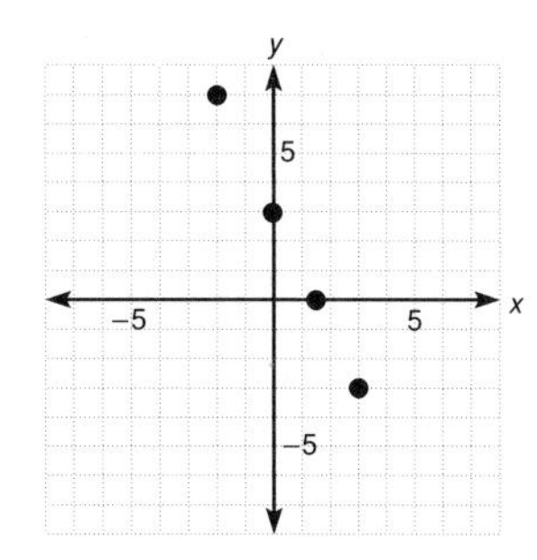

103. (0,4), (−3,7), (2,2), (3,1)

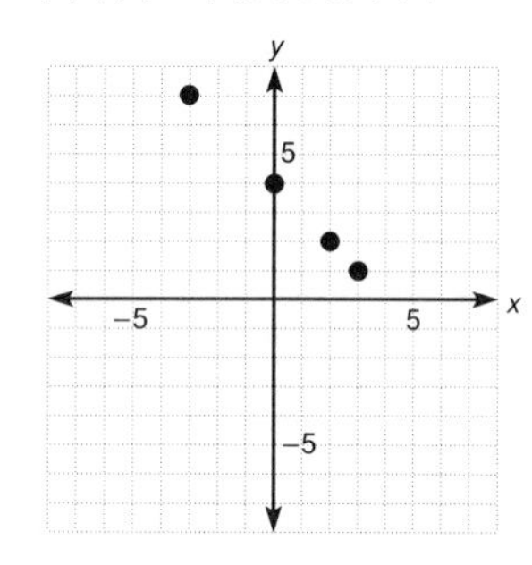

Solutions to trial exercise problems

1. $y = 3x - 1$; (1,2), (−1,−4), (2,3)

Solution:

(a) (1,2)	(b) (−1,−4)	(c) (2,3)
$2 = 3(1) - 1$	$-4 = 3(-1) - 1$	$3 = 3(2) - 1$
$2 = 3 - 1$	$-4 = -3 - 1$	$3 = 6 - 1$
$2 = 2$	$-4 = -4$	$3 = 5$
(true)	(true)	(false)

Therefore (1,2) and (−1,−4) are solutions and (2,3) is not a solution.

7. $3x = 2y$; (2,3), (3,2), (0,0)

Solution:

(a) (2,3)	(b) (3,2)	(c) (0,0)
$3(2) = 2(3)$	$3(3) = 2(2)$	$3(0) = 2(0)$
$6 = 6$	$9 = 4$	$0 = 0$
(true)	(false)	(true)

So (2,3) and (0,0) are solutions and (3,2) is not a solution.

9. $x = -4$; (−4,1), (4,2), (−4,−4)

Solution:

Since we can write $x = -4$ as $x = 0 \cdot y - 4$, then $x = -4$ for *any* value of y. Therefore (−4,1) and (−4,−4) are solutions but (4,2) is not since, substituting 4 for x, $4 = -4$ is false.

15. $y = 3 - 2x$; $x = 3$, $x = -4$, $x = 0$

Solution:

When $x = 3$, $y = 3 - 2(3) = 3 - 6 = -3$ (3,−3)

When $x = -4$, $y = 3 - 2(-4) = 3 + 8 = 11$ (−4,11)

When $x = 0$, $y = 3 - 2(0) = 3 - 0 = 3$ (0,3)

20. $x + 4y = 0$; $x = -4$, $x = 8$, $x = 0$

Solution:

When $x = -4$, $-4 + 4y = 0$
$$4y = 4$$
$$y = 1 \quad (-4,1)$$

When $x = 8$, $8 + 4y = 0$
$$4y = -8$$
$$y = -2 \quad (8,-2)$$

When $x = 0$, $0 + 4y = 0$
$$4y = 0$$
$$y = 0 \quad (0,0)$$

23. $y = 5$; $x = 1$, $x = -6$, $x = 0$

Solution:

We write $y = 5$ as $y = 0 \cdot x + 5$, so for *every* value of x, $y = 5$. Therefore (1,5), (−6,5), (0,5).

28. $x = -3y + 1$; $y = -1$, $y = 2$, $y = 0$

Solution:

When $y = -1$, $x = -3(-1) + 1 = 3 + 1 = 4$ (4,−1)

When $y = 2$, $x = -3(2) + 1 = -6 + 1 = -5$ (−5,2)

When $y = 0$, $x = -3(0) + 1 = 0 + 1 = 1$ (1,0)

31. $x = 1$; $y = -2$, $y = 7$, $y = 0$

Solution:

Since $x = 1$ can be written $x = 0 \cdot y + 1$, then $x = 1$ for *any* value of y, so we have (1,−2), (1,7), (1,0).

36. Given $y = 240 - 2x$, when

a. $y = 200$, then $200 = 240 - 2x$
$$-40 = -2x$$
$$20 = x \qquad (20,200)$$

b. $y = 0$, then $0 = 240 - 2x$
$$-240 = -2x$$
$$120 = x \qquad (120,0)$$

c. $y = 210$, then $210 = 240 - 2x$
$$-30 = -2x$$
$$15 = x \qquad (15,210)$$

43. **48.** **53.**

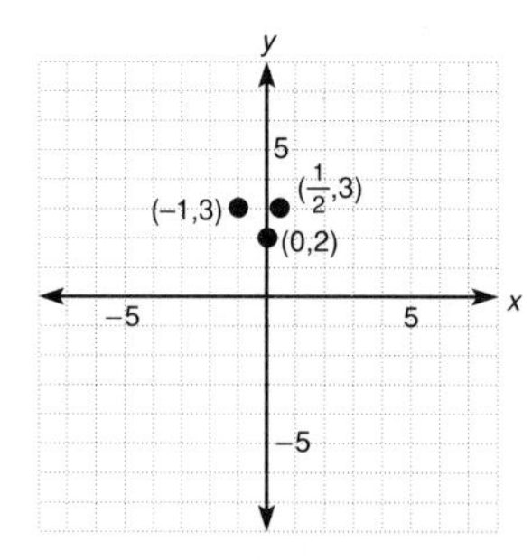

58. $C(-2,0)$ **63.** $H(-4,4)$ **64.** $I(7,0)$ **68.** (4,−1) lies in quadrant IV. **73.** $\left(-\frac{5}{2}, -\frac{3}{4}\right)$ lies in quadrant III.

82. (−3,3), (−3,0), (−3,−2), (−3,5), (−3,−5)
They lie on a vertical straight line.

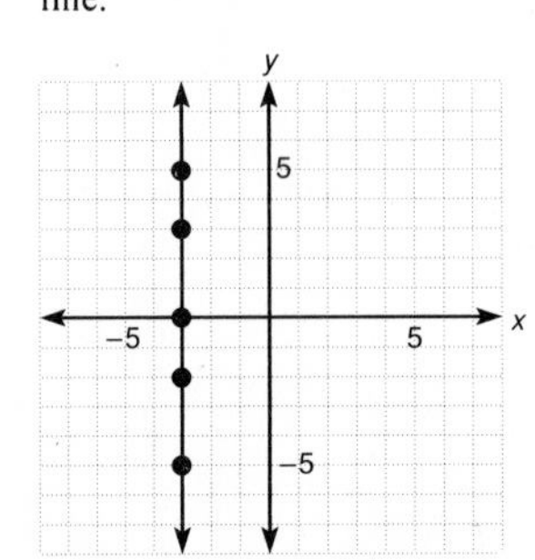

86.

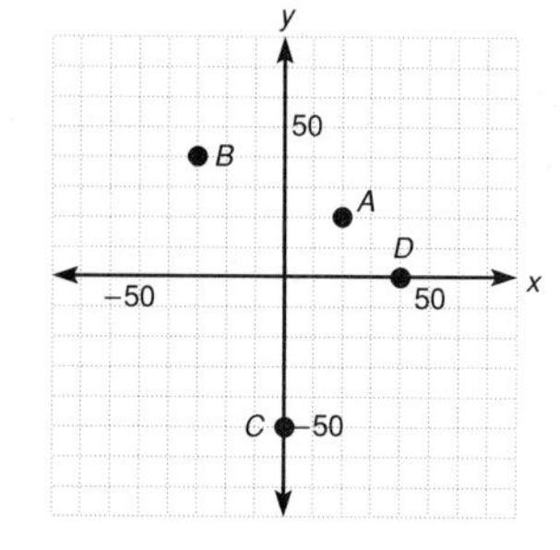

91. A (Tues., 70), the temperature was 70° on Tuesday.
B (Wed., 80), the temperature was 80° on Wednesday.
C (Fri., 90), the temperature was 90° on Friday.
D (Sat., 85), the temperature was 85° on Saturday.

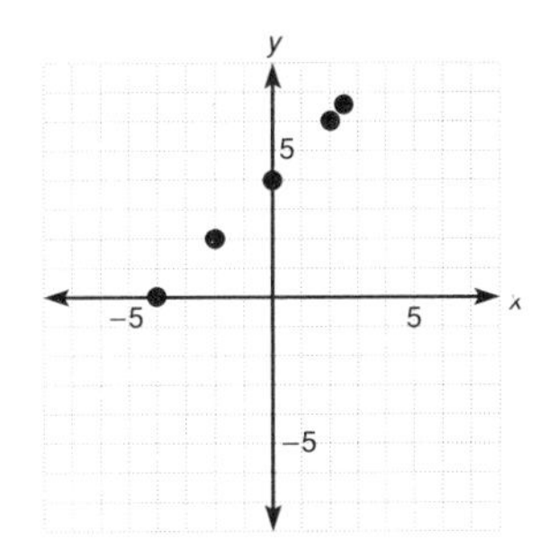

93. When $x = 0, y = 0 + 4 = 4 \quad (0,4)$
When $x = -4, y = -4 + 4 = 0 \quad (-4,0)$

When $x = \frac{5}{2}, y = \frac{5}{2} + 4 = \frac{13}{2} \quad \left(\frac{5}{2}, \frac{13}{2}\right)$

When $x = 2, y = 2 + 4 = 6 \quad (2,6)$

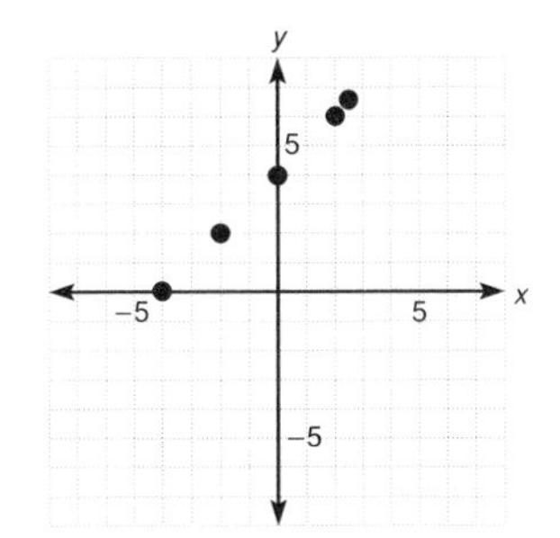

Review exercises

1. $2x^2 + 4x$ **2.** $\frac{x+3}{2x+1}$ **3.** $\frac{x+10}{(x-2)(x+1)}$ **4.** -8 **5.** x^3
6. 0

Exercise 7–2

Answers to odd-numbered problems

1. x-intercept, $(-2,0)$; y-intercept, $(0,4)$ **3.** x-intercept, $\left(-\frac{1}{3},0\right)$; y-intercept, $(0,1)$ **5.** x-intercept, $(3,0)$; y-intercept, $(0,2)$
7. x-intercept, $\left(\frac{11}{2},0\right)$; y-intercept, $\left(0,\frac{11}{5}\right)$ **9.** x-intercept, $(0,0)$; y-intercept, $(0,0)$ **11.** x-intercept, $(0,0)$; y-intercept, $(0,0)$
13. x-intercept, $(4,0)$; y-intercept, $(0,2)$ **15.** x-intercept, $(3,0)$; y-intercept, $\left(0,-\frac{3}{2}\right)$ **17.** x-intercept, $(-2,0)$; y-intercept, $(0,6)$

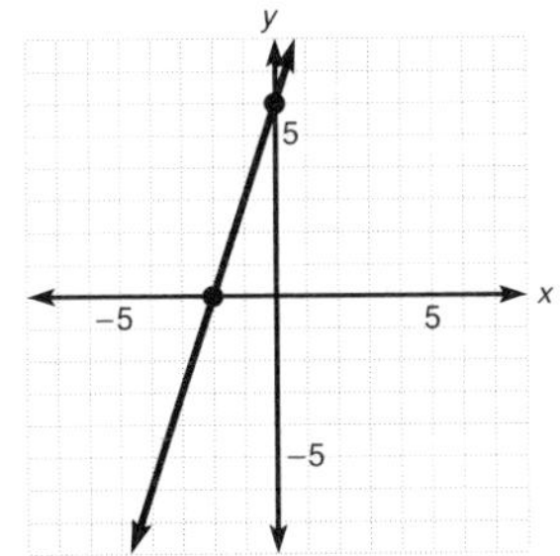

19. x-intercept, $(2,0)$; y-intercept, $(0,-2)$

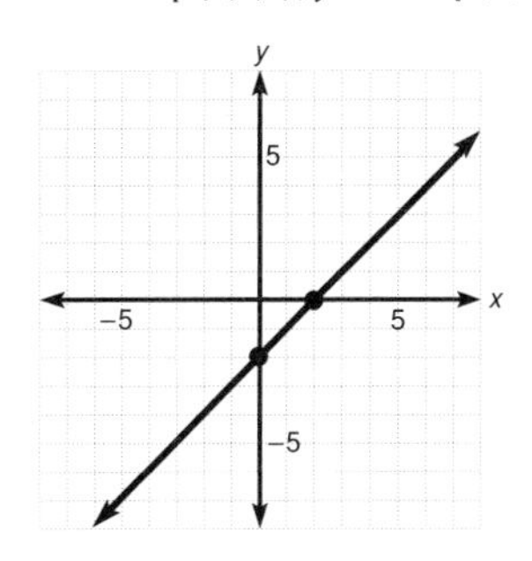

21. x-intercept, $(4,0)$; y-intercept, $(0,-8)$

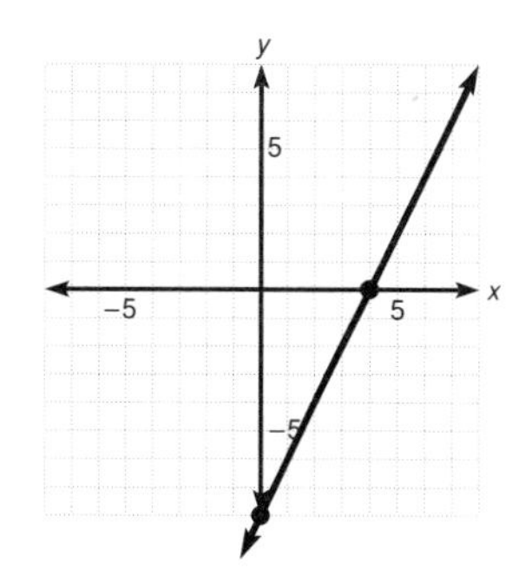

23. x-intercept, $(0,0)$; y-intercept, $(0,0)$; $(1,1)$

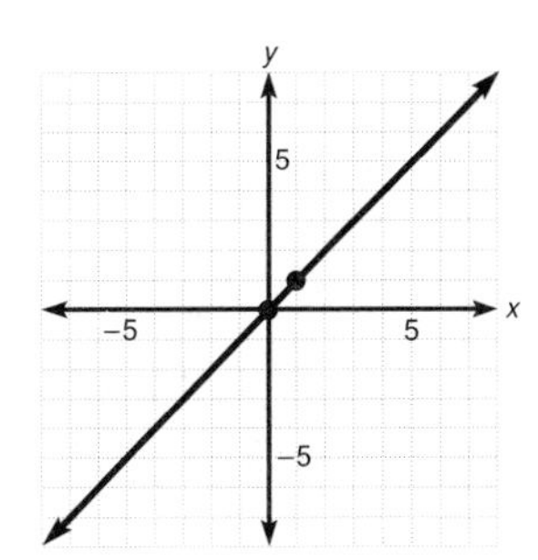

25. x-intercept, $(0,0)$; y-intercept, $(0,0)$; $(1,-2)$

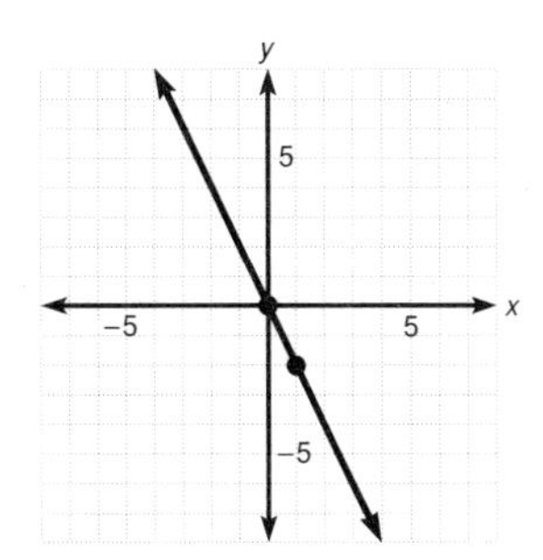

27. x-intercept, (0,0); y-intercept, (0,0); (1,2)

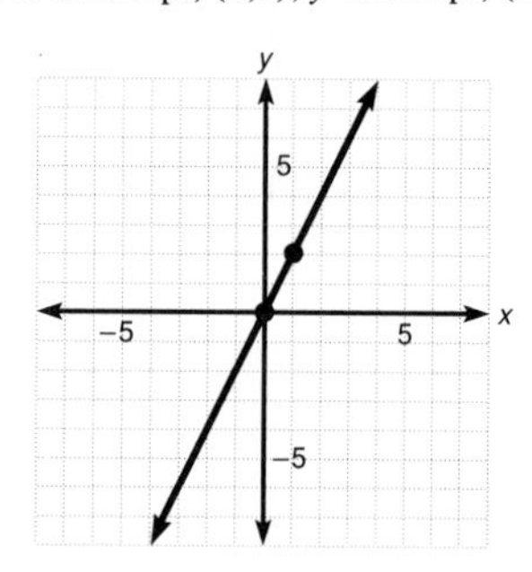

29. x-intercept, (3,0); y-intercept, (0,−4)

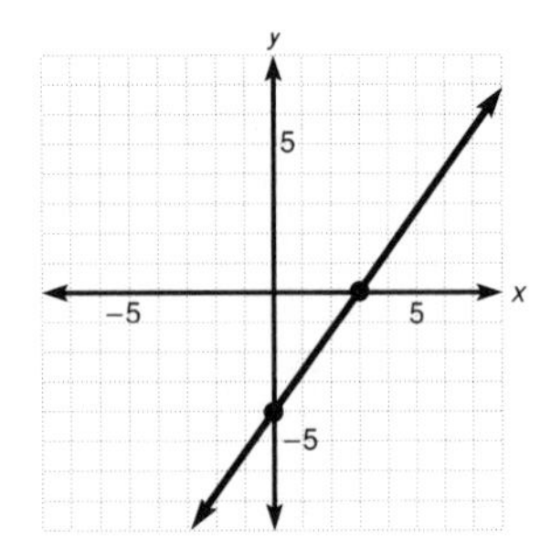

31. x-intercept, (2,0); y-intercept, (0,5)

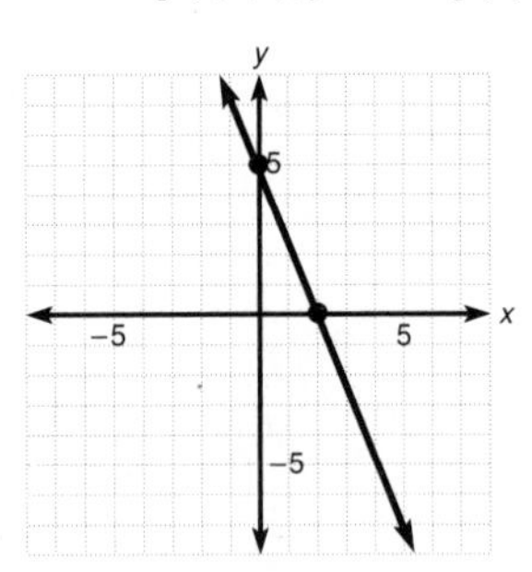

33. x-intercept, (6,0); y-intercept, (0,−5)

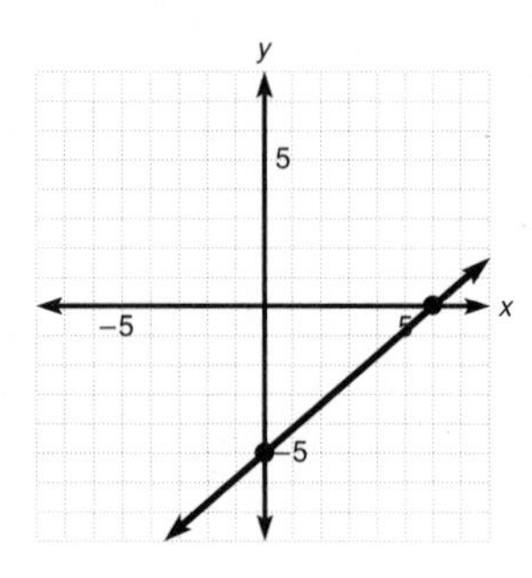

35. no x-intercept; y-intercept, (0,6)

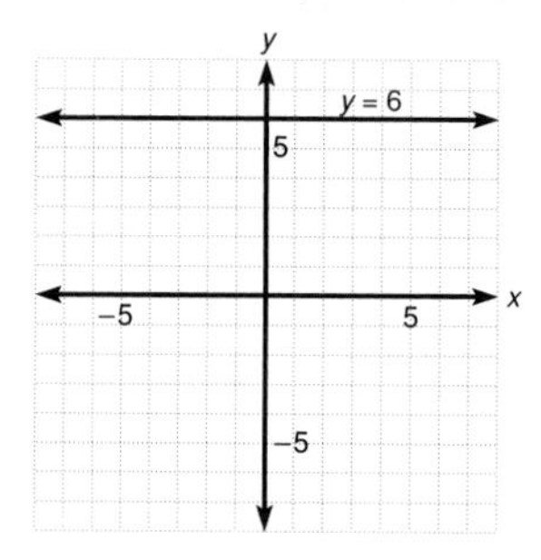

37. x-intercept, (5,0); no y-intercept

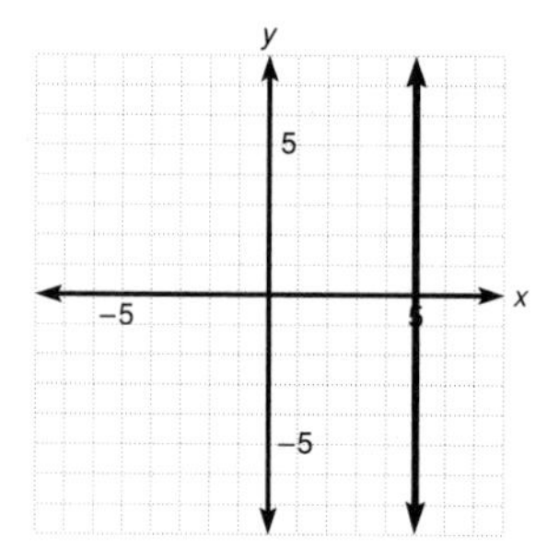

39. x-intercept, (0,0); All real numbers are y-intercepts.

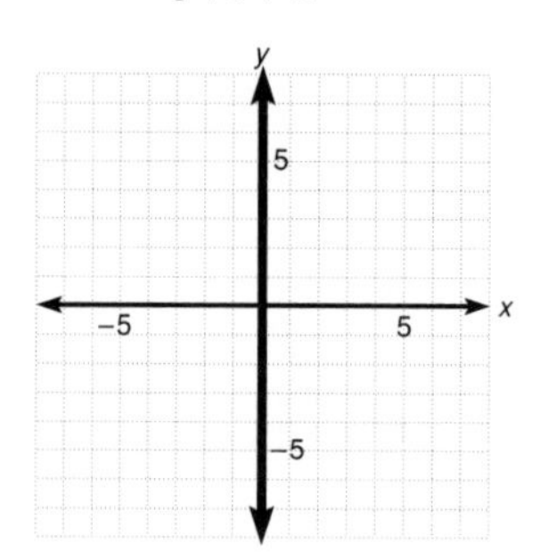

41. $y = 2x - 7$ **43.** $y = \frac{4}{3}x + 3$ **45.** $y = -\frac{7}{3}x + \frac{10}{3}$

47. $y = \frac{1}{5}x + \frac{7}{5}$ **49.** $y = \frac{8}{5}x - \frac{14}{5}$

51. **a.** $y = -2x + 5$ **b.** $y = -2x$ **c.** $y = -2x - 3$

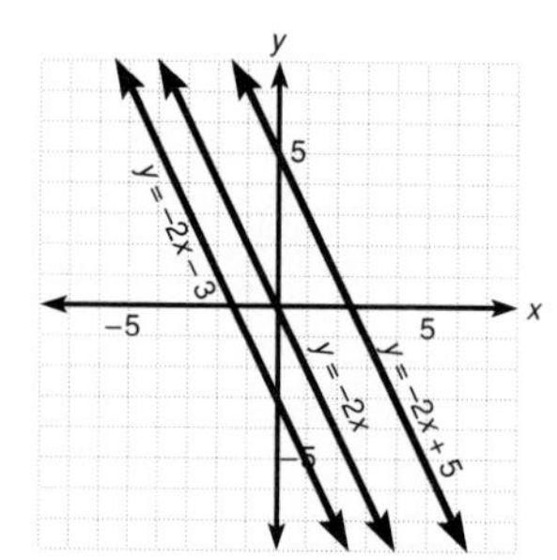

53. $y = 2x - 3$; x-intercept $\left(\frac{3}{2},0\right)$; y-intercept, $(0,-3)$

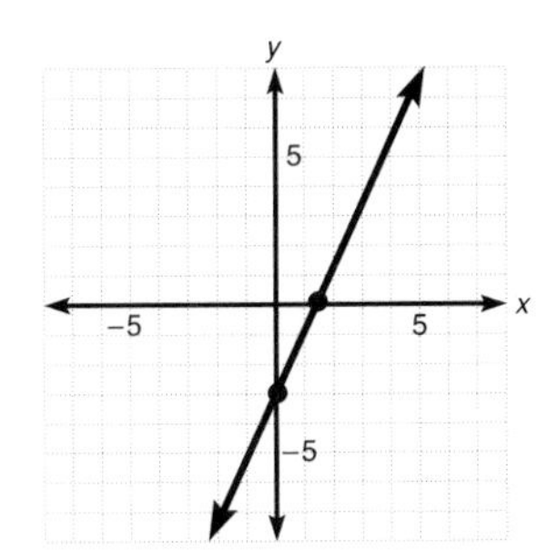

55. $3y - 2x = 6$; x-intercept, $(-3,0)$;
y-intercept, $(0,2)$

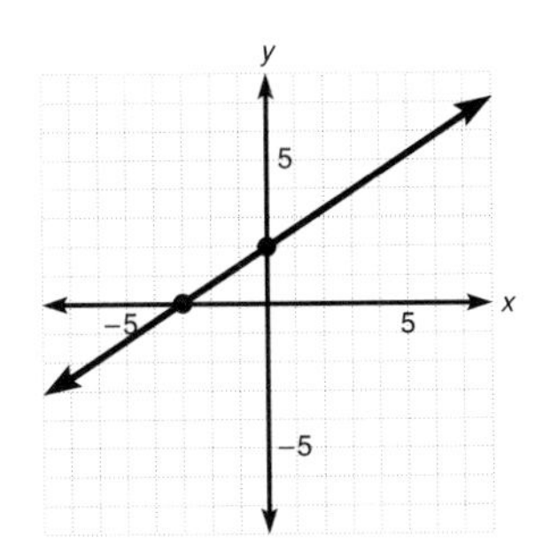

Solutions to trial exercise problems

1. $y = 2x + 4$
Let $x = 0$,
then $y = 2(0) + 4 = 0 + 4 = 4$ $(0,4)$ is the y-intercept.
Let $y = 0$,
then $0 = 2x + 4$, $2x = -4$
$x = -2$ $(-2,0)$ is the x-intercept.

5. $2x + 3y = 6$
Let $x = 0$, then
$2(0) + 3y = 6$
$0 + 3y = 6$
$3y = 6$
$y = 2$ $(0,2)$ is the y-intercept.
Let $y = 0$, then
$2x + 3(0) = 6$
$2x + 0 = 6$
$2x = 6$
$x = 3$ $(3,0)$ is the x-intercept.

12. $y - 4x = 0$
Let $x = 0$, then
$y - 4(0) = 0$
$y - 0 = 0$
$y = 0$ $(0,0)$ is the y-intercept.
Let $y = 0$, then
$0 - 4x = 0$
$-4x = 0$
$x = 0$ $(0,0)$ is the x-intercept.

17. $y = 3x + 6$

x	y	
0	6	y-intercept
−2	0	x-intercept
−1	3	checkpoint

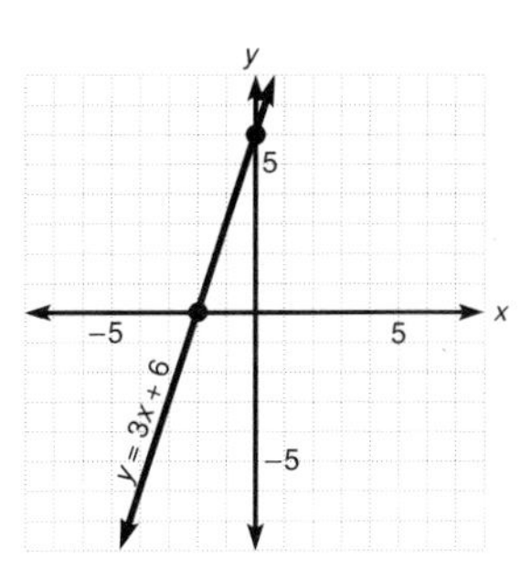

30. $3x - 2y = 6$

x	y	
0	−3	y-intercept
2	0	x-intercept
1	$-\frac{3}{2}$	checkpoint

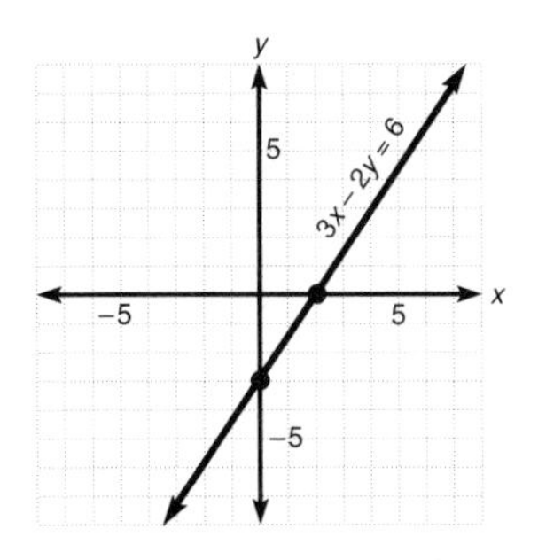

41. $y - 2x + 7 = 0$
$y - 2x = -7$ Add -7 to both members.
$y = 2x - 7$ Add $2x$ to both members.

43. $3y - 4x = 9$
$3y = 4x + 9$ Add $4x$ to both members.
$y = \frac{4x + 9}{3}$ Divide both members by 3.
$y = \frac{4}{3}x + 3$ Divide by 3.

51. When $b = 5$, $y = -2x + 5$
When $b = 0$, $y = -2x$
When $b = -3$, $y = -2x - 3$

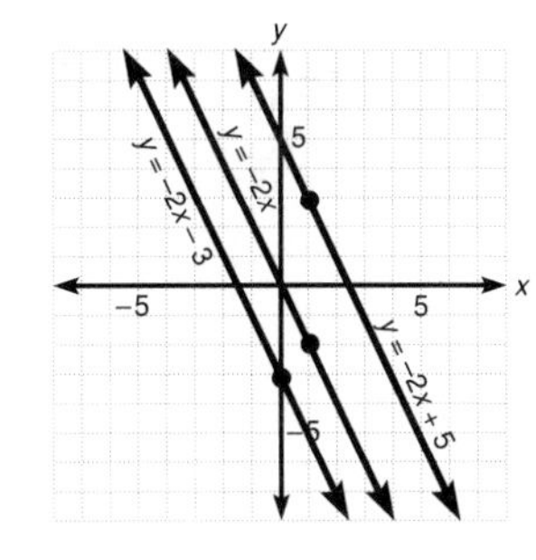

54. $x = y + 4$

x	y	
0	−4	y-intercept
4	0	x-intercept
1	−3	checkpoint

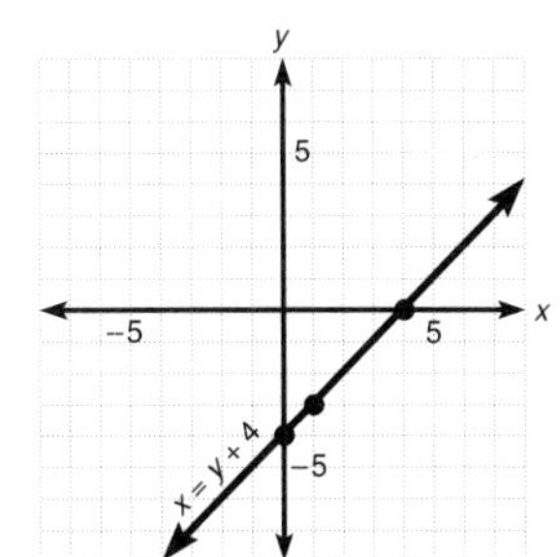

Review exercises

1. $\frac{6}{5}$ **2.** $\frac{1}{x+4}$ **3.** $\frac{x-4}{x+5}$ **4.** $-\frac{1}{5}$ **5.** $y = -3x + 6$
6. 4 and 12; -12 and -4

Exercise 7–3

Answers to odd-numbered problems

1. $m = -\frac{1}{2}$

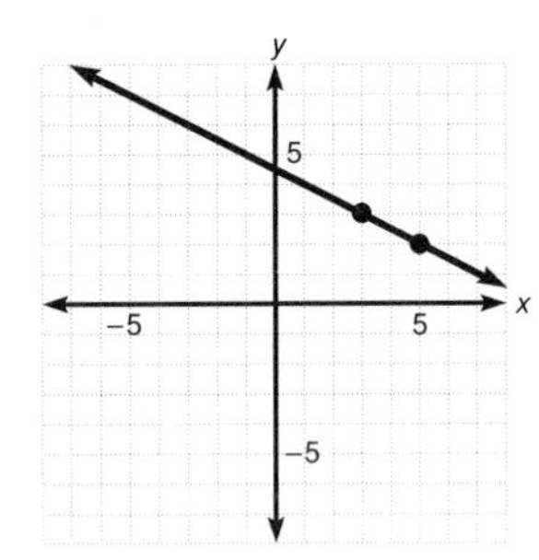

3. $m = \frac{2}{3}$

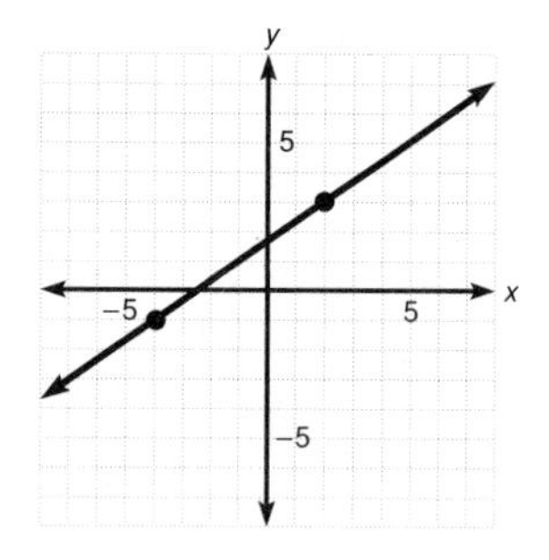

5. $m = 0$

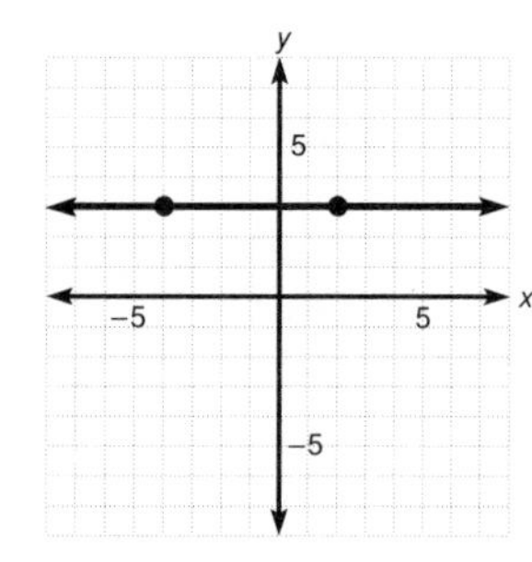

7. m is undefined.

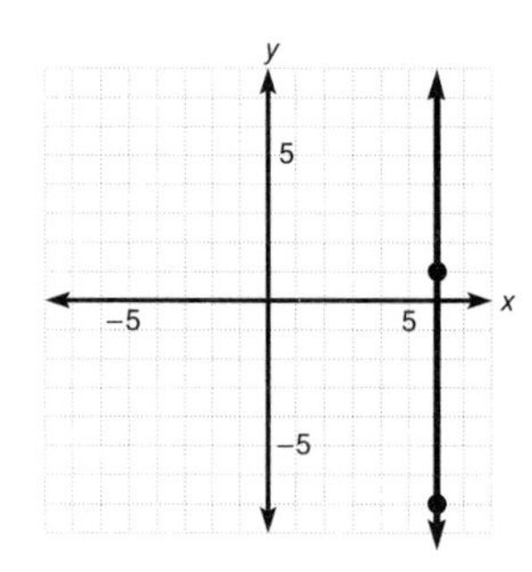

9. $m = \frac{2}{3}$ **11.** undefined **13.** $m = 1$ **15.** undefined

17. $m = \frac{1}{7}$ **19.** $m = -\frac{3}{4}$ **21.** $m = \frac{10}{11}$

23. $(-5,3), (4,-2); m = -\frac{5}{9}$ **25.** $(3,5), (-2,-6); m = \frac{11}{5}$

27. $(6,5), (-3,-5); m = \frac{10}{9}$ **29.** $m = \frac{2}{3}$ **31.** $m = \frac{2}{3}$

33. $m = 4$ **35.** $m = \frac{15}{2}$

Solutions to trial exercise problems

1. $(5,2), (3,3); m = \frac{3-2}{3-5} = \frac{1}{-2} = -\frac{1}{2}$

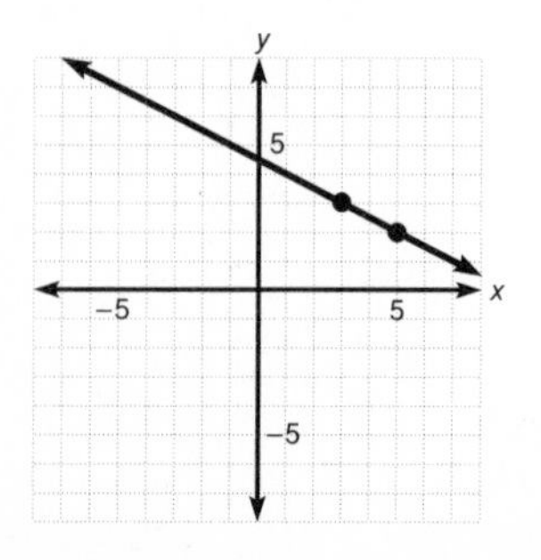

6. $(4,-4), (2,-4); m = \frac{-4-(-4)}{4-2} = \frac{0}{2} = 0$

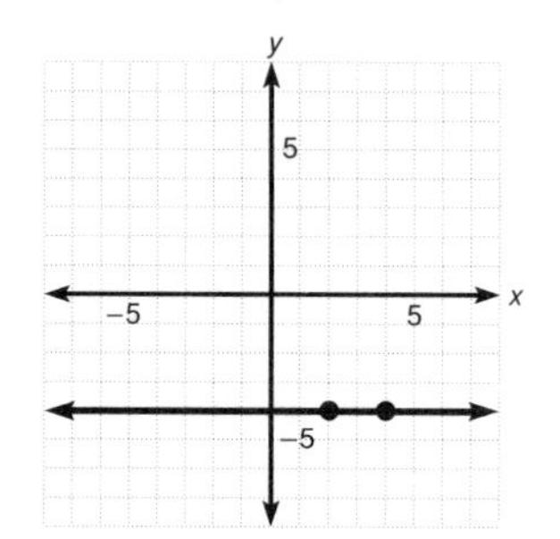

15. $(0,7), (0,-8); m = \frac{7-(-8)}{0-0} = \frac{15}{0}$ slope is undefined

23. Using points $(-5,3)$ and $(4,-2)$, $m = \frac{3-(-2)}{-5-4} = \frac{5}{-9} = -\frac{5}{9}$

29. $m = \frac{8}{12} = \frac{2}{3}$

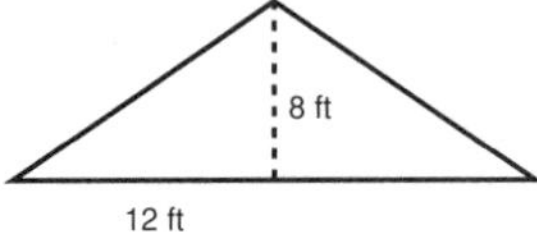

33. $m = \frac{1{,}000}{250} = \frac{4}{1} = 4$

Review exercises

1. -9 **2.** -13 **3.** $-\frac{1}{8}$ **4.** $\frac{b^5}{a^5}$ **5.** x **6.** 13
7. $y = -3x - 1$ **8.** $y = x - 2$

Exercise 7–4

Answers to odd-numbered problems

1. $y = 4x - 1$ **3.** $y = -2x - 1$ **5.** $5y - 3x = 29$
7. $6y + 7x = 35$ **9.** $4y - 5x = -13$ **11.** $y = -3$
13. $x = 1$ **15.** $x - 2y = 0$ **17.** $3x + 8y = 7$
19. $-x + y = 4$ **21.** $x + 7y = -6$ **23.** $-8x + 3y = 24$
25. $8x + 5y = 0$ **27.** $y = -x + 2$; $m = -1$; y-intercept, 2
29. $y = -3x - 2$; $m = -3$; y-intercept, -2
31. $y = -\frac{2}{5}x + 2$; $m = -\frac{2}{5}$; y-intercept, 2
33. $y = \frac{7}{2}x + 2$; $m = \frac{7}{2}$; y-intercept, 2 **35.** $y = \frac{9}{2}x - 3$; $m = \frac{9}{2}$; y-intercept, -3 **37.** $y = \frac{8}{9}x - \frac{1}{9}$; $m = \frac{8}{9}$; y-intercept, $-\frac{1}{9}$

39. $m = 2$; $b = -4$

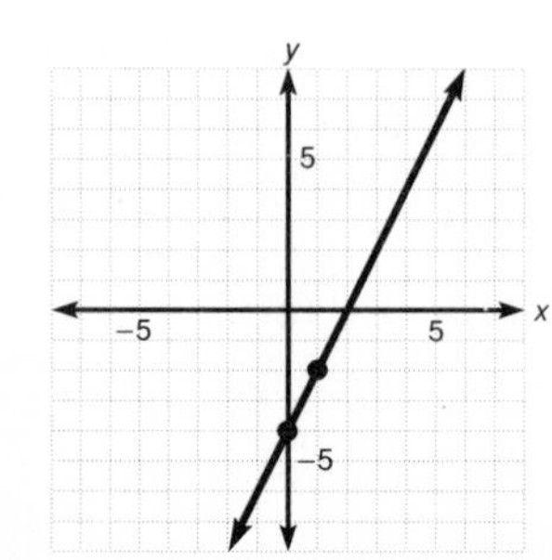

41. $m = -5$; $b = 2$

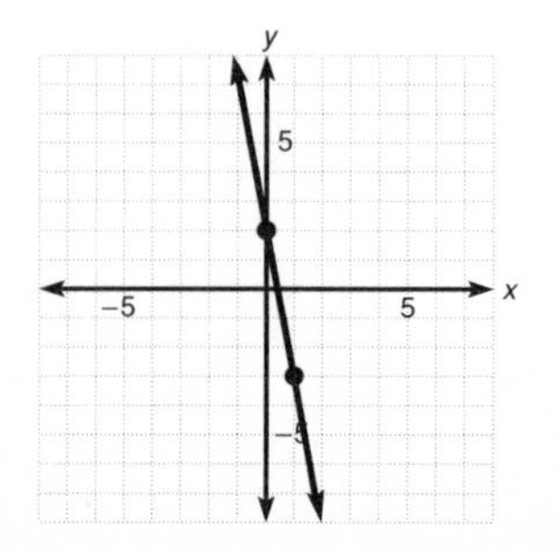

43. $m = \frac{2}{3}$; $b = -1$

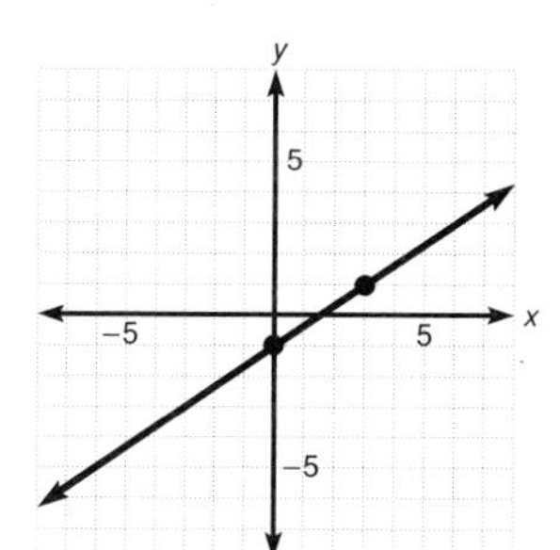

45. $m = -2$; $b = -3$

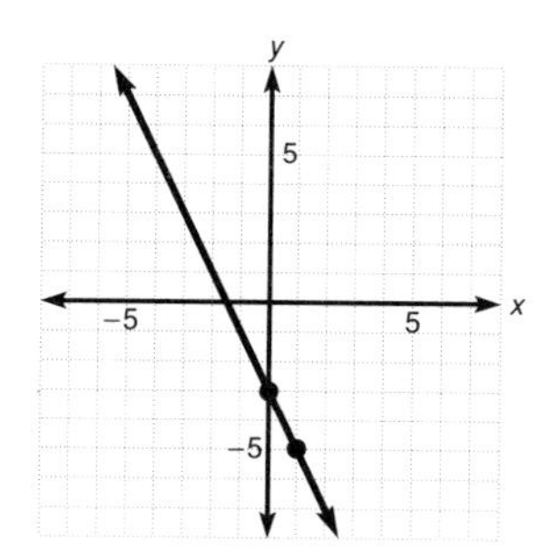

47. $m = \frac{5}{2}$; $b = -3$

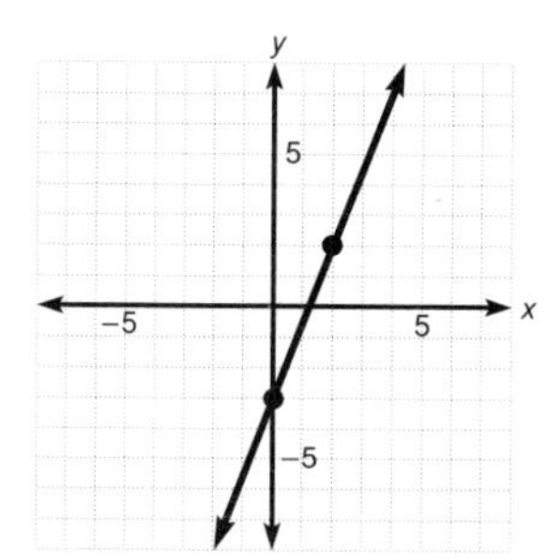

49. $m = -\frac{4}{3}$; $b = -3$

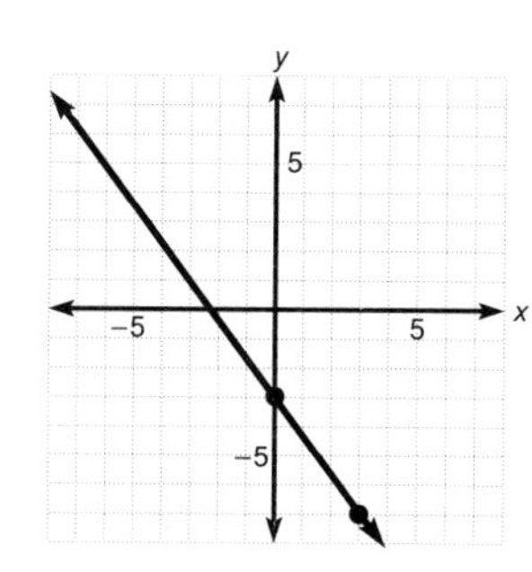

51. $(4,3)$; $m = \frac{3}{4}$

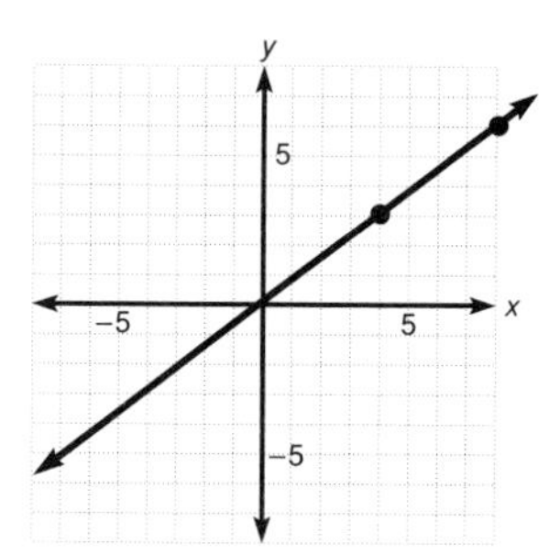

53. $(-1,2)$; $m = -\frac{1}{3}$

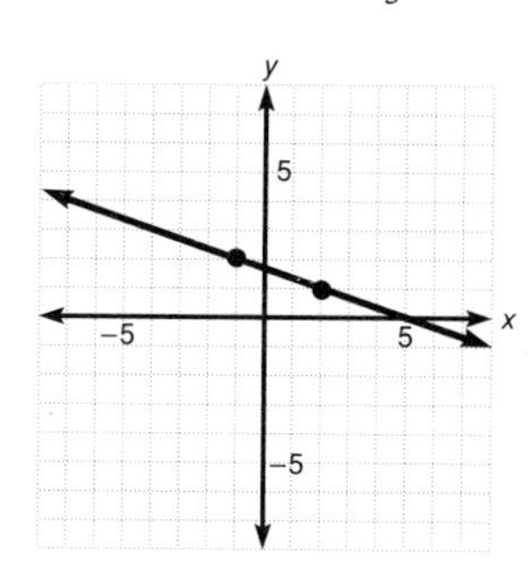

55. $(-4,-5)$; $m = 4$

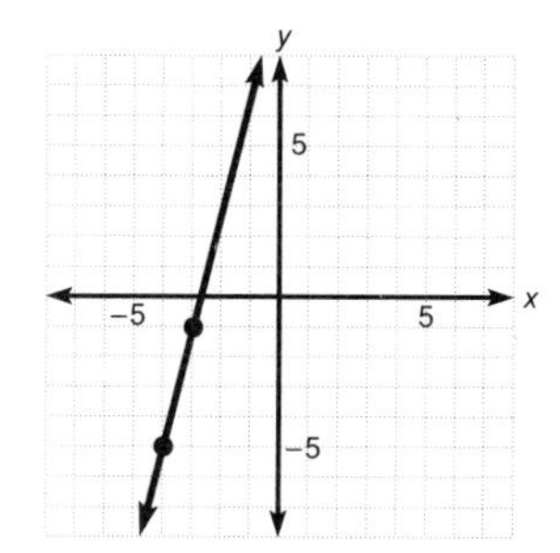

57. $(0,3)$; $m = -1$

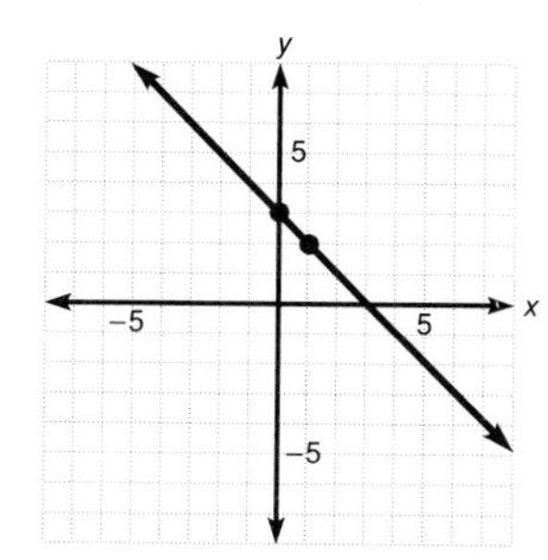

59. $(5,6)$; $m = 0$

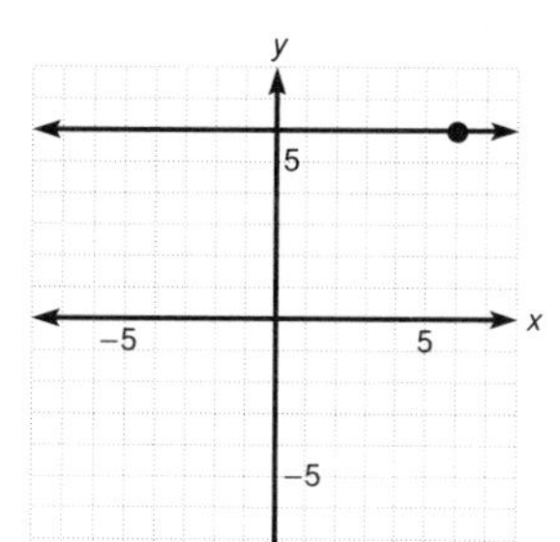

61. $(-1, -1)$; undefined slope

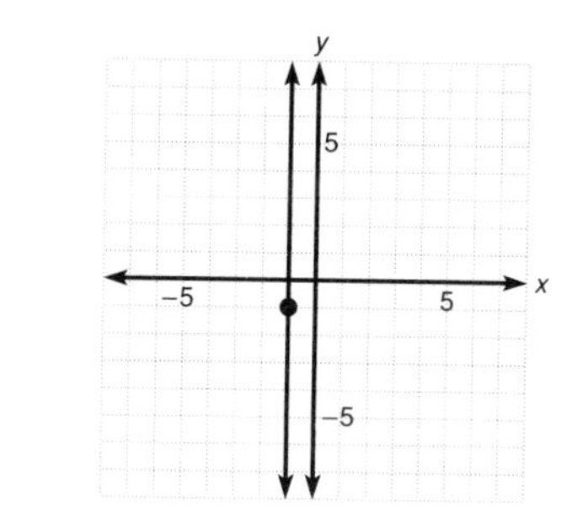

63. $y = 4x - 1$ **65.** $y = -\frac{5}{3}x + 2$ **67.** $y = -1$

69. $m_1 = -1$, $m_2 = -1$; parallel **71.** $m_1 = -1$, $m_2 = 1$; perpendicular **73.** $m_1 = 3$, $m_2 = -3$; neither **75.** $m_1 = -5$, $m_2 = -\frac{10}{3}$; neither **77.** $m_1 = 4$, $m_2 = 4$; parallel **79.** $m_1 = 0$, $m_2 =$ undefined; perpendicular **81.** $(-6,2)$, $(7,-3)$; $5x + 13y = -4$ **83.** $(3,-2)$, $(-4,-2)$; $y = -2$ **85.** $(1,-4)$, $(-6,4)$; $8x + 7y = -20$

Solutions to trial exercise problems

1. $(1,3)$; $m = 4$

Using $y - y_1 = m(x - x_1)$

$$y - 3 = 4(x - 1)$$
$$y - 3 = 4x - 4$$
$$y = 4x - 1$$

8. $(-3,0)$; $m = -\frac{5}{8}$

$$y - 0 = -\frac{5}{8}[x - (-3)]$$
$$8y = -5(x + 3)$$
$$8y = -5x - 15$$
$$8y + 5x = -15$$

15. $(2,1)$ and $(6,3)$

$$m = \frac{3 - 1}{6 - 2} = \frac{2}{4} = \frac{1}{2}$$

Using $y - y_1 = m(x - x_1)$ and point $(2,1)$

$$y - 1 = \frac{1}{2}(x - 2)$$
$$2y - 2 = 1(x - 2)$$
$$2y - x = 0$$

23. $(0,8)$ and $(-3,0)$

$$m = \frac{8 - 0}{0 - (-3)} = \frac{8}{3}$$

Using $y - y_1 = m(x - x_1)$ and point $(0,8)$

$$y - 8 = \frac{8}{3}(x - 0)$$
$$3y - 24 = 8x$$
$$3y - 8x = 24$$

29. $3x + y = -2$ $\qquad m = -3$

$y = -3x - 2$ $\qquad b = -2$

31. $2x + 5y = 10$

$$5y = -2x + 10$$
$$y = \frac{-2x + 10}{5} \qquad m = -\frac{2}{5}$$
$$y = -\frac{2}{5}x + 2 \qquad b = 2$$

39. $y = 2x - 4;$
$m = 2$ and $b = -4$
$\left(\textit{Note: } m = \dfrac{2}{1} = \dfrac{\text{rise}}{\text{run}}\right)$

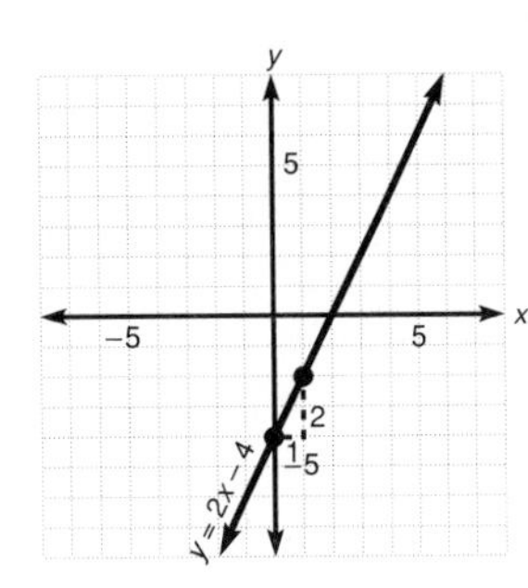

47.
$$5x - 2y = 6$$
$$-2y = -5x + 6$$
$$y = \frac{-5x + 6}{-2}$$
$$y = \frac{5}{2}x - 3$$
$$m = \frac{5}{2} \text{ and } b = -3$$

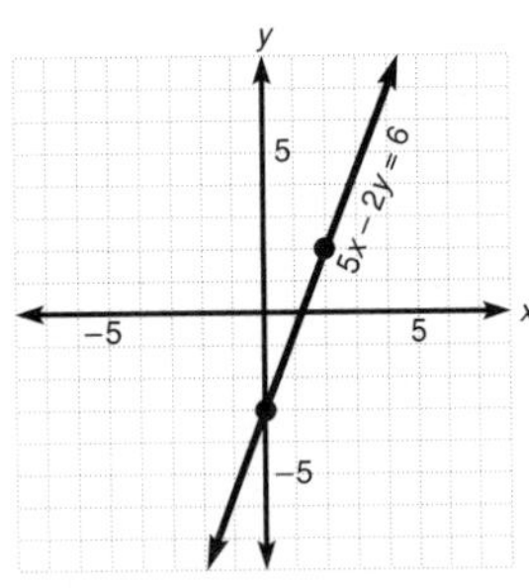

51. $(4,3);\ m = \dfrac{3}{4}$

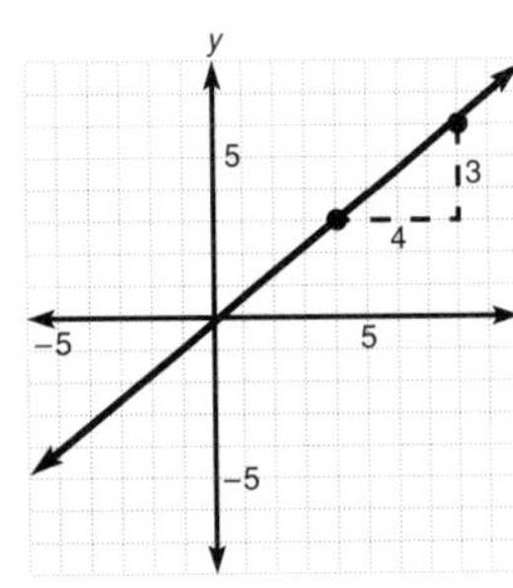

56. $(2,-3);\ m = 2 = \dfrac{2}{1}$

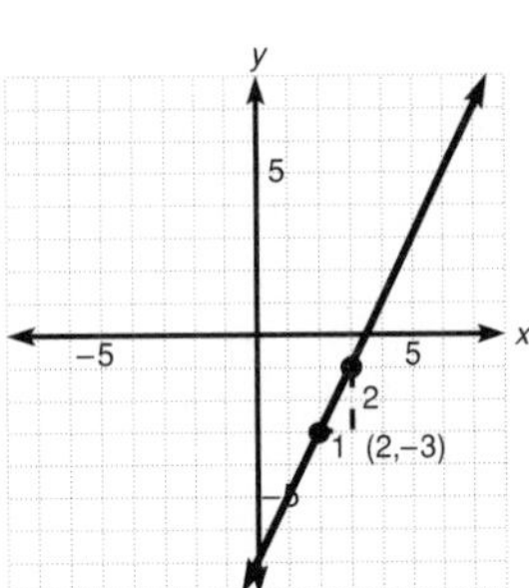

59. $(5,6);\ m = 0$

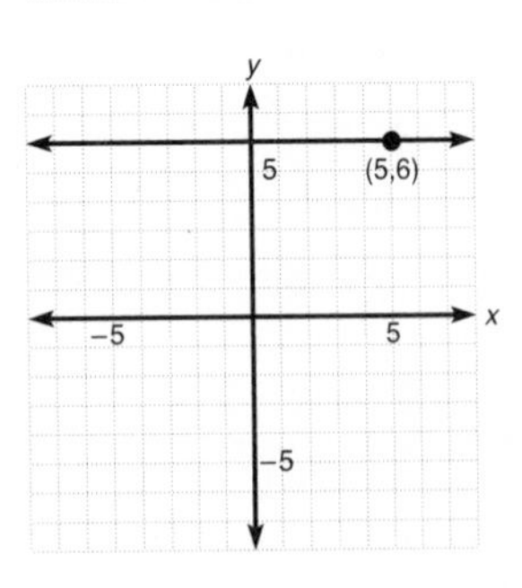

69. $x + y = 4$
$x + y = -7$
Then $y = -x + 4 \quad m = -1$
$y = -x - 7 \quad m = -1$
The slopes are equal so the lines are *parallel.*

74.
$$x + 2y = 5$$
$$-6x + 3y = -1$$
Now $x + 2y = 5$
$$2y = -x + 5$$
$$y = -\frac{1}{2}x + \frac{5}{2}$$
$$m_1 = -\frac{1}{2}$$

$$-6x + 3y = -1$$
$$3y = 6x - 1$$
$$y = 2x - \frac{1}{3}$$
$$m_2 = 2$$

The lines are *perpendicular* since $m_1 m_2 = -\dfrac{1}{2} \cdot 2 = -1.$

81. It appears the line passes through points $(7,-3)$ and $(-6,2)$.
Then $m = \dfrac{2 - (-3)}{-6 - 7} = \dfrac{5}{-13} = \dfrac{-5}{13}$
and $y - (-3) = -\dfrac{5}{13}(x - 7)$
$$y + 3 = -\frac{5}{13}(x - 7)$$
$$13y + 39 = -5x + 35$$
$$5x + 13y = -4$$

Review exercises

1. $41\frac{1}{2}$ ft and $58\frac{1}{2}$ ft **2.** $0, -3, 9, \frac{9}{3}, \frac{0}{4}$ **3.** $\{-3\}$ **4.** $\left\{\frac{1}{7}\right\}$
5. $\{10\}$ **6.** $(3x - 2)(x + 2)$ **7.** Not factorable
8. $8(y + 2x)(y - 2x)$

Exercise 7–5

Answers to odd-numbered problems

1. $x < 4$

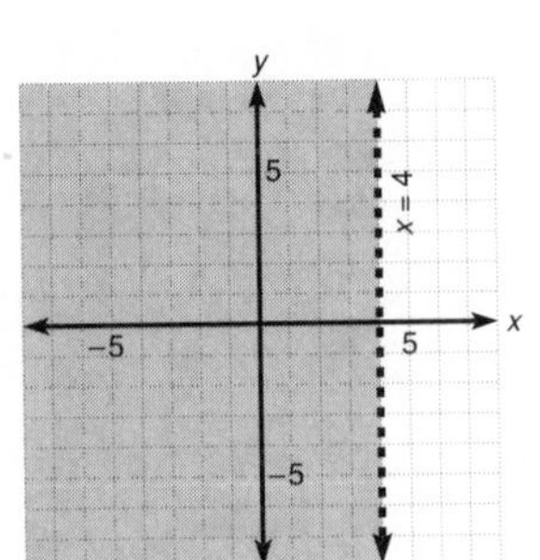

3. $x + y \geq 4$

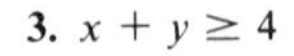

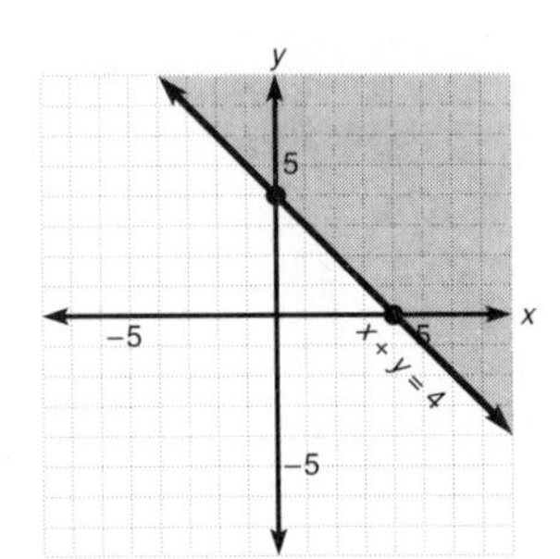

5. $3x - y \leq 4$

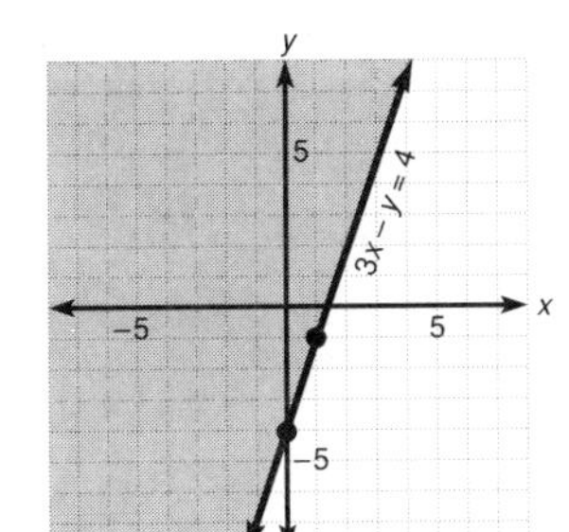

7. $3x + 5y \geq 0$

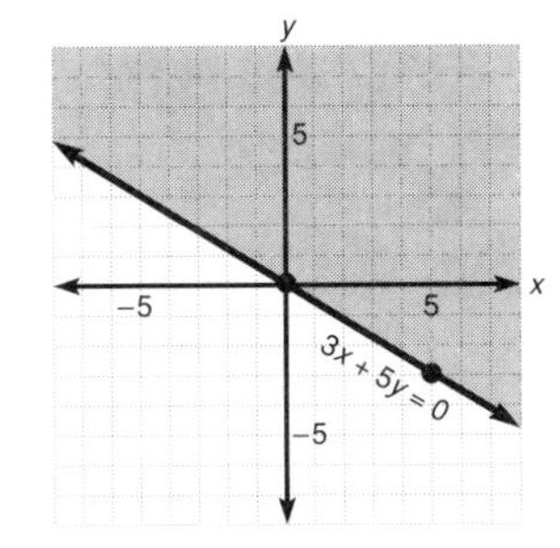

21. $5x - 4y \leq 20$

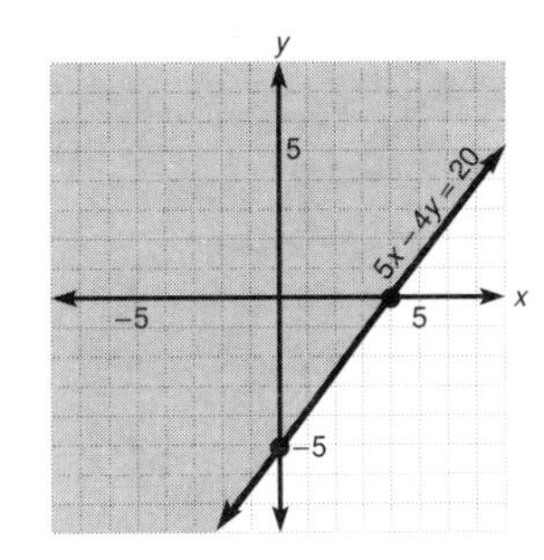

23. $x > 3$

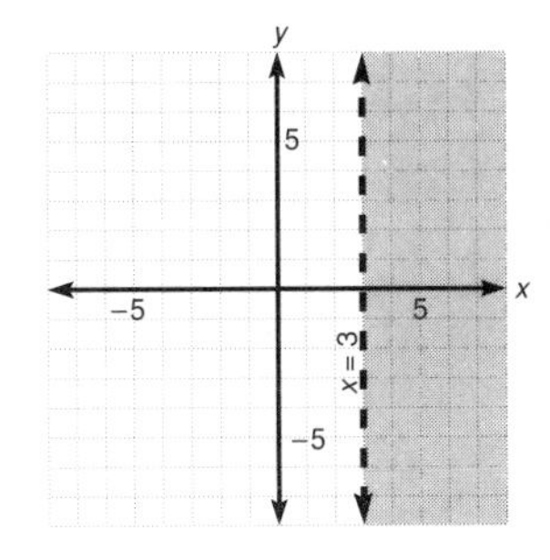

9. $2y \leq 3x$

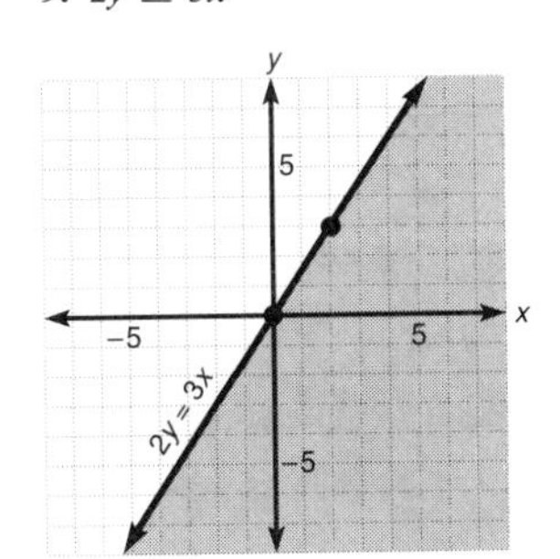

11. $x + y \leq -2$

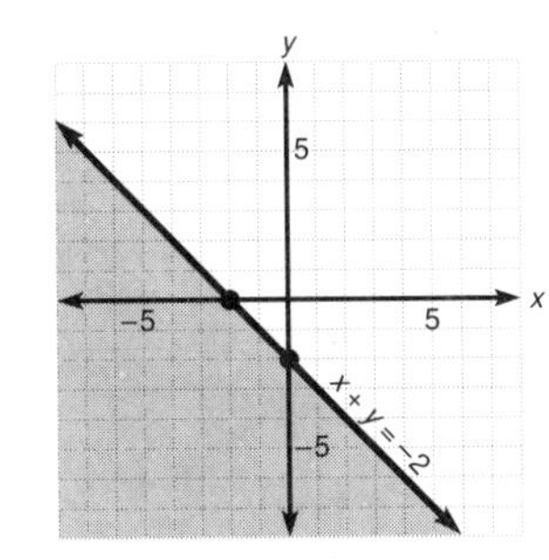

25. $x \leq -2$

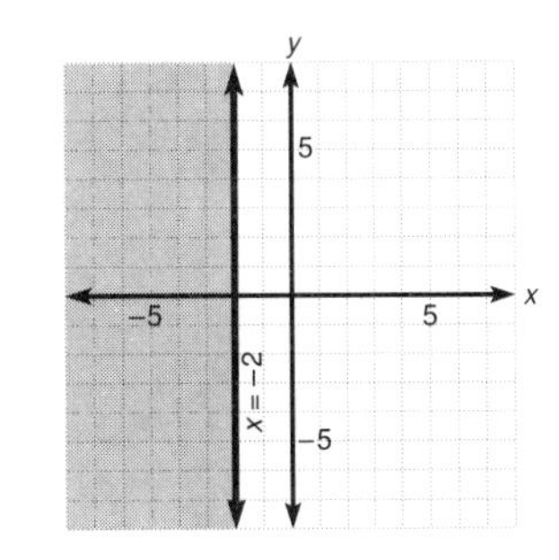

27. $x + 4 > 0$

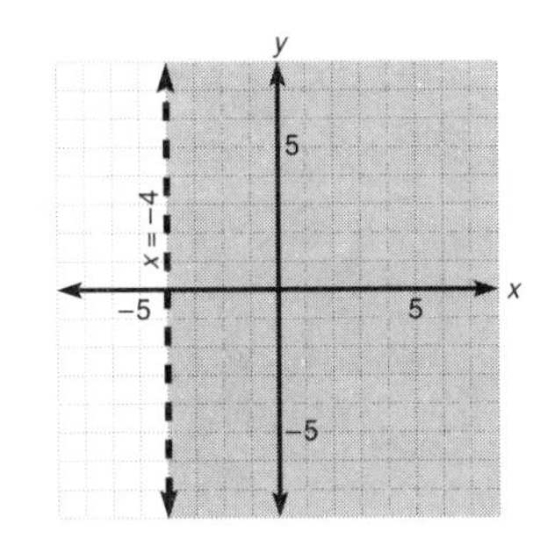

13. $x - y < 6$

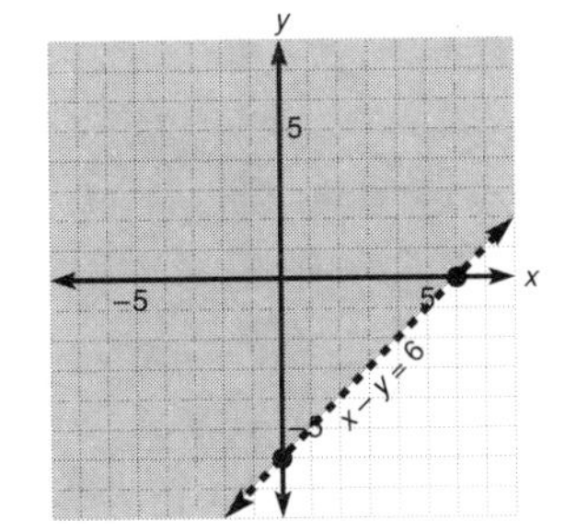

15. $x + 3y > 1$

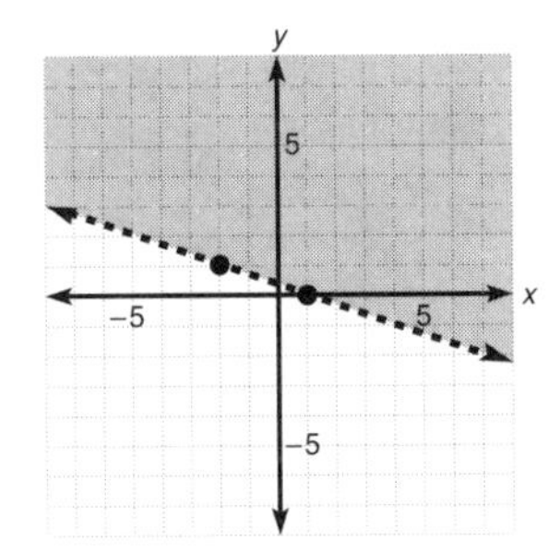

29. $y + 3 \leq 5$

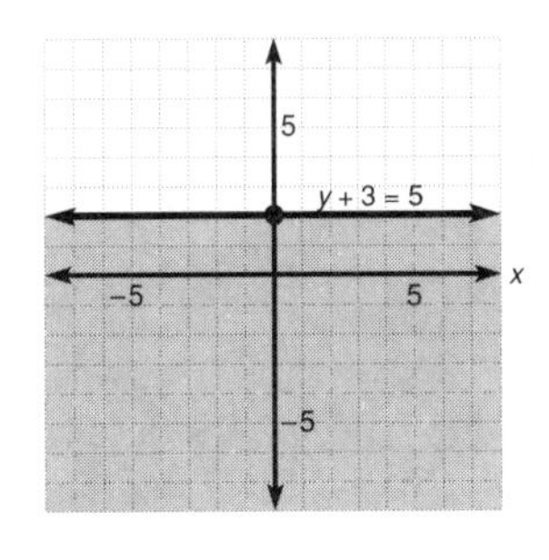

31. $y > x - 2$

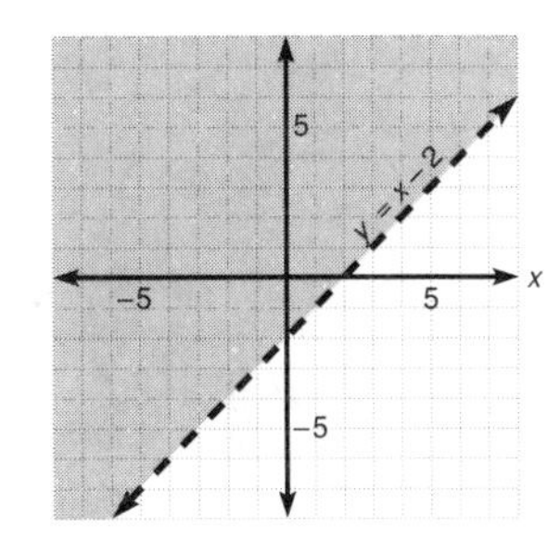

17. $y \geq 4x + 1$

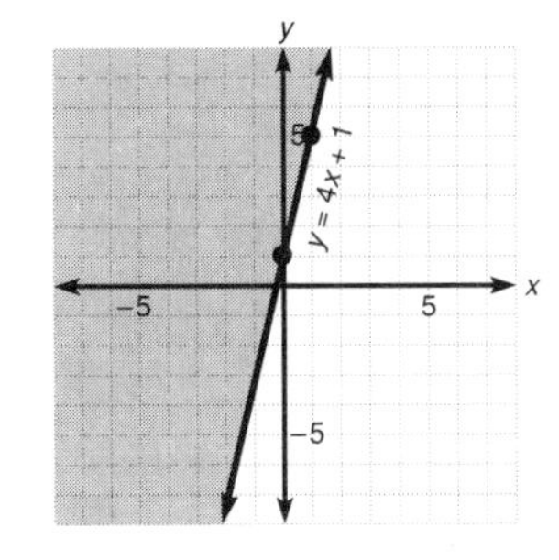

19. $3x - 5y < -15$

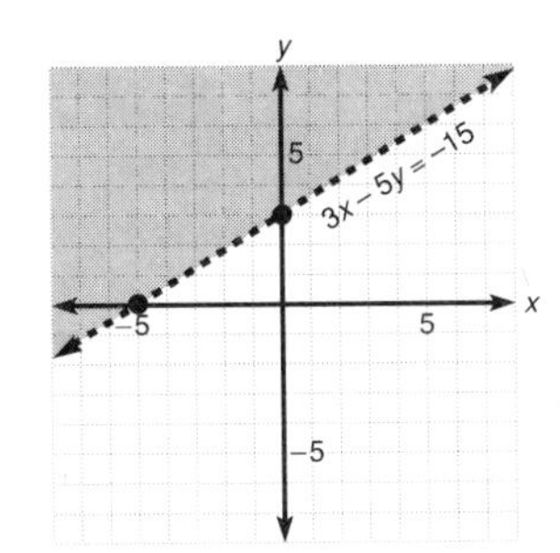

33. $x < -y$

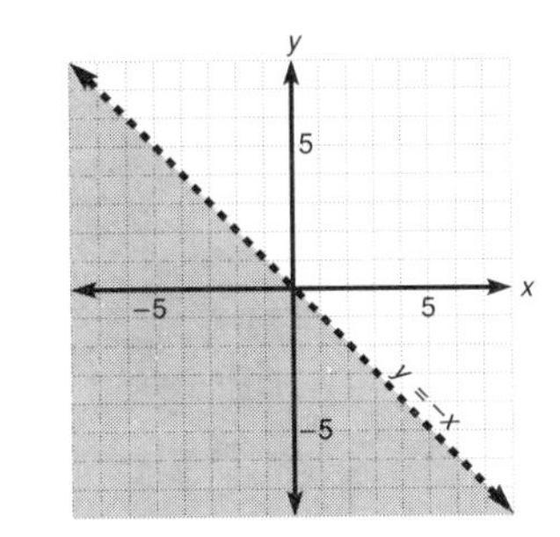

35. $x \geq 3y$

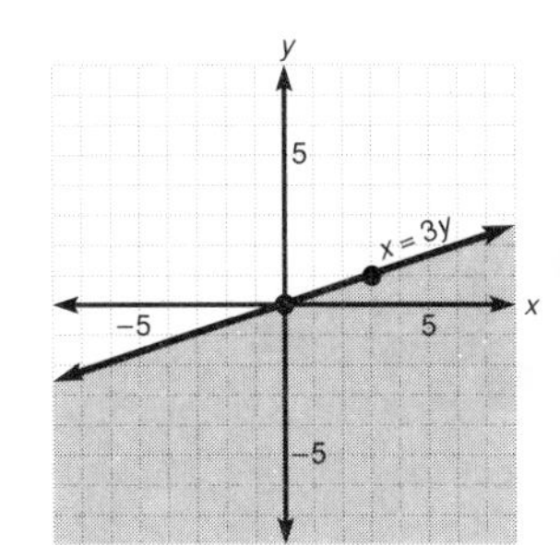

37. $8x - 3y \geq 0$

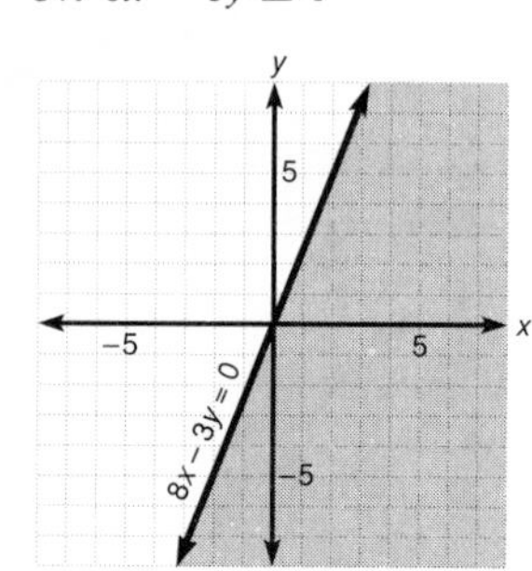

39. $3x + 2y < 0$

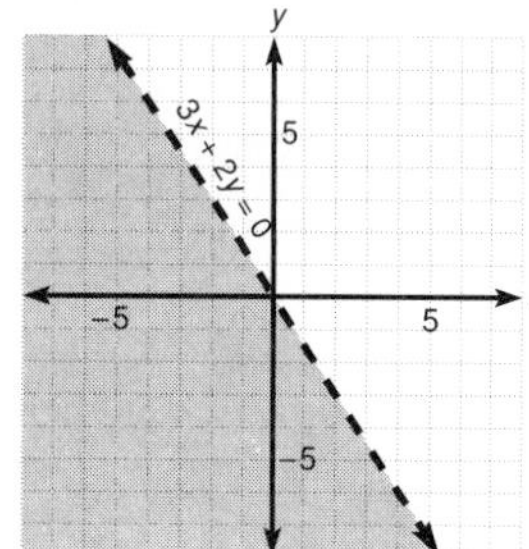
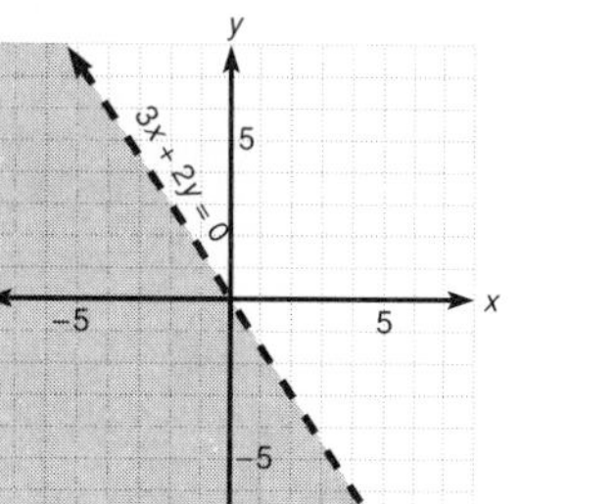

41. a. $x > 0$ and $y > 0$
b. $x < 0$ and $y > 0$
c. $x < 0$ and $y < 0$
d. $x > 0$ and $y < 0$

Solutions to trial exercise problems

1. $x < 4$
Graph $x = 4$ in a dashed line.
Using (0,0),
$0 < 4$ (true)
Shade half-plane containing the origin.

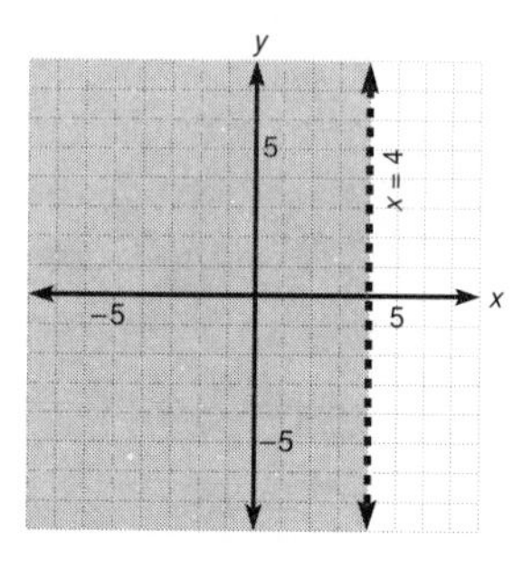

6. $2x - 4y > 8$
Graph $2x - 4y = 8$ in a dashed line.
Using (0,0),
$2(0) - 4(0) > 8$
$0 - 0 > 8$
$0 > 8$ (false)
Shade half-plane that does not contain the origin.

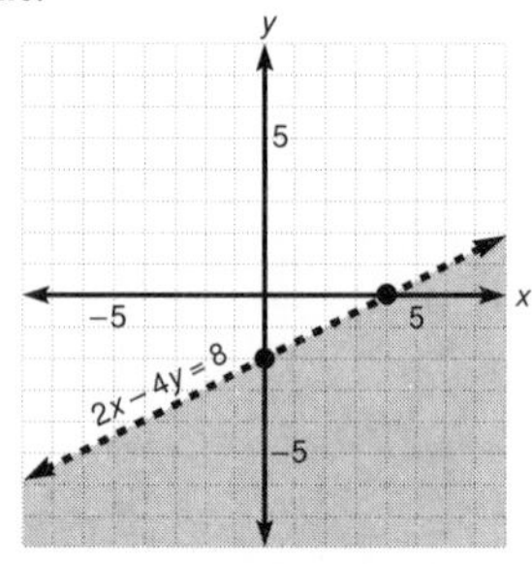

8. $y - 2x < 0$
Graph $y - 2x = 0$ in a dashed line.
Using point (3,0),
$0 - 2(3) < 0$
$-6 < 0$ (true)
Shade half-plane containing point (3,0).

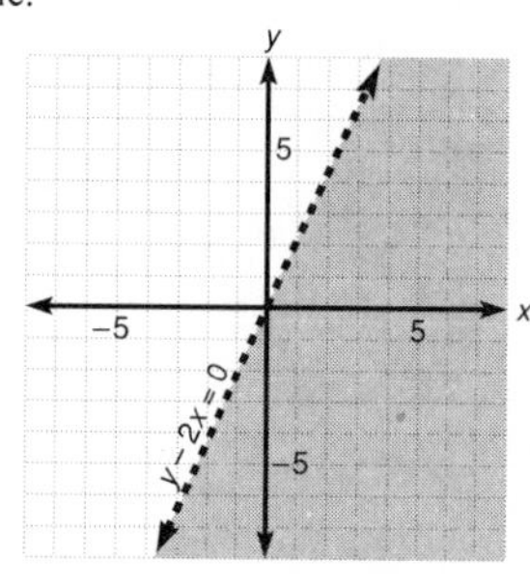

14. $x - y \geq -5$
Graph $x - y = -5$ in a *solid* line.
Using (0,0),
$0 - 0 \geq -5$
$0 \geq -5$ (true)
Shade half-plane containing the origin.

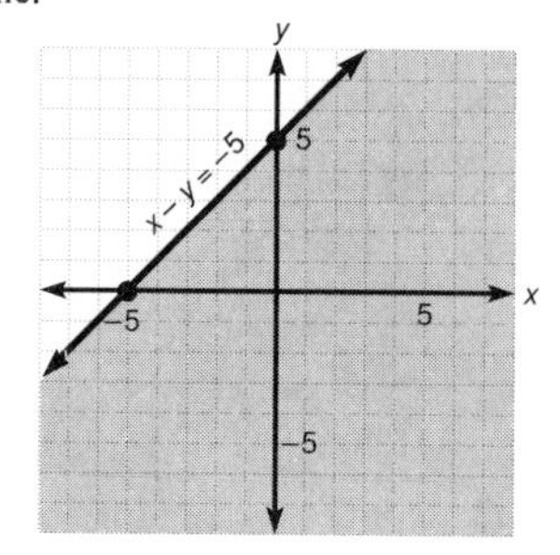

23. $x > 3$
Graph $x = 3$ in a dashed line.
Using (0,0),
$0 > 3$ (false)
Shade half-plane *not* containing the origin.

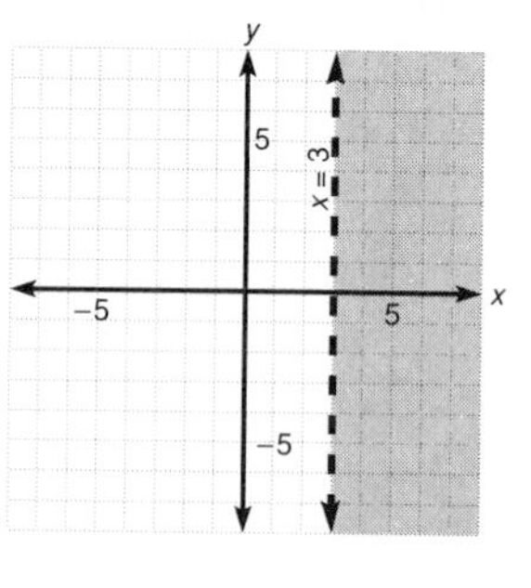

35. $x \geq 3y$
Graph $x = 3y$ in a solid line.
Using point (1,1),
$1 \geq 3$ (false)
Shade the half-plane that does not contain the point (1,1).

Review exercises

1. all real numbers except 3 **2.** all real numbers except -5 and 2 **3.** all real numbers **4.** $y = \frac{5}{2}$ **5.** $x = 13$ **6.** $x + 3y = 7$ **7.** $3x - 4y = 8$ **8.** $x = -3$ **9.** $y = -3$

Exercise 7–6

Answers to odd-numbered problems

1. domain $= \{-5,-3,-1,0,1,3,5\}$, range $= \{-16,-10,-4,-1,2,8,14\}$ **3.** domain $= \{0,1,2,3,4,5\}$, range $= \left\{-\frac{1}{3},\frac{1}{3},1,\frac{5}{3},\frac{7}{3},3\right\}$ **5.** domain $= \{-10,-5,0,5,10\}$, range $= \{-4\}$ **7.** $f(2) = 4, f(0) = 0, f(-1) = -2$ **9.** $f(5) = 10, f(-5) = 0, f(0) = 5$ **11.** $f(-3) = -4$, $f\left(-\frac{5}{3}\right) = 0, f(1) = 8$ **13.** $f(-7) = 3, f(0) = 3, f(0.41) = 3$

15. $(-3,-14), \left(-\frac{2}{5},-1\right), (0,1), (4,21), \left(\frac{1}{5},2\right)$

17. $(6,5), (-3,-1), (0,1), \left(\frac{3}{2},2\right), \left(-\frac{3}{2},0\right)$

19. $(-1,0), (-10,0), (0,0), \left(\frac{8}{3},0\right), (2,0)$ **21.** $f(100) = 212$, $f(0) = 32, f(25) = 77, f(-10) = 14$ **23.** $h(2) = 58$, $h(3) = 87, h(4) = 116$, domain = {positive real numbers} **25.** $g(2) = 0.14, g(3) = 0.21, g(1.2) = 0.084$
27. $g(N) = 0.7N, g(5) = 3.5, g(10) = 7, g(3) = 2.1$
29. $h(n) = \frac{n}{4} + 40, h(50) = 52.5, h(12) = 43, h(120) = 70$
31. $g(h) = 10.5\,h, g(40) = 420, g(48) = 504, g(28) = 294$
33. $f(C,M) = C + M; f(25,7) = 32, f(32,11) = 43, f(146,27) = 173$

Solutions to trial exercise problems

3. $2x - 3y = 1; x \in \{0,1,2,3,4,5\}$
The domain is $\{0,1,2,3,4,5\}$ and since when we solve for y we get $y = \frac{2x-1}{3}$, then when

$x = 0, y = \frac{2(0)-1}{3} = -\frac{1}{3}$

$x = 1, y = \frac{2(1)-1}{3} = \frac{1}{3}$

$x = 2, y = \frac{2(2)-1}{3} = 1$

$x = 3, y = \frac{2(3)-1}{3} = \frac{5}{3}$

$x = 4, y = \frac{2(4)-1}{3} = \frac{7}{3}$

$x = 5, y = \frac{2(5)-1}{3} = 3$

Then the range is $\left\{-\frac{1}{3},\frac{1}{3},1,\frac{5}{3},\frac{7}{3},3\right\}$.

7. $f(x) = 2x; f(2), f(0), f(-1)$
$f(2) = 2(2) = 4, f(0) = 2(0) = 0, f(-1) = 2(-1) = -2$

12. $f(s) = 5s + 2; f(-2), f(3), f\left(-\frac{2}{5}\right)$
$f(-2) = 5(-2) + 2 = -10 + 2 = -8$
$f(3) = 5(3) + 2 = 15 + 2 = 17$
$f\left(-\frac{2}{5}\right) = 5\left(-\frac{2}{5}\right) + 2 = -2 + 2 = 0$

13. $f(x) = 3; f(-7), f(0), f(0.41)$
$f(-7) = 3, f(0) = 3, f(0.41) = 3$

16. $f(x) = 7 - 2x; f(-4), f\left(-\frac{1}{2}\right), f(0), f\left(\frac{1}{2}\right), f(1.7)$
$f(-4) = 7 - 2(-4) = 7 + 8 = 15 \quad (-4,15)$
$f\left(-\frac{1}{2}\right) = 7 - 2\left(-\frac{1}{2}\right) = 7 + 1 = 8 \quad \left(-\frac{1}{2},8\right)$
$f(0) = 7 - 2(0) = 7 - 0 = 7 \quad (0,7)$
$f\left(\frac{1}{2}\right) = 7 - 2\left(\frac{1}{2}\right) = 7 - 1 = 6 \quad \left(\frac{1}{2},6\right)$
$f(1.7) = 7 - 2(1.7) = 7 - 3.4 = 3.6 \quad (1.7,3.6)$

21. $f(C) = \frac{9}{5}C + 32$

$f(100) = \frac{9}{5}(100) + 32 = 9(20) + 32 = 180 + 32 = 212$

$f(0) = \frac{9}{5}(0) + 32 = 0 + 32 = 32$

$f(25) = \frac{9}{5}(25) + 32 = 9(5) + 32 = 45 + 32 = 77$

$f(-10) = \frac{9}{5}(-10) + 32 = 9(-2) + 32 = -18 + 32 = 14$

26. Using $F_f = \mu N$, where $\mu = 0.5$, $F_f = 0.5N$ and we define function h by $h(N) = 0.5N$. Then
$h(100) = 0.5(100) = 50$
$h(50) = 0.5(50) = 25$
$h(25) = 0.5(25) = 12.5$

Review exercises

1. $y = -2x + 7$ **2.** $y = \frac{3x-8}{4}$ **3.** $y = \frac{3}{5}$

4. Point of intersection is (2,0).

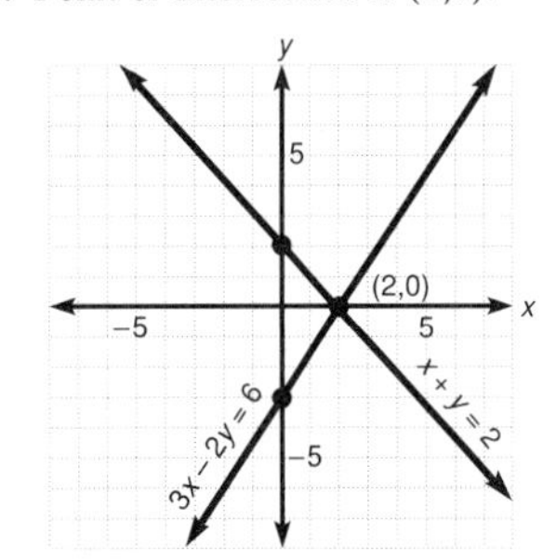

5. $x = 10$ **6.** $y = 2$

Chapter 7 review

1. $(-1,1), (0,4), (4,16)$ **2.** $(-2,-1), \left(0,\frac{1}{3}\right), (1,1)$
3. $(-7,-3), (0,-3), (5,-3)$ **4.** $(-3,15), (0,0), (3,-15)$
5. $(7,-2), (1,0), (-14,5)$ **6.** $\left(\frac{9}{4},-1\right), \left(\frac{7}{4},0\right), \left(\frac{1}{4},3\right)$
7. $(1,-8), (1,0), (1,2)$ **8.** $(-2,-3), (0,0), \left(\frac{2}{3},1\right)$
9. $(1,5)$ **10.** $(4,-4)$ **11.** $(-1,-6)$ **12.** $(0,-4)$
13. $\left(2,\frac{1}{2}\right)$ **14.** $\left(5,-\frac{2}{3}\right)$

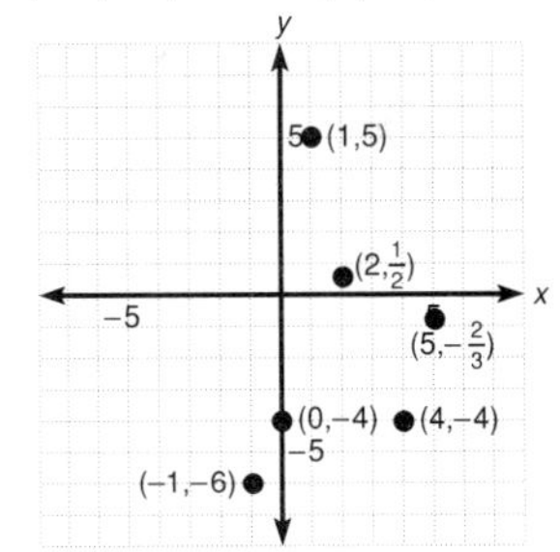

15. $(-1,-2)$ **16.** $(1,1)$ **17.** $(5,1)$ **18.** $(0,3)$ **19.** $(-2,1)$
20. $(3,-4)$ **21.** $y = 3x + 4$; $(2,10), (0,4), \left(\frac{4}{3},8\right), \left(-\frac{2}{3},2\right)$

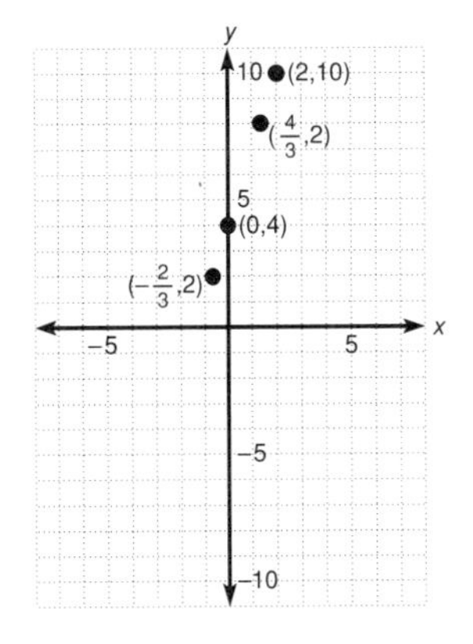

22. $y = -2x + 5$; $(-2,9)$, $\left(\frac{5}{2},0\right)$, $(0,5)$, $(3,-1)$

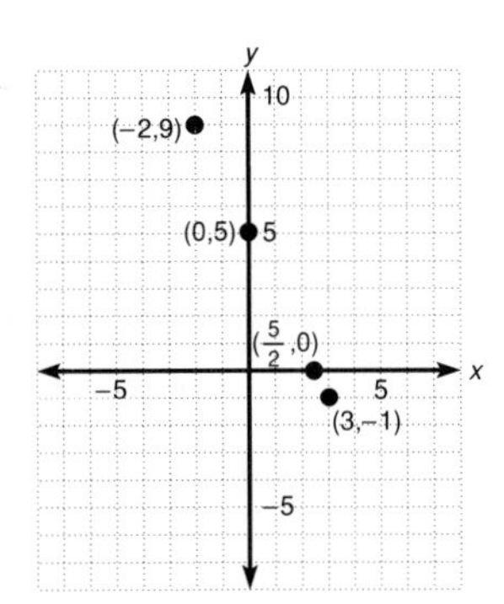

23. $y = x^2 + 2$; $(1,3)$, $(-1,3)$, $(0,2)$, $(2,6)$

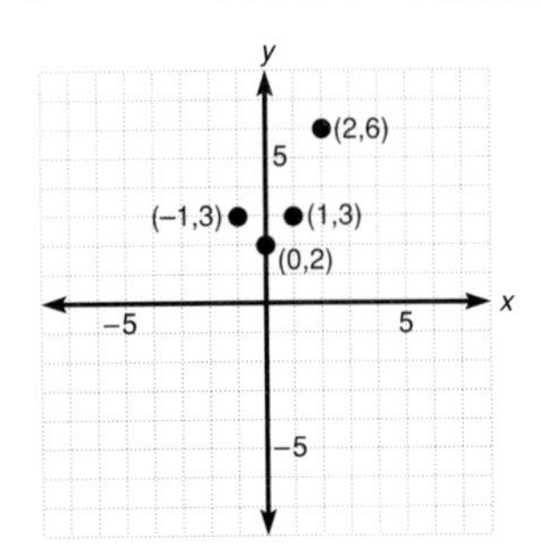

24. $y = x^2 - 16$; $(4,0)$, $(3,-7)$, $(0,-16)$, $(-4,0)$

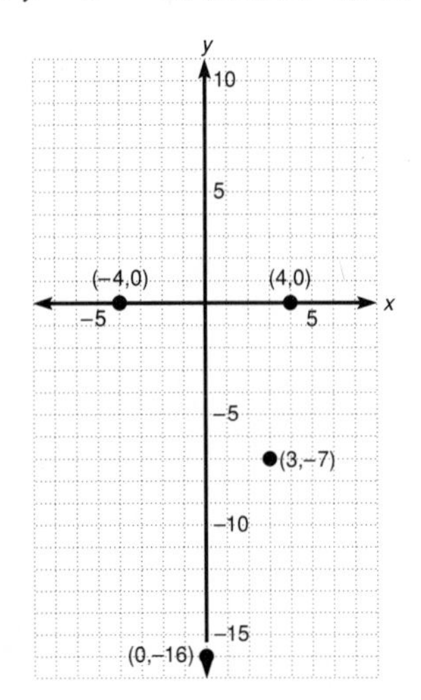

25. $y = 3x + 5$; x-intercept, $\left(-\frac{5}{3},0\right)$; y-intercept, $(0,5)$

26. $y = -4x$; x-intercept, $(0,0)$; y-intercept, $(0,0)$

27. $y + 2 = 0$; x-intercept, (none); y-intercept, $(0,-2)$

28. $x - 6 = 0$; x-intercept, $(6,0)$; y-intercept, (none)

29. $2x - 7y = 4$; x-intercept, $(2,0)$; y-intercept, $\left(0,-\frac{4}{7}\right)$

30. $4y - x = 0$; x-intercept, $(0,0)$; y-intercept, $(0,0)$

31.

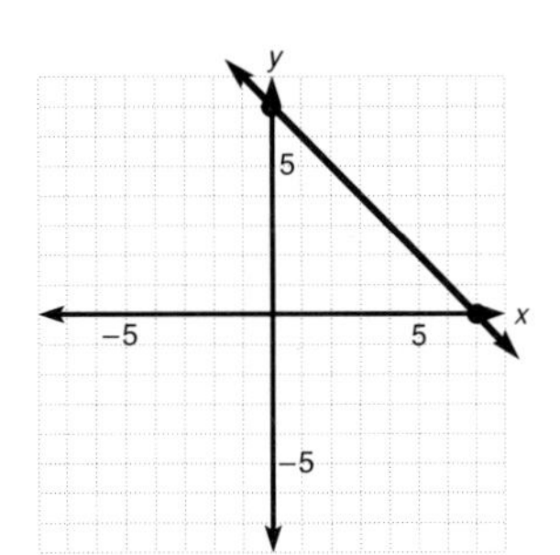

32.

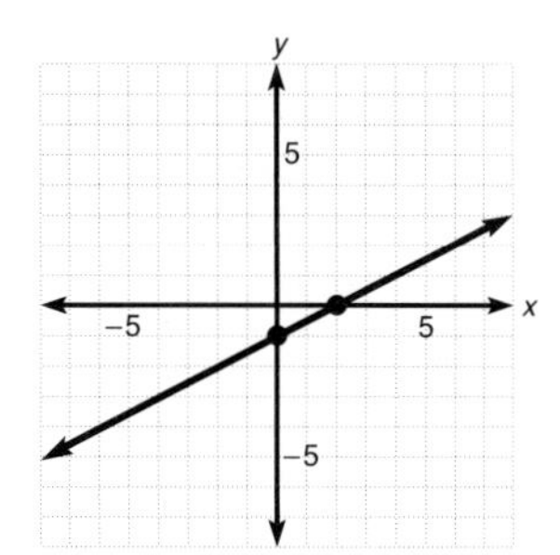

33.

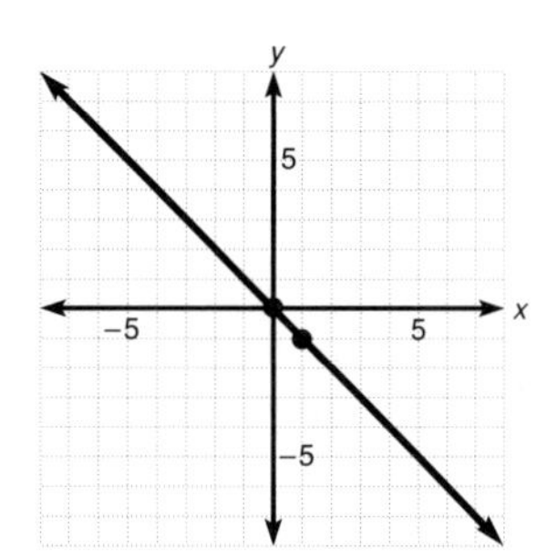

34.

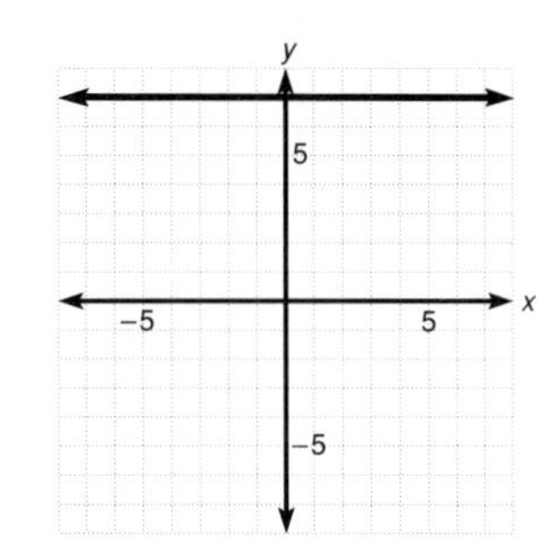

35.

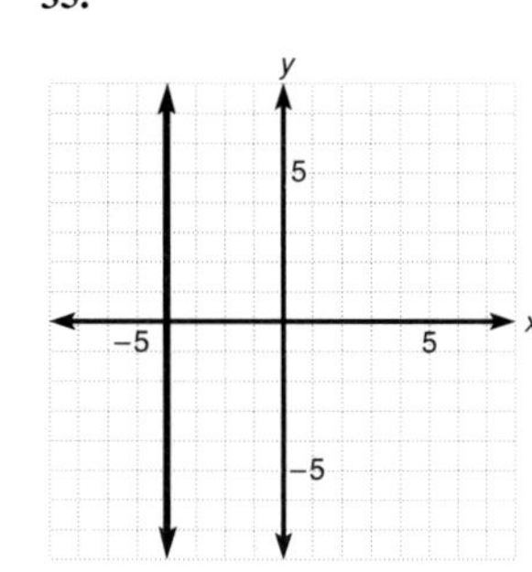

36.

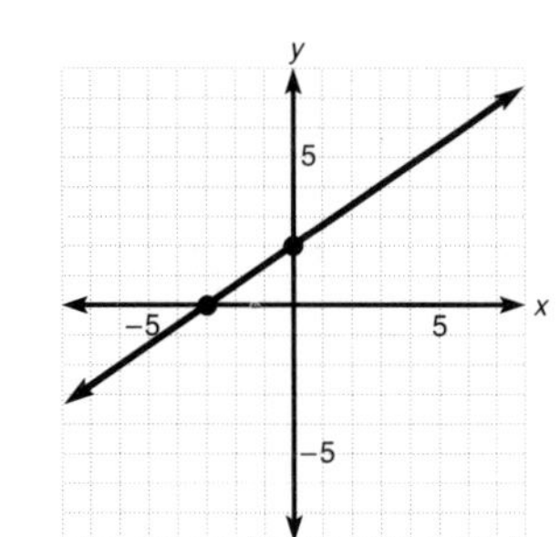

37.

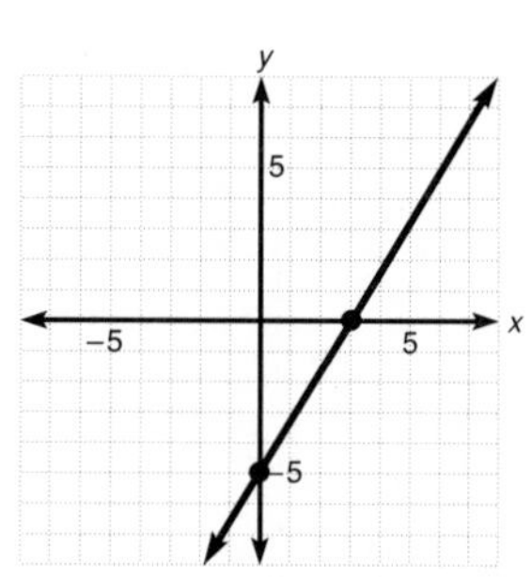

38. $(-4,3)$ and $(1,0)$; $m = -\frac{3}{5}$ **39.** $(-5,3)$ and $(-5,-1)$; m = undefined **40.** $(1,-2)$, $(5,-2)$; $m = \frac{0}{4} = 0$

41. $(-4,-4)$, $(1,1)$; $m = 1$ **42.** $y = \frac{3}{4}x - 2$; $m = \frac{3}{4}$; $b = -2$

43. $y = -\frac{4}{3}x + \frac{2}{3}$; $m = -\frac{4}{3}$; $b = \frac{2}{3}$

44. $y = \frac{8}{3}x + \frac{1}{3}$; $m = \frac{8}{3}$; $b = \frac{1}{3}$ **45.** $y = 5x - 10$; $m = 5$; $b = -10$
46. $m = -1$ passing through $(0,-1)$; $y + x = -1$
47. $m = 1$ passing through $(2,0)$; $y - x = -2$
48. $m = \frac{150}{1{,}050} = \frac{3}{21} = \frac{1}{7}$ **49.** $y + 4x = 21$
50. $-x + 3y = 12$
51. $y - 2 = \frac{1}{8}(x - 5)$; $8y - 16 = x - 5$; $x - 8y = -11$
52. $-4x + 3y = 12$
53. $m = 3$; $b = 4$

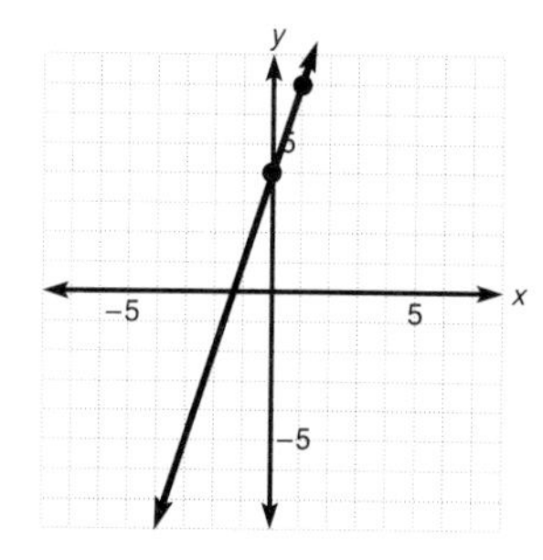

54. $m = 2$; $b = -5$

55. $m = \frac{3}{4}$; $b = 2$

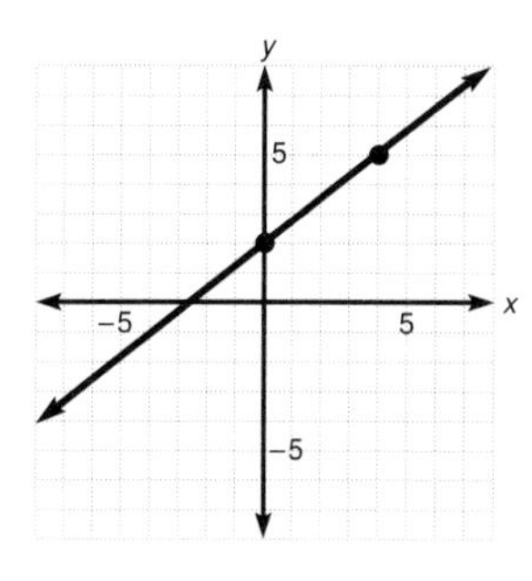

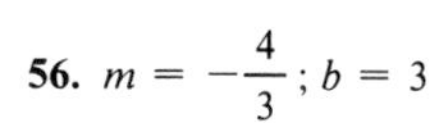
56. $m = -\frac{4}{3}$; $b = 3$

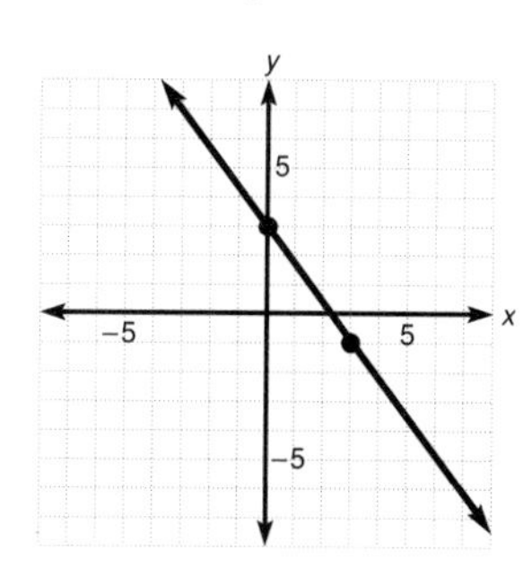

57. $x \le 2y + 1$

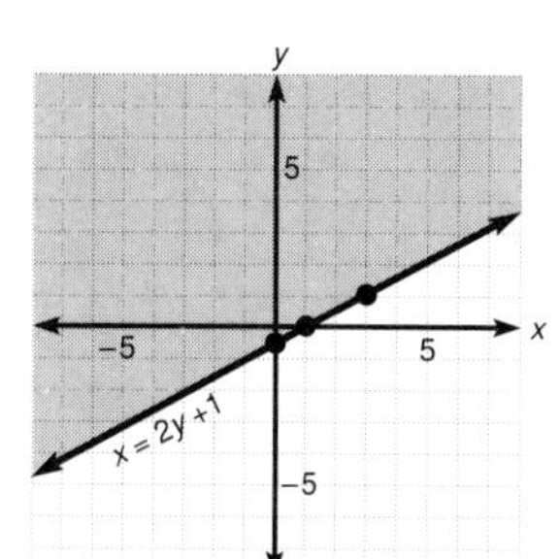

58. $y > 3x - 2$

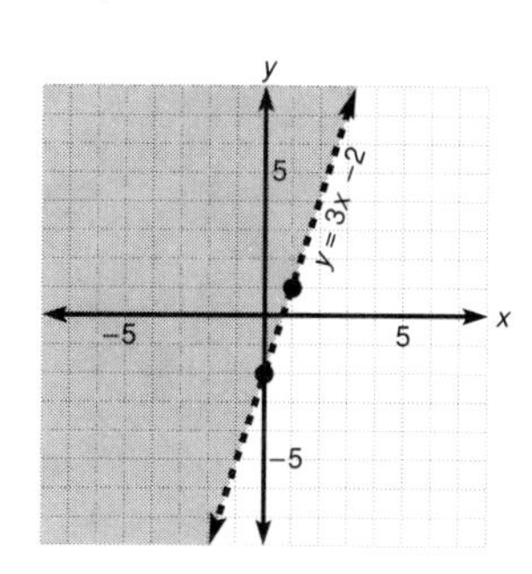

59. $2x - y \ge 4$

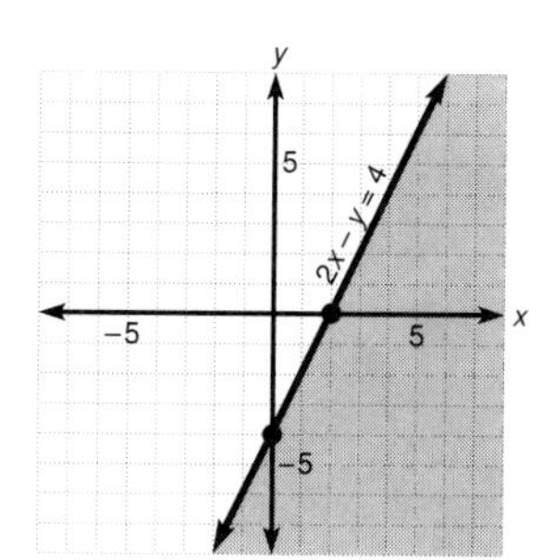

60. $y \le 5$

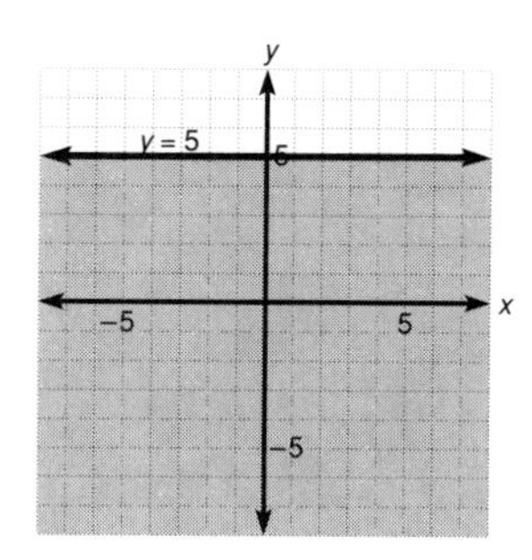

61. $x < -1$

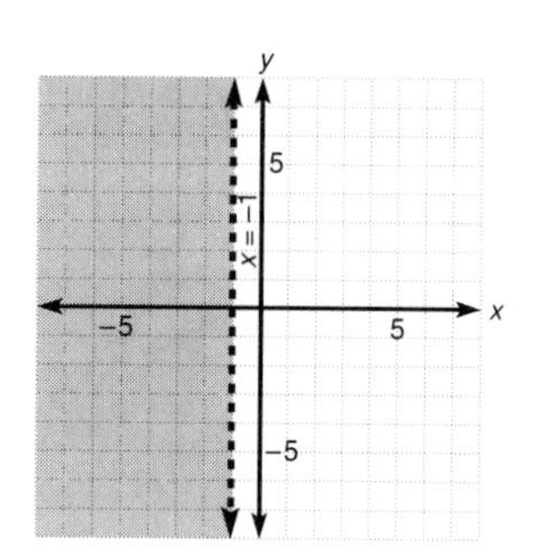

62. $2y - 5x > 0$

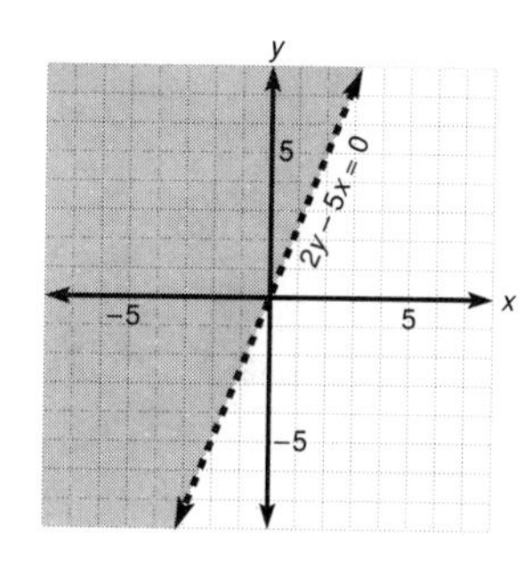

63. $3x + 2y \ge 9$

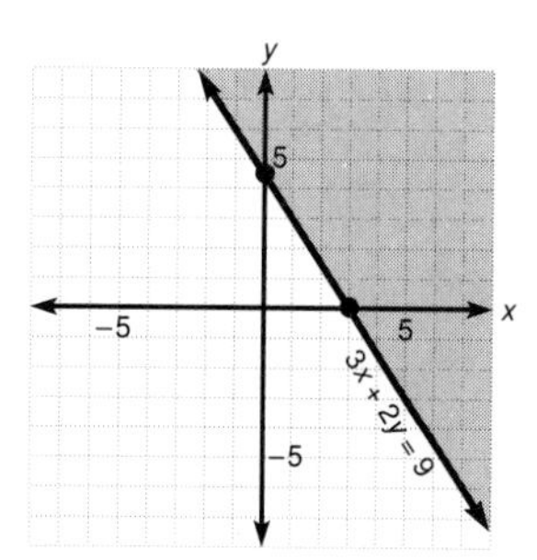

64. $5x - 2y < 10$

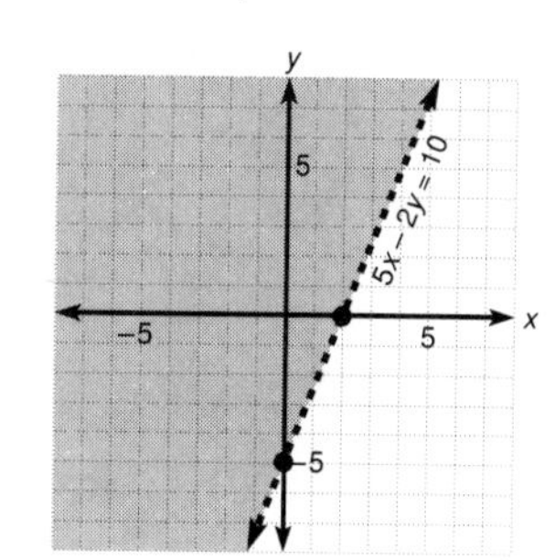

65. domain $= \{-5,-3,-1,0,1,3,5\}$, range $= \{14,8,2,-1,-4,-10,-16\}$
66. domain $= \{-4,-1,0,2,3\}$, range $= \left\{\frac{7}{4},1,\frac{3}{4},\frac{1}{4},0\right\}$
67. domain $= \{-7,-3,0,2,6\}$, range $= \left\{\frac{-7}{2},\frac{-3}{2},0,1,3\right\}$
68. $f(-1) = -4, f(0) = 0, f(4) = 16$; $(-1,-4), (0,0), (4,16)$
69. $f(-2) = 6, f(0) = 4, f(2) = 2$; $(-2,6), (0,4), (2,2)$
70. $f(-4) = -23, f(0) = -3, f(6) = 27$; $(-4,-23), (0,-3), (6,27)$ **71.** $f(-8) = 5, f(0) = 5, f(9) = 5$; $(-8,5), (0,5), (9,5)$
72. $f(s) = 6s$, $f(3) = 18$, $f(14) = 84$; domain $=$ {positive real numbers}

Chapter 7 cumulative test

1. 25 **2.** $-27x + 36$ **3.** $\left\{\frac{8}{3}\right\}$ **4.** $\left\{\frac{60}{17}\right\}$ **5.** $\left\{-\frac{5}{3}\right\}$
6. $\left\{-\frac{2}{3},2\right\}$ **7.** $-1 < x \le 4$ **8.** $(7x + 1)(x - 1)$

9. $y^6(y^2 + y - 1)$ **10.** $p(p^3 - 4)(p + 1)$ **11.** $16(a + b)(a - b)$ **12.** $t(t - 7)(t - 1)$ **13.** $-7(5y + 4z)$ **14.** $16y^2 - 24xy + 9x^2$ **15.** $20x^2 - 3x - 56$ **16.** $25 - \frac{9}{16}y^2$ **17.** $x^3 - 3x^2y + 3xy^2 - y^3$ **18.** $15y^5 + 18y^4 + 32y^3 + 66y^2 - 7y + 56$ **19.** $\frac{5x^2 + 32x}{(x - 7)(x + 7)(x + 6)}$ **20.** $\frac{a^2 - 2a + 1}{(a + 5)(a - 2)}$ **21.** $\frac{6}{y + 3}$ **22.** $\frac{1}{x^9}$ **23.** $\frac{1}{8^8}$ **24.** $-\frac{125}{a^9}$ **25.** 7.76×10^{-6} **26.** $3x^2 - 8x + 20 + \frac{-45}{x + 2}$ **27.** $\frac{2x^2 - x + 9}{3x + 4}$ **28.** x-intercept, $\frac{7}{2}$; y-intercept, -7

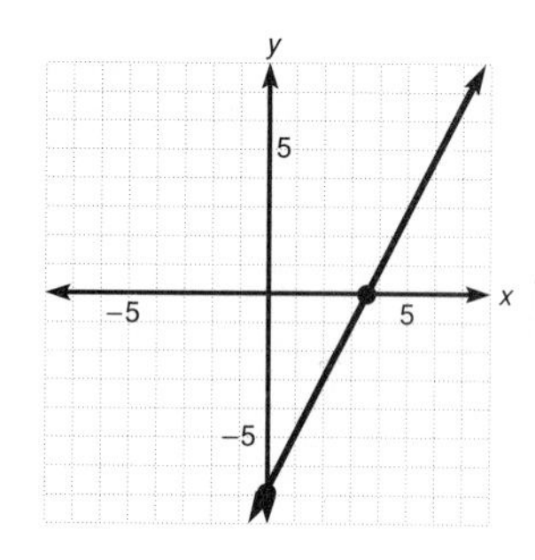

29. x-intercept, 7; y-intercept, none **30.**

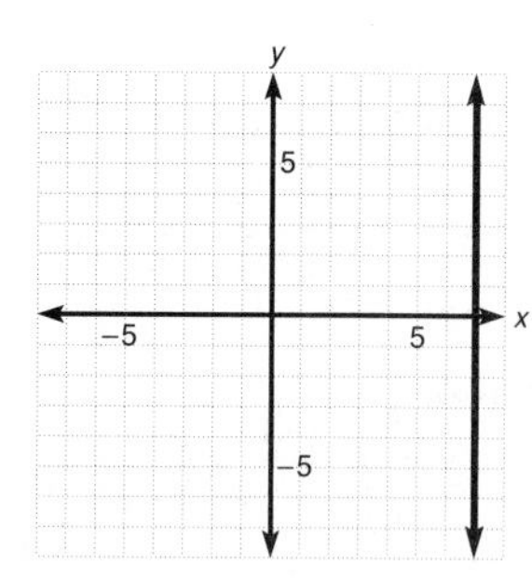

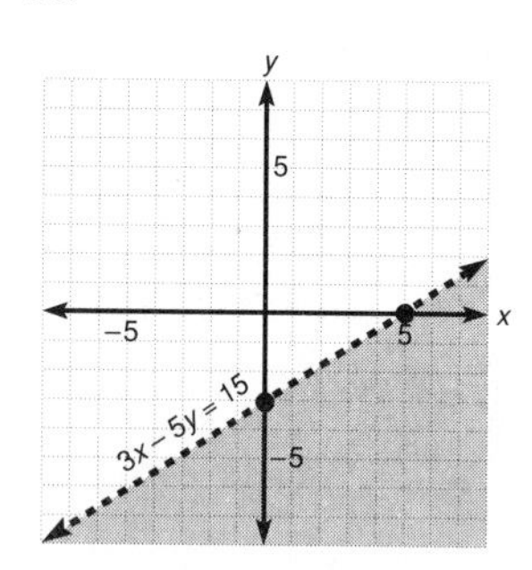

31. $m = \frac{7}{3}$ **32.** $3x - 4y = -13$ **33.** $3x + 5y = 27$ **34.** parallel **35.** neither **36.** $f(0) = -2, f(-3) = -11, f(4) = 10$ **37.** 136 **38.** 12 **39.** legs are 5 inches and 12 inches

Chapter 8

Exercise 8–1

Answers to odd-numbered problems

1. yes **3.** yes **5.** yes **7.** no **9.** yes **11.** no **13.** solution: (3,2)

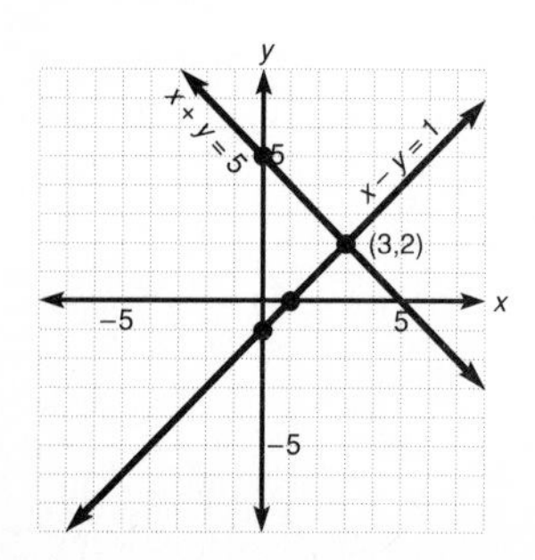

15. solution: (3,−1)

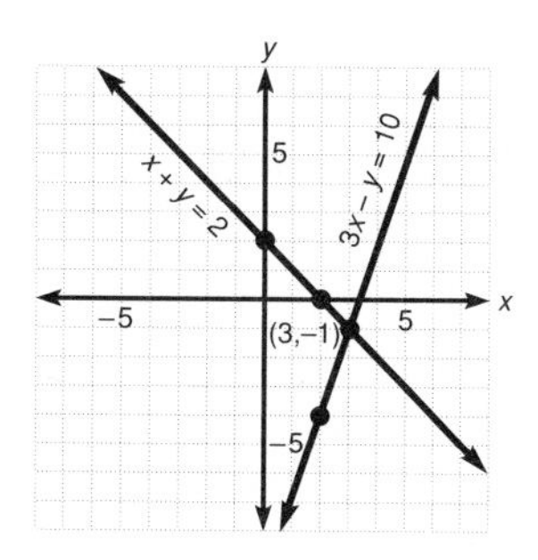

17. solution: (−1,−5)

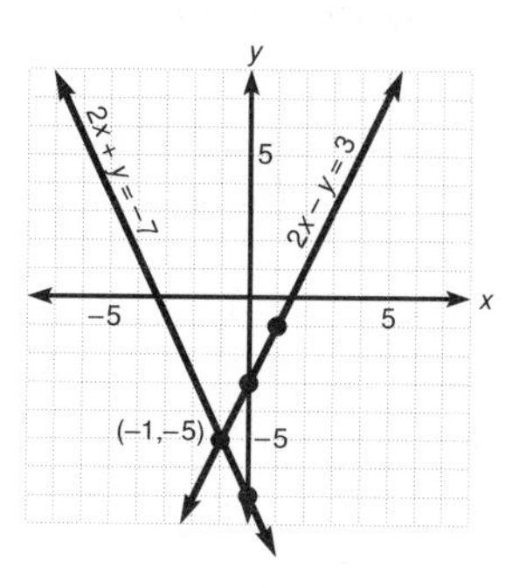

19. solution: (2,3)

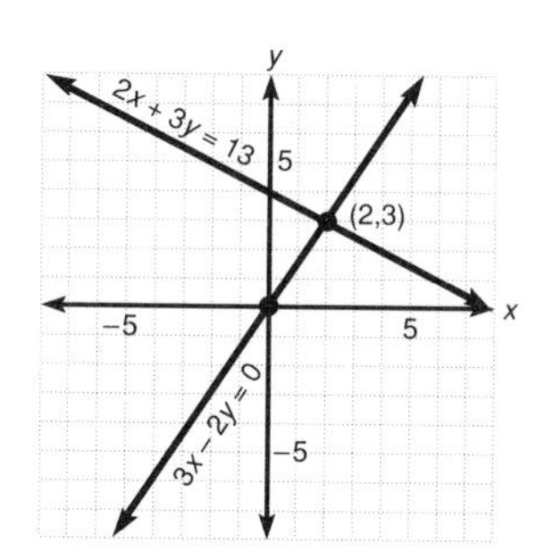

21. parallel lines (the slopes of both lines are the same) ∴ *inconsistent* with no solution

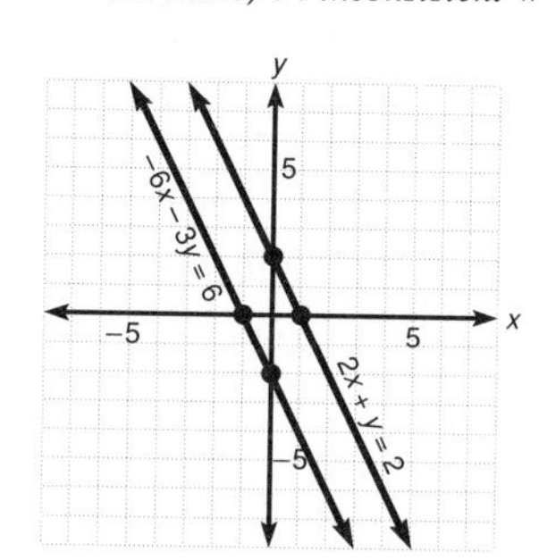

23. solution: (−1,0)

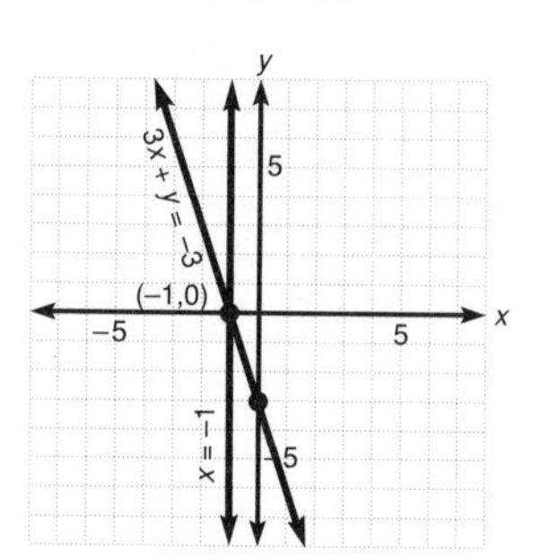

25. inconsistent; no solution

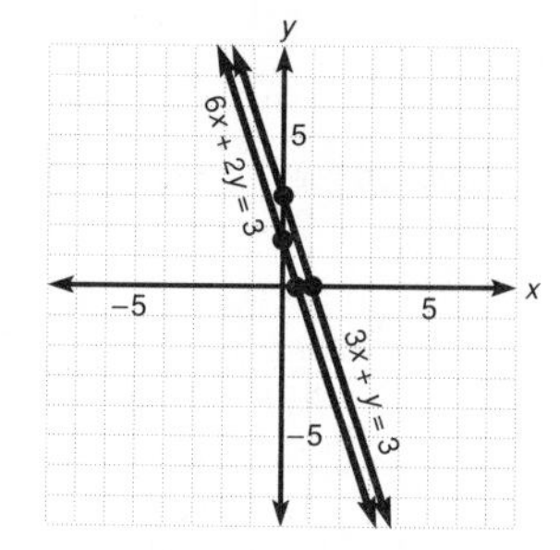

27. Both equations define the same line ∴ *dependent* with an unlimited number of solutions

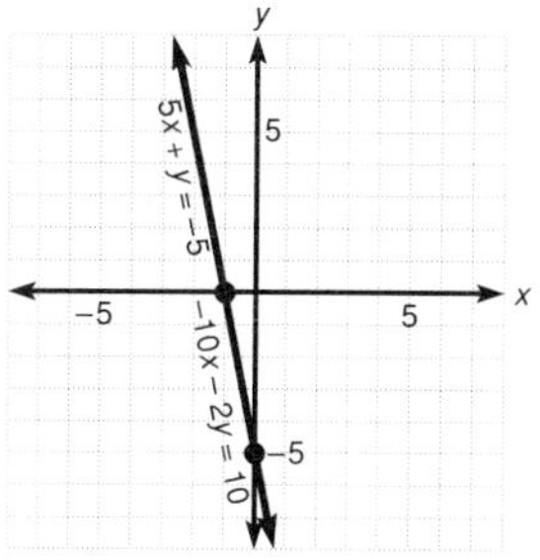

29. solution: $\left(\frac{12}{5},\frac{6}{5}\right)$

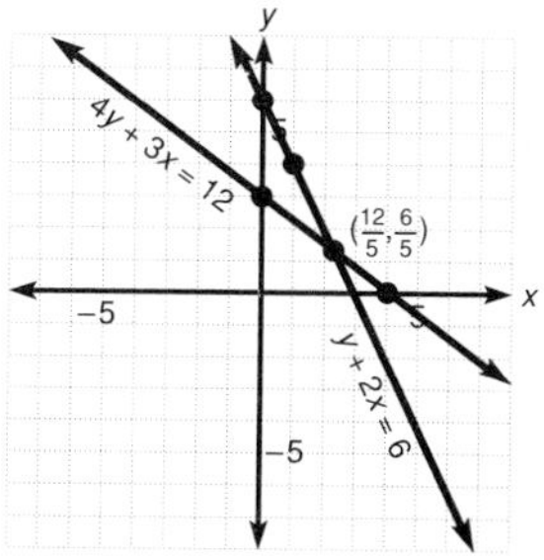

Solutions to trial exercise problems

5. $2x + y = 2$
$6x - y = 22$ (3,−4)
Let $x = 3$ and $y = -4$.
$2(3) + (-4) = 2$ $6(3) - (-4) = 22$
$6 - 4 = 2$ $18 + 4 = 22$
$2 = 2$ (true) $22 = 22$ (true)
Therefore (3,−4) is a simultaneous solution.

10. $y = 4x - 3$
$y = -1$ (−1,−1)
Let $x = -1$ and $y = -1$.
$-1 = 4(-1) - 3$ $y = -1$
$-1 = -4 - 3$ $-1 = -1$ (true)
$-1 = -7$ (false)
Therefore (−1,−1) is *not* a simultaneous solution.

15. $3x - y = 10$
$x + y = 2$
$a_1 = 3, b_1 = -1, c_1 = 10$
$a_2 = 1, b_2 = 1, c_1 = 2$
Since $\frac{3}{1} \neq \frac{-1}{1}$ the system is independent and consistent with simultaneous solution (3,−1).

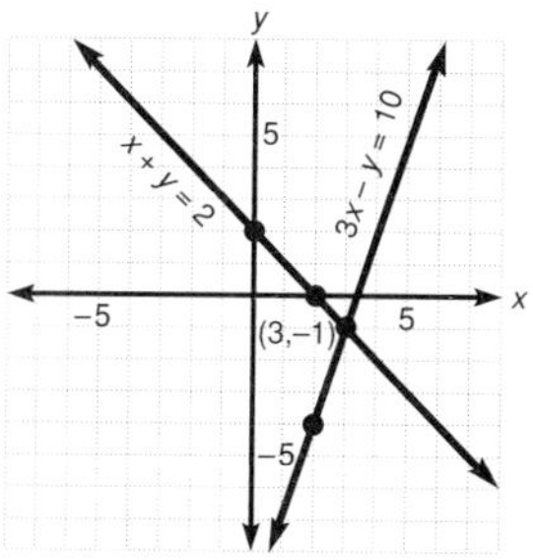

27. The system is *dependent* so all points in the line represent simultaneous solutions. (*Note:* If we divide the members of $-10x - 2y = 10$ by −2, we get equation $5x + y = -5$.)

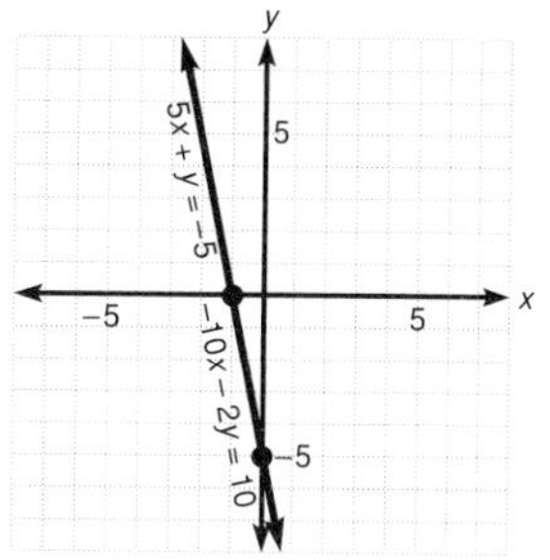

Review exercises

1. $\left\{\frac{4}{5}\right\}$ **2.** $3x^2 - 10x + 30 + \frac{-89}{x+3}$

3. $y \leq 5$

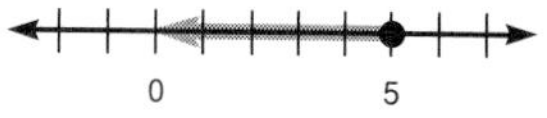

4. $-3 < x \leq 3$

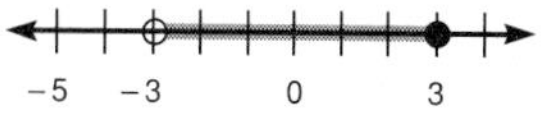

5. x^4 **6.** $x^5 - 2x^4 + x^3$ **7.** $4x^2 - 4xy + y^2$

Exercise 8–2

Answers to odd-numbered problems

1. {(3,2)} **3.** $\left\{\left(\frac{5}{3},-5\right)\right\}$ **5.** inconsistent (no solution)
7. $\left\{\left(\frac{5}{3},2\right)\right\}$ **9.** {(−2,3)} **11.** {(−1,3)} **13.** $\left\{\left(\frac{29}{10},-\frac{3}{5}\right)\right\}$
15. dependent (unlimited number of solutions) **17.** {(1,−5)}
19. dependent (unlimited number of solutions) **21.** $\left\{\left(\frac{17}{21},-\frac{11}{21}\right)\right\}$
23. $\left\{\left(\frac{22}{21},\frac{13}{21}\right)\right\}$ **25.** {(5,−2)} **27.** $\left\{\left(\frac{4}{5},\frac{9}{5}\right)\right\}$ **29.** inconsistent (no solution) **31.** {(−3,7)} **33.** (7,3) **35.** $y = 5x$
37. $y = x + 20$ **39.** $x + y = 14{,}000$

Solutions to trial exercise problems

6. $-x + y = 6$
$x + 4y = 4$

$$\begin{array}{r} -x + y = 6 \\ \underline{x + 4y = 4} \\ 5y = 10 \\ y = 2 \end{array}$$

Add the equations.

Substitute 2 for y in $-x + y = 6$. $-x + (2) = 6$
$-x = 4$ so $x = -4$
The solution is (−4,2).
The solution set is {(−4,2)}.

10. $x - y = 3$ Multiply equation one by 3.
$\underline{2x + 3y = 11}$
$3x - 3y = 9$
$\underline{2x + 3y = 11}$
$5x \quad = 20$ Add the equations.
$x = 4$
Substitute 4 for x in $2x + 3y = 11$.
$2(4) + 3y = 11$
$8 + 3y = 11$
$3y = 3$
$y = 1$
The solution is (4,1).
The solution set is {(4,1)}.

19. $-6x + 3y = 9$ $\qquad$ $-6x + 3y = 9$

$2x - y = -3$ $\quad$ Multiply by 3. $\quad$ $6x - 3y = -9$

$0 = 0$

The system is dependent so an unlimited number of solutions.

27. $\frac{1}{2}x + \frac{1}{3}y = 1$ $\quad$ Multiply by 6. $\quad$ $3x + 2y = 6$

$\frac{2}{3}x - \frac{1}{4}y = \frac{1}{12}$ $\quad$ Multiply by 12. $\quad$ $8x - 3y = 1$

(*Note:* We have just cleared denominators.)

$3x + 2y = 6$ $\quad$ Multiply by 3. $\quad$ $9x + 6y = 18$

$8x - 3y = 1$ $\quad$ Multiply by 2. $\quad$ $16x - 6y = 2$

Add: $\quad$ $25x = 20$

$x = \frac{20}{25} = \frac{4}{5}$

Substitute $\frac{4}{5}$ for x in $\frac{1}{2}x + \frac{1}{3}y = 1$.

$\frac{1}{2}\left(\frac{4}{5}\right) + \frac{1}{3}y = 1$

$\frac{2}{5} + \frac{1}{3}y = 1$

$\frac{1}{3}y = \frac{3}{5}$

$y = \frac{9}{5}$

The solution is $\left(\frac{4}{5},\frac{9}{5}\right)$.

The solution set is $\left\{\left(\frac{4}{5},\frac{9}{5}\right)\right\}$.

34. $x + (0.4)y = 3.4$ $\quad$ Multiply by 10. $\quad$ $10x + 4y = 34$

$(0.6)x - (1.4)y = 0.4$ $\quad$ Multiply by 10. $\quad$ $6x - 14y = 4$

(*Note:* The coefficients are now integers.)

$10x + 4y = 34$ $\quad$ Multiply by 3. → $\quad$ $30x + 12y = 102$

$6x - 14y = 4$ $\quad$ Multiply by -5. → $\quad$ $-30x + 70y = -20$

$82y = 82$

$y = 1$

Substitute 1 for y in $x + 0.4y = 3.4$.

$x + (0.4)(1) = 3.4$

$x + 0.4 = 3.4$

$x = 3$

The solution is (3,1). The solution set is $\{(3,1)\}$.

35. Since the length, y, is 5 times the width, x, then $y = 5x$.

38. Since each auto travels 3 hours, then A travels $3x$ miles and B travels $3y$ miles. They are 500 miles apart after 3 hours, so $3x + 3y = 500$.

Review exercises

1. $x = \frac{1}{3}$ **2.** $2x - 3y = -20$ **3.** $4x - 7y = -35$ **4.** $4(2x + y)(2x - y)$ **5.** $(3x - 2)^2$ **6.** $(5y + 4)(y - 2)$ **7.** $\{1\}$

Exercise 8–3

Answers to odd-numbered problems

1. $\{(3,-1)\}$ **3.** $\{(-1,3)\}$ **5.** $\{(-5,-1)\}$ **7.** $\{(5,2)\}$ **9.** $\left\{\left(-\frac{5}{3},-\frac{17}{3}\right)\right\}$ **11.** $\{(4,5)\}$ **13.** $\{(5,8)\}$ **15.** $\left\{\left(\frac{7}{5},\frac{1}{5}\right)\right\}$ **17.** $\left\{\left(\frac{2}{13},-\frac{14}{13}\right)\right\}$ **19.** no solution (inconsistent) **21.** dependent (unlimited solutions) **23.** $\{(-1,0)\}$ **25.** $\left\{\left(-\frac{11}{13},-\frac{17}{13}\right)\right\}$ **27.** $\left\{\left(\frac{17}{20},\frac{1}{10}\right)\right\}$ **29.** $\{(4,-2)\}$ **31.** dependent (unlimited solutions) **33.** $\left\{\left(\frac{13}{4},\frac{7}{4}\right)\right\}$ **35.** inconsistent (no solution) **37.** $\left\{\left(5,\frac{21}{5}\right)\right\}$ **39.** $\left\{\left(\frac{28}{3},2\right)\right\}$ **41.** $\left\{\left(-\frac{7}{3},-\frac{26}{3}\right)\right\}$ **43.** $\left\{\left(0,\frac{1}{2}\right)\right\}$ **45.** $\{(3,-4)\}$ **47.** $\left\{\left(\frac{12}{13},-\frac{33}{13}\right)\right\}$ **49.** $\left\{\left(-\frac{76}{13},\frac{60}{13}\right)\right\}$ **51.** $\left\{\left(\frac{20}{17},-\frac{56}{17}\right)\right\}$ **53.** $x = 2y$ **55.** $x + y = 80$

Solutions to trial exercise problems

3. $-2x + 5y = 17$

$y = 2x + 5$

Substitute $2x + 5$ for y in the first equation.

$-2x + 5(2x + 5) = 17$

$-2x + 10x + 25 = 17$

$8x + 25 = 17$

$8x = -8$

$x = -1$

Substitute -1 for x in $y = 2x + 5$.

$y = 2(-1) + 5$

$y = -2 + 5$

$y = 3$

The solution is $(-1,3)$.

The solution set is $\{(-1,3)\}$.

17. $5x - 3y = 4$

$x + 2y = -2$ $\qquad$ $x = -2y - 2$

Substitute $-2y - 2$ for x in $5x - 3y = 4$.

$5(-2y - 2) - 3y = 4$

$-10y - 10 - 3y = 4$

$-13y - 10 = 4$

$-13y = 14$

$y = -\frac{14}{13}$

Substitute $-\frac{14}{13}$ for y in $x = -2y - 2$.

Then $x = -2\left(\frac{-14}{13}\right) - 2 = \frac{28}{13} - \frac{26}{13} = \frac{2}{13}$

The solution is $\left(\frac{2}{13},\frac{-14}{13}\right)$.

The solution set is $\left\{\left(\frac{2}{13},-\frac{14}{13}\right)\right\}$.

35. $-3x - 2y = 6$

$6x + 4y = 5$

Solving for y, $-3x - 2y = 6$. (Multiply by -1.)

$3x + 2y = -6$

$2y = -6 - 3x$

$y = -3 - \frac{3}{2}x$

Substitute $-\frac{3}{2}x - 3$ for y in $6x + 4y = 5$.

$6x + 4\left(-\frac{3}{2}x - 3\right) = 5$

$6x - 6x - 12 = 5$

$-12 = 5$ (false)

The system is *inconsistent* so there is no solution.

42. $y = 3x - 4$
$y = 3x + 5$
Then $3x - 4 = 3x + 5$
$-4 = 5$ (false)
The system is *inconsistent* so there is no solution.

45. $x - 3 = 0$
$3x + 2y = 1$
Solving $x - 3 = 0$ for x, we have $x = 3$.
Substitute 3 for x in $3x + 2y = 1$.
$3(3) + 2y = 1$
$9 + 2y = 1$
$2y = -8$
$y = -4$
The solution is $(3,-4)$.
The solution set is $\{(3,-4)\}$.

48. $\frac{2x}{3} - \frac{y}{2} = 1$ Multiply by 6. $4x - 3y = 6$

$3x + \frac{y}{4} = 2$ Multiply by 4. $12x + y = 8$

Now $4x - 3y = 6$ $\quad 4x - 3y = 6$
$12x + y = 8$ Multiply by 3. $36x + 3y = 24$
$40x = 30$
$x = \frac{3}{4}$

Substitute $\frac{3}{4}$ for x in the equation $3x + \frac{y}{4} = 2$.

Then $3\left(\frac{3}{4}\right) + \frac{y}{4} = 2$

$\frac{9}{4} + \frac{y}{4} = 2$ Multiply by 4.
$9 + y = 8$
$y = -1$

The solution is $\left(\frac{3}{4},-1\right)$.

The solution set is $\left\{\left(\frac{3}{4},-1\right)\right\}$.

53. The first current, x, is twice as much as the second current, y, so $x = 2y$.

Review exercises

1. 16 ft and 26 ft **2.** 4 yd and 9 yd **3.** $b = \frac{2A - hc}{h}$ **4.** $\frac{2}{9}$

5. $\left\{\frac{5}{8}\right\}$ **6.** $-1 < x \leq 1$

Exercise 8–4

Answers to odd-numbered problems

1. $w = 4$ meters; $\ell = 14$ meters **3.** $w = 14$ meters; $\ell = 36$ meters **5.** $w = 11$ feet; $\ell = 20$ feet **7.** \$14,000 at 9%; \$4,000 at 7½% **9.** \$22,000 at 7%; \$8,000 at 9% **11.** \$13,300 at 9%; \$7,700 at 13% **13.** 4 feet and 8 feet **15.** 28 amperes and 52 amperes **17.** 38 teeth and 26 teeth **19.** 47 volts; 79 volts **21.** 19 ohms; 5 ohms **23.** 28 suits at \$125; 12 suits at \$185 **25.** 16 volts; 11 volts **27.** 1st gear—6 teeth; 2nd gear—17 teeth **29.** 80 cubic centimeters of 4%; 40 cubic centimeters of 10% **31.** 9 ounces of 80%; 3 ounces of 60% **33.** 87.5% **35.** $\frac{600}{7}$ centiliters of 3%; $\frac{40}{7}$ centiliters of 38% **37.** 25 mph; 75 mph **39.** Jane, $2\frac{1}{2}$ mph; Jim, 3 mph **41.** wind, 69 mph; airplane in still air, 391 mph **43.** $m = 3$, $b = -4$; $y = 3x - 4$ **45.** $m = -2$, $b = -1$; $y = -2x - 1$ **47.** $m = \frac{4}{5}$, $b = -4$

Solutions to trial exercise problems

1. The perimeter P of a rectangle, by formula, is found by $P = 2\ell + 2w$, where ℓ is the length and w is the width of the rectangle. From "the perimeter of a rectangle is 36 meters," we get $2\ell + 2w = 36$. From "the length is 2 meters more than three times the width," we get $\ell = 3w + 2$.
We solve the system $2\ell + 2w = 36$
$\ell = 3w + 2$
Using substitution, substitute $3w + 2$ for ℓ in the first equation.
$2(3w + 2) + 2w = 36$
$6w + 4 + 2w = 36$
$8w + 4 = 36$
$8w = 32$
$w = 4$
Then $\ell = 3w + 2$
$\ell = 3(4) + 2$
$\ell = 12 + 2$
$\ell = 14$
The solution of the system is $(\ell,w) = (14,4)$. The rectangle has length $\ell = 14$ m and width $w = 4$ m.

6. Let $x =$ the amount invested at 8% interest and $y =$ the amount invested at 6% interest. From "Phil has \$20,000, part of which he invests at 8% interest and the rest at 6%," we get $x + y = 20{,}000$. From "if his total income from the two investments was \$1,460," we get $0.08x + 0.06y = 1{,}460$. Multiply by 100 to clear decimal points: $8x + 6y = 146{,}000$.
We solve the system $x + y = 20{,}000$
$8x + 6y = 146{,}000$
Multiply the first equation by -6. $(-6)x + (-6)y = -6(20{,}000)$
$-6x - 6y = -120{,}000$
Add. $8x + 6y = 146{,}000$
$2x = 26{,}000$
$x = 13{,}000$
Substitute 13,000 for x in $x + y = 20{,}000$:
$13{,}000 + y = 20{,}000$, $y = 7{,}000$. Phil invested \$13,000 at 8% and \$7,000 at 6% interest.
Check: $(0.08)13{,}000 = \$1{,}040.00$
$(0.06)7{,}000 = \$\ 420.00$
Add to get income. $= \$1{,}460.00$

12. Let $x =$ the length of the longer piece of pipe and $y =$ the length of the shorter piece of pipe. From "a piece of pipe is 19 feet long," we get $x + y = 19$. From "one piece is 5 feet longer than the other piece," we get $x = y + 5$.
Solve the system: $x + y = 19$
$x = y + 5$
Use substitution: Substitute $y + 5$ for x in $x + y = 19$.
$(y + 5) + y = 19$
$2y + 5 = 19$
$2y = 14$
$y = 7$
Substitute 7 for y in $x + y = 19$.
$x + 7 = 19$
$x = 12$
The pipe is cut into pieces 12 feet and 7 feet long.

27. Let x = the number of teeth on the first gear and y = the number of teeth on the second gear. From "three times the number of teeth on a first gear is 1 more than the number of teeth on a second gear," we get $3x = y + 1$. From "five times the number of teeth on the first gear is 4 less than twice the number of teeth on the second gear," we get $5x = 2y - 4$.

Then $3x = y + 1$
$5x = 2y - 4$ is the system.

Rewrite the system.

$$\begin{array}{lll} 3x - y = 1 & \text{Multiply by } -2. \rightarrow & -6x + 2y = -2 \\ 5x - 2y = -4 & & \underline{\;\;5x - 2y = -4} \\ & \text{Add:} & -x \qquad\;\; = -6 \\ & & \qquad\quad x = 6 \end{array}$$

Substitute 6 for x in $3x - y = 1$. $3(6) = y + 1$
$18 = y + 1$
$17 = y$

The first gear has 6 teeth and the second gear has 17 teeth.

29. Let x = the number of cubic centimeters (cc) of the 10% solution and y = the number of cc of the 4% solution.

Then $x + y = 120$ cc and $(0.10)x + (0.04)y = (0.06)120$.

Multiply the second equation by 100: $10x + 4y = 6(120)$, so $10x + 4y = 720$. Solve the system.

$$\begin{array}{lll} x + y = 120 & \text{Multiply by } -4. & -4x - 4y = -480 \\ 10x + 4y = 720 & & \underline{10x + 4y = \quad 720} \\ & \text{Add.} & 6x \qquad = \quad 240 \\ & & \qquad x = \quad 40 \end{array}$$

Substitute 40 for x in $x + y = 120$.
$40 + y = 120$
$y = 80$

Therefore the mechanic must have 40 cc of 10% solution and 80 cc of 4% solution.

37. Let x = the speed of the faster car and y = the speed of the slower car. Then using "if they drive toward each other they will meet in 1 hour," $t = 1$ for each car, so $x + y = 100$. Using "if they drive in the same direction they will meet in 2 hours," $t = 2$ for each car and the faster car travels 100 miles farther, so $2x - 2y = 100$.

We now have the system $x + y = 100$
$2x - 2y = 100$

Multiply the first equation by 2.

$$\begin{array}{l} 2x + 2y = 200 \\ \underline{2x - 2y = 100} \\ 4x \qquad = 300 \\ \quad x = 75 \end{array}$$

Then since $x + y = 100$, $75 + y = 100$, $y = 25$. The average speeds are 75 mph and 25 mph.

40. Let x = the speed of the boat in still water and y = the speed of the current. Using $d = rt$ and (a) "a boat can travel 24 miles downstream in 2 hours," we get $2x + 2y = 24$; using $d = rt$ and (b) "and 16 miles upstream in the same length of time," we get $2x - 2y = 16$.

We have the system of equations

$$\begin{array}{l} 2x + 2y = 24 \\ \underline{2x - 2y = 16} \\ 4x \qquad = 40 \\ \quad x = 10 \end{array}$$

Then substituting 10 for x, $2(10) + 2y = 24$
$20 + 2y = 24$
$2y = 4$
$y = 2$

Therefore the boat travels at 10 mph in still water and the current is traveling at 2 mph.

43. Using $y = mx + b$, substitute
(a) 1 for x and -1 for y $\quad -1 = (1)m + b$
(b) 2 for x and 2 for y $\quad 2 = 2m + b$

$$\begin{array}{lll} \text{Then } m + b = -1 & \text{Multiply by } -1. & -m - b = 1 \\ \quad\;\; 2m + b = 2 & & \underline{2m + b = 2} \\ & & m \qquad = 3 \end{array}$$

Substitute 3 for m in $m + b = -1$.
$3 + b = -1$
$b = -4$

An equation of the line is $y = 3x - 4$.

Review exercises

1. $16x^4y^6$ **2.** $\frac{y}{x^3}$ **3.** $\frac{7}{12}$ **4.** $x^2 - 5x + 4$
5. $12x^4 + 8x^3 - 12x^2$ **6.** $9y^2 - 4$ **7.** $16x^2 - 40xy + 25y^2$
8. $\{-1\}$ **9.** $\{-2,5\}$

Exercise 8–5

Answers to odd-numbered problems

1.

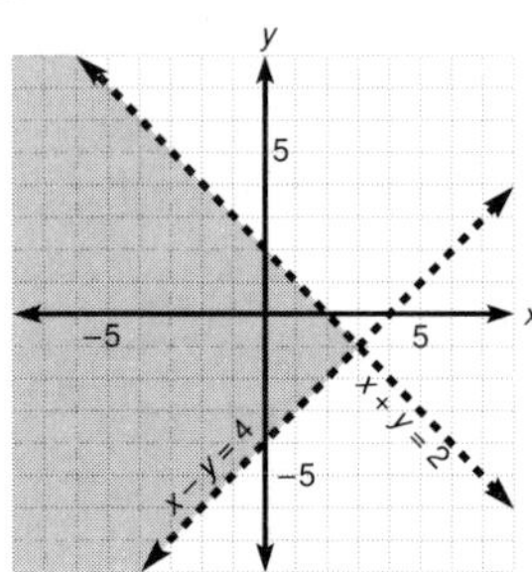

3.

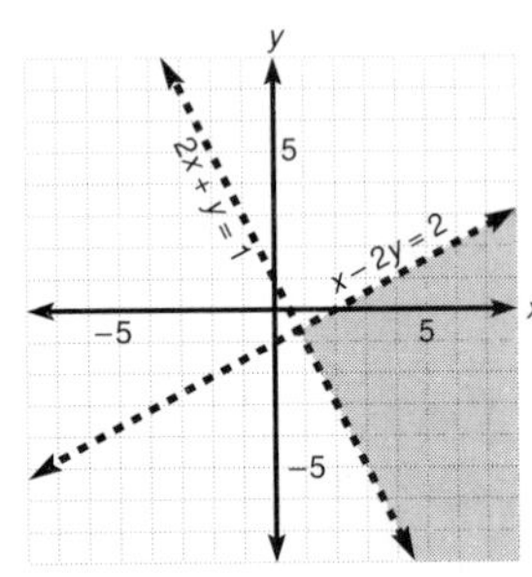

5.

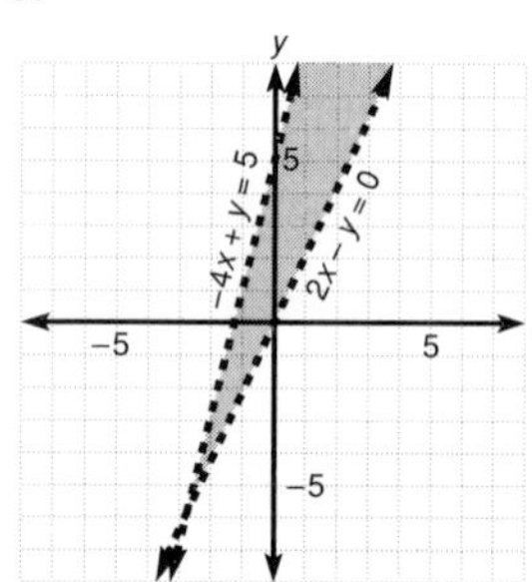

7.

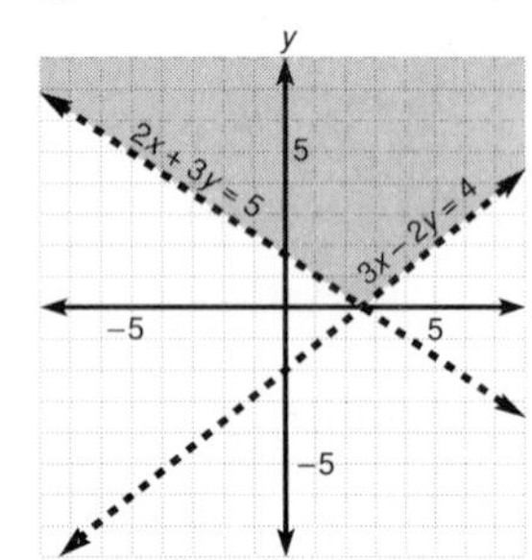

9.

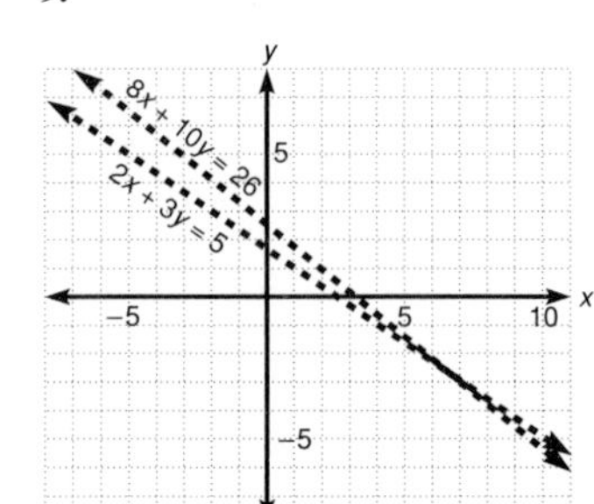

11.

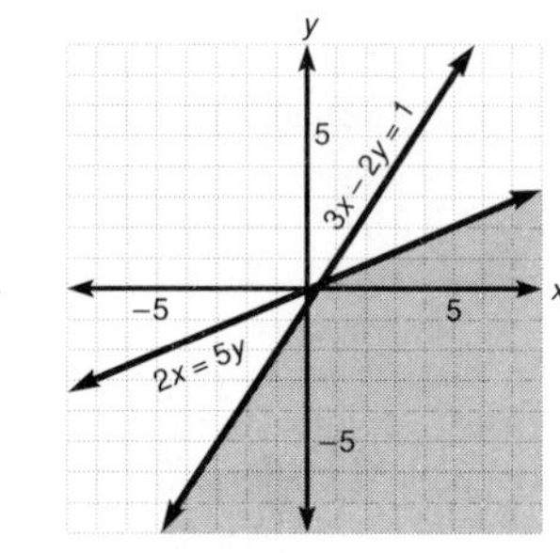

13.

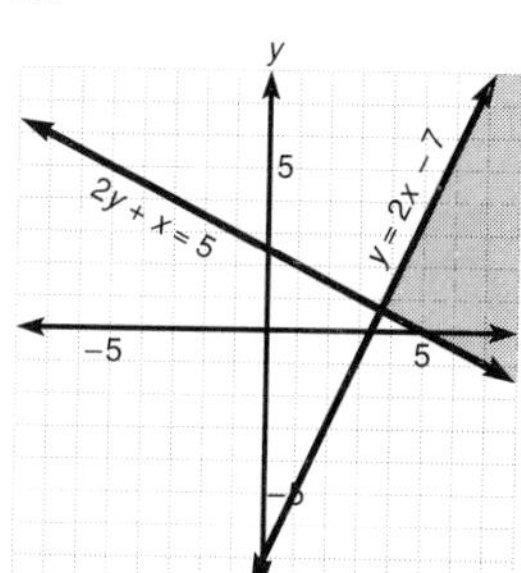

15.

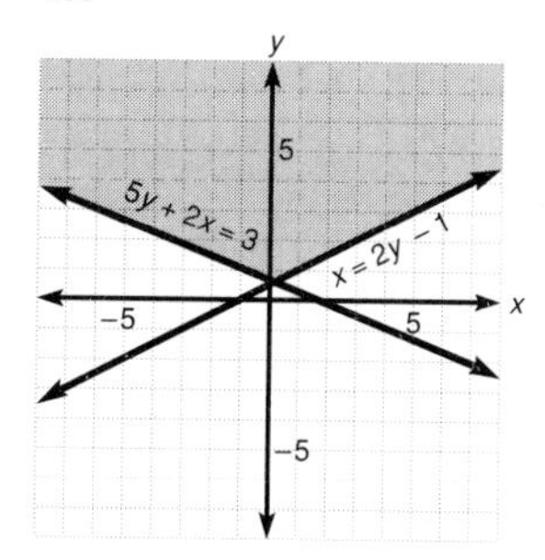

17.

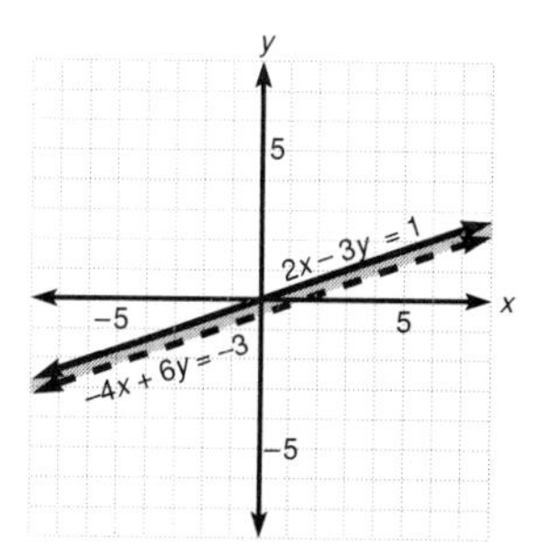

19.

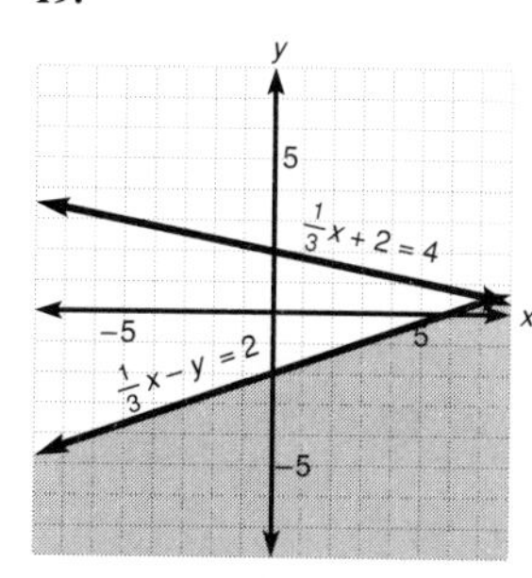

21.

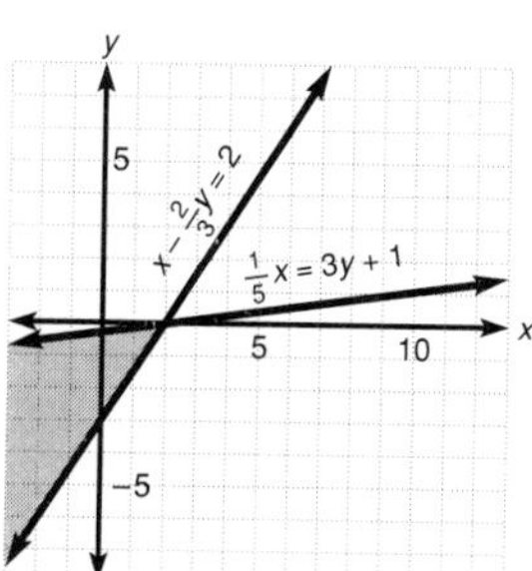

23.

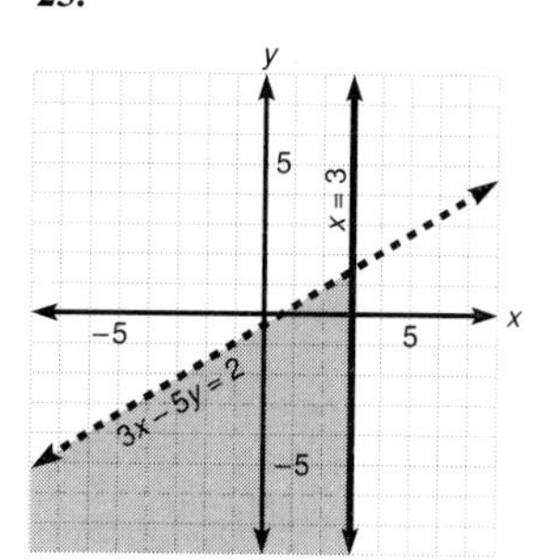

25.

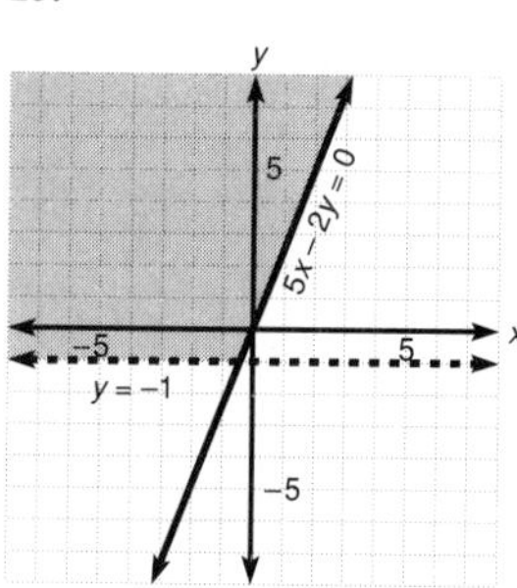

27.

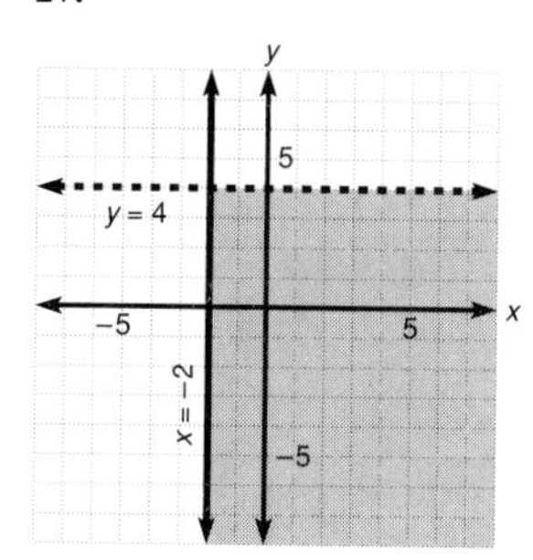

29.

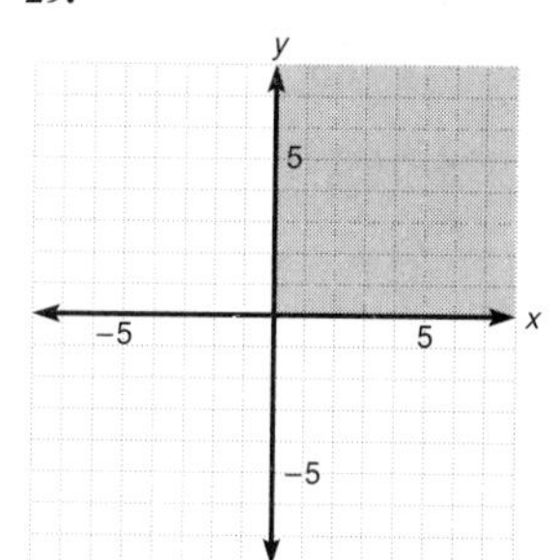

31. a. $x < 0$ **b.** $x > 0$
$y < 0$ $\quad y < 0$

Solutions to trial exercise problems

6. $5x - y < 5$
$2x + y \geq 1$

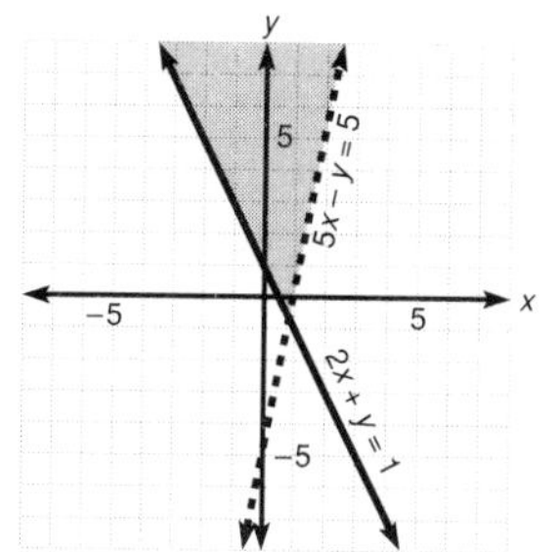

11. $3x - 2y \geq 1$
$2x \geq 5y$

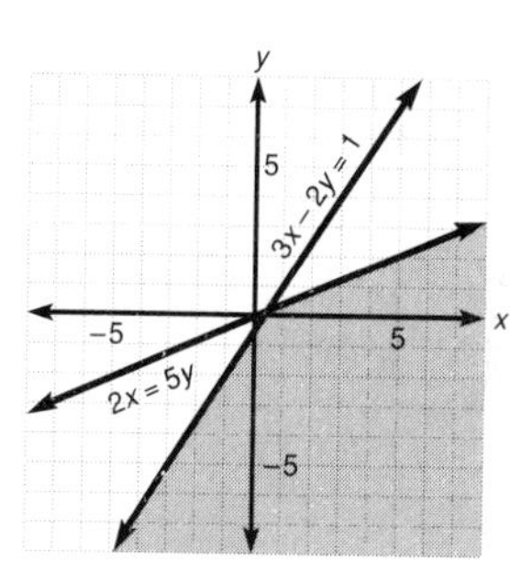

22. $\frac{3}{2}y - 4x \geq 2$
$\frac{1}{3}y > 1 - x$
(*Note:* Clear fractions.)
$3y - 8x \geq 4$
$y > 3 - 3x$

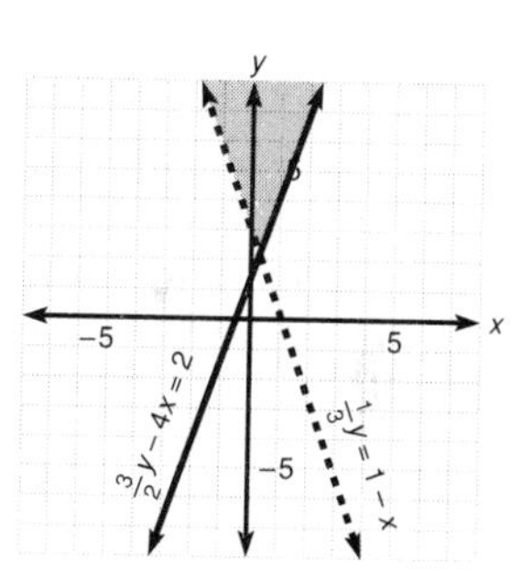

25. $5x - 2y \leq 0$
$y > -1$

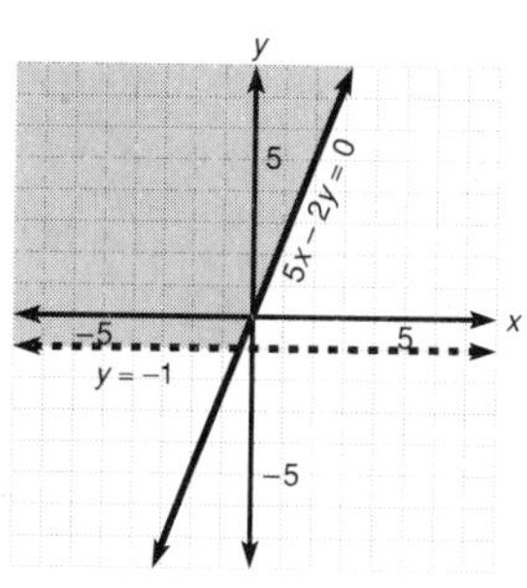

Review exercises

1. $2x^2 + 3x + 7$ **2.** $4x^2 - 11x - 3$ **3.** $25y^2 - 20y + 4$ **4.** $49z^2 - 1$ **5.** a^4 **6.** xy^2 **7.** $-y^9$ **8.** $x(5x - 3)$ **9.** $(2x - 1)(2x + 5)$ **10.** $(3y - 5)^2$

Chapter 8 review

1. $\{(3,1)\}$ **2.** $\{(2,4)\}$ **3.** $\{(4,2)\}$ **4.** $\{(-2,6)\}$ **5.** inconsistent; no solution **6.** dependent; unlimited number of solutions **7.** $\{(3,0)\}$ **8.** $\{(7,2)\}$ **9.** $\{(-1,1)\}$ **10.** $\{(1,-1)\}$

11. $\{(0,2)\}$ **12.** $\{(5,-3)\}$ **13.** $\{(5,2)\}$ **14.** inconsistent; no solution **15.** $\{(2,1)\}$ **16.** dependent; unlimited number of solutions **17.** $\{(5,3)\}$ **18.** $\{(5,-3)\}$ **19.** inconsistent; no solution **20.** $\left\{\left(\frac{24}{11},\frac{5}{11}\right)\right\}$ **21.** $\{(3,1)\}$ **22.** $\{(7,2)\}$ **23.** dependent; unlimited number of solutions **24.** $\left\{\left(\frac{19}{11},-\frac{17}{11}\right)\right\}$ **25.** $\ell = 25$; $w = 7$ **26.** \$11,000 at 14% profit; \$7,000 at 23% loss **27.** 1st gear, 12 teeth; 2nd gear, 54 teeth **28.** 30 milliliters at 60%; 90 milliliters at 80%

29. $x + y \geq 5,\ 2x - y < 3$

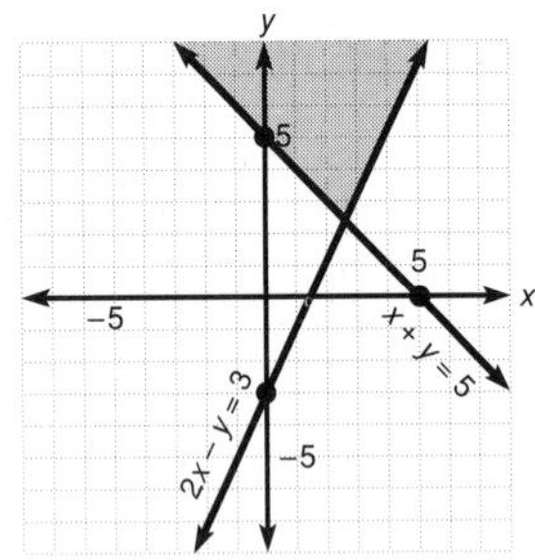

30. $4x - 3y < 1,\ 2x + 5y \leq 2$

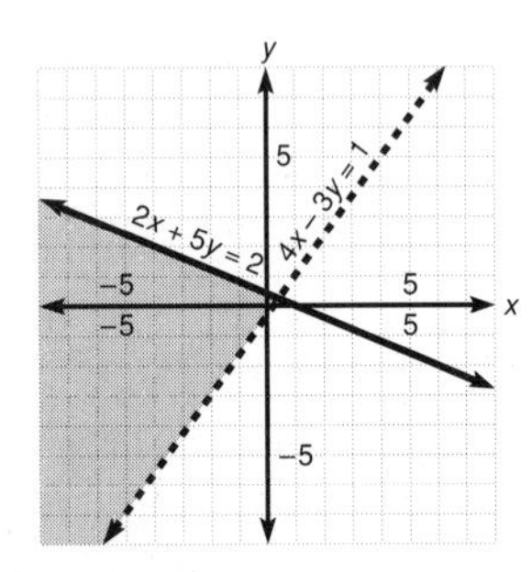

31. $4y - 3x \geq 2,\ y \leq 1 - 2x$

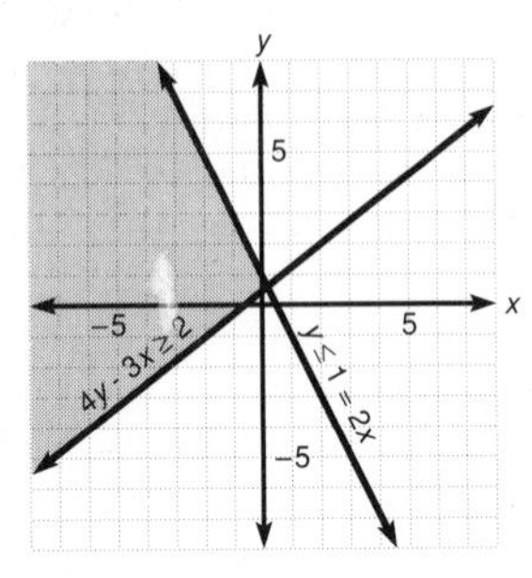

32. $2y \leq 4 - x,\ 5y - x > 2$

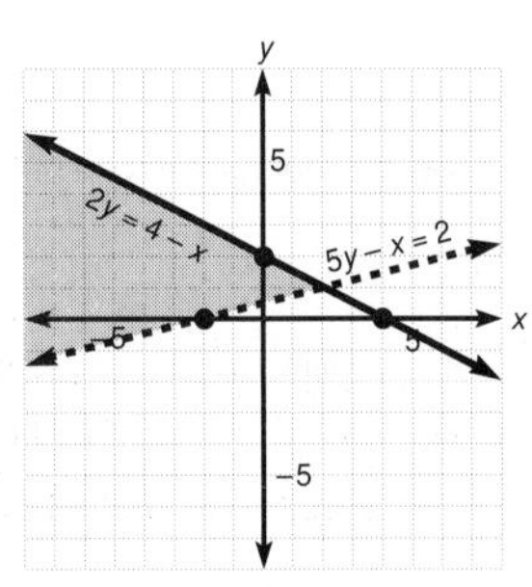

33. $x > 4y + 3,\ y \leq 4$

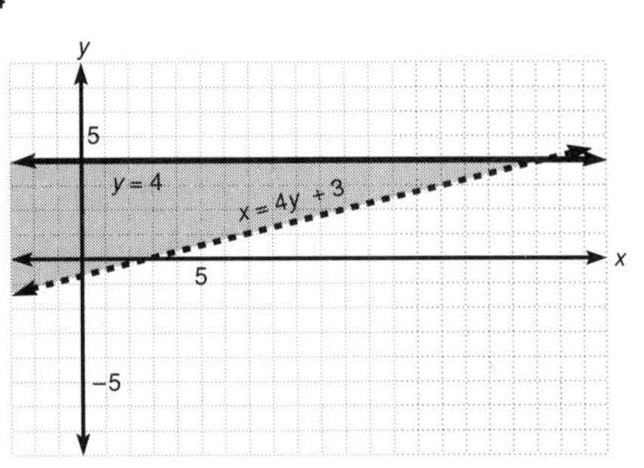

34. $x > 1,\ 5y - x \leq 0$

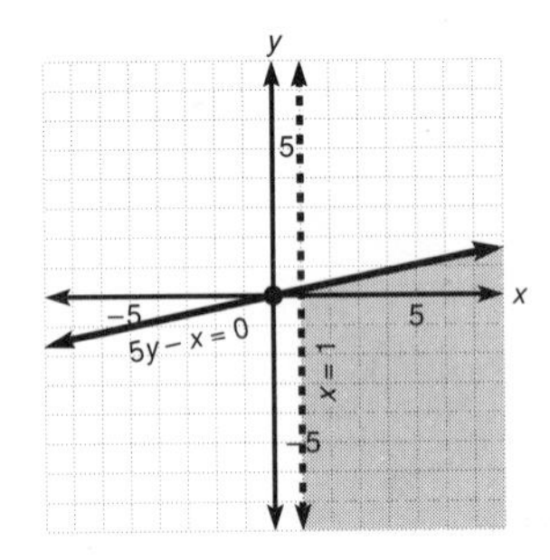

Chapter 8 cumulative test

1. $-\frac{1}{64}$ **2.** $\frac{4}{3}$ **3.** 1 **4.** $\frac{1}{25}$ **5.** 16 **6.** $-\frac{1}{3}$
7. $3a(a^2 + 5a + 9)$ **8.** $(x + 11)(x - 11)$ **9.** $(a - 5)(x + y)$
10. $(3a + b)(a - 2b)$ **11.** $7x(x + 1)(x - 1)$
12. $(5b - 2)(b - 4)$ **13.** $\frac{y^2}{2x}$ **14.** $\frac{32 - 3x}{x(x - 9)}$ **15.** $81x^4y^6z^6$
16. $9x^2y - 14y^2$ **17.** $\frac{a - 7}{7}$ **18.** $25x^2 - 30xy + 9y^2$
19. $2y + 6$ **20.** $\left\{\frac{15}{8}\right\}$ **21.** $\{0,-9\}$ **22.** $\left\{-\frac{11}{36}\right\}$
23. $\left\{\frac{3}{7},-1\right\}$ **24.** 6 **25.** $\frac{4}{5}$ **26.** $-\frac{1}{3}$ **27.** $\{(4,2)\}$
28. $\{(1,-2)\}$ **29.** $\left\{\left(\frac{14}{17},\frac{1}{17}\right)\right\}$ **30.** inconsistent; no solution
31. $\left\{\left(\frac{5}{8},1\right)\right\}$ **32.** $\{(0,8)\}$ **33.**

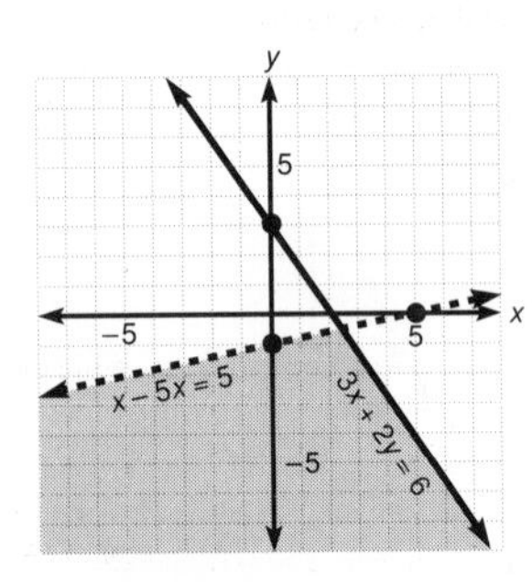

34. 6, 19 **35.** 42, 37 **36.** 5, 6 or −6, −5 **37.** 0 or 5

38. 1 or 2 **39.** $\frac{20}{9}$ hours or $2\frac{2}{9}$ hours

40. length = 13 inches; width = 7 inches

Chapter 9

Exercise 9–1

Answers to odd-numbered problems

1. 10 **3.** 2 **5.** -12 **7.** -11 **9.** 11 **11.** -4 **13.** 4.243 **15.** 6.403 **17.** -7.211 **19.** 24 amperes **21.** 13 units **23.** 5 meters **25.** 13 inches **27.** 8 yards **29.** 20 millimeters **31.** 2 **33.** 5 **35.** -2 **37.** 3 **39.** -3 **41.** 2 **43.** -3 **45.** 1 **47.** -1 **49.** 9 units

Solutions to trial exercise problems

19. $I = \sqrt{\frac{\text{watts}}{\text{ohms}}} = \sqrt{\frac{(\quad)}{(\quad)}} = \sqrt{\frac{1{,}728}{3}} = \sqrt{576} = 24$
Answer: 24 amperes **31.** $\sqrt[3]{8} = 2$, since $2 \cdot 2 \cdot 2 = 2^3 = 8$. **39.** $-\sqrt[4]{81} = -3$, since $3 \cdot 3 \cdot 3 \cdot 3 = 3^4 = 81$ and the negative sign indicates that we want the negative root. **45.** $\sqrt[10]{1} = 1$, since $1^{10} = 1$. 1 raised to any power is 1 and the nth root of 1 is 1.

Review exercises

1. 3^2 **2.** $2^2 \cdot 3$ **3.** 2^3 **4.** $2^3 \cdot 5$ **5.** $2 \cdot 5^2$ **6.** 3^4 **7.** 2^6 **8.** 2^4

Exercise 9–2

Answers to odd-numbered problems

1. 4 **3.** $2\sqrt{5}$ **5.** $3\sqrt{5}$ **7.** $4\sqrt{2}$ **9.** $4\sqrt{5}$ **11.** $7\sqrt{2}$ **13.** $a^3\sqrt{a}$ **15.** $2ab\sqrt{b}$ **17.** $3ab^2\sqrt{3ab}$ **19.** $3\sqrt{2}$ **21.** 15 **23.** $2\sqrt{15}$ **25.** $5\sqrt{15}$ **27.** $5\sqrt{3}$ **29.** $5x\sqrt{3}$ **31.** $4b\sqrt{3a}$ **33.** 13 feet **35.** 24 mph **37.** $2\sqrt[5]{2}$ **39.** $2\sqrt[3]{3}$ **41.** $b^2\sqrt[3]{b^2}$ **43.** y^3 **45.** $b\sqrt[3]{4a^2}$ **47.** $2ab\sqrt[3]{2ab^2}$ **49.** $3ab^3\sqrt[3]{3a^2b^2}$ **51.** $b\sqrt[3]{b}$ **53.** a **55.** $3ab\sqrt[3]{2b}$ **57.** $3b\sqrt[4]{3a^3}$ **59.** $4a^5b^3\sqrt[3]{3b}$ **61.** 2 inches **63.** 2 inches

Solutions to trial exercise problems

15. $\sqrt{4a^2b^3} = \sqrt{4 \cdot a^2 \cdot b^2 \cdot b} = \sqrt{4}\sqrt{a^2}\sqrt{b^2}\sqrt{b} = 2ab\sqrt{b}$
19. $\sqrt{6}\sqrt{3} = \sqrt{18} = \sqrt{9 \cdot 2} = \sqrt{9}\sqrt{2} = 3\sqrt{2}$
42. $\sqrt[3]{x^9} = \sqrt[3]{x^3 \cdot x^3 \cdot x^3} = \sqrt[3]{x^3}\sqrt[3]{x^3}\sqrt[3]{x^3} = x \cdot x \cdot x = x^3$
54. $\sqrt[3]{5a^2b}\,\sqrt[3]{75a^2b^2} = \sqrt[3]{375a^4b^3} = \sqrt[3]{125 \cdot 3 \cdot a^3 \cdot a \cdot b^3}$
$= \sqrt[3]{125}\sqrt[3]{a^3}\sqrt[3]{b^3}\sqrt[3]{3a} = 5ab\sqrt[3]{3a}$
61. $h = \sqrt[3]{\frac{12I}{b}} = \sqrt[3]{\frac{12(\quad)}{(\quad)}} = \sqrt[3]{\frac{12(2)}{(3)}} = \sqrt[3]{\frac{24}{3}} = \sqrt[3]{8}$
$= \sqrt[3]{2^3} = 2$ Answer: $h = 2$ inches

Review exercises

1. $\frac{7}{8}$ **2.** $-4x^2y$ **3.** $\frac{2y+10}{y+1}$ **4.** 5 **5.** 3 **6.** 4 **7.** 6 **8.** x

Exercise 9–3

Answers to odd-numbered problems

1. $\frac{3}{5}$ **3.** $\frac{5}{7}$ **5.** $\frac{\sqrt{3}}{2}$ **7.** $\frac{8}{a}$ **9.** $\frac{\sqrt{2}}{2}$ **11.** $\frac{2\sqrt{7}}{7}$ **13.** $\frac{\sqrt{15}}{15}$ **15.** $\frac{2\sqrt{3}}{15}$ **17.** $\sqrt{2}$ **19.** $\frac{5\sqrt{2}}{2}$ **21.** $\frac{x\sqrt{y}}{y}$ **23.** $\frac{\sqrt{x}}{x}$ **25.** a^2 **27.** $\sqrt{41}$ meters **29.** 12 feet **31.** $\frac{1}{2}$ **33.** $\frac{3}{5}$ **35.** $\frac{a^2\sqrt[3]{3}}{b}$ **37.** $\frac{\sqrt[5]{a^4}}{b^2}$ **39.** $\frac{ab^2\sqrt[4]{bc}}{c^3}$ **41.** $\frac{\sqrt[5]{x^3y^2}}{z^3}$ **43.** $\frac{\sqrt[3]{20}}{5}$ **45.** $\frac{2\sqrt[4]{5}}{5}$ **47.** $\frac{x\sqrt[3]{y}}{y}$ **49.** $b\sqrt[3]{a}$ **51.** $\frac{a\sqrt[3]{bc^2}}{bc}$ **53.** $\frac{\sqrt[3]{a^2bc^2}}{bc}$ **55.** $\sqrt[3]{a^2b}$ **57.** 3 units

Solutions to trial exercise problems

15. $\sqrt{\frac{4}{75}} = \frac{\sqrt{4}}{\sqrt{75}} = \frac{\sqrt{2^2}}{\sqrt{3 \cdot 5^2}} = \frac{2}{5\sqrt{3}} \cdot \frac{\sqrt{3}}{\sqrt{3}} = \frac{2\sqrt{3}}{5 \cdot 3} = \frac{2\sqrt{3}}{15}$
35. $\sqrt[3]{\frac{3a^6}{b^3}} = \frac{\sqrt[3]{3a^6}}{\sqrt[3]{b^3}} = \frac{a^2\sqrt[3]{3}}{b}$ **39.** $\sqrt[4]{\frac{a^4b^9}{c^{11}}} = \frac{\sqrt[4]{a^4b^9}}{\sqrt[4]{c^{11}}}$
$= \frac{ab^2\sqrt[4]{b}}{c^2\sqrt[4]{c^3}} \cdot \frac{\sqrt[4]{c^1}}{\sqrt[4]{c^1}} = \frac{ab^2\sqrt[4]{bc}}{c^2\sqrt[4]{c^4}} = \frac{ab^2\sqrt[4]{bc}}{c^2 \cdot c} = \frac{ab^2\sqrt[4]{bc}}{c^3}$
45. $\sqrt[4]{\frac{16}{125}} = \frac{\sqrt[4]{16}}{\sqrt[4]{125}} = \frac{\sqrt[4]{2^4}}{\sqrt[4]{5^3}} = \frac{2}{\sqrt[4]{5^3}} \cdot \frac{\sqrt[4]{5^1}}{\sqrt[4]{5^1}} = \frac{2\sqrt[4]{5^1}}{\sqrt[4]{5^4}} = \frac{2\sqrt[4]{5}}{5}$
49. $\frac{ab}{\sqrt[3]{a^2}} \cdot \frac{\sqrt[3]{a^1}}{\sqrt[3]{a^1}} = \frac{ab\sqrt[3]{a}}{\sqrt[3]{a^3}} = \frac{ab\sqrt[3]{a}}{a} = b\sqrt[3]{a}$
51. $\sqrt[3]{\frac{a^3}{b^2c}} = \frac{\sqrt[3]{a^3}}{\sqrt[3]{b^2c}} = \frac{a}{\sqrt[3]{b^2c}} \cdot \frac{\sqrt[3]{b^1c^2}}{\sqrt[3]{b^1c^2}} = \frac{a\sqrt[3]{bc^2}}{\sqrt[3]{b^3c^3}} = \frac{a\sqrt[3]{bc^2}}{bc}$
57. $r = \sqrt[3]{\frac{3V}{4\pi}} = \sqrt[3]{\frac{3(\quad)}{4(\quad)}} = \sqrt[3]{\frac{3(113.04)}{4(3.14)}} = \sqrt[3]{\frac{339.12}{12.56}}$
$= \sqrt[3]{27} = \sqrt[3]{3^3} = 3$ Answer: 3 units

Review exercises

1. $6x$ **2.** $4y$ **3.** $8ab$ **4.** $5xy$ **5.** $x^2 - 9$ **6.** $x^2 - y^2$

Exercise 9–4

Answers to odd-numbered problems

1. $9\sqrt{3}$ **3.** $10\sqrt{5}$ **5.** $6\sqrt{3}$ **7.** $3\sqrt{7}$ **9.** $3\sqrt{a}$ **11.** $8\sqrt{a}$ **13.** $7\sqrt{xy}$ **15.** $8\sqrt{a} + 2\sqrt{ab}$ **17.** $4\sqrt{xy} + 3\sqrt{y}$ **19.** $7\sqrt{2}$ **21.** $\sqrt{3}$ **23.** $17\sqrt{7}$ **25.** $7\sqrt{2}$ **27.** $2\sqrt{3} + 8\sqrt{2}$ **29.** $7\sqrt{2a}$ **31.** $-\sqrt{x}$ **33.** $17\sqrt{2a}$ **35.** $2\sqrt{2a} + 6\sqrt{3a}$ **37.** 14 units **39.** 8.1 units by 8.1 units **41.** $6\sqrt[5]{2}$ **43.** $5\sqrt[3]{2}$ **45.** $3\sqrt[3]{3} + 10\sqrt[3]{2}$ **47.** $5\sqrt[4]{x^3}$ **49.** $\sqrt[3]{x^2y}$ **51.** $2a\sqrt[3]{b^2}$

Solutions to trial exercise problems

14. $3\sqrt{x} + 2\sqrt{y} - \sqrt{x} = 3\sqrt{x} - \sqrt{x} + 2\sqrt{y} = 2\sqrt{x} + 2\sqrt{y}$
18. $\sqrt{20} + 3\sqrt{5} = \sqrt{2^2 \cdot 5} + 3\sqrt{5} = 2\sqrt{5} + 3\sqrt{5} = 5\sqrt{5}$
29. $\sqrt{50a} + \sqrt{8a} = \sqrt{2 \cdot 5^2 \cdot a} + \sqrt{2^3 \cdot a} = 5\sqrt{2a} + 2\sqrt{2a}$
$= 7\sqrt{2a}$ **37.** $h = b + s$, where $s = 6$ units and $b = \sqrt{c^2 - s^2}$
$= \sqrt{(10)^2 - (6)^2} = \sqrt{100 - 36} = \sqrt{64} = 8$. Then $h = b + s$
$= (8) + (6) = 14$. Answer: 14 units
43. $\sqrt[3]{16} + \sqrt[3]{54} = \sqrt[3]{2^4} + \sqrt[3]{2 \cdot 3^3} = 2\sqrt[3]{2} + 3\sqrt[3]{2} = 5\sqrt[3]{2}$
50. $\sqrt[3]{x^6y} + 2x^2\sqrt[3]{y} = x^2\sqrt[3]{y} + 2x^2\sqrt[3]{y} = (1 + 2)x^2\sqrt[3]{y} = 3x^2\sqrt[3]{y}$

Review exercises

1. $6x^2 - 3xy$ **2.** $2a^4 - 2a^2b^2$ **3.** $x^2 - 2x + 1$ **4.** $y^2 - 1$ **5.** $4x^2 - 1$ **6.** $x^2 + 5xy + 6y^2$ **7.** $x^2 - 2xy + y^2$ **8.** $a^2 + 4ab + 4b^2$

Exercise 9–5

Answers to odd-numbered problems

1. $3\sqrt{2} + 3\sqrt{3}$ **3.** $\sqrt{6} + \sqrt{14}$ **5.** $6\sqrt{6} - 3\sqrt{22}$ **7.** $5\sqrt{3} - 5\sqrt{2}$ **9.** $14\sqrt{5} - 42\sqrt{2}$ **11.** $3a + \sqrt{ab}$ **13.** $14 + 7\sqrt{2}$ **15.** $12 - 25\sqrt{a} + 12a$ **17.** 1 **19.** -2 **21.** $9 + 4\sqrt{5}$ **23.** $x + 2\sqrt{xy} + y$ **25.** $x - y$ **27.** $x^2y - z$ **29.** $4x + 4y\sqrt{x} + y^2$ **31.** $11 + \sqrt{3}$ **33.** $\sqrt{a} - 3\sqrt{b}$ **35.** $\frac{-\sqrt{2} + 3}{7}$ **37.** $\frac{-14 + 7\sqrt{7}}{3}$ **39.** $\sqrt{6} + \sqrt{3}$ **41.** $\frac{6\sqrt{3} + 3\sqrt{5}}{7}$ **43.** $\frac{-3 - \sqrt{5}}{2}$ **45.** $\frac{a + 2b\sqrt{a} + b^2}{a - b^2}$

Solutions to trial exercise problems

5. $3\sqrt{2}(2\sqrt{3} - \sqrt{11}) = 3\sqrt{2}\cdot 2\sqrt{3} - 3\sqrt{2}\cdot\sqrt{11}$ $= 6\sqrt{6} - 3\sqrt{22}$ **15.** $(3 - 4\sqrt{a})(4 - 3\sqrt{a})$ $= 3\cdot 4 - 3\cdot 3\sqrt{a} - 4\sqrt{a}\cdot 4 + 4\sqrt{a}\cdot 3\sqrt{a}$ $= 12 - 9\sqrt{a} - 16\sqrt{a} + 12\cdot a = 12 - 25\sqrt{a} + 12a$ **17.** $(\sqrt{3} + \sqrt{2})(\sqrt{3} - \sqrt{2})$ Conjugates, therefore $= (\sqrt{3})^2 - (\sqrt{2})^2 = 3 - 2 = 1$ **21.** $(2 + \sqrt{5})^2$ $= (2 + \sqrt{5})(2 + \sqrt{5}) = 2\cdot 2 + 2\sqrt{5} + 2\sqrt{5} + \sqrt{5}\sqrt{5}$ $= 4 + 4\sqrt{5} + 5 = 9 + 4\sqrt{5}$ **26.** $(2\sqrt{a} - \sqrt{b})(2\sqrt{a} + \sqrt{b})$ Conjugates, therefore $= (2\sqrt{a})^2 - (\sqrt{b})^2 = 2^2(\sqrt{a})^2 - b$ $= 4a - b$ **38.** $\dfrac{6}{3 - \sqrt{6}} = \dfrac{6}{3 - \sqrt{6}}\cdot\dfrac{3 + \sqrt{6}}{3 + \sqrt{6}} = \dfrac{6(3 + \sqrt{6})}{(3)^2 - (\sqrt{6})^2}$

$= \dfrac{6(3 + \sqrt{6})}{9 - 6} = \dfrac{6(3 + \sqrt{6})}{3} = 2(3 + \sqrt{6}) = 6 + 2\sqrt{6}$

43. $\dfrac{1 + \sqrt{5}}{1 - \sqrt{5}} = \dfrac{1 + \sqrt{5}}{1 - \sqrt{5}}\cdot\dfrac{1 + \sqrt{5}}{1 + \sqrt{5}} = \dfrac{(1)^2 + \sqrt{5} + \sqrt{5} + (\sqrt{5})^2}{(1)^2 - (\sqrt{5})^2}$

$= \dfrac{1 + 2\sqrt{5} + 5}{1 - 5} = \dfrac{6 + 2\sqrt{5}}{-4} = \dfrac{2(3 + \sqrt{5})}{-4}$

$= \dfrac{-(3 + \sqrt{5})}{2} = \dfrac{-3 - \sqrt{5}}{2}$

Review exercises

1. 64 **2.** 32 **3.** 9 **4.** $\dfrac{1}{27}$ **5.** $\dfrac{1}{x^6}$ **6.** $8a^6b^3$

7. $\dfrac{y^3}{x}$ **8.** $\dfrac{1}{x^9}$

Exercise 9–6

Answers to odd-numbered problems

1. 6 **3.** a^2 **5.** 2 **7.** -3 **9.** 9 **11.** 27 **13.** $\dfrac{1}{5}$ **15.** $\dfrac{1}{8}$

17. $-\dfrac{1}{2}$ **19.** 4 **21.** $2^{5/6}$ **23.** x **25.** $c^{3/4}$ **27.** 2 **29.** $2^{1/6}$

31. $a^{3/5}$ **33.** $y^{1/6}$ **35.** $a^{1/3}$ **37.** $x^{2/3}$ **39.** $c^{1/6}$ **41.** $a^{1/3}$

43. $\dfrac{1}{x^{1/6}} = \dfrac{x^{5/6}}{x}$ **45.** $\dfrac{1}{y^{1/4}} = \dfrac{y^{3/4}}{y}$ **47.** $8a^3$ **49.** $4a^4b^2$

51. $b^{1/2}c^{1/4}$ **53.** $a^{1/2}b^{2/3}$ **55.** 81 **57.** 49 mph **59.** 24 miles **61.** Take the square root 3 times, that is, the square root of the square root of the square root is the eighth root.

Solutions to trial exercise problems

3. $(a^6)^{1/3} = a^{6/1\cdot 1/3} = a^2$ **7.** $(-27)^{1/3} = \sqrt[3]{(-27)} = -3$

9. $(27)^{2/3} = (\sqrt[3]{27})^2 = (3)^2 = 9$ **17.** $(-8)^{-1/3} = \dfrac{1}{(-8)^{1/3}}$

$= \dfrac{1}{\sqrt[3]{-8}} = \dfrac{1}{-2} = -\dfrac{1}{2}$ **35.** $(a^{2/3})^{1/2} = a^{2/3\cdot 1/2} = a^{1/3}$

39. $(c^{-1/4})^{-2/3} = c^{(-1/4)\cdot(-2/3)} = c^{1/6}$ **51.** $\dfrac{b^{3/4}c^{1/2}}{b^{1/4}c^{1/4}}$

$= b^{3/4 - 1/4}c^{1/2 - 1/4} = b^{2/4}c^{2/4 - 1/4} = b^{1/2}c^{1/4}$

Review exercises

1. 7 **2.** x **3.** $x + 1$ **4.** $x^2 + 2x + 1$ **5.** $x^2 - 4x + 4$ **6.** $\{-2,3\}$ **7.** $\{2,8\}$ **8.** $\{-1,0\}$

Exercise 9–7

Answers to odd-numbered problems

1. $\{16\}$ **3.** $\{81\}$ **5.** $\{11\}$ **7.** $\{43\}$ **9.** $\{12\}$ **11.** $\{5\}$ **13.** $\{9\}$ **15.** $\{36\}$ **17.** $\emptyset$ **19.** $\emptyset$ **21.** $\{4\}$ **23.** $\{4\}$ **25.** $\{3\}$ **27.** $\{2\}$ **29.** $\{5\}$ **31.** $\{3\}$ **33.** $\{16\}$ **35.** $\{9\}$ **37.** $\{2\}$ **39.** $\left\{-\dfrac{3}{4}\right\}$ **41.** $\emptyset$ **43.** $\{3\}$ **45.** $\{4\}$ **47.** $\{8\}$ **49.** $\{12\}$ **51.** $\{2\}$ **53.** 19 **55.** 24 **57.** 4 **59.** 5 **61.** 24 feet **63.** 108 feet **65.** 100

Solutions to trial exercise problems

17.
$$\begin{aligned} \sqrt{x} + 7 &= 5 \\ \sqrt{x} &= -2 \\ (\sqrt{x})^2 &= (-2)^2 \\ x &= 4 \end{aligned}$$

Check:
$$\begin{aligned} \sqrt{4} + 7 &= 5 \\ 2 + 7 &= 5 \\ 9 &= 5 \text{ (false)} \end{aligned}$$

Therefore no solution and the solution set is $\emptyset$.

21.
$$\begin{aligned} \sqrt{2x + 1} &= \sqrt{x + 5} \\ (\sqrt{2x + 1})^2 &= (\sqrt{x + 5})^2 \\ 2x + 1 &= x + 5 \\ x + 1 &= 5 \\ x &= 4 \end{aligned}$$

Check:
$$\begin{aligned} \sqrt{2(4) + 1} &= \sqrt{(4) + 5} \\ \sqrt{8 + 1} &= \sqrt{9} \\ \sqrt{9} &= \sqrt{9} \\ 3 &= 3 \text{ (true)} \end{aligned}$$

$\{4\}$

39.
$$\begin{aligned} \sqrt{x^2 + 1} &= x + 2 \\ (\sqrt{x^2 + 1})^2 &= (x + 2)^2 \\ x^2 + 1 &= (x + 2)(x + 2) \\ x^2 + 1 &= x^2 + 2x + 2x + 4 \\ x^2 + 1 &= x^2 + 4x + 4 \\ 1 &= 4x + 4 \\ -3 &= 4x \\ -\frac{3}{4} &= x \end{aligned}$$

Check:
$$\begin{aligned} \sqrt{\left(-\frac{3}{4}\right)^2 + 1} &= \left(-\frac{3}{4}\right) + 2 \\ \sqrt{\frac{9}{16} + 1} &= -\frac{3}{4} + \frac{8}{4} \\ \sqrt{\frac{9}{16} + \frac{16}{16}} &= \frac{5}{4} \\ \sqrt{\frac{25}{16}} &= \frac{5}{4} \\ \frac{\sqrt{25}}{\sqrt{16}} &= \frac{5}{4} \\ \frac{5}{4} &= \frac{5}{4} \text{ (true)} \end{aligned}$$

$\left\{-\dfrac{3}{4}\right\}$

43.
$$\begin{aligned} \sqrt{x + 6} &= x \\ (\sqrt{x + 6})^2 &= (x)^2 \\ x + 6 &= x^2 \\ 0 &= x^2 - x - 6 \\ 0 &= (x - 3)(x + 2) \\ x - 3 &= 0 \text{ or } x + 2 = 0 \\ x &= 3 \text{ or } \quad x = -2 \end{aligned}$$

Check:

$\sqrt{(3)+6} = (3)$
$\sqrt{9} = 3$
$3 = 3$ (true)
$\sqrt{(-2)+6} = (-2)$
$\sqrt{4} = -2$
$2 = -2$ (false)

$\{3\}$

47. $\sqrt{x-4} = x-6$
$(\sqrt{x-4})^2 = (x-6)^2$
$x-4 = (x-6)(x-6)$
$x-4 = x^2-6x-6x+36$
$x-4 = x^2-12x+36$
$0 = x^2-13x+40$
$0 = (x-8)(x-5)$
$x-8 = 0$ or $x-5 = 0$
$x = 8$ or $x = 5$

Check:

$\sqrt{(8)-4} = (8)-6$
$\sqrt{4} = 2$
$2 = 2$ (true)
$\sqrt{(5)-4} = (5)-6$
$\sqrt{1} = -1$
$1 = -1$ (false)

$\{8\}$

57. Let x = the number.
$\sqrt{x+12} = x$
$(\sqrt{x+12})^2 = (x)^2$
$x+12 = x^2$
$0 = x^2-x-12$
$0 = (x-4)(x+3)$
$x-4 = 0$ or $x+3 = 0$
$x = 4$ or $x = -3$

Check:

$\sqrt{(4)+12} = (4)$
$\sqrt{16} = 4$
$4 = 4$ (true)
$\sqrt{(-3)+12} = (-3)$
$\sqrt{9} = -3$
$3 = -3$ (false)

Hence the number is 4.

Review exercises

1. $(x+2)(x-2)$ **2.** $(x+3)(x+6)$ **3.** $(x+2)(x-5)$ **4.** $(x-3)^2$ **5.** 9 **6.** 7 **7.** 11 **8.** $\{-8,8\}$

Chapter 9 review

1. 9 **2.** 5 **3.** -3 **4.** -7 **5.** $2\sqrt{10}$ **6.** $3ab\sqrt{2b}$ **7.** $2\sqrt{7}$ **8.** $6\sqrt{5}$ **9.** $\frac{4\sqrt{17}}{17}$ **10.** $\frac{\sqrt{14}}{6}$ **11.** $\frac{\sqrt{ab}}{b}$ **12.** $\frac{\sqrt{xy}}{y^2}$ **13.** $\frac{\sqrt{ab}}{b}$ **14.** $\frac{2\sqrt{xy}}{y}$ **15.** $7\sqrt{7}$ **16.** $8\sqrt{2}$ **17.** $3\sqrt{5}$ **18.** $24\sqrt{3}$ **19.** $\sqrt{2a}$ **20.** $17\sqrt{x}$ **21.** $\sqrt{15}-\sqrt{21}$ **22.** $2\sqrt{35}+2\sqrt{15}$ **23.** $8-2\sqrt{7}$ **24.** $39-12\sqrt{3}$ **25.** $8+2\sqrt{15}$ **26.** $4a-b$ **27.** $-\sqrt{3}-2$ **28.** $\frac{-\sqrt{6}+4}{5}$ **29.** $\frac{\sqrt{a}-b}{a-b^2}$ **30.** $\frac{\sqrt{xy}-x}{y-x}$ **31.** $\frac{a^2-a\sqrt{b}}{a^2-b}$ **32.** $\frac{-11-6\sqrt{2}}{7}$ **33.** 6 **34.** 4 **35.** -2 **36.** $\frac{1}{4}$ **37.** a **38.** $b^{13/12}$ **39.** $a^{1/4}$ **40.** $a^{9/8}$ **41.** $8a^3b^6$ **42.** $a^{3/2}b^{1/2}$ **43.** $\{64\}$ **44.** $\{53\}$ **45.** $\{4\}$ **46.** $\{1\}$ **47.** $\{2\}$ **48.** $\{3\}$ **49.** $\{-2\}$ **50.** $\{4,3\}$

Chapter 9 cumulative test

1. 45 **2.** x^7 **3.** x^6 **4.** $2x^2+2x+13$ **5.** $8a^4b^3-12a^3b^4+16a^2b^5$ **6.** $2a^3b^2$ **7.** -36 **8.** $9a^2-6ab+b^2$ **9.** $\frac{a-3}{3a+6}$ **10.** $16\sqrt{3}$ **11.** -2 **12.** $\frac{x+3}{x-1}$ **13.** $\frac{x\sqrt{x}+\sqrt{xy}}{x^2-y}$ **14.** $3xy^2\sqrt[3]{3xz}$ **15.** $25x^2-y^2$ **16.** $2x-4y$ **17.** $2a^3b^3(3b-b^2+4a^2)$ **18.** $(5c+d)(5c-d)$ **19.** $(2x-1)(x+4)$ **20.** $(y^2+2z)(y^2-2z)$ **21.** $(2x+1)(3x+4)$ **22.** $(x+7)(x-4)$ **23.** $\left\{-\frac{1}{3}\right\}$ **24.** $\{-3,3\}$ **25.** $\{-6\}$ **26.** $\{12\}$ **27.** $\left\{\frac{19}{12}\right\}$ **28.** $\left\{-1,-\frac{1}{2}\right\}$ **29.** $1 < x < 8$ **30.** $x > \frac{7}{2}$ **31.** -3 **32.** $y = 4x-2$; slope is 4; y-intercept is $(0,-2)$ **33.** $\left(\frac{7}{8},\frac{3}{8}\right)$ **34.** 14, 56 **35.** 1,350 **36.** 16, 18 **37.** 21 feet by 27 feet

Chapter 10

Exercise 10–1

Answers to odd-numbered problems

1. $\{-5,3\}$ **3.** $\left\{-\frac{3}{2},2\right\}$ **5.** $\{-2,2\}$ **7.** $\{-8,8\}$ **9.** $\{-\sqrt{11},\sqrt{11}\}$ **11.** $\{2\sqrt{5},-2\sqrt{5}\}$ **13.** $\{-\sqrt{3},\sqrt{3}\}$ **15.** $\{-4\sqrt{2},4\sqrt{2}\}$ **17.** $\{-3,3\}$ **19.** $\{-\sqrt{6},\sqrt{6}\}$ **21.** $\{-5\sqrt{2},5\sqrt{2}\}$ **23.** $\{-2\sqrt{2},2\sqrt{2}\}$ **25.** $\{-2\sqrt{2},2\sqrt{2}\}$ **27.** $\{-\sqrt{2},\sqrt{2}\}$ **29.** $\left\{-\frac{\sqrt{6}}{5},\frac{\sqrt{6}}{5}\right\}$ **31.** $\{-\sqrt{11},\sqrt{11}\}$ **33.** $\{-4,0\}$ **35.** $\{-1,9\}$ **37.** $\{-3-\sqrt{6},-3+\sqrt{6}\}$ **39.** $\{9-3\sqrt{2},9+3\sqrt{2}\}$ **41.** $\{-5+4\sqrt{2},-5-4\sqrt{2}\}$ **43.** $\{-a-6,-a+6\}$ **45.** $\{6-a,6+a\}$ **47.** $\{p-q,p+q\}$ **49.** $\left\{-\frac{1}{2},\frac{7}{2}\right\}$ **51.** 5 meters **53.** 2 feet **55.** 9, -9 **57.** 0,9 **59.** 7 inches, 14 inches **61.** length = 24 meters; width = 6 meters **63.** 4 and 8 **65.** $5\sqrt{2}$ centimeters

Solutions to trial exercise problems

11. $a^2 = 20$

Extract the roots.
$a = \sqrt{20}$ or $a = -\sqrt{20}$
$a = \sqrt{4\cdot 5}$ or $a = -\sqrt{4\cdot 5}$
$a = 2\sqrt{5}$ or $a = -2\sqrt{5}$
$\{2\sqrt{5},-2\sqrt{5}\}$

18. $5x^2 = 75$

Divide each member by 5.
$x^2 = 15$
Extract the roots.
$x = \sqrt{15}$ or $x = -\sqrt{15}$
$\{\sqrt{15},-\sqrt{15}\}$

25. $\frac{3}{4}x^2 - 6 = 0$

Multiply each member by 4.
$3x^2 - 24 = 0$
Add 24 to each member.
$3x^2 = 24$
Divide each member by 3.
$x^2 = 8$
Then $x = \sqrt{8}$ or $x = -\sqrt{8}$
$x = 2\sqrt{2}$ or $x = -2\sqrt{2}$
$\{2\sqrt{2}, -2\sqrt{2}\}$

33. $(x + 2)^2 = 4$
Extract the roots.
$x + 2 = \pm 2$
Add -2 to each member.
$x = -2 \pm 2$
So $x = -2 + 2$ or $x = -2 - 2$
$x = 0$ or -4
$\{0, -4\}$

39. $(x - 9)^2 = 18$
Extract the roots.
$x - 9 = \pm\sqrt{18} = \pm 3\sqrt{2}$
Add 9 to each member.
$x = 9 + 3\sqrt{2}$ or $x = 9 - 3\sqrt{2}$
$\{9 + 3\sqrt{2}, 9 - 3\sqrt{2}\}$

44. $(x - a)^2 = 50$
$x - a = \sqrt{50} = 5\sqrt{2}$ or
$x - a = -\sqrt{50} = -5\sqrt{2}$
Add a to each member.
$x = a + 5\sqrt{2}$ or $x = a - 5\sqrt{2}$
$\{a + 5\sqrt{2}, a - 5\sqrt{2}\}$

58. Let n = the number. Then n^2 = the square of the number and $8n$ = eight times the number.
The equation is
$$2n^2 - 8n = 0$$
$$2n(n - 4) = 0$$
$$2n = 0 \text{ or } n - 4 = 0$$
$$n = 0 \text{ or } n = 4$$
The number n is 0 or 4.

61. Using $A = \ell w$, let ℓ = length of the rectangle. Then $\frac{1}{4}\ell$ = width of the rectangle.

The equation is $\ell \cdot \frac{1}{4}\ell = 144$
$$\frac{1}{4} \cdot \ell^2 = 144$$
$$\ell^2 = 576$$
$$\ell = \pm\sqrt{576} = \pm 24$$
Since length cannot be negative, then $\ell = 24$ meters and

$w = \frac{1}{4}(24) = 6$ meters.

Review exercises

1. $x^2 - 4x + 4$ **2.** $9z^2 + 12z + 4$ **3.** $(x + 9)^2$
4. $(3y + 5)^2$ **5.** $\dfrac{3x^2 - 7x}{(x + 2)(x - 2)}$ **6.** $\dfrac{1}{(x - 2)(x + 3)}$

Exercise 10–2

Answers to odd-numbered problems

1. $x^2 + 10x + 25; (x + 5)^2$ **3.** $a^2 - 12a + 36; (a - 6)^2$
5. $x^2 + 24x + 144; (x + 12)^2$ **7.** $y^2 - 20y + 100; (y - 10)^2$
9. $x^2 + x + \frac{1}{4}; \left(x + \frac{1}{2}\right)^2$ **11.** $x^2 - 7x + \frac{49}{4}; \left(x - \frac{7}{2}\right)^2$
13. $x^2 + \frac{1}{2}x + \frac{1}{16}; \left(x + \frac{1}{4}\right)^2$ **15.** $s^2 - \frac{1}{5}s + \frac{1}{100}; \left(s - \frac{1}{10}\right)^2$ **17.** $y^2 + \frac{2}{3}y + \frac{1}{9}; \left(y + \frac{1}{3}\right)^2$
19. $m^2 - \frac{2}{5}m + \frac{1}{25}; \left(m - \frac{1}{5}\right)^2$ **21.** $a^2 - \frac{3}{2}a + \frac{9}{16}; \left(a - \frac{3}{4}\right)^2$ **23.** $\{-7, -1\}$ **25.** $\{-2, 6\}$ **27.** $\{1, 3\}$
29. $\left\{\frac{1 - \sqrt{5}}{2}, \frac{1 + \sqrt{5}}{2}\right\}$ **31.** $\left\{\frac{5 - \sqrt{17}}{2}, \frac{5 + \sqrt{17}}{2}\right\}$
33. $\{2 - \sqrt{85}, 2 + \sqrt{85}\}$ **35.** $\left\{\frac{-21 - \sqrt{401}}{2}, \frac{-21 + \sqrt{401}}{2}\right\}$
37. $\left\{-\frac{3}{2}, 1\right\}$ **39.** $\left\{-3, -\frac{1}{2}\right\}$ **41.** $\left\{-\frac{1}{2}, \frac{3}{2}\right\}$ **43.** $\left\{\frac{2}{3}, \frac{3}{2}\right\}$
45. $\left\{\frac{-1 - \sqrt{13}}{2}, \frac{-1 + \sqrt{13}}{2}\right\}$ **47.** $\left\{-\frac{2}{3}, \frac{1}{2}\right\}$
49. $\{3 - \sqrt{5}, 3 + \sqrt{5}\}$ **51.** $\left\{\frac{1 - \sqrt{57}}{4}, \frac{1 + \sqrt{57}}{4}\right\}$
53. $\left\{\frac{-1 - \sqrt{29}}{2}, \frac{-1 + \sqrt{29}}{2}\right\}$ **55.** $\left\{\frac{-5 - \sqrt{17}}{4}, \frac{-5 + \sqrt{17}}{4}\right\}$
57. 12 inches; 8 inches **59.** ℓ = 15 millimeters; w = 7 millimeters
61. 17 inches by 9 inches **63.** $\frac{11}{2}$ meters by $\frac{7}{2}$ meters
65. 14 rods by 6 rods **67.** w = 3 inches

Solutions to trial exercise problems

3. $a^2 - 12a$
Square one-half of the coefficient of a, -12.
$$\left[\frac{1}{2}(-12)\right]^2 = (-6)^2 = 36$$
Then $a^2 - 12a + 36 = (a - 6)^2$

9. $x^2 + x$
Square one-half of the coefficient of x, 1.
$$\left[\frac{1}{2}(1)\right]^2 = \left(\frac{1}{2}\right)^2 = \frac{1}{4}$$
So $x^2 + x + \frac{1}{4} = \left(x + \frac{1}{2}\right)^2$

13. $x^2 + \frac{1}{2}x$

Square one-half of the coefficient of x, $\frac{1}{2}$.
$$\left[\frac{1}{2}\left(\frac{1}{2}\right)\right]^2 = \left(\frac{1}{4}\right)^2 = \frac{1}{16}$$
So $x^2 + \frac{1}{2}x + \frac{1}{16} = \left(x + \frac{1}{4}\right)^2$

16. $x^2 - \frac{3}{8}x$

Square one-half of the coefficient of x, $-\frac{3}{8}$.
$$\left[\frac{1}{2}\left(-\frac{3}{8}\right)\right]^2 = \left(-\frac{3}{16}\right)^2 = \frac{9}{256}$$
So $x^2 - \frac{3}{8}x + \frac{9}{256} = \left(x - \frac{3}{16}\right)^2$

29. $u^2 - u - 1 = 0$

Add 1 to each member.

$u^2 - u = 1$

Add $\left[\frac{1}{2}(-1)\right]^2 = \left(-\frac{1}{2}\right)^2 = \frac{1}{4}$ to each member.

$u^2 - u + \frac{1}{4} = 1 + \frac{1}{4}$

Then $\left(u - \frac{1}{2}\right)^2 = \frac{5}{4}$ Then

$u - \frac{1}{2} = \pm\sqrt{\frac{5}{4}} = \pm\frac{\sqrt{5}}{2}$

so $u = \frac{1}{2} \pm \frac{\sqrt{5}}{2} = \frac{1 \pm \sqrt{5}}{2}$

Then $u = \frac{1 + \sqrt{5}}{2}$ or $u = \frac{1 - \sqrt{5}}{2}$

$\left\{\frac{1 + \sqrt{5}}{2}, \frac{1 - \sqrt{5}}{2}\right\}$

36. $3x^2 + 6x = 3$

Divide each term by 3.

$x^2 + 2x = 1$

Then $\left[\frac{1}{2}(2)\right]^2 = (1)^2 = 1$, so add 1 to each member.

$x^2 + 2x + 1 = 1 + 1$

So $(x + 1)^2 = 2$. Then

$x + 1 = \pm\sqrt{2}$

so $x = -1 \pm\sqrt{2}$

$x = -1 + \sqrt{2}$ or $x = -1 - \sqrt{2}$

$\{-1 + \sqrt{2}, -1 - \sqrt{2}\}$

37. $2x^2 + x - 3 = 0$

Add 3 to each member to get $2x^2 + x = 3$.

Now divide each term by 2.

$x^2 + \frac{1}{2}x = \frac{3}{2}$

Add $\left[\frac{1}{2}\left(\frac{1}{2}\right)\right]^2 = \left(\frac{1}{4}\right)^2 = \frac{1}{16}$ to each member. Then

$x^2 + \frac{1}{2}x + \frac{1}{16} = \frac{3}{2} + \frac{1}{16}$

$\left(x + \frac{1}{4}\right)^2 = \frac{25}{16}$

Extract the roots.

$x + \frac{1}{4} = \sqrt{\frac{25}{16}}$ or $x + \frac{1}{4} = -\sqrt{\frac{25}{16}}$

So $x = -\frac{1}{4} \pm \frac{5}{4}$ and we have

$x = -\frac{1}{4} + \frac{5}{4} = \frac{4}{4} = 1$ or $x = -\frac{1}{4} - \frac{5}{4} = \frac{-6}{4} = -\frac{3}{2}$.

$\left\{1, -\frac{3}{2}\right\}$

48. $4 - x^2 = 2x$

Add x^2 to each member.

$4 = x^2 + 2x$ or $x^2 + 2x = 4$

Then add $\left[\frac{1}{2}(2)\right]^2 = (1)^2 = 1$ to each member.

$x^2 + 2x + 1 = 4 + 1$

so $(x + 1)^2 = 5$

Extract the roots.

$x + 1 = \sqrt{5}$ or $x + 1 = -\sqrt{5}$

Then $x = -1 + \sqrt{5}$ or $x = -1 - \sqrt{5}$

$\{-1 + \sqrt{5}, -1 - \sqrt{5}\}$

53. $(x + 3)(x - 2) = 1$

Perform the indicated multiplication in the left member.

$x^2 + x - 6 = 1$

Add 6 to each member to get $x^2 + x = 7$.

Add $\left[\frac{1}{2}(1)\right]^2 = \left(\frac{1}{2}\right)^2 = \frac{1}{4}$ to each member.

Thus $x^2 + x + \frac{1}{4} = 7 + \frac{1}{4}$

and $\left(x + \frac{1}{2}\right)^2 = \frac{29}{4}$

Then $x + \frac{1}{2} = \pm\sqrt{\frac{29}{4}} = \pm\sqrt{\frac{29}{2}}$

so $x = -\frac{1}{2} \pm \frac{\sqrt{29}}{2} = \frac{-1 \pm \sqrt{29}}{2}$

Then $x = \frac{-1 + \sqrt{29}}{2}$ or $x = \frac{-1 - \sqrt{29}}{2}$

$\left\{\frac{-1 + \sqrt{29}}{2}, \frac{-1 - \sqrt{29}}{2}\right\}$

59. By "a surface of a rectangular solid has a width w that is 8 millimeters shorter than its length ℓ," we get ℓ = the length of the part and
$\ell - 8$ = the width of the part. Then given area $A = 105$ square millimeters, and using $A = \ell w$, $\ell(\ell - 8) = 105$, then
$\ell^2 - 8\ell = 105$.

Add $\left[\frac{1}{2}(-8)\right]^2 = (-4)^2 = 16$ to both members.

$\ell^2 - 8\ell + 16 = 105 + 16$

$(\ell - 4)^2 = 121$

$\ell - 4 = \pm\sqrt{121} = \pm 11$

so $\ell - 4 = 11$ or $\ell - 4 = -11$

Then

$\ell = 4 + 11$ or $\ell = 4 - 11$

$\ell = 15$ $\qquad$ $\ell = -7$

Since a rectangle must have positive length, -7 is ruled out. So the length $\ell = 15$ millimeters and the width $\ell - 8 = 7$ millimeters.

Review exercises

1. $\sqrt{9} = 3$ **2.** $\sqrt{45} = 3\sqrt{5}$ **3.** $\{(2,0)\}$

4.

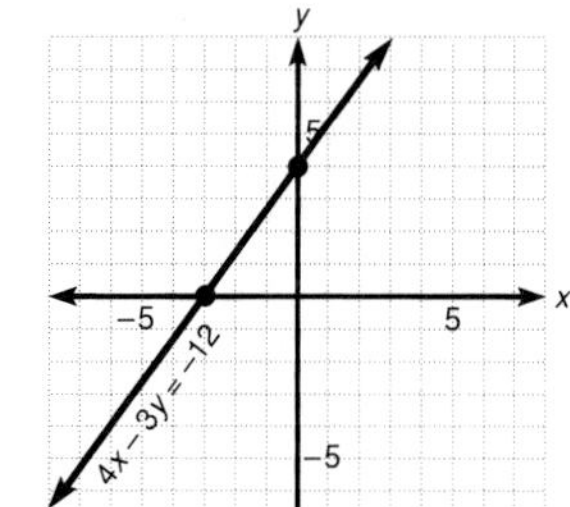

5. 9 dozen

Exercise 10–3

Answers to odd-numbered problems

1. $5x^2 - 3x + 8 = 0$; $a = 5, b = -3, c = 8$

3. $6z^2 + 2z - 1 = 0$; $a = 6, b = 2, c = -1$

5. $4x^2 - 2x + 1 = 0$; $a = 4, b = -2, c = 1$

7. $x^2 + 3x = 0$; $a = 1, b = 3, c = 0$

9. $5x^2 - 2 = 0$; $a = 5, b = 0, c = -2$

11. $p^2 + 3p - 4 = 0$; $a = 1, b = 3, c = -4$

13. $x^2 + 2x - 9 = 0$; $a = 1, b = 2, c = -9$
15. $8m^2 - 3m - 2 = 0$; $a = 8, b = -3, c = -2$ **17.** $\{1,2\}$
19. $\{1\}$ **21.** $\{-5,5\}$ **23.** $\{-\sqrt{2},\sqrt{2}\}$ **25.** $\{0,3\}$ **27.** $\left\{0,\frac{9}{5}\right\}$
29. $\left\{\frac{9+\sqrt{65}}{2},\frac{9-\sqrt{65}}{2}\right\}$ **31.** $\{-1-\sqrt{7}, -1+\sqrt{7}\}$
33. $\{4+\sqrt{15}, 4-\sqrt{15}\}$ **35.** $\left\{\frac{5-\sqrt{97}}{6},\frac{5+\sqrt{97}}{6}\right\}$
37. $\left\{\frac{-9-\sqrt{57}}{6},\frac{-9+\sqrt{57}}{6}\right\}$ **39.** $\left\{-\frac{5}{3},2\right\}$ **41.** $\{-4\}$
43. $\left\{\frac{5}{2}\right\}$ **45.** $\left\{-\frac{3}{2}\right\}$ **47.** $\left\{\frac{1-\sqrt{22}}{3},\frac{1+\sqrt{22}}{3}\right\}$
49. $\{-1-\sqrt{7},-1+\sqrt{7}\}$ **51.** $\left\{\frac{3-\sqrt{105}}{6},\frac{3+\sqrt{105}}{6}\right\}$
53. $\left\{\frac{5-\sqrt{85}}{10},\frac{5+\sqrt{85}}{10}\right\}$ **55.** $\left\{\frac{3-\sqrt{41}}{4},\frac{3+\sqrt{41}}{4}\right\}$
57. $\left\{-\frac{3}{2},3\right\}$ **59.** $\left\{\frac{-1-\sqrt{3}}{3},\frac{-1+\sqrt{3}}{3}\right\}$ **61. a.** 2 seconds
b. $\sqrt{6}$ seconds $\approx$ 2.5 sec **c.** $\frac{\sqrt{30}}{2}$ seconds $\approx$ 2.74 sec
63. a. $-20 + 20\sqrt{11}$ **b.** $\frac{-15+5\sqrt{329}}{2}$
65. $b = 9\sqrt{10}$ inches; $h = 3\sqrt{10}$ inches **67.** 6 millimeters; 8 millimeters **69.** $-1+\sqrt{7}$; $1+\sqrt{7}$ **71.** $5\sqrt{19}$ feet
73. -16 and -14 **75.** 7 or $\frac{1}{7}$

Solutions to trial exercise problems

7. $x^2 = -3x$
Add $3x$ to each member. Then
$x^2 + 3x = 0$ and
$a = 1, b = 3, c = 0.$

12. $2x(x-9) = 1$
Perform the indicated multiplication.
$2x^2 - 18x = 1$
Add -1 to each member.
$2x^2 - 18x - 1 = 0$, so
$a = 2, b = -18, c = -1.$

18. $y^2 + 6y + 9 = 0$
Here $a = 1, b = 6$, and $c = 9$ so

$$y = \frac{-6 \pm \sqrt{(6)^2 - 4(1)(9)}}{2(1)} = \frac{-6 \pm \sqrt{36-36}}{2} = \frac{-6 \pm \sqrt{0}}{2} = \frac{-6}{2} = -3$$

$\{-3\}$

21. $x^2 - 25 = 0$
We can write this
$x^2 + 0x - 25 = 0$, so
$a = 1, b = 0, c = -25.$

$$\text{Then } x = \frac{-0 \pm \sqrt{0^2 - 4(1)(-25)}}{2(1)}$$

$$x = \frac{\pm\sqrt{100}}{2}$$

so $x = \pm\frac{10}{2}$ Then
$x = 5$ or $x = -5$
$\{5,-5\}$

26. $x^2 = 4x$
Add $-4x$ to each member and write the equation as
$x^2 - 4x + 0 = 0$
So $a = 1, b = -4, c = 0,$

$$\text{and } x = \frac{-(-4) \pm \sqrt{(-4)^2 - 4(1)(0)}}{2(1)} = \frac{4 \pm \sqrt{16}}{2} = \frac{4 \pm 4}{2}$$

$$\text{Then } x = \frac{4+4}{2} = \frac{8}{2} = 4 \text{ or } x = \frac{4-4}{2} = \frac{0}{2} = 0$$

$\{0,4\}$

36. $4t^2 = 8t - 3$
Add $3 - 8t$ to each member.
$4t^2 - 8t + 3 = 0$
So $a = 4, b = -8, c = 3,$

$$\text{and } t = \frac{-(-8) \pm \sqrt{(-8)^2 - 4(4)(3)}}{2(4)} = \frac{8 \pm \sqrt{64-48}}{8} = \frac{8 \pm \sqrt{16}}{8}$$

$$\text{so } t = \frac{8 \pm 4}{8}$$

$$\text{Then } t = \frac{8+4}{8} = \frac{12}{8} = \frac{3}{2} \text{ or } t = \frac{8-4}{8} = \frac{4}{8} = \frac{1}{2}$$

$\left\{\frac{1}{2},\frac{3}{2}\right\}$

54. $2x^2 - \frac{7}{2} + \frac{x}{2} = 0$
Multiply by the LCD, 2.
$4x^2 - 7 + x = 0$
Then write in standard form.
$4x^2 + x - 7 = 0$
Then $a = 4, b = 1$, and $c = -7.$

$$\text{Thus } x = \frac{-1 \pm \sqrt{1^2 - 4(4)(-7)}}{2(4)} = \frac{-1 \pm \sqrt{1+112}}{8} = \frac{-1 \pm \sqrt{113}}{8}$$

$$\text{So } x = \frac{-1+\sqrt{113}}{8} \text{ or } x = \frac{-1-\sqrt{113}}{8}$$

$\left\{\frac{-1+\sqrt{113}}{8},\frac{-1-\sqrt{113}}{8}\right\}$

60. a. Using $s = vt + \frac{1}{2}at^2$, replace s with 8, v with 3, and a with 4.

$$8 = 3t + \frac{1}{2}(4)t^2$$
$$8 = 3t + 2t^2$$
$$2t^2 + 3t - 8 = 0$$
$$t = \frac{-3 \pm \sqrt{3^2 - 4(2)(-8)}}{2(2)}$$
$$= \frac{-3 \pm \sqrt{9 + 64}}{4}$$
$$= \frac{-3 + \sqrt{73}}{4}$$
$$t = \frac{-3 + \sqrt{73}}{4} \approx 1.39 \text{ or } t = \frac{-3 - \sqrt{93}}{4} \text{ (reject)}$$

66. Using $a^2 + b^2 = c^2$, replace a with x, b with $x + 14$, and c with $x + 16$.

$$x^2 + (x + 14)^2 = (x + 16)^2$$
$$x^2 + x^2 + 28x + 196 = x^2 + 32x + 256$$
$$2x^2 + 28x + 196 = x^2 + 32x + 256$$
$$x^2 - 4x - 60 = 0$$
$$x = \frac{-(-4) \pm \sqrt{(-4)^2 - 4(1)(-60)}}{2(1)}$$
$$= \frac{4 \pm \sqrt{16 + 240}}{2}$$
$$= \frac{4 \pm \sqrt{256}}{2}$$
$$= \frac{4 \pm 16}{2}$$

Then $x = \frac{4 + 16}{2} = \frac{20}{2} = 10$ or

$x = \frac{4 - 16}{2} = \frac{-12}{2} = -6$ (reject).

Thus, $x = 10$.

74. Let n = the first odd positive integer. Then $n + 2$ = the next consecutive odd positive integer.

The equation is then $n(n + 2) = 143$

$$n^2 + 2n = 143$$
$$n^2 + 2n - 143 = 0$$
$$(n + 13)(n - 11) = 0$$

Then $n = -13$ and $n + 2 = -11$ or $n = 11$ and $n + 2 = 13$. Reject -13 and -11 since we want positive integers. Thus 11 and 13 are two consecutive odd positive integers.

Review exercises

1. $3x^2 + x + 5$ **2.** $5y^2 + 33y - 14$ **3.** $16z^2 - 9$ **4.** $9x^2 - 30x + 25$ **5.** $\{-2, 2\}$ **6.** $\left\{-\frac{1}{2}, 4\right\}$ **7.** $\left\{\frac{1 - \sqrt{41}}{2}, \frac{1 + \sqrt{41}}{2}\right\}$ **8.** $\frac{2x - 8}{(x + 2)(x - 2)(x - 3)}$

Exercise 10–4

Answers to odd-numbered problems

1. $9 + 0i$ **3.** $0 + 4i$ **5.** $0 + 5i$ **7.** $4 + 4i$ **9.** $4 + i$ **11.** $1 - 4i$ **13.** $5 - 2i$ **15.** $-9 - 2i\sqrt{7}$ **17.** $-12 + 6i$ **19.** $10 + 11i$ **21.** 41 **23.** $-33 + 56i$ **25.** $\frac{15}{13} + \frac{10}{13}i$ **27.** $\frac{1}{17} + \frac{13}{17}i$ **29.** $\frac{17}{25} - \frac{19}{25}i$ **31.** $\{-2 + 4i, -2 - 4i\}$ **33.** $\left\{\frac{-1 + i\sqrt{7}}{2}, \frac{-1 - i\sqrt{7}}{2}\right\}$ **35.** $\left\{\frac{3 + i\sqrt{11}}{2}, \frac{3 - i\sqrt{11}}{2}\right\}$ **37.** $\left\{\frac{-1 + i\sqrt{31}}{4}, \frac{-1 - i\sqrt{31}}{4}\right\}$ **39.** $\left\{\frac{-1 + i\sqrt{19}}{2}, \frac{-1 - i\sqrt{19}}{2}\right\}$ **41.** $b^2 - 4ac = 24$; two distinct irrational solutions **43.** $b^2 - 4ac = 0$; one rational solution **45.** $b^2 - 4ac = 9$; two distinct rational solutions **47.** $b^2 - 4ac = 5$; two distinct irrational solutions

Solutions to trial exercise problems

7. $4 + 2\sqrt{-4} = 4 + 2(2i) = 4 + 4i$

14. $(1 - \sqrt{-4}) - (3 + \sqrt{-9})$
$$= (1 - 2i) - (3 + 3i) = 1 - 2i - 3 - 3i$$
$$= (1 - 3) + (-2i - 3i)$$
$$= -2 + (-5i)$$
$$= -2 - 5i$$

23. $(4 + 7i)^2$
$$= 4^2 + 2(4)(7i) + (7i)^2 = 16 + 56i + 49i^2 = 16 + 56i + 49(-1)$$
$$= 16 + 56i - 49$$
$$= -33 + 56i$$

28. $\frac{1 + i}{2 - i} \cdot \frac{2 + i}{2 + i} = \frac{(1 + i)(2 + i)}{2^2 + 1^2} = \frac{2 + 3i + i^2}{4 + 1} = \frac{2 + 3i + (-1)}{5}$
$$= \frac{1 + 3i}{5} = \frac{1}{5} + \frac{3}{5}i$$

38. $3y^2 - 2y + 3 = 0$. Here $a = 3$, $b = -2$, and $c = 3$.
$$y = \frac{-(-2) \pm \sqrt{(-2)^2 - 4(3)(3)}}{2(3)} = \frac{2 \pm \sqrt{4 - 36}}{6}$$
$$= \frac{2 \pm \sqrt{-32}}{6}$$
$$= \frac{2 \pm 4\sqrt{-2}}{6}$$
$$= \frac{2 \pm 4i\sqrt{2}}{6}$$
$$= \frac{2(1 \pm 2i\sqrt{2})}{6}$$
$$= \frac{1 \pm 2i\sqrt{2}}{3}$$

The solution set is $\left\{\frac{1 + 2i\sqrt{2}}{3}, \frac{1 - 2i\sqrt{2}}{3}\right\}$.

47. $(x + 4)(x + 3) = 1$

$x^2 + 7x + 12 = 1$, and then we have $x^2 + 7x + 11 = 0$. Then $b^2 - 4ac = (7)^2 - 4(1)(11) = 49 - 44 = 5$; two distinct irrational solutions.

Review exercises

1.

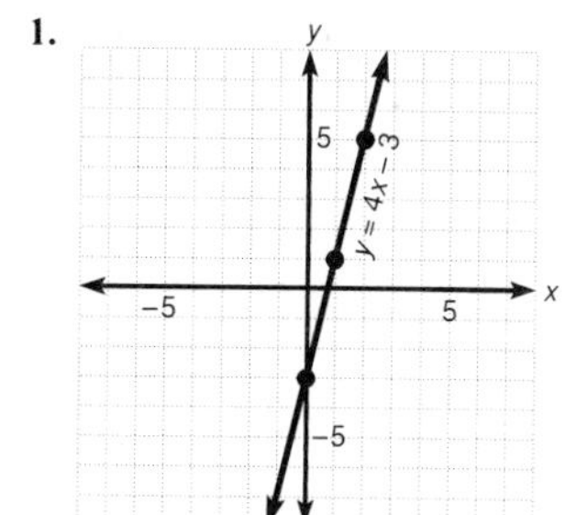

2.

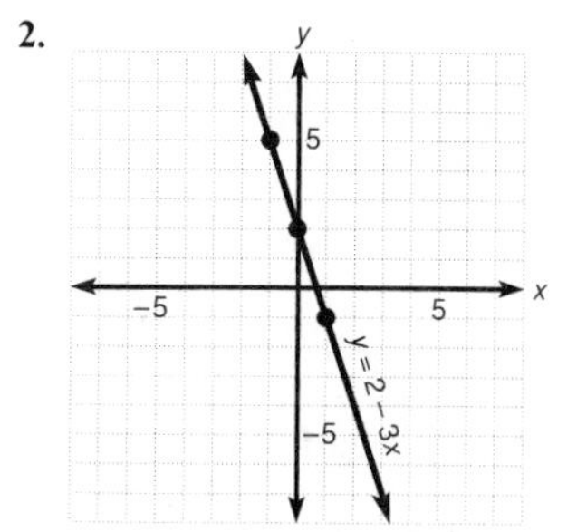

3. $\left(\frac{4}{3}, \frac{5}{3}\right)$ **4.** $8x - 3y = 17$ **5.** $\frac{3x + 3}{4x}$

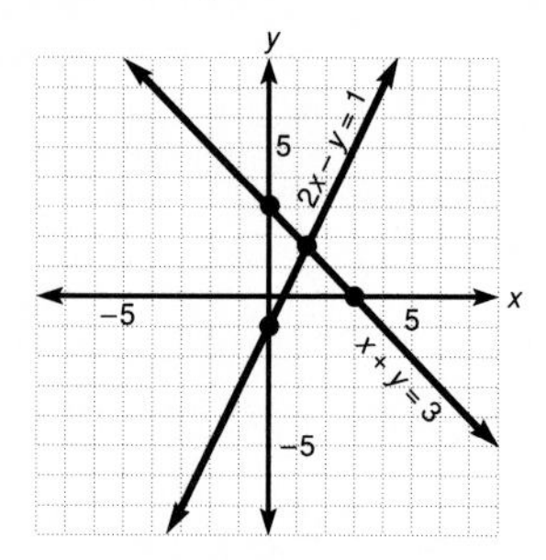

Exercise 10–5

Answers to odd-numbered problems

1. $(-1,-6)$, $(0,-4)$, $(3,14)$ **3.** $(-3,38)$, $(0,5)$, $(2,23)$ **5.** $(-6,198)$, $(0,0)$, $(6,162)$ **7.** $(-3,44)$, $(0,-1)$, $(5,124)$ **9.** $(-2,-4)$, $(0,4)$, $\left(\frac{3}{4}, \frac{13}{16}\right)$ **11.** y-intercept, -16; x-intercepts, 4 and -4 **13.** y-intercept, 8; x-intercepts, 4 and 2 **15.** y-intercept, 12; x-intercepts, -2 and -6 **17.** y-intercept, 5; x-intercepts, $\sqrt{5}$ and $-\sqrt{5}$ **19.** y-intercept, 9; x-intercept, -3 **21.** y-intercept, 5; x-intercepts, none **23.** y-intercept, 6; x-intercepts, none **25.** y-intercept, -16; x-intercept, 4 **27.** y-intercept, 1; x-intercepts, -1 and $-\frac{1}{2}$ **29.** y-intercept, 6; x-intercepts, -2 and $\frac{3}{2}$ **31.** $(0,-16)$; $x = 0$ **33.** $(3,-1)$; $x = 3$ **35.** $(-4,-4)$; $x = -4$ **37.** $(0,5)$; $x = 0$ **39.** $(-3,0)$; $x = -3$ **41.** $(0,5)$; $x = 0$ **43.** $(-2,2)$; $x = -2$ **45.** $(4,0)$; $x = 4$ **47.** $\left(-\frac{3}{4}, -\frac{1}{8}\right)$; $x = -\frac{3}{4}$ **49.** $\left(-\frac{1}{4}, \frac{49}{8}\right)$; $x = -\frac{1}{4}$

51.

x	y	
± 1	-15	arbitrary points
± 2	-12	
± 3	-7	
-4	0	x-intercepts
4	0	
0	-16	y-intercept; vertex

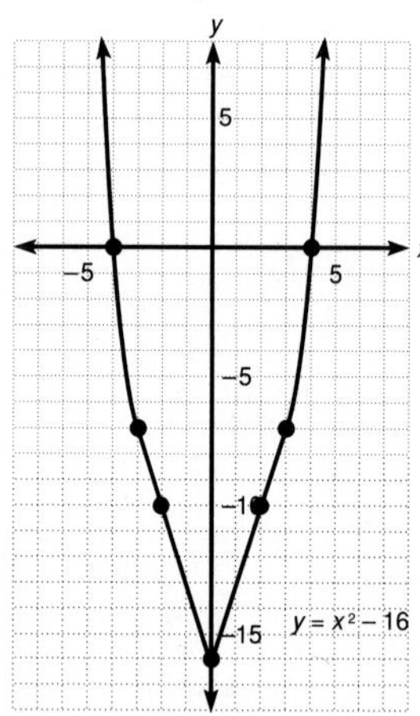

53.

x	y	
1	3	arbitrary points
5	3	
6	8	
3	-1	vertex
2	0	x-intercepts
4	0	
0	8	y-intercept

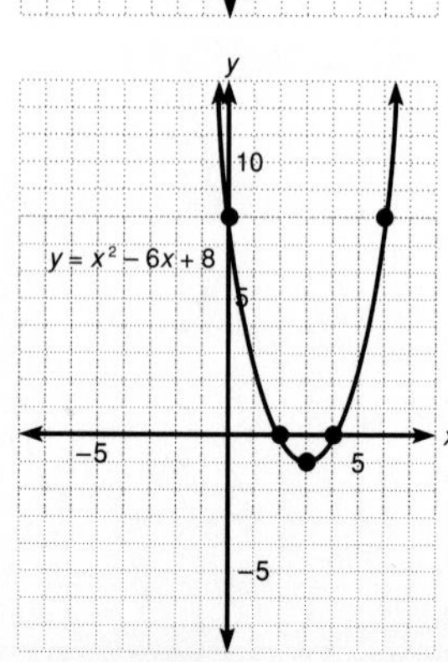

55.

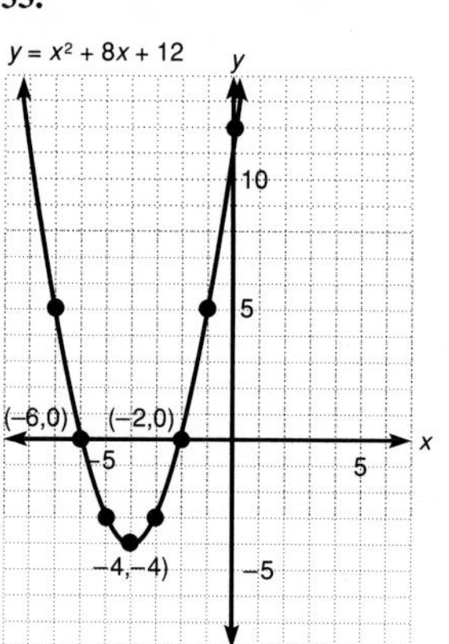

57.

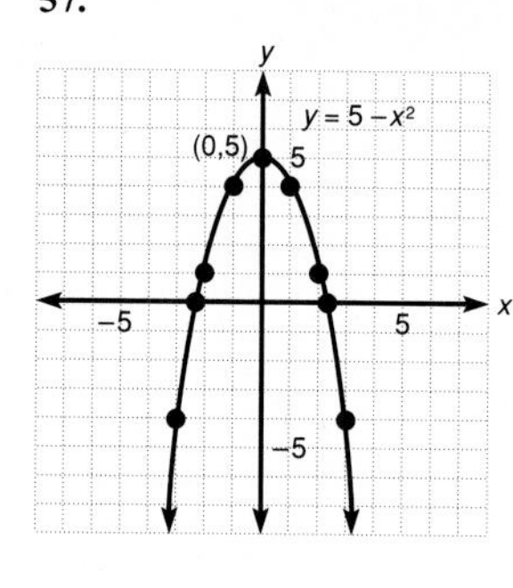

59.

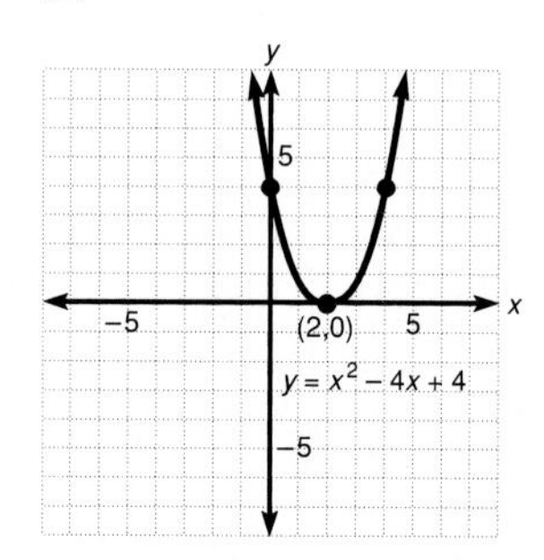

61.

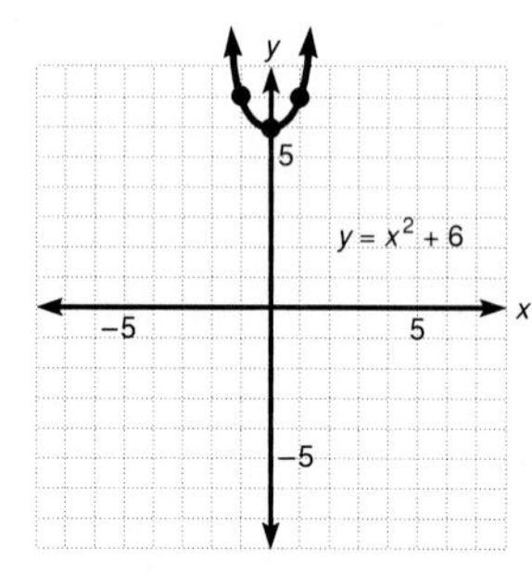

63.

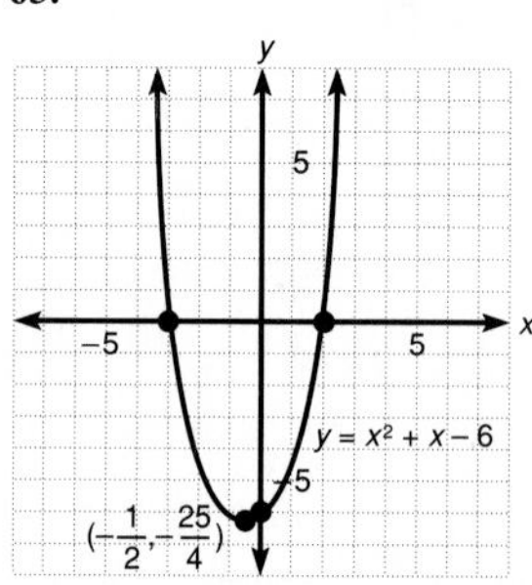

65.

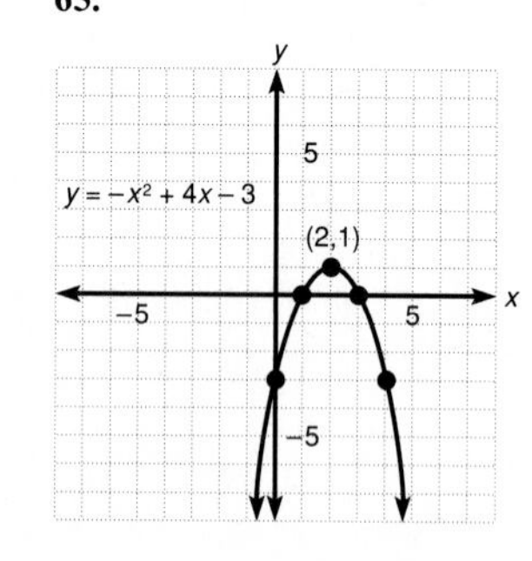

67.

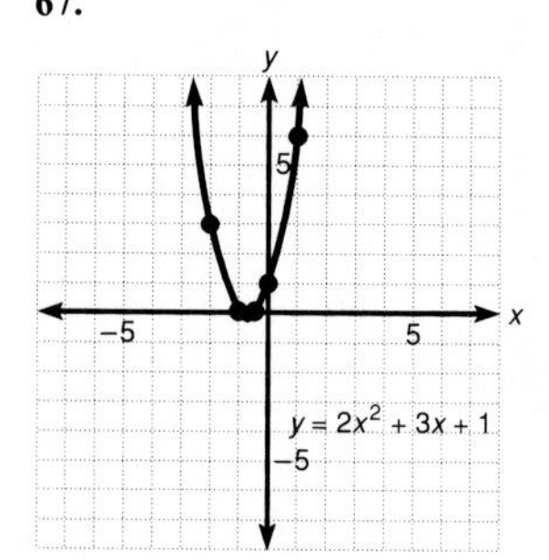

69.

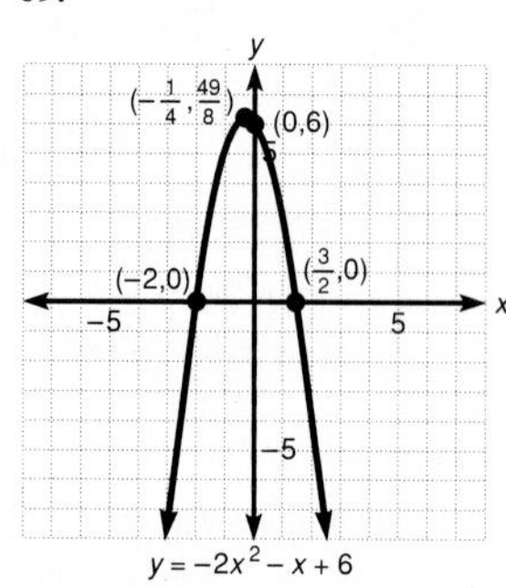

71.

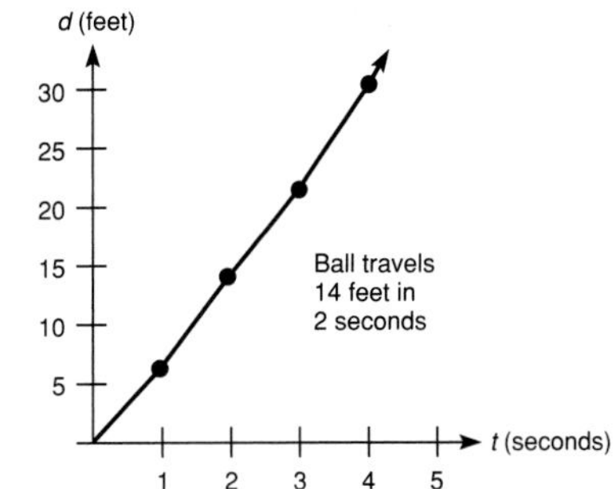

73.

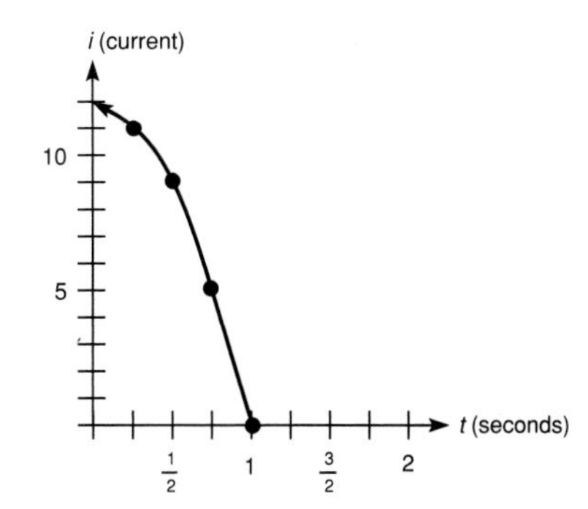

75.

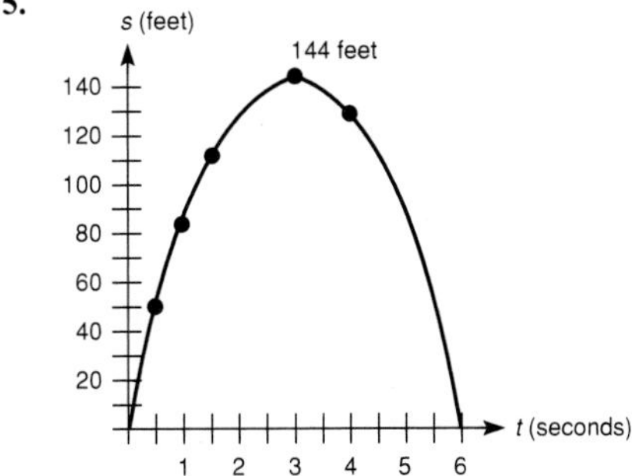

77.

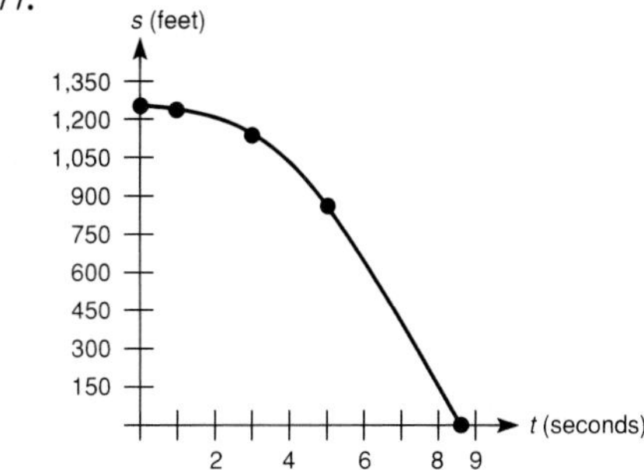

Solutions to trial exercise problems

3. $h(x) = 4x^2 + x + 5$

$h(-3) = 4(-3)^2 + (-3) + 5$
$= 4(9) - 3 + 5$
$= 36 - 3 + 5 = 38$; $(-3,38)$

$h(0) = 4(0)^2 + 0 + 5$
$= 0 + 0 + 5$
$= 5$; $(0,5)$

$h(2) = 4(2)^2 + 2 + 5$
$= 4(4) + 2 + 5$
$= 16 + 2 + 5$
$= 23$; $(2,23)$

13. $y = x^2 - 6x + 8$

Let $x = 0$, then $y = 0^2 - 6(0) + 8 = 8$.
Let $y = 0$, then $0 = x^2 - 6x + 8$. Factor the right member.
$0 = (x - 4)(x - 2)$
so $x = 4$ or $x = 2$
The y-intercept is 8 and the x-intercepts are 4 and 2.

17. $y = 5 - x^2$

Let $x = 0$, then $y = 5 - 0^2 = 5$.
Let $y = 0$, then $0 = 5 - x^2$. Add x^2 to each member.
$x^2 = 5$
Extract the roots.
$x = \sqrt{5}$ or $x = -\sqrt{5}$ so the y-intercept is 5 and the x-intercepts are $\sqrt{5}$and $-\sqrt{5}$.

27. $y = 2x^2 + 3x + 1$

Let $x = 0$, then $y = 2(0)^2 + 3(0) + 1 = 1$.
Let $y = 0$, then $0 = 2x^2 + 3x + 1$. Factor the right member.
$0 = (2x + 1)(x + 1)$
then $2x + 1 = 0$ or $x + 1 = 0$

so $x = -\dfrac{1}{2}$ or $x = -1$

The y-intercept is 1 and the x-intercepts are $-\dfrac{1}{2}$ and -1.

33. $y = x^2 - 6x + 8$

Here $a = 1$ and $b = -6$

so $x = \dfrac{-b}{2a} = -\dfrac{-6}{2(1)} = 3$

then $y = (3)^2 - 6(3) + 8$
$= 9 - 18 + 8$
$= -1$
The vertex is at $(3,-1)$. Axis of symmetry is $x = 3$.

37. $y = 5 - x^2$

Here $a = -1$ and $b = 0$

so $x = \dfrac{-b}{2a} = \dfrac{-0}{2(-1)} = 0$

then $y = 5 - 0^2 = 5$
Therefore the vertex is at $(0,5)$. Axis of symmetry is $x = 0$.

47. $y = 2x^2 + 3x + 1$

Here $a = 2$ and $b = 3$ so $x = \dfrac{-b}{2a} = \dfrac{-3}{2(2)} = -\dfrac{3}{4}$ Then

$$y = 2\left(-\frac{3}{4}\right)^2 + 3\left(-\frac{3}{4}\right) + 1$$
$$= 2\left(\frac{9}{16}\right) - \frac{9}{4} + 1$$
$$= \frac{9}{8} - \frac{18}{8} + 1$$
$$= -\frac{9}{8} + 1 = -\frac{1}{8}$$

The vertex is at $\left(-\frac{3}{4}, -\frac{1}{8}\right)$. Axis of symmetry is $x = -\frac{3}{4}$.

53. $y = x^2 - 6x + 8$

x	y	
0	8	y-intercept
4	0	x-intercepts
2	0	
3	−1	vertex
1	3	arbitrary points
5	3	
6	8	

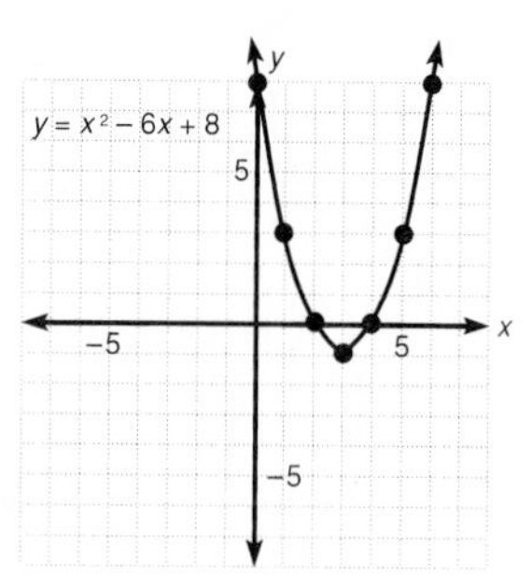

57. $y = 5 - x^2$

x	y	
0	5	y-intercept; vertex
$\sqrt{5}$	0	x-intercepts
$-\sqrt{5}$	0	
1	4	arbitrary points
2	1	
−2	1	
−1	4	

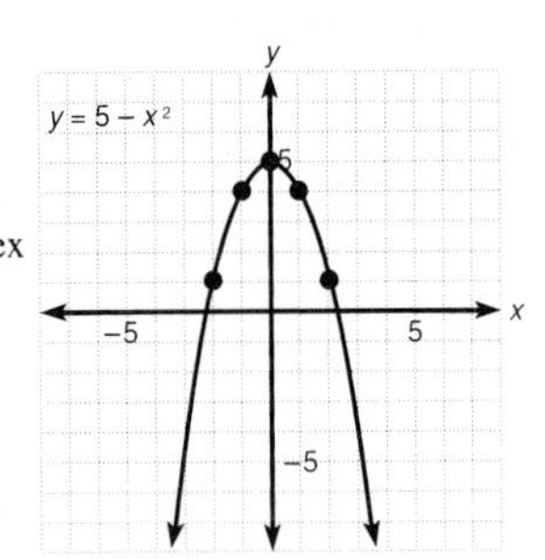

67. $y = 2x^2 + 3x + 1$

x	y	
0	1	y-intercept
$-\frac{1}{2}$	0	x-intercepts
−1	0	
$-\frac{3}{4}$	$-\frac{1}{8}$	vertex
1	6	arbitrary points
−2	3	
−3	10	

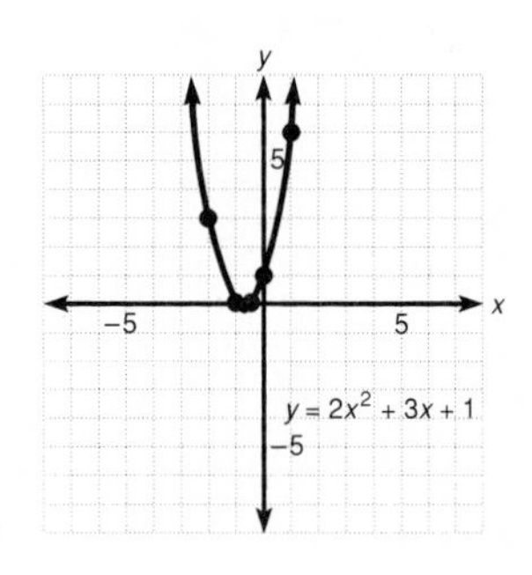

71. $d = 6t + \dfrac{t^2}{2}$ (*Note:* We must choose $t \geq 0$.)

t	d
0	0
1	$\frac{13}{2}$ or $6\frac{1}{2}$
2	14
3	$\frac{45}{2}$ or $22\frac{1}{2}$
4	32

$t = 2$ seconds when $d = 14$ feet

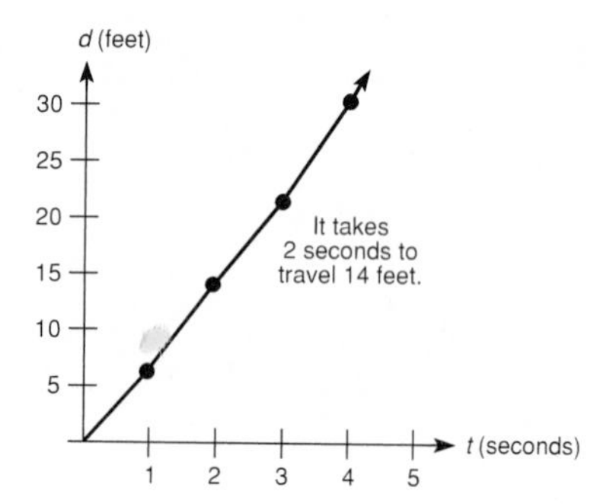

Chapter 10 review

1. $\{10, -10\}$ **2.** $\{-5, 5\}$ **3.** $\{\sqrt{2}, -\sqrt{2}\}$ **4.** $\{\sqrt{6}, -\sqrt{6}\}$
5. $\{2, -2\}$ **6.** $\{2\sqrt{3}, -2\sqrt{3}\}$ **7.** $\{4, -4\}$ **8.** $\{2\sqrt{3}, -2\sqrt{3}\}$
9. $\left\{\frac{3}{2}\sqrt{11}, -\frac{3}{2}\sqrt{11}\right\}$ **10.** $\{-6, 7\}$ **11.** $\left\{1, \frac{4}{3}\right\}$
12. $\{3 + \sqrt{5}, 3 - \sqrt{5}\}$ **13.** $\{5 + \sqrt{21}, 5 - \sqrt{21}\}$
14. $\left\{\frac{4 + \sqrt{14}}{2}, \frac{4 - \sqrt{14}}{2}\right\}$ **15.** $\left\{\frac{3 + 2\sqrt{6}}{3}, \frac{3 - 2\sqrt{6}}{3}\right\}$
16. $\left\{\frac{-5 + \sqrt{41}}{2}, \frac{-5 - \sqrt{41}}{2}\right\}$ **17.** $\left\{\frac{11 + \sqrt{101}}{2}, \frac{11 - \sqrt{101}}{2}\right\}$
18. $\left\{-\frac{3}{4}, 1\right\}$ **19.** $\left\{\frac{1 + \sqrt{13}}{2}, \frac{1 - \sqrt{13}}{2}\right\}$
20. $\left\{\frac{3 + \sqrt{41}}{8}, \frac{3 - \sqrt{41}}{8}\right\}$ **21.** $\left\{-2, \frac{5}{3}\right\}$
22. $w = 2$, meters, $\ell = 8$ meters **23.** $\{1 + \sqrt{6}, 1 - \sqrt{6}\}$
24. $\{-2 + 2\sqrt{3}, -2 - 2\sqrt{3}\}$ **25.** $\left\{\frac{5}{2}, -1\right\}$
26. $\left\{\frac{-7 + \sqrt{145}}{6}, \frac{-7 - \sqrt{145}}{6}\right\}$ **27.** $\left\{\frac{3\sqrt{2}}{2}, \frac{-3\sqrt{2}}{2}\right\}$
28. $\left\{0, -\frac{7}{4}\right\}$ **29.** $\left\{\frac{1 + \sqrt{13}}{3}, \frac{1 - \sqrt{13}}{3}\right\}$
30. $\left\{\frac{4 + \sqrt{34}}{6}, \frac{4 - \sqrt{34}}{6}\right\}$ **31.** 6 inches, 9 inches
32. $5 - i$ **33.** $4 + 9i$ **34.** $15 - 16i$ **35.** $3 + 21i$
36. 52 **37.** $16 - 30i$ **38.** $\frac{3}{2} + \frac{3}{2}i$ **39.** $\frac{4}{5} + \frac{8}{5}i$

40. $\frac{7}{10} - \frac{1}{10}i$ **41.** $\frac{26}{25} - \frac{7}{25}i$ **42.** $\{-2 + i\sqrt{3}, -2 - i\sqrt{3}\}$
43. $\left\{\frac{1 + i\sqrt{79}}{8}, \frac{1 - i\sqrt{79}}{8}\right\}$ **44.** $\{i\sqrt{7}, -i\sqrt{7}\}$
45. $\left\{\frac{-3 + 2i}{2}, \frac{-3 - 2i}{2}\right\}$ **46.** $b^2 - 4ac = 0$; one rational solution **47.** $b^2 - 4ac = 25$; two distinct rational solutions **48.** $b^2 - 4ac = -56$; two distinct complex solutions **49.** $b^2 - 4ac = 37$; two distinct irrational solutions **50.** $f(-5) = 35, f(0) = -5, f(1) = -7$ **51.** $g(-1) = -3, g(0) = 4, g(3) = 1$
52. $h(-3) = 30, h(0) = 0, h(4) = 72$
53. $f(-4) = -36, f(0) = 12, f(2) = 0$
54. y-intercept, -12; x-intercepts, $6, -2$; vertex, $(2, -16)$; $x = 2$
55. y-intercept, 1; x-intercepts, $1, \frac{1}{5}$; vertex, $\left(\frac{3}{5}, -\frac{4}{5}\right)$; $x = \frac{3}{5}$
56. y-intercept, 8; x-intercepts, $2, -4$; vertex, $(-1, 9)$; $x = -1$
57. y-intercept, 2; x-intercepts, $1, -\frac{2}{3}$; vertex, $\left(\frac{1}{6}, \frac{25}{12}\right)$; $x = \frac{1}{6}$
58. y-intercept, 0; x-intercepts, $0, \frac{2}{5}$; vertex, $\left(\frac{1}{5}, -\frac{1}{5}\right)$; $x = \frac{1}{5}$
59. y-intercept, 0; x-intercepts, $0, \frac{1}{3}$; vertex, $\left(\frac{1}{6}, \frac{1}{12}\right)$; $x = \frac{1}{6}$
60. y-intercept, -8; x-intercepts, $\sqrt{2}, -\sqrt{2}$; vertex, $(0, -8)$; $x = 0$
61. y-intercept, 9; x-intercepts, $3, -3$; vertex, $(0, 9)$; $x = 0$
62. y-intercept, 2; x-intercepts, none; vertex, $(0, 2)$; $x = 0$
63. y-intercept, 3; x-intercepts, none; vertex, $(-1, 2)$; $x = -1$
64. $4\frac{1}{2}$ sec **65.** $93\frac{1}{3}$ ft

Final examination

1. $>$ **2.** 42 **3.** 0 **4.** -52 **5.** x^6 **6.** $\frac{1}{2x^5}$ **7.** $-12x^3y^5$
8. $\frac{y^6}{27x^3}$ **9.** 1 **10.** $7x^2 - 8y^2$ **11.** $-3y$ **12.** $y^2 - 81$
13. $49z^2 - 42zw + 9w^2$ **14.** $5x^3 + 17x^2 - 11x + 4$
15. $5y - 3x^3 + xy$ **16.** $4y + 1 + \frac{-2}{2y - 1}$ **17.** $\{7\}$
18. $\{12, -1\}$ **19.** $\left\{2, \frac{1}{3}\right\}$ **20.** $\left\{-\frac{8}{3}\right\}$ **21.** $\left\{0, \frac{3}{2}\right\}$
22. $3x(x - 2y + 3)$ **23.** $(a - 7)(a + 3)$
24. $(2x - 1)(2x - 5)$ **25.** $(3a + 8)(3a - 8)$
26. $(2a + b)(3x - y)$ **27.** $(x - 5)^2$ **28.** 11 and 12 or -12 and -11 **29.** $\frac{x + 1}{x + 2}$ **30.** $\frac{x - 2}{12}$ **31.** $\frac{14}{x - 6}$
32. $\frac{x^2 - 6x + 14}{(x + 5)(x - 5)}$ **33.** $\frac{5y + 4}{4y - 6}$ **34.** $x = \frac{16}{5}$ **35.** $\frac{21}{40}$ or 21:40
36. $2x + y = 1$ **37.** $m = \frac{2}{3}$; $b = -3$ **38.** $f(2) = 17, (2, 17)$; $f(0) = -1, (0, -1)$; $f(-1) = -1, (-1, -1)$ **39.** $\{(-19, -11)\}$
40. length = 12 feet; width = 5 feet
41. $-\sqrt{3}$ **42.** $\sqrt{6} + 3$ **43.** 13 **44.** $11 - 4\sqrt{7}$
45. $\frac{9 + 3\sqrt{5}}{4}$ **46.** -3 **47.** 8 **48.** $\{-1, 0\}$

49.

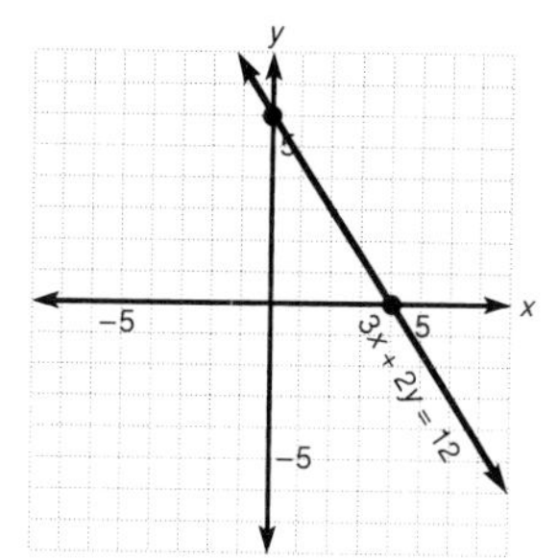

50.

x	y	
-1	0	x-intercepts
6	0	
0	-6	y-intercept
$\frac{5}{2}$	$-\frac{49}{4}$	vertex
1	-10	arbitrary points
2	-12	
3	-12	
4	-10	

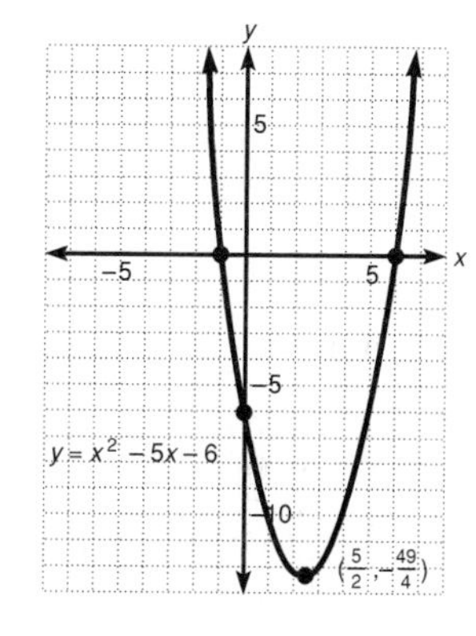

51. $\left\{\frac{7 + \sqrt{97}}{8}, \frac{7 - \sqrt{97}}{8}\right\}$ **52.** $1 - 19i$ **53.** $17 + 9i$ **54.** 65
55. $-24 + 70i$ **56.** $10 + 2i$ **57.** 6 **58.** $\frac{27}{37} - \frac{23}{37}i$
59. $-\frac{6}{13} + \frac{15}{13}i$

Index